AF864070

Lecture Notes in Electrical Engineering

Volume 877

The book series *Lecture Notes in Electrical Engineering* (LNEE) publishes the latest developments in Electrical Engineering - quickly, informally and in high quality. While original research reported in proceedings and monographs has traditionally formed the core of LNEE, we also encourage authors to submit books devoted to supporting student education and professional training in the various fields and applications areas of electrical engineering. The series cover classical and emerging topics concerning:

- Communication Engineering, Information Theory and Networks
- Electronics Engineering and Microelectronics
- Signal, Image and Speech Processing
- Wireless and Mobile Communication
- Circuits and Systems
- Energy Systems, Power Electronics and Electrical Machines
- Electro-optical Engineering
- Instrumentation Engineering
- Avionics Engineering
- Control Systems
- Internet-of-Things and Cybersecurity
- Biomedical Devices, MEMS and NEMS

For general information about this book series, comments or suggestions, please contact leontina.dicecco@springer.com.

To submit a proposal or request further information, please contact the Publishing Editor in your country:

China

Jasmine Dou, Editor (jasmine.dou@springer.com)

India, Japan, Rest of Asia

Swati Meherishi, Editorial Director (Swati.Meherishi@springer.com)

Southeast Asia, Australia, New Zealand

Ramesh Nath Premnath, Editor (ramesh.premnath@springernature.com)

USA, Canada:

Michael Luby, Senior Editor (michael.luby@springer.com)

All other Countries:

Leontina Di Cecco, Senior Editor (leontina.dicecco@springer.com)

**** This series is indexed by EI Compendex and Scopus databases. ****

Brijesh Mishra · Manish Tiwari
Editors

VLSI, Microwave and Wireless Technologies

Select Proceedings of ICVMWT 2021

Springer

Editors
Brijesh Mishra
Department of Electronics
and Communication Engineering
Shambhunath Institute of Engineering
and Technology
Prayagraj, Uttar Pradesh, India

Manish Tiwari
Department of Electronics
and Communication Engineering
Manipal University Jaipur
Jaipur, India

ISSN 1876-1100 ISSN 1876-1119 (electronic)
Lecture Notes in Electrical Engineering
ISBN 978-981-19-0311-3 ISBN 978-981-19-0312-0 (eBook)
https://doi.org/10.1007/978-981-19-0312-0

This Springer imprint is published by the registered company Springer Nature Singapore Pte Ltd.
The registered company address is: 152 Beach Road, #21-01/04 Gateway East, Singapore 189721, Singapore

Preface

Research on VLSI and wireless technologies has lately advanced at a rapid pace. The compactness of the gazettes, approaches to providing high data rates to the masses are transforming and expanding in a way that is beyond our imagination. As electronic devices get compact, engineers and designers are faced with the growing challenge of keeping up with the need to optimize processing speed within a shrinking form factor also the challenges in providing uninterrupted data and broadband communications have not changed. Our mission as a technical community is to understand these challenges and find ways to mitigate them. This includes the development and management of appropriate channels, novel devices, new protocols, efficient networks, and their integration. Keeping in view the amalgamation of these issues, the proceedings of the International Conference on VLSI and Microwave and Wireless technologies (ICVMWT 2021) is being presented herewith.

The conference (ICVMWT 2021) was held in the campus of Madan Mohan Malaviya University of Technology Gorakhpur, Uttar Pradesh, in virtual mode during 20–21 March 2021.

A total of 200 participants including the invited speakers, contributing authors, researchers, and attendees participated in the conference. The participants were explored to a broad range of topics critical to our society and industry in the related areas. The conference provided an opportunity to exchange ideas among global leaders and experts from academia and industry on topics like optical signal processing and networking, photonic communication systems and networks, all-optical systems, microwave photonics, software-defined and cognitive radio, signal processing for wireless communications, antenna systems, spectrum management and regulatory issues, vehicular communications, wireless sensor networks, machine-to-machine communications, cellular Wi-Fi integration, etc.

Apart from high-quality contributed papers presented by the authors from all over the country and abroad, the conference participants also witnessed the informative demonstrations and technical sessions from the industry as well as invited talks from renowned experts aimed at advances in these areas. The overall response to the conference was quite encouraging. A large number (184) of research papers were received for consideration for publication in the conference proceedings. After a rigorous editorial and review process and oral presentation, 79 papers were selected for inclusion in the conference proceedings. We are confident that the papers presented in these proceedings shall provide a platform for young as well as experienced professionals to generate new ideas and networking opportunities.

The editorial team members would like to extend gratitude and sincere thanks to all contributed authors, reviewers, panellists, organizing committee members, volunteers, and the session chair for paying attention to the quality of the publication. We are thankful to our sponsors (TEQIP-III) for generously supporting this event and institutional partners (J. K. Institute of Applied Physics and Technology University of Allahabad and Manipal University Jaipur) for providing motivation, support, and encouragement duringdifferent stages. At last, we pay the highest regard to the Technical Education Quality Improvement Programme of Government of India (TEQIP-III) for extending support for the financial management of the ICVMWT 2021.

Best wishes from

Prayagraj, Uttar Pradesh, India — Dr. Brijesh Mishra
Jaipur, India — Prof. Manish Tiwari

Acknowledgements

The editors wish to extend heartfelt acknowledgement to Prof. J. P. Pandey, the Hon'ble Vice-Chancellor, MMM University of Technology, for his motivation and constant encouragement for the organization of ICVMWT 2021 in the university campus. We are also thankful to Prof. R. K. Chauhan, HoD, ECED, MMM University of Technology, for his constant support during the different stages of this big event. We must thank Dr. Dharmendra Kumar and Dr. B. P. Pandey for their initiative and efforts for the successful organization of this conference.

We also pay our best regards to the faculty members from institutional partners (J. K. Institute of Applied Physics and Technology University of Allahabad and Manipal University Jaipur) for extending their enormous assistance during the conference-related assignments, especially to Prof. Rajeev Singh, Dr. Sudhanshu Kumar Jha, and Dr. Shudhakar Singh, University of Allahabad; Dr. Dinesh Yadav and Dr. Tarun Kr. Dubey, Manipal University Jaipur.

The editors are heartily thankful to contributing authors, esteemed reviewers for their timely response, members of various organizing committees, and production staff whose diligent work put shape to the ICVMWT 2021 proceedings. We especially thank our dedicated reviewers for their volunteering efforts to check the manuscript thoroughly to maintain the technical quality and for useful suggestions.

We also acknowledge the financial support received from our esteemed sponsors Technical Education Quality Improvement Programme of Government of India (TEQIP-III).

Most sincerely, we extend our gratitude to Springer for agreeing to be our publishing partner.

Lastly, we are heartily thankful to our student volunteers like Vijay, Gyan, Anuj Shukla, Shobhit gupta, Abhishek Singh, Swarnim Dixit, Shivani Singh, Arpan Bansal, Sarthak Srivastava, Priyanshi chauharwal, Minal Mishra, Prianshu parasar, and Deepanshu katiyar who have continuously involved from day one and did very great job; especially, they managed all the virtual mode sessions in a very efficient manner.

Dr. Brijesh Mishra
Prof. Manish Tiwari

Our Reviewers

1. Dr. Hemant Kumar, Jaypee Institute of Information Technology, Noida, hkumarbhatt@gmail.com
2. Dr. Bajrang Bansal, Jaypee Institute of Information Technology, Noida, bajrangbnsl@gmail.com
3. Dr. Kuldeep Bandera, Jaypee Institute of Information Technology, Noida, kuldeep_gec2000@yahoo.co.in
4. Dr. Yogesh Kumar, Jaypee Institute of Information Technology, Noida, bhyogesh@gmail.com
5. Dr. Neeraj Kumar Mishra, Bharat Institute of Engineering Technology, Hyderabad, neeraj.mishra3@gmail.com
6. Dr. Abhishek Tripathi, NIT Raipur, abhi.sanutripathi@gmail.com
7. Dr. Sunil pandey, Intel Bangalore, sunilpandeyiert@gmail.com
8. Dr. Nilesh Anand Srivastava, MITS Andhra Pradesh, nilesh.iert@gmail.com
9. Dr. Yatindra Gaurav, IERT Prayagraj, UP, ygaurav2000@gmail.com
10. Dr. Arvind Pandey, BIT Gorakhpur, arvindmknk@gmail.com
11. Dr. Saiyed Salim Sayeed, Dr. B. R. Ambedkar Institute of Technology, Port Blair saiyedsalimsayeed@gmail.com
12. Dr. Akhilesh Kumar Pandey, IERT, akhileshjkit@gmail.com
13. Dr. Vivek Singh, Department of Electronics and Communication, SIET, Prayagraj, UP vivek.10singh@gmail.com
14. Dr. Satyendra Kumar Mourya, Birla Institute of Technology and Science Pilani, Pilani Campus, Rajasthan, India, satyendra.mourya@pilani.bits-pilani.ac.in
15. Dr Vanya Arun, IILM Greater Noida, vanyaarun@gmail.com
16. Dr. Gaurav Kumar Sharma, GLA University, Mathura, gauravkr.sharma@gla.ac.in
17. Dr. Arun Kishor Johar, MNIT Jaipur, Johar.arun@gmail.com
18. Dr. Ashish Ranjan, National Institute of Technology Manipur, ashish.ism@rediffmail.com
19. Dr. Bhanu Pratap Singh, Lakshmi Narain College of Technology, Bhopal, bhanupprataps@lnct.ac.in

20. Dr. Mayur Agarwal, College of Technology, G. B. Pant University of Agriculture and Technology Pantnagar, agrmayur@gmail.com
21. Dr. Nikhil Deep Gupta, Centre for VLSI and Nanotechnology, VNIT Nagpur, nikhildeepgupta@cvn.vnit.ac.in
22. Dr. Amitesh Kumar, IIT(ISM), Dhanbad, amitesh.ism@gmail.com
23. Dr. M. T. Beig, Shree Guru Gobind Singh Tricentenary University, mirzatanweer@gmail.com
24. Dr. Anurag Upadhyay, Rajkiya Engineering College Azamgarh, uanurag89@gmail.com
25. Dr. Ashish Kumar, JECRC Engineering College And Research Centre, Jaipur Ashishkumar.ece@jecrc.ac.in
26. Dr. Md Azharuddin, Aliah University, Kolkata, WB, md.azharuddin@aliah.ac.in
27. Dr. Mohd Naseem, Baba Ghulam Shah Badshah University, Rajouri, J&K, mohdnaseem@bgsbu.ac.in
28. Dr. Ajay Kumar, National Institute of Technology Jamshedpur, ajay.ece@nitjsr.ac.in
29. Dr. Chandra Mohan Singh Negi, Banasthali Vidyapith, echandru2001@yahoo.com
30. Dr. Bal Chand Nagar, National Institute of Technology Patna, balchandnagar@nitp.ac.in
31. Dr. Anita Garhwal, Ton Duc Thang University, HCMC, Vietnam, anitagarhwal@gmail.com
32. Dr. Aanjey Mani Tripathi, Meerut Institute of Engineering and Technology, Meerut, aanjey.mani@miet.ac.in
33. Dr. Sumit Saha, NIT Rourkela, sahas@nitrkl.ac.in
34. Dr. Gaurav Baranwal, Department of Computer Science, Banaras Hindu University, Varanasi, gaurav.vag@gmail.com
35. Dr. Rajeev Arya, NIT Patna, rajeev.arya@nitp.ac.in
36. Dr. Naveen Mishra, Vellore Institute of Technology, Vellore, naveenmishra.ece@gmail.com
37. Dr. Nikhil Sharma, LNMIIT, Jaipur, nikhil.sharma@lnmiit.ac.in
38. Dr. Ashutosh Kumar Singh, I.E.T., Dr. Ram Manohar Lohia Avadh University, Ayodhya, aksinghelectronics@gmail.com
39. Dr. Manisha Bhatt, Govind Ballabh Pant Institute of Engineering and Technology, bhatt.manisha1@gmail.com
40. Dr. Mohammad Aneesh, Veer Bahadur Singh Purvanchal University Jaunpur, Aneeshau14@gmail.com
41. Dr. Shashwat Pathak, MIET Meerut, pathak.shashwat@gmail.com
42. Dr. Narendra Yadava, BIT Gorakhpur, narendrayadava5@gmail.com
43. Dr. Sanjay Kumar Biswash, NIIT University, skbismu@gmail.com
44. Dr. Sneha Sharma, Dayananda Sagar University, Bangalore, sneha.ece@gmail.com
45. Dr. Anurag Mishra, Iswar Saran PG College, University of Allahabad, Prayagraj, mishraanurag31@gmail.com

46. Dr. Mohd Gulman Siddiqui, Banasthali Vidyapith Rajasthan, mohdgulman@gmail.com
47. Dr. Bhubon Chandra Mech, IIT Gandhinagar, bhubon1126@gmail.com
48. Dr. Nishit Malviya, IIIT Ranchi, nishit.malviya@gmail.com
49. Dr. Anand Sharma, MNNITA, Prayagraj, UP, anandsharma2312@gmail.com
50. Dr. Ashok Kumar, NIT Srinagar, ashok1987.in@gmail.com
51. Dr. Mourina Ghosh, Indian Institute of Information Technology, Guwahati, mourina@iiitg.ac.in
52. Dr. Pratosh Kumar Pal, MMMUT, Gorakhpur, UP, pkpec@mmmut.ac.in
53. Dr. Chandra Bhan Jha, SIET, Prayagraj, UP, cbhan.1965@gmail.com
54. Dr. Subodh Wariya, IET, Lucknow, UP, swairya@gmail.com
55. Dr. Vimal Kumar Mishra, JUIT, Noida UP, vimalk.mishra@jiit.ac.in
56. Dr. Satish Chandra, MMMUT, Gorakhpur, UP, scece@mmmut.ac.in

Contents

Compact Circular Polarized Square Slot Antenna for ISM Band Applications 1
Sunidhi Dubey, Saurabh Singh, Sudhanshu Verma, and Abhay Krishna Yadav

Miniaturization Capability of the Pixel Antenna for Nanosatellite Footprints 9
Milind T. Themalil, Satya Singh, Divanshu Jain, Dinesh Yadav, Mohamad Rammal, and Mayank Sharma

Wideband Antenna Design for 5G Communications 17
Amruta S. Dixit and Sumit Kumar

SIW Slot Antenna Array for 5G Applications 25
Sayyed Arif Ali, Mohd Wajid, and Muhammad Shah Alam

Compact Multiple Input Multiple Output Antenna with High Isolation for UWB Applications 37
Pankaj Kumar Keshri, Sanjay Sahu, and Richa Chandel

Survey on Services and Support in Cloud Computing Environment 45
Abhishek Kumar Pandey and Sarvpal Singh

Human Authentication Using Score Level Fusion of Face and Palm Print Biometrics 55
Rohit Srivastava and Dhirendra Kumar Sharma

Design of 16-Bit Vedic Multiplier Using Modified Logic Gates and BEC Technique 65
K. Rupa Lakshmi, Kandula Bala Sindhuri, K. Mani Kumar, and N. Udaya Kumar

A Tunable Resistorless Floating Inductance Simulator Using MO-DXCCTA 73
Yumnam Shantikumar Singh, Ashish Ranjan, Shuma Adhikari, and Benjamin A. Shimray

Design and Implementation of All Optical Processing Units Together Performing Arithmetic and Logical Functions 83
Vanya Arun, Kapil Deo Bodha, Awadhesh K. Maurya, and Ashutosh K. Singh

Design of Fruit-V2 and Fruit-80 Stream Ciphers Using Basic Version 95
Vamsi Krishna Guntapudi and Dheeraj Kumar Sharma

Software Implementation of Plantlet Stream Cipher Using Verilog Hardware Description Language 107
Goundla Sandhya and Dheeraj Kumar Sharma

An Implementation of Tunable All-Pass Filter Employing CCDDCC 119
Rajkumar Mumusana, Yumnam Shantikumar Singh, and Ashish Ranjan

A Cost-Efficient QCA RAM Cell for Nanotechnology Applications 127
Divya Tripathi and Subodh Wairya

An Impact of Efficient Backoff Algorithm in MANETs 139
Dhirendra Kumar Sharma and Rohit Srivastava

Heterogeneous Energy-Efficient Clustering Protocol for Wireless Sensor Networks 149
Sandip K. Chaurasiya, Arindam Biswas, and P. K. Bandyopadhyay

High-performance 3–2 Compressor Using Efficient XOR-XNOR in Nanotechnology 159
Anum Khan and Subodh Wairya

Area-Efficient and Fast Computing Chebyshev Type-I Filter Design 171
Ajay Kumar Yadav and Sangeeta Singh

Analog and Digital Applications of 4-T Based Memristor Emulator 183
Semba Walli, Jyoti Garg, and Subodh Wairya

Perfomance Analysis of Text Extraction from Complex Degraded Image Using Fusion of DNN, Steganography, and AGSO 195
Digvijay Pandey and Subodh Wairya

Simulation-Based Analysis of AlGaN/GaN Gate All Around Field Effect Transistor (AlGaN/GaN GAA-FET) 205
Yogesh Kumar Verma, Varun Mishra, Rohit Gurjar, Rajeev Kumar Chauhan, and Santosh Kumar Gupta

Electrical and RF Performance Enhancement of Pocket Doped Junctionless Transistor Using Dual Metal Gate 215
Priyansh Tripathi, Narendra Yadava, and R. K. Chauhan

Improvement of Leakage Current in Double Pocket FDSOI 22 nm Transistor Using Gate Metal Arrangement 227
Aryan Kannaujiya, Narendra Yadava, Mangal Deep Gupta, and Rajeev Kumar Chauhan

Beat Frequency Detection on Boolean Chaotic Oscillator for True Random Number Generation on FPGA 237
R. Sivaraman, N. Adithya, Siva Janakiraman, and Amirtharajan Rengarajan

Femtocell-Based Interference Reduction Methodology for Self-organized Networks: A Green Network Prospective 247
Arshita Batra, Simran Wasnik, and Sanjay Kumar Biswash

Heterodielectric-Based Gate Oxide Stack Engineering in FDSOI Structure with Enhanced Analog Performance 253
Aditya Kushwaha, Narendra Yadava, Mangal Deep Gupta, and R. K. Chauhan

Numerical Study of Various ETL Materials for an Efficient Lead-Free Perovskite Solar Cell 265
Km Shivani Bhardwaj, Shambhavi Rai, Vandita Rao, Pooja Lohia, and D. K. Dwivedi

High Sensitive Surface Refractive Index Sensor Using Multiple Vertical Silicon Nanowire-Based PIC Waveguide 273
Rukmani Singh and Vishnu Priye

A Compact Slot and Notch Loaded Microstrip Antenna for Wireless Applications 283
Ramesh Kumar Verma and Akhilesh Kumar

Real-time Face mask Detection Using Deep Learning and MobileNet V2 297
Akshay Verma and Manish Kumar Srivastava

A Novel TD-UTD Coefficients for Evaluation of Diffraction and Transmission from Dielectric Wedges 307
Vinod Kumar, Sanjay Soni, and N. S. Raghava

Quad-Band MIMO Antenna for Radar and Satellite Applications 317
Amrees Pandey, Aditya Kumar Singh, Sweta Singh, Rajeev Singh, and Brijesh Mishra

Comparative Study Between Silver and Gold Metal Film-Based Surface Plasmon Resonance Biosensor with Platinum Diselenide 329
Rashmi Tiwari, Sachin Singh, Pooja Lohia, and D. K. Dwivedi

Mobile Application on Drowsiness Detection When Driving Car 337
Akshat Singhal and Sunil Kumar

Recent Advancement in Long-Period Fiber Grating (LPFG) 347
Pragya Mishra, Sachin Singh, Pooja Lohia, and D. K. Dwivedi

Novel Tri-band Microstrip Bandpass Filter with Stub Loaded in Circular Ring Resonator 357
Prashant Pandey, Arvind Kumar Pandey, and R. K. Chauhan

Automatic Detection of Counterfeit and Genuine News Article 367
Vipula Yadav and Manish Kumar Srivastava

Twitter Sentiment Analysis Using Machine Learning 379
Ayesha Quazi and Manish Kumar Srivastava

Specific Pose Detection Through Efficient Neural Network 391
Harsh Malik, Mansi Malik, Siddhant Malik, Vimal Kumar, and Mukesh Rawat

Tri-forked (Trishul Shape) Triple-band MIMO Antenna for Satellite Applications 401
Aditya Kumar Singh, Amrees Pandey, Sweta Singh, and Rajeev Singh

Effect of Packet Size on the Performance of Wireless Ad Hoc Network with Various Data Rates 411
Alok Singh, Saurabh Sharma, and Pramod Narayan Tripathi

Performance of Wireless Ad Hoc Network with All the Data Rate of IEEE 802.11b Using NS-3 421
Saurabh Sharma, Alok Singh, and Pramod Narayan Tripathi

Impact of Parasitic Component on Bandwidth of Printed Dipole Antenna 431
Harshit Kumar Dhameniya, Ramesh Kumar Verma, Dheeraj Tripathi, and Shadab Azam Siddique

Charge Selective Layer Optimization of a Double Perovskite Solar Cell by Numerical Simulation 445
Shambhavi Rai, B. K. Pandey, and D. K. Dwivedi

An Improved Pseudo-Domino Technique for Low-Power Applications 455
Shivangi Pandey and Manish Kumar

Memristor Emulator Circuits an Emerging Technology with Applications ... 467
Jyoti Garg, Aishita Verma, and Subodh Wairya

Sensitivity Enhancement of Graphene and Blue Phosphorene/Mos_2 Heterostructure-Based SPR Biosensor Using Gold (Au) Metal Layer: Theoretical Insight ... 481
Sachin Singh, Pravin Kumar Singh, A. K. Sharma, Pooja Lohia, and D. K. Dwivedi

Recent Development of Hardware-Based Random Number Generators on FPGA for Cryptography ... 489
Mangal Deep Gupta and R. K. Chauhan

Ergodic Secrecy Capacity Analysis Over Composite Weibull/Inverse Gamma Fading Channel ... 501
Vipin Kumar Upaddhyay, Puspraj Singh Chauhan, and Sanjay Kumar Soni

IoT-Based Efficient Parking System Using IntelliP ... 509
Dheeraj Tiwari and Rajan Mishra

Study of Area-delay and Energy Efficient Multi-operand Binary Tree Adder ... 519
Km Jyoti Singh and Manish Kumar

Source/Drain Engineered Silicon-on-Insulator Transistor with Improved Analog Performance ... 531
Shailesh Shankar Upadhyay, Narendra Yadava, Mangal Deep Gupta, and R. K. Chauhan

Design of Partial Product Generator Circuit for Approximate Radix-8 Booth Multiplier with Lower Delay ... 543
Pooja Singh and Manish Kumar

Power Optimisation of UAV-Based Relay for Selective Decoding and Forward Protocol Co-operation in Wireless Communication ... 553
Deepak Kumar, Satish Chandra, and Seemanti Saha

AUC Analysis for Generalised Bessel K Fading Model ... 563
Manisha Bhatt, Sanjay Kumar Soni, and Sandeep Bhatt

Analysis of Erroneous SIC and CSI on the Outage Performance of Cooperative NOMA Systems ... 569
P. Bachan, Aasheesh Shukla, and Atul Bansal

Study the Impact of ZrO_2 High-k Dielectrics Gate Material on FD-SOI and PD-SOI MOSFET ... 579
Astha Srivastava, Narendra Yadava, Mangal Deep Gupta, and R. K. Chauhan

Permanent Magnet Synchronous Motor (PMSM) Drive Using Multi-Objective Genetic Algorithm (MOGA) Technique 587
Deepti Yadav, Arunima Verma, and Frank Tittel

Study of Structural and Electronic Properties of 2D WSe_2 Monolayer 599
Arunima Singh and Bramha P. Pandey

Four-Port Compact MIMO Antenna for 5.8 GHz Applications 605
Akhilesh Kumar, Shadab Azam Siddique, Sanjay Kumar Soni, Kritika Singh, and Brijesh Mishra

Energy Band Gap Tuning Using 633 nm Laser Exposed on $(GeS_2)_{85}(Sb_2S_3)_{15}$ Thin Films: An Optical Analysis Using Spectroscopic Techniques 617
Rajnish Raj, Pooja Lohia, and D. K. Dwivedi

A Compact Circular Ultra-Wideband MIMO Antenna Sensor (CUMAS) Probe for Breast Cancer Detection 625
Ankit Kumar Gupta, Praveen Kumar Rao, and Rajan Mishra

A Novel Framework for Recommendation of Various Communication Networks Using Multi-criteria 635
Pradeep Kr. Tiwari, Animesh Tripathi, Gaurav Srivastava, Shiv Prakash, and Narendra Kr. Shukla

CDBA-Based MOS-C Quadrature Oscillator 647
Shekhar Suman Borah and Mourina Ghosh

Impact of Mixed Interlayer Thickness on Performance of Organic Light-Emitting Diodes 655
Pooja Maurya and Brijesh Kumar

Physical Properties Resemblance of Optical Material $ZnGeN_2$ with GaN Under Different Higher Pressures 665
Satish Chandra, Deepak Kumar, Rukmani Singh, Ritesh Kumar, and Virendra Kumar

Photonic Generation of Arbitrary Microwave Signal Based on Tunable Optoelectronic Oscillator 675
Shivangi Yadav, Ritesh Kumar, and Pooja Lohia

Triple Band Semi-Fractal-Based Antenna for Wireless Applications in S, C and X-Bands 685
Vinay Singh, Brijesh Kumar Maurya, Praveen Kumar Rao, and Vipin Kumar Upaddhyay

Delay and Dispersion Investigation of Optical Components for Microwave Photonic Filter 693
Ritesh Kumar, Yadvendra Singh, Sanjeev Kumar Raghuwanshi, Satish Chandra, and Danish Nadeem

Enhanced Performance of $MoSe_2$-CIGS Solar Cell with ZnSe Buffer Layer 705
Jyoti Singh, Uzair Alam, and Anupam Sahu

Study and Implementation of Smart Water Supply Management Model for Water Drain Region in India 711
Prashant Pandey, Ashish Ranjan Mishra, P. K. Verma, and Ravi P. Tripathi

Improvement of Electrical Characteristics for Nanoscale Single-Gate FDSOI Using Gate Oxide Engineering 723
Aditya Kushwaha, Shailesh Shankar Upadhyay, Narendra Yadava, Mangal Deep Gupta, and R. K. Chauhan

Improved Digital Performance of Modified Source-Drain FDSOI by Optimization of Buried Oxide Properties 735
Sudhir Rai, Mangal Deep Gupta, Saurabh Kumar, and R. K. Chauhan

Silicon on Insulator-Based Ultra-Small Micro-Ring Resonator for Temperature Sensing 747
Rukmani Singh, Rupam Srivastava, Y. K. Prajapati, and Dharmendra Kumar

Study of Sum-Rate Performance of Cooperative Cognitive Radio Networks Using MIMO 757
Champa Tanga, Joyatri Bora, Md. Anwar Hussain, and Karik Lego

Novel Power Gated (PG) and Sleep Body Bias (SBB) 6T CNTFET-Based SRAM Design for Ultra-Low-Power Application 767
Hemant Kumar, Balwinder Singh, Subodh Srivastava, Gulman Siddiqui, and Shekhar Yadav

Tuning of White-Colour Gamut Using Strain Adapting Interfacial Layer in 'Giant' Colloidal Core–shell Quantum Dot 783
Anupam Sahu and Dharmendra Kumar

The Effect of Finite Ground Plane on Wideband Slim Rectangular Dielectric Resonator Antenna 791
Sovan Mohanty and Baibaswata Mohapatra

Design and Computational Simulation of CZTS-Based Solar Cell Using SCAPS-1-D Software 801
Avijeet Rai and B. K. Panday

About the Editors

Dr. Brijesh Mishra received his M. Tech. (Electronics Engineering) and Ph.D. (RF & Microwaves) degrees from the University of Allahabad in 2012 and 2018, respectively. Dr. Brijesh Mishra worked as an Assistant Professor in the Department of Electronics and Communication Engineering at Shambhunath Institute of Engineering and Technology (SIET) during 2012–2013 and again from 2017–2018, and as an Assistant Professor (NPIU-MHRD) in the Department of Electronics and Communication Engineering at Madan Mohan Malaviya University of Technology during 2018–2021. He is currently working as an Associate Professor in the Department of Electronics and Communication Engineering at the Shambhunath Institute of Engineering and Technology (SIET). He has published 01 patent, and more than 45 research papers in journals of international repute, international conferences, and book chapters. Dr. Brijesh Mishra has completed two projects funded by the NPIU & the World Bank. He is the recipient of awards like Excellence in Performance and Outstanding Contributions. He has been a potential reviewer of more than 20 SCI & Scopus indexed journals, organizing track chair, and organizing secretary for several conferences. He is a member of IEEE, ISTE, IE(I), IETE, IAENG and IFERP. His research interests include modeling, simulation, and fabrication of RF and microwave devices and their applications.

Manish Tiwari received his Ph.D. in Electronics and Communication Engineering (ECE) in Photonics from Malaviya National Institute of Technology (MNIT), Jaipur. He is currently Professor and Head of the ECE department at Manipal University, Jaipur. He was Visiting Researcher at City University, London, in 2010 and 2011, and Tsinghua University, Beijing, China, in 2016. His current research interests include micro/nano-structured photonic devices, fiber optics, numerical modeling, nonlinear optics, and photonic crystal fibers. He has published over 100 research papers in reputed journals and conferences. He has also authored/edited over 10 books and got 15 book chapters published to his credit.

Compact Circular Polarized Square Slot Antenna for ISM Band Applications

Sunidhi Dubey, Saurabh Singh, Sudhanshu Verma, and Abhay Krishna Yadav

Abstract The presented paper aims at a compact circularly polarized (CP) square slot antenna (CCPSSA) for ISM band applications. The proposed antenna design is composed of *L*-shaped patch and wide square slot with asymmetric rectangular patches, excited by 50 Ω micro-strip line feed. Two of the rectangular patches are present at the opposite corners on the ground that are responsible for generating two orthogonal modes, and other strips are located at the top and side of *L*-shaped patch to enhance the axial ratio bandwidth (ARBW). The proposed CCPSSA achieved 10 dB return loss of 66.9% (2.05–4.11 GHz), and the axial ratio (AR) < 3-dB bandwidth is 39.18% (2.36–3.51 GHz). The proposed CCPSSA radiates with bidirectional radiation pattern (left-hand circular polarization (LHCP) in negative z-direction and right-hand circular polarization (RHCP) in positive z-direction) at 2.45 GHz.

Keywords Circularly polarized · Slot antenna · Axial ratio · ISM band

1 Introduction

In recent times, printed slot antennas are highly concerned because of desirable properties of less bulk, light profile, low cost with greater flexibility, wide bandwidth, and easy fabricating process used for circular polarization antennas in modern communication systems, such as RFID, WLAN, and GPS [1]. An antenna is said to be circularly polarized when the plane of polarization rotates in a helical pattern forming a complete revolution in each wavelength. The circular polarization is termed as left hand when the rotation is clockwise (LHCP), and it is termed as right hand when it rotates anti-clockwise (RHCP). The CP signals are good at penetration and bending around obstacles and are more repellent to signal distortion due to cold weather conditions for reflection, absorption, multi-path, and LOS, and also, it is much effective to establish and maintain communication links [2].

S. Dubey (✉) · S. Singh · S. Verma · A. K. Yadav
Department of Electronics and Communication, Madan Mohan Malaviya University of Technology, Gorakhpur, UP, India
e-mail: dubeysunidhi1211@gmail.com

B. Mishra and M. Tiwari (eds.), *VLSI, Microwave and Wireless Technologies*, Lecture Notes in Electrical Engineering 877,
https://doi.org/10.1007/978-981-19-0312-0_1

There are various circularly polarized printed antennas that are reported in literature [3–10] such as in [3] a coupled cylindrical DRA of dimension 60 × 60 × 1.6mm^3 fed by an off-centered micro-strip line feed which is responsible for CP generation. The DRA is radiated with 10-dB return loss bandwidth (BW) of 6.2% (2359–2510 MHz) and 3-dB ARBW of 3.48% (2399–2484 MHz). In [4], the CP is obtained by matching two perpendicular side of square ring having a chain of micro-strip-line-feed configuration. The size of antenna is 54 × 54 × 1.6 mm^3 which radiates with ARBW greater than 6%. A metal strip of T-shape, ground protruded, is operated to attain circular polarized radiation [5]. The size of an antenna is 70 × 70 × 1.6 mm^3 radiated with axial ratio BW that can extend up to 18%. In [6], a new design of stair-shaped slot antenna of 60 × 40 × 0.8 mm^3 is introduced fed by CPW. By varying, the dimension of the slot CP is generated and wide band performance of CP is obtained. The 3-dB AR bandwidth is of 31.2% (2.30–3.15 GHz). In [1], a micro-strip line feed with L-shaped radiation patch square slot and dual rectangular strip antenna of dimension 60 × 60 × 1 mm^3 is introduced. On tuning the rectangular strips, a dual-band CP radiation is attained. The AR bandwidths obtained are 37.4% (2.50–3.65 GHz) and 16.3% (5.02–5.91 GHz). Dual flipped-L strip about two opposite corners of overall dimension 25 × 25 × 0.8 mm^3 is aimed to obtain circular polarized radiation [11] fed by (CPW) co-planar waveguide. A broadband axial ratio (AR) bandwidth > 38.2% is achieved. To attain a CP wave in [12], S-shaped slot antenna of 54 × 54 × 1 mm^3 is introduced in the ground for RHCP radiation. The reflection coefficient obtained is 104% (1.78–5.64 GHz), and a 3-dB AR bandwidth is 58.6% (2.85–5.21 GHz). The overall literature survey depicts that the antennas are large in size, narrow bandwidth, low ARBW, and low gain.

In this paper, a compact circularly polarized square slot antenna is presented, which consists of an *L*-shaped radiating patch with wide square slot having unequal rectangular patches at the opposite corners excited by a 50 Ω micro-strip line feed. Two of the rectangular patches are present at the opposite corners on the ground that are responsible for generating two orthogonal modes, and other strips are located at the top and side of *L*-shaped patch which are used to enhance the axial ratio. The designing and simulation of proposed antenna are done on HFSS software.

2 Antenna Design

The designed geometry of CCPSSA is illustrated in Fig. 1. FR4 is used as substrate of dielectric constant 4.4, loss tangent of 0.02 and 0.8 mm thickness to print the antenna. This antenna is composed of *L*-shaped radiation patch fed by 50 Ω micro-strip line feed, a wide square slot antenna having asymmetric rectangular patches to realize the circularly polarized radiation. The compact size of antenna is 40 × 40 × 0.8 mm^3. The micro-strip line feed ($L_1 \times W_1$) is punched on the top layer of the substrate, i.e., FR4. The *x*-axis strip of L-shaped patch consists ($K_1 \times L_2$), and the *y*-axis strip is ($K_2 \times L_3$). The ground plane is imprinted on the other edge of antenna. The dual rectangular strips ($S_1 \times S_2$, $S_3 \times S_4$) are at the inverse corner of the ground, and

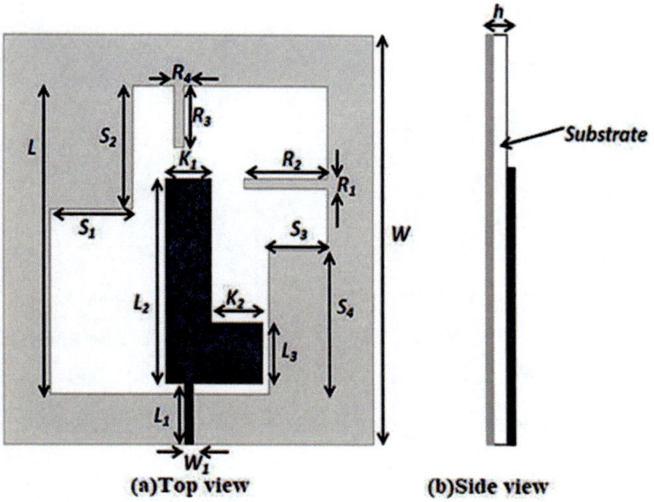

Fig. 1 Antenna Geometry of proposed CCPSSA: ($W = 40.0$, $L = 30.0$, $L_1 = 6.0$, $L_2 = 20.0$, $L_3 = 6.0$, $K_1 = 5.0$, $K_2 = 5.5$, $S_1 = 9.0$, $S_2 = 12.0$, $S_3 = 7.3$, $S_4 = 14.0$, $R_1 = 1.0$, $R_2 = 10.0$, $R_3 = 6.0$, $R_4 = 1.0$, $W_1 = 1.0$, $h = 1.0$, units: mm)

other two strips are ($R_1 \times R_2$, $R_3 \times R_4$) at the side and top of the patch connected with ground. In the designing of the wide square slot circular polarized antenna, four steps are depicted to explain the evolution steps in Fig. 2. Antenna 1 comprises only a straight micro-strip ($K_1 \times L_2$). Ant.2 adds another end ($K_2 \times L_3$) connected with the straight micro-strip ($K_1 \times L_2$). Ant. 3 comprises two rectangular strips ($S_1 \times S_2$, $S_3 \times S_4$) at the inverse corner of the ground. Ant. 4 comprises two additional strips, i.e., ($R_1 \times R_2$, $R_3 \times R_4$).

The graph of impedance bandwidths and Axial Ratio bandwidths of all the four antennas are illustrated in Fig. 3. The IBW of Ant.1 is bad as it does not achieve 10 dB return loss and 3-dB ARBW is not achieved. Further adding another strip (K_2

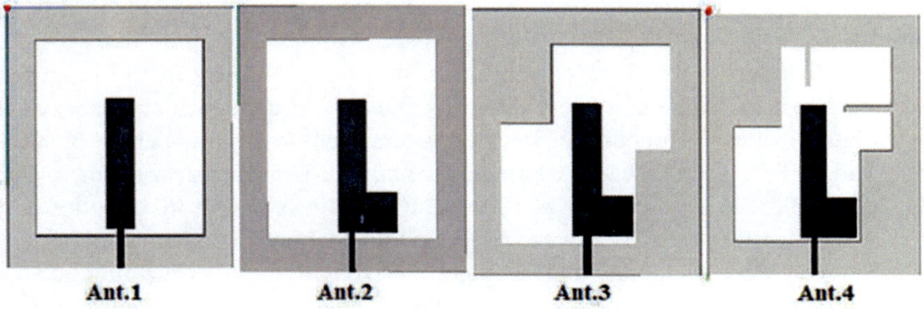

Fig. 2 Stages to realize the designed antenna (1–4)

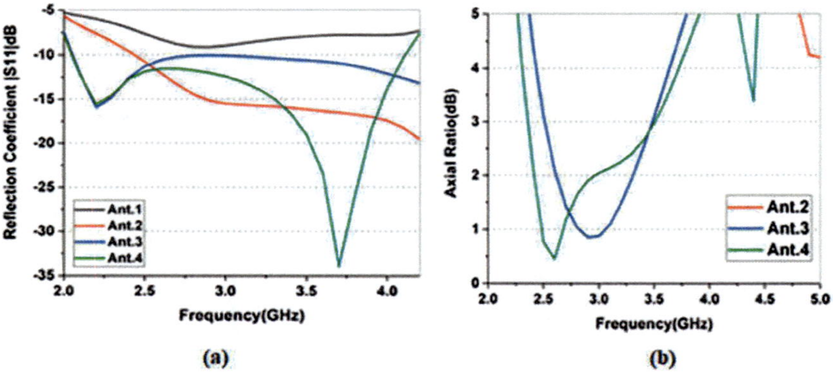

Fig. 3 Performances of simulated **a** reflection coefficient $|S_{11}|$ and **b** axial ratio of four antennas

$\times$ L_3) in Ant.2, the return loss is enhanced but ARBW is not attained. When the two rectangular strips ($S_1 \times S_2$, $S_3 \times S_4$) are added, it improves the IBW and ARBW in Ant.3. By adding ($R_1 \times R_2$, $R_3 \times R_4$) rectangular strips, it greatly improves the impedance bandwidth and increases the ARBW. Furthermore, a circular polarized antenna is achieved.

3 Results and Discussion

The simulated results of L-shaped patch square slot antenna are analyzed on HFSS software version 16.2, having solution frequency of 4 GHz. The antenna's performance having 10 dB return loss is depicted in Fig. 4 (reflection coefficient vs. frequency graph) which covers a wide frequency band. The simulated result is 66.9% (2.05–4.11 GHz). The resonance peak is obtained at 3.7 GHz which is near to 4 GHz which is desirable to solution frequency. Figure 5 shows the simulation result of axial ratio curve of the proposed antenna. The resultant 3-dB AR bandwidth is 39.18% (2.36–3.51 GHz) and covering 2.45 GHz that can be utilized for ISM band applications.

Figure 6 shows the flow of current distribution at 2.45 GHz of the proposed antenna. The distribution of current shows a centered frequency which is also used for radiation pattern. The current flows at different phase angles such as 0°, 90°, 180°, and 270°. It depicts that at 0° the flow of current is toward downward, at 90° flow of current is in $+x$-direction, at 180° the current flows in upward direction and at 270° the direction of current is $-x$-direction which shows the phenomena of CP radiation which is right hand circularly polarized (RHCP) at positive z-direction.

Figure 7 illustrates the simulated gain of the proposed antenna which shows that the peak gain of proposed CCPSSA is 4.3 dBi at 3.7 GHz. The simulated result of radiation patterns of the proposed antenna at 2.45 GHz is depicted in Fig. 8. The

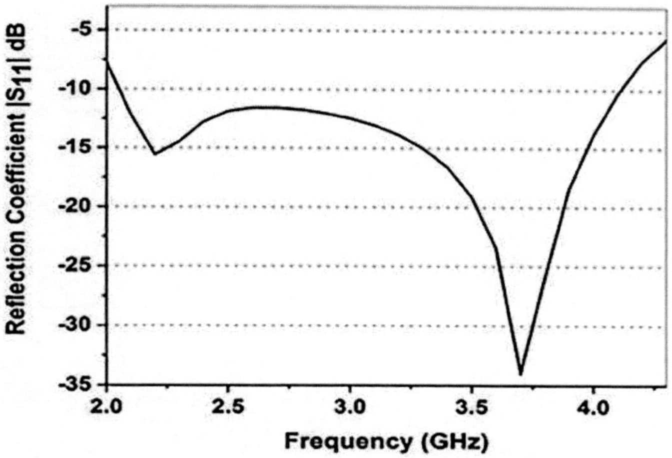

Fig. 4 Simulated reflection coefficient $|S_{11}|$ of proposed CCPSSA

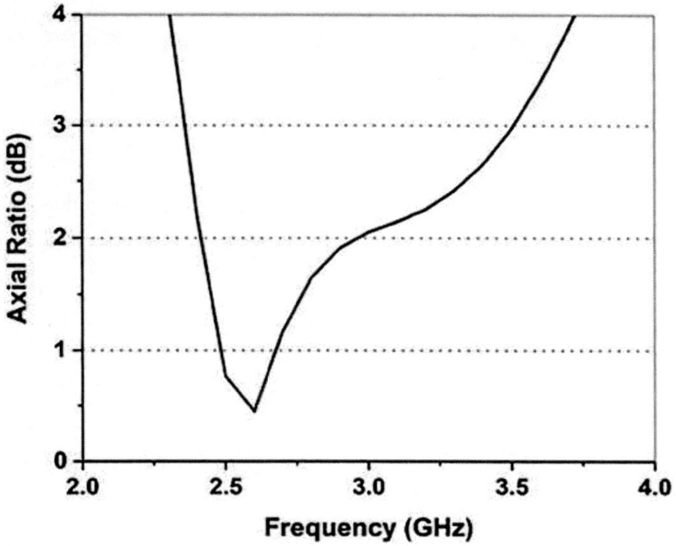

Fig. 5 Simulated axial ratio of proposed CCPSSA

radiation pattern is obtained in xoz and yoz planes and radiates LHCP in negative z-direction and RHCP in positive z-direction. It is bidirectional in nature which shows that it is circularly polarized.

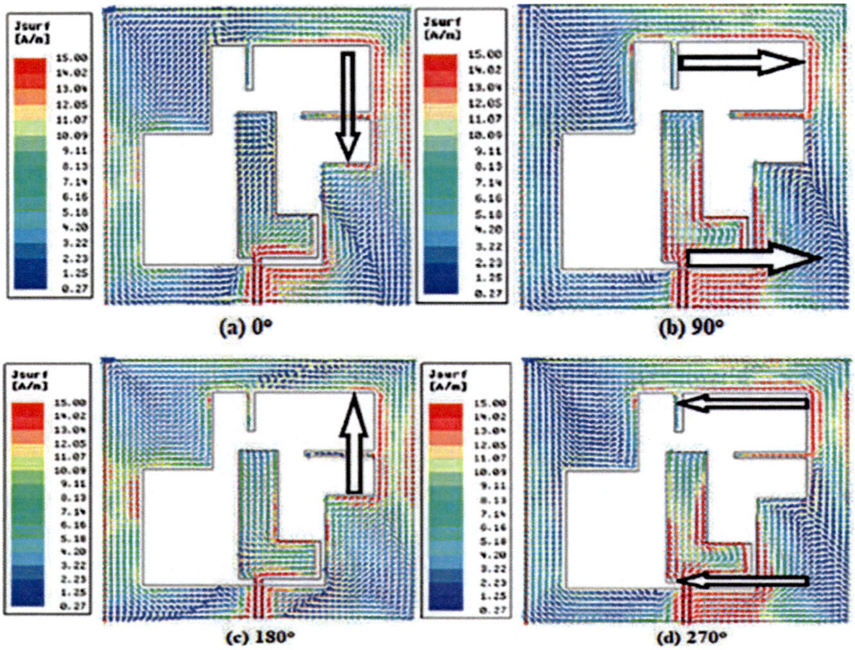

Fig. 6 Distribution of current in proposed CCPSSA at 2.45 GHz

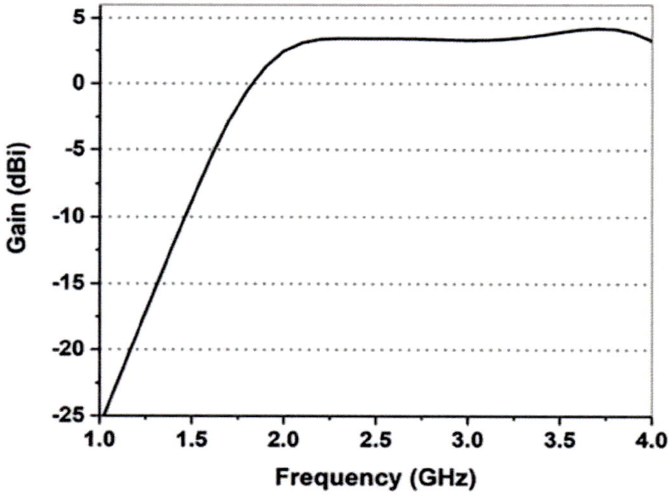

Fig. 7 Simulated antenna gain of proposed CCPSSA

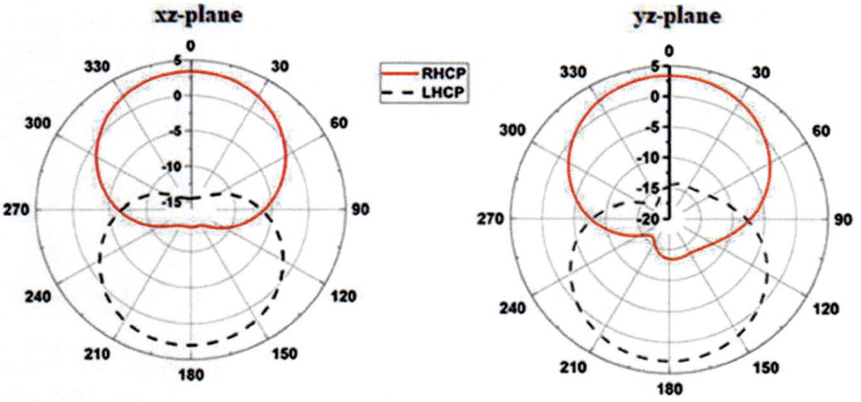

Fig. 8 Radiation patterns of proposed CCPSSA at 2.45 GHz

Table 1 Comparative study of proposed CCPSSA with reference antennas

References	IBW (GHz)	ARBW (GHz)
Zhao et al. [7]	7.95% (2.391–2.589)	0.33% (2.393–2.401)
Hao et al. [8]	17.4.0% (3.05–3.63)	5.4% (3.25–3.43)
Ahdi Rezaeieh et al. [9]	40% (4.8–7.2)	31.84% (5.15–7.1)
Mousavi et al. [10]	30% (1.41–1.91)	32% (1.43–1.98)
Proposed work	66.9% (2.05–4.11)	39.18% (2.36–3.51)

Table 1 shows the comparison of radiation characteristics of the designed antenna with reference antennas. In contrast, it is observed that the proposed antenna is having a miniaturized structure with wide impedance bandwidth and axial ratio bandwidth.

4 Conclusion

A compact antenna with simple structure of size $40 \times 40 \times 0.8$ mm^3 is presented in this paper. The proposed antenna design is composed of *L*-shaped patch and wide square slot with asymmetric rectangular patches, excited by 50 Ω micro-strip line feed. Two of the rectangular patches are present at the opposite corners on the ground that are responsible for generating two orthogonal modes, and other strips are located at the top and side of *L*-shaped patch to enhance the axial ratio bandwidth. The proposed CCPSSA achieved 10 dB return loss of 66.9% (2.05–4.11 GHz), and the axial ratio <3-dB bandwidth is 39.18% (2.36–3.51 GHz). The peak gain of proposed

CCPSSA is 4.3dBi at 3.7 GHz. The proposed CCPSSA radiates with bidirectional radiation pattern (LHCP in negative z-direction and RHCP in positive z-direction) at 2.45 GHz. The proposed CCPSSA is well suited for ISM band application for the circular polarized radiation.

References

1. Xu R, Li J, Wei K (2016) Dual-band dual-sense circularly polarized square slot antenna with simple structure. Electronics Lett 52(8):578–580
2. Gao S, Luo Q, Zhu F (2014) Antenna theory-analysis and design. John Wiley & Sons, Ltd
3. Lin C-C, Sun J-S (2013) Circularly-polarized dielectric resonator antenna fed by off-centered micro strip line for 2.4-GHz ISM band applications. IEEE Antennas Wireless Propag Lett 1536-1225 (c) 2013 IEEE
4. Row J-S (2005) The design of a squarer-ring slot antenna for circular polarization. IEEE Trans Antennas Propag 53(6):1967–1972
5. Sze J-Y, Wong K-L, Huang C-C (2003) Coplanar waveguide-fed square slot antenna for broadband circularly polarized radiation. IEEE Trans Antennas Propag 51(8):2141–2144
6. Wang C-J, Chen C-H (2009) CPW-fed stair-shaped slot antennas with circular polarization. IEEE Trans Antennas Propag 57(8)
7. Zhao Z, Liu F, Ren J, Liu Y, Yin Y (2020) Dual-sense circularly polarized antenna with a dual-coupled line. IEEE Antennas Wireless Propag Lett 1536-1225 (c) 2020 IEEE
8. Hao S, Chen Q, Li J, Xie J (2020) A high-gain circularly polarized slotted patch antenna. IEEE Antennas Wireless Propag Lett
9. Ahdi Rezaeieh S, Abbosh A, Antoniades MA (2013) Compact CPW-fed planar monopole antenna with wide circular polarization bandwidth. IEEE Antennas Wireless Propag Lett 12
10. Mousavi P, Miners B, Basir O (2010) Wideband L-shaped circular polarized monopole slot antenna. IEEE Antennas Wireless Propag Lett 9
11. Pourahmadazar J, Mohammadi S (2011) Compact circularly-polarised slot antenna for UWB applications. Electron Lett 47(15):837–838
12. Li G, Zhai H, Li T, Li L, Liang C (2013) CPW-fed s-shaped slot antenna for broadband circular polarization. IEEE Antennas Wireless Propag Lett 12

Miniaturization Capability of the Pixel Antenna for Nanosatellite Footprints

Milind T. Themalil, Satya Singh, Divanshu Jain, Dinesh Yadav, Mohamad Rammal, and Mayank Sharma

Abstract Present day dimension of satellites is constantly being scaled down; hence, there is a stronger need for the antennas placed on them to be highly miniaturized. The emergence of small satellites needs new antenna designs compatible with footprints scaled down highly comparative to that of traditional satellites for similar earth coverage operation. The phenomenon of transverse evanescent modes of the inner cavity field and their wide band capability (Monediere et al. in Int J Antennas Propagat 2014 [1]) allows the pixel antenna design to be a feasible approach for miniaturization on such scale. The antenna proposed in this paper is redesigned to be used as a lone pixel, but it can be arranged as an array in the agile reconfigurable radiating matrix format, hence being termed as pixel antennas (Jecko and Arnaud in FERMAT, 20, 2017 [2, 3]). The objectivity of this present work is to establish the miniaturization capability of this pixel antenna by designing a miniaturized planar antenna for a small satellite specification, implanted on a nanosatellite and to realize the communication uplink/downlink links in right hand circularly polarized in the upper UHF band.

M. T. Themalil (✉) · S. Singh · D. Jain
JK Lakshmipat University, Mahindra SEZ, Jaipur 302026, India
e-mail: milind.themalil@jklu.edu.in

S. Singh
e-mail: satyasingh@jklu.edu.in

D. Jain
e-mail: divanshujain@jklu.edu.in

D. Yadav
Manipal University Jaipur, Jaipur, Rajasthan, India
e-mail: dinesh.yadav@jaipur.manipal.edu

M. Rammal
ITHPP, 46500 Thegra, France

M. Sharma
Celestial Space Technologies GmbH, Amtsgericht Nürnberg, Nürnberg HRB 37759, Germany
e-mail: mayank@celestialspacetechnologies.com

B. Mishra and M. Tiwari (eds.), *VLSI, Microwave and Wireless Technologies*,
Lecture Notes in Electrical Engineering 877,
https://doi.org/10.1007/978-981-19-0312-0_2

Keywords Pixel antennas · Pre- and post-miniaturize · Low lateral profile · Ellipticity · Axial ratio · Realized gain

1 Introduction

Each unit of the small satellites, like CubeSats, is of the geometric volume of 10 cm^3; hence, surmounted antennas must be redesigned that can scale down to these small surface areas. The objective of this present work is to establish a new procedure to miniaturize from a traditional wide band antenna, implementing the design of the pixel antenna for installation on a small satellite (Fig. 1) according to the following specifications:

- UHF link: Downlink: 400 MHz
  Uplink: 465 MHz
- Return loss ≤ -10 dB
- Footprint: 20 cm × 20 cm
- Circular polarization; ellipticity ≈10%

2 Pre-miniaturized Pixel Antenna

The low-profile plane pixel antenna (shown in Fig. 2) has a profile λc/10 and transverse dimension of λc/2 × λc/2 where wavelength λc corresponds to the center frequency of the antenna. It exhibits circular polarization [4–6]. The ellipticity is obtained by exciting a patch (inside a cavity) (Fig. 2) fed by four ports [7].

The antenna exhibits bi-band behavior (Fig. 3), and its dimensions of 34 cm × 34 cm must be miniaturized to be compatible with the small satellite footprint, which is the aim of this paper.

As far as the characteristics of this antenna are concerned, the pixel exhibits a dual-band nature [8–11] in radiation characteristics (Fig. 4). The realized gain is obtained at the desired adapted uplink/downlink frequencies.

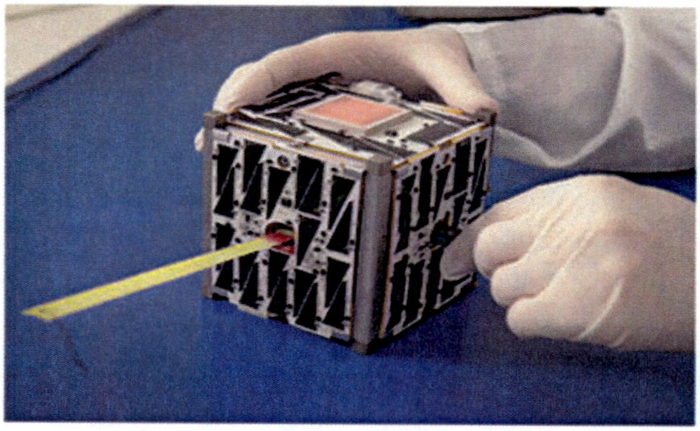

Fig. 1 A nanosatellite

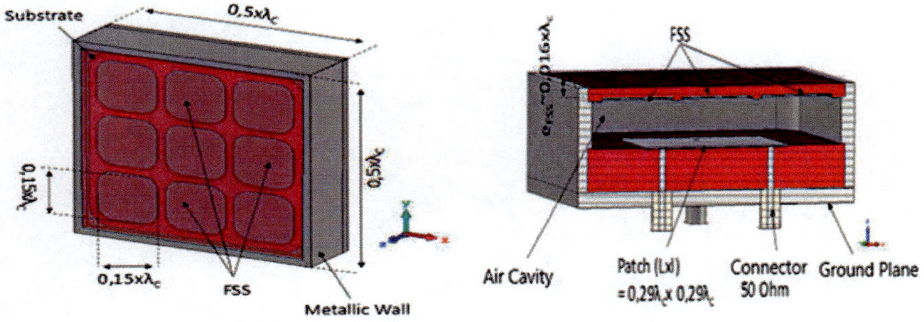

Fig. 2 Wide band pixel antenna excited by four ports

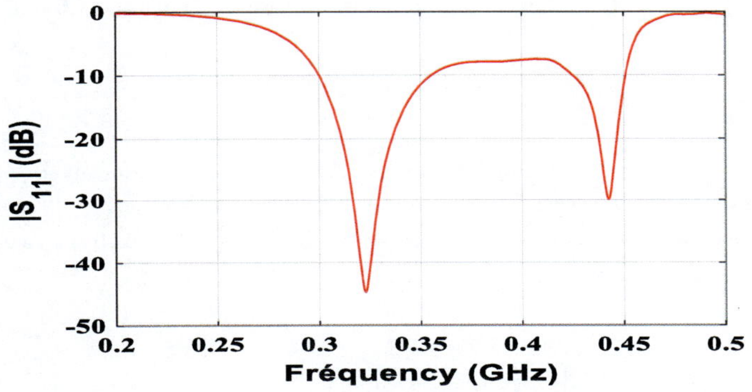

Fig. 3 Return loss of the pre-miniaturized pixel antenna

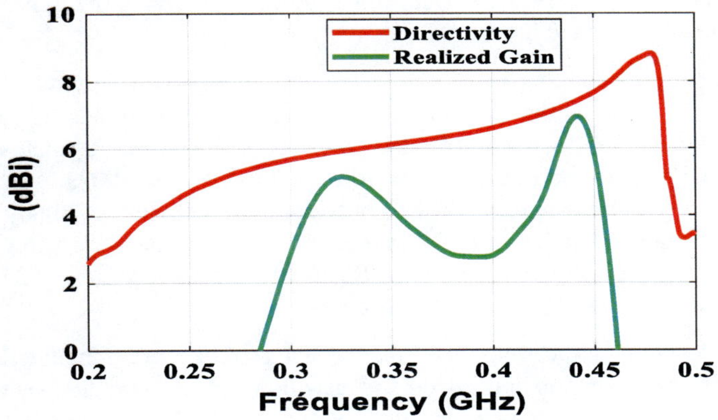

Fig. 4 Directivity and realized gain obtained against frequency

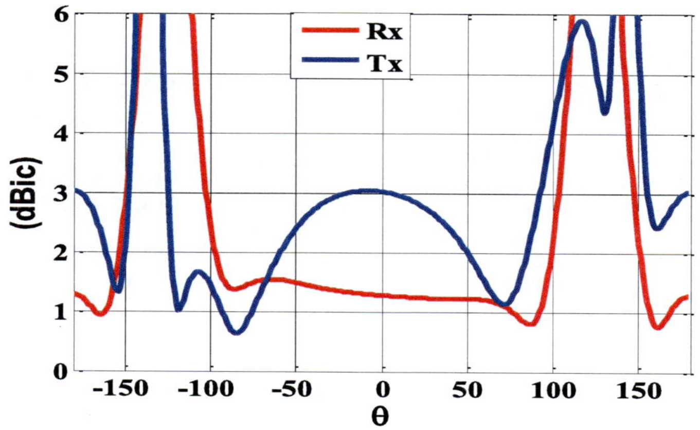

Fig. 5 Ellipticity evolution against elevation Theta

Figure 5 shows the rate of ellipticity for the uplink and downlink frequencies; axial ratio level lesser than 2.9 dB.

Ellipticity evolution is of the range of 1.55 dBiC in the axis for the downlink and 3.1 dBiC for the uplink.

3 Post-miniaturized Pixel Antenna

Few modifications in design are necessary before miniaturizing the dimensions of the antenna as mentioned below:

1. Replacing the exciting patch substrate with a higher permittivity material to secure an exciting element with dimensional geometry not exceeding that of the structure after it is miniaturized. Arlon is preferred as the patch substrate material having a permittivity $\varepsilon r = 10.2$ and loss tan $\delta = 0.0023$.
2. The periodic arrangement of the FSS on top of the pixel (Fig. 6) is highly reduced to keep enough number of FSS periodic pattern in the antenna structure having miniaturized dimensional geometry. Rogers 4003 substrate ($\varepsilon r = 3.37$ and tan $\delta = 0.0026$) is used for the FSS substrate.

These design modifications alter the behavior of the pre-miniaturized pixel antenna. The phenomenon in the pixel antenna allowing miniaturization is the presence of lateral evanescent transverse fields inside the cavity [2, 12] making it feasible to reduce the lateral dimensional geometry of the antenna without altering its functioning indiscriminately, in terms of directivities, intrinsic gains, realized gains, and elliptic axial ratios.

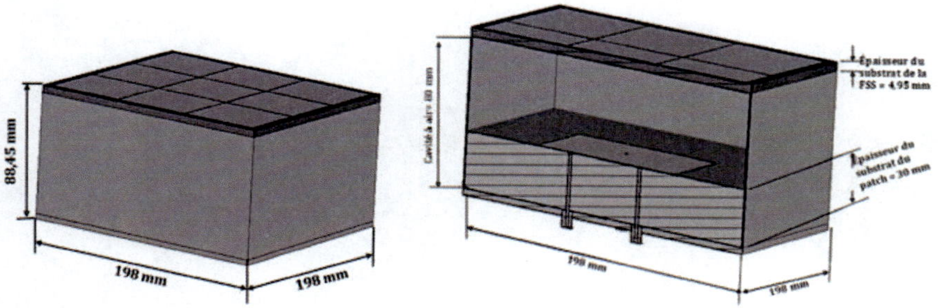

Fig. 6 Post-miniaturized pixel antenna

For small satellite applications, because the two downlink and uplink bands are in close proximity, the design solution employed for this current viability study is not two bands (since the impedance matched peak values are much separated in axis), but the first singular band matched on the first adapted value is optimized to provide a maximum realized gain for the downlink and uplink frequencies [13–15]. The miniaturized pixel antenna thus obtained is shown in Fig. 6

Figure 7 demonstrates that the new behavior of directivities and realized gains against bands give optimally similar result due to the low dielectric lossy tangents. The gains adapt at these curved graphs at the desired maximal impedance matching frequency nearer to the downlink frequency than to that of the uplink frequency since the minimal gains restricted by the small satellite required specifics must be prominent in those regions, respectively.

Fig. 7 Gains plotted against frequency for the post-miniaturized pixel antenna

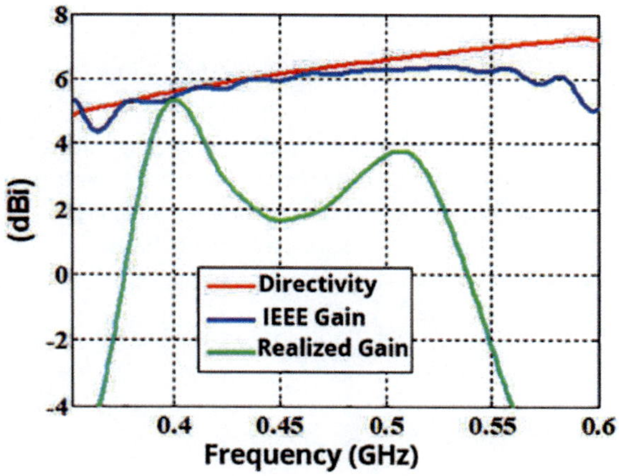

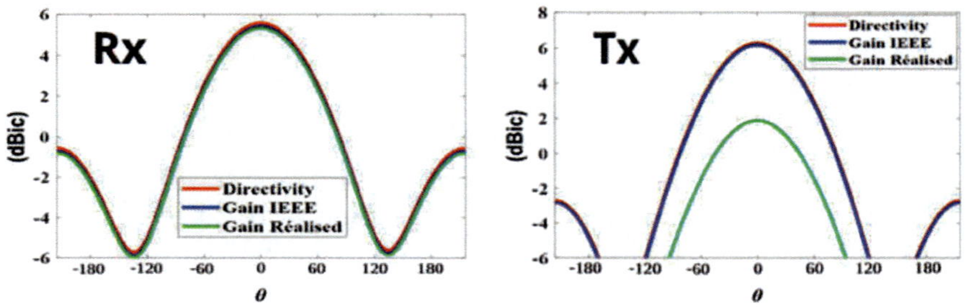

Fig. 8 Gains against elevation θ at center frequency **a** in downlink **b** in uplink

4 Downlink and Uplink Frequencies Results

4.1 Downlink and Uplink Frequencies Characteristic Results

Since the downlink/uplink bands are narrower, the computed result is taken for the center frequency, for downlink 400 MHz and for uplink 465 MHz. The evolution of the computed result against function of elevation angle θ for the different gains is presented in Fig. 8a and b. For the downlink, the realized gain is 5.34 dBi in the axis which secures a gain of 0.75 dBi for $\theta = \pm 60°$, and for uplink, the realized gain in the axis is 1.87 dBi equivalent to -3.15 dBi for $\theta = \pm 60°$. The realized gain for uplink can be optimized by employing an electronic band gap material to shorten the separation between the two lower matched peak values (Fig. 7).

4.2 Ellipticity of Downlink and Uplink

Ellipticity evolution of the axial ratio against elevation θ (Fig. 9) is plotted for the downlink and uplink frequencies. Ellipticity evolution is of the range of 1.6 dB in the axis at 60°, and minimum value is near 0.5 dB.

Considering the closer proximity of the downlink and uplink frequencies, it is necessary to construct a circuit to obtain circular polarization thus covering these two bands together [7].

5 Conclusions

This paper establishes that the pixel antenna design is an excellent approach for miniaturization to conform to ever decreasing satellite area footprints at the same time retaining its applications in terms of gains and elliptic axial ratio. The specifications for a small satellite were employed to exemplify this property and to provide a

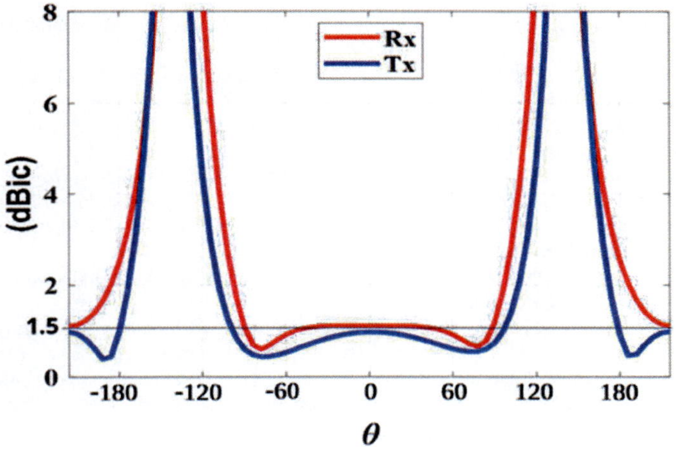

Fig. 9 Axial ratio against elevation θ in downlink and uplink

practicality that can be enhanced by enlarging the band or employing a bi-band. So work is being continued presently to obtain Isoflux patterns by using multiple pixel antennas. A challenge, however, is to obtain a sharper miniaturization.

Acknowledgements The authors thank the Institute of Engineering and Technology of JK Lakshmipat University, Jaipur, for their laboratory facilities and library. The authors also thank XLIM laboratory, University of Limoges, France, TU Berlin, and Celestial Space Technologies GmbH for measurement facilities and other resources.

References

1. Monediere T, Jecko B, Rammal M (2014) Agile radiating electromagnetic band gap antenna. Int J Antennas Propag
2. Jecko B, Arnaud E (2017) The ARMA concept: AESA and ARMA technologies for agile antenna. FERMAT 20
3. Themalil M, Majed M, Rammal M, Martinod E, Jecko B (2019) Pixel antenna for implantation on CubeSat. ICEAA – IEEE—APWC 2019 (also in IEEE Xplore), Granada, Spain, 9–13 September 2019
4. Themalil M, Majed M, Rammal M, Martinod E, Jecko B (2019) Pixel antenna for implantation on Argos CubeSat, 21st National Microwave Day (Journee National Microonde). Caen, France, JNM 2019:14–17
5. Milind Thomas JR, Bhaskar G (2016) Design of circularly polarized T—stub coupled microstrip patch antenna. Asian J Appl Sci 4(4):883–888
6. Roy J, Thomas M (2007) Investigations on new proximity coupled dual frequency microstrip antenna for wireless communication. Microw Rev 13(1):12–15
7. Siblini A, Taam HA, Jecko B, Rammal M, Belion A (2017) ARMA feeding for Isoflux coverage from microsatellite. ASTE J Adv Engg Tech 2(6):63–69

8. Roy J, Thomas M (2007) Compact microstrip antennas for next generation wireless communication. Int J Microw Sc Tech 2007(10):1–4
9. Roy J, Thomas M (2007) Broadband microstrip antennas for HIPERLAN/2. Rom J Inf Sci Technol 10(4):355–360
10. Roy J, Thomas M, Jayanta G (2008) Miniaturized circularly polarized microstrip antenna. Int J Microw Opt Technol 3(4):467–473
11. Roy J, Thomas M (2008) Design of circularly polarized microstrip antenna for WLAN. Prog Electromagnet Res PIER M, 3:79–90
12. Jecko B, Majed M, Chreim H, Siblini A, Ab Taam H, Andrieu J, Lalande M, Martinod E (2018) Agile radiating surfaces. FERMAT 30
13. Milind Thomas JR, Bhaskar G (2010) Design of proximity coupled microstrip antenna for wireless communication. ACTA Tech J 55(2):131–138
14. Thomas M, Jayanta G, Roy J (2006) Performance of compact broadband proximity-coupled microstrip antennas for HIPERLAN. Int J Trans Electron Comm Comput Netw 1(4):217–223
15. Thomas M, Chelani V, Sukumar S, Ramesh R, Singh S (2012) Parametric investigation on slot loaded microstrip antenna for wireless communication. Int Jo Comput Intell Tel Syst 2(1)

Wideband Antenna Design for 5G Communications

Amruta S. Dixit and Sumit Kumar

Abstract In this paper, the antenna is designed for the 5G communication application. It is a microstrip patch antenna which is called an antipodal Vivaldi antenna. The proposed antenna dimensions are 10 mm × 5 mm, and it is designed on RO 4003 substrate of height 0.8 mm. It is a wideband antenna that operates from 20.72 GHz to above 45 GHz with a resonance frequency at 31.77 GHz. The antenna gain is almost stable, and the highest gain is 5.22 dBi. The designed antenna is very appropriate for various 5G communication devices.

Keywords Wideband · Antipodal Vivaldi antenna (AVA) · Compact · 5G communication

1 Introduction

The 5G technology is the new emerging technology in the area of wireless communication. This 5G technology is the best solution for increasing demand for high data rate, faster communication, and device-to-device communication [1]. The important 5G communication specifications are as follows [2]:

- Mobility should be greater than or equal to 500 km/h
- Spectrum efficiency should be up to 9 bits/s/Hz
- Connection density should be high which is equal to 1 million/km^2
- Enhanced data rate from 2 to 20 Gbps
- Frequency range 1 (below 6 GHz) and frequency range 2 (above 6 GHz).

Out of these specifications of 5G communication, to satisfy frequency range specification, we need an antenna with large bandwidth and which can operate at high frequencies. This requirement of 5G technology can be satisfied by antipodal Vivaldi antenna (AVA) [3]. The AVA is the next version of the Vivaldi antenna which was

A. S. Dixit · S. Kumar (✉)
Symbiosis Institute of Technology (SIT) Symbiosis International (Deemed University)(SIU), Lavale, Pune, India
e-mail: er.sumitkumar21@gmail.com

B. Mishra and M. Tiwari (eds.), *VLSI, Microwave and Wireless Technologies*,
Lecture Notes in Electrical Engineering 877,
https://doi.org/10.1007/978-981-19-0312-0_3

designed by Dr. Gazit in 1988 [4], and Dr. Gibson invented the Vivaldi antenna in 1979 [5]. This AVA and Vivaldi antennas are inherently wideband and high operating frequency antennas. The Vivaldi antenna has two exponential flares on one side, and the other side consists of a feed line, whereas AVA has two flares on the substrate's different sides. In AVA, one flare acts as a radiator while the other acts as a ground.

The antenna performance can be enhanced by using different performance enhancement techniques like corrugations, metamaterial, metasurface, balanced AVA, "substrate integrated waveguide (SIW)," "coplanar waveguide (CPW)," multilayer, array, and "multiple input multiple output (MIMO)" [6]. Metamaterials are designed on the substrate and by forming different shapes or structures of the copper layer [7]. Next, the metasurfaces are a single layer of metamaterial unit cells [8]. Both metamaterial and metasurface structures are complex in design. Next, corrugations improve bandwidth and directivity of an antenna [9]. SIW balances the current distribution of the antenna, but its arrangement on an antenna is a daunting task [10]. The multilayer, array, and MIMO performance enhancement techniques improve the antenna gain significantly, but they are best for large size devices [11–14]. MIMO antenna requires additional mutual coupling reduction techniques [15–17].

This paper presents a simple and miniaturized AVA design for 5G application devices. It consists of two exponential-shaped flares that are identical and mirror images of each other. Mostly in AVA, the ground patch has an exponential extension on the lower side of the substrate, whereas in the presented AVA design, the ground patch has a microstrip line structure to improve its lower cut-off frequency. The paper explains the antenna design in Sect. 2, simulated results in Sect. 3, and conclusion in the last Sect. 4.

2 AVA Design

The proposed antenna design is shown in Fig. 1. Figure 1a is the conventional AVA which contains exponential curves at the bottom side of the ground, and Fig. 1b is the proposed AVA with no exponential curves at the bottom side of the ground. Both antennas are designed on Roger's RO 4003 substrate whose thickness is 0.8 mm, and its permittivity is 3.55. The ground and radiator patches are consist of exponential flares at the upper side and microstrip lines at the lower sides. Dimensions of the AVA are given in Table 1.

The reflection coefficient results of both designs are shown in Fig. 2. It is depicted in Fig. 2 that the −10 dB bandwidth of conventional AVA (CAVA) is from 24.55 to 45 GHz, whereas the −10 dB bandwidth of the proposed AVA is from 20.72 to 45 GHz. Hence, the proposed antenna provides more bandwidth as compared to the CAVA. Also, the resonance frequency of CAVA is 32.88 GHz, and the resonance frequency of the proposed AVA is 31.77 GHz.

Next, Fig. 3a shows the smith chart, and Fig. 3b shows the input impedance plot. Both figures depict that the input impedance of the proposed antenna is near to

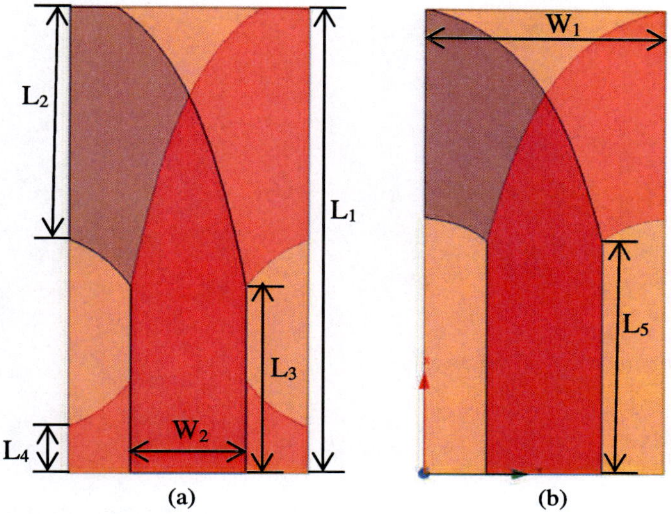

Fig. 1 Design of AVA

Table 1 Proposed AVA dimensions

Parameter	L_1	L_2	L_3	L_4	L_5	W_1	W_2
Size (mm)	10	5	4	1	5	5	2.4

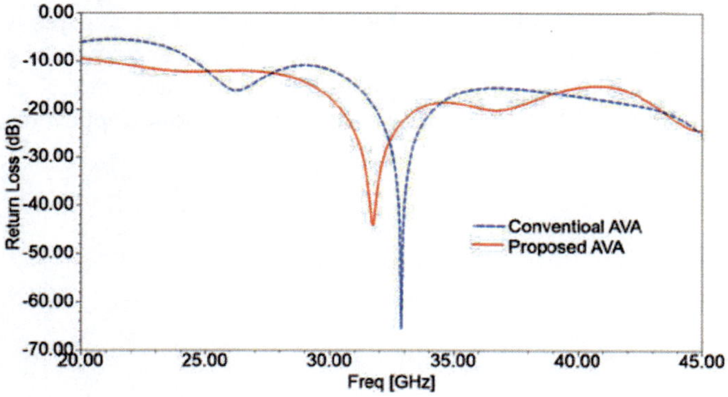

Fig. 2 Reflection coefficient

50 Ω over the operating frequency range of 20.72–45 GHz. Hence, good impedance matching is achieved in the proposed antenna.

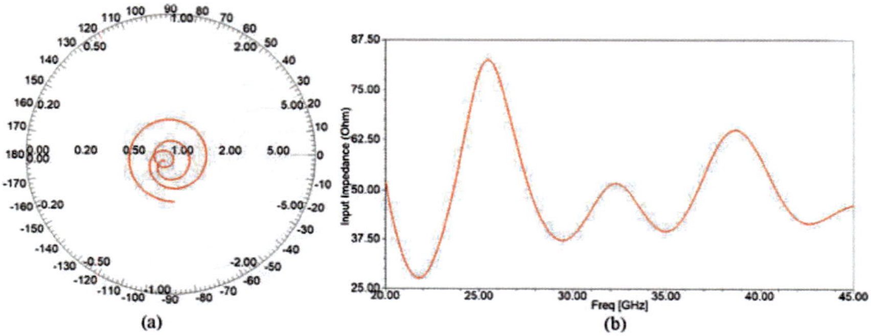

Fig. 3 Smith chart and input impedance

3 Results

The antenna is designed by using HFSS version 2020 R2. The gain versus frequency graph is shown in Fig. 4. The gain range is 0.6–5.2 dBi. The maximum gain is obtained at 34.72 GHz. Further, the electric field distribution is shown in Fig. 5. It shows that the maximum radiation is in the end fire direction, and hence, AVA is the end fire antenna. Also, the proposed AVA radiates maximum energy in the end fire direction and the corresponding electric field is in the range of 142.44–7000 V/m (Fig. 5).

The elevation and azimuth planes at four frequencies are plotted in Figs. 6 and 7 respectievely which show that the proposed antenna is the directional antenna. Also, the radiation patterns are almost stable. Further, the antenna is compared with other antennas from the literature, and it is presented in Table 2. The antennas presented in [18, 19] are of very large size, and their gain is also moderate. Next, in [20, 21], antenna gain is good but their size is large for 5G mobile applications. Out of

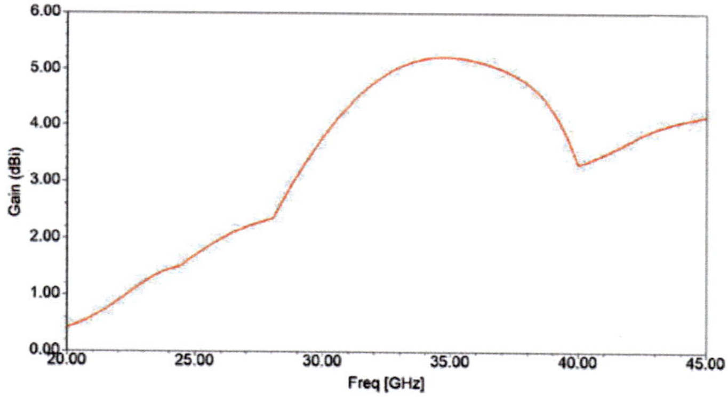

Fig. 4 Gain versus frequency plot

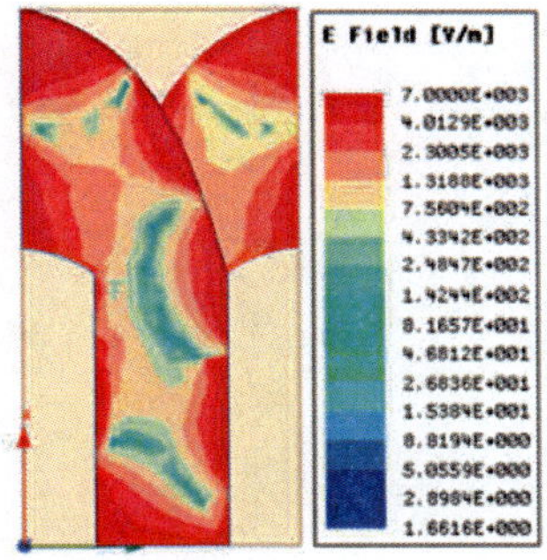

Fig. 5 Electric field distribution

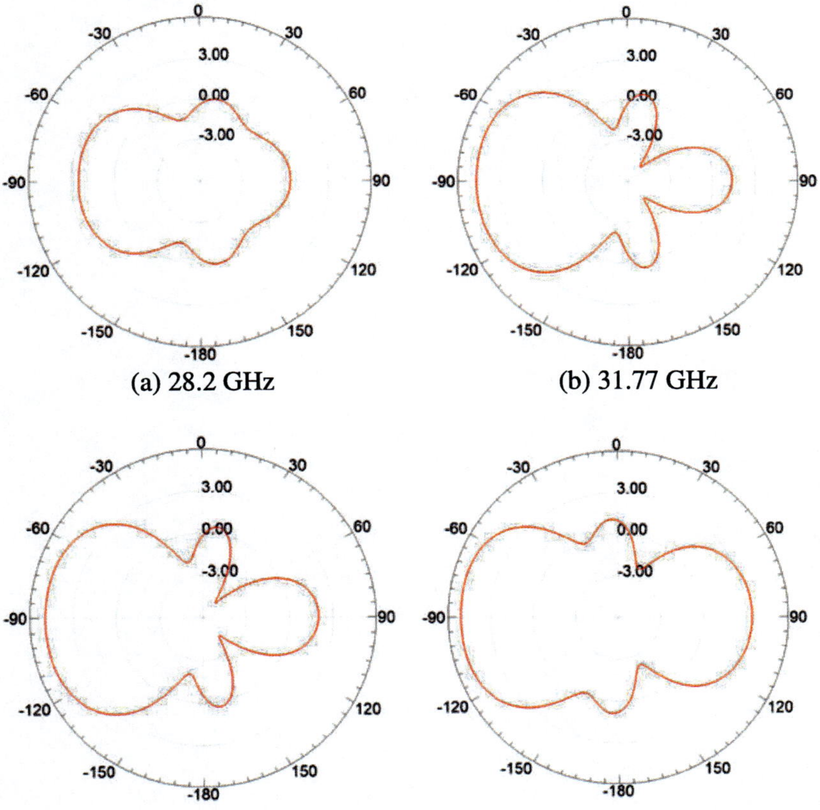

(a) 28.2 GHz

(b) 31.77 GHz

(c) 32.5 GHz

(d) 36.38 GHz

Fig. 6 Elevation planes

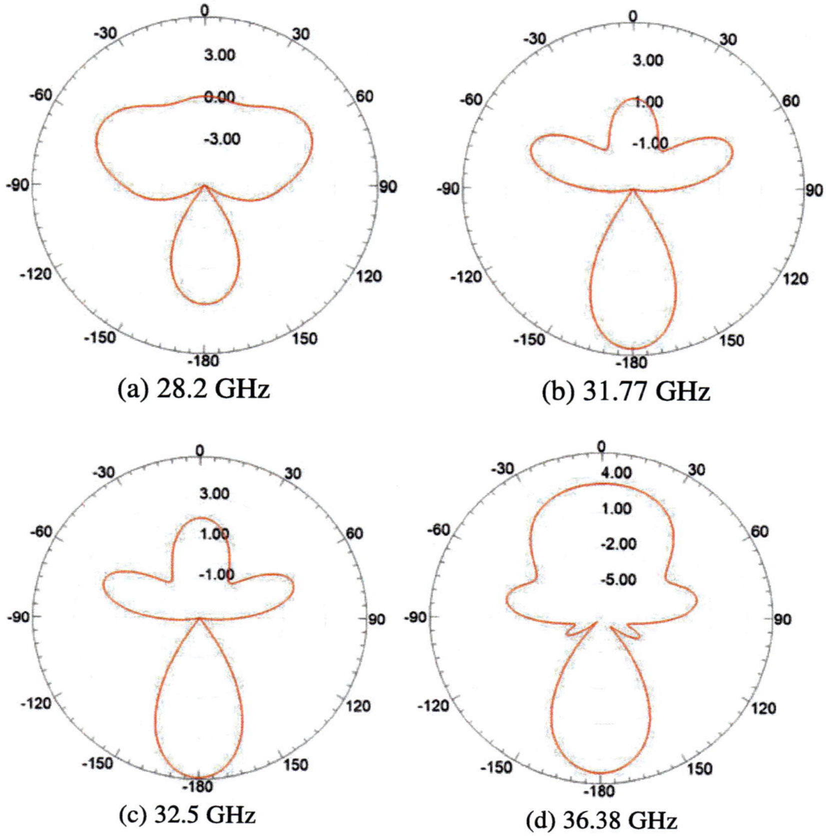

Fig. 7 Azimuth planes

Table 2 Comparison of the proposed AVA with other antennas for 5G applications

Ref. No.	Dimensions (mm^3)	Gain (dBi)	Frequency (GHz)
[18]	55 × 110 × 1.6	4.6–5.1	27.7–28.7 37.3–38.6
[19]	158 × 77.8 × 0.381	5.8–7.6	25–40 1.8–2.6
[20]	40 × 60 × 0.508	8–15	3.4–40
[21]	64 × 64 × 0.254	3–12	4–50
Proposed work	10 × 5 × 0.8	0.6–5.2	20.72–45

these antennas, the proposed wideband antenna is very compact with moderate gain. Hence, it is very suitable for 5G mobile applications.

4 Conclusion

The paper presented a compact and wideband AVA for 5G communication applications. Its design is very simple, and it is easy to fabricate. Mainly, the structure of the bottom patch is modified to reduce the lower cut-off frequency from 24.55 to 20.72 GHz. Also, it can operate from 20.72 to 45 GHz with a moderate gain of 0.6–5.2 dBi. The proposed antenna has enhanced two important parameters of an antenna which are bandwidth and dimensions. Hence, the proposed AVA is suitable to integrate in 5G mobile devices.

References

1. Dixit AS, Shevada LK, Raut HD, Malekar RR, Kumar S (2020) Fifth generation antennas: a bibliometric survey and future research directions. Libr Philos Pract 2020:1–24
2. Kumar S, Dixit AS, Malekar RR, Raut HD, Shevada LK (2020) Fifth generation antennas: a comprehensive review of design and performance enhancement techniques. IEEE Access 8:163568–163593
3. Zhu S, Liu H, Wen P (2019) A new method for achieving miniaturization and gain enhancement of Vivaldi antenna array based on anisotropic metasurface. IEEE Trans Antennas Propag 67(3):1952–1956
4. Gazit E (1988) Improved design of the Vivaldi antenna. IEE Proc H Microwaves Antennas Propag 135:89–92
5. Gibson PJ (1979) The Vivaldi aerial. In: 9th European microwave conference, vol 1, pp 101–105
6. Dixit AS, Kumar S (2020) A survey of performance enhancement techniques of antipodal Vivaldi antenna. IEEE Access 8:45774–45796
7. Raut HD, Shevada L, Malekar R, Kumar S (2021) High gain wideband antennas for 5G applications: a review. In: Ranganathan G et al (eds) Inventive communication and computational technologies. Lecture notes in networks and systems. Lecture notes in networks and systems, vol 145. Springer, Singapore, pp 777–787
8. Hussain S, Qu SW, Zhou WL, Zhang P, Yang S (2020) Design and fabrication of wideband dual-polarized dipole array for 5G wireless systems. IEEE Access 8:65155–65163
9. Dixit AS, Kumar S (2020) The enhanced gain and cost-effective antipodal Vivaldi antenna for 5G communication applications. Microwave Opt Technol Lett 1–10
10. Deckmyn T, Cauwe M, Vande Ginste D, Rogier H, Agneessens S (2019) Dual-band (28, 38) GHz coupled quarter-mode substrate-integrated waveguide antenna array for next-generation wireless systems. IEEE Trans Antennas Propag 67(4):2405–2412
11. Dixit AS, Kumar S (2020) A miniaturized antipodal Vivaldi antenna for 5G communication applications. In: 7th International conference on signal processing and integrated networks, pp 800–803
12. Hussain N, Jeong MJ, Abbas A, Kim TJ, Kim N (2020) A metasurface-based low-profile wideband circularly polarized patch antenna for 5G millimeter-wave systems. IEEE Access 8:22127–22135
13. Chen Q et al (2019) Single ring slot-based antennas for metal-rimmed 4G/5G smartphones. IEEE Trans Antennas Propag 67(3):1476–1487

14. Bhadoria B, Kumar S (2018) A novel omnidirectional triangular patch antenna array using Dolph Chebyshev current distribution for C-band applications. Prog Electromagn Res M 71:75–84
15. Shevada L, Raut HD, Malekar R, Kumar S (2021) Comparative study of different beamforming techniques for 5g: a review. In: Ranganathan G et al (eds) Inventive communication and computational technologies. Lecture notes in networks and systems. Lecture notes in networks and systems, vol 145, pp 589–595
16. Malekar RR, Shevada LK, Raut HD, Dixit AS, Kumar S (2020) MIMO antenna for fifth generation mm-wave applications: a bibliometric survey. Libr Philos Pract 1–21
17. Shevada LK, Raut HD, Malekar RR, Dixit AS, Kumar S (2021) A bibliometric survey on ultra wideband multiple input multiple output antenna with improved isolation. Libr Philos Pract 1–20
18. Marzouk HM, Ahmed MI, Shaalan AA (2020) A novel dual-band 28/38 GHz AFSL MIMO antenna for 5G smartphone applications. J Phys Conf Ser 1447(1):1–7
19. Al Abbas E, Ikram M, Mobashsher AT, Abbosh A (2019) MIMO antenna system for multi-band millimeter-wave 5G and wideband 4G mobile communications. IEEE Access 7:181916–181923
20. Moosazadeh M, Kharkovsky S (2015) A compact high-gain and front-to-back ratio elliptically tapered antipodal Vivaldi antenna with trapezoid-shaped dielectric lens. IEEE Antennas Wirel Propag Lett 15:552–555
21. Bai J, Shi S, Prather DW (2011) Modified compact antipodal Vivaldi antenna for 4–50-GHz UWB application. IEEE Trans Microwave Theory Technol 59(4):1051–1057

SIW Slot Antenna Array for 5G Applications

Sayyed Arif Ali, Mohd Wajid, and Muhammad Shah Alam

Abstract A substrate-integrated waveguide (SIW) slot antenna array is reported in this paper for 5G applications. The proposed broadside antenna array has ten linearly arranged slots optimized to operate at the lower 5G band of 3.5 GHz. A prototype of the antenna in a unit configuration is measured to verify the frequency of operation. The proposed antenna achieved a relative bandwidth of 3.7% with 12.3 dBi of gain, and the beam scanning range from −60° to +60° in the azimuth plane is demonstrated. The 3-dB beamwidth of the main lobe of 15° is realized, thus restricting the first side-lobe level (SLL1) below −14 dB in the H-plane. The front-to-back ratio (FTBR) of nearly 24.83 dB and mutual coupling ≤−20 dB) were verified. The volume percentage reduction of almost 68% compared to similar work reported in the literature justifies its compact size and low cost.

Keywords Enhanced mobile broadband (eMBB) · Fifth generation (5G) · Substrate-integrated waveguide (SIW) · Slot antenna array · Beam scanning

1 Introduction

The fifth-generation (5G) mobile opened up several possibilities like enhanced mobile broadband (eMBB), massive machine-type communications (mMTC), and ultra-reliability and low latency communication (URLLC) services [1, 2]. With the increasing demand of data rate, particularly in the present pandemic scenario (i.e., COVID-19), eMBB services are in great need which offer peak data rates up to

S. A. Ali (✉) · M. Wajid
ZH College of Engineering & Technology, Aligarh Muslim University, Aligarh, India
e-mail: sayyedarifali@zhcet.ac.in

M. Wajid
e-mail: mohdwajid@zhcet.ac.in

M. S. Alam
Electrical Engineering Department, College of Engineering, Imam Mohammad Ibn Saud Islamic University, Riyadh 11432, Saudi Arabia
e-mail: malam@imamu.edu.sa

B. Mishra and M. Tiwari (eds.), *VLSI, Microwave and Wireless Technologies*,
Lecture Notes in Electrical Engineering 877,
https://doi.org/10.1007/978-981-19-0312-0_4

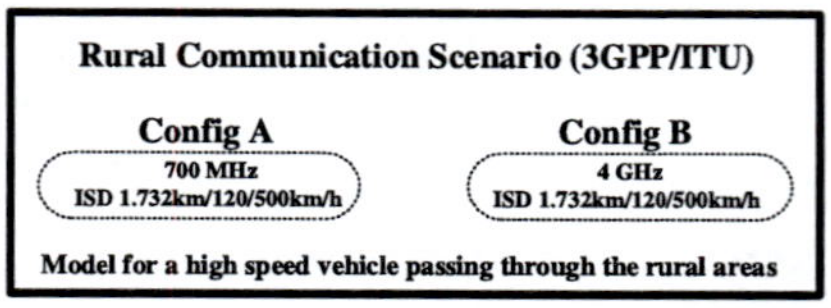

(a)

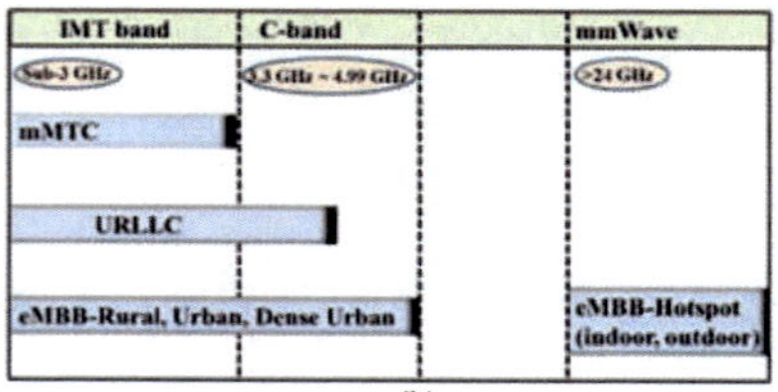

(b)

Fig. 1 **a** Rural communication scenario in various configurations (3GPP/ITU) [4] and **b** various frequency bands of operation for 5G [5]

20 gigabits per second (Gbps), as well as supporting high mobility reaching up to 500 km/h [1, 2]. A new version of 5G, called 5Gi, is proposed for India, mainly serving the country's rural population and providing high data rate connectivity [3]. The proposed 5Gi model for India differs from what is adopted by 3GPP/ITU [4] to serve the rural masses as shown in Fig. 1a, where inter-site distance (ISD) has been increased from 1.732 to 6 km keeping in view the geographical condition in the country.

Several frequency bands are used for 5G [5], as shown in Fig. 1b. As shown in Fig. 1b, sub-3 GHz, C-band (3.3 ~ 4.99 GHz), and mmWave (i.e., 28 GHz) are the possible options [1, 2]. More specifically, sub-3 GHz and C-bands are used for mobile broadband and the Internet of Things (IoT) to serve rural, urban, and dense urban populations for fulfilling their communication requirement. Thus, various opportunities exist to design and develop antennas operating in these bands to deliver high data rate services. However, the C-band, also called the FR1 band, provides broader coverage but lower data rate capability, which is more particularly suited to serve rural masses.

In contrast, the mmWave band delivers higher data rates but limited coverage area. Therefore, the MIMO concept, particularly in sub-THz and C-band, is used to enhance data rate capability. By utilizing beam-forming and steering in the antenna design, signal connectivity is improved [6]. Furthermore, compact antenna size is another critical concern due to the limited space available on integrated circuits to reduce costs.

The substrate-integrated waveguide (SIW) technique, which is planar, is mainly used for developing low-cost antenna design. Therefore, this work is utilized to develop a compact size antenna operating in the lower 5G band, i.e., C-band (i.e., 3.3 ~ 4.99 GHz). Furthermore, it improves isolation with other radio circuit elements by using vias in the walls [7]. To realize a single antenna element, the SIW mainly utilizes either microstrip patch [8–10] or slot [11–13]. However, slots are a better option in the array design to realize compact and high gain antenna. There are several papers reported in the literature using the SIW targeting mainly the mmWave and satellite bands [14–17]. However, the SIW potential to realize antenna in the lower 5G band is not fully established. Therefore, in this work, the SIW slot antenna is proposed for the 5G operating at 3.5 GHz, and its array configuration to realize high gain is presented, and performances are investigated. A prototype of the antenna in a unit configuration is developed, and return loss (S_{11}) is measured to verify its operating frequency. The volume percentage reduction of nearly 68% has been achieved compared to similar work reported in the literature [13], which justifies its compact size and low cost of the proposed antenna. The result achieved more specifically fan-beam radiation patterns with beam-steering from $-60°$ to $+60°$, first side-lobe level (SLL1) of around −14 dB, high gain of 12.30 dBi, and high isolation (≤ -20 dB) confirm the proposed antenna suitability of 5G base station terminal [5, 6, 13]. The paper is organized as follows: Sect. 2 provides details on the antenna design. The results are discussed in Sect. 3, and finally, conclusions are drawn in Sect. 4.

2 Antenna Design

2.1 Substrate-Integrated Waveguide (SIW)

For antenna design, first, a SIW structure is realized on a dielectric substrate with copper covering at the top and bottom layer, acting as a ground plane. In this structure, two rows of metalized circular holes of uniform diameter d are called via to short these ground planes. There are several vias in each row separated by a constant distance p called pitch. The two rows of via form the waveguide's narrow walls, whereas the top and bottom conductors covering form the broad walls. The separation between the two rows of vias is denoted by a, called width, which is decided by the cut-off frequency f_c. The effective width a_{eff} of the dielectric-filled rectangular waveguide is expressed as [18, 19]:

$$a_{\text{eff}} = a - \frac{d^2}{0.95 \times p}, \quad \text{for } p \leq 4d \tag{1}$$

The validity of (1) lies within ±5% of accuracy [19]. Therefore, a more accurate expression for a_{eff} was suggested in [20] as given below:

$$a_{\text{eff}} = a - 1.08\frac{d^2}{p} + 0.1\frac{d^2}{a}, \quad \text{for } p < 3d \text{ and } d < \frac{w}{5} \tag{2}$$

While designing the waveguide, it is to be ensured that $p \leq 2d$minimizes the electromagnetic radiation leakage from the vias [21, 22]. Since creating vias in SIW involves mechanical drilling in the substrate, there should be no more than 20 vias (i.e., $p \geq \frac{\lambda_c}{20}$) in a single cut-off wavelength λ_c to ensure the SIW mechanical strength. Furthermore, the pitch is used larger or equal to d, i.e., $p \geq d$ for the physical realization of SIW. Since SIW is a periodic structure with vias created at regular length, the system is prone to the bandgap effect, which is expressed as [22]:

$$\beta p = \pi \tag{3}$$

where β is the propagation constant, and the p value in (3) is carefully chosen so that the first bandgap exists beyond the upper cut-off frequency, thus ensuring the mono-mode bandwidth operation of SIW. By considering the upper cut-off frequency equal to $2f_c$, the corresponding propagation constant β is given by:

$$\beta = \sqrt{k_0^2 - k_c^2} \tag{4}$$

where k_0 and k_c are wavenumbers corresponding to the operating and cut-off wavelength, respectively. By solving (4) for $k_0 = 2k_c$, the condition to avoid the bandgap is achieved at the operating frequency when $p < \frac{\lambda_c}{4}$. By using the above-stated design rules, the SIW is designed, as shown in Fig. 2a, b. The SIW is developed using FR4 substrate, which has a height (h) of 1.6 mm, dielectric constant (ε_r) of 4.4 and a loss tangent of 0.002.

Figure 2a shows the modeled SIW for f_c equal to 2.5 GHz, and its associated design parameters are shown in Fig. 2b. Using the optimized SIW parameters given in Table 1, the electric field patterns in TE101 mode are plotted, as shown in Fig. 2c. The display of various higher-order modes with their respective cut-off frequencies that is generated inside the SIW which is demonstrated in Fig. 3a. The S-parameters of the SIW (S_{11} and S_{21}), as plotted in Fig. 3b, suggest that the SIW pass band begins from the 2.5 GHz and extends up to the 10 GHz band. The reflection coefficient (S_{11}) within the passband is sufficiently low (<−10 dB), which justifies the suitability of SIW for the proposed antenna design.

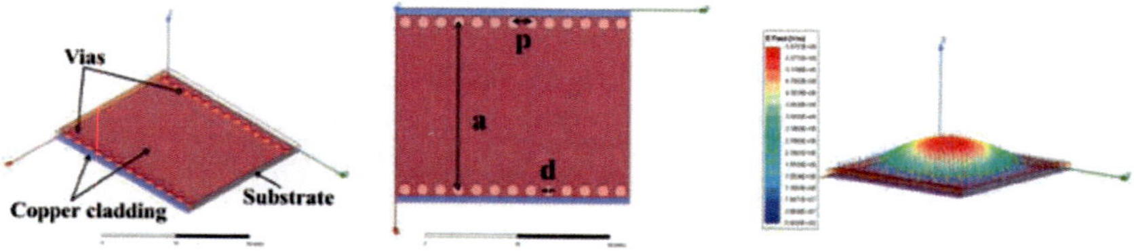

(a) (b) (c)

Fig. 2 SIW design parameters and electric field patterns **a** geometry, **b** design parameters, and **c** electric field patterns in TE101 mode

Table 1 Optimized parameters of the SIW

Parameter	a	p	d	h
Value (mm)	30.6	3.0	2.0	1.6

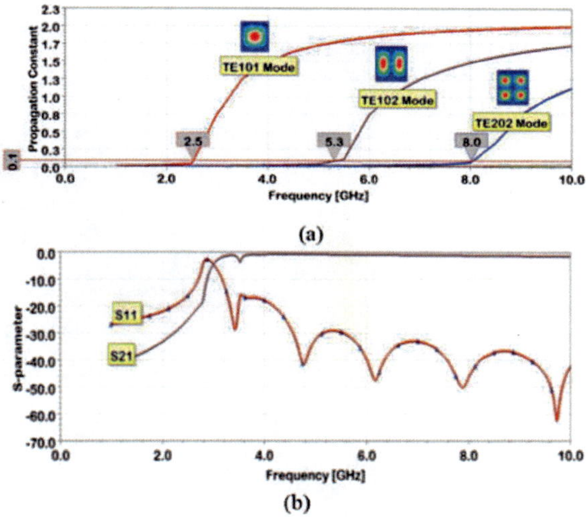

Fig. 3 **a** Different modes of propagation and **b** plot of S_{21} and S_{11} parameters in the substrate-integrated waveguide

2.2 *SIW Slot Antenna*

The tapered microstrip feed [23] is used to excite SIW, which is shorted at one end to form a cavity (see Fig. 4a). As shown in Fig. 4a, the W_t is a critical parameter of the tapered feed, which mainly decides the various wave propagation modes in the SIW. The taper transition dimension W_t is related by the following transcendental equations [23]:

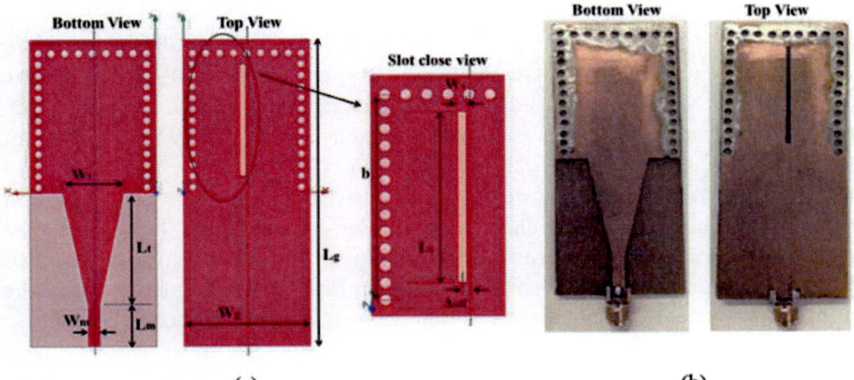

(a) (b)

Fig. 4 **a** Tapered feed geometry of the SIW slot antenna and **b** prototype

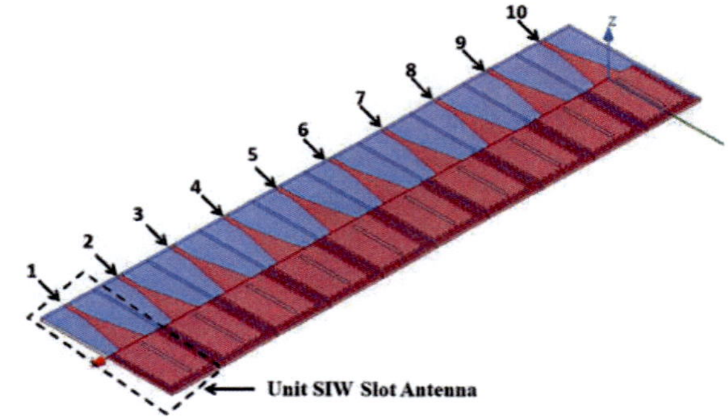

Fig. 5 SIW slot antenna array in 1 × 10 configuration

Table 2 Optimized dimensions of the SIW slot antenna

Parameter	b	W_g	L_g	W_m	L_m	W_t	L_t	W_s	L_s	Δ_{off}
Value (mm)	36	33	82.4	3	11.4	16	30	1.5	30	1

$$\left(\frac{1}{w}\right) = \begin{cases} \left(\frac{60}{\eta \times h}\right) \ln\left(8\frac{h}{W_t} + 0.25\frac{W_t}{h}\right), & \text{for } \frac{W_t}{h} < 1 \\ \frac{120\pi}{(\eta \times h)\left\{\frac{W_t}{h} + 1.393 + 0.667 \ln\left(\frac{W_t}{h} + 1.444\right)\right\}}, & \text{for } \frac{W_t}{h} > 1 \end{cases} \tag{5}$$

$$\left(\frac{1}{w}\right) = \left(\frac{4.38}{a}\right) \times \mathrm{e}^{-\left\{0.63 \times \frac{\varepsilon_r}{\frac{\varepsilon_r+1}{2} + \frac{\varepsilon_r-1}{2}\left(\frac{1}{\sqrt{1+12\frac{h}{W_t}}}\right)}\right\}} \tag{6}$$

where $\eta = \sqrt{\frac{\mu_0}{\varepsilon_0}}$ and h and ε_r are substrate parameters. For a given SIW width a, (5) and (6) are used and solve for W_t. As one end of the SIW is shorted, the standing waves will be developed inside the structure. In order to radiate out the wave, a slot of half-wavelength size is cut, placed at quarter wavelength apart from the shorted end, thus to realize a SIW slot antenna. The slot is slightly offset from the central line to interrupt the current's full path to maximize the antenna's radiation performance. The designed antenna is shown in Fig. 4a, and its prototype, developed by using chemical etching, is demonstrated in Fig. 4b. The shorting vias are created by a drilling machine and soldered, as shown in Fig. 5b. Table 2 summarizes the optimized SIW slot antenna parameters, and the results are discussed in Sect. 3.

2.3 *SIW Slot Array Antenna*

The antenna array having ten linearly arranged slots (i.e., 1 × 10) optimized to operate at the lower 5G band of 3.5 GHz is developed as shown in Fig. 5. These antenna units are uniformly spaced at half-wavelength. The measured and simulated performances of the proposed antenna are discussed in section III. The antenna simulation study is carried out using Ansys HFSS [24].

3 Results and Discussion

The SIW slot antenna is used as a unit element to construct the array. First, the unit element is optimized to ensure its operation in TE_{101} mode at 3.5 GHz, as shown in Fig. 6a. The slot is slightly offset by Δ_{off} from the center axis to interrupt the current's maximum path to optimize the radiation performance, i.e., when more current is interrupted, more will be the radiation. Figure 6b compares the simulated and measured S_{11} using Agilent FieldFox vector network analyzer N9923A for the proposed antenna. The antenna achieves a relative bandwidth (BW) determined when $S_{11} = -10$ dB of 3.7% with frequency span from 3.44 to 3.56 GHz, which is sufficient to operate at the lower 5G band. A good match is observed between the measured and simulated S_{11} as shown in Fig. 6b. The proposed antenna's radiation performances are demonstrated in Fig. 6b–d. As observed from Fig. 6c, the slot radiates in a broadside direction with a peak gain of 2.5 dBi in the unit configuration in both the E and H planes, with an excellent front-to-back ratio (FTBR) of 19.3 dB.

The optimized unit slot antenna element is used to develop a 1 × 10 slot array for 5G applications. The array gain achieved is 12.3 dBi, as shown in Fig. 7a, b. The beam-steering capability of the proposed array is demonstrated as shown in Fig. 8. As observed from Fig. 8, the beam can scan from $-60°$ to $+60°$ in the azimuth plane. The 3-dB beamwidth of the main lobe of 15° is achieved, which restricts SLL1 ≤ -14 dB in H-plane. The FTBR of nearly 24.83 dB and mutual coupling ≤ -20 dB have been demonstrated as shown in Figs. 7 and 9, respectively. The volume percentage reduction by almost 68% compared to similar work reported in the literature [13] has been observed, which justifies the proposed antenna's compactness and low cost. After comparing with the documented paper [13], the proposed antenna is found compact in size (see Table 3) and planar, which are essential features required to integrate with the other radio frequency circuit. The proposed antenna performs better in terms of beam-steering range, FTBR, SLL1, and isolation performances needed to operate for the 5G base station terminal.

4 Conclusion

A 1 × 10 SIW slot antenna array is proposed in this work to operate at 3.5 GHz for a 5G application. A prototype of the proposed antenna in-unit antenna element configuration is developed to justify the frequency of operation. The array exhibits a high gain of 12.3 dBi with beam-steering capability in the azimuth plane to cover the

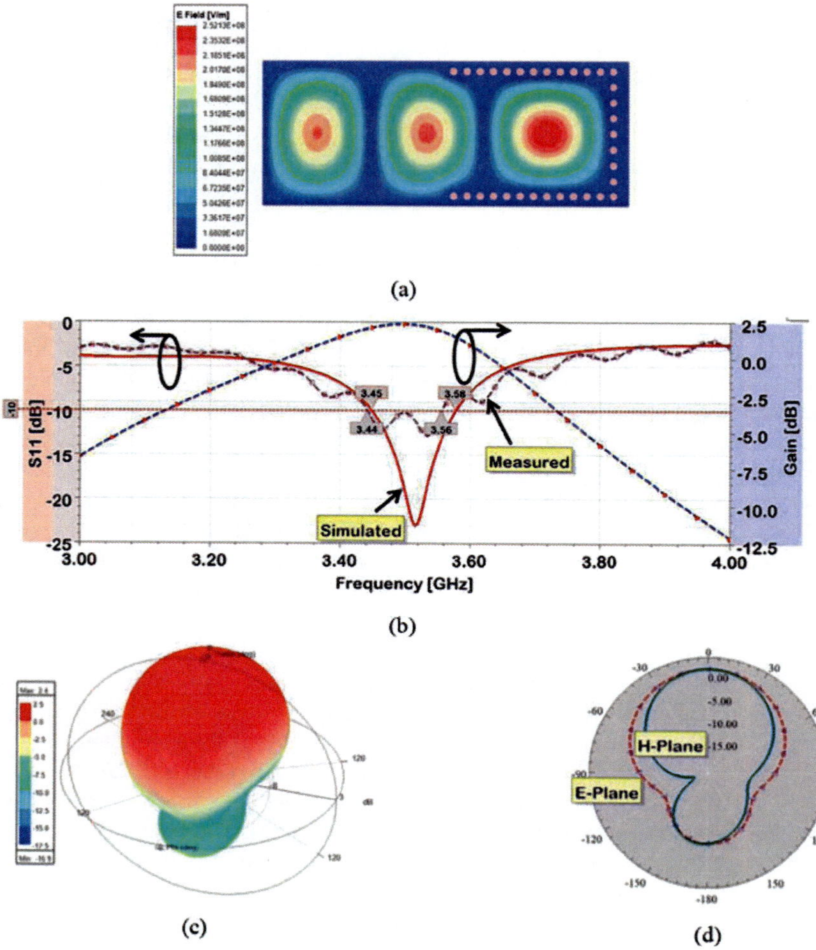

Fig. 6 SIW slot antenna performance **a** Eigenmode TE101 at 3.5 GHz, **b** reflection coefficient (S_{11}) and gain, **c** 3D radiation pattern, and **d** 2D radiation pattern

−60° to +60° scanning range. Its compact size, planar design features justify its integration with other radio circuit elements. The proposed antenna covers the frequency band from 3.46 to 3.59 GHz with ≤−20 dB isolation. Thus different reported results justify the proposed antenna suitability for the 5G base station terminal.

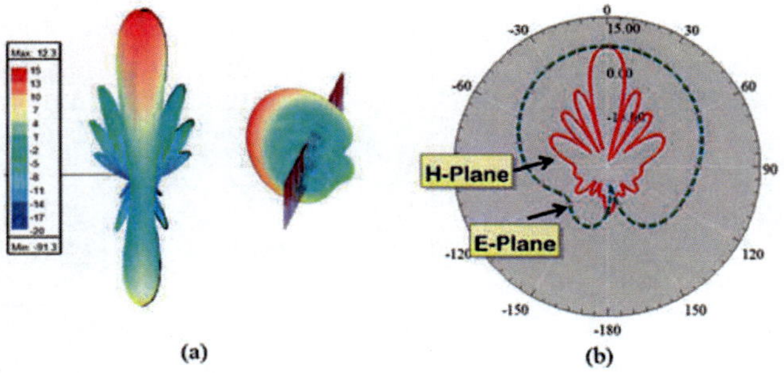

Fig. 7 Radiation pattern of 1 × 10 slot antenna array **a** 3D radiation pattern, and **b** 2D radiation pattern

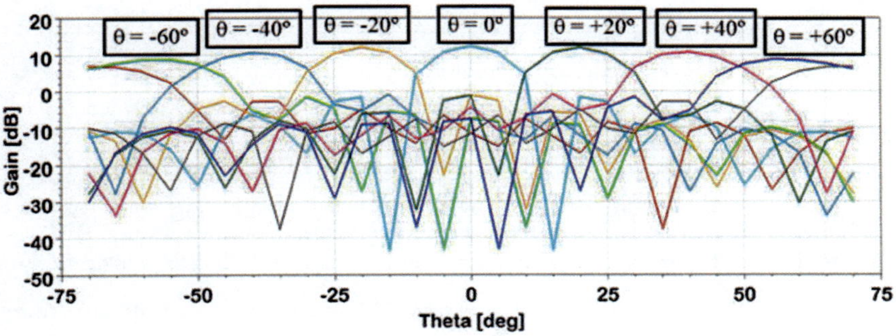

Fig. 8 Slot antenna array beam steers from −60° to +60° in the azimuth plane

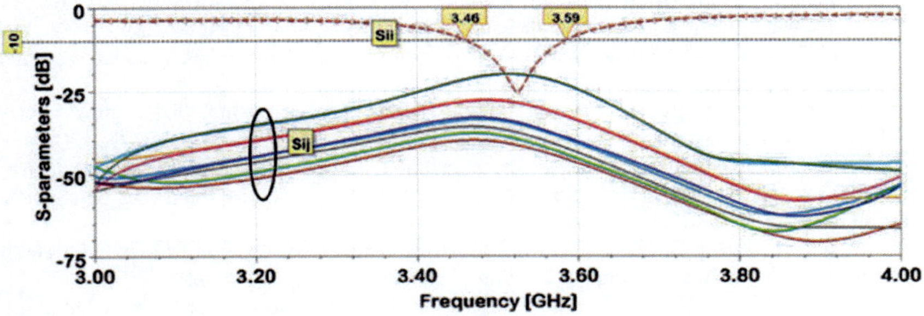

Fig. 9 Reflection coefficient (S_{ii}) and mutual coupling (S_{ij}) of 1 × 10 slot antenna array

Table 3 Comparison of performance with the work reported in the literature

Parameter	Array configuration	BW (MHz)	Gain (dBi)	Scanning range	FTBR (dB)	SLL1 (dB)	Isolation (dB)	Unit element volume (mm^3)
[13]	1 × 15	970	19.28	±30°	21.79[a]	– 13.3	≤−18	13,601
Present work	1 × 10	130	12.3	±60°	24.83[b]	– 14	≤−20	4351

[a]With extra ground plane
[b]Without any extra ground plane

Acknowledgements This work is supported by the Ministry of Education and Ministry of Minority Affairs, Govt. of India through UGC under MANF Ph.D. scheme, Grant no. MANF-2018-19-RAJ-95315.

References

1. Ali SA, Wajid M, Alam MS (2020) Antenna design challenges for 5G—assessing future direction. In: Enabling technologies for next generation wireless communications, CRC Press, Taylor & Francis Group, UK, ch 9, pp 149–175 [Online]. Available: https://doi.org/10.1201/9781003003472
2. Mukundan C (2020) Understanding 5G new radio (NR) release 15/16 standards. In: Industry workshop 26A: keysight technologies, IMS2020
3. TSDSI Homepage (2020) https://tsdsi.in/tsdsis-5g-radio-interface-technology-5gi-approved-by-sg5-of-itu-r-as-part-of-upcoming-itu-r-recommendation-m-imt-2020-specs/. Last accessed 21 Dec 2020
4. Aniruddhan S (2020) Systems and circuits for 5G new radio. In: Online workshop on microwave and antenna applications in 5G (2020). Source: https://youtu.be/gMTIKF115LM
5. Yong W (2018) Brings 5G vision into reality. In: Industry workshop: Huawei, 2nd international conference 5G India (2018)
6. Sligar A et al (2020) Ansys 5G antenna solutions. White paper Ansys Inc. (2019). https://www.ansys.com/resource-library/white-paper/5g-antenna-solutions, Last accessed 21 Dec 2020
7. Wu K (2001) Integration and interconnect techniques of planar and non-planar structures for microwave and millimeter-wave circuits-current status and future trends. In: Microwave conference (APMC'01), pp 411–416
8. Li JF, Wu DL, Zhang G, Wu YJ, Mao CX (2019) Compact Dual-Polarized Antenna for Dual-Band Full-Duplex Base Station Applications. IEEE Access 7:72761–72769
9. Ciydem M, Miran EA (2020) Dual Polarization Wideband Sub-6 GHz Suspended Patch Antenna for 5G Base Station. IEEE Antennas and Wireless Propagation Letters 19(7):1142–1146
10. El Hasnaoui Y, Mazri T (2020) Study, design, and simulation of an array antenna for base station 5G. In: International conference on intelligent systems and computer vision (ISCV). IEEE, Morocco, pp 1–5
11. Wu YM, Wong SW, Lin JY, Zhu L, He Y, Chen FC (2018) A Circularly Polarized Cavity-Backed Slot Antenna with Enhanced Radiation Gain. IEEE Antennas and Wireless Propagation Letters 17(6):1010–1014

12. Yong WY, Haddadi A, Bagheri A, Emanuelsson T, Glazunov AA (2019) Wideband cavity-backed slot subarray fed by gap ridge waveguide for 5G mmWave base station. In: International symposium on antennas and propagation (ISAP). IEEE, Xi'an, China, pp 1–3
13. Chen Y, Elsherbeni AZ, Alharbi K, Aldhaheri R (2020) Cavity backed patch-slot antenna for lower band 5G communications. In: XXXIIIrd general assembly and scientific symposium of the International Union of Radio Science. IEEE, Rome, Italy, pp 1–3
14. Kumar A, Chaturvedi D, Raghavan S (2018) SIW cavity-backed Circularly Polarized Square Ring Slot Antenna with Wide Axial-Ratio Bandwidth. AEU-International Journal of Electronics and Communications 94:122–127
15. Kumar A (2020) Wideband Circular Cavity-Backed Slot Antenna with Conical Radiation Patterns. Microwave and Optical Technology Letters 62(6):2390–2397
16. Xie JJ, Chen Z (2020) Wideband Dual-Polarized SIW cavity-backed Patch Antenna with Multimode Characteristics. Progress in Electromagnetics Research C 103:237–249
17. Jing P, Li B, Wu B, Zhang N, Xu Y, Lv X (2018) A millimeter wave SIW slot antenna array for 5G communication. In: International conference on microwave and millimeter wave technology (ICMMT). IEEE, Chengdu, pp 1–3
18. Che W, Deng K, Wang D, Chow YL (2008) Analytical Equivalence between Substrate-Integrated Waveguide and Rectangular Waveguide. IET microwaves, antennas & propagation 2(1):35–41
19. Cassivi Y, Perregrini L, Arcioni P, Bressan M, Wu K, Conciauro G (2002) Dispersion Characteristics of Substrate Integrated Rectangular Waveguide. IEEE Microwave and Wireless components letters 12(9):333–335
20. Xu F, Wu K (2005) Guided-wave and Leakage Characteristics of Substrate Integrated Waveguide. IEEE Transactions on Microwave Theory and Techniques 53(1):66–73
21. Bozzi M, Georgiadis A, Wu K (2011) Review of Substrate-Integrated Waveguide Circuits and Antennas. IET Microwaves, Antennas & Propagation 5(8):909–920
22. Deslandes D, Wu K (2006) Accurate Modeling, Wave Mechanisms, and Design Considerations of a Substrate Integrated Waveguide. IEEE Transactions on Microwave Theory and Techniques 54(6):2516–2526
23. Deslandes D (2010) Design equations for tapered microstrip-to-substrate integrated waveguide transitions. In: IEEE MTT-S international microwave symposium, pp 704–707
24. Ansys® Electronics desktop, HFSS-v20, Ansys Inc. www.ansys.com

Compact Multiple Input Multiple Output Antenna with High Isolation for UWB Applications

Pankaj Kumar Keshri, Sanjay Sahu, and Richa Chandel

Abstract A miniature multiple-input multiple-output antenna that comprised of four elements is presented in this literature. The antenna is design with FR4 substrate whose dimension is 38 mm × 38 mm × 1.6 mm. The antenna consists of four fork-shaped radiator out of two placed on top and two placed on bottom of the substrate. The isolation is improved by placing antenna element orthogonal which gives dual polarization. Further, for more isolation improvement, a decoupling structure has been used between partial slotted ground planes. The bandwidth is improved by slotted ground structure and multiple slit cut on radiator. The impedance bandwidth of the proposed antenna is found to be 3.1–11 GHz which covers whole UWB range. The characteristics of the antenna has been described in terms of return loss less than −10 dB, isolation parameter less than −20 dB up to 8 GHz, remaining frequency band less than −16 dB and realized gain from 0.5 to 6.4 dB. The performance of MIMO antenna has been measured in terms of isolation, ECC and diversity gain. The proposed antenna is suitable for various portable devices.

Keyword Envelope correlation coefficient (ECC) · Ultra-wide band (UWB) · MIMO antenna

1 Introduction

Nowadays, due to fast development in wireless communication every wireless device required high data speed along with large bandwidth. The UWB technology full-filled the above demand with low power spectral density, low cost and low power consumption. The unlicenced frequency range 3.1–10.6 GHz for UWB was decided

P. K. Keshri (✉) · S. Sahu
Lovely Professional University, Phagwara, Punjab, India
e-mail: pankaj.05ec@gmail.com

R. Chandel
University Institute of Technology, Himachal Pradesh University, Shimla, India

B. Mishra and M. Tiwari (eds.), *VLSI, Microwave and Wireless Technologies*,
Lecture Notes in Electrical Engineering 877,
https://doi.org/10.1007/978-981-19-0312-0_5

by FCC in 2002 [1]. The UWB technology is suitable for short distance communication but multipath fading is big problem with this technology. This issue resolved by implementing MIMO antenna along with UWB technology. The high polarization diversity is achieved in MIMO system because of placing multiple antenna in transmitter and receiver side. In MIMO, multiple signals are transmitted simultaneously and receiver received required signal correctly in the presence of any obstacle. However, in a MIMO system, the high possibility of mutual interaction between antenna elements is increased by arranging several antennas in a limited space. The performance of MIMO antennas is degraded due to mutual coupling. The mutual coupling could be improved by placing antenna elements far apart, reducing radiator size, or making antenna elements orthogonal. In recent years, there are several technique used to reduce mutual coupling like defected ground structure [2–5], tree-like structure [6], electromagnetic bandgap structure (EBG) which absorbed interference [7, 8], metamaterial structure [9] and various types of structure used on ground like F-shaped, T-shaped or inverted L-shaped stub [10–12]. The isolation is also improved by using complementary split ring resonator on ground [13] and neutralization line between feed line [14, 15].

In this literature, a miniature MIMO antenna whose dimension 38 $mm^2 \times 38$ mm^2 is proposed. The miniaturization and bandwidth of antenna are improved by introducing multiple slits on radiator and cutting slot at optimum position on partial ground plane. The isolation is enhanced by placing antenna orthogonally, and high diversity is achieved by placing two antenna elements placed on both side of the substrate. Placing antenna orthogonally increases polarization diversity and achieves dual polarization. Furthermore, isolation is improved by addition of decoupling structure on ground plane and top surface between two partial slotted ground. It acts as a filter to restrict the flow of current from one antenna element to other. The impedance bandwidth of designed antenna is 7.9 GHz (3.1–11 GHz) which satisfies the UWB range decided by FCC. The proposed design is suitable for recent wireless devices.

2 MIMO Design

The two-dimensional layout of proposed antenna is presented in Fig. 1 and has four identical elements which are orthogonally placed. The overall dimension of antenna is very compact 38 mm × 38 mm.The proposed antenna is optimized and designed using high frequency structure simulator (HFSS) software. The substrate material FR4 (dielectric constant 4.4 and loss tangent 0.02) with 1.6 mm thick is used for designing the antenna. The optimized dimension of various parameters labelled in antenna structure is presented in Table 1.

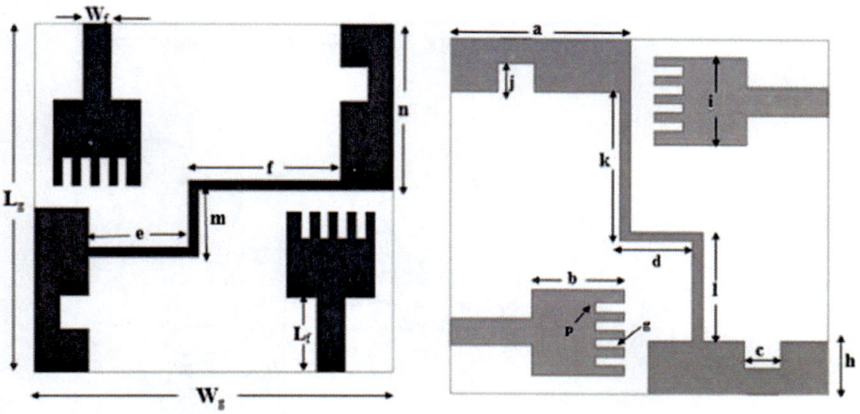

(a) Top (b) Bottom

Fig. 1 Two-dimensional layout of proposed antenna

Table 1 Optimized dimension (mm) of various parameters depicted in Fig. 1

Parameter	W_g	W_f	a	b	c	d	e	f	g	p
Dimension (mm)	38	3	18	9	3.8	7.5	11	16.4	3	1
Parameter	L_g	L_f	h	i	j	k	l	m	n	
Dimension (mm)	38	8.5	5.8	9	3	16.4	12	7.5	3	

3 Results and Discussions

Figure 2 represents the return loss at various steps. In step 1, all elements are placed orthogonally and it is resonating at 5.6, 7.6 and 10.6 GHz. But isolation is very poor up to 8 GHZ as shown in Fig. 3, which has more value than desired value −15 dB, and for higher frequency, it is less than −15 dB. In step 2, a strip was placed on top of the substrate which adds one lower resonant frequency and also improves isolation. Further, for more reduction of mutual coupling, strip was added on ground as well as on top surface. It can be seen in Fig. 3, and isolation is less than −20 dB up to 7.8 GHz and at higher frequency less than −16 dB. The isolation has measured when port 1 excited and other port deactivated.

Fig. 4 shows a two-dimensional surface current distribution in which the current density is highly coupled in Fig. 4a, c without the use of any decoupling mechanism. This mutual coupling is reduced by adding decoupling element between two partial slotted ground plane on top and bottom surface. The function of decoupling element is to restrict the flow of current to other element. It acts as filter to reject the current and divert in some other direction. It can be seen in Fig. 5, the realized maximum gain of antenna is 6.4 dB, and radiation efficiency varied from 28 to 70%. The antenna gain is maximum at operating frequency. The two-dimensional radiation pattern in

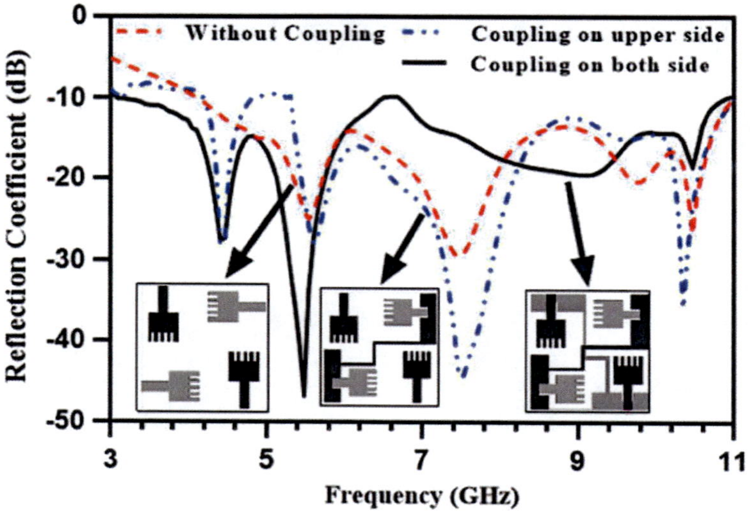

Fig. 2 Reflection coefficient at various stages

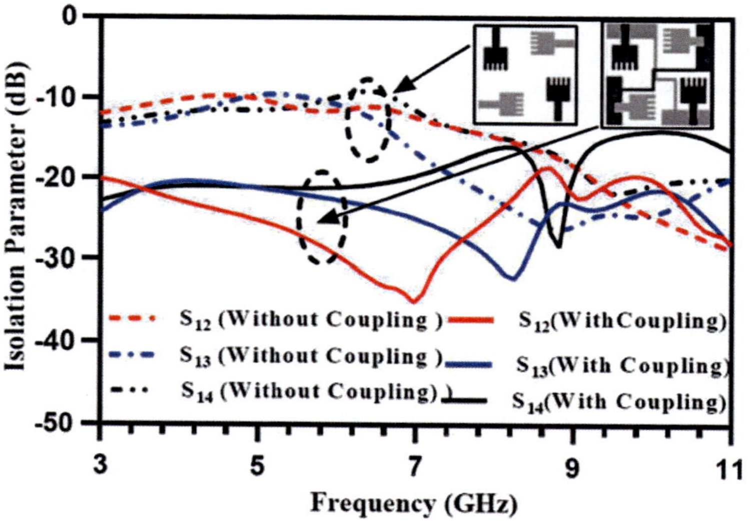

Fig. 3 $S_{12}/S_{13}/S_{14}$ at various stages when port 1 excited

xz (E-plane) and yz (H-plane) direction at 4.3 and 5.4 GHz is presented in Fig. 6. It can be seen that radiation pattern is directional in E-plane and close to omnidirection in H-plane (Fig. 7).

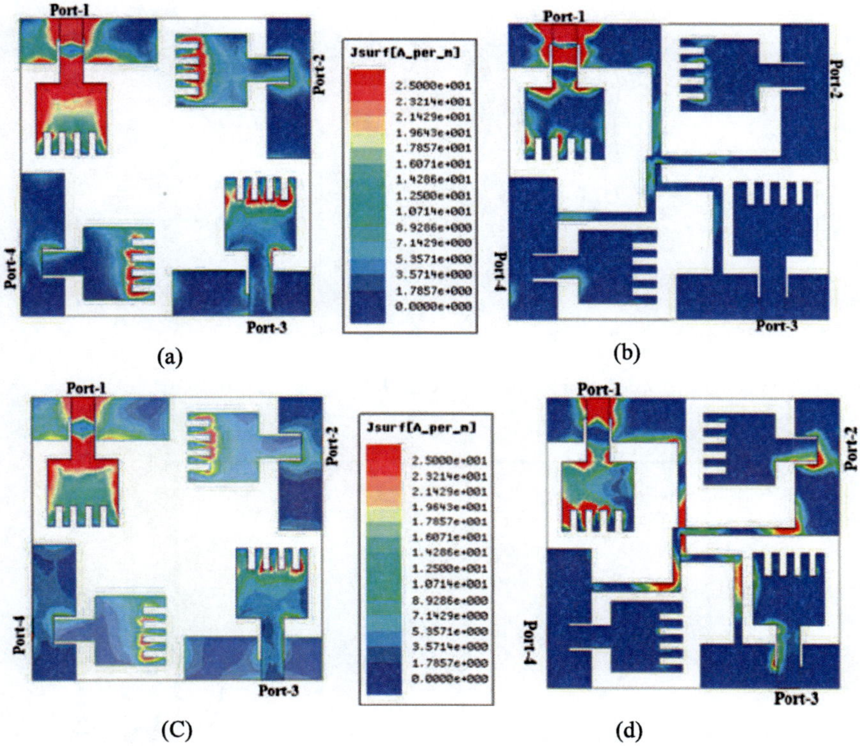

Fig. 4 Two-dimensional current distribution at operating frequency 4.3 GHz in **a**, **b** and at 5.4 GHz in **c**, **d**

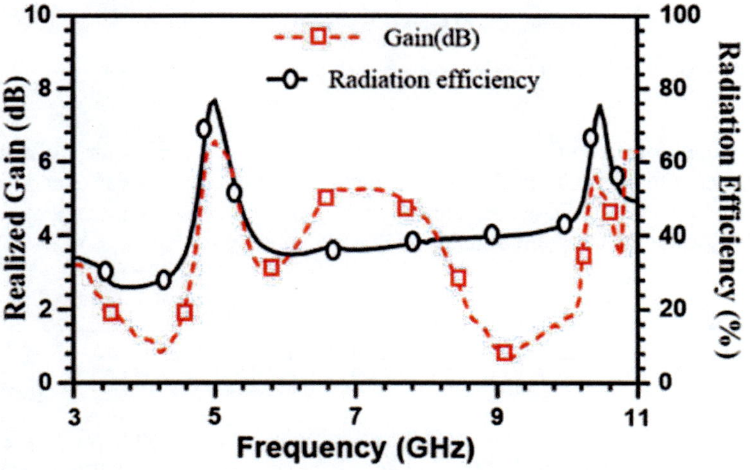

Fig. 5 Realized gain and radiation efficiency

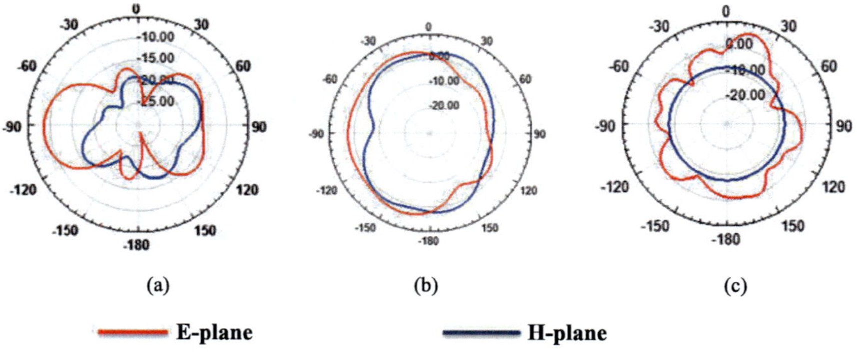

Fig. 6 Two-dimensional radiation pattern **a** 4.3 GHz, **b** 5.4 GHz and **c** 10.4 GHz

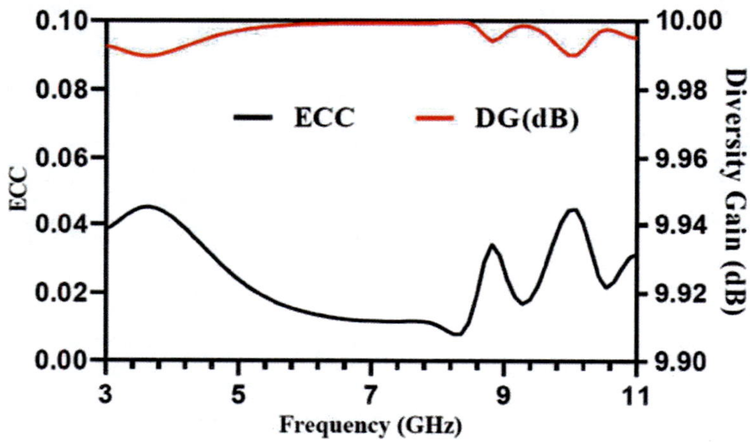

Fig. 7 Simulated ECC and DG of proposed antenna

Envelope correlation and diversity gain are important parameter for deciding MIMO performance. The ECC value decides how much antenna correlated to each other. The lower the ECC value, the better the isolation between antenna elements. The ECC value for four-port antenna can be obtained by [16].

$$\rho_e = \frac{\left|S_{11}^* S_{12} + S_{21}^* S_{22} + S_{13}^* S_{32} + S_{14}^* S_{42}\right|^2}{\left(1 - |S_{11}|^2 - |S_{21}|^2 - |S_{31}|^2 - |S_{41}|^2\right)\left(1 - |S_{12}|^2 - |S_{22}|^2 - |S_{32}|^2 - |S_{42}|^2\right)} \quad (1)$$

The ECC value of proposed antenna is very low 0.047 as compared to practical accepted value. It signifies that the suggested antenna will work well with wireless devices. The diversity gain of antenna is directly calculated from ECC by [11].

$$DG = 10\sqrt{1 - |\rho_e|^2} \tag{2}$$

The diversity gain of antenna is very close to practical accepted value. It has more than 9.9 dB in whole UWB range.

4 Conclusion

In present literature, a novel miniature quad port antenna is described. The high bandwidth 3.1–11 GHz is achieved due to partial ground and multiple slits on radiator. The isolation is improved by adding strip on ground and top surface along with placing antenna element in orthogonal orientation. In most of the bands, the antenna achieved less than −20 dB isolation, a lower ECC value, and excellent efficiency and gain. The antenna is good candidates for various wireless devices.

References

1. Revision of part 15 of the commission's rules regarding ultra-wideband transmission systems first report and order. Federal Communications Commission, Washington, DC, USA, Tech. Rep. FCC 02.V48, 2002
2. Zhu F-G, Xu J-D, Xu Q (2009) Reduction of mutual coupling between closely-packed antenna elements using defected ground structure. Electron Lett 45(12):601–602
3. Khandelwal MK, Kanaujia BK, Kumar SS (2017) Defected ground structure: fundamentals, analysis, and applications in modern wireless trends. Int J Antennas Propag 2017:22. ID 2018527
4. Zhu J, Feng B, Peng B, Deng L, Li S (2016) A dual notched band MIMO slot antenna system with Y-shaped defected ground structure for UWB applications. Microw Opt Technol Lett 58(3):626–630
5. Srivastava G, Mohan A (2016) Compact MIMO slot antenna for UWB applications. IEEE Antennas Wirel Propag Lett 15:1057–60
6. Zhang S, Ying Z, Xiong J, He S (2009) Ultrawideband MIMO/diversity antennas with a tree-like structure to enhance wideband isolation. IEEE Antenna Wirel Propag Lett 8:1279–1282
7. Ghosh S, Tran T-N, Le-Ngoc T (2014) Dual-layer EBG-based miniaturized multi-element antenna for MIMO systems. IEEE Trans Antennas Propag 62(8):3985–3997
8. Jaglan N, Gupta SD, Thakur E, Kumar D, Kanaujia BK, Srivastava S (2018) Triple band notched mushroom and uniplanar EBG structures based UWB MIMO/diversity antenna with enhanced wideband isolation. Int J Electron Commun (AEÜ) 90:36–44
9. Ketzaki DA, Yioultsis TV (2013) Metamaterial-based design of planar compact MIMO monopoles. IEEE Trans Antennas Propag 61(5):2758–2766
10. Kang L, Li H, Wang X, Shi X (2015) Compact offset microstripfed MIMO antenna for band-notched UWB applications. IEEE Antenna Wirel Propag Lett 14:1754–1757
11. Chandel R, Gautam AK, Rambabu K (2018) Tapered fed compact UWB MIMO-diversity antenna with dual band-notched characteristics. IEEE Trans Antennas Propag 66(4):1677–84
12. Iqbal A, Saraereh OA, Ahmad AW, Bashir S (2018) Mutual coupling reduction using F-shaped stubs in UWB-MIMO antenna. IEEE Access 6:2755–2759

13. Khan MS, Capobianco AD, Asif SM, Anagnostou DE, Shubair RM, Braaten BD (2017) A compact CSRR-enabled UWB diversity antenna. IEEE Antennas Wirel Propag Lett 16:808–812
14. Kayabasi A, Toktas A, Yigit E, Sabanci K (2018) Triangular quad-port multi-polarized UWB MIMO antenna with enhanced isolation using neutralization ring. AEU Int J Electron Commun 85(4):7–53
15. Zhang S, Pedersen GF (2016) Mutual coupling reduction for UWB MIMO antennas with a wideband neutralization line. IEEE Antennas Wirel Propag Lett 15:166–169
16. Yussuf AA, Paker S (2019) Design of a compact quad-radiating element MIMO antenna for LTE/Wi-Fi application. AEU-Int J Electron Commun 111:152983–152902

Survey on Services and Support in Cloud Computing Environment

Abhishek Kumar Pandey and Sarvpal Singh

Abstract Cloud computing serves a large portion of enterprises and intellectuals with the facility of computation as per their demand and proves to be successful. There are multiple services being provided by numerous cloud service providers (CSPs) along with widespread range of framework for software. Also, the client possesses some requirement of application and capabilities of system that are not outsourced. Thus, making it difficult to fit everything under one cloud environment. In this paper, we explain the services and platforms provided by cloud service providers. We also showcase a study, comparative in nature, detailing the services by open-source-based cloud computing environment. The paper gives justifications to support which cloud environment a client is comfortable with according to the requirement and resources of client.

Keywords IaaS cloud computing environment · PaaS cloud computing environment · SaaS cloud computing environment

1 Introduction

Cloud computing is a model for supplying sources, namely hardware and software, over the network and is widely spread. It utilizes pay per use billing model [1] to ensure provisioning of resources on demand. In the aforesaid provisioning, it unites the principles of grid computing and utility computing. Elasticity and scalability of infrastructure and ability to share same application with multiple customers are the prominent characteristics of cloud computing.

A. K. Pandey (✉) · S. Singh
Department of Information Technology and Computer Application, MMMUT, Gorakhpur, UP, India
e-mail: abhi_007cs@yahoo.co.in

S. Singh
e-mail: spsingh@mmmut.ac.in

B. Mishra and M. Tiwari (eds.), *VLSI, Microwave and Wireless Technologies*, Lecture Notes in Electrical Engineering 877,
https://doi.org/10.1007/978-981-19-0312-0_6

Categorization of cloud computing is as follows: IaaS supplying hardware facilities, PaaS providing prebooted with OS server facilities and SaaS providing applications user-centric in nature. The approach of cloud deployment also differs. CSP owns and set up public cloud, whereas private clouds are controlled only by single client. Applications based on outdated technology can be integrated by small and medium enterprises with applications based on cloud.

Cloud computing is defined as fifth utility in paper [2]. Comparison of cloud computing environment is given in paper [3–6]. Private cloud performance evaluation has been described in paper [7]. Discussion to evaluate clouds based on frameworks and parameters is done in paper [8, 9]. PaaS CSPs are detailed in paper [10]. Paper [11] showcases contrast between open-source clouds based on different service providers. But all of the aforesaid studies fail to give deep understanding of selection of cloud environment based upon requirement of client. Also, most of the studies earlier focus on IaaS.

In this paper, we have provided some part of the study of PaaS service providers as well. The factors that client must consider for the selection of cloud service include budget, expertise in IT, sensitivity of data and application and availability of infrastructure. The same can be used to choose appropriate CSP and cloud deployment model. The aim of this paper is to study various widespread cloud computing environments and as per the requirements of the client to suggest a suitable cloud environment. Services provided under IaaS and PaaS are discussed in Sects. 2 and 3. Recommendations with justifications for appropriate cloud environment are given in Sect. 5.

2 Infrastructure as a Service (IaaS) Cloud Computing Environment

Hardware resources including memory, CPU and bandwidth of networks are provided by IaaS. CSP owns and manages the resources. Resources are rented to the clients. There is flexibility to client for choosing framework, OS. By choosing IaaS, client gets to invest in other fields rather than infrastructure management. IaaS services can be classified as:

2.1 Computational Services

Resources that are essential, like CPU and memory, for execution of program, are provided by computational services in many ways as logical units. Hypervisor layer hides underlying hardware and introduces server virtualization. The host hardware is divided into virtual machines as logical units. Each virtual machine acts as an independent machine. Hypervisor isolates the VM which share the hardware of host machine. Virtualization-based instances can be opted by clients having single-tiered software application. Some CSP uses VMWare ESXi, and some uses open-source hypervisor.

For example, GCP [12] uses kernel VM. XenServer is supported by AWS [13]. A less complex option other than VM is the containerization-based approach's practice. It divides user space in containers but uses host OS. OS instance is attached to a particular application in each container, thus, making it lighter than VM.

Docker [14], a runtime-based container service, is supported by most of the open-source cloud environments. Clients who develop granulated microservices-based applications can choose this.

Very few CSPs provide bare metal-based IaaS, like IBM. Flexibility is harnessed by the client but the client also gets the responsibility of administration of IT.

Preemptive VMs are offered at low cost but when the scarcity of resources is encountered, and instances having strict SLA require resources, then, preemption happens. This may lead the user to lose the work. Non-critical applications may use this.

Eucalyptus gives descent API with services of AWS, so it is easy to develop hybrid cloud. CSPs support migration of legacy applications on cloud, making them more scalable.

2.2 *Network Services*

Network resources consist of interface of network, bandwidth and some traffic dealing components, mainly middleware. Virtual networking is provided over physical networking, among VMs, by the service providers. VPN is used to establish network traffic isolation between different organization's VM. A private network is created for the virtual machine that the organization of client uses. Machines can access each other only if they fall under same VPN, else they cannot access each other. This is used to ensure the isolation of application and data of various organizations in a shareable environment. Clients need VPN to establish secure inter-application communication if they work with workflow-based applications.

Content delivery network (CDN) is another major service that CSPs provide whose principle is to cache the web content near client on network edge point. This minimizes delay in loading web content. When the client sends request the first time, it is sent to the actual server. The server then responds whether the response can be cached or not. If the response is cacheable, then it is cached by the network edge node. So, if the same request is sent again, then, instead of the server giving the response, cache is used to send the response, this is how the response time is improved. Content delivery network benefits web applications with static content.

2.3 Storage Services

There are two types of storage provided by IaaS. They are block storage facility and object storage facility. In block storage facility, the structure of data is not considered and is stored in raw format. Block storage breaks up data into blocks and then stores these blocks as separate pieces, each with a unique identifier. A block is transferred only if it contains the part of file being requested rather than transferring the whole file. Some examples of block type storage are AWS's elastic block storage, GCP's persistent and OpenStack's cinder service. Cold storage service is used to archive least frequently used data like videos and images. They offer low prices but have higher latency.

Another one is the object storage. Object storage saves data as of particular type like table of a database or a file; it is used when we need to store massive unstructured data. Some examples of object storage are OpenStack's [15] swift, AWS's S3 or GCP's cloud storage. SAN-based block storage/NAS-based file system-based storage is offered by some CSPs like IBM, VMWare [16] and Microsoft.

3 Platform as a Service (PaaS) Cloud Computing Environment

Software frameworks, in PaaS, are used for easy deployment of applications by pretending them as an instance on hardware. The responsibility of application configuration and its installation is taken by the CSP. This ensures the concentration of client developer is only on development of application. In PaaS, the CSP like APP engine or Google can themselves own the hardware, or other IaaS providers might rent them like cloud foundry on IBM [17] datacenters.

Linux distributions and version of windows are supported by most of the CSPs. A vast range of languages used for programming like Ruby, Java, PHP, python, etc., are also supported. Mobile applications are also supported by providers, which use the features of cloud computing like massiveness and scalability. Google cloud platform [12] provides firebase framework, which is one among the examples.

CSPs provide database services for both structured and unstructured databases. Transactional data are managed by MySQL, SQL Server and PostgreSQL, which fall under the category of structured or relational databases. Big data, images and videos that are collected from different sources are stored using unstructured databases or NoSQL databases like MongoDB, Cassandra, HBase, etc. The increase in popularity of big data has introduced the use of unstructured databases. Latency of I/O operations can be reduced by using Redis, an example of in-memory databases.

Areas like IoT, ML and big data are witnessing an increment of demand of development of applications in the past one decade. Many CSPs keep this in mind and they provide platforms for the development and deployment of applications related to these fields. Frameworks like cloud-based Hadoop cluster are special framework which facilitates parallel and distributed processing which are map-reduce based. To process stream data, spark integrated with Hadoop is provided.

Many service providers provide ML-based APIs to develop ML-based application. APIs for image classification, NLP, text analysis are provided.

IoT cloud is a new concept which is the integration of IoT and cloud computing. For the development of IoT-oriented applications, software development kits of various languages are provided by many CSP. These include embedded C, Java, .net, etc. Message queuing protocols, for example, AMQP or MQTT are supported to ensure fault tolerance and reliability.

4 Software as a Service (SaaS) Cloud Computing Environment

User-centric applications are provided over the network by SaaS that are ready to use. Here, importance is given to user-based customization, convenient use, look and feel, etc. In SaaS, application development, its deployment and publication are the responsibility of SaaS providers. Frameworks and infrastructure that are related to an owner or ownership may be used by SaaS. For supplying the services, Facebook, LinkedIn, government offices and Twitter, that are social networking giants, use their own datacenters. SaaS providers that are small and medium-sized, guarantee scalability, availability and robustness by utilizing public IaaS and PaaS without straight investment in the aforesaid technologies. Public cloud service providers host the websites that offer courses and tourism. SaaS service model offers a vast range of services. Web-based applications are a majority of most of the applications. The management of exceeding application requests, upgradation of platform and security threats that are network based should be ensured by the SaaS provider.

5 Key Findings-IaaS and PaaS Cloud Computing Environments

As an IaaS client, one should note the following key points regarding CSPs.

Computational, network and storage resources are provided by AWS, GCP and Azure which provide huge service range. Bare metal servers are provided by IBM as part of IaaS. This supplies the client with the flexibility of choosing middleware components and hypervisors. This is useful for the client having skilled IT staff but low infrastructure. Deployment of private or dedicated cloud can be recommended on VMWare IaaS, but it is not feasible economically being licensed.

From the perspective of open-source IaaS providers, reasonable rates are provided by OpenStack. Client enjoys flexibility with full control, parallelly, it requires highly skilled IT staff. Eucalyptus can be a choice if in the future, hybrid cloud with AWS exists together. Other than OpenStack, no open-source IaaS support CDN services and cold storage.

As a PaaS client, one should keep the following pointed noted.

GCP and AWS are equally considerable for their offerings. Vast range of programming languages and database support are provided by both. They provide IoT-related machine learning application development frameworks. Microsoft Azure [18] supports most programming languages and database over PaaS, but has limited support for machine learning and IoT-related application. Third party cloud foundry framework is the base for IBM PaaS service. If PaaS cloud has to be built on third party infrastructure, then cloud foundry and open shift online can be considered as they support many languages and databases. If amalgamated with GCP or AWS or third party software is used, they can be used for machine learning, IoT domain services.

A chart representing the recommendation for various cloud environments according to the requirement set of the client is given below.

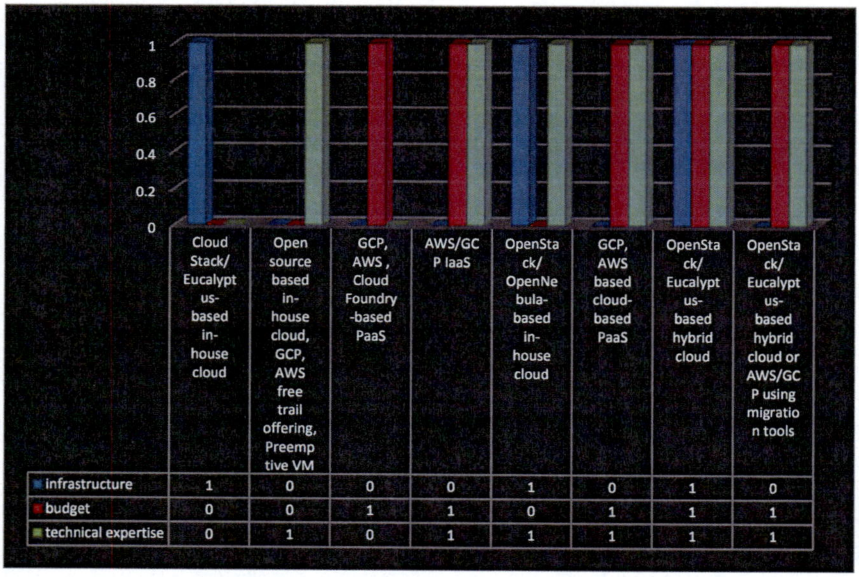

Here, 0 means the corresponding parameter for consideration is Irrelevant, whereas 1 represents that the parameter for consideration is Relevant.

6 Conclusion and Future Plans

If the client has infrastructure only as system parameter, for example, academic projects, then cloud stack [19]/eucalyptus-based in-house cloud is recommended as it is open source and easy to install.

If the client has technical expertise as system parameter, for example, academic projects, research projects, then open-source-based in-house cloud, GCP, AWS free trial offerings, preemptive virtual machine are recommended as they are economical and setup can be done on commodity hardware.

If the client has budget only as system parameter, for example, SME, startups, then GCP, AWS, Microsoft, cloud foundry-based PaaS services are recommended as they have less IT overhead due to ready to use instances.

If the client has budget and technical expertise as system parameter, for example, SME, startup, then AWS/GCP IaaS services are recommended as they provide scalable infrastructure on public cloud.

If the client has infrastructure and technical expertise as system parameter, for example, research projects, then OpenStack/OpenNebula [20]-based in-house cloud is recommended as they provide complex process to install but flexible framework.

If the client has budget and technical expertise as system parameter, with special domains such as big data, machine learning, IoT, for example, research projects SME, then GCP, AWS cloud-based PaaS services are recommended as they offer rich APIs and high-quality library-based service.

If the client has infrastructure, budget and technical expertise as system parameter, with sensitive data and critical applications, for example, research projects SMEs, then OpenStack/eucalyptus-based hybrid cloud is recommended as the private cloud is under client' control.

If the client has budget and technical expertise as system parameter, with multiple legacy applications, for example, old corporates or organizations, then OpenStack/eucalyptus-based hybrid cloud or AWS/GCP using migration tools is recommended as hybrid cloud is cost effective.

Scalability, availability and supply of network-based resources on demand are ensured by cloud computing. Various services are provided by service providers under IaaS, PaaS and SaaS models. Services such as computing, storage and network resources are provided by IaaS CSPs, whereas the services such as offering development and deployment environment for software are provided by PaaS CSPs. Frameworks for new technology are also provided by PaaS.

As the CSP provides a widespread range of services, then, it becomes easier for the user to choose from. Clients can easily choose an appropriate CSP as per their requirement. We have given the justification for recommending different clouds to client according to our study. Open-source cloud frameworks can be easily used for academic and research projects due to being economical. Although, they have limitation of providing limited support to special purpose applications like machine learning, IoT and big data. They can be amalgamated with third party tools. Google cloud platform and Amazon web services can be chosen more adequately at an enterprise level as they provide services that are highly rich and comprehensive. They serve as public proprietary cloud providers. For hosting non-critical applications, applications, proprietary CSPs have introduced preemptive instances that are economical.

The study of geographical distribution of datacenters is intended as a future plan. The technical strategies of resources provisioning are also planned to be studied. This will further extend to dead balancing, auto-scaling to ensure scalability, availability and reliability.

References

1. Buyya R, Yeo CS, Venugopal S, Broberg J, Brandic I (2009) Cloud computing and emerging IT platforms: vision, hype, and reality for delivering computing as the 5th utility. Futur Gener Comput Syst 25:599–616. https://doi.org/10.1016/j.future.2008.12.001
2. Ahmed M, Litchfield AT (2018) Taxonomy for identification of security issues in cloud computing environments. J Comput Inf Syst 58:79–88. https://doi.org/10.1080/08874417.2016.1192520
3. Butt SA, Tariq MI, Jamal T, Ali A, Diaz Martinez JL, De-La-Hoz-Franco E (2019) Predictive variables for agile development merging cloud computing services. IEEE Access 7:99273–99282. https://doi.org/10.1109/ACCESS.2019.2929169
4. Dhivya R, Devi R, Shanmugalakshmi R (2016) Parameters and methods used to evaluate cloud service providers: a survey. In: International conference on computer communication and informatics (ICCCI). IEEE, pp 1–5. https://doi.org/10.1109/ICCCI.2016.7479947
5. Varma MK, Choi E (2016) Comparative study of various platform as a service frameworks. Int J Cloud Comput Serv Archit 6:23–34. https://doi.org/10.5121/ijccsa.2016.6103
6. von Laszewski G, Diaz J, Wang F, Fox GC (2012) Comparison of multiple cloud frameworks. In: IEEE fifth international conference on cloud computing. IEEE, pp 734–741. https://doi.org/10.1109/CLOUD.2012.104
7. Lou P, Yuan L, Hu J, Yan J, Fu J (2018) A comprehensive assessment approach to evaluate the trustworthiness of manufacturing services in cloud manufacturing environment. IEEE Access 6:30819–30828. https://doi.org/10.1109/ACCESS.2018.2837664
8. Sun L, Dong H, Hussain FK, Hussain OK, Chang E (2014) Cloud service selection: state-of-the-art and future research directions. J Netw Comput Appl 45:134–150. https://doi.org/10.1016/j.jnca.2014.07.019
9. Casas P, Schatz R (2014) Quality of experience in cloud services: survey and measurements. Comput Networks 68:149–165. https://doi.org/10.1016/j.comnet.2014.01.008
10. Noshy M, Ibrahim A, Ali HA (2018) Optimization of live virtual machine migration in cloud computing: a survey and future directions. J Netw Comput Appl 110:1–10. https://doi.org/10.1016/j.jnca.2018.03.002
11. Mateo-Fornes J, Solsona-Tehas F, Vilaplana-Mayoral J, Teixido-Torrelles I, Rius-Torrento J (2019) CART, a decision SLA model for SaaS providers to keep QoS regarding availability and performance. IEEE Access 7:38195–38204. https://doi.org/10.1109/ACCESS.2019.2905870
12. Tang M, Dai X, Liu J, Chen J (2017) Towards a trust evaluation middleware for cloud service selection. Futur Gener Comput Syst 74:302–312. https://doi.org/10.1016/j.future.2016.01.009
13. Bilecki LF, Fiorese A (2017) A trust reputation architecture for cloud computing environment. In: IEEE/ACS 14th international conference on computer systems and applications (AICCSA). IEEE, pp 614–621. https://doi.org/10.1109/AICCSA.2017.31
14. Wang Y, Wen J, Zhou W, Tao B, Wu Q, Tao Z (2019) A cloud service selection method based on trust and user preference clustering. IEEE Access 7:110279–110292. https://doi.org/10.1109/ACCESS.2019.2934153

15. Singh A, Chatterjee K (2017) A multi-dimensional trust and reputation calculation model for cloud computing environments. In: ISEA Asia security and privacy (ISEASP). IEEE, pp 1–8. https://doi.org/10.1109/ISEASP.2017.7976983
16. Rajganesh N, Ramkumar T (2016) A review on broker based cloud service model. J Comput Inf Technol 24:283–292. https://doi.org/10.20532/cit.2016.1002778
17. Anjana PS, Badiwal P, Wankar R, Kallakuri S, Rao CR (2019) Cloud service provider evaluation system using fuzzy rough set technique. In: IEEE international conference on service-oriented system engineering (SOSE). IEEE, pp 187–18709. https://doi.org/10.1109/SOSE.2019.00033
18. Nagarajan R, Thirunavukarasu R (2018) A review on intelligent cloud broker for effective service provisioning in cloud. In: Second international conference on intelligent computing and control systems (ICICCS). IEEE, pp 519–524. https://doi.org/10.1109/ICCONS.2018.8662953
19. Noor TH, Sheng QZ, Yao L, Dustdar S, Ngu AHH (2016) CloudArmor: supporting reputation-based trust management for cloud services. IEEE Trans Parallel Distrib Syst 27:367–380. https://doi.org/10.1109/TPDS.2015.2408613
20. Garg SK, Versteeg S, Buyya R (2013) A framework for ranking of cloud computing services. Futur Gener Comput Syst 29:1012–1023. https://doi.org/10.1016/j.future.2012.06.006

Human Authentication Using Score Level Fusion of Face and Palm Print Biometrics

Rohit Srivastava and Dhirendra Kumar Sharma

Abstract Human authentication has always been of significant concern for research. Different methodologies and approaches have been proposed in the last few years to accurately authenticate humans on different biometric characteristics. For the accurate authentication process, there is a requirement of a template, which can identify humans from a given set of images. For the generation of templates, various biometric features have been tested and different methodologies have been applied. Out of these methods, the most prominent biometric features identified are face and palm print. These biometric traits have given the maximum authentication rate when applied as a single biometric. In this paper, a combination framework is proposed in which principal component analysis (PCA) and linear binary pattern (LBP) are applied on both face and palm print to generate a unique score that is used to authenticate the human. For validation of the framework two different databases, ORL (Olivetti Research Laboratory) and PolyU (Hong Kong Polytechnic University) are used. The framework achieved an accuracy of 99.8%, which is far better as compared to the unimodal system.

Keywords Multimodal biometric · PCA · LBP · ORL · PolyU · Face recognition · Palm recognition

1 Introduction

The biometric framework is utilized for authentication and acknowledgment framework. The framework utilizes the data of an individual to recognize dependent information, which is novel and explicit to their natural characteristics. The biometric is utilized in businesses, associations, and governments for security reasons. A biometric framework is ordered into unimodal/single modular and multimodal. The unimodal utilize a single characteristic, for example, a face, finger, and palm print for the acknowledgment [1]. The unimodal do not satisfy the degree of execution and

R. Srivastava (✉) · D. K. Sharma
School of Computer Science, Department of Cybernetics, UPES, Dehradun, India
e-mail: rts080185@gmail.com

B. Mishra and M. Tiwari (eds.), *VLSI, Microwave and Wireless Technologies*,
Lecture Notes in Electrical Engineering 877,
https://doi.org/10.1007/978-981-19-0312-0_7

has less exactness due to which the multimodal framework is an assurance to create a superior outcome. The multimodal uses more characteristics, for example, finger–face, face–palm print, and palm–retina print. The multimodal biometric framework shows a few favorable circumstances in numerous angles when contrasted with that of a unimodal biometric framework because of different sources. Multimodal biometric has amazing precision and accuracy than unimodal [2]. The multimodal biometric framework combination methods consolidate at least two distinctive biometric modalities (feature, matching score, or decision) into another space to produce another combination esteem [3].

Data combination is a significant technique in numerous images preparing the application. The combination can be separated into two classes, which are before matching and after matching. In multimodal biometric, combination is a significant stage to deliver higher segregation power in the component space. The combined features now and then has high separation power and required component determination calculation to lessen the dimensionality. A wide range of combination images can be actualized in the spatial or recurrence space. The term of spatial area alludes to the typical space of the image. The information image straightforwardly works at pixel level depends on pixels' position understanding. The recurrence area term alludes to the change from spatial space to another space. The destinations of this exploration center to an improvement of combination procedure to upgrade the most elevated precision of the acknowledgment framework in the spatial space for face and palm print biometric.

2 Proposed Methodology

2.1 Preprocessing

Image preprocessing is the most important stage in the biometric acknowledgment framework. The pre-handling process incorporates changing example images into grayscale, upgrade image quality, edge identification, and ROI trimming [4]. During the image catching procedure, numerous elements influence the information, for example, brightening, differentiation, etc. It responds to the acknowledgment framework exactness, instantaneity, and strength [5, 6]. Before the preparation procedure, the information must be handled and converted into the appropriate organization, for example, size, grayscale, and direction. A trimming strategy is then required for the palm print image to catch the focal point of the palm image (Fig. 1).

Local Binary Pattern (LBP): Feature extraction gives the effect to the presentation and exactness of the framework. The element vector of the example images removed utilizing the particular strategy permits the framework to perceive the extraordinary example in the element space [7–9]. Feature extraction dependent on LBP is the basic structure that speaks to the histogram and has shown predominant outcome on

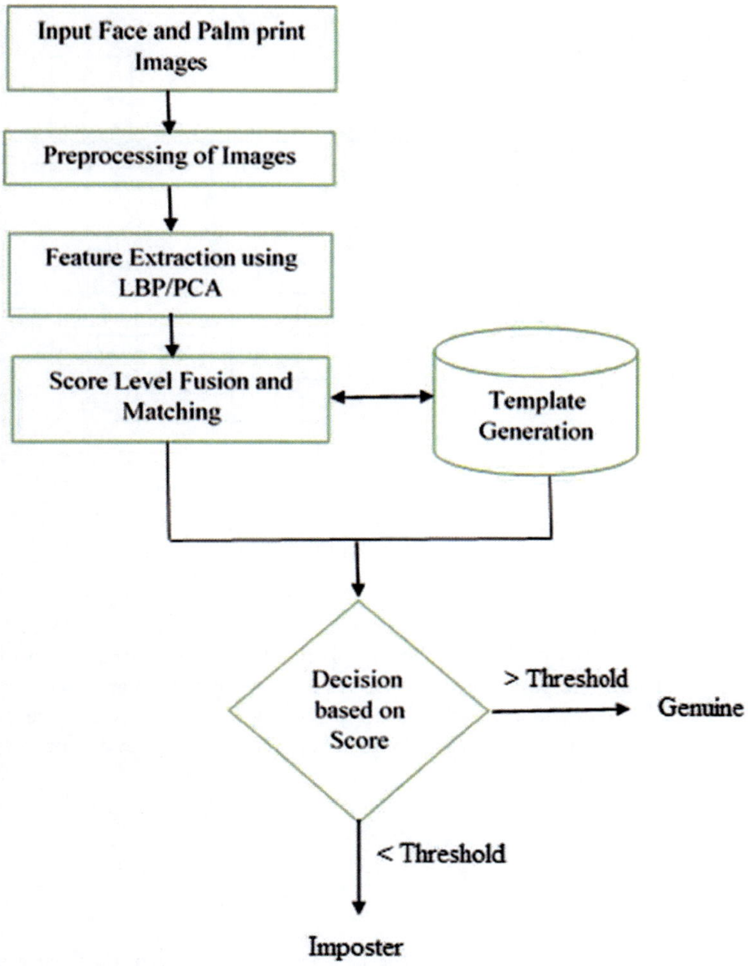

Fig. 1 Proposed framework

the component arrangement execution. LBP separates the image on the pixel esteem with it's neighbourhood in a grayscale image.

The pixel estimation of the neighborhood is deducted with the middle pixel esteem. The deduction is performed utilizing a limit and incentive. The edge esteem utilizes the standard if neighborhood pixel higher than focus pixel offered 1 to that area, and in any case given 0 [10]. The aftereffect of the limit is changed over into a decimal incentive to supplant the middle worth. The outcomes for all pixels in the images speak to the histogram of LBP.

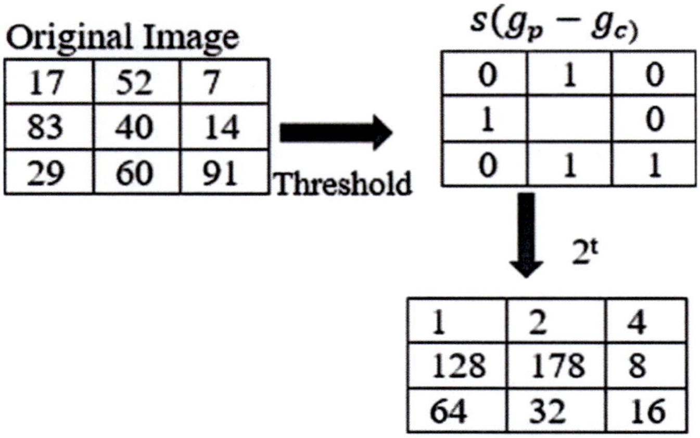

Fig. 2 LBP process

$$\mathrm{LBP}_{\mathrm{p1\ p2}} = \sum_{\mathrm{p=0}} S(g_p - g_c)2^{\mathrm{t}} \tag{1}$$

In the Eq. 1, g_p and g_c are the pixel power of the middle pixel with range R, and P is the absolute number of the neighboring pixel. LBP process appeared in Fig. 2.

In this procedure, the palm print images are isolated into 8 squares and the LBP window is 3 × 3 pixel. The clockwise strategy is applied to create all an incentive in each square to include the extraction process. The LBP surface can be portrayed by utilizing a histogram of 256 codes.

2.2 *Principal Component Analysis (PCA)*

The yield of LBP feature extraction is the new vector speaking to the histogram in 256 measurements. Straight projection technique, for example, principal component analysis (PCA) [3, 11–13], is a reasonable and viable answer for diminishing the element of this feature. The quantity of head parts in terms of information measurement is not exactly the first information. The main part is arranged dependent on the most noteworthy Eigen esteem speaking from the most significant information to less significant information. The most significant information in the eigenvector is situated in initial barely any columns of the grid [14].

The calculation of the most significant eigenvector is executed in both the preparation stage and the testing stage as follows:

- Prepare the wellspring of information with a similar size.
- Rearrange the 2D grid into a 1D framework.

- Calculate a mean of information and normalization utilized zero scores (Z-score) method, as in:

$$Z_S = D{-}a \tag{2}$$

- In the equation, D is 1D grid information and a is mean (normal vector of information).
- After normalization utilizing Z-score, at that point covariance is determined as:

$$\text{Cov} = Z^t \times Zs \tag{3}$$

- Calculate the Eigenvector of covariance grid above as in:

$$[V, \text{d}] = \text{Eig}(\text{cov}) \tag{4}$$

- Calculate the principal components of data as shown in the equation, where Zs is Z-score data and V is variance.

$$Pc = (Zs \times V) \tag{5}$$

In this process, PCA is used to reduce the feature vector dimensionality by choosing a relevant high variance of data. PCA observing correlated data then produces the uncorrelated data by removing all redundant and noise that exist in the data.

2.3 *Fusion Method*

In multimodal biometric, the combination is utilized to expand segregation power in the element space. There are four phases of combination, which are sensor level, feature level, score level, and decision level. Few strategies of combination have been utilized in each level, for example, whole principle, weight sum rule [1], and rule. In these trials, the combination at matching score level is performed utilizing weight aggregate standard.

3 Experimental Results

The calculation in this process is demonstrated utilizing MATLAB (R2014a). The investigation of each face and palm print is discussed in detail in the following section.

3.1 ORL Database

The investigation of face images is led utilizing Olivetti Research Laboratory (ORL), Cambridge. In the ORL database, there are 40 subjects comprising 10 unique images of each subject. The image is captured on various levels, outward appearance, changing the lighting, and detail in facial (wearing glasses and not wearing glasses). A sample image from the ORL database is presented in Fig. 3.

Every pixel comprises 256 Gy levels. The process of feature extraction is as follows:

1. The preparing information is arbitrarily chosen from 40 subjects. They have a few varieties.
2. Read the preparation image and preprocess every one of them.
3. Crop each image into 78 × 81 pixels and convert it into a single vector.
4. PCA strategy is utilized to extricate the neighborhood feature from the yield of preprocessing stage and develop the element vector as follows:

    - Transform 2D information into a 1D long vector.

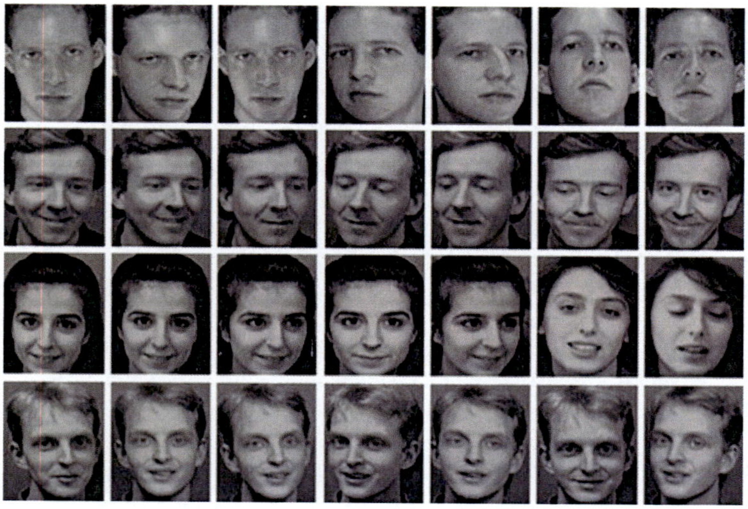

Fig. 3 Sample ORL face images

- The entire of preparing information is standardized with the Z-score (zero scores) procedure.
- Calculate the mean image.
- Find the covariance lattice of preparing information.
- Compute the Eigenvalue and Eigenvector of the preparation information. Each Eigenvector connects to Eigenvalue.

5. The figuring of preparing and testing coordinating score utilizing Euclidean separation classifier.

3.2 PolyU Database

The palm print images are taken from the PolyU (Hong Kong Polytechnic University) database. The images comprise 250 palm images gathered from the volunteer. The images are separated into 55 females and 195 males. In this examination, 40 subjects of palm print images are utilized and each subject comprises 10 unique images. The size of each image is 191 $\times$ 131 pixels, and every one of them is reshaped into 80 $\times$ 80 pixels. A sample image from the PolyU database for palm print is depicted in Fig. 4.

In this work, a similar strategy as face investigation was additionally applied on palm print images. For the palm print image, the nearby element is extricated utilizing the LBP technique, and afterward, the PCA strategy is utilized to decrease include measurements.

LBP is utilized to remove the palm print image surface. The histograms are built from a surface element. The test result of the palm print utilizing LBP-PCA is 89% acknowledgment rates.

Fusion of two modalities: A mix of palm print and face matching scores from the past investigation can accomplish a better outcome. Weight aggregate method is

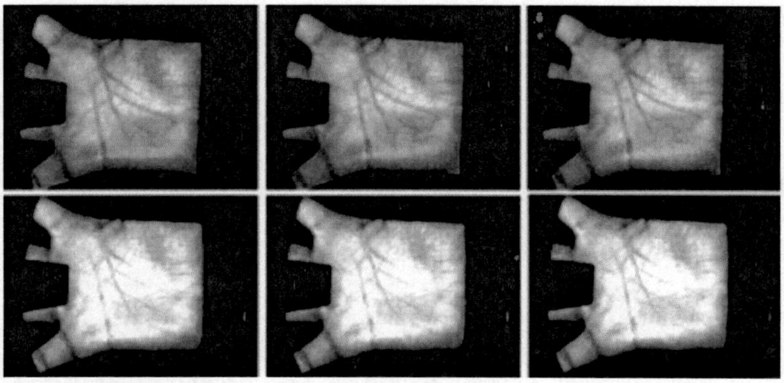

Fig. 4 Sample Palm print images from PolyU database

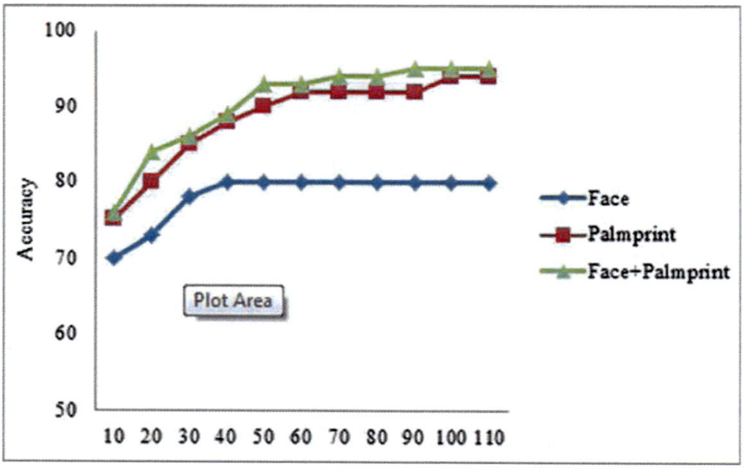

Fig. 5 Accuracy plot of both modalities

utilized to coordinate both methodologies at matching score by utilizing this standard: new_score = a.face + fJ.palm, where a + fJ = 1. The accuracy of the trial result is depicted in Fig. 5.

Face acknowledgment precision utilizing PCA produces 98.5% acknowledgment rates, while palm print acknowledgment is 99%. The palm print acknowledgment produces a better outcome contrasted with face acknowledgment after utilizing the proposed technique. The combination of the two modalities has the most noteworthy exactness of 99.8%. The combination can increment the acknowledgment rates contrasted with the normal precision of the face and palm print. The examination between face, palm print, and the combination appears in Fig. 6.

4 Conclusion

A framework design is proposed based on the combination of Face and Palm print Biometrics which gives high accuracy in Human authentication. Score level fusion strategy is applied for generating the final scores from the input images of Face and Palm print. The input images are preprocessed, and then a combined strategy of PCA (principal component analysis) and LBP (linear binary pattern) is applied on the processed image to remove surface features and reduce feature level calculations. The Gaussian dissemination of feature vector in the element space produces a Euclidean separation classifier to compute the matching score. Matching based on weighted sum rule is used to compute the matching score. If the value of the score generated is above threshold, then the authentication is successful and the user is said to be Genuine otherwise Imposter. The rate of accuracy obtained by the combination of

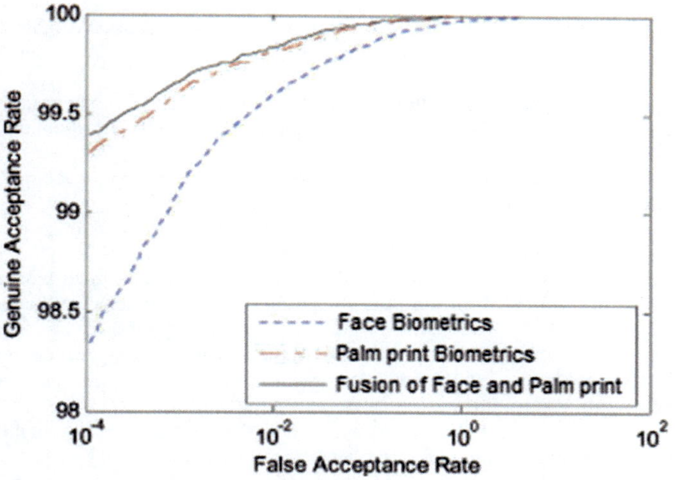

Fig. 6 Comparison of face, Palm print, and their fusion

face and palm print biometrics is 99.8% which is extremely high compared to the single use of face and palm print alone as unimodal biometrics.

References

1. Kumar Y, Nigam A, Tiwari K, Gupta P (2015) A mechanized multimodal biometric framework and combination. In: IEEE workshop computational intelligence biometrics identity management CIBIM, January, vol 2015, pp 16–21
2. Mondal A, Kaur A (2016) Relative study of feature level and decision level fusion in multimodal biometric recognition of face, ear, and iris. 5(5):822–842
3. Morris C, Rajesh RS (2015) An epic and improved spatial space combination strategy utilizing simple—PCA strategies. In: 2014 international conference communications networks technology ICCNT 2014, vol 2015, March, pp 90–94
4. Bounneche MD, Boubchir L, Bouridane A, Nekhoul B, Ali-Cherif A (2016) Multi-phantom palmprint acknowledgment dependent on arranged multiscale log-Gabor channels. Neurocomputing 205:274–286
5. Jain AK, Nandakumar K, Ross A (2016) 50 years of biometric research: accomplishments, difficulties, and openings. Pattern Recognit Lett 79:80–105
6. Zuo W et al (2017) Palmprint recognition based on complete direction representation. IEEE Trans Image Process 26(9):4483–4498
7. Xiao-Qiang Y, Xiao-bing Z (2017) Face recognition research based on the fusion of layered LBP feature. In: IEEE workshop computational intelligence biometrics identity management CIBIM, pp 75
8. Ali MMH, Yannawar PL, Gaikwad AT (2017) ScienceDirect multi-algorithm of Palmprint recognition system based on fusion of local binary pattern and two-dimensional locality preserving projection. Proc Comput Sci 115:482–492

9. Skillet Z, Wu X, Li Z (2019) Focal pixel choice procedure dependent on neighborhood dim worth circulation by utilizing angle data to upgrade LBP for surface order. Expert Syst Appl 120:319–334
10. Srivastava R (2020) Score-level multimodal biometric authentication of humans using retina, fingerprint, and finger vein. Int J Appl Evol Comput (IJAEC) 11(3). https://doi.org/10.4018/IJAEC.2020070102
11. Luo YT et al (2016) Neighborhood line directional example for Palmprint acknowledgment. Pattern Recognit 50:26–44
12. Meher B, Agrawal S, Panda R, Abraham A (2019) A study on area-based image combination strategies. Inf Combination 48:119–132
13. Moutafis P, Kakadiaris IA (2015) Rank-based score standardization for multi-biometric score combination. In: 2015 IEEE international symposium technology homeland security, HST 2015, pp 1–6
14. Moi H, Asmuni H, Hassan R, Othman RM (2014) Master systems with applications multimodal biometrics: weighted score level combination dependent on non-perfect iris and face images. Expert Syst Appl 41(11):5390–5404
15. Khodadoust J, Khodadoust AM, Li X, Kumari S (2018) Plan and usage of a multibiometric framework dependent available's characteristics. Expert Syst Appl 97:303–314
16. Jain MP, An Alad Manoj Peter M (2016) Combination of PalmPrint and Iris for multimodal biometric. Int J Scient Res Publ 6(5):221–227
17. Srivastava R, Srivastava P (2019) A multimodal based approach for face and unique mark based combination for confirmation of human. Int J Bus Anal 6(3). https://doi.org/10.4018/IJBAN.2019070102
18. Srivastava R (2020) A score level fusion approach for multimodal biometric fusion. Int J Scient Technol Res. ISSN: 22778616
19. Zhou LF, Du YW, Li WS, Mi JX, Luan X (2018) Posture strong face acknowledgment with Huffman-LBP improved by divide-and-rule procedure. Pattern Recognit 78:43–55
20. Modak SKS, Jha VK (2019) Multibiometric combination technique and its applications: a survey. Inf Combination 49:174–204

Design of 16-Bit Vedic Multiplier Using Modified Logic Gates and BEC Technique

K. Rupa Lakshmi, Kandula Bala Sindhuri, K. Mani Kumar, and N. Udaya Kumar

Abstract Higher throughput arithmetic operations are important to achieve the desired performance in many real-time signal and image processing applications. Thus, for the design of an efficient arithmetic unit, the key functioning elements are multipliers, adders, and the basic logic gates for which the performance parameters which decide the complete performance of the arithmetic unit are area, power dissipation, and speed. Research has been going through the past few decades to design efficient VLSI arithmetic circuits by using logic optimization and circuit optimization techniques to further reduce area, power, and delay. The multiplier design which is initially developed for area optimizations is an array multiplier in which the main disadvantage is the worst-case delay. To overcome this disadvantage, many methodologies came into existence in which Vedic methodology has acquired prominence because of fast computations. It has been found that Vedic multiplier adopting Urdhva Tiryagbhyam is one of the most effective multipliers with minimum delay for multiplication of all types of numbers, either large or small. Though the speed of this multiplier is high, area occupancy is more. Thus, Binary to Excess-one Converter (BEC) technique is employed in this multiplier to reduce area. To further improve the performance, a 16×16 Vedic multiplier employing BEC adders and modified logic gates is developed and coded in Verilog HDL and further synthesized using Vivado 2017.2. Synthesis results disclose that the 16-bit Vedic multiplier using BEC technique employing Modified AND gate and Modified OR gate architecture accomplishes power savings of 0.13% compared to a Vedic multiplier using BEC technique.

K. Rupa Lakshmi (✉) · K. Bala Sindhuri · K. Mani Kumar · N. Udaya Kumar
Department of E.C.E, S.R.K.R Engineering College, Bhimavaram, India
e-mail: rupalakshmikallepalli@gmail.com

K. Bala Sindhuri
e-mail: kbsinduri@srkrec.ac.in

N. Udaya Kumar
e-mail: nuk@srkrec.ac.in

B. Mishra and M. Tiwari (eds.), *VLSI, Microwave and Wireless Technologies*,
Lecture Notes in Electrical Engineering 877,
https://doi.org/10.1007/978-981-19-0312-0_8

Keywords Vedic multiplier · Urdhva tiryagbhyam · BEC technique · Modified logic gates

1 Introduction

In applications like multimedia signal processing and data mining, adders and multipliers form the key components. Throughput [1], power dissipation, and propagation delay are the important factors to choose a suitable design for these applications. Designing a multiplier involves various steps which are done by adopting various techniques like logic optimization technique and circuit optimization technique in which logic optimization technique plays the crucial role in digital electronics as it helps in reducing the chip area which affects the cost of the chip. The survey carried out to find the efficient multiplier elevates the problems associated in the design of multiplier are carry propagation delay in the partial product addition and more area because of greater number of full adders used in the computations. Through this survey, it has been understood that the multipliers designed using Vedic Mathematics helps to reduce the area, speed, and power [1–3]. Partial product generation and their addition play an important role in determining the delay and speed of the multiplication process. In [4], multiplication is done by following the Urdhva Tiryagbhyam principle using ripple carry adders (RCA) for delay reduction. Time required to propagate a carry through the adder determines the speed of addition, and from the literature survey, it is understood that the parallel addition in carry select adder (CSLA) helped to reduce the propagation delay. Usage of two independent RCAs in CSLA architecture leads to increase in delay and area [5]. To reduce the number of gates used and to further increase the speed of addition, RCA in the CSLA architecture is replaced with BEC [5–7]. Mohan shobha et al. [8] designed a BEC-based hierarchy multiplier using carry select adder (CSLA).

Multiplexers are also used as one of the methods to reduce the number of integrated circuit packages which further helped to reduce area and power [9]. Logic gates are designed using multiplexers, and they are used in different combinations to check the performance of the multiplier. Modified arithmetic and logical units (AND, OR XOR, half adder, and full adder) replaced the basic units which lead to decrease in power consumption and delay with slight increase in the number of Look-Up Tables (LUTs).

2 Preliminaries

The main ideology is to replace the existing logic gates with modified logic gates using multiplexers and compare the performance of the Vedic multiplier, i.e., multiplication is done by using efficient carry select adders using BEC technique, modified logic gates, and Vedic algorithm.

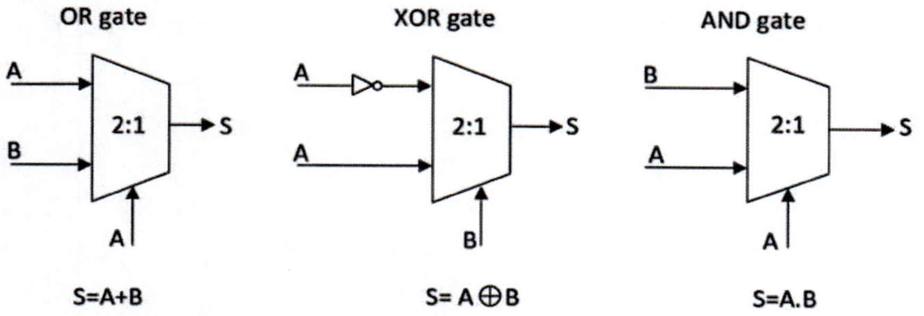

Fig. 1 Basic logic gates designed using 2:1 multiplexer

2.1 *Modified Logic Gates*

Multiplexers are used in the design of all the basic logic gates. Usage of multiplexers reduces the number of computations performed. The Fig. 1 shows the modified logic gates using 2:1 multiplexer. These gates are used in the VLSI architecture for Vedic multiplier employing BEC technique in CSLA adder.

2.2 *BEC*

In regular CSLA, there is RCA with $C\text{in}$= '0' and RCA with $C\text{in}$=1.The actual $C\text{in}$ from the previous sector selects one of the two RCAs. Hence, the number of full adders used is large which leads to larger area and more delay. As an alternative, using Binary to Excess-one converter (BEC) instead of the RCA with $C\text{in}$=1 reduces the area and power consumption of the regular CSLA. The Fig. 2 shows the modified 16-bit CSLA architecture using BEC adders [7, 8]. This adder along with the modified gates is used in the VLSI architecture for Vedic multiplier employing BEC technique.

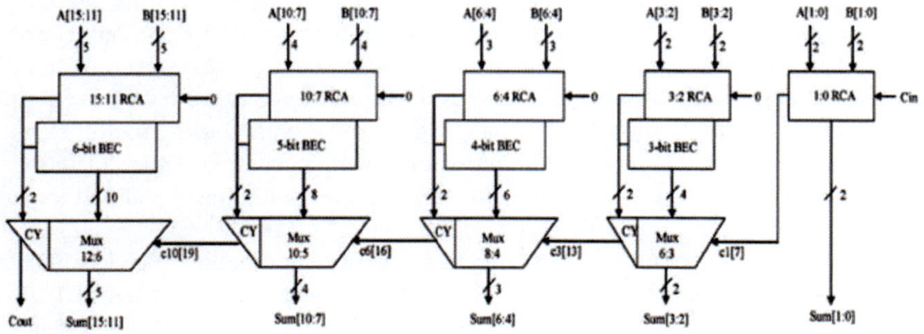

Fig. 2 Modified 16-bit CSLA architecture with RCA $C\text{in}$ = '1' replaced by BEC

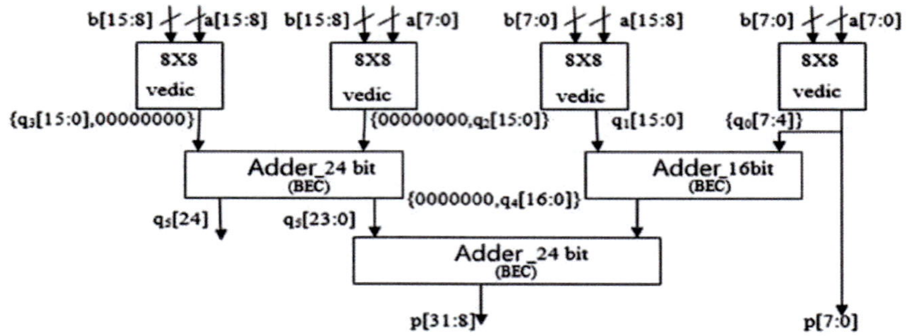

Fig. 3 Vedic 16-bit multiplier architecture using BEC adder and modified logic gates

3 Architecture

The 16-bit Vedic multiplier is implemented using four 8-bit Vedic multipliers and a 16-bit BEC adder with a couple of 24-bit BEC adders and modified logic gates as discussed in preliminaries section. Here, partial product generation and additions are done concurrently in a vertical and crosswise manner. The partial products generated are added using Urdhva Tiryagbhyam algorithm and BEC technique. Vedic algorithm greatly reduces the number of partial products generated leading to fast multiplication process. The VLSI architecture for 16-bit Vedic multiplier using modified logic gates and BEC technique is shown in the Fig. 3. The 16-bit adder, i.e., Adder_16 used in Fig.3, is designed by using the architecture shown in Fig.2, and further, 24-bit adder, i.e., Adder_24bit, is also designed by extending the architecture shown in Fig.2.

4 Simulation Results

G.C.Ram et al. [10] designed Vedic multiplier using BEC technique employing basic logic gates such as NOT, AND, XOR, and OR gates. The VLSI architectures employing these gates consume more area and power because of a greater number of transistors used in the design. To further reduce the area and power consumption of full adder, Jiang et al. [11] designed "A Novel Multiplexer-Based Low-Power Full Adder." In the similar fashion, the basic logic gates are modified using multiplexers to further improve the performance in terms of power consumption and delay; the existing basic gates in the BEC architecture are replaced with these modified logic gates.

The architecture is modified using these seven different combinations like BEC architecture using Modified AND gate (MA); BEC architecture using Modified OR gate (MR); BEC architecture using Modified XOR gate (MX); BEC architecture using Modified AND gate and Modified OR gate (MAMR); BEC architecture using

Modified XOR gate and Modified OR gate (MXMR); BEC architecture using Modified AND gate and Modified XOR gate (MAMX); and BEC architecture using Modified OR gate, Modified XOR gate, and Modified AND gate (MAMR MX).

The developed VLSI architectures such as Vedic multiplier using BEC employing MA, MR, MX, MAMR, MX MR, MAMX, and MAMXMR are coded in Verilog HDL. Simulation and synthesis are carried out using Vivado 2017.2 software on an I5 processor with 8 GB RAM and 64-bit operating system. Inputs to the 4-bit Vedic multiplier using BEC technique employing MAMR architecture are a [3:0], b [3:0]. The output c [7:0] obtained for different combinations of inputs is shown in Fig. 5. The Fig. 4 shows the simulation results of all the mentioned combinations of high-speed architecture of 4-bit Vedic multiplier.

Inputs to the 8-bit Vedic multiplier using BEC technique employing MAMR architecture are a [7:0], b [7:0]. The output c [15:0] obtained for different combinations of inputs is shown in Fig. 6. Figure 7 shows the performance characteristics of 8-bit Vedic multiplier by replacing logic gates with revised logic gates in the BEC adders.

The architecture of 16-bit Vedic multiplier using BEC technique employing MAMR combination is checked using different inputs which is shown in Fig. 8

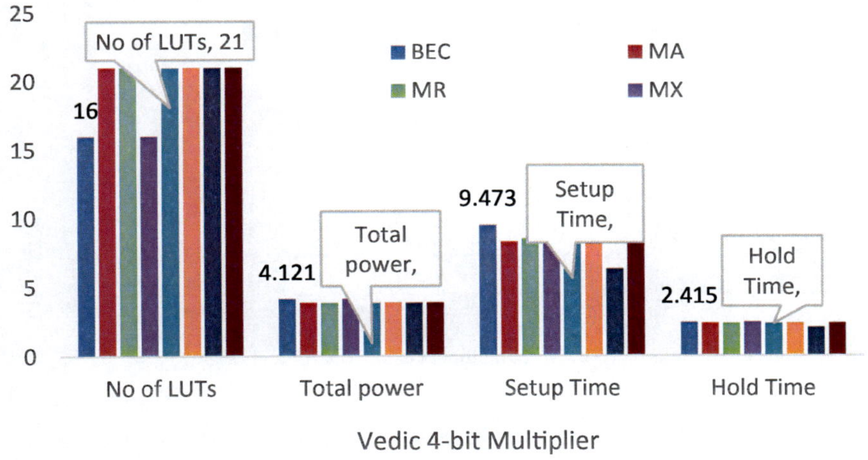

Fig. 4 Performance characteristics of Vedic 4-bit multiplier

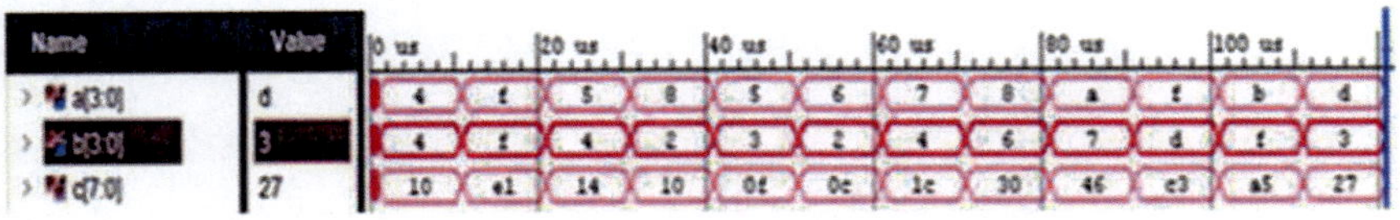

Fig. 5 Simulation results of 4-bit Vedic multiplier using MAMR architecture

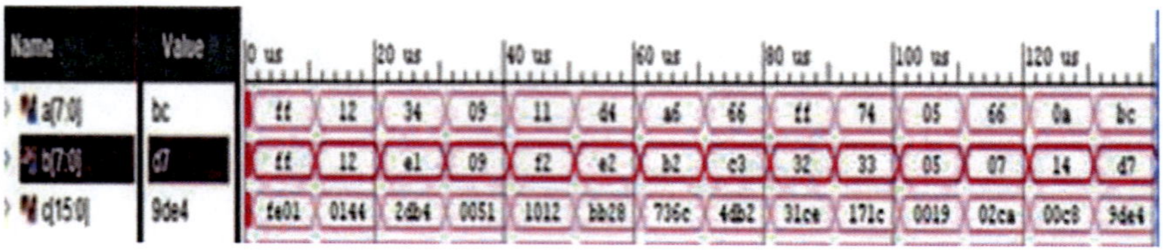

Fig. 6 Simulation results of 8-bit Vedic multiplier using MAMR architecture

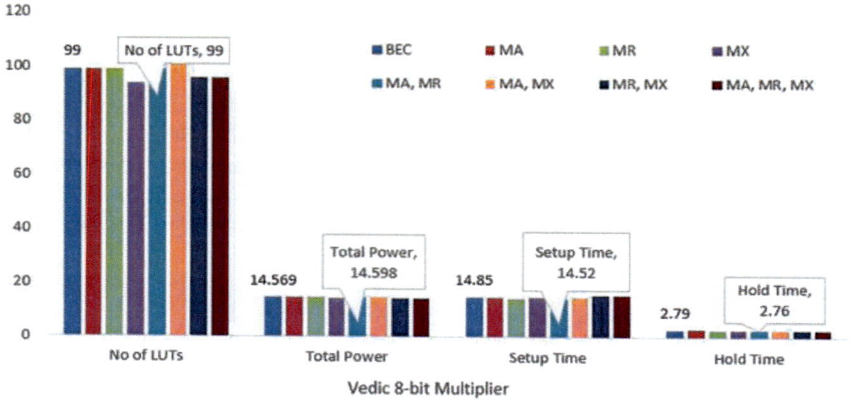

Fig. 7 Performance characteristics of Vedic 8-bit multiplier

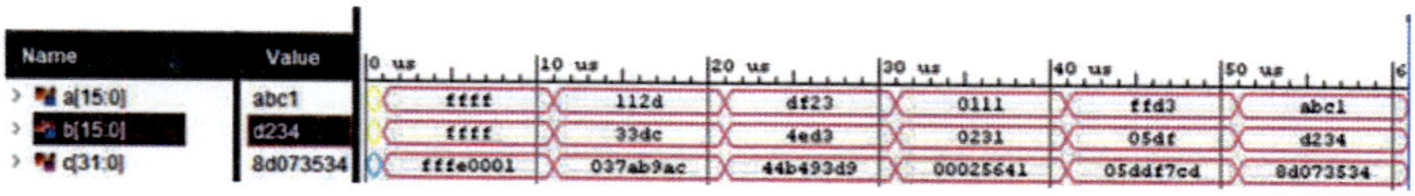

Fig. 8 Simulation results of 16-bit Vedic multiplier using MAMR architecture

where a [15:0], b [15:0] are inputs and c [31:0] is the output. All the seven combinations of modified architecture are simulated, and the power consumption, number of LUTs, setup, and hold timings are observed. Figure 9 gives the detail performance of the Vedic 16-bit multiplier.

5 Comparison Results

The architecture of Vedic multiplier using BEC adders operates on a larger number of transistors. Table 1 gives the power consumption, number of LUTs, setup, and hold timing of 4-bit, 8-bit, and 16-bit Vedic multiplier using BEC technique.

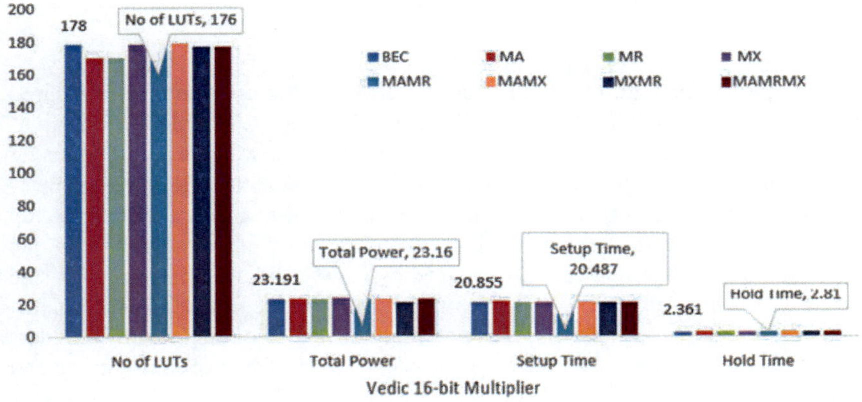

Fig. 9 Performance characteristics of Vedic 16-bit multiplier

Table 1 Parameters of BEC architecture

BEC	No of LUTs	Power	Setup time	Hold time
Vedic 4	16	4.121	9.47	2.41
Vedic 8	99	14.569	14.85	2.79
Vedic 16	178	23.191	20.85	2.36

The proposed architecture of Vedic multiplier with modified logic gates using multiplexers ensured significant change in power consumption and delay in almost all the mentioned combinations. Of all the proposed modified combinations, MAMR architecture has reduced the power consumption and delay timings irrespective of the bit size. Table 2 gives the power consumption, number of LUTs, setup, and hold timings of the Vedic multiplier using BEC technique employing MAMR architecture. The developed VLSI architecture of Vedic 4-bit multiplier using BEC technique employing MAMR has reduced power consumption, setup, and hold time by 7.23, 10.3, and 2.89%, respectively. While in the case of 8-bit Vedic multiplier using MAMR architecture, 2.22% reduction in setup time and 1.07% reduction in hold time have been observed. Modified MAMR architecture of Vedic 16-bit multiplier using BEC technique led to a decrease of 0.133% and 1.832% in power consumption and setup time, respectively.

Table 2 Parameters of MAMR architecture

MAMR	No of LUTs	Power (W)	Setup time	Hold time
Vedic 4	21	3.823	8.49	2.345
Vedic 8	99	14.598	14.52	2.76
Vedic 16	176	23.160	20.487	2.81

6 Conclusion

In this paper, Vedic multiplier with modified BEC adders using multiplexer-based logic gates (i.e., MA, MR, MX, MAMR, MXMR, MAMX, MAMXMR) is simulated, and comparison is done among them. From the results, Vedic multiplier using BEC technique employing Modified AND gate and Modified OR gate architecture (MAMR) is more significant in the reduction of power consumption and setup time. Power consumption and setup time of Vedic multiplier using BEC technique employing MAMR architecture are reduced by 0.13% and 1.832% for 16-bit size and by 7.23% and 10.3% for 4-bit size multipliers when compared to Vedic multiplier with BEC adders, respectively. Hence, implementation of modified BEC adders in Vedic multipliers improved the performance in terms of power consumption and delay.

References

1. Bathija RK, Meena RS, Sarkar S (2012) Low power high speed 16*16-bit multiplier using Vedic mathematics. Int J Comput Appl
2. Jie LS, Rusian SH (2017) A 2x2 bit Vedic multiplier with different adders in 90nm CMOS technology. AIP Conf Proc 1883:020017
3. Gadakh SN, Khade A (2016) Design and optimization of 16×16 Bit multiplier using Vedic mathematics. In: 2016 international conference on automatic control and dynamic optimization techniques (ICACDOT). Pune, pp 460–464
4. Rao KD, Gangadhar C, Korrai PK (2016) FPGA implementation of complex multiplier using minimum delay Vedic real multiplier architecture. In: 2016 IEEE Uttar Pradesh section international conference on electrical, computer and electronics engineering (UPCON). Varanasi, pp 580–584
5. Natarajan PB, Ghosh S, Karthik R (2017) Low power high performance carry select adder, pp 601–603
6. Rao PR, Halle P (2013) An efficient carry select adder with less delay and reduced area application. Int J Eng Trends Technol (IJETT) 4(9)
7. Ramkumar B, Kittur HM (2012) Low-power and area-efficient carry select adder. IEEE Trans. Very Large Scale Integr (VLSI) Syst 20(2):371–375
8. Shoba M, Nakkeeran R (2017) Energy and area efficient hierarchy multiplier architecture based on Vedic mathematics and GDI logic. Eng Sci Technol Int J 20(1)
9. Venkatachalam S, Ko S (2017) Design of power and area efficient approximate multipliers. IEEE Trans Very Large-Scale Integ (VLSI) Syst 25(5):1782–1786
10. Ram GC, Lakshmanna YR, Rani DS, Sindhuri KB (2016) Area efficient modified Vedic multiplier. In: 2016 international conference on circuit, power and computing technologies (ICCPCT). Nagercoil, pp 1–5
11. Jiang Y, Al-Sheraidah A, Wang Y, Sha Ea, Chung J-G (2004) A novel multiplexer-based low-power full adder. IEEE Trans Circ Syst II: Exp Briefs 51(7):345–348

A Tunable Resistorless Floating Inductance Simulator Using MO-DXCCTA

Yumnam Shantikumar Singh, Ashish Ranjan, Shuma Adhikari, and Benjamin A. Shimray

Abstract A tunable resistorless floating positive inductance simulator using two MO-DXCCTA active elements and one grounded capacitor has been presented. The proposed design uses only one grounded capacitor which makes it easy to fabricate. Additionally, the design circuit does not require any component matching. The non-idealities and parasitic effects of the proposed floating inductance simulator are also studied. To test the functionality of the proposed floating inductance circuit, it is used in a band-pass filter and second-order low-pass filter. The functionality test is carried out using 0.18 μm TSMC parameters in OrCAD PSpice software. To support the proposed floating inductance simulator, a few simulation results are established.

Keywords MO-DXCCTA · Inductance simulator · Active Filter

1 Introduction

An inductor is an essential component in the analog circuit design. A tunable filter circuit design in a communication system, inductor, capacitor, and resistor is used. The physical inductors have many applications like LC oscillators, for parasitic element cancelation, impedance matching, phase shifters, etc. [1–3]. However, the use of physical inductor has some limitation constraint due to bulky size, lack of tunable, occupied more area, not suitable for fabrication, low-quality factor, etc. From these constraints, it becomes an interesting research area for making active inductors using several active building blocks (ABBs). Various grounded/floating active inductance simulators using ABBs such as CC-CFA [4, 5], MO-CCCCTA [6], M-CDTA [7], CCCII [8], CDTA [9], VD-DIBA [10], VDTA [11], ZC-VDTA

Y. S. Singh (✉) · A. Ranjan
Department of Electronics and Communication Engineering, National Institute of Technology Manipur, Imphal, India
e-mail: yshantikumar99@gmail.com

S. Adhikari · B. A. Shimray
Department of Electrical Engineering, National Institute of Technology Manipur, Imphal, India
e-mail: benjaminshimray@ieee.org

B. Mishra and M. Tiwari (eds.), *VLSI, Microwave and Wireless Technologies*, Lecture Notes in Electrical Engineering 877,
https://doi.org/10.1007/978-981-19-0312-0_9

[12], CCCDTA [13], CFTA [14], ZC-CFCCC [15], FTFNTA [16, 17], VDDDD [18], DXCCCII [19], DXCCTA [20], MO-DXCCTA [21], CCCFTA [22], and VDBA [23] have been investigated.

Recently, several ABBs have been introduced in [24]. Among the reported ABBs, MO-DXCCTA [21] is also one of them. Since MO-DXCCTA has the electronically tunable property due to an operational transconductance amplifier (OTA) at the Z+ and Z− terminals and transferring of both current and voltage in its relevant ports, it is also suitable for designing active filters [25–27] or inductance simulators.

In this paper, we proposed a tunable floating positive active inductance simulator using two MO-DXCCTA active elements with a grounded capacitor. This paper is described as a short study of the proposed floating active inductor using MO-DXCCTA in Sect. 2. The non-idealities and parasitic effects are given in Sect. 3, respectively. The functionality and possible test are presented in Sect. 4 and Sect. 5 using PSpice parameter of 0.18 μm technology, respectively. Lastly, brief study reports are given in Sect. 6.

2 The Inductance Simulator

The circuit symbol and internal CMOS structure of MO-DXCCTA are shown in Fig. 1a, b. The terminal characteristic of the MO-DXCCTA is given as:

$$\begin{bmatrix} I_Y \\ V_{X\pm} \\ I_{Z\pm} \\ I_{1O\pm} \\ I_{2O\pm} \end{bmatrix} = \begin{bmatrix} 0 & 0 & 0 & 0 & 0 \\ \pm 1 & 0 & 0 & 0 & 0 \\ 0 & 1 & 0 & 0 & 0 \\ 0 & 0 & \pm g_{m1} & 0 & 0 \\ 0 & 0 & 0 & \pm g_{m2} & 0 \end{bmatrix} \begin{bmatrix} V_Y \\ I_{X\pm} \\ V_{Z+} \\ V_{Z-} \\ V_O \end{bmatrix} \tag{1}$$

where g_m tends to transconductance parameter and it can be controlled with the bias current (I_B) present in [21].

Figure 2 shows the proposed floating positive inductance simulator. In a straightforward routine analysis using terminal characteristics of Eq. (1) in Fig. 2, the input impedance value is given as:

$$Z_{\text{in}}(s)|_{\text{Floating}} = \frac{1}{Y_{\text{in}}} = \frac{V_{\text{in}} - V_o}{I_{\text{in}}} = \frac{sC}{g_{m1}g_{m2}} = sL_{\text{eq}} \tag{2}$$

Hence, the equivalent inductance value is obtained as:

$$L_{\text{eq}} = \pm\frac{C}{g_{m1}g_{m2}} \tag{3}$$

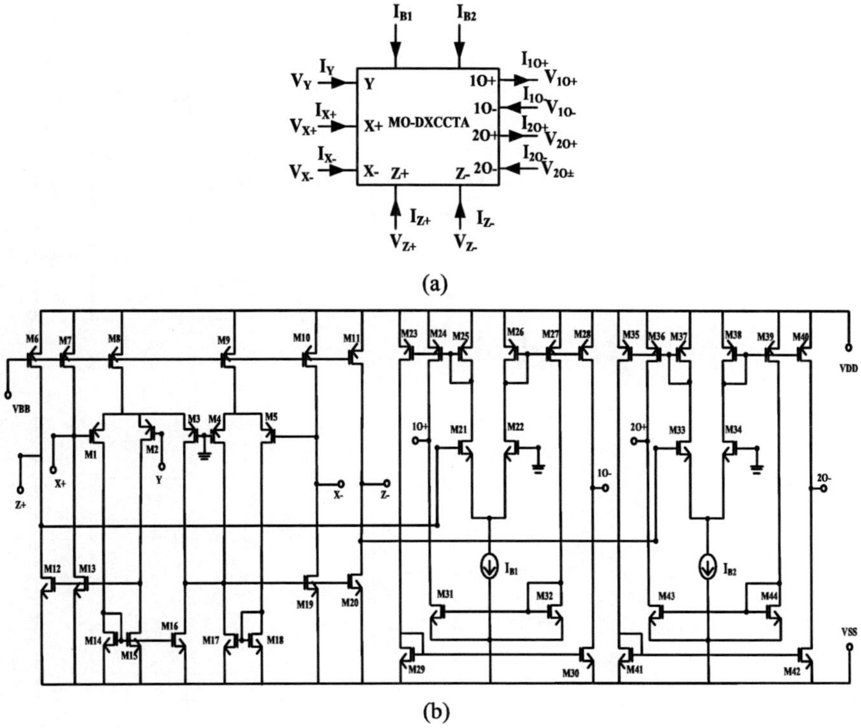

Fig. 1 MO-DXCCTA a Electrical symbol, b Internal structure

From Eq. (3), it provides the electronically tunable through a transconductance (g_{m1}, g_{m2}).

3 Non-ideal Analysis

The deviation from the theoretical performance is due to the presence of non-ideality constraints in the active element. The frequency-dependent constraints such as current transfer gain (α), voltage transfer gain (β), and transconductance transfer accuracy (γ) of the MO-DXCCTA, the non-ideality terminal characteristics of MO-DXCCTA are given as follows:

$$I_Y = 0,\ V_Y = \beta V_{X\pm},\ I_{X\pm} = \alpha I_{Z\pm},\ I_{1O\pm} = \pm\gamma g_m V_{Z+},\ I_{2O\pm} = \pm\gamma g_m V_{Z-}$$

A straight forward analysis with the non-ideality constraint in MO-DXCCTA provides a new expression for the presented simulator as:

Fig. 2 The proposed floating positive AI

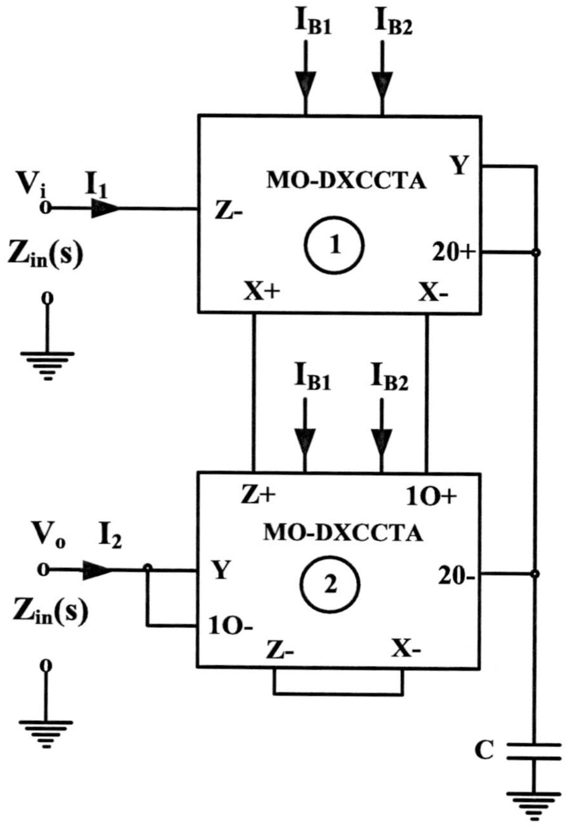

$$Z_{\text{in}}|_{\text{Floating}} = \frac{V_{\text{in}} - V_o}{I_{\text{in}}} = \frac{sC\alpha\beta}{\gamma^2 g_{m1} g_{m2}} \tag{4}$$

In the active inductor circuit, the practical input impedance is a supplementary utility of the frequency-dependent parameters that differ from the actual frequency response. The subsequently non-ideality word comes from the parasitic element of $R_p||C_p$ in each terminal (X, Y, Z, and O) which is presented in Fig. 3.

When considering the parasitic impedances present inside the MO-DXCCTA-based floating positive active inductor, the admittance function in terms of parasitic is given as:

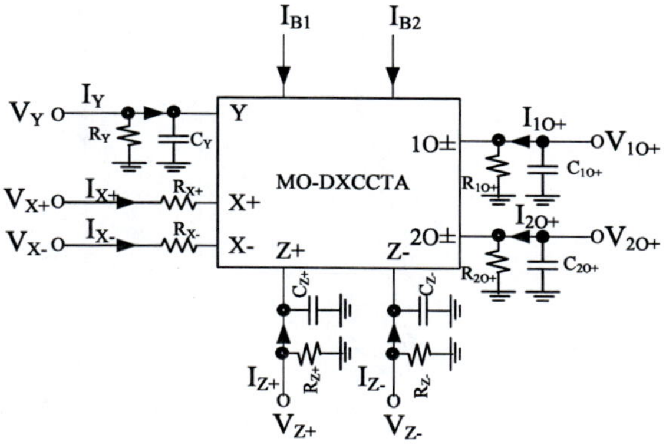

Fig. 3 MO-DXCCTA with parasitic elements

$$
\left.\begin{array}{l}
Y_{\text{in}}(s)|_{\text{Floating}} = \left\{ sC_{P1} + \dfrac{1}{R_{P1}} + \dfrac{g_{m1}g_{m2}}{\left(s(C + C_{P2}) + \frac{1}{R_{P2}}\right) * \left(sC_{P4} + \frac{1}{R_{P4}}\right)} \right\} \\
\text{if } C >> C_{P2} \text{ and } \dfrac{1}{R_{P2}} \\
Y_{\text{in}}(s)|_{\text{Floating}} = \left\{ sC_{P1} + \dfrac{1}{R_{P1}} + \dfrac{g_{m1}g_{m2}}{sC * \left(sC_{P4} + \frac{1}{R_{P4}}\right)} \right\}
\end{array}\right\} \tag{5}
$$

where $R_{P1} = R_{Z-}, C_{P1} = C_{Z-}, R_{P2} = R_{20-}||R_{20+}||R_Y$, $C_{P2} = C_{20-}||C_{20+}||C_Y$, $R_{P3} = R_{X-}$, $C_{P3} = C_{X-}||C_{1O-}$, $R_{P4} = R_{Z+}||R_Y$, and $C_{P4} = C_Y||C_{Z+}$.

4 Computer Results

The proposed floating inductance simulator is confirmed in PSpice through CMOS parameter of [21]. The aspect ratio, supply voltage, and bias voltage are used from the author presented in [21]. Figure 4a shows the transient response for floating positive inductance simulator by choosing $I_{B1} = 50\ \mu\text{A}$, $I_{B2} = 100\ \mu\text{A}$, and $C = 100$ nF that gives the inductance value $L_{\text{eq}} = 115.81\ \mu\text{H}$, for 1 MHz frequency in signal. Figure 4b shows the frequency responses of the floating positive inductance simulator, $L_{\text{eq}} = 115.81\ \mu\text{H}$, and also shows the tunable properties through bias currents. The operating frequency range is found as 1 kHz–1 MHz.

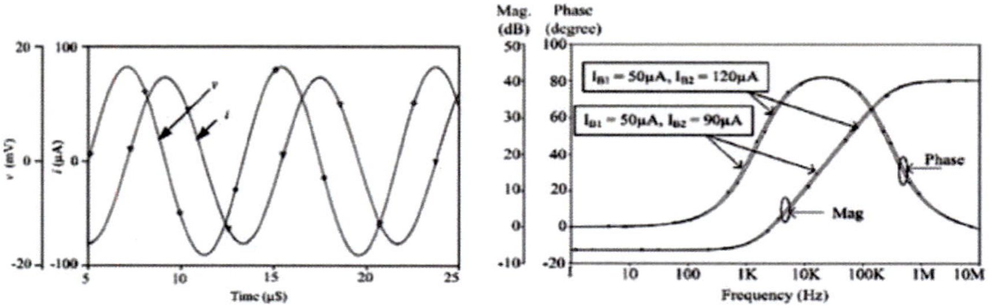

(a) (b)

Fig. 4 Positive AI response: **a** Time domain. **b** Frequency domain

5 Inductance Simulator Applications

For verification, the floating positive inductance simulator is utilized in a simple RLC band-pass filter (BPF) with standard passive components as $R_L = 1\ \text{k}\Omega$, $C =$ 1 nF, and $L_{\text{eq.}} = \{0.456\ \mu\text{H}, 0.527\ \mu\text{H}, 0.608\ \mu\text{H}\}$ which is shown in Fig. 5a. The frequency response of the filter for different L_{eq} as 0.456 μH ($I_{B1} = 50\ \mu\text{A}$, $I_{B2} =$ 160 μA), 0.527 μH ($I_{B1} = 50\ \mu\text{A}$, $I_{B2} = 120\ \mu\text{A}$), and 0.608 μH ($I_{B1} = 50\ \mu\text{A}$, $I_{B2} = 90\ \mu\text{A}$) using proposed floating positive AI circuit is depicted in Fig. 5b with different quality factor (Q) as 2.9, 4.6, and 10.2, respectively.

The transfer function of the simple RLC band-pass filter using the proposed inductance simulator is given as:

$$\frac{V_o(s)}{V_i(s)} = \frac{sR_LC}{s^2L_{\text{eq}}C + sR_LC + 1} \tag{7}$$

From Eq. (7), the natural frequency (ω_o) and Q_o can be expressed as:

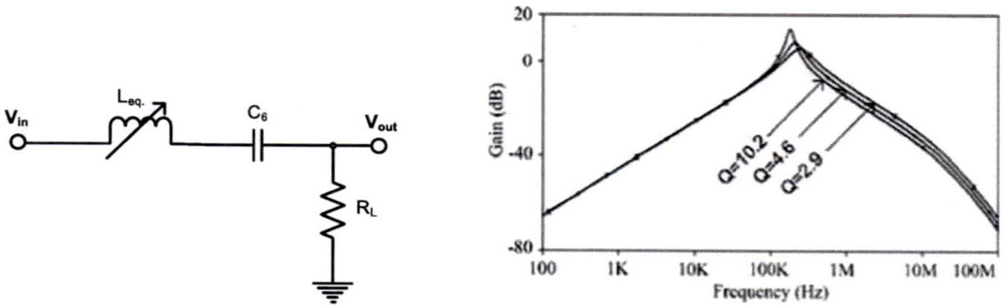

(a) (b)

Fig. 5 **a** RLC band-pass filter. **b** Frequency response

$$\omega_o = \frac{1}{\sqrt{L_{\text{eq}}C}} \tag{8}$$

$$Q_o = \frac{1}{R_L}\sqrt{\frac{L_{\text{eq}}}{C}} \tag{9}$$

$$\text{BW} = \frac{R_L}{L_{\text{eq}}} \tag{10}$$

In addition to test for floating active inductor circuit, it is utilized in a fourth-order low-pass filter (LPF) which is shown in Fig. 6a. The Fig. 6a is performed using standard passive elements as $R_S = 200\ \Omega$, $R_L = 1\ \text{k}\Omega$, $L_{\text{eq.}} = 120\ \mu\text{H}$ ($I_{B1} = 50\ \mu\text{A}$, $I_{B2} = 93.13\ \mu\text{A}$), and $C_7 = 1$ nF. The frequency responses of the fourth-order LPF is shown in Fig. 6b. The transient responses of Fig. 6a for 200 kHz and 1 MHz input frequencies and 5 mVp-p sinusoidal input signals are shown in Fig. 7a, b.

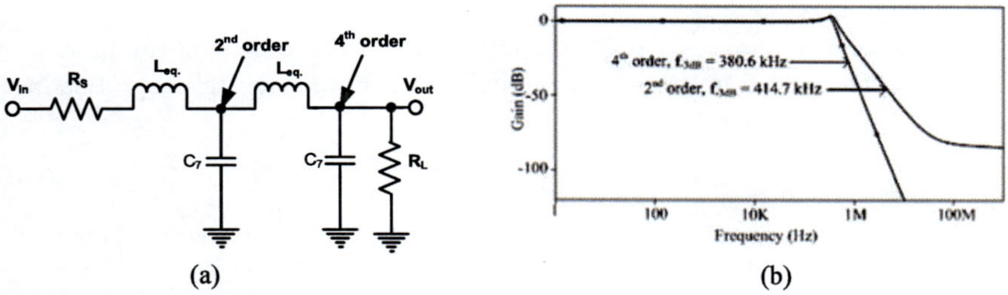

Fig. 6 Low-pass filter: **a** fourth-order LPF circuit. **b** Frequency response

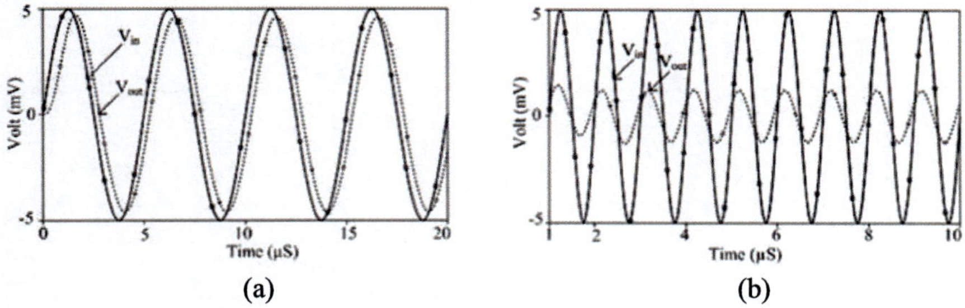

Fig. 7 Time domain responses of second-order low-pass filter at: **a** 200 kHz and, **b** 1 MHz

6 Comparison Study

As shown in Table 1, a short comparison study of the proposed floating active inductor design with the available one is investigated in terms of passive element count (active and passive), passive element matching constraint, type of simulator, operating frequency, and power utilization. Table 1 provides the following essential potential:

(i) The inductance simulator uses two number of active elements similar to [2, 4, 11, 17, 18, 20], and other uses more than two active elements [5, 8–10]
(ii) The proposed simulator uses single passive element similar to [4, 11], and other uses two or three passive elements [1, 3, 17, 18, 20].
(iii) The simulator has no passive element matching constraint and is electronically tunable like [9–11, 17].
(iv) All passive elements are grounded in proposed simulator as similar to [4, 5, 8–11]; however, some simulator uses floating passive elements [1, 3, 14, 17, 18, 20].
(v) The simulator has capable of integrated circuit (IC) conversion similar to [4, 11], and other simulators have difficulty with IC conversion due to floating passive elements used [1, 20].

Table 1 Comparison of proposed work with the available works

References	No. of ABBs used	No. of passive element used	Type of simulator	Frequency range (Hz)	The technology used (μm)	Power dissipation (mW)	Power supply (±Volt)
[1]	1	3	F	10–400 k	0.35	NA	1.5
[3]	2	3	F	NA	IC	NA	12
[4]	2	1	F	100–5 M	ALA400	0.449	1.5
[5]	3	1	F	NA	ALA400 + 0.35	NA	1.5
[8]	3	1	F	NA	ALA400	4.98	2.5
[9]	3	1	F	Up to 1 M	0.18	NA	2.5
[10]	3	1	F	Up to 1 M	IC	NA	1
[11]	2	1	F	Up to 10 M	0.18	NA	0.9
[14]	3	1	F	NA	ALA400	NA	1.5
[17]	2	3	F	100–1 M	0.18	8.59	1.65
[18]	2	2	F	3 k–1.56 M	IC	NA	5
[20]	2	2	F	1 k–10 M	0.25	1.15	0.75
Pro	2	*1*	*F*	*1 k–1 M*	*0.18*	*2.96*	*1.25*

* *Notes NA* = Not Available

7 Conclusions

In this paper, we proposed an electronically tunable new floating positive inductance simulator utilizing two MO-DXCCTA active elements with one grounded capacitor. The inductance simulator is free from matching constraints and suitable for monolithic IC design. The inductance simulator is verified through the 0.18 μm Taiwan Semiconductor Manufacturing Company (TSMC) SPICE parameter in OrCAD software. The proposed simulator can be used as an alternative circuit in analog signal processing applications. An application of second-order BPF and fourth-order LPF is well supported by the proposed simulator.

References

1. Ibrahim MA, Minaei S, Yuce E, Herencsar N, Koton J (2012) Lossy/lossless floating/grounded inductance simulation using one DDCC. Radioengineering 21(1):3–10
2. Yuce E (2006) Floating inductance, FDNR, and capacitance simulation circuit employing only grounded passive elements. Int J Electron 93(10):679–688
3. Metin B, Cicekoglu O (2006) A novel floating lossy inductance realization topology with NICs using current conveyors. IEEE Trans Circuits Syst II Express Briefs 53(6):483–486
4. Srisakultiew S, Silapan P, Siripruchyanun M (2009) Novel floating and grounded inductance simulators based on CC-CFAs. In: The seventh PSU engineering conference, pp 21–22
5. Singh A, Jain MK, Wairya S (2019) Novel lossless grounded and floating inductance simulators employing a grounded capacitor based in CC-CFA. J Circuits, Syst, Comput 28(6):1950093
6. Pandey N, Bazaz R, Manocha R (2011) MO-CCCCTA-based floating positive and negative inductors and their applications. J Electr Comput Eng, p 8
7. Prasad D, Haseeb Z, Mainuddin Akram Md. W. Realization of resistorless floating inductor using modified CDTA. Indian J Pure Appl Phys 57(1)
8. Yuce E, Minaei S, Cicekoglu O (2006) Resistorless floating immittance function simulators employing current controlled conveyors and a grounded capacitor. Electr Eng 88:519–525
9. Prasad D, Bhaskar DR, Singh AK (2010) New grounded and floating simulated inductance circuits using current differencing transconductance amplifiers. Radioengineering 19:194–198
10. Prasad D, Bhaskar DR, Pushkar KL (2011) Realization of new electronically controllable grounded and floating simulated inductance circuits using voltage differencing differential input buffered amplifier. Act Passiv Electron Compon 101432
11. Prasad D, Bhaskar DR (2012) Grounded and floating inductance simulation circuits using VDTAs. Circuits Syst 3:342–347
12. Guney A, Kuntman H (2014) New floating inductance simulator employing a single ZC-VDTA and one grounded capacitor. In: 9th IEEE international conference on design & technology of integrated systems in nanoscale era; Santorini, Greece. IEEE, New York, NY, USA, pp 9–10
13. Siripruchyanun M, Jaikla W (2008) CMOS current-controlled current differencing transconductance amplifier and applications to analog signal processing. AEU-Int J Electron Commun 62:277–287
14. Li Y (2012) A series of new circuits based on CFTAs. AEU—Int J Electron Commun 66:587–592
15. Singh A, Pragati K, Senani R (2018) Electronically tunable grounded/floating inductance simulators using Z-copy CFCCC. Turk J Electr Eng Comput Sci 26:1041–1055
16. Tarunkumar H, Singh YS, Ranjan A (2019) An active inductor employing a new four terminal floating transconductance amplifier (FTFNTA). Int J Electron 1362–3060:2–21

17. Singh R, Prasad D (2020) Comment floating simulated inductance circuits using FTFNTAs. Int J Electron. https://doi.org/10.1080/00207217.2020.1726495
18. Jaikla W, Sotner R, Khateb F (2019) Design and analysis of floating inductance simulators using VDDDAs and their applications. Int J Electron Commun. https://doi.org/10.1016/j.aeue.2019.152937
19. Tekin SA, Ercan H, Alci MA (2014) versatile active block: DXCCCII and tunable applications. Radioengineering 23:1130–1139
20. Tangsrirat W (2018) Actively floating lossy inductance simulators using voltage differencing buffered amplifiers. IETE J Res. https://doi.org/10.1080/03772063.2018.1433082
21. Navnit K, Vista J, Ranjan A (2019) A tuneable active inductor employing DXCCTA: grounded and floating operation. Microelectron J 90:1–11. https://doi.org/10.1016/j.mejo.2019.05.014
22. Singh Y S, Ranjan A, Adhikari S, Shimray B A (2020) Dual mode grounded active inductor employing MO-DXCCTA. In: 4th International conference on electronics, communication and aerospace technology (ICECA), pp 368–373. https://doi.org/10.1109/ICECA49313.2020.9297654
23. Siriphot D, Maneewan S, Jaikla W (2013) Single active element based electronically controllable grounded inductor simulator. In: IEEE 6th Biomedical engineering international conference, vol 23, no 3. IEEE, Krabi, Thialand New York, NY, USA, pp 1–4
24. Roongmuanpha N, Pukkalanun T, Tangsrirat W (2017) Resistorless realization of grounded lossy series inductor with two VDBAs and a grounded capacitor. ICCIP, Tokyo, Japan, pp 24–26. https://doi.org/10.1145/3162957.3163047
25. Biolek D, Senani R, Biolkova V, Kolka Z (2008) Active elements for analog sig-nal processing: classification, review, and new proposals. Radioengineering 17(4):15–32
26. Ranjan A, Perumalla S, Kumar R, Vista J, Yumnam S (2019) Second order universal filter using four terminal floating nullor (FTFN). J Circuits, Syst Comput 28:1950091
27. Rana P, Ranjan A (2021) Odd-and even-order electronically controlled wave filter employing differential difference transconductance amplfier (DDTA). Int J Electron. https://doi.org/10.1080/00207217.2020.1870737

Design and Implementation of All Optical Processing Units Together Performing Arithmetic and Logical Functions

Vanya Arun, Kapil Deo Bodha, Awadhesh K. Maurya, and Ashutosh K. Singh

Abstract Two different optical processing units have been designed in the proposed manuscript. They are integrated together to get different arithmetic and logical functions like half adder, NOT, NOR, XOR and AND gates. The designed photonic circuit is based on photonic interactions in the active region of semiconductor optical amplifier (SOA) which forms the heart of the design. This manuscript also discusses in detail two important non-linear attributes of semiconductor optical amplifier (SOA) which are cross gain modulation, (XGM) and cross phase modulation, (XPM). Both are utilized to obtain the desired functions. Input signals to the designed photonic circuit are non-return to zero (NRZ) pulses with varied envelope types such as on–off, on–off exponential, on–off ramp and raised cosine. The performance of the designed circuit is analyzed with parameters like extinction ratio (ER) and quality factor (QF). The reported results show that the quality factor remains unaffected at 63.01 dB for various pulse type but extinction ratio is highest for on–off ramp type which is 15.45 dB.

Keywords Optical processing unit (OPU) · Semiconductor optical amplifier (SOA) · Optical gates · Non-linearity · Extinction ratio (ER) · Quality factor (QF)

1 Introduction

Since the advent of the Internet until its maturity stage, generation and transportation of data has increased many folds. To fulfill the requirement of transporting huge amount of data, optical fibers are used, which can provide large bandwidth and

V. Arun (✉) · A. K. Maurya
IILM-CET, Greater Noida, India
e-mail: vanyaarun@gmail.com

K. D. Bodha
GCET, Greater Noida, India

A. K. Singh
FET, MJP Rohilkhand University, Bareilly, India

B. Mishra and M. Tiwari (eds.), *VLSI, Microwave and Wireless Technologies*, Lecture Notes in Electrical Engineering 877,
https://doi.org/10.1007/978-981-19-0312-0_10

fast speed. However, for signal processing of optical data, optoelectronic conversions were necessary, which had their drawbacks of high power consumption, heat generation and added noise from electronic circuitry for very high data rates. In some of the literature and recent researches it has been found that the electronic transfer of data even between intra processing nodes consumes a substantial portion of power [1, 2]. Data is transported (long distances) in the optical domain, but if it is also processed in optical domain the need for optoelectronic conversion will be eliminated. Basic building blocks of all optical signal processing are optical logic gates. Any photonic processing circuitry can be modeled using optical logic gates. The main component in designing circuits of optical logic gates [3, 4], optical flip flops [5–7], optical shift registers [8, 9], optical buffer [10], wavelength converter [11], and optical adder subtractor [12, 13] is SOA. Non-linearity of its active region such as XGM, XPM and four-wave mixing (FWM) are utilized in designing certain processing circuits. The main advantage of fabricating SOA in a photonic chip are its integration capabilities, stability, compactness, low latency and low switching energy and its independence from polarization and wavelength. The phenomenon of XGM arises when the gain saturates in the SOA's active region. It means the carrier density of the active length has reached its peak. Now this peaked carrier density will amplify any signal propagating through its active region. The amplified signal leaves behind decreased carrier density or reduced gain in the region. Any other signal propagating after the first signal in this region will experience less gain. Therefore due to propagation of one signal, gain for the other signal has been modulated or changed [14]. The other important phenomenon of XPM is also linked with the carrier density in the active region of SOA. With the variation in the carrier density the refractive index changes and hence the phase of the pulse traveling through it changes [15].

Many researchers in the recent years have designed photonic circuitries capable of performing functions like binary to gray code conversion [16], Adder [17, 18], Full Adder/Subtractor [19], NOT, OR and XOR [20] gates with different methods like utilizing the polarization rotation of the beam in highly non-linear fiber [21], by utilizing the non-linear phenomenon of four-wave mixing of SOA [22], with the help of ultrafast non-linear interferometers (UNI) gates [23], by using Mach-Zehnder interferometer (MZI) arrangement [24] and many more such as designing of universal gate like NAND gate [25]. These different photonic arithmetic and logical units when integrated together in a photonic circuit becomes a potential processing unit capable of performing many important functions. This manuscript successfully reports the integrated photonic processing unit capable of performing functions of half adder, NOT, XOR, AND and NOR gates. Two differ optical processing units (OPU-*A* and OPU-*B*) are designed as shown in Figs. 1 and 2. The components used in the design are all optical and are identical in design parameters. The OPU-*A* is implementing NOR gate function and requires three inputs (input signal 1, input signal 2 and probe pulse), whereas OPU-*B* is implementing NOT gate and requires only two inputs (input signal 1 and probe pulse).

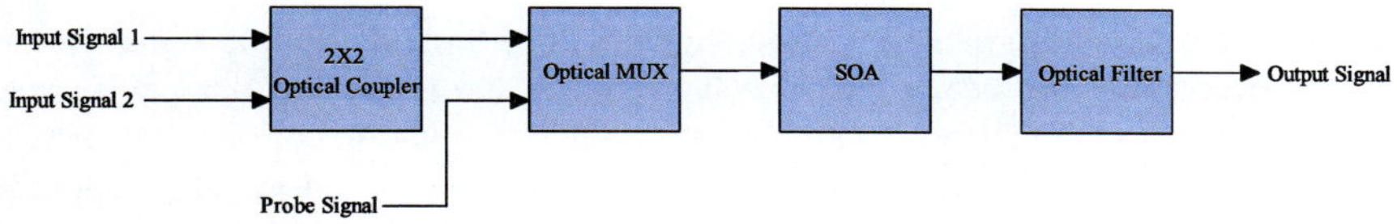

Fig. 1 Optical implementation of OPU-*A*

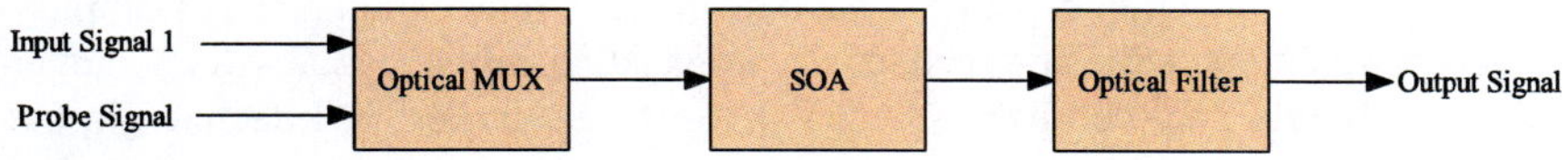

Fig. 2 Optical implementation of OPU-*B*

2 Principle of Operation and Model Formulation

The designed OPU-*A* and OPU-*B* with their input and output scheme is shown in Figs. 1 and 2. Total of five OPU's are integrated into a photonic circuit to make the proposed design. The input optical signals are *A* and *B*. They are treated as binary bits in the computation process. There are three probe signals given at various inputs to the designed optical circuit. The inputs of OPU-*A*-1 are optical signals *A*, *B* and probe signal 1 as shown in Fig. 3. Two inputs (*A* and B) are at wavelength 1555 nm. They are given to 2X2 optical coupler. The coupler couples the two input hence at is output port binary signal $A + B$ at 1555 nm comes out. Then this signal ($A + B$ at 1555 nm) is multiplexed with a low power probe signal named PROBE 1 at 1552 nm. The output of the optical multiplexer (mux) gives the binary output as $A + B +$ PROBE1. It is worth mentioning here that whenever two signals will be

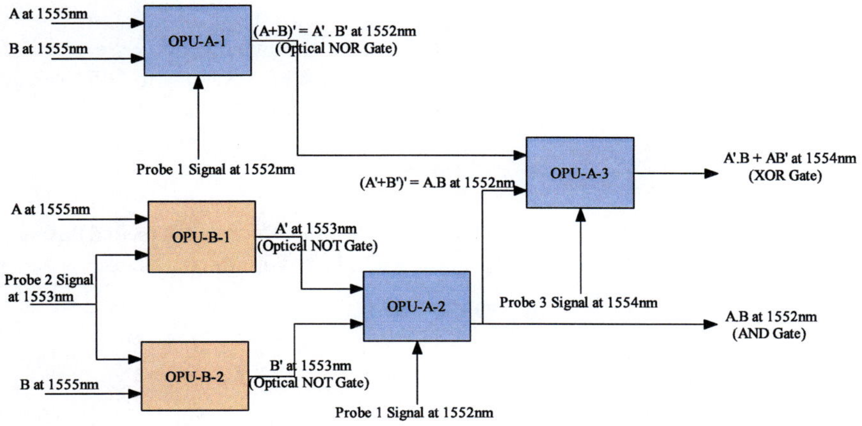

Fig. 3 Integrated photonic circuit implementing half adder, NOT gate and NOR gate

multiplexed in this design, higher power signal will be termed as pump signal and a lower power signal will be termed as probe signal. $A + B +$ PROBE1 signal is then given to SOA, which behaves like an inverter and a wavelength converter. It writes the invert of $A + B$ on PROBE1. Hence $\overline{A + B}$ at 1552 nm is its output. This happens because the high power pump signal ($A + B$) saturates the gain of SOA when binary 1 (presence of pump signal $A + B$) is transmitted across the active region of SOA. It means that it stimulates the free carriers present in the active region of SOA to return to their conduction or valance band as the case may be and thereby amplify the pump signal. Now the simultaneously propagating probe signal will encounter less or negligible gain because the population of free carriers has declined in the active region hence a binary 0 is imprinted on the probe signal till there is a binary 1 on the pump signal. So the inverted signal now has the new wavelength of the probe signal. After SOA the output will contain both the wavelengths 1555 nm as well as the 1552 nm. The filter with the pass band of 1552 nm will allow the inverted pump signal at probe1 wavelength, i.e., 1552 nm to pass. At filter binary output $\overline{A + B} = \overline{A} \cdot \overline{B}$ at wavelength 1552 nm will be obtained which is intended output of NOR gate and can be seen in Fig. 3.

Inputs A and B are then given to two different OPU-B as shown in Fig. 3 with PROBE2 at 1553 nm. After passing through SOA's present in different OPU-B structures, $\overline{A}$ and $\overline{B}$ at 1553 nm wavelength is obtained at the outputs which are output of NOT gate. These outputs obtained from NOT gate ($\overline{A}$ and $\overline{B}$) are given as inputs to the OPU-A-2. As shown in Fig. 1 the 1st component of OPU-A-2 is a 2X2 optical coupler which will couple its two input and will give the Boolean function of $\overline{A} + \overline{B}$. This signal is multiplexed with PROBE1 at 1552 nm and passed through SOA and optical filter. At the output port of filter, $\overline{\overline{A} + \overline{B}}$ is obtained which according to De Morgan's law can be written as:

$$\overline{\overline{A} + \overline{B}} = \overline{\overline{A}} \cdot \overline{\overline{B}} = A \cdot B \tag{1}$$

This is the output of Boolean function of AND gate which forms the carry of the half adder. Now the outputs of AND gate and NOR gate is given to OPU-A-3 as shown in Fig. 3. The output of 2X2 optical coupler of OPU-A-3 is:

$$\overline{A} \cdot \overline{B} + A \cdot B \tag{2}$$

This signal in (2) is multiplexed with PROBE3 at 1554 nm and then passed through SOA and filter whose pass band is centered at 1554 nm with the output as:

$$\overline{\overline{A} \cdot \overline{B} + A.B} \tag{3}$$

According to De Morgan's law:

$$= \overline{(\overline{A} \cdot \overline{B})} \cdot \overline{(A \cdot B)} \tag{4}$$

$$= \left(\overline{\overline{A}} + \overline{\overline{B}}\right) \cdot \left(\overline{A} + \overline{B}\right) \tag{5}$$

$$= (A + B) \cdot \left(\overline{A} + \overline{B}\right) \tag{6}$$

$$= \overline{A} \cdot B + A \cdot \overline{B} \tag{7}$$

Equation (7) is the output of OPU-*A*-3 and the output of the proposed design of the Exclusive-OR gate as well. This also forms the Sum of the half adder.

3 Theoretical Analysis

The carrier density *N* is an important parameter for determining the refractive index of SOA's active region. It is in this region that all the processing will take place. When a signal propagates across the active region, it gets amplified due to the carriers present in the region. When the carrier density of the SOA's active region is high enough it means that the gain has saturated, this saturated gain of the region will give rise to XGM [26].

$$\frac{\mathrm{d}N}{\mathrm{dt}} = \frac{J}{\mathrm{ed}} - R(N) - v_g.g(N).N_{\mathrm{ph}} \tag{8}$$

$$g(N) = \frac{\Gamma.\sigma_g}{V}(N - N_0) \tag{9}$$

where J is the current injection density, $R(N)$ is recombination rate, v_g is the group velocity of light, $g(N)$ is the material gain coefficient, N_{ph} is photon density, electron charge is denoted by e and the thickness of the active layer is denoted by, N_{ph} is photon density. Γ is confinement factor, differential gain is represented by σ_g, active region's volume is denoted by v and the carrier density for transparency is N_0. Gain per unit length of SOA is written as:

$$g = \Gamma.g(N) - \alpha \tag{10}$$

where loss coefficient α is taken per unit length of the active region. As the signal travels along the active region, it experiences an increasing gain which is exponential g and is described as:

$$G_s = e^{g.L} \tag{11}$$

For such gain as described in Eq. (11) length of the active region of SOA becomes very important. The width of the active region of SOA in the proposed design is kept as 50 nm and length as 100 nm. The parameter $g(N)$ (per unit length) which is material gain coefficient of the region is dependent on the density of the carriers and is the property of the material used. The material gain coefficient $g(N)$ can be written mathematically as:

$$g(N) = \frac{\mathrm{d}g}{\mathrm{d}N}(N - N_{\mathrm{tran}}) - K(\lambda - \lambda_p)^2 = S(N - N_{\mathrm{tran}})^2 - K(\lambda - \lambda_p)^2 \quad (12)$$

In Eq. (12) $\mathrm{d}g/\mathrm{d}N$ is termed as differential gain, and it characterizes the slope of material gain coefficient $g(N)$ when N is the carrier density of the medium used. N_{tran} is the carrier density at transparency. K is the parameter that defines the spectral gain curvature. λ_p is the wavelength at which the gain has peaked. The entire gain and processing of the designed gate is dependent on the material gain coefficient $g(N)$.

4 Results and Discussion

The inputs and the output plots of the designed photonic circuit is shown in Fig. 4a–h. The bitwise input and output are shown in Table 2. There are 00000 pre bits and 0000000 post bits and no operation is done on pre and post bits; hence it is not shown in Table 2. However they can be seen in the plots that is why starting bits are 00000 and ending bits are 0000000 for every plot. The input binary bits for all the three probe signals are the same. Table 1 shows the different logical operation obtained from the designed circuit.

Table 1 can be verified from plots in Fig. 4a–h. From the plots, it can be inferred that the intended results are obtained and the proposed photonic circuit.

Figure 4a–c shows the input data for binary optical signals A, B and all the probe signals. The X-axis shows the time, and Y-axis shows the power in watt (W). Figure 4d–h are the output of designed optical circuit (as shown in Fig. 3) giving a precise, logical expressions depending upon the input given to each one of them (Table 1). The output has been taken after the optical receiver; hence here y-axis shows the signal voltage. The simulation result fulfills the aim of this paper of designing all optical logic gates which are cable of performing Boolean functions.

The designed model is tested for various pulse type like an on–off, on–off ramp and raised cosine. The modulation type is NRZ for all the pump and probe pulses. The quality factor of the designed circuit can be calculated as [27]:

$$Q = \frac{\overline{P_1} - \overline{P_0}}{\sigma_1 + \sigma_0} \quad (13)$$

where $\overline{P_1}$ and $\overline{P_0}$ are the mean value of output powers of binary signal '1' and output binary signal '0', parameters σ_1 and σ_0 are the standard deviation of all 1 bits and 0 bits. Extinction ratio of the proposed optical design can be calculated as [28]:

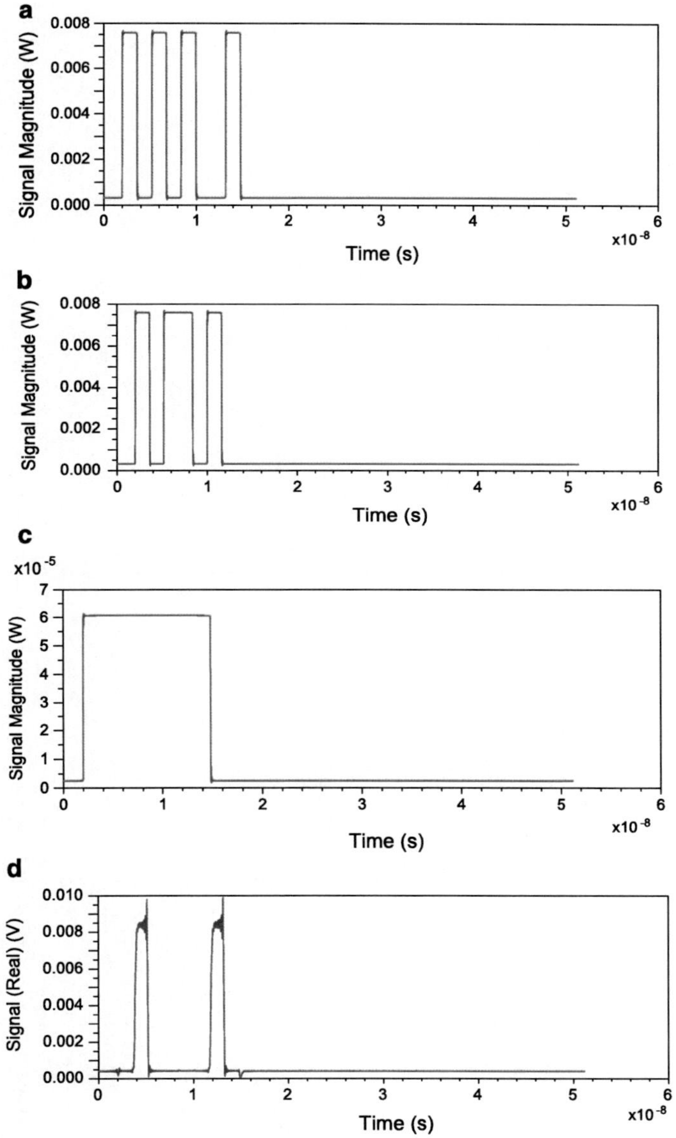

Fig. 4 **a** Plot of input binary signal *A* (Input), **b** Plot of input binary signal *B* (Input), **c** Plot of input binary probe signal (Input), **d** Plot of output binary signal of OPU-*A*-1/*A* NOR *B*, **e** Plot of output binary signal of OPU-*B*-1/NOT *A*, **f** Plot of output binary signal of OPU-*B*-2/NOT *B*, **g** Plot of output binary signal of OPU-*A*-2/*A* AND *B*/Carry, **h** Plot of output binary signal of OPU-*A*-3/*A* XOR *B*/Sum

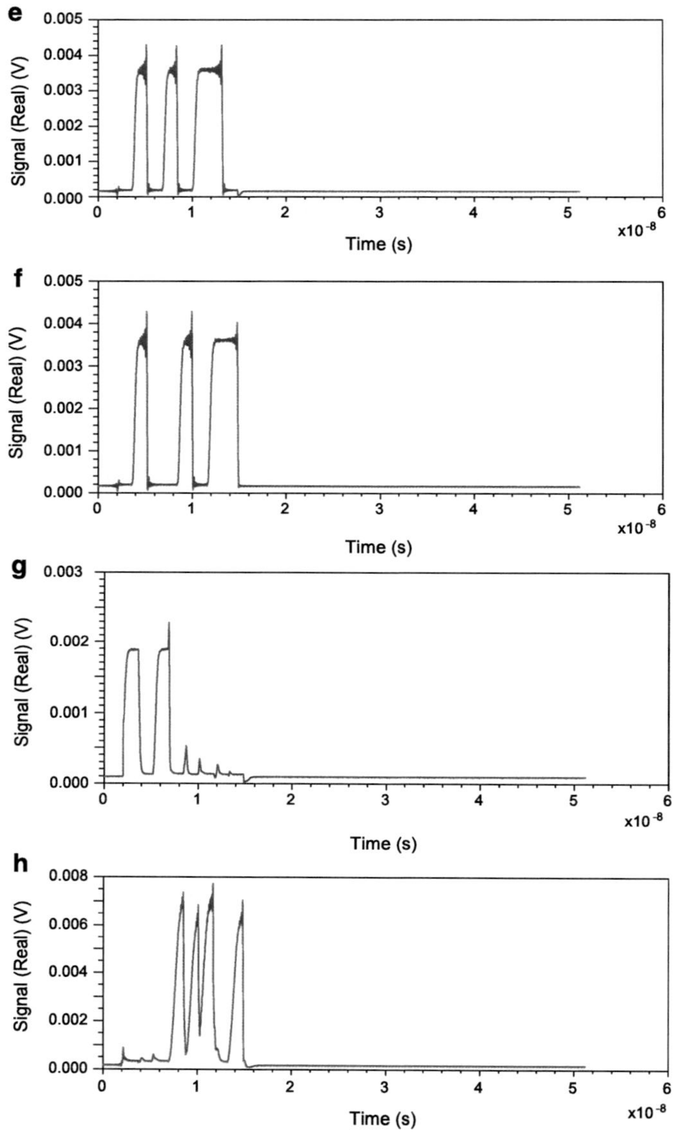

Fig. 4 (continued)

Table 1 Logical operations of the designed photonic circuit

S. No	OPU	Inputs to OPU			Output of OPU
1	OPU-*A*-1	A	B	Probe 1	$\overline{A+B}=\overline{A}.\overline{B}$
2	OPU-*B*-1	A	–	Probe 2	$\overline{A}$
3	OPU-*B*-2	B	–	Probe 2	$\overline{B}$
4	OPU-*A*-2	$\overline{A}$	$\overline{B}$	Probe 1	$A.B$ (Carry)
5	OPU-*A*-3	$\overline{A}.\overline{B}$	$A.B$	Probe 3	$A \oplus B$(Sum)

Table 2 Input and output bits of the proposed design

Input/Output	Bits						
Input (*A*)	0	1	0	1	0	0	1
Input (*B*)	0	1	1	0	1	0	0
Input (PROBE)	1	1	1	1	1	1	1
Output of OPU-*A*-1 (*A* NOR *B*)	1	0	0	0	0	1	0
Output of OPU-*B*-1(NOT *A*)	1	0	1	0	1	1	0
Output of OPU-*B*-2 (NOT *B*)	1	0	0	1	0	1	1
Output of OPU-*A*-2 (*A* AND *B*/Carry)	0	1	0	0	0	0	0
Output of OPU-*A*-3 (*A* XOR *B*/SUM)	0	0	1	1	1	0	1

$$\mathrm{ER(d}B) = 10\log_{10}\left(P^{1}_{\min}/P^{0}_{\max}\right) \tag{14}$$

$P^{1}{}_{\min}$ is the minimum power when binary bit is high or 1 and $P^{0}{}_{\max}$ is the maximum power obtained when binary bit is low or 0 is transmitted. Greater the extinction ratio better is the performance. Table 3 compares the two parameters for different types of signal.

Table 3 shows that the quality factor is the same for all the signal types, whereas there is a slight difference in the extinction ratios. The on–off ramp has the highest extinction ratio while raised cosine has its minimum value.

Table 3 Extinction ratio and Q-factor for different signal type

S. No	Signal type	Extinction ratio (ER) in dB	Q-factor (QF) in dB
1	on–off	15.38	63.01
2	on–off exponential	15.37	63.01
3	on–off ramp	15.45	63.01
4	Raised cosine	15.32	63.01

5 Conclusion

In this manuscript, an integrated photonic circuit is proposed which is capable of performing half adder function and logical functions like NOT, NOR, AND and XOR. The photonic circuit is designed successfully with satisfactory extinction ratio and quality factor. It is now time that electronic circuitry is replaced by photonic circuitry to eliminate optoelectronic conversions. To compete with the electronic devices, optical devices should be designed such that the consumption of power per bit and the switching time should be low enough. All optical signal processing is the future.

Acknowledgements This research is funded by AKTU Lucknow (U.P.) as award of grant under "Collaborative Research Innovation Program (CRIP)" funding through TEQIP-III of AKTU reference no: AKTU/Dean-PGSR/2019/ CRIP/46.

References

1. Notomi M, Shinya A, Nozaki K et al (2011) Low-power nanophotonic devices based on photonic crystals towards dense photonic network on chip. IET Circuits Devices Syst 5:84. https://doi.org/10.1049/iet-cds.2010.0159
2. Leung SN (2018) Thermally conductive polymer composites and nanocomposites: processing-structure-property relationships. Compos Part B Eng
3. Kaur S, Kaler RS, Kamal TS (2015) All-optical binary full adder using logic operations based on the nonlinear properties of a semiconductor optical amplifier. J Opt Soc Korea. https://doi.org/10.3807/JOSK.2015.19.3.222
4. Kotiyal S, Thapliyal H, Ranganathan N (2013) Mach-Zehnder interferometer based design of all optical reversible binary adder
5. Bakopoulos P, Vyrsokinos K, Fitsios D et al (2012) All-optical T-flip-flop using a single SOA-MZI-based latching element. IEEE Photonics Technol Lett. https://doi.org/10.1109/LPT.2012.2187779
6. Li W, Ma S, Hu H, Dutta NK (2012) All optical latches using quantum-dot semiconductor optical amplifier. Opt Commun. https://doi.org/10.1016/j.optcom.2012.07.085
7. Pitris S, Vagionas C, Kanellos GT et al (2016) Optical static RAM cell using a monolithically integrated InP Flip-Flop and wavelength-encoded signals. In: 2016 Optical fiber communications conference and exhibition, OFC 2016
8. Shcherbakov MR, Liu S, Zubyuk VV et al (2017) Ultrafast all-optical tuning of direct-gap semiconductor metasurfaces. Nat Commun. https://doi.org/10.1038/s41467-017-00019-3
9. Chattopadhyay T, Reis C, André P, Teixeira A (2012) Theoretical analysis of all-optical clocked D flip-flop using a single SOA assisted symmetric MZI. Opt Commun. https://doi.org/10.1016/j.optcom.2011.12.103
10. Fu S, Shum P, Zhang L et al (2006) Design of SOA-based dual-loop optical buffer with a 3 × 3 collinear coupler: guideline and optimizations. J Light Technol. https://doi.org/10.1109/JLT.2006.876092
11. Matsuura M, Raz O, Gomez-Agis F et al (2011) Ultrahigh-speed and widely tunable wavelength conversion based on cross-gain modulation in a quantum-dot semiconductor optical amplifier. Opt Express. https://doi.org/10.1364/oe.19.00b551

12. Datta K, Chattopadhyay T, Sengupta I (2015) All optical design of binary adders using semiconductor optical amplifier assisted Mach-Zehnder interferometer. Microelectron J. https://doi.org/10.1016/j.mejo.2015.06.020
13. Chauhan C, Kaushik BK, Kumar S (2019) Design of optical reversible hybrid adder-subtractor device using Mach-Zehnder interferometers for WDM applications. J Opt Commun. https://doi.org/10.1515/joc-2019-0031
14. Contestabile G (2014) Wavelength conversion of PAM signals by XGM in SOAs. In: Optics InfoBase Conference Papers
15. Tao Z, Yan W, Liu L et al (2011) Simple fiber model for determination of XPM effects. J Light Technol. https://doi.org/10.1109/JLT.2011.2107728
16. Kumari A, Pal A, Singh A, Sharma S (2020) All-optical binary to gray code converter using non-linear material based MIM waveguide. Optik (Stuttg). https://doi.org/10.1016/j.ijleo.2019.163449
17. Tan H, Liu G, Yang H et al (2017) Light-gated memristor with integrated logic and memory functions. ACS Nano. https://doi.org/10.1021/acsnano.7b05762
18. Ying Z, Dhar S, Zhao Z et al (2018) Electro-optic ripple-carry adder in integrated silicon photonics for optical computing. IEEE J Sel Top Quantum Electron. https://doi.org/10.1109/JSTQE.2018.2836955
19. Kumar PK, Rao PP, Kakarla HK (2017) Optimal design of reversible parity preserving new Full adder/Full subtractor. In: Proceedings of 2017 11th international conference on intelligent systems and control, ISCO 2017
20. Pal A, Ahmed MZ, Swarnakar S (2021) An optimized design of all-optical XOR, OR, and NOT gates using plasmonic waveguide. 53:84. https://doi.org/10.1007/s11082-021-02732-2
21. Yu C, Christen L, Luo T et al (2005) All-optical XOR gate using polarization rotation in single highly nonlinear fiber. IEEE Photonics Technol Lett 17:1232–1234. https://doi.org/10.1109/LPT.2005.846918
22. Deng N, Chan K, Chan CK, Chen LK (2006) An all-optical XOR logic gate for high-speed RZ-DPSK signals by FWM in semiconductor optical amplifier. IEEE J Sel Top Quantum Electron 12:702–707. https://doi.org/10.1109/JSTQE.2006.876603
23. Yang X, Manning RJ, Hu W (2010) Simple 40Gbit/s all-optical XOR gate. Electron Lett 46:222–223. https://doi.org/10.1049/el.2010.3039
24. Sonarita G, Bharata H, Asnawati R (2014) Peningkatan Kemampuan Komunikasi Matematis Melalui Model Pembelajaran Kooperatif Tipe Three-Step Interview. J Pendidik Mat Unila 2:1–4. https://doi.org/10.14042/j.cnki.32.1309.2013.06.005
25. Singh A, Pal A, Singh Y, Sharma S (2019) Design of optimized all-optical NAND gate using metal-insulator-metal waveguide. Optik (Stuttg). https://doi.org/10.1016/j.ijleo.2019.01.098
26. Arun V, Singh AK, Shukla NK, Tripathi DK (2016) Design and performance analysis of SOA–MZI based reversible toffoli and irreversible AND logic gates in a single photonic circuit. Opt Quantum Electron 48:1–15. https://doi.org/10.1007/s11082-016-0711-y
27. Wang C, Burek MJ, Lin Z et al (2014) Integrated high quality factor lithium niobate microdisk resonators. Opt Express. https://doi.org/10.1364/oe.22.030924
28. Shen W, Hu C, Huo S et al (2019) Black phosphorus nano-polarizer with high extinction ratio in visible and near-infrared regime. Nanomaterials. https://doi.org/10.3390/nano9020168

Design of Fruit-V2 and Fruit-80 Stream Ciphers Using Basic Version

Vamsi Krishna Guntapudi and Dheeraj Kumar Sharma

Abstract Nowadays, technology has improved a lot. The use of computing power for reducing human power and time has increased. The use of low-resources devices like WSN, RFID, and IIOT is also increased. Xilinx FPGA solutions provide real-time processing and flexibility. For all these, the main challenging problem is the confidentially with limited resources. eSTREAM portfolio was started to discover the efficient lightweight stream ciphers. Many lightweight stream ciphers came into existence like Sprout. Sprout was introduced to achieve all the goals but researchers have proven that Sprout was insecure, i.e., it cannot withstand the attacks. Fruit-v2 is introduced with improved features. It has proven that is very secure and ultra-lightweight. This paper discusses the design of the basic version of Fruit-v2 and Fruit-80. The main objective is to obtain a high frequency so that the speed is achieved.

Keywords Stream Cipher · Ultra-lightweight · LFSR · NLFSR · Basic Version · Frequency

1 Introduction

The development of cryptographic algorithms for achieving high throughput and optimized area consumption is required. There are two cryptographic algorithms [1]: symmetric ciphers and asymmetric ciphers. For encryption and decryption, the secret key used is the same in the symmetric ciphers. In asymmetric ciphers, the public-key is used for encryption and the secret key is used for decryption. Symmetric ciphers have more demand compared to asymmetric ciphers because they have optimized consumed area and throughput.

V. K. Guntapudi (✉)
Embedded System Design, National Institute of Technology, Kurukshetra, India
e-mail: vamsiguntapudi@gmail.com

D. K. Sharma
Department of ECE, National Institute of Technology, Kurukshetra, India
e-mail: sharma986dheeraj@nitkkr.ac.in

B. Mishra and M. Tiwari (eds.), *VLSI, Microwave and Wireless Technologies*,
Lecture Notes in Electrical Engineering 877,
https://doi.org/10.1007/978-981-19-0312-0_11

In symmetric algorithms, there are two ciphers, block ciphers and stream ciphers [2]. The block ciphers encrypt the data in blocks, and the stream ciphers encrypt the data in a stream of bits.

In cryptology, the European Network of Excellence has started eSTREAM project in 2004 [1]. The newly discovered stream ciphers, which qualify the three levels, are provided in the eSTREAM project. The eSTREAM project has been classified into two portfolios: portfolio-1 and portfolio-2. Portfolio-1 is software-oriented, and portfolio-2 is hardware-oriented.

The demand for lightweight and highly secured stream ciphers has led to the implementation of many stream ciphers. Design of Grain-v1 [3], Mickey 2.0 [4], Lizard [5], Grain 128a [6], Plantlet [7], and Sprout [8] is implemented. The implementation can be done in the basic version. In the basic version, the separate shift register is used to load the key, and initialization vector (IV) is loaded directly into two shift registers. With this basic version, high frequency is achieved. Using Verilog Hardware Description Language [9], the code is simulated on Xilinx ISE design Suite, version 14.7 [10].

In the eSTREAM project, three ciphers have been introduced. They are Grain-v1, MICKEY 2.0, and Trivium. These stream ciphers consist of nonlinear feedback shift register (NLFSR), linear feedback shift register (LFSR), counter, and a filter or Boolean function. The NLFSR is used for protecting the cipher against attacks. The LFSR is used for obtaining the maximum length and avoiding the all-zero state.

Grain-v1 was proved insecure. Later few successors of Grain-v1 are discovered but they are also proved insecure. Grain128a was better than earlier. But in these ciphers, the internal state is longer. But there is a restriction on sources, i.e., lightweight stream ciphers are having great demand. To withstand the attacks, the length of the secret key should be half of the length of the internal state. Now the problem is to withstand the attacks with the smaller internal state. A new stream cipher sprout was introduced which has a small internal state. But sprout was proved insecure by the researchers. Then with some improvements, Fruit-v2 was introduced informally. Fruit-v2 is having a shorter internal state, i.e., 80 bits, so it is an ultra-lightweight stream cipher. It is also withstanding the attacks from attackers. Later, with some modifications, Fruit-80 was implemented. In this paper, design architecture of Fruit-v2 and Fruit-80 has been discussed in Sect. 2. Appendix A consists of Verilog code for Fruit-v2, and in the same way, Fruit-80 can be written. Section 3 consists of simulation results, and conclusion is written in Sect. 4.

2 Design Architecture

2.1 Fruit-v2

Fruit-v2 consists of one LFSR, one NLFSR, and one counter. The LFSR length is 43 bits, i.e., $d_{\mathrm{t}} \ldots d_{(\mathrm{t}+42)}$, NLFSR length is 37 bits, i.e., $m_{\mathrm{t}} \ldots m_{(\mathrm{t}+36)}$, and the counter length is 15 bits, i.e., $C^0{}_t \ldots C^{14}{}_t$. The initialization vector (IV) is of length 70 bits, i.e., $v_0 \ldots v_{69}$. Secret key is of length 80 bits, i.e., $K_0 \ldots K_{79}$. IV should be different for different keys, i.e., IV should be uniquely produced. The block diagram is shown in Fig. 1.

The counter has two separate counters. The first counter (C1) has the first 7 bits ($C^6{}_t - C^0{}_t$, where $C^6{}_t$ is the LSB) and is used for the round key function. It is incremented by one for every cycle. The second counter (C2) has the remaining 8 bits ($C^{14}{}_t - C^7{}_t$, where $C^{14}{}_t$ is the LSB) and is used for initialization and generation of keystreams. It counts from all zeros to all ones and it repeats.

The sprout is insecure. The main reason for it is that the key is not involved in the NLFSR update function. The attacker can attack the key easily. If attacker knows some bits of the key, it is very easy to get the secret key. So, the round key function should satisfy two important factors. The key should be involved in the internal state update function, and it has to be lightweight.

As it is well known that the round key function of Fruit-v2 is involved in the update function, Fruit-v2 is more secure. Due to this, it will be a difficulty for attackers to get the secret key. Since the length of the round key function is 80 bits, it shows

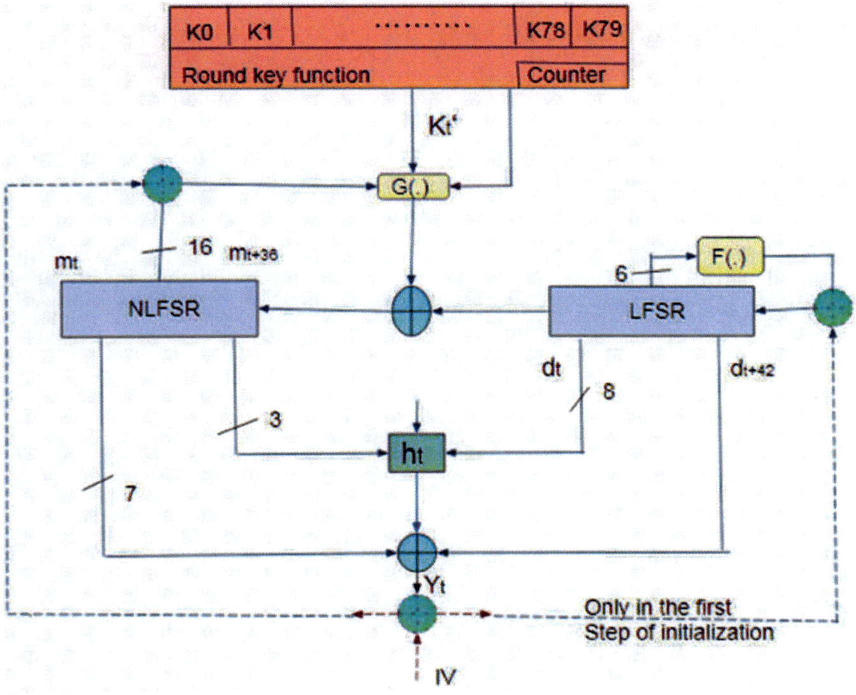

Fig. 1 Fruit-v2 block diagram

that it is lightweight. It has been already discussed that the C1 counter is used in the round key function. Now, it will be explored that how it can be used.

Round key function can be obtained as follows in every cycle.

$$K_t' = K_P \cdot K_{(Q+32)} + K_{(R+64)} \cdot K_S + K_{(T+16)} + K_{(U+48)} \tag{1}$$

where $P = (C^0{}_tC^1{}_tC^2{}_tC^3{}_tC^4{}_t)$, $Q = (C^5{}_tC^6{}_tC^0{}_tC^1{}_tC^2{}_t)$, $R = (C^3{}_tC^4{}_tC^5{}_tC^6{}_t)$, $S = (C^0{}_tC^1{}_tC^2{}_tC^3{}_t)$, $T = (C^4{}_tC^5{}_tC^6{}_tC^0{}_tC^1{}_t)$, $U = (C^2{}_tC^3{}_tC^4{}_tC^5{}_tC^6{}_t)$.

NLFSR update function can be written as:

$$\begin{aligned} m_{(t+37)} = {} & K_t' + d_t + C_t^3 + m_t + m_{(t+20)} \\ & + m_{(t+12)} \cdot m_{(t+3)} + m_{(t+14)} \cdot m_{(t+25)} + m_{(t+5)} \cdot m_{(t+23)} \\ & \cdot m_{(t+31)} + m_{(t+8)} \cdot m_{(t+18)} + m_{(t+28)} \cdot m_{(t+30)} \cdot m_{(t+32)} \cdot m_{(t+34)} \end{aligned} \tag{2}$$

The last bit of NLFSR is updated with the above equation. It is using the LSB bit of LFSR, G function, and 1 bit from the counter, where G function is having 15 bits from NLFSR and $K_t{}'$.

LFSR update function can be written as:

$$d_{(t+43)} = d_t + d_{(t+8)} + d_{(t+18)} + d_{(t+23)} + d_{(t+28)} + d_{(t+37)} \tag{3}$$

It is a primitive polynomial. By using a primitive polynomial, the sequence generated is of the maximum length. Here for LFSR, the primitive polynomial is used for feedback. So, the maximum length is obtained. The other disadvantage of sprout was all-zero state. In all-zero state, the cipher will be locked in that state. This problem was not there in Fruit-v2 because, in the second state of initialization, initialization was done in such a way that the all-zero states is avoided.

Pre-output function can be written as shown below:

$$\begin{aligned} h_t = {} & d_{(t+6)} \cdot d_{(t+15)} + d_{(t+1)} \cdot d_{(t+22)} + m_{(t+35)} \cdot d_{(t+27)} \\ & + m_{(t+1)} \cdot m_{(t+24)} + m_{(t+1)} \cdot m_{(t+33)} \cdot m_{(t+42)} \end{aligned} \tag{4}$$

Here, 8 bits from LFSR and 3 bits from NLFSR are used in this function.

The output function can be written as follows:

$$\begin{aligned} Y_t = {} & h_t + m_t + m_{(t+7)} + m_{(t+13)} + m_{(t+19)} + m_{(t+24)} \\ & + m_{(t+29)} + m_{(t+36)} + m_{(t+38)} \end{aligned} \tag{5}$$

Here, 1 bit from LFSR, 7 bits from NLFSR, and pre-output function are used in this function.

IV is extended to 130 bits by padding 59 zeros and 1 one:

$$IV' = 1000000000v_0v_1v_2v_3 \ldots v_{66}v_{67}v_{68}v_{69}000 \ldots 00 \tag{6}$$

The initialization phase is discussed below:

In the first step of initialization, NLFSR is loaded with the first 37 bits of the secret key and the LFSR is loaded with the remaining bits of the secret key (K_0 to m_0, K_1 to m_1, …, K_{36} to m_{36}, K_{37} to d_1, …, K_{79} to d_{42}). The counter is set to zero and incremented by one for each clock cycle. The output is feedback to both NLFSR and LFSR by XOR operation with IV bits for 130 clock cycles.

In the second step of initialization, to prevent all-zero states in LFSR, the counter C1 is initialized as ($C^0{}_{130} = m_{130}$, $C^1{}_{130} = m_{131}$, …, $C^4{}_{130} = m_{134}$, $C^5{}_{130} = m_{135}$, $C^6{}_{130} = d_{130}$). Now, the feedback is disconnected and does not take any output till 80 clock cycles. After 210 clock cycles, the keystream is produced at the output.

2.2 *Fruit-80*

Fruit-80 consists of one NLFSR, one LFSR, and one counter. The NLFSR length is 37 bits, i.e., $m_t \ldots m_{(t+36)}$, LFSR length is 43 bits, i.e., $d_t \ldots d_{(t+43)}$, and the counter length is 7 bits, i.e., $C^0{}_t \ldots C^6{}_t$. The secret key is of length 80 bits, i.e., $K_0 \ldots K_{79}$, and initialization vector (IV) is of length 70 bits, i.e., $v_0 \ldots v_{69}$. IV should be different for a different key. The block diagram is shown in Fig. 2.

Counter bits are used in round key function as shown below:

$$p = (C_t^0C_t^1C_t^2C_t^3), q = (C_t^1C_t^2C_t^3C_t^4C_t^5), r = (C_t^2C_t^3C_t^4C_t^5C_t^6)$$

The round key function is used in the NLFSR update function and as well as in the pre-output function.

$$\begin{aligned} K_t' = {} & K_p \cdot K_{(q+16)} \cdot K_{(r+48)} + K_p \cdot K_{(q+16)} \\ & + K_{(q+16)} \cdot K_{(r+48)} + K_p \cdot K_{(r+48)} + K_{(q+16)} \end{aligned} \tag{7}$$

$$\begin{aligned} K_t^* = {} & K_p \cdot K_{(q+16)} + K_{(q+16)} \cdot K_{(r+48)} + K_p \cdot K_{(r+48)} \\ & + K_p + K_{(q+16)} + K_{(r+48)} \end{aligned} \tag{8}$$

NLFSR update function can be written as:

$$\begin{aligned} m_{(t+37)} = {} & K_t' + d_t + m_t + m_{(t+20)} + m_{(t+12)} \cdot m_{(t+3)} \\ & + m_{(t+14)} \cdot m_{(t+25)} + m_{(t+5)} \cdot m_{(t+23)} \cdot m_{(t+31)} \\ & + m_{(t+8)} \cdot m_{(t+18)} + m_{(t+28)} \cdot m_{(t+30)} \cdot m_{(t+32)} \cdot m_{(t+34)} \end{aligned} \tag{9}$$

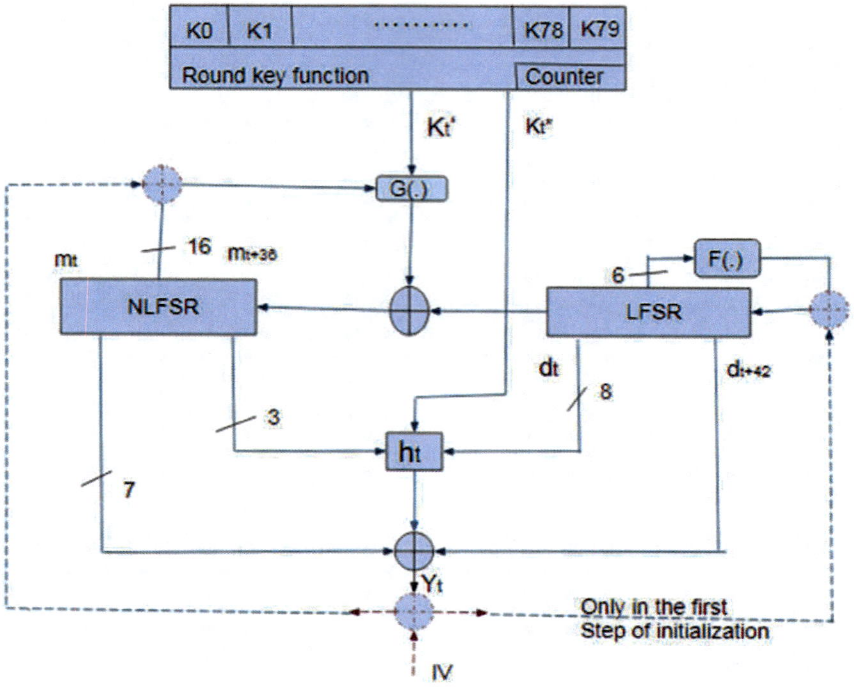

Fig. 2 Fruit-80 block diagram

The MSB bit of the NLFSR is updated with the above equation. The bits involved are LSB bit from the LFSR and G function, where G function is taking 16 bits from NLFSR and K_t'.

LFSR update function can be written as

$$d_{(t+43)} = d_t + d_{(t+8)} + d_{(t+18)} + d_{(t+23)} + d_{(t+28)} + d_{(t+37)} \tag{10}$$

The MSB bit of the NLFSR is updated with the above equation. It is a linear function of 6 bits from LFSR.

Pre-output function can be written as

$$\begin{aligned} h_t = {} & K_t^* \cdot \left(m_{(t+36)} + m_{(t+19)}\right) + d_{(t+6)} \cdot d_{(t+15)} \\ & + d_{(t+1)} \cdot d_{(t+22)} + m_{(t+35)} \cdot d_{(t+27)} \\ & + m_{(t+1)} \cdot m_{(t+24)} + m_{(t+1)} \cdot m_{(t+33)} \cdot m_{(t+42)} \end{aligned} \tag{11}$$

Unlike Fruit-v2, round key function is also involved in pre-output function. It is having 7 bits from NLFSR, 5 bits from LFSR, and K_t^*.

Output function can be written as

$$Y_t = h_t + m_t + m_{(t+7)} + m_{(t+19)} + m_{(t+29)} + m_{(t+36)} + d_{(t+38)} \quad (12)$$

In output function, the bits involved are LFSRs 1 bit, NLFSRs 5 bits, and pre-output function.

In the initialization phase, there are three phases. IV is extended to 80 bits by padding 9 zeros and 1 one.

$$IV' = 1000000000v_0v_1 \ldots v_{68}v_{69} \quad (13)$$

In the first step, NLFSR is initialized with the starting 37 bits of secret key, i.e., K_0 to m_0, K_1 to m_1, … K_{36} to m_{36} and LFSR is initialized with the remaining bits of secret key, i.e., K_{37} to d_0, K_{38} to d_1, … K_{79} to d_{42}. NLFSR and LFSR are fed back with the output bits XOR with the IV' bits at each clock cycle for 80 clock cycles.

In the second step, counter is initialized with NLFSR and LFSR bits as shown. ${C^0}_{80} = m_{80}$, ${C^1}_{80} = m_{81}$, …, ${C^5}_{80} = m_{85}$, ${C^6}_{80} = d_{80}$.

In the final step, i.e., the third step, the feedback is removed for 80 clock cycles and the output is also not taken for these 80 clock cycles. After completing the 160 clock cycles, the output stream is taken at the output.

3 Simulation Results

In this paper, the software implementation was performed by writing code in Verilog HDL, on Xilinx ISE Design suite, version 14.7, and the results are shown below.

3.1 Fruit-v2

Simulation results of the basic version of Fruit-v2 are given below. Simulation result for initialization phase is shown in Fig. 3. Simulation result for key generation, i.e., the output is ready to take at this point of time, is shown in Fig. 4. In the below figures, where lf[0:42] is LFSR which is of 43 bits, nf[0:36] is NLFSR which is of 37 bits, count1 is the first counter, count2 is the second counter, clk is the clock signal and rst is the reset signal. Clock signal has the time period of 10 ns.

When rst becomes active at positive-edge clock NLFSR and LFSR is initialized with key bits in both Fruit-v2 and Fruit-80. This is shown in hexadecimal format as shown below. The first step of initialization phase is executed for 130 clock cycles, i.e., the values of count2 from 0 to 129, and then second step for 80 clock cycles, i.e., the values of count2 from 130 to 209. After 210 clock cycles, i.e., from 210th clock

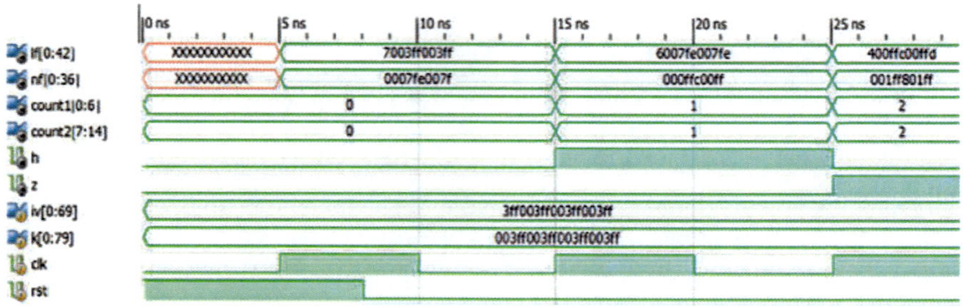

Fig. 3 Simulation output of initialization phase Fruit-v2

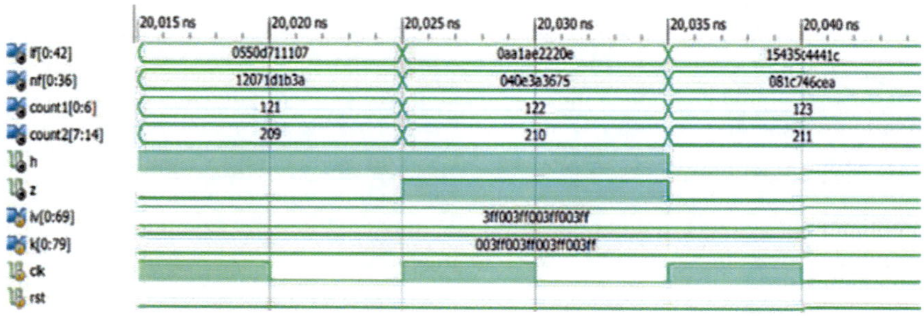

Fig. 4 Simulation output of key generation phase Fruit-v2

cycle, the key bit is taken as output. Both the initialization phase and key generation phase are shown below.

3.2 *Fruit-80*

The simulation output of basic version of Fruit-80 is shown below. Simulation result for initialization phase of Fruit-80 is shown in Fig. 5, and simulation result for keystream generation is shown in Fig. 6. The first step of initialization phase is executed for 80 clock cycles, i.e., the values of count1 from 0 to 79, and in the second step, count1 is initialized as discussed in Sect. 2 and it is executed for next 80 clock cycles. After 160 clock cycles, the key bit 'z' is ready to take as output.

4 Conclusion

In this paper, the software implementation of basic version of Fruit-v2 and Fruit-80 is performed using Verilog hardware description language on Xilinx ISE Design

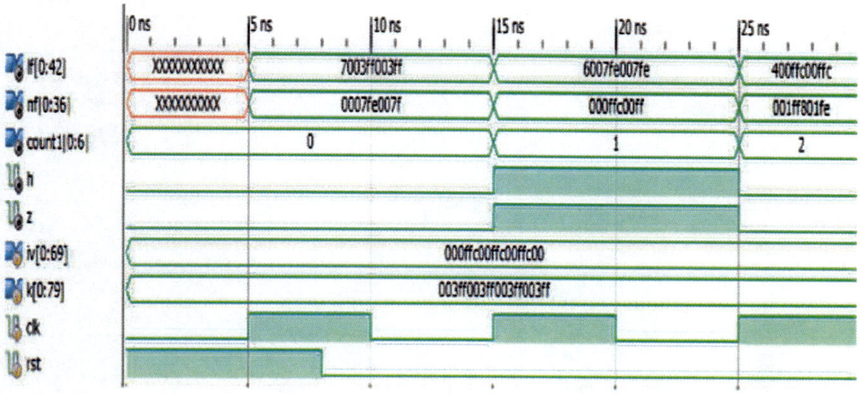

Fig. 5 Simulation output for initialization phase of Fruit-80

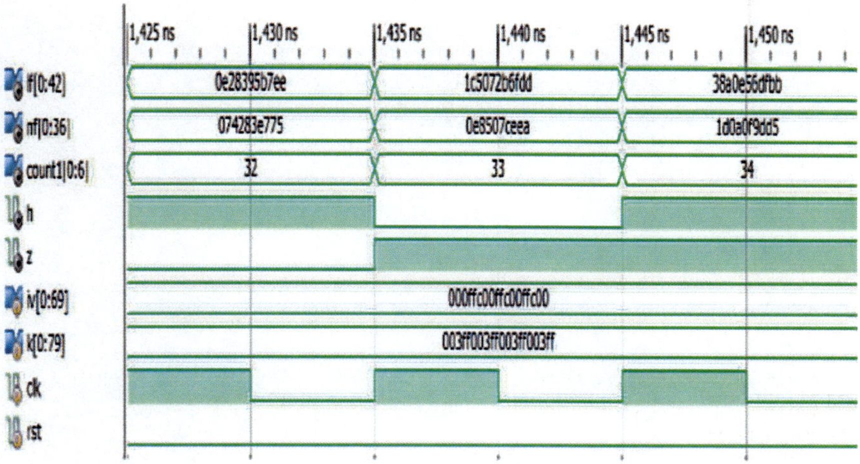

Fig. 6 Simulation output of key generation phase of Fruit-80

suite. The device used on Xilinx is xc7a100t-3csg324. Using the basic version, the maximum frequency obtained for Fruit-v2 is 404.989 MHz and the area is 99 slices. The maximum frequency obtained for Fruit-80 using basic version is 373.31 MHz and the area is 92 slices. Fruit-v2 has high speed compared to Fruit-80. But it can be seen that Fruit-v2 is having more area compared to Fruit-80. We get the speed at the cost of area. So, basic version is chosen when high speed is the main criteria.

Appendix A: Program Code

A.1 Fruit-v2 Design code:

```
module fruitv2(output reg [0:42]lf,                    //LFSR
                          output reg [0:36]nf,           //NLFSR
                          output reg [0:6]count1=7'b0,   // first counter
                          output reg[7:14]count2=8'b0,   // second counter
                          input [0:69]iv,                // initialization vector
                          input [0:79]k,                 // secret key
                          input clk,                     // clock signal
                          input rst,                     // synchronous reset
                          output reg h=0,                // pre-output function
                          output reg z=0);               // output function
reg kt;          // round key function
wire [0:4]s,y,q,r;  // counter bits used in round key function
wire[0:3]u,p;
wire [0:129]iv1;
reg i;
assign iv1 = {1'b1,9'b0,iv[0:69],50'b0}; // extending IV to 130 bits
assign fl=lf[0]^lf[8]^lf[18]^lf[23]^lf[28]^lf[37]; // LFSR update function
assign
nfb=kt^lf[0]^count1[3]^nf[0]^nf[10]^nf[20]^(nf[12]*nf[3])^(nf[14]*nf[25])^(nf[5]*
    nf[23]*nf[31])^(nf[8]*nf[18]*nf[28]*nf[30]*nf[32]*nf[34]);//NLFSR update function

//initializing counter bits to use in round key function
assign s ={count1[0],count1[1],count1[2],count1[3],count1[4]};
assign y ={count1[5],count1[6],count1[0],count1[1],count1[2]};
assign u ={count1[3],count1[4],count1[5],count1[6]};
assign p ={count1[0],count1[1],count1[2],count1[3]};
assign q ={count1[4],count1[5],count1[6],count1[0],count1[1]};
assign r ={count1[2],count1[3],count1[4],count1[5],count1[6]};

//counter block
always@(posedge clk )
begin
if(rst)
     begin
     count1<=7'b0;
     count2<=8'b0;
     end
//second step of initialization
else if(count2==130)
     begin
     count1[0]<=nf[0];
     count1[1]<=nf[1];
     count1[2]<=nf[2];
     count1[3]<=nf[3];
     count1[4]<=nf[4];
     count1[5]<=nf[5];
     count1[6]<=lf[0];
     count2<=count2+1'b1;
     end
else
     begin
     count1<=count1+1'b1;
     count2<=count2+1'b1;
     end
end

//round-key function block
always@(posedge clk or posedge rst)
begin
if(rst)
     kt<=1'b0;
else
     kt<=(k[s]*k[y+32])^(k[u+64]*k[p])^k[q+16]^k[r+48];
end
```

```
//LFSR and NLFSR block
always@(posedge clk )
if(rst)
     begin
     nf[0:36]<=k[0:36];      //initializing key bits to NLFSR
     lf[0:42]<=k[37:79];     //initializing key bits to LFSR
     end
// initialization phase
else if(count2<=129)
     begin
     nf[0:35]<=nf[1:36];
     nf[36]<=z^iv1[i]^nfb;
     lf[0:41]<=lf[1:42];
     lf[42]<=z^iv1[i]^f1;
     end
//second step of initialization
else if(count2==130)
     lf[0]<=1'b1;
// key-generation phase starts after 80 clock cycles
else if(count2>129)
     begin
     nf[0:35]<=nf[1:36];
     nf[36]<=nfb;
     lf[0:41]<=lf[1:42];
     lf[42]<=f1;
     end
end

//used for selecting IV bits for first step of initialization phase
always@(posedge clk or posedge rst)
if(rst)
     i<=1'b0;
else if(count2<130)
     i<=i+1'b1;

//pre-output function block
always@(posedge clk or posedge rst)
if(rst)
    h<=0;
else
    h<=(lf[6]*lf[15])^(lf[1]*lf[22])^(nf[35]*lf[27])^(lf[11]*lf[33])^(nf[1]
         *nf[33]*lf[42]);

//output function block
always@(posedge clk or posedge rst)
if(rst)
    z<=0;
else
    z<=h^nf[0]^nf[7]^nf[13]^nf[19]^nf[24]^nf[29]^nf[36]^lf[38];

endmodule
```

A.2 Fuit-v2 Testbench Code:

```
module testbench;

      // Inputs
      reg [0:69] iv;
      reg [0:79] k;
      reg clk, rst;

      // Outputs
      wire [0:42] lf;
      wire [0:36] nf;
      wire [0:6] count1;
      wire [7:14] count2;
      wire h, z;

      // Instantiate the Unit Under Test (UUT)
      fruitv2 uut (.lf(lf), .nf(nf), .count1(count1), .count2(count2), .iv(iv),
             .k(k), .clk(clk), .rst(rst), .h(h), .z(z));

      initial begin
             // Initialize Inputs
             clk = 0;
             iv =
70'b1111111111000000000011111111110000000000111111111100000000001111111111;
             k =
80'b00000000001111111111000000000011111111110000000000111111111100000000001111111111;
              rst = 1;
             #8rst=0;
             end

             initial begin
              // clock signal
             forever #5clk=~clk;
             // Wait 100 ns for global reset to finish
             #100;
             end
endmodule
```

References

1. eSTREAM, ECRYPT Stream Cipher Project. http://www.ecrypt.eu.org/stream
2. Li B, Liu M, Lin D (2020) FPGA implementations of Grain v1, Mickey 2.0, Trivium, Lizard and Plantlet. Microprocess Microsyst 78:103210
3. Hell M, Johansson T, Maximov A, Meier W (2008) The grain family of stream ciphers. In: New stream cipher designs. Springer, Heidelberg, pp 179–190
4. Babbage S, Dodd M (2006) The stream cipher MICKEY 2.0. ECRYPT Stream Cipher
5. Hamann M, Krause M, Meier W (2017) LIZARD-A lightweight stream cipher for power-constrained devices, IACR Trans Symmetr Cryptol :45–49
6. Agren M, Hell M, Johansson T, Meier W (2011) Grain-128 a: a new version of Grain-128 with optional authentication. Int J Wirel Mob Comput 5(1):48–59
7. Mikhalev V, Armknecht F, Muller C (2016) On ciphers that con-tinuously access the non-volatile key. IACR Trans Symmetr Cryptol :52–79
8. Armknecht F, Mikhalev V (2015) On lightweight stream ciphers with shorter internal states. In: International workshop on fast software encryption. Springer, Heidelberg, pp 451–470
9. Palnitkar S (1996) Verilog HDL: a guide to digital design and synthesis, Prentice Hall. https://dl.amobbs.com/bbs_upload782111/files_33/ourdev_585395BQ8JA.pdf
10. Xilinx ISE 14.7 Software Manuals and help. http://www.xilinx.com/support/documentation/sw_manuals/xilinx14_1/ise_tutorial_ug695.pdf

Software Implementation of Plantlet Stream Cipher Using Verilog Hardware Description Language

Goundla Sandhya and Dheeraj Kumar Sharma

Abstract Nowadays, use of resources like area, energy, or power is restricted. The fewer resources should be used as much as possible. Thus, there is a lot of scope for lightweight stream ciphers. In this paper, software implementation of plantlet stream cipher has been performed using Verilog hardware description language on the Xilinx ISE design suite. One approach to obtain a lightweight stream cipher is to store the cipher key in non-volatile memory. This cipher key is used not only for the initialization process but it is also used during the encryption/decryption process. Plantlet stream cipher was proposed to improve the design of sprout that has minimum area compared to all other stream ciphers. In this paper, plantlet stream cipher is compared with two other ciphers. It is also a secured stream cipher to overcome the time-memory-data attacks. The main application of plantlet stream cipher is for security and lightweight constrained devices. In the plantlet, a double-layer LFSR is used such that the area is minimized and an all-zero state is avoided.

Keywords Stream cipher · Ultra-lightweight · EEPROM · eSTREAM · Keystream · LFSR · NLFSR

G. Sandhya (✉)
VLSI Design, National Institute of Technology, Kurukshetra, India
e-mail: goundla_31911212@nitkkr.ac.in

D. K. Sharma
Department of ECE, National Institute of Technology, Kurukshetra, India
e-mail: sharma986dheeraj@nitkkr.ac.in

B. Mishra and M. Tiwari (eds.), *VLSI, Microwave and Wireless Technologies*,
Lecture Notes in Electrical Engineering 877,
https://doi.org/10.1007/978-981-19-0312-0_12

1 Introduction

The cryptographic algorithms are of two types [1–3]. First one is symmetric ciphers, and the second one is asymmetric ciphers. In symmetric ciphers, the encryption and decryption can be done using the same secret key, and in asymmetric ciphers, the data is encrypted using a public key and decrypted using a secret key. The cryptographic algorithms can provide communication with security.

As it has been seen by researchers that many lightweight stream ciphers like sprout [1], E0 that they are very insecure for attacks. So, researchers have performed many modifications and introduced plantlet stream cipher. In plantlet stream cipher, a double-layer linear feedback shift register (LFSR) has been used to ensure higher security and also using lesser area. In these types of stream ciphers, the key is used not only for initialization but also for the keystream generation phase to result in high security and to reduce the area consumption.

The plantlet stream cipher consists of one LFSR, one nonlinear feedback shift register (NLFSR), and a counter. The double-layer LFSR is also used to avoid the all-zero state in the register. The plantlet has a goal to use the non-volatile key approach. The main requirements of newly proposed stream ciphers are about the area and to access key from non-volatile memory. In the first criteria, the cipher consumes less area, and then, the remaining area is used for integrating other functionalities. The cipher consumes more power if the area is large.

Keeping in mind these requirements, designers proposed a lightweight stream cipher named sprout [1]. Sprout consists of one LFSR and one NLFSR, each consisting of a length of 40 bits. The size of keystream bits is 80 bits. The length of the IV is 80 bits. Sprout can produce a maximum length of keystream bits of 2^{40} using the same IV [2]. It accesses the key from non-volatile memory. It is a small stream cipher. But sprout cannot withhold time-memory-data tradeoff (TMDTO) attacks.

Due to the drawback of sprout, many proposals are done for implementing new lightweight stream ciphers. Fruit-v1 is firstly proposed in the series of Fruit stream ciphers. Fruit accessed the key from volatile memory at random positions. Fruit-v1 is attacked by the divide-and-conquer attack and correlation attack. After Fruit-v1, later Fruit-80 is proposed. Fruit-80 can withstand the attacks. Some modifications are done to sprout, and the plantlet stream cipher is proposed. The plantlet consists of one LFSR of 61 bits and one NLFSR of 40 bits. The plantlet stream cipher can withstand the TMDTO attacks.

Various techniques are developed to design lightweight stream cipher. One approach is to use the key for initialization and as well as during the encryption/decryption process. The key accessed is stored in non-volatile memory. The key is continuously accessed using the updating function. The area is reduced by storing the key in non-volatile memory.

There are three different categories of non-volatile memory. The first category is to have a memory that is programmed by manufacturers and cannot be changed later. In lightweight constrained devices, MROM is used. The second category is also same as first category but the difference is the time taken for programming. In second category, PROM is used. And the third category uses EEPROM. It is rewritable and provides a lot of flexibility.

In plantlet stream cipher, EEPROM was used to store and access keys. The user can rewrite the key. Though it is expensive, it can be used because it improves the level of security. EEPROM can be used in two different ways. In first way, it is used as an external EEPROM which can communicate via standard interfaces. And in second manner, EEPROM is integrated into the design itself. The first approach is cheaper, but it is affected by side-channel attack. Therefore, the second approach is generally used. In this manner, the security is improved by using non-volatile memory, and the flexibility is also improved.

The key is accessed during both the initialization phase and key generation phase. It is accessed by sequential reading rather than selective reading. Due to the use of sequential reading, high throughput is achieved. Sequential reading can be of two types, i.e., wrap case and no-wrap case. Plantlet uses wrap case, i.e., it points to the first bit of key, once it completes reading the last bit of key. Plantlet requires computation of $2^{46.26}$ encryptions for key recovery attack [2].

In this paper, the software implementation of plantlet stream cipher is performed in Verilog hardware description language [4]. The software tool used is Xilinx ISE design suite, version 14.7 [5]. In Sect. 2, plantlet architecture has been discussed. Applications of plantlet stream cipher have been discussed in Sect. 3. Verilog code has been written in appendix. Results and comparison table have been given in Sect. 4. Finally, conclusion has been discussed in Sect. 5.

2 Plantlet Architecture

Plantlet consists of one NLFSR, one LFSR, a round key function, a counter, and an output function. The length of LFSR is 61 bits. The length of NLFSR is 40 bits. The length of the counter is 9 bits. The first 7 bits of the counter are used for counting 0–79 clock cycles, and on the 80th clock cycle, it resets. The last 2 bits are used to count the number of cycles completed by the initialization phase (Fig. 1).

The round key size is 80 bits. The initialization vector (IV) size is 90 bits. The first 40 bits of IV are initialized to NLFSR, i.e., $s_0 = \mathrm{iv}_0$, $s_1 = \mathrm{iv}_1$, …, $s_{39} = \mathrm{iv}_{39}$. The remaining 50 bits of IV are initialized to first 50 bits of LFSR, i.e., $b_0 = \mathrm{iv}_{40}$, $b_1 = \mathrm{iv}_1$, …, $b_{49} = \mathrm{iv}_{79}$. The last 11 bits of LFSR are initialized with '0's and '1's, i.e., $b_{50} = b_{51} = \cdots = b_{58} = 1$, $b_{59} = 0$, $b_{60} = 1$.

In the first 320 clock cycles, the keystream bits are produced in initialization phase. The output is fed back to LFSR and NLFSR inputs. In the initialization phase, the first 60 bits of LFSR are updated, while b_{60} is not utilized in the initialization phase; hence, it remains at '1.' In this phase, it uses the primitive polynomial $F_i(X)$.

$$F_i(X) = X^{60} + X^{54} + X^{43} + X^{34} + X^{20} + X^{14} + 1 \tag{1}$$

The update function of LFSR during the initialization phase is

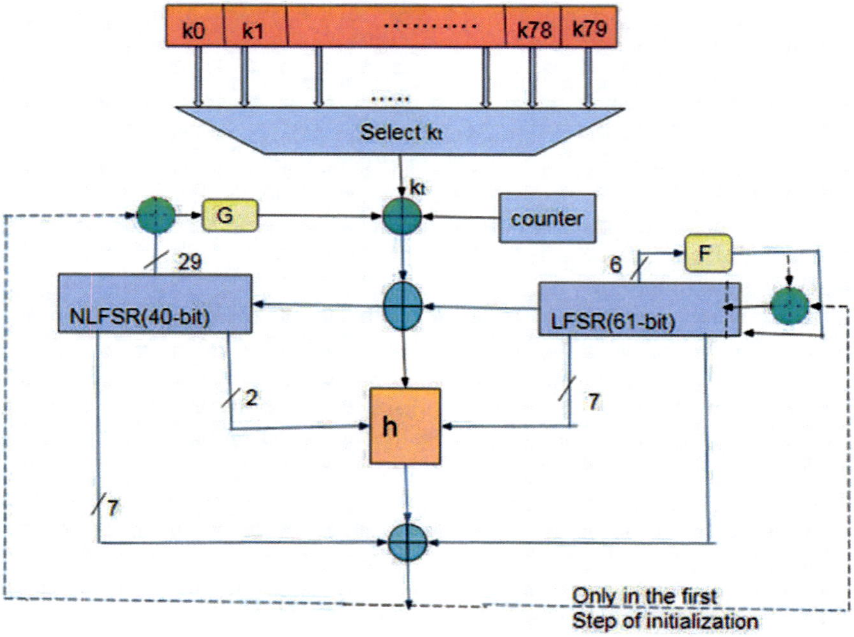

Fig. 1 The block diagram of plantlet

$$b_{60}^{t+1} = 1 \tag{2}$$

$$b_{59}^{t+1} = b_{54}^{t} + b_{43}^{t} + b_{34}^{t} + b_{20}^{t} + b_{14}^{t} + b_{0}^{t} + y^{t} \tag{3}$$

$$b_{i}^{t+1} = b_{i+1}^{t}, \quad \text{for} \;\; 0 \leqq \mathrm{i} \leqq 58 \tag{4}$$

The update function for NLFSR is

$$s_{39}^{t+1} = g(s_t) + k^t + b_0^t + C_4^t \tag{5}$$

$$\begin{aligned} s_{39}^{t+1} = {} & k^t + b_0^t + C_4^t + s_0^t + s_{13}^t + s_{19}^t + s_{35}^t + s_{16}^t s_{18}^t + s_{22}^t s_{24}^t \\ & + s_{26}^t s_{32}^t + s_{33}^t s_{36}^t + s_{37}^t s_{38}^t + s_{10}^t s_{11}^t s_{12}^t + s_{27}^t s_{30}^t s_{31}^t + y^t \end{aligned} \tag{6}$$

$$s_i^{t+1} = s_i + b^t, \;\; 0 \leqq \mathrm{i} \leqq 39 \tag{7}$$

After 320 clock cycles, the key generation phase starts. In this phase, it uses 61-bit LFSR. To reduce area, it reutilizes the registers of the initialization phase. This is known as double-layer LFSR. Due to this, all-zero state is avoided and high period is achieved. In this phase, the 61 bits are updated. It uses the primitive polynomial $F_k(X)$.

$$F_k(X) = X^{61} + X^{54} + X^{43} + X^{34} + X^{20} + X^{14} + 1 \quad (8)$$

The update function of LFSR during the key generation phase is

$$b_{60}^{t+1} = b_{54}^t + b_{43}^t + b_{34}^t + b_{20}^t + b_{14}^t + b_0^t \quad (9)$$

$$b_i^{t+1} = b_{i+1}^t, \quad 0 \leqq i \leqq 59 \quad (10)$$

The update function for NLFSR is

$$s_{39}^{t+1} = g(s_t) + k^t + b_0^t + C_4^t \quad (11)$$

$$\begin{aligned} s_{39}^{t+1} = {} & k^t + b_0^t + C_4^t + s_0^t + s_{13}^t + s_{19}^t + s_{35}^t + s_{39}^t \\ & + s_2^t s_{25}^t + s_3^t s_5^t + s_7^t s_8^t + s_{14}^t s_{21}^t + s_{16}^t s_{18}^t + s_{22}^t s_{24}^t \\ & + s_{26}^t s_{32}^t + s_{33}^t s_{36}^t s_{37}^t s_{38}^t + s_{10}^t s_{11}^t s_{12}^t + s_{27}^t s_{30}^t s_{31}^t \end{aligned} \quad (12)$$

$$s_i^{t+1} = s_i + b^t, \quad 0 \leqq i \leqq 39 \quad (13)$$

The round key function is as follows:

$$k^t = k_{(t \bmod 80)}, \quad t \geqq 0 \quad (14)$$

The output function is as follows:

$$y^t = h(X) + b_{30}^t + s_1^t + s_6^t + s_{15}^t + s_{17}^t + s_{23}^t + s_{28}^t + s_{34}^t \quad (15)$$

where $h(X) = s^t{}_4 b^t{}_6 + b^t{}_8 b^t{}_{10} + b^t{}_{32} b^t{}_{17} + b^t{}_{19} b^t{}_{23} + s^t{}_4 b^t{}_{32} b^t{}_{38}$ is the pre-output function.

3 Applications

There are many applications of stream ciphers that are used for security such as RFID (radio frequency identification) and also in SIM cards and mobile phones. Plantlet is used in applications where the requirements are as the first one is it is used for light weight constrained devices. And the second one is it is used for security purposes.

In lightweight constrained devices, the key is accessed directly from non-volatile memory, and the area is reduced. In this, it consists of a smaller inner state length, i.e., 109 bits, and it also uses a double-layer LFSR. Due to this, the area is minimized and higher throughput is also achieved. Plantlet was designed to overcome the attacks faced by sprout. The attacks are as follows.

3.1 Merge and Sieving Technique

In this attack, the sprout is attacked by the idea of observing some possible states. Using the sieving technique, the observed keystream bits are applied. The states which do not contain the observed bits are discarded. Not only that but some of the key bits are also recovered. This is due to the reason that the key is a nonlinear function in the update function. Plantlet uses a round key function, and this selects one key and uses in update function at every clock cycle. The internal state is increased. Plantlet can withstand these attacks.

3.2 Tradeoff Attacks

The attack against sprout for key recovery is given below. This is based on the sequences in which the key is not used. In plantlet, key is used in the update function at every clock cycle. There is no possibility of these attacks in plantlet.

3.3 All-Zero LFSR State

In keystream generation phase, the LFSR gets into all-zero states. So, it is possible to recover the key by attacking. This is not possible in plantlet because it uses double-layer LFSR. Due to this, all the zero state is avoided. We never get an all-zero state in this. So these attacks cannot affect the plantlet.

4 Results of Simulation

Simulation was performed on Xilinx ISE design suite, version 14.7. Code was written in Verilog HDL language that has been given in appendix. The simulation results for the initialization phase are shown in Fig. 2 and for the keystream generation phase are shown in Fig. 3. The comparison of different stream ciphers like plantlet, sprout [1], and Fruit-80 [6] is shown in Table 1. Area is given in gate equivalent (GE). From

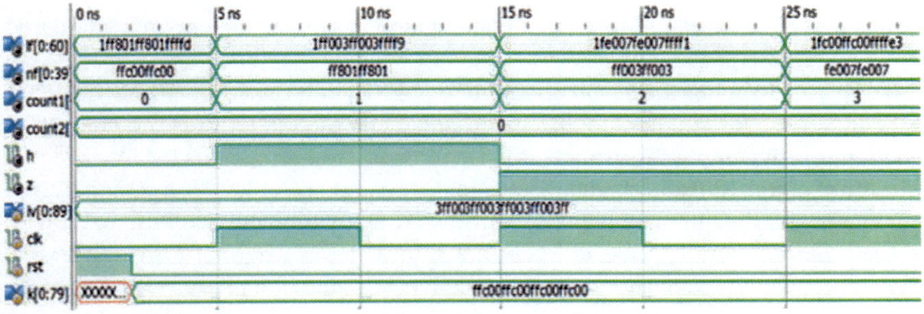

Fig. 2 Simulation output of initialization phase

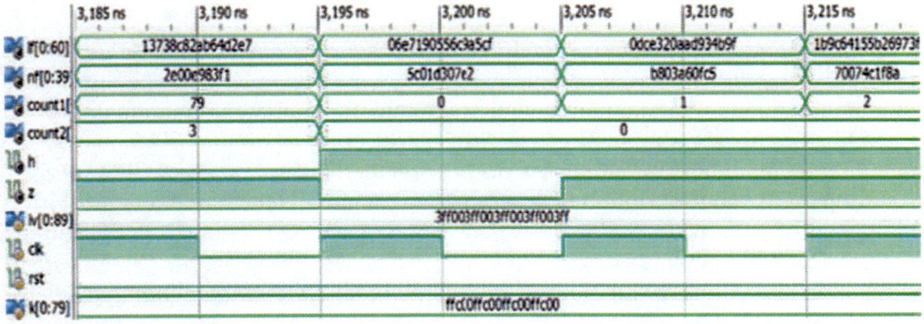

Fig. 3 Simulation output of keystream generation phase

Table 1 Comparative study of different stream ciphers

Parameters	Plantlet	Sprout	Fruit-80
Size of NLFSR (bits)	40	40	37
Size of LFSR (bits)	61	40	43
Length of IV (bits)	90	80	70
Internal state (bits)	109	89	87
Length of key stream (bits)	80	80	80
Number of clocks cycles for initialization	320	320	160
Data limitation (bits)	2^{30}	2^{40}	2^{43}
Area (GE)	807	810	960
Throughput (kb/s)	100	100	100

the above table, one can observe that plantlet stream cipher has less area compared to other stream ciphers.

In Figs. 2 and 3, lf is the 61-bit double-layer LFSR, nf is the 40-bit NLFSR, count1 is the first 7 bits of counter, count2 is the last 2 bits of counter, h is the pre-output function, z is the output function, iv is the initialization vector, clk is the clock signal with time period 10 ns, rst is the reset signal, and k is secret key. Reset signal used is active-high asynchronous signal. The NLFSR and LFSR bits are initialized with IV bits. The bits of NLFSR and LFSR are shown in hexadecimal format. The bits of count1 and count2 are shown in decimal. The initialization phase is for 320 clock cycles, i.e., count1 is counting 0–79 for 4 cycles which is shown by count2 starting from zero. The keystream is generated after 320 clock cycles. The keystream bit 'z' is taken from this clock cycle onwards.

5 Conclusion

In this paper, a study of plantlet stream cipher and its comparison with other stream ciphers has been done. Further, the software implementation of plantlet stream cipher in Verilog hardware description language is performed. Plantlet stream cipher simulation results for initialization and key generation phase are shown. The applications of plantlet stream ciphers are also discussed. So, plantlet stream cipher can withstand the attacks. Due to the double-layer LFSR, all-zero state is avoided and area is also minimized. From the table, it is shown that plantlet stream cipher has minimum area compared to other.

Appendix: Program Code

Verilog Code

```
module lfsr(output reg [0:60] lf,           //61-bit double-layer LFSR
            output reg [0:39] nf,           //40-bit NLFSR
            output reg [0:6]count1=6'b0,    //first 7-bits of counter
            output reg [7:8]count2=2'b00,   //last 2-bits of counter
            output reg h=0,                 //pre-output function
            output reg z=0,                 //output function
            input [0:89]iv,                 //IV
            input clk,                      //clock signal
            input rst,                      //reset signal
            input [0:79]k                   //secret key
    );

wire [0:8] count; //7-bit counter
reg [0:79] kt;    //round key function


//LFSR feedback function
assign f1=lf[54]^lf[43]^lf[34]^lf[20]^lf[14]^lf[0]^z;//initialization phase
assign f2=lf[54]^lf[43]^lf[34]^lf[20]^lf[14]^lf[0];//keystream generation phase

//NLFSR feedback function
assign
nt=nf[0]^nf[13]^nf[19]^nf[35]^nf[39]^(nf[2]*nf[25])^(nf[3]*nf[5])^(nf[7]*nf[8])^(nf[14
]*nf[21])^(nf[16]*nf[18])^(nf[22]*nf[24])^(nf[26]*nf[32])^(nf[33]*nf[36]*nf[37]*nf[38]
)^(nf[10]*nf[11]*nf[12])^(nf[27]*nf[30]*nf[31])^z;//initialization phase
assign
nt1=nf[0]^nf[13]^nf[19]^nf[35]^nf[39]^(nf[2]*nf[25])^(nf[3]*nf[5])^(nf[7]*nf[8])^(nf[1
4]*nf[21])^(nf[16]*nf[18])^(nf[22]*nf[24])^(nf[26]*nf[32])^(nf[33]*nf[36]*nf[37]*nf[38
])^(nf[10]*nf[11]*nf[12])^(nf[27]*nf[30]*nf[31]);//keystream generation phase

assign count={count1, count2}; //concatenating counter bits

//LFSR block
always@(posedge clk or posedge rst)
//initialization of LFSR
if(rst)
     begin
     lf[0:49]<=iv[40:89];
     lf[50:58]<={9{1'b1}};
     lf[59]<=1'b0;
     lf[60]<=1'b1;
     end
//initialization phase
else if(count<=463)
     lf<={lf[1:59],f1,1'b1};
//keystream generation phase
else
     lf<={lf[1:60],f2};

//NLFSR block
always @(posedge clk or posedge rst)//asynchronous reset signal
//initializing NLFSR bits with IV bits
if(rst)
     nf[0:39]<=iv[0:39];
//initialization phase
else if(count<=463)
     begin
     nf<={nf[1:39],nf[39]};
     nf[39]<=nt^k[kt]^lf[0]^count[4];
     end
//keystream generation phase
else
     begin
     nf<={nf[1:39],nf[39]};
     nf[39]<=nt1^k[kt]^lf[0]^count[4];
     end
```

```
//counter block
always @(posedge clk)
if(count1==79)
      begin
      count1<=0;
      count2<=count2+1;
      end
else
      count1<=count1+1;

//round key function block
always@(posedge clk or posedge rst)
if(rst||count1==79)
      kt<=0;
else
      kt<=kt+1;


//pre-output function
always@(posedge clk or posedge rst)
if(rst)
      h<=0;
else
h<=(nf[4]*lf[6])^(lf[8]*lf[10])^(lf[32]*lf[17])^(lf[19]*lf[23])^(nf[4]*lf[32]*nf[38]);

//output function
always@(posedge clk or posedge rst)// entire output function
if(rst)
      z<=0;
else
      z<=h^lf[30]^nf[1]^nf[6]^nf[15]^nf[17]^nf[23]^nf[28]^nf[34];

endmodule
```

Testbench Code

```
Module testbench;
        // Inputs
        reg [0:89] iv;
        reg clk;
        reg rst;
        reg [0:79] k;

        // Outputs
        wire [0:60] lf;
        wire [0:39] nf;
        wire [0:6] count1;
        wire [7:8] count2;
        wire h;
        wire z;

        // Instantiate the Unit Under Test (UUT)
        lfsr uut (
                .lf(lf),
                .nf(nf),
                .count1(count1),
                .count2(count2),
                .h(h),
                .z(z),
                .iv(iv),
                .clk(clk),
                .rst(rst),
                .k(k)
        );
initial
begin
// Initialize Inputs
clk=0;
iv=
90'b1111111111000000000011111111110000000000111111111100000000001111111111000000000011
11111111;
k =
80'b11111111110000000000111111111100000000001111111111000000000011111111110000000000;
rst = 1;
#2 rst=0;
end
initial
    begin
        forever #5clk=~clk; //clock signal
        // Wait 100 ns for global reset to finish
        #100
    end

endmodule
```

References

1. Banik S (2015) Some results on sprout. In: Biryukov A, Goyal V (eds) Progress in cryptology—INDOCRYPT 2015—16th international conference on cryptology in India, Bangalore, India, 6–9 Dec 2015, Proceedings, vol 9462 of Lecture Notes in Computer Science. Springer, pp 124–139
2. Mikhalev V, Armknecht F, Muller C (2017) On ciphers that continuously access non-volatile key. ToSc 2016(2):52–79

3. Li B, Liu M, Lin D (2020) FPGA implementations of Grain v1, Mickey 2.0, Trivium, Lizard and Plantlet. Microprocess Microsyst 78:103210
4. Palnitkar S (1996) Verilog HDL: a guide to digital design and synthesis. Prentice Hall
5. Xilinx ISE 14.7 Software manuals and help. http://www.xlinx.com/support/documentation/sw_manuals/xilinx14_1/ise_tutorial_ug695.pdf
6. Amin Ghafari V, Hu H (2018) Fruit-80: a secure ultra-lightweight stream cipher for constrained environments. Entropy 20(3):180

An Implementation of Tunable All-Pass Filter Employing CCDDCC

Rajkumar Mumusana, Yumnam Shantikumar Singh, and Ashish Ranjan

Abstract This research paper contains voltage mode first-order filter structure using single current controlled differential difference current conveyor (CCDDCC) with an external capacitor. The proposed design offers a tunable property. The all-pass filter is free from external resistor and suitable for low sensitivity. The workability test of the proposed APF is verified by using PSPICE simulation in which CCDDCC is implemented with 0.25 μm TSMC CMOS process parameter.

Keywords Current controlled differential difference current conveyor (CCDDCC) · First-order active filter · All-pass filter (APF)

1 Introduction

All-pass filter plays a vital role in signal processing circuit and its major capability is widely accepted for delay equalizers or phase correctors. A first-order all-pass filter (APF) is one of the simplest order filter function among all higher order APFs. This traditional tunable first-order APF is suitable for shifting the input signal phase from 0° to 180° or vice versa with constant amplitude [1]. In recent times, analogue circuit design has fulfilled the flip sides of conventional op-amp-based circuitry design that includes low power dissipation property, wide dynamic range, bandwidth enhancement and many more [2]. A wide range of APF circuit design is enriched in the literatures as

(i) Inverting-type second generation current conveyor (ICCII) [3]
(ii) Second generation current conveyor (CCII) [4]
(iii) Current controlled differential difference current conveyor (CCDDCC) [5]
(iv) Current controlled differential difference current conveyor transconductance amplifier (CCDDCCTA) [6].

R. Mumusana · Y. S. Singh · A. Ranjan (✉)
Department of Electronics and Communication Engineering, National Institute of Technology Manipur, Imphal, India
e-mail: ashish.ism@rediffmail.com

B. Mishra and M. Tiwari (eds.), *VLSI, Microwave and Wireless Technologies*, Lecture Notes in Electrical Engineering 877,
https://doi.org/10.1007/978-981-19-0312-0_13

Among the above mentioned active blocks-based APFs, a new block named as current controlled differential difference current conveyor (CCDDCC) has a special feature with a flavour of second generation current conveyor (CCII) and differential difference current conveyor (DDCC). In this research article, a simple first-order APF using single CCDDCC active block with a used of external capacitor is designed. The heart of the APF is CCDDCC which is integrated with 0.25 μm TSMC CMOS parameter in this design. Lastly, a short comparison of proposed work with the reported one is given in Table 1.

2 Circuit Descriptions

The core of the APF uses CCDDCC as a proposed active block. The equivalent circuit symbol and MOS-based internal structure of CCDDCC are shown in Fig. 1a, b, respectively. The aspect ratio of the transistor used in the CCDDCC active element is given in Table 2. The port behaviour is characterized in terms of the following matrix form as

$$\begin{bmatrix} V_X \\ I_{Y1} \\ I_{Y2} \\ I_{Y2} \\ I_Z \end{bmatrix} = \begin{bmatrix} R_X & 1 & -1 & 1 & 0 \\ 0 & 0 & 0 & 0 & 0 \\ 0 & 0 & 0 & 0 & 0 \\ 0 & 0 & 0 & 0 & 0 \\ \pm 1 & 0 & 0 & 0 & 0 \end{bmatrix} \begin{bmatrix} I_X \\ V_{Y1} \\ V_{Y2} \\ V_{Y2} \\ V_Z \end{bmatrix} \tag{1}$$

The most significant parameter is R_X that designates the intrinsic series input resistance of the conveyor at X port which is electronically tunable by bias current I_B, can be expressed as a function of transconductance (g_m) as

$$R_x = \frac{1}{g_{m19} + g_{m20}} \tag{2}$$

Finally, the theoretical investigation is tested with PSPICE simulation to justify the viability of the APF design. The key design for the implementation of first-order APF using CCDDCC and an external capacitor is shown in Fig. 2.

By utilizing the port characteristics of CCDDCC and per family of Kirchoff's law, Fig. 2 results the following voltage transfer function for APF as:

$$\frac{V_{out}}{V_{in}} = \frac{sCR_X - 1}{sCR_X + 1} \tag{3}$$

The phase angle of the inverting all-pass filter can be obtained as:

$$\psi_{inverting} = -2\tan^{-1}(\omega CR_x) \tag{4}$$

Table 1 Comparison of proposed work with the reported one

References	No. of ABBs # type of ABB	No. of passive elements	Free from matching	Operating frequency	CMOS count	Electronically tunable	Power consumption (mW)
[3]	1#ICCII	2/3	Yes	370 kHz	16	No	NA
[4]	1#CCII	4	Yes	500 kHz	18	No	1.6
[5]	1#CCDDCC	3	No	4.54 MHz	29	Yes	NA
[6]	1#CCDDCCTA	1	Yes	1.28 MHz	31	Yes	NA
[7]	1#DVCC	2	Yes	1.59 MHz	32	No	NA
[8]	2#DVCC	2	Yes	397.89 kHz	32	No	NA
[9]	2#CCII±	5	No	200 kHz	AD844	No	NA
[10]	3#OTRA	9	Yes	100 kHz	14	Yes	NA
Prop.	**1#CCDDCC**	**1**	**Yes**	**800 kHz**	**29**	**Yes**	**1.5**

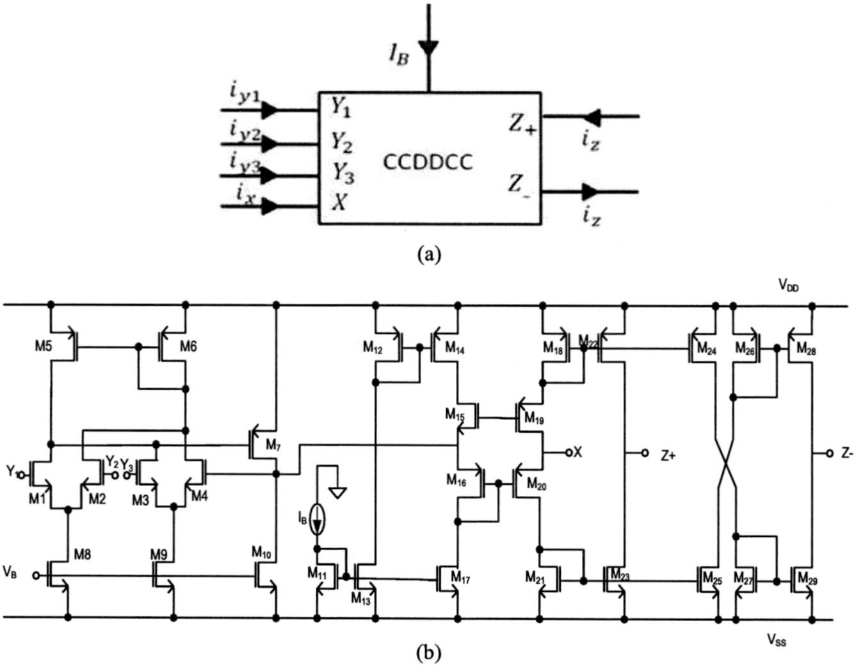

Fig. 1 CCDDCC: **a** circuit symbol **b** internal MOS structure

Table 2 *W*/*L* ratio of the CCDDCC active element

Transistor	*W*/*L* ratio (μm)
M1–M4	1/0.25
M15, M19	5/0.25
M16, M20	8/0.25
All NMOS	3/0.25
All PMOS	5/0.25

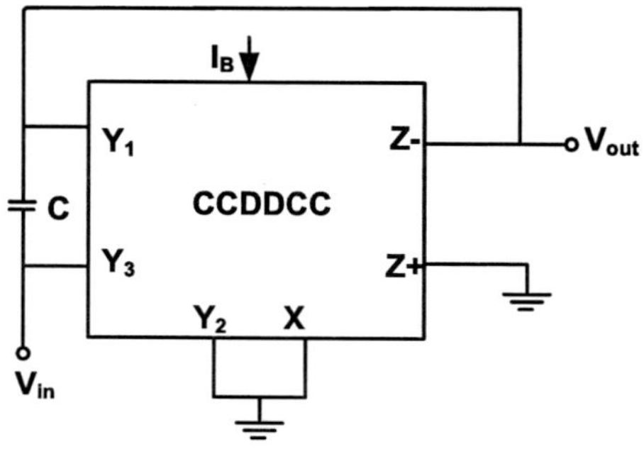

Fig. 2 Proposed first-order APF

In addition, an active and passive sensitivity of cut-off frequency are observed as:

$$S_{C}^{\omega_0} = S_{R_X}^{\omega_0} = -1 \tag{5}$$

Here, the low value of sensitivity is investigated which is suitable for filter design.

Moreover, the ideal transfer function of the ALP is deviated due to the non-ideal and parasitic elements present in the CCDDCC active element. Therefore, the modified transfer function of the proposed ALP with the voltage gain (β) and current gain (α) is characterized as:

$$\frac{V_o(s)}{V_i(s)} = \frac{sCR_X - \alpha(s)\beta(s)}{sCR_X + \alpha(s)\beta(s)} \tag{6}$$

3 Simulation Results

To verify the theoretical observation of the proposed first-order APF, a complete schematic representation of the CCDDCC-based APF circuit is simulated using PSPICE. It is integrated with 0.25 μm TSMC CMOS process parameter, with $V_{DD} = -V_{SS} = 1.25$ V and $V_B = -0.55$ V. The APF is designed for a frequency of 800 kHz by using capacitor $C = 1$ pF and biased current $I_B = 450$ μA. Figure 3 shows the transient response with input and output voltage waveforms. Moreover, the Lissajous pattern is also observed in Fig. 4. Finally, the phase response and its Fourier frequency curve are observed in Figs. 5 and 6, respectively. Figure 7 shows the different frequency responses of APF, and it supports the electronically tunable

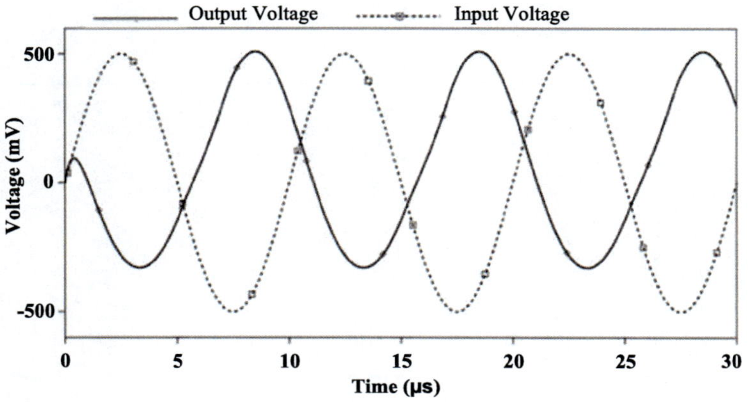

Fig. 3 Transient response of proposed APF

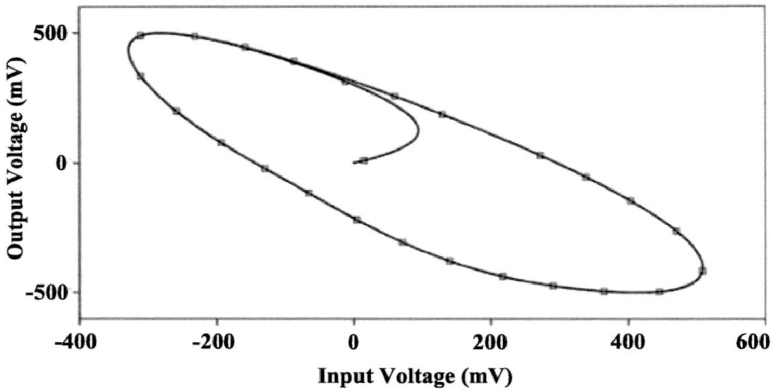

Fig. 4 APF Lissajous pattern (V_{out} vs. V_{in})

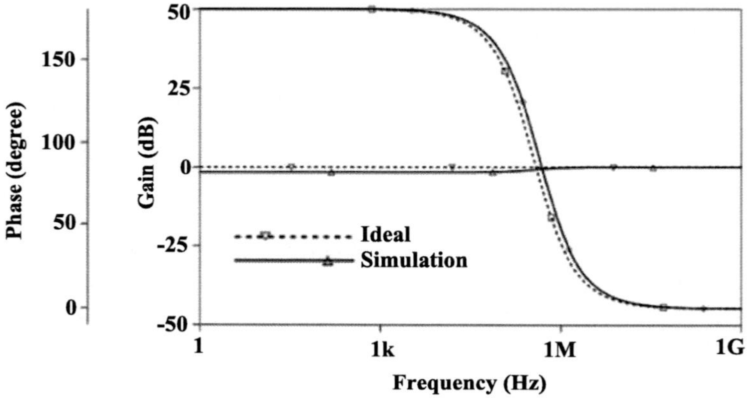

Fig. 5 Proposed APF frequency response

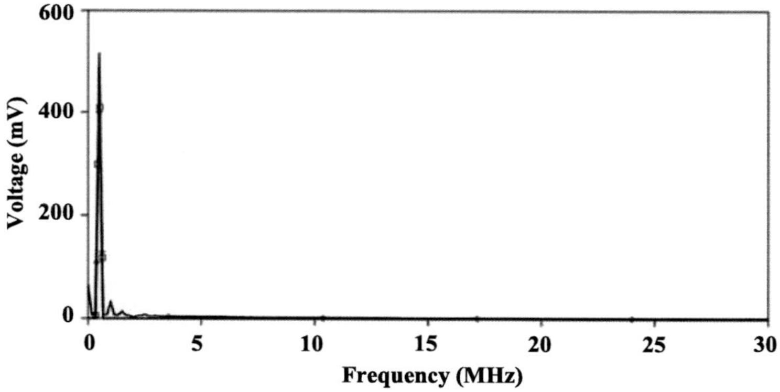

Fig. 6 Fourier response of APF

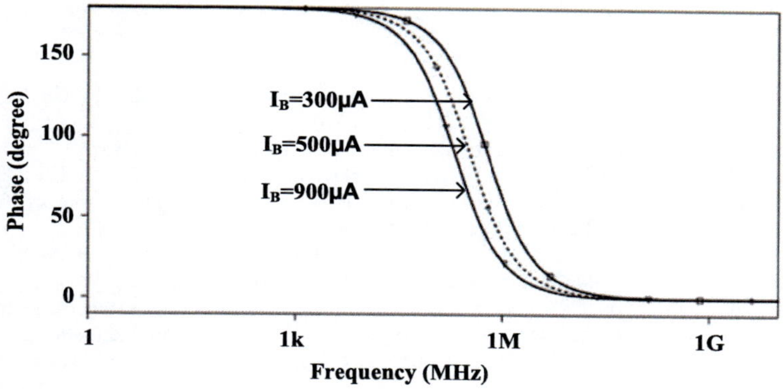

Fig. 7 Tunable phase response of proposed APF circuit

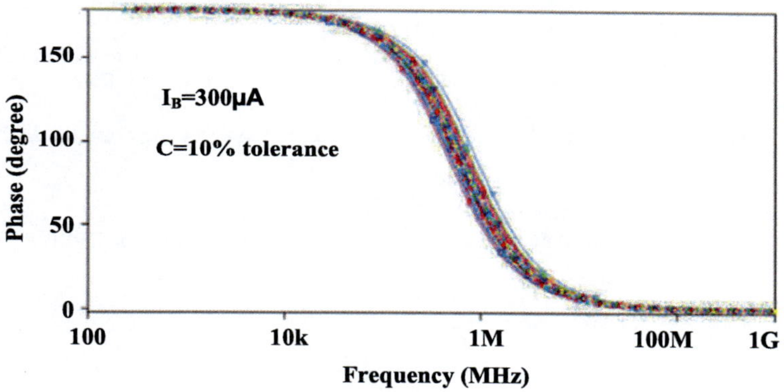

Fig. 8 Monte Carlo analysis phase response of APF circuit

of proposed APF circuit. Lastly, Monte Carlo analysis of phase response with 10% tolerance in capacitance value is shown in Fig. 8.

4 Conclusion

A simple first-order APF topology using CCDDCC is designed with a single capacitor. The proposed design also fulfils the requirement of inbuilt tunability of cut-off frequency by using intrinsic series resistance (R_X) of CCDDCC. Simulation results follow the theoretical investigation and are suitable for the design of lower-order filter with low sensitivity values.

References

1. Lear E, John P (1966) Active all-pass filter using a differential operational amplifier. Electron Lett 2(4):134–135. https://doi.org/10.1049/el:19660107
2. Fidler JK (1991) Analogue IC design: the current-mode approach, vol 37, no 1
3. Ibrahim MA, Kuntman H, Ozcan S, Suvak O, Cicekoglu O (2004) New first-order inverting-type second-generation current conveyor-based all-pass sections including canonical forms. Electr Eng 86(5):299–301. https://doi.org/10.1007/s00202-003-0205-3
4. Sharma S, Rajput SS, Mangotra LK, Jamuar SS (2004) Applications of new FGMOS based CCII in low voltage analog filters. Indian J Eng Mater Sci 11(5):397–400
5. Ranjan A, Paul SK (2012) A generalized current mode first order filter using single current controlled differential difference current conveyor. In: CODEC 2012—5th International conference on computers and devices for communication, vol 3, no 2, pp 7–10. https://doi.org/10.1109/CODEC.2012.6509211
6. Kushwaha AK (2017) First order multifunction filters based on CCDDCCTA. In: Proceedings of the 2nd international conference on 2017 devices for integrated circuit, DevIC 2017, vol 25, no 4, pp 71–75. https://doi.org/10.1109/DEVIC.2017.8073909
7. Chen HP (2012) DVCC-based first-order filter with grounded capacitor. Int J Inf Electron Eng 2(1):50–54. https://doi.org/10.7763/ijiee.2012.v2.54.
8. Horng JW (2010) DVCCs based high input impedance voltage-mode first-order allpass, highpass and lowpass filters employing grounded capacitor and resistor. Radioengineering 19(4):653–656
9. Shah NA, Iqbal SZ, Parveen B (2004) Simple first-order multifunction filter. Indian J Pure Appl Phys 42(11):854–856
10. Banerjee K, Ranjan A, Paul SK (2012) New first order multifunction filter employing operational transresistance amplifier. In: CODEC 2012—5th International conference on computers and devices for communication, vol 3, pp 3–5. https://doi.org/10.1109/CODEC.2012.6509218

A Cost-Efficient QCA RAM Cell for Nanotechnology Applications

Divya Tripathi and Subodh Wairya

Abstract Quantum-dot cellular automata (QCA) is an inventive nano-level computation which suggests less dimension and less power consumption, with more speed and premeditated as an amplification to the scaling obstacle with CMOS methodology. One of the newest and rising nanotechnologies used today is QCA based on the repulsion of Coulomb. The enhancement of RAM is expanding with each passing day. This can be primarily due to the advancement of handheld, portable devices such as smart phones and laptops. Cache memory is utilized within the microprocessor to get data quicker. In this paper, cell optimization and realization of the QCA XOR gate are suggested, and further this QCA XOR gate is used to design 2:1 QCA MUX and these QCA structures will be utilized to design interesting and ideal designs of QCA RAM. The proposed QCA RAM contains less number of cells and lesser cost as related to its best previous QCA layouts. The comprehensive analysis of the proposed design suggests that the proposed RAM cell architecture is 61.53% cost efficient and 68.18% area decrement as compared to its best previous existing layout. The suggested circuits are executed and tested with the QCA designer simulation environment. The simulation outcomes illustrate that the suggested architecture outperforms in comparison with best counterpart designs in terms of quantum cell count, area and latency.

Keywords QCA · MUX · Memory · RAM

1 Introduction

Quantum-dot cellular automata (QCA) suggests higher device density, less power consumption and rapid switching speed structures [1, 2]. Traditional CMOS has

D. Tripathi (✉) · S. Wairya
Electronics & Communication Engineering, Institute of Engineering & Technology, Lucknow,, India
e-mail: divyatripathi.cest@gmail.com

Dr. APJ. Abdul Kalam Technical University, Lucknow, India

B. Mishra and M. Tiwari (eds.), *VLSI, Microwave and Wireless Technologies*, Lecture Notes in Electrical Engineering 877,
https://doi.org/10.1007/978-981-19-0312-0_14

conquered our fabrication in recent years and prove to be an improved alternative than previous automation methods. Quantum computing, molecular computing and power efficient nanocomputing are few evolutionary nanotechnologies implemented based on Coulomb repulsion. Presently, circuit designers are leading toward an advanced applied science dependent on electron polarization is named as quantum-dot cellular automata, which gives productive results about decrement and quantum cost to the excessive low horizon [3]. The QCA design tool has been tailored to simulate and verify physical results. QCA-based layouts are best for the boundaries of density (1012 device/cm^2), frequency or speed (range of terahertz) and energy dissipation(100 W/cm^2) [4].

Numerous RAM cells are utilized to design a cluster of RAM cells to store the information. Memory covers an expansive region of the framework on the chip. Scaling of transistor measurements is the requirement of modern innovation. It makes a difference in diminishing the region of the cell which comes about in diminishment of measure of SRAM and progresses integration thickness, but it increments leakage current [5]. Each cell has an issue of leakage. SRAM array, which comprises numerous cells, gets to be an enormous source of leakage current. The issue of data stability in the SRAM cell emerges due to voltage scaling and device dimension in a novel methodology [6]. The RSNM is SNM at the read mode is weakened due to bungle in transistor and a reduction in voltage supply. This issue can be settled by utilizing a separate read circuit [7]. This read circuit increment and the amount of transistors within the cell which effects in an increment in the space of the SRAM cell. Within the examined operation, existing data in the cell are perused out utilizing sense amplifiers. To move forward SNM and decrease in area of SRAM, different SRAM cells are outlined [8]. The SNM of the SRAM cell can too improve by using proper sizing the ratio of the transistor. Proper sizing of transistor improves data stability in both read and write operation. Read–write circuit is required in memory design which can change the data on memory array at the time of write mode and retain data in read mode. SRAM is static due to its retaining information property as long as the power supply is applied without needing periodic refreshment [9]. These are the methods through which data stability can improve during read and write mode of operation. Data stability of the cell in read mode and hold mode is a very main restriction in advanced technology. So, this paper is based on cell optimization, and the realization of QCA XOR and multiplexer which is further used in designing of the QCA RAM cell.

1.1 Paper Contribution

The major implementation of this article is as follows:

(a) Designing of a cost-efficient XOR (N8) gate using QCA.
(b) Designing of an efficient 2:1 MUX (N16) using a proposed cost-efficient XOR gate in QCA.

(c) Designing of QCA RAM cell using a proposed cost-efficient XOR (N8) and 2:1 MUX (N16) gate.
(d) The proposed architectures are tabulated based on quantum cell count, area, latency and the quantum cost that affirms the suggested designs have a lesser arena and way better quantum cost as related to its past best counterpart.

1.2 Paper Organization

The layout of the article is organized in the following manner. Section 2 deals with a brief overview of QCA. Section 3 shows the proposed architectures, the QCA XOR gate, QCA, 2:1 MUX and QCA RAM cell. Section 4 explained simulation results and discussion of the proposed architecture and comparison to its previous best counterparts. Finally, Sect. 5 draws the conclusion of the research work.

2 An Overview of Quantum-Dot Cellular Automata (QCA)

QCA is an arrangement that is put together as a course of action of quantum cells inside that each and every cell has transmitted data electrostatically by its neighboring cells. QCA technology has built-in capabilities for digital circuitry without taking any Boolean function [10, 11]. Every cell is involved by a pair of electrons. The Coulombic interchange among electrons could make two unmistakable cells phases with distinctive charge courses of action. Coulumbia's repulsion of the electrons to involve the furthermost dots in a QCA cell which resembles the nethermost energy state owned of the circuitry [5]. This field for an electron is implied by a dot in the cube cell. Quantum mechanical tunneling obstructions are utilized to couple the stages so that electrons could move with them and determine the structural phase. Columbia's repugnance compels the electrons to include the furthermost dots in a QCA cubicles which takes after the near most energy state governed of the circuitry (Fig. 1).

2.1 Majority Gate

The positioning of the majority gates is advertised in Fig. 2. The outcome of a QCA cell wanders, managing to the calculations of a QCA cell within the medium of the gate. The majority gate plays a critical part to build the OR/AND gates [4, 10].

$$M(P, R, Q) = F = (P * R) + (P * Q) + (R * Q) \quad (1)$$

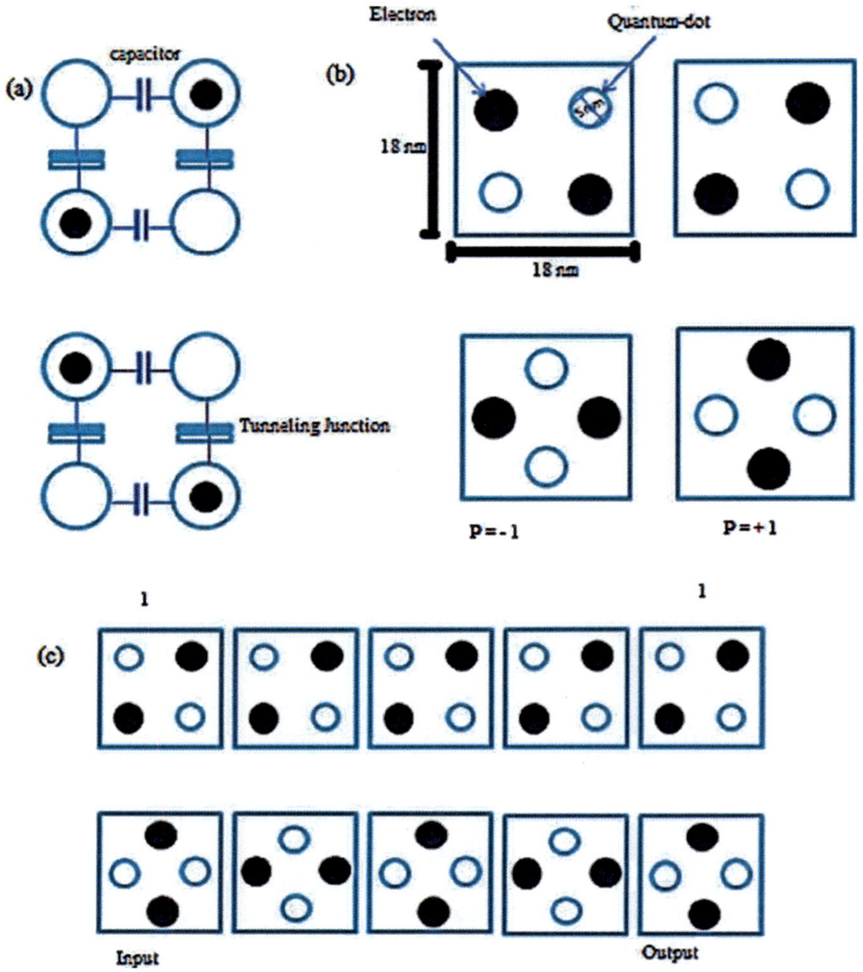

Fig. 1 QCA cells patterns **a** functional diagram **b** cells polarization **c** QCA wire

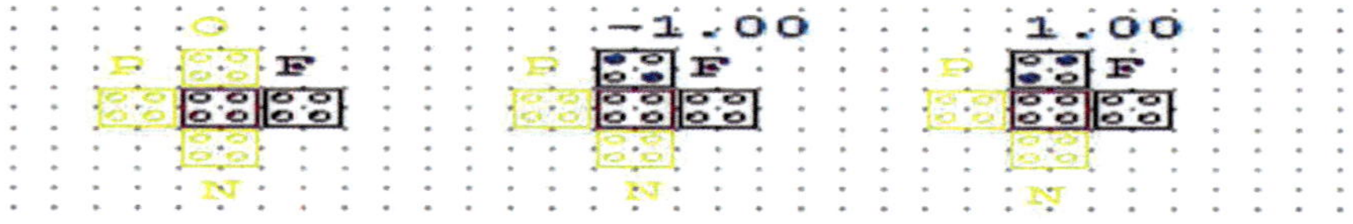

(a) Majority Gate (b) QCA AND Gate (c) QCA OR Gate

Fig. 2 Basic QCA logic gates architecture

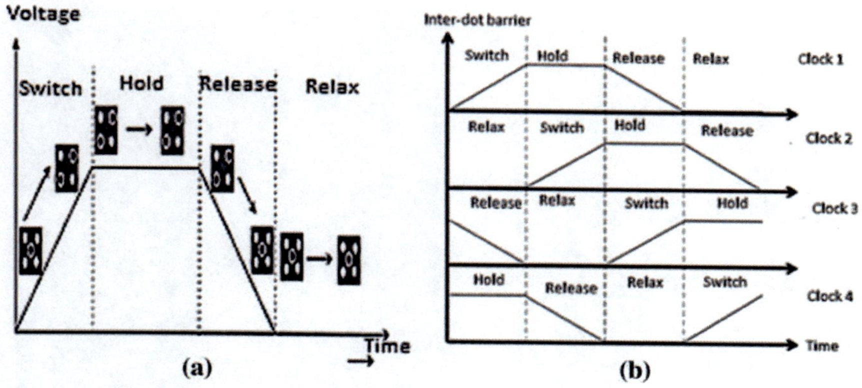

Fig. 3 Clocking concept in QCA **a** clock zone of a QCA cell **b** clocking in four phase

where P, R and Q are three inputs of the majority gate, and F is the one outcome of the majority gate. Fundamental digital logic gates are made of the majority gates. If the polarization of the majority gate is -1, then it becomes AND gate and if the polarization is $+1$, then it becomes OR gate QCA architecture.

2.2 Clocking in QCA

QCA clocking gives the capability to create and invalidate the metastable stage [5–7, 11]. The clocking in QCA innovation is not comparable as they are standard CMOS circuitry, and the QCA clocking courses of action contain fourfold stages: switch (unpolarized cells decided by few inputs and ended up polarized subordinate on their neighbors' polarization), control (cells consist within the same binary state so that which could be apply as per the input to other cells), release (obstruction is taken down and cells stay non-polarized) and relax (cell remains non-polarized) [5, 6] (Fig. 3).

3 Proposed QCA-Based Logic Circuit Architectures

3.1 Proposed QCA XOR Gate

An exclusive-OR (XOR) gate is a base of every contemporary Ics and ALU. The exclusive-OR (XOR) gate is vital to suggest designing of any complex circuitry. In literature, various QCA designer creates XOR gates that are explained [5–9, 12] (Fig. 4).

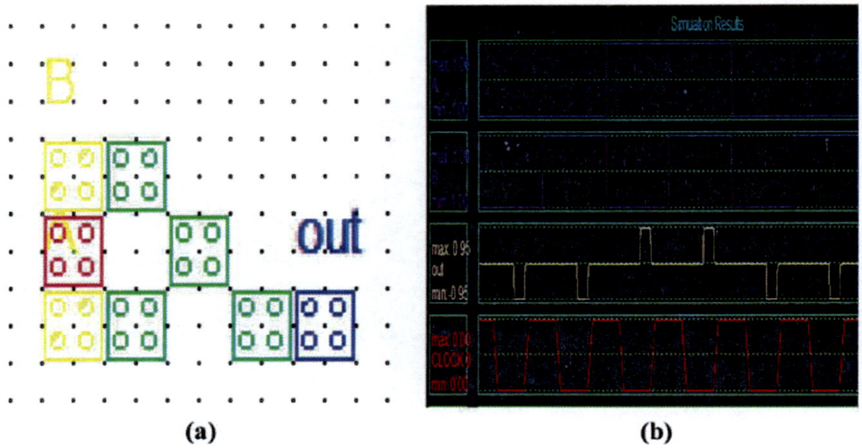

Fig. 4 Proposed QCA XOR gate (N8) **a** layout **b** simulation waveform

3.2 Proposed QCA MUX (N16)

The suggested 2:1 MUX layout has only 16 cells and consists of 0.03 μm^2 area that is smaller as related to past existing designs [6]. The cost-efficient 2:1 multiplexer block diagram is shown in Fig. 5.

Previously, various designs of 2:1 QCA multiplexer were introduced [6, 13–18]. The proposed design consists of 16 numbers of quantum cells and area is 0.21 μm^2 and latency is 0.5 clocking cycles as presented in Fig. 6.

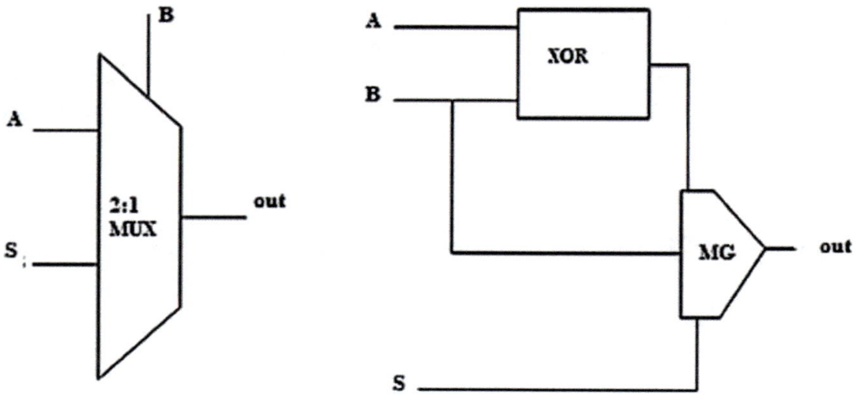

Fig. 5 Block diagram of proposed MUX(N16)

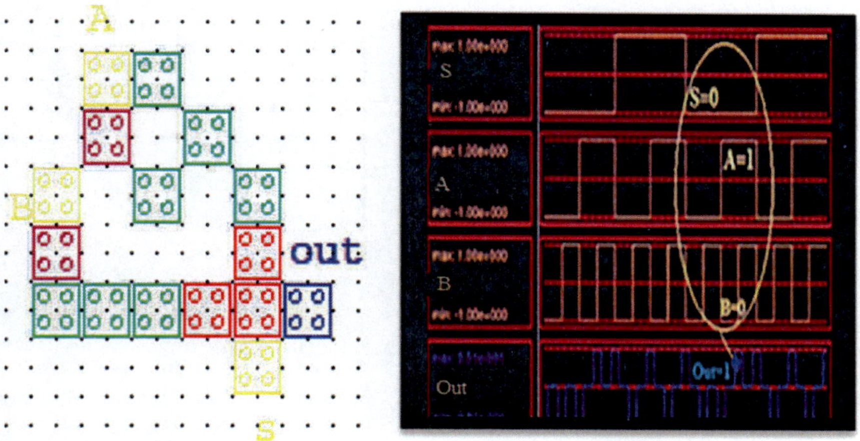

(a) (b)

Fig. 6 Proposed QCA 2:1 MUX(N16) **a** layout **b** simulation waveform

3.3 Proposed QCA RAM Cell

The abbreviation of electronics circuitry is the key for semiconductor industries. Along with the advancements in Si-based electronics, many alternative technologies are being developed at the same time. SRAM also fulfills the need of battery operated devices like wireless sensors and biomedical devices, where the lifetime of battery and power consumption is a very important criterion. Many smart devices are multi-functional and require less power dissipation [4, 5, 10, 11, 19]. In VLSI, thousands of devices are fabricated on a single chip which has a smaller surface area [6]. Low power in portable devices becomes a primary factor [7]. Numerous SRAM cells are utilized to plan a cluster of SRAM cells to store the huge amount of data. Memory covers a large area of the system on chip. Scaling of the transistor dimension is the need for the new technology. It helps in reducing the area of the cell which results in reduction of size of SRAM and improves integration density.

The proposed QCA RAM architecture is shown in Fig. 12, which contains 57 numbers of quantum cells in an area of 0.07 μm^2. Flag sel is utilized as a particular line for the primary multiplexer, and the flag R/W (read/write) is utilized as a selective line for second multiplexer as shown in the block diagram of RAM in Fig. 7.

The total operation of proposed QCA slam design is clarified in Table 3. As per the table, when R/W is '0', the output of the circuit will be uncharged. The output of the circuit distinguishes the impact of input and set/reset signals by setting R/W = '1'. Additionally, making W is '1' and set is '1' gives an input flag to be exchanged to the output, and for the select line sel is '0' the output gets charged as per set/reset flag (Figs. 8 and 9).

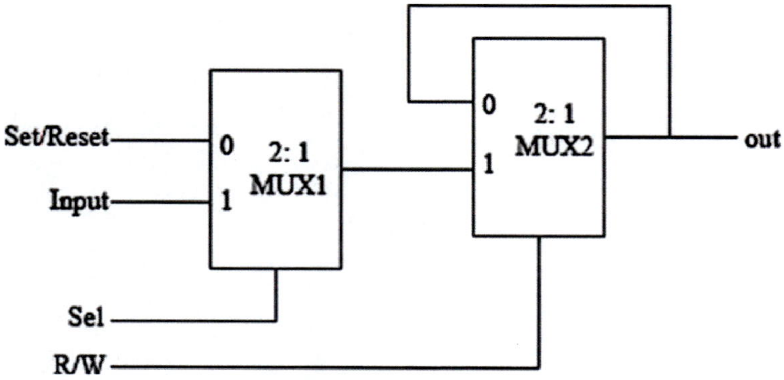

Fig. 7 Schematic design of the memory cell

Fig. 8 QCA implementation of proposed memory cell

4 Simulation Result and Discussion

QCA designer tool is taken for simulation of the QCA layout of proposed designs. To assess the effectiveness and adaptability of proposed QCA layouts is to an extent and comparable with its best previous existing counterpart in parameters of quantum cell count, area, latency and quantum cost is presented in tabular form in Tables 1, 2 and 3. QCA could be a new methodology for optimization of the area by reducing the quantum cell for a nanoscale circuitry design that is suitable to the design of

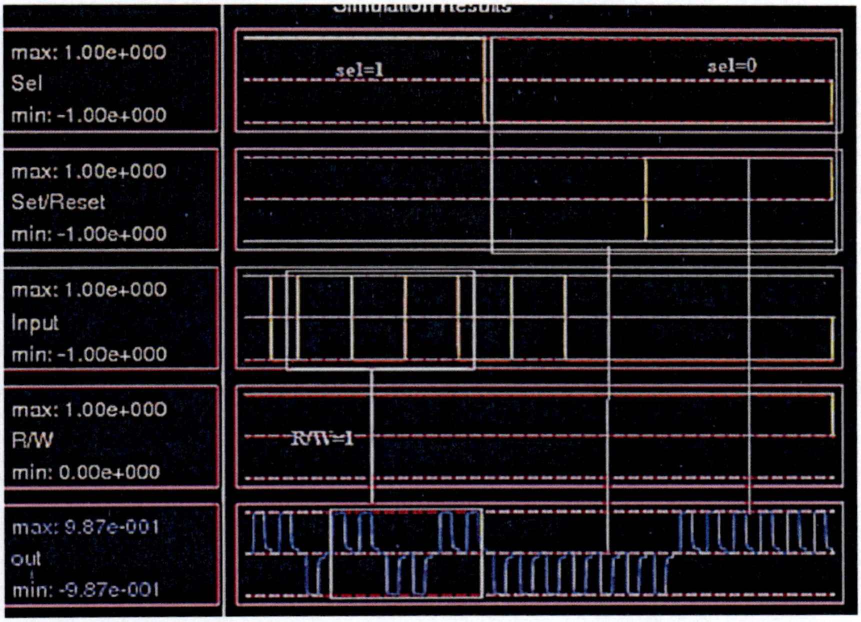

Fig. 9 Simulation waveform of proposed QCA RAM cell

Table 1 Performance comparison of proposed QCA XOR gate

QCA XOR gate	No. of cells	Area (μm^2)	Latency (clocking cycles)	Quantum cost (area * latency)
[5]	60	0.011	1.5	0.016
[6]	51	0.092	2.0	0.184
[7]	29	0.041	0.25	0.010
[8]	28	0.035	0.75	0.026
[9]	12	0.021	0.05	0.001
[12]	14	0.034	0.50	0.017
Proposed	8	0.012	0.25	0.003

exceedingly versatile digital logic cost which has been measured and matched and shows that our proposed designs gained less QCA cells and area also. The proposed QCA digital logic gate is executed and simulated on the QCA designer simulation environment with all the default parameters [20]. The functionality and performance of the proposed designs are better as related with the best existing previous design.

Table 2 Performance comparison of proposed QCA 2:1 multiplexer

QCA MUX designs	Number of cells	Area (μm^2)	Latency clocking cycle	Quantum cost (area * latency)
[21]	27	0.028	3.00	0.084
[22]	26	0.019	2.00	0.038
[23]	23	0.022	3.00	0.066
[24]	23	0.022	2.00	0.044
[25]	43	0.044	4.00	0.176
[5]	36	0.041	4.00	0.164
[6]	22	0.037	1.25	0.046
[13]	19	0.031	0.75	0.023
Proposed	16	0.021	0.50	0.010

Table 3 Performance comparison of proposed QCA memory cell

QCA memory cell designs	Number of cells	Area (μm^2)	Latency clocking cycle	Quantum cost (area * latency)
[15]	233	0.31	1.73	0.53
[16]	173	0.25	1.50	0.37
[17]	158	0.16	2.00	0.32
[18]	100	0.11	3.00	0.33
[26]	63	0.07	2.00	0.14
[27]	109	0.13	1.75	0.05
[28]	75	0.06	1.50	0.09
[29]	71	0.09	1.25	0.11
[20]	177	0.22	0.60	0.13
Proposed	57	0.07	0.75	0.05

5 Conclusion

Nanotechnology has been the center of current analysis about the growth of nano-electronics circuitry. As digital logic gates are rudimentary for maximum digital circuitry, requiring higher speed, less complex and least zone designs are imperative. The morphological characteristics of this efficient design ease highly dense circuitry execution. This paper offered an optimum approach to implementing all proposed circuits. QCA is a worthy and reliable alternative to CMOS technology at a nanoscale as of less power consumption, higher speed and higher density devices. The suggested plans have further reduced the area, which illustrate to be a finest overseeing when considered broad circuits like processors, ALU, etc. In addition, the usual environment of QCA rationale encourages the reversal work impartial, thereby heightening the speedy function within the terahertz extend. In this article, efficient and optimum

QCA XOR N8 and MUX N16 proposed and further, we suggested a QCA RAM by using 3 MUX circuitry and these suggested models were utilized in the QCA RAM layout in a QCA designer simulation environment. The comprehensive analysis of the proposed design is suggested that the proposed RAM cell architecture is 61.53% cost-efficient and 68.18% area decrement and 67.77% reduction in quantum cell count as compared to its best previous existing layout. It is observed that the proposed plan has been displayed with Set/Reset capacity, and on the off chance that we need to concern the highlight for RAM cells, the suggested plan is ideal in parameter of the number of cells and possesses range and quantum fetches related to its best past layouts.

References

1. Taugaw PD, Lent CS (1994) Logical devices implemented using QCA. J Appl Phys 75:1818–1825
2. Shafi MA, Bahar AN, Bhuiyan MMR, Shamim SM (2018) The average output polarization dataset for signifying the temperature influence for QCA designed reversible logic circuits, vol 409. Elsevier Inc., pp 42–48
3. Palli B, Hemant (2012) Design, implementation and test of novel quantum-dot cellular automata FPGAs for the beyond CMOS era. Thesis and dissertations, 260
4. Niemier MT (2004) Designing digital systems in QCA. M.S. thesis, University of Notre Dame
5. Singh G, Sarin RK, Raj B (2016) A novel robust exclusive-or function implementation in QCA nanotechnology with energy dissipation analysis. J Comput Electron 15:455–465
6. Bahar AN, Waheed S, Hossain N, Asaduzzaman M (2017) A novel 3-input XOR function implementation in quantum dot-cellular automata with energy dissipation analysis. Alexandria Eng J 57:729–738
7. Chabi A, Roohi A, Khademolhosseini H, Sheikhfaal S, Angizi S, Navi K, DeMara RF (2017) Towards Ultra-efficient QCA reversible circuits. Microprocess Microsyst 49:127–138
8. Fengbin D, Guangjun X, Yongqiang Z, Fei P, Hongjun L (2017) A novel design and analysis of comparator with XNOR gate for QCA. Microprocess Microsyst 55:131–135
9. Tripathi D, Wairya S (2020) Energy efficient code converter for nanotechnology applications. J Crit Rev 7:2916–2925
10. Hashemi S, Farazkish R, Navi K (2013) New quantum dot cellular automata cell arrangements. J Comput Theor Nanosci 10:798–809
11. Chabi A, Sayedsalehi S, Angizi S, Navi K (2014) Efficient QCA exclusive-or and multiplexer circuits based on a nanoelectronic-compatible designing approach. Hindawi Publishing Corporation International Scholarly, 9
12. Wang L, Xie G (2020) A novel XOR/XNOR structure for modular design of QCA circuits. Trans Circuits Syst II:1–5
13. Chabi AM, Roohi A, Khademolhosseini H, Sheikhfaal S, Angizi S, Navi K, DeMara RF (2017) Towards ultra-efficient QCA reversible circuits. Microprocess Microsyst 49:127–138
14. Tripathi D, Wairya S (2020) An efficient energy ripple carry adder for nanotechnology applications. In: 3rd international conference on VLSI, communication & signal processing (3rd VCAS-2020), MNNIT, Allahabad
15. Vankamamidi V, Ottavi M, Lombardi F (2005) A line-based parallel memory for QCA implementation. IEEE Trans Nanotechnol 4(6):690–698
16. Taskin B, Hong B (2006) Dual-phase line-based QCA memory design. In: Proceedings of the IEEE conference on nanotechnology, Cincinnati, OH, U.S.A., pp 302–305

17. Walus K, Vetteth A, Jullien GA, Dimitrov VS (2003) RAM design using quantum-dot cellular automata. In: Proceedings of the nanotechnology conference and trade show, vol 2, pp 160–163
18. Dehkordi MA, Shamsabadi AS, Ghahfarokhi BS, Vafaei A (2011) Novel RAM cell designs based on inherent capabilities of quantum-dot cellular automata. Microelectron J 42:701–708
19. Sridharan K, Pudi V (2015) Design of arithmetic circuits in quantum dot cellular automata nanotechnology. In: Studies in computational intelligence, 599
20. Singh R, Sharma DK (2020) QCA based RAM design using a resilient reversible gate with improved performance. J Circuits Syst Comput 29
21. Sen B, Goswami M, Mazumdar S, Sikdar BK (2015) Towards modular design of reliable quantum-dot cellular automata logic circuit using multiplexers. Comput Electr Eng 45:42–54
22. Khosroshahy MB, Moaiyeri MH, Angizi S, Bagherzadeh N, Navi K (2017) Quantum-dot cellular automata circuits with reduced external fixed inputs. Microprocess Microsyst 50:154–163
23. Ahmadpour SS, Mosleh M (2018) A novel fault-tolerant multiplexer in QCA technology. J Supercomput 74:4696–4716
24. Xingjun L, Zhiwei S, Hongping C (2020) A new design of QCA based nanoscale multiplexer and its usage in communications. Int J Commun Syst 33:11–12
25. Tripathi D, Sana, Wairya S (2020) Cell optimization and realization of MGDI and QCA based combinational logic circuits for nanotechnology applications. In: 2020 17th IEEE India Council international conference (INDICON).IEEE Delhi Section, Netaji Subhas University, New Delhi, India
26. Hashemi S, Navi K (2012) New robust QCA D flip-flop and memory structures. Microelectron J 43:929–940
27. Azimi S, Angizi S, Moaiyeri MH (2018) Efficient and robust SRAM cell design based on quantum-dot cellular automata. ECS J Solid State Sci Technol 7(3):38–45
28. Khosroshahy MB, Moaiyeri MH, Navi K, Bagherzadeh N (2017) An energy and cost efficient majority-based RAM cell in quantum-dot cellular automata. Results Phys 7:3543–3551
29. Sasamal TN, Singh AK, Ghanekar U (2019) Design and implementation of QCA D-flip flops and RAM cell using majority gates. J Circuits Syst Comput 28(5)

An Impact of Efficient Backoff Algorithm in MANETs

Dhirendra Kumar Sharma and Rohit Srivastava

Abstract Generally, in mobile ad hoc networks, node failure and link failure are identified in mobility scenario. Here, we have considered failure during channel access, in this case, when two mobile nodes are sending data at the same time such type of scenario leads to packet collision when mobile nodes are unable to adjust the contention window. By backoff tuning mechanism, a mobile has an attempt to retransmit the lost packets. Existing backoff algorithms have better results in certain network scenario. In this paper, effective backoff algorithm proposed to regulate the retransmission of delayed packet in high traffic. We have received significant improvement in performance by proposed method and its utility in mobile ad hoc network (MANET).

Keywords Backoff algorithm · Collisions · Fairness index · Minimum contention window (CW_{min}) · Maximum contention window (CW_{max}) · Contention resolution algorithm

1 Introduction

Collision avoidance in mobile ad hoc network (MANET) becomes more challenging when all mobile nodes are transmitting simultaneously [1, 2]. IEEE 802.11 DCF standard has CSMA/CA and backoff algorithm concepts to cope such type of issues. Backoff algorithm is used estimate the waiting time (backoff time; t_{backoff}) for retransmission of lost or unacknowledged packets. Actually, backoff algorithm uses the contention resolution algorithm which became significant for certain network configurations as noisy channel and congested network. Besides these facts, increase in hidden node scenario causes the high collision probability which rapidly decreases the network performance. Here, backoff tuning may give excessive attempts for successful transmission of packets which may lead to increase delay. To illustrate

D. K. Sharma (✉) · R. Srivastava
School of Computer Science, University of Petroleum and Energy Studies, Bidholi, Dehradun, India
e-mail: dk.sharma@ddn.upes.ac.in

B. Mishra and M. Tiwari (eds.), *VLSI, Microwave and Wireless Technologies*,
Lecture Notes in Electrical Engineering 877,
https://doi.org/10.1007/978-981-19-0312-0_15

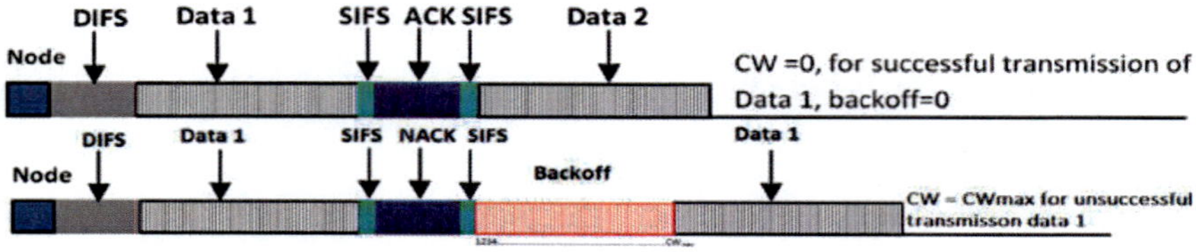

Fig. 1 IEEE 802.11 DCF backoff tuning

the backoff algorithm functionality, we have considered an example in Fig. 1 which shows the traditional backoff process.

In MANETs, distributed coordinated function (DCF) is a basic medium access control mechanism which uses retransmission for drop packets, this approach becomes ineffective in case of high traffic and node density [4]. An efficient DCF algorithm shows performance enhancement when mobile nodes are moving [5]. Here, existing algorithm became less effective in frequent link failure, so CSMA/CA uses its effective strategy for data frame transfer. When a mobile node senses the medium is free, then it starts the data transmission otherwise waiting for transmission. For successful reception of packet, CSMA/CA protocol uses a positive acknowledgment (ACK) scheme. After a short time interval, the receiving node sends the ACK packet to sender node. In case of unsuccessful delivery of ACK packet, sender arranges the retransmission of packets in Fig. 1 [5–7].

In shared wireless medium, collisions of packets are reduced by adjustment of contention window for backoff estimation. In the paper, contention window threshold used to adjust the backoff time for retransmission of dropped packets [3]. BEB algorithm works better in low dense region, where traffic density is low. As well as, when node density increases in the medium, a node increases the backoff slot number, hence the waiting time increases.

In the papers [5–12], authors have the MAC protocols that use the binary exponential backoff (BEB) algorithm for contention resolution. However, in congested and noisy scenario, backoff parameter does not work to avoid collision mechanism. Here, such type of scenario increases the collision probability and degrades the channel utilization [10]. In case of low-level congestion in the system, small backoff window is considered. When level of congestion increases, then small backoff value failed to handle the packet loss. In mobile ad hoc network, mobile node is free to transmit the data packet at any time for this frequent calling of backoff incurs but in this scenario, minimum throughput achieved and increases the end-to-end delay. For practical scenario, wireless links are unreliable and noisy so maximum path loss occurs in the channel due to fading and interference causes the bit errors. During the initialization when first transmission attempt occurs, backoff timer takes the minimum contention window (CW_{min}). For every unsuccessful transmission, the contention window (CW) size becomes double until it reaches to maximum backoff window size (CW_{max}). When the CW set to CW_{max} until it is reset for successful transmission, it decreases linearly or multiplicatively or exponentially. ACK frames are sent by destination when it successfully received [8–13, 16–18].

2 Proposed Method

In the IEEE 802.11 wireless local area networks (WLANs), network nodes experiencing collisions on the shared channel need to backoff for a random period of time, which is uniformly selected from the contention window (CW). This contention window is dynamically controlled by the binary exponential backoff (BEB) algorithm [2, 3, 8]. The BEB scheme suffers from a fairness problem that shows low throughput under high traffic load. To eliminate this fairness problem, an enhanced version of this algorithm as enhanced binary exponential backoff algorithm (EBEB) for successful transfer of data packets. In EBEB, constant counter variable used to maintain the contention window [12]. The counter variable accounts for several successful attempts. Our proposed method is the modification of EBEB algorithm for improvement of MANET performance. We have assumed two separate counters, namely C_1 and C_2 are used to encounter both the successful transmission state and unsuccessful transmission state, respectively. In case of EBEB algorithm, only the successful transmission state was taken into consideration which causes the ambiguity. EBEB algorithm works only for successful transmission but in our proposal, we have checked for unsuccessful transmission by applying a condition to check the delivery of data frame. In case of unsuccessful transmission, the counter C_2 is initialized to counts the no. of unsuccessful delivery of data frame. And before adjustment of contention window, we have compared the last two values of both counter C_1 and C_2. In order to get rid of this ambiguity, both the cases are taken into account in our proposed method. Figure 2 shows that in flowchart, there are two separate flows for the successful and unsuccessful transmission by the node. The successful transmission is same as that of EBEB algorithm [12], and the difference lies in the case of unsuccessful transmission. Here, first the value of unsuccessful counter C_2 is compared with the maximum threshold value, and then further compared with the value of counters C_1. According to the higher value among the two, the further flow chart is carried out as shown in Fig. 2 (Table 1).

3 Simulation

We are using network simulator (ns-2.34) [13] for implementation of algorithms in proposed method. IEEE 802.11 WLAN mac.cc, *mac.h* and *timer.cc* files are studied for implementation of proposed method. At network layer, we have used existing AODV routing protocol for simulation in different scenario [14, 15]. We compared proposed method results with that of existing EBEB method [12] (Fig. 3 and Table 2).

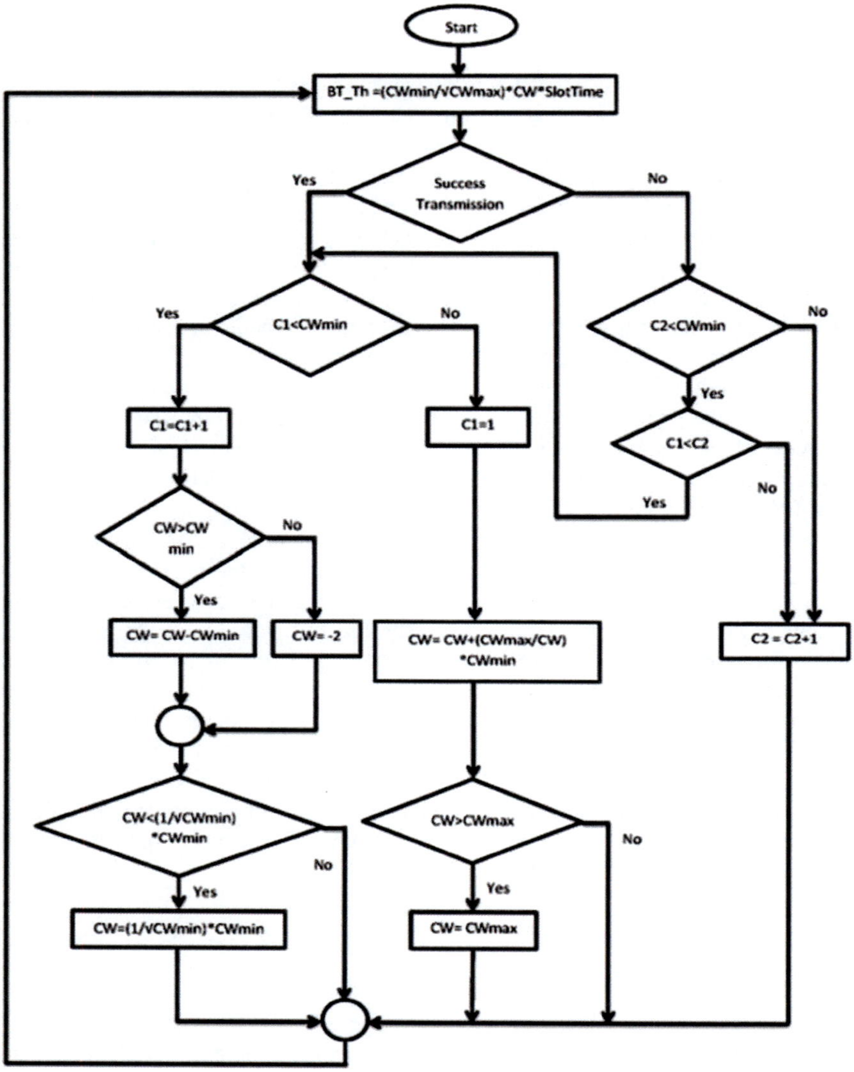

Fig. 2 Adjustment of contention window

4 Results and Discussion

Performance metrics

We have done the simulation to check the functionality of proposed method. Results are shown in Figs. 4 and 5 that show significant improvement in performance metric when network traffic increases.

Table 1 Notations

Symbols	Description
BT_Th	Threshold backoff time
CW	Default contention window
CW_{min}	Minimum contention window
CW_{max}	Maximum contention window
C_1	Counter for successful attempt
C_2	Counter for unsuccessful attempt
t_{slot}	Contention window slot time
N	Number of slots
C	Number of collisions
W	Wait or backoff states
T	Transmission state

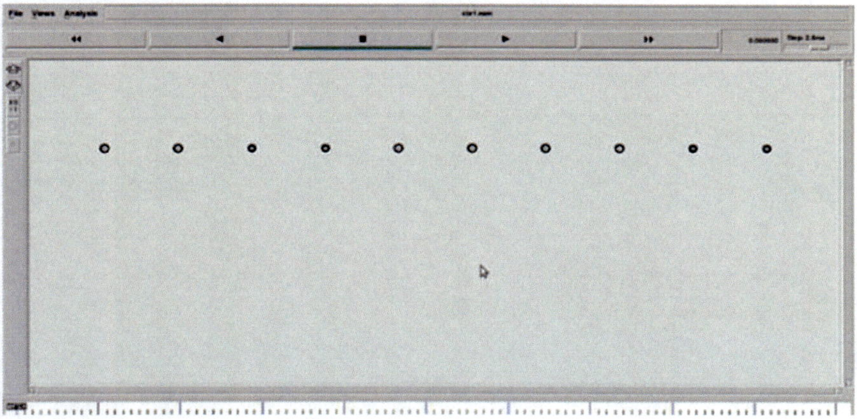

Fig. 3 Simulation snapshot of 10 node chain topology

Scenario 1

In first scenario, we have increased number of hidden terminals in a network (Table 3).

As the hop length increases, then throughput decreases because increasing the hop length increases the number of active nodes and hidden node scenarios. Here, proposed method gives better throughput than EBEB algorithm.

Scenario 2

When increasing the network traffic means, number of source destination pairs in the network (Table 4).

Table 2 Simulation parameter

Parameter	Value
Simulation time	600 s
Topology size	2000*1000 m^2
Number of mobile nodes	10 (chain topology)
MAC type	MAC 802.11
Radio propagation model	Two ray ground
Range	250 m
Size of packet	512 Bytes
Transmitter power	0.281 W
Receiver threshold	$7.69113*10^{-8}$ W
Traffic type	CBR
CBR rate	1Mbps
Promiscuous mode	Enable
Hello interval	1500 ms
Speed	5–10 m/s

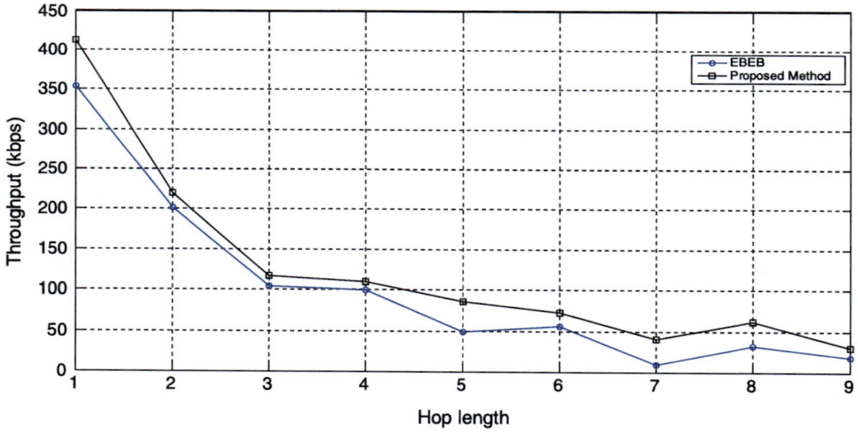

Fig. 4 Throughput versus hop length

As the number of active nodes in the network increases, throughput increases because the increase in traffic results more collision. Proposed method gives better throughput with respect to EBEB algorithm.

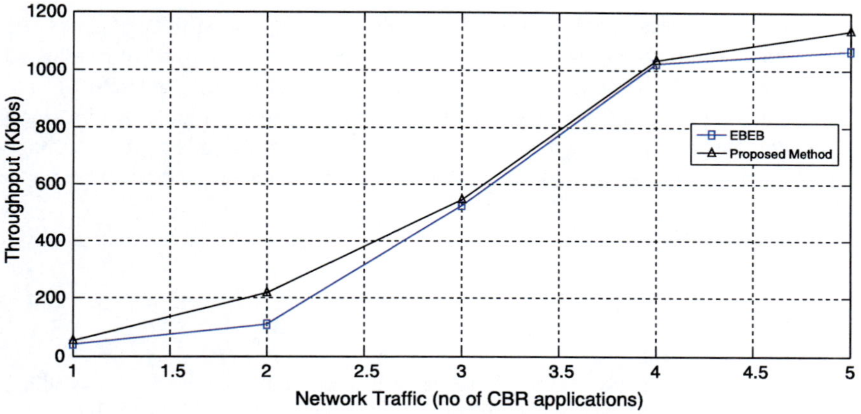

Fig. 5 Throughput versus network traffic

Table 3 Results 1

Hop length	EBEB method	Proposed method	% Improvement
1	353.61	412.28	16.59
2	201.6	219.33	8.79
3	103.59	116.61	12.56
4	98.95	109.42	10.58
5	48.83	86.2	76.53
6	56.24	72.31	28.57
7	8.85	41.3	36.66
8	31.71	61.98	95.45
9	18	30.04	66.88

Table 4 Results 2

Traffic pairs	EBEB	Proposed method	% Improvement
1	41.07	51.88	26.32091551
2	109.91	216.62	97.08852698
3	525.8	545.23	3.695321415
4	1021.88	1032.44	1.033389439
5	1068.55	1135.81	6.294511254

5 Conclusions

We proposed an algorithm to enhance the performance of backoff algorithm by modification of contention window by use of counter mechanism. We identified that mobile node with less contention window (CW) have high possibility of accessing the shared medium among competing mobile nodes. Here, proposed work adjusted CW size because of C_2 counter to monitor the unsuccessful attempt, and before transmission, it compared to C_1 counter. When channel becomes available by observing C_2 counter, we find that rapid decrease in backoff time, which leads to get advantage by proposed scheme. We got significant improvement in performance parameters in terms of throughput.

References

1. Bensaou B, Wang Y, Ko CC (2000) Fair medium access in 802.11 based wireless ad-hoc networks. In: Mobile and ad hoc networking and computing, Boston, MA, pp 99–106
2. Li T, Tang T, Chang C (2009) A new backoff algorithm for IEEE 802.11 distributed coordination function. Fuzzy Syst Knowl Discov. In: FSKD '09, Tianjin, pp 455–459
3. Krishna CR, Chakrabarti S, Datta D (2004) A modified backoff algorithm for IEEE 802.11 DCF-based MAC protocol in a mobile ad hoc network. In: IEEE TENCON 2004, pp 664–667
4. Kwak B-J, Song N-O, Miller LE (2005) Performance analysis of exponential backoff. IEEE/ACM Trans Netw 13(2):343–355
5. Xia Q, Hamdi M (2006) Contention window adjustment for IEEE 802.11 WLANs: a control-theoretic approach. In: IEEE international conference on communications Istanbul, pp 3923–3928
6. Bih-Hwang L, Hui-Cheng L (2008) Analysis of adaptive control method in IEEE 802.11 and IEEE 802.11 e wireless LANs. IEICE Trans Commun 91(3):862–870
7. Guang L, Assi CM, Benslimane A (2008) Enhancing IEEE 802.11 random backoff in selfish environments. IEEE Trans Veh Technol 57(3):1806–1822
8. Manaseer S, Masadeh M (2009) Pessimistic backoff for mobile ad hoc networks. In: International conference on information technology (ICIT'09)
9. Deng DJ, Chao HC, Chen HH (2010) Slow start backoff algorithm for ad-hoc wireless networks. In: IEEE global telecommunications conference (GLOBECOM 2010), pp 1–5
10. Laddomada M, Mesiti F, Mondin M, Daneshgaran F (2010) On the throughput performance of multirate IEEE 802.11 networks with variable-loaded stations: analysis, modeling, and a novel proportional fairness criterion. IEEE Trans Wirel Commun 9(5):1594–1607
11. Zhu YH, Tian XZ, Zheng J (2011) Performance analysis of the binary exponential backoff algorithm for IEEE 802.11 based mobile ad hoc networks. In: IEEE international conference on communications (ICC), Kyoto, pp 1–6
12. Al-Hubaishi M, Alahdal T, Alsaqour R, Berqia A, Abdelhaq M, Alsaqour O (2014) Enhanced binary exponential backoff algorithm for fair channel access in the IEEE 802.11 medium access control protocol. Int J Commun Syst 27(12): 4166–4184
13. McCanne S, Floyd S, Fall K, Varadhan K (1997) Network simulator ns-2, pp 1059–1068
14. Sharma DK, Patra AN, Kumar C (2017) P-AODV: a priority based route maintenance process in mobile ad hoc networks. Wirel Pers Commun 95:4381–4402
15. Sharma DK, Patra AN, Kumar C (2016) An improvement in performance of mobile ad hoc networks using modified route maintenance. Comput Electr Eng 56:300–312
16. Srivastava R, Srivastava P (2019) A multimodal based approach for face and unique mark based combination for confirmation of human. Int J Bus Anal 6(3)
17. Srivastava R (2020) A score level fusion approach for multimodal biometric fusion. Int J Sci Technol Res. ISSN 22778616

18. Sharma DK, Patra AN, Kumar C (2014, March) AODV-OF: An overhearing based reactive routing protocol for Mobile Ad hoc Networks. In: 2014 International Conference on Computing for Sustainable Global Development (INDIACom) (pp. 674–679). IEEE

Heterogeneous Energy-Efficient Clustering Protocol for Wireless Sensor Networks

Sandip K. Chaurasiya, Arindam Biswas, and P. K. Bandyopadhyay

Abstract With the wide variety of applications involving different types of sensor nodes, heterogeneous wireless sensor Network (HWSN) has attained a lot of popularity. However, alike wireless sensor network (WSN), performance of the nodes in HWSN also succumbs due to scarcity of power. And hence, the researchers are consistently aiming to achieve energy-efficient network operations. And in this attempt, clustering has been found as a significant tool. In this work, two popular routing schemes for homogeneous WSN- energy-efficient protocol with static clustering (EEPSC) and an enhanced energy-efficient protocol with static clustering (E^3PSC) have been deployed in the heterogeneous environment in order to figure out their performance; then, a new clustering-based scheme—heterogeneous energy-efficient clustering protocol for wireless sensor network (HEECP) has been proposed. The suitability of the HEECP is established through an extensive set of experiments conducted in MATLAB with respect to the network parameters—network lifetime and network throughput.

Keywords Clustering · Energy efficiency · Energy heterogeneity · Network lifetime

1 Introduction

With the rapid growth in technological advancement, wireless sensor network (WSN) has been evolved as a great tool for many of the applications viz. Habitat monitoring,

S. K. Chaurasiya (✉)
Department of Cybernetics, School of Computer Science, University of Petroleum and Energy Studies (UPES), Energy Acres Building, Dehradun, Uttarakhand 248007, India
e-mail: sandipchaurasiya@gmail.com

A. Biswas
School of Mines, Kazi Nazrul University, Asansol, West Bengal, India

P. K. Bandyopadhyay
Department of Electronics and Communication Engineering, DIET Murshidabad, Baharampur, West Bengal, India

B. Mishra and M. Tiwari (eds.), *VLSI, Microwave and Wireless Technologies*,
Lecture Notes in Electrical Engineering 877,
https://doi.org/10.1007/978-981-19-0312-0_16

environment monitoring, disaster management, industrial monitoring, and patient monitoring, etc. [1]. Based on the intended application, WSN may comprised of a variety of nodes with different ability and power, resulting into a new variant popularly called as heterogeneous wireless sensor network (HWSN). However, alike WSN, performance of HWSN might succumb due to scarcity of power as nodes once depleted energy might not get replaced or recharged. To overcome this insufficiency, clustering, i.e., grouping of nodes based on some common attribute, has been investigated accordingly also in the HWSN, and a number of works have already addressed this issue viz. [2–6]. However, all the aforementioned schemes are based on dynamic clustering in which clusters are formed repeatedly and hence consuming a lot of network energy in the cluster formation. Contrary to the dynamic clustering, static clustering-based routing schemes viz. [7], Chaurasiya et al. [8], and [9] save the significant amount of network energy from being consumed in repeated cluster formation and make that available for the other important network activities like sensing, transceiving, and data aggregation.
Hereby, in this work, the existing schemes of EEPSC [7], & E^3PSC [8] have been examined in the heterogeneous environment, and the E^3PSC has been modified accordingly under the title, heterogeneous energy-efficient clustering protocol for wireless sensor network (HEECP) to compete with the new and changed context of HWSN. In the subsequent Sect. 2, the proposed scheme-HEECP has been discussed in detail. Section 3 discusses the simulation environment and simulation results, and finally, Sect. 4 concludes the entire work.

2 HEECP-Heterogeneous Energy-Efficient Clustering Protocol for Wireless Sensor Network

This section discusses the proposed scheme, heterogeneous energy-efficient clustering protocol for wireless sensor network (HEECP) in detail. The HEECP is basically an extension of the scheme—an enhanced energy-efficient protocol with static clustering (E^3PSC) in order to ensure even load distribution among the nodes deployed with varying initial energy. The central idea in the E^3PSC is to select the node with the highest residual energy located near to the mean position in the concerned cluster as cluster head (CH). However, contrary to the homogeneous WSN where nodes are deployed with the same amount of initial energy, nodes deployed in the HWSN might vary in terms of their initial energy; and hence, spatial distribution of the nodes must accommodate their energy values, i.e., the energy distribution. HEECP proceeds with this idea and computes the center of cluster energy taking both spatial distribution and energy distribution into account. Moreover, alike E^3PSC, the network operation starts with the formulation of distance-based static clusters.

2.1 Network Model

The following heterogeneous network model has been adopted for the network operation:

1. The sensor nodes are deployed randomly and with definite initial energy.
2. The sensor nodes are static in nature, i.e., they cannot change their respective location once deployed.
3. The sensor nodes are facilitated by power control feature, i.e., they can adjust their transmission power as and when required.
4. The sensor nodes might be featured with any initial energy according to the nature of application, i.e., the network can be deployed with n-level of energy heterogeneity.
5. The base station is fixed and has been located far beyond the sensor nodes. (However, it can be situated at any place as per the application requirement.
6. The network adopts the continuous data flow model for periodically sensing and transmitting the information.

In this work, three-level HWSN has been considered in an attempt to be aligned with [3–6]. In the three-level HWSN, three types of nodes—normal, advanced, and super nodes—are deployed in the network in different proportion defined by the network administrator. For example, if a total of N number of nodes are being deployed, then m might be the fraction of non-normal nodes, i.e., the $m.N$ nodes would be deployed as advanced and super nodes and there would be $1 - m.N$ normal nodes. Moreover, if m_0 be the fraction of super nodes, then it can be easily computed that there would be $m.m_0.N$ super nodes and $(1 - m.m_0.N)$ number of advanced nodes. Also, if the initial energy of the normal nodes is $E - 0$, then that of advanced nodes & super nodes are $E_0 + a.E_0$ & $E_0 + b.E_0$, respectively, where a and b are the energy multiplier for the advanced and super nodes to be defined by the network administrator.

In addition to the aforementioned network model, first-order radio model [2–8, 10] has been adopted here for the measurement of energy consumption in each sensor activity viz. transmission, reception, and data aggregation as follows:

$$E_{\mathrm{TX}}(l, d) = \begin{cases} E_{\mathrm{elec}} * l + \varepsilon_{fs} * l * d^2, d \leq d_0 \\ E_{\mathrm{elec}} * l + \varepsilon_{mp} * l * d^4, d > d_0 \end{cases} \quad (1)$$

$$E_{\mathrm{RX}}(l) = E_{\mathrm{elec}} * l \quad (2)$$

$$E_{\mathrm{DA}}(l) = \varepsilon_{\mathrm{da}} * l \quad (3)$$

Here, $E_{\mathrm{TX}}(l, d)$ is the energy consumption by the node in order to send l-bits of data over a distance of d meters; $E_{\mathrm{RX}}(l)$ is the energy consumed by the node in receiving l-bits of message; and $E_{\mathrm{DA}}(l)$ is the energy required for the aggregation of l-bits of message. ε_{da} is the energy required to aggregate a single bit information. ε_{fs} and ε_{mp} are the energy required to run the amplification circuitry in free space and multipath fading models, respectively, i.e., when the distance to be traveled is $\leq d_0$ and $> d_0$, where d_0 is the threshold distance adapted by the first-order radio model based on ε_{fs} and ε_{mp} as follows:

$$d_0 = \sqrt{\frac{\varepsilon_{fs}}{\varepsilon_{mp}}} \tag{4}$$

2.2 *Working and Operation*

In the beginning of network operation, distance-based static clusters are formed as in [7, 8]. Afterward, some distinguished nodes are selected to carry out the high energy-consuming network operations and then, data exchange among the nodes and the sink happen. Accordingly, the entire network operation in HEECP is divided into number of rounds where each round is comprised of three different phases—setup phase, responsible node selection phase, and steady state phase. Each of these three phases are being explained below.

Phases of Operation

A. **SetUp Phase**: Alike [7, 8], k number of distance-based clusters are formed in this phase where k is defined by the network-administrator. In order to achieve this, the base station (BS) broadcasts $(k-1)$ different messages with different transmission power. The sensor nodes deployed in the network if listen to ith broadcasts, they set their cluster identity to i accordingly where $i = 1, 2, 3, 4, \ldots (k-1)$ and inform the BS via JOIN-REQ message. More illustratively, nodes listening to the 1st broadcast will set their cluster-id to 1, nodes listening to the 2nd one would set their cluster-id equal to 2 and so on. At last, the nodes who would have not heard any of the broadcasts after a predetermined time period would set their cluster-id equal to k and respond to BS via the JOIN-REQ message. In order to prevent collisions while sending their respective JOIN-REQ message to the BS, nodes employ CSMA. Once the clusters are formed, the BS prepares a TDMA schedule and randomly select a distinguished node to act as temporary cluster head (TCH) for each of the clusters formed. After performing all these computation, the BS communicates a 2-tuple, $< \mathbf{TDMA_i^j}, \mathbf{TCH_i} >$ to every jth node of the ith cluster. Once all the nodes receive their respective 2-tuple, setup phase is complete. Moreover, it is to be noted here that pursuing the static clustering, the setup phase will be called only once, i.e., at the beginning of network operation.

B. **Responsible Node Selection Phase**: This phase mainly focuses on the selection of TCH for the next round and CH for the current round. Here, temporary cluster head (TCH) is a node whose main responsibility is to select CH for the next round based on the input received from the member nodes of the same cluster whereas cluster head (CH) is the node which receives data from the member nodes, aggregate those, and communicate such aggregated data to the base station on behalf of the entire cluster. At the beginning of this round, every member node sends its current energy status along with the respective coordinates to the TCH. Based on the information received, TCH locates the center of cluster energy (CoCE) which is nothing but a point in the cluster that embarks the mean position in relation to the residual energy of the nodes. In this process, it locates a point say $(X_{\text{CoCE}_i}, Y_{\text{CoCE}_i})$ for the ith cluster which is taken as weighted mean of location defined as follows:

$$X_{\text{CoCE}_i} = \frac{\sum_{j=1}^{m} x_i^j * \text{RE}_i^j}{\sum_{j=1}^{m} \text{RE}_i^j}$$
$$Y_{\text{CoCE}_i} = \frac{\sum_{j=1}^{m} y_i^j * \text{RE}_i^j}{\sum_{j=1}^{m} \text{RE}_i^j} \quad (5)$$

here, m is the number of nodes in the ith cluster and RE_i^j is the residual energy of the jth node in the ith cluster. Once the center of cluster energy is determined by the TCH in the concerned cluster, closest node to this is chosen to be the cluster head for the current round and node with the least amount of energy is chosen as TCH for the next round. Such selected responsible nodes are then informed to each member node in the cluster.

C. **Steady State Phase**: This phase defines the data transmission between the clusters and the base station (BS). In the beginning of this phase, member nodes send the collected data to the respective CH in their predefined time slots. Thereafter, cluster heads send their data to the BS directly. In order to save energy, member nodes in a cluster turn their radio off until their time slots arrive; however, the radio of various cluster heads is always on.

3 Simulation Results and Analysis

In this section, all the results obtained through an extensive set of experiments have been discussed in detail. To measure the performance of the proposed scheme, two different yardsticks have been chosen here, network stability, and network throughput.

Network stability is measured in terms of network lifetime and network energy consumption. Here, network lifetime is defined as the time until the last node dies in the network, i.e., the time when there is no more alive nodes in the network, whereas network throughput refers to the number of packet delivered to the base station.

3.1 Simulation Environment

MATLAB has been used here to simulate the network operation of the schemes—EEPSC, E^3PSC, & HEECP. To model the energy consumption by the nodes, first-order radio model has been used as explained in the previous section. The standard set of parameters used in the simulation is listed in Table 1.

3.2 Results and Discussion

1. The first set of experiments demonstrates the performance comparison of the schemes with respect to the chosen yardstick network stability. In the following Fig. 1, the same has been summarized. Figure 1a briefs the outperformance of HEECP over EEPSC and E^3PSC in terms of network lifetime. Similarly, Fig. 1b and c describes the pattern of energy consumption for the different schemes in the network. Figure 1b clearly demonstrates that any point in time, network employing HEECP is left with more residual network energy when compared with the remaining two schemes- [7, 8]. Moreover, Fig. 1c depicts explicitly that the average energy per node in the network following HEECP is always greater than that in the network following EEPSC and E^3PSC.
2. In the second set of experiments, performance of the proposed scheme, HEECP has been measured in terms of packet delivery to the base station with respect to

Table 1 Parameters used in the simulation

Parameter	Parameter's value
Network area	100×100 m^2
Base station's position	(50, 150 m)
Number of nodes deployed in the network (N)	100
Initial energy of the normal nodes	1.0 J
Energy multiplier for the advanced nodes (a)	2.0
Energy multiplier for the super nodes (b)	3.0
Proportional factor for non-normal nodes (m)	0.8
Proportional factor for super nodes (m_0)	0.6
Size of data message	4000 bits
Energy consumed in data aggregation (ε_{da})	5 nJ/bits/signal
Energy consumed in the transceivers' circuitry (E_{elec})	50 nJ/bit
Amplification factor in free space model (ε_{fs})	10 pJ/bit/m^2
Amplification factor in multipath fading model (ε_{mp})	0.0013 pJ/bit/m^4

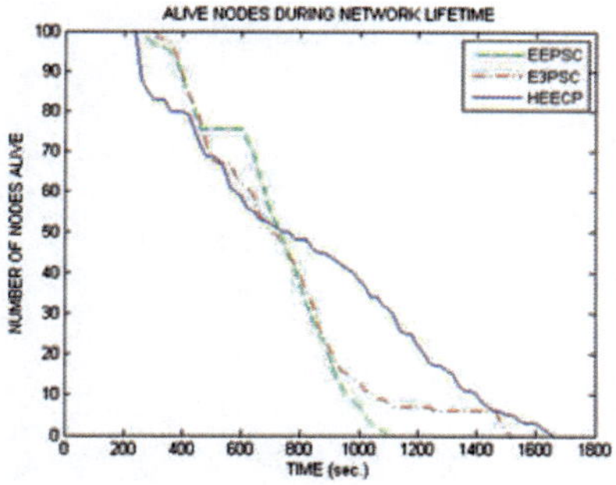

(a) Alive nodes in the n/w over time

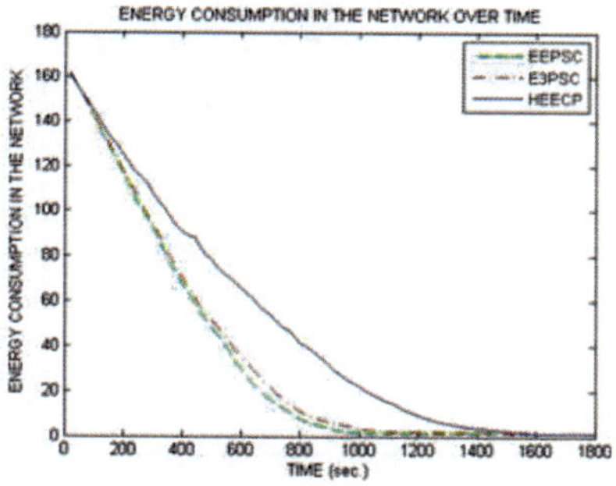

(b) Energy consumption over time

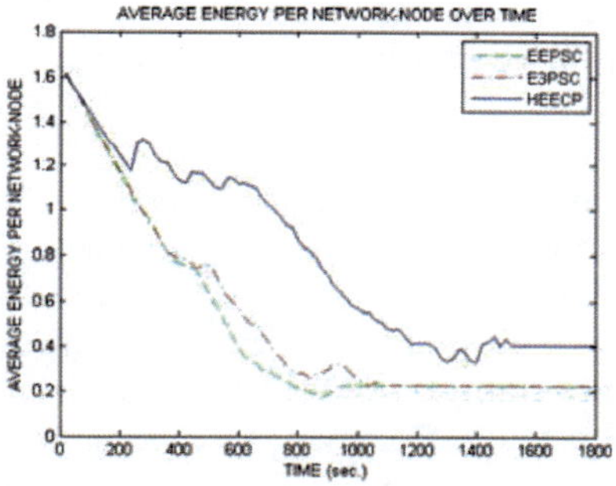

(c) Average residual energy per node in the n/w over time

Fig. 1 Network stability- EEHCP [7], E^{3}PSC [8], and HEECP in three-level HWSN

the [7, 8].Fig. 2a confirms that the overall number of packet delivered to the base station in HEECP (18513 packets) is higher than that in EEPSC (13648 packets) and E^{3}PSC (14926 packets). Similarly, Fig. 2b implies that the higher number of data packets can be delivered to the base station in HEECP while consuming less amount of network energy in comparison to EEPSC and E^{3}PSC.

Thus, from the obtained results as described in Figs. 1 and 2, the suitability of the proposed scheme, HEECP can easily be concluded.

4 Conclusion and Future Works

In this work, two of the popular static clustering-based routing protocols, EEPSC and E^{3}PSC, have been examined in the context of heterogeneous environment and

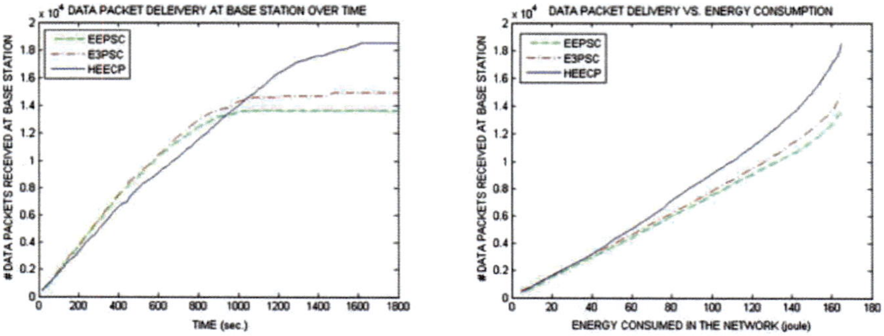

(a) Data Packet Delivery at Base Station over Time (b) Data Packet Delivery over Network Energy Consumption

Fig. 2 Network throughput- EEHCP [7], E^3PSC [8], and HEECP in 3-level HWSN

a new scheme, HEECP, has been introduced as a derivation of E^3PSC in order to further improve the performance especially in heterogeneous networks. HEECP not only considers the spatial distribution of the deployed nodes but also the energy distribution of the nodes is brought into consideration while formulating the cluster heads in the network, hence ensures even load distribution among the nodes leading to improved network performance. Moreover, since the cluster head selection process does not count on the types of participating nodes, i.e., normal, advanced, or super, etc., the proposed scheme can be extended to accommodate any level of energy heterogeneity.

As a future work of this, mobility enabled wireless sensor network might be investigated.

References

1. Dumka A, Chaurasiya SK, Biswas A, Mandoria HL (2019) A complete guide to wireless sensor networks: from inception to current trends, 1st edn. CRC Press, Boca Raton, Florida, USA
2. Qing L, Zhu Q, Wang M (2006) Design of a distributed energy-efficient clustering algorithm for heterogeneous wireless sensor networks. Comput Commun 29:2230–2237. https://doi.org/10.1016/j.comcom.2006.02.017
3. Parul Saini, Ajay. K. Sharma (2010) E-DEEC-enhanced distributed energy efficient clustering scheme for heterogeneous WSN. In: 2010 1st international conference on parallel, distributed and grid computing (PDGC-2010), pp 205–210. https://doi.org/10.1109/PDGC.2010.5679898
4. Javaid N, Qureshi TN, Khan AH, Iqbal A, Akhtar E, Ishfaq M (2013) EDDEEC: enhanced developed distributed energy-efficient clustering for heterogeneous wireless sensor networks. Proced Comput Sci 19:914–919. https://doi.org/10.1016/j.procs.2013.06.125
5. Singh S, Chand S, Kumar B (2013) 3-Level heterogeneity model for wireless sensor networks. Int J Comp Netw Info Secur 5(4):40–47. https://doi.org/10.5815/ijcnis.2013.04.06

6. Singh S, Malik A, Kumar R (2017) Energy efficient heterogeneous DEEC protocol for enhancing lifetime in WSNs. Eng Sci Technol Int J 20:345–353. https://doi.org/10.1016/j.jestch.2016.08.009
7. Zahmati AS, Abolhassani B, Shirazi AAB, Bahitiari AS (2007) An energy-efficient protocol with static clustering for wireless sensor networks. Int J Electron Circ Syst 1(2):135–138. https://doi.org/10.5281/zenodo.1329188
8. Chaurasiya SK, Pal T, Bit SD (2011) An enhanced energy-efficient protocol with static clustering for WSN. In: Proceedings IEEE Xplore, international conferenece of information networking (ICOIN), Kuala Lumpur, Malaysia, pp 58–63. https://doi.org/10.1109/ICOIN.2011.5723134
9. Chaurasiya SK, Sen J, Chatterjee S, Bit SD (2012) EBLEC: an energy-balanced lifetime enhancing clustering for WSN. In: Proceeding IEEE Xplore, 14th international conference on advanced communication and technology-2012, PyeongChang, Korea (South), pp 189–194, INSPEC Accession Number: 12656578
10. Mishra S, Chaurasiya SK (2015) Cluster based coverage enhancement for directional sensor networks. In: 2015 1st international conference on next generation computing technologies (NGCT), IEEE, pp 212–216

High-performance 3–2 Compressor Using Efficient XOR-XNOR in Nanotechnology

Anum Khan and Subodh Wairya

Abstract In this paper, two new 3–2 compressor architecture topologies are proposed which have low power and high speed using both CMOS and CNTFET technology. A 3–2 compressor topology involves XOR-XNOR module and 2:1 multiplexer module. The performance of recent XOR-XNOR circuits is analyzed in terms of transistor count, power dissipation, delay, power-delay product (PDP), and energy delay product (EDP). The superiority of pass transistor logic-based multiplexer is established over transmission gate-based multiplexer is validated by simulation results. Thus, high-performance 3–2 compressor is implemented using 10 T XOR-XNOR gate and PTL-based 2:1 multiplexer. In order to prove effectiveness, the performance of the proposed 3–2 compressor design 1 is compared with the 3–2 compressor design 2 in terms of transistor count, power dissipation, delay, power-delay product (PDP), and energy delay product (EDP). Performances of all circuits are justified using Cadence Virtuoso Analog Environment in 45 nm gpdk technology and 10 nm Stanford CNTFET model at 0.6–1.4 V supply.

Keywords 3–2 compressor · CNTFET · XOR-XNOR circuit · Full adder

1 Introduction

Modern technology requires high-performance devices on the go. The designers have the main task of optimizing the existing circuitry to low-power, high-speed, and low-area design. However, there is always some tradeoff between area, power, and speed. Hence, the designer's methodology must be based on the application's need and targets. Apart from that the continuous trend of scaling down VLSI technology has lead to leakage and reliability problems such as short channel effect, high-leakage current, and interconnect problems. CNTFET is the most probable substitutes for traditional MOSFET. The CNTFET device-based circuits portray

A. Khan (✉) · S. Wairya
Department of Electronics and Communication Engineering, Institute of Engineering and Technology, Lucknow, India
e-mail: anumkhan0902@gmail.com

B. Mishra and M. Tiwari (eds.), *VLSI, Microwave and Wireless Technologies*, Lecture Notes in Electrical Engineering 877,
https://doi.org/10.1007/978-981-19-0312-0_17

advantages with their better control over the device channel, reduced leakage current, and reduced short channel effects [1]. Thus in this paper, the most efficient topologies are implemented using the 10 nm CNTFET model.

In the digital multiplier, the generation of partial products is an integral part. These partial products add significantly to the overall area, power, and delay. The compressors are used to reduce the latency of this step. Therefore, compressors are critical to multiplier circuits that greatly influence the whole multiplier performance [2]. In larger systems, the compressors are used repeatedly. Therefore, research is required toward compressors with improved design with the lowest transistor count and reduced delay performance. Several compressor topologies like 3–2, 4–2, 5–2, 7–2, 5–3, 5–4, etc exist. This paper focuses only on 3–2 compressors. A 3–2 compressor is essentially a 1-bit full adder, but its performance in terms of delay and power is better than a conventional full adder.

The paper is organized in five sections which include: Sect. 1 is the introduction, Sect. 2 discusses conventional 3–2 compressor designs which include the XOR-XNOR topologies and 2:1 multiplexer topology. Two proposed 3–2 compressor design is depicted in Sect. 3. Simulation results are in Sect. 4. Lastly, a conclusion is drawn in the last Sect. 5.

2 3–2 Compressor Design

Several conventional 3–2 compressor designs have been proposed in the literature [3–7]. It basically comprises XOR-XNOR modules and 2:1 multiplexers. Figure 1 shows conventional block diagram of 3–2 compressor using XOR-XNOR topology. As discussed earlier, the 3–2 compressor is inherently a full adder having 3 inputs × 1, × 2, and × 3, and two outputs are carry and sum.

As it is clear from Fig. 1, to implement high-performing compressors, it is important to implement efficient XOR-XNOR gate and 2:1 multiplexers.

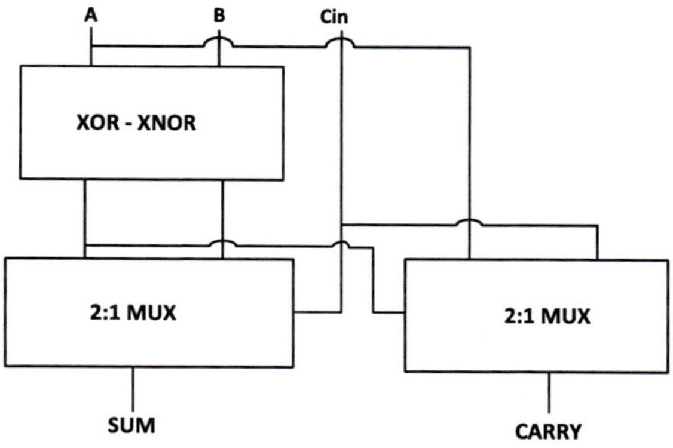

Fig. 1 3–2 compressor block diagram

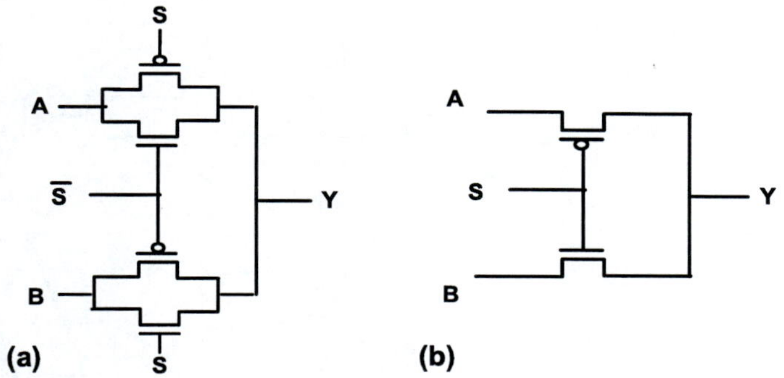

Fig. 2 **a** 2:1 multiplexer using transmission gates, **b** 2:1 multiplexer using pass transistor logic

2.1 2:1 MULTIPLEXER Design

A multiplexer (mux) is a data selector that chooses from input signals and sends the selected input to an individual output line [8]. The selection is decided by the input known as select lines (S). A 2:1 multiplexer has 2 inputs, one output, and one select line. Figure 2 depicts the 2:1 multiplexer having A and B as inputs, S as select line, and Y as output.

Figure 2a shows the architecture of multiplexer using transmission gates (TG). It uses two transmission gates and one inverter. Thus, the total transistor count is 6; on the other hand, pass transistor logic-based multiplexer requires just 2 transistors at the cost of little voltage degradation. The topology of pass transistor logic (PTL)-based multiplexer is shown in Fig. 2b.

2.2 XOR-XNOR Gate Implementation

Several XOR-XNOR topologies have been proposed in [8–20], but in this paper, the focus has been on XOR-XNOR gate having low-transistor count and better performance. Figure 3 shows a simultaneous XOR-XNOR circuit having just six transistors [8, 9] proposed by Radhakrishnan et al.

It employs two complementary transistors (PMOS and NMOS) to restore the output voltage level which is affected due to degradation. Two complementary feedback transistors are used in this design to regain the weak logic in complementary output nodes when both the inputs are same logic. The drawback of this circuit is that it has slow response.

Naseri et al. [16] presented design of the XOR- XNOR circuit having 12 transistors. This circuit depicted in Fig. 4 has low-power consumption and gives better delay performance than its counterparts. But it requires an additional NOT gate which increases the transistor count by 2.

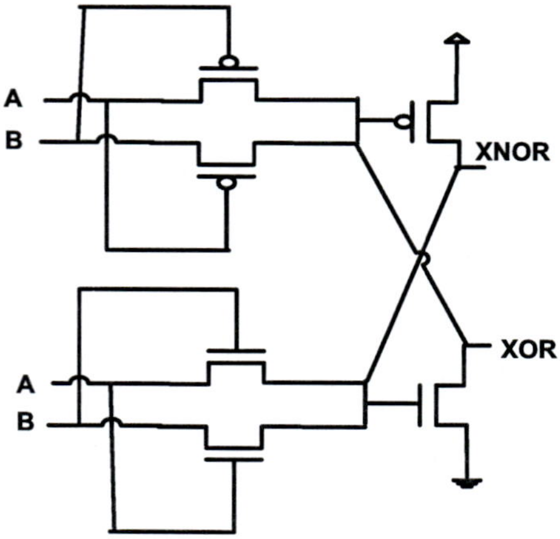

Fig. 3 6 T XOR-XNOR circuit

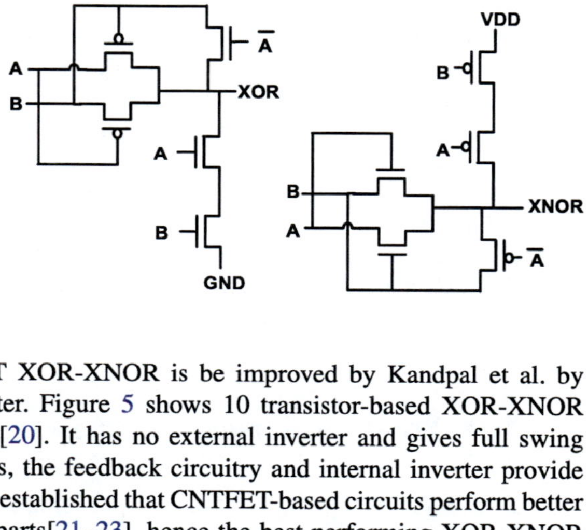

Fig. 4 12 T XOR-XNOR circuit

The performance of 10 T XOR-XNOR is be improved by Kandpal et al. by removing the external inverter. Figure 5 shows 10 transistor-based XOR-XNOR proposed by Kandpal et al. [20]. It has no external inverter and gives full swing output. For all the transitions, the feedback circuitry and internal inverter provide full swing output. It has been established that CNTFET-based circuits perform better than their MOSFET counterparts[21–23], hence the best performing XOR-XNOR topology has been implemented using 10 nm Stanford CNTFET model [24].

3 Proposed 3–2 Compressor Design Architecture

Compressors are essential components of fast digital multipliers used in digital signal processing. In order to implement a highly efficient compressor, all its modules must be high performing, therefore most efficient XOR-XNOR topology is selected from the above implemented circuits. For multiplexer, instead of transmission gate-based multiplexer having 4 transistors, a pass transistor logic-based multiplexer is used.

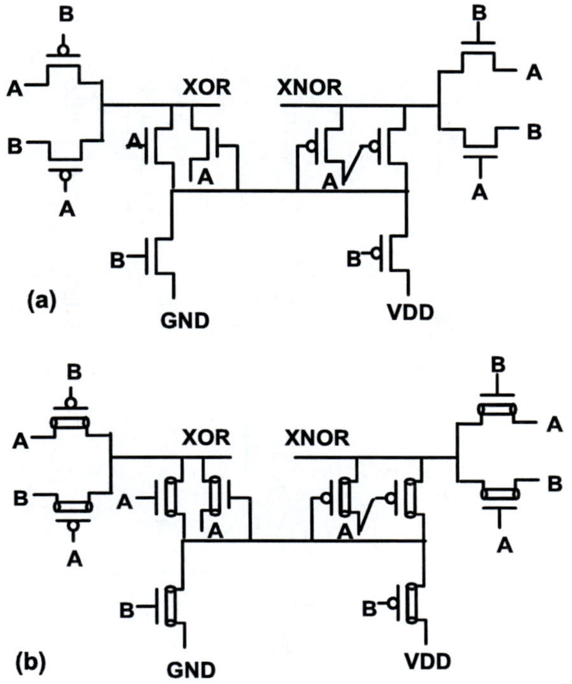

Fig. 5 10 T XOR-XNOR **a** MOSFET-based, **b** CNTFET-based

The performance of 6 T XOR-XNOR topology was unsatisfactory as it gives high delay, for that reason, it has been not included in 3–2 compressor design.

Figure 6 shows architecture topology of 3–2 compressor using 12 T XOR-XNOR and two PTL multiplexers. The total transistor count for the circuit is 16.When compared independently, the performance of 10 T XOR-XNOR is better than 12 T XOR-XNOR.

Figure 7 shows proposed circuit of 3–2 compressor based on the block diagram depicted in Fig. 1. The XOR-XNOR module is filled by 10 T XOR-XNOR, and both the multiplexers are pass transistor logic-based. The next section discusses the performance of each implemented circuits.

4 Results and Discussion

All the implemented circuits are validated using Cadence Virtuoso at room temperature, using 45 nm gpdk technology and 10 nm CNTFET Stanford model. A typical XOR-XNOR circuit's simulation waveform is shown in Fig. 8, where A and B are inputs, and XOR and XNOR are outputs.

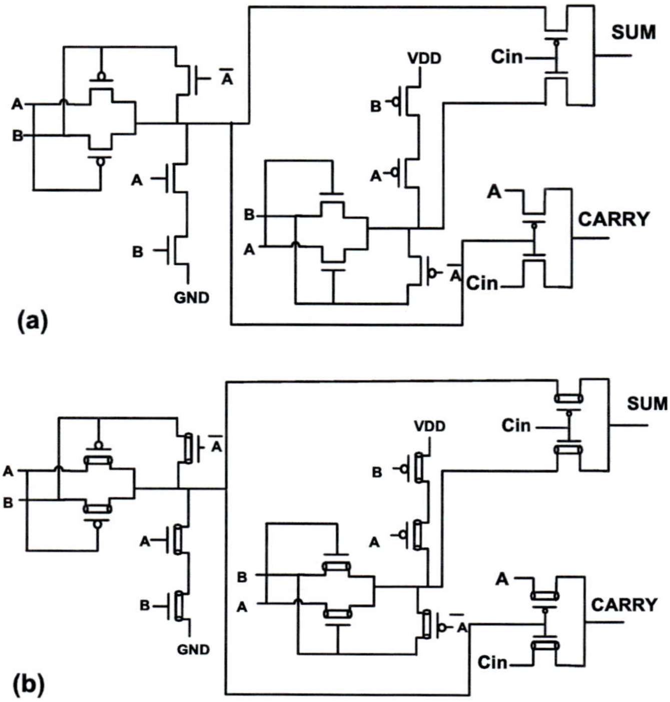

Fig. 6 3–2 compressor using 12 T XOR-XNOR and PTL multiplexer **a** CMOS-based, **b** CNTFET-based

Voltage variation of all implemented XOR-XNOR circuit is done from 0.6–1.4 V to observe the circuit performance. The performances of XOR-XNOR circuits are shown in Table 1. The PDP variation observed due to voltage variation of 6 T XOR-XNOR, 12 T XOR-XNOR, and 10 T XOR-XNOR is plotted as a graph in Fig. 9.

From Table 1 and Fig. 9, it is observed that 6 T XOR-XNOR gives very large PDP, and therefore, it is worst performing. On comparing 12 T XOR-XNOR and 10 T XOR-XNOR, the latter gives better performance up to 97% and gives the finest performance by substantial amount. Hence, the process corner variation is done to check the performance at different process corners as summarized in Table 2.

After establishing 10 T XOR-XNOR circuit topology having superior performance than other implemented XOR-XNOR circuits, the same circuit is implemented using 10 nm CNTFET. The results obtained are shown in Table 3.Voltage variation is done from 0.6–1.4 V, and delay, power, and power-delay product are noted.

From the results in Table 3, it is concluded that CNTFET-based circuit provides up to 83% PDP improvement. The next important module of 3–2 compressor is a multiplexer. The two different topologies of multiplexers are using transmission gates and pass transistor logic. These circuits are analyzed at identical conditions having

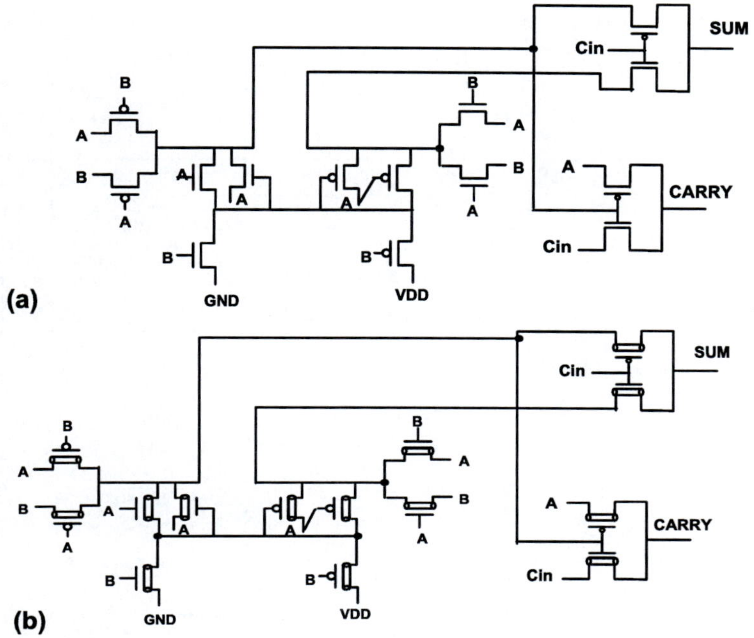

Fig. 7 3–2 compressor design 2 **a** CMOS-based circuit, **b** CNTFET-based circuit

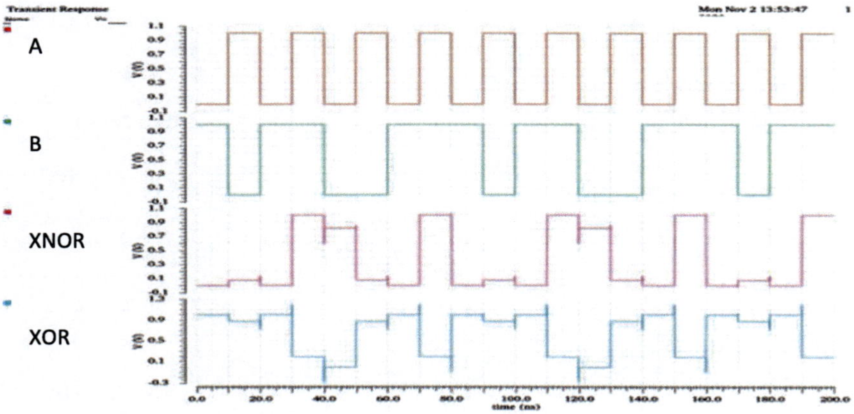

Fig. 8 Simulation waveforms for XOR-XNOR circuits

Table 1 Performance analysis of XOR-XNOR circuits with voltage variation of 0.6 V-1.4 V

	6 T XOR-XNOR			12 T XOR-XNOR			10 T XOR-XNOR		
Supply voltage (V)	Power (uW)	Delay (fs)	PDP (10^{-21} J)	Power (uW)	Delay (fs)	PDP (10^{-21} J)	Power (uW)	Delay (fs)	PDP (10^{-21} J)
0.6	FAIL	FAIL	FAIL	0.04	5500	239.8	0.0241	335.68	8.08
0.8	2.66	26,165	69,598.9	5.82	3300	19,209.3	3.587	276.5	991.80
1	22.3	8084	180,273.2	27.9	816	22,766.4	29.89	184.2	5505.73
1.2	56.82	4533	257,565.1	64.24	439	28,201.36	76.21	112.06	8540.09
1.4	102.9	3062	315,079.8	114	375	42,750	143.094	59.05	8449.70

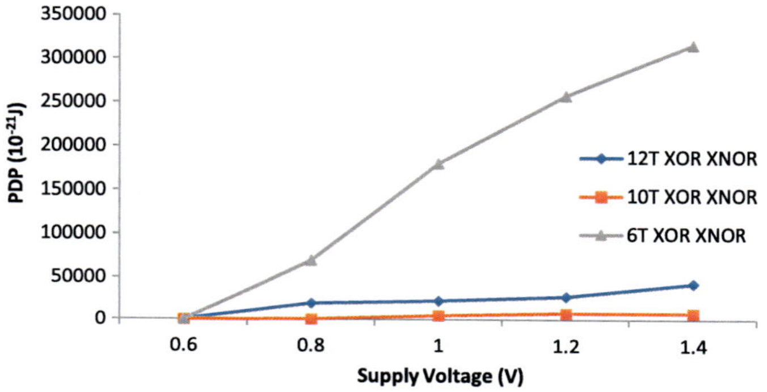

Fig. 9 PDP comparison of implemented XOR-XNOR circuits

Table 2 Process variation of 10 T XOR-XNOR topology

	Power (uW)	Delay (fs)	PDP(10^{-21} J)	EDP(10^{-18}Js)
TT	29.94	80.28	2403.50	192.95
FF	30.04	80.61	2421.52	195.19
SS	22.89	82.91	1897.81	157.34
FS	30.135	74.2	2236.017	165.91
SF	29.029	89.09	2586.194	230.40

Table 3 CNTFET 10 T XOR-XNOR topology

Supply voltage (V)	Power (uW)	Delay (fs)	PDP(10^{-21} J)	EDP(10^{-18}Js)
0.6	0.020	273.7	5.59	1.32
0.8	0.037	281.6	10.42	2.934
1	0.66	285.8	187.71	53.64
1.2	5.30	270.3	1433.67	409.7
1.4	13.23	257.2	3402.75	875.18

a 1 V input supply at room temperature. Figure 10 shows simulation waveform of a typical 2:1 multiplexer having S as select line, A and B as two inputs, and OUT is the output line. Table 4 gives the performance analysis of both PTL and TG based 2:1 multiplexer by analysing both CMOS and CNTFET implementation.

From the results obtained in Table 4 PTL-based mux shows better performance than TG MUX by 65%, CNTFET implementation gives performance improvement by 36 and 89% for PTL mux and TG mux, respectively.

Now, the results of two proposed 3–2 compressor are contained in Table 5, wherein all parameters are evaluated over the voltage variation of 0.6–1.4 V. Figure 11 shows typical output waveform of 3–2 compressor having X1, X2, and X3 as three inputs, SUM and Carry as two outputs. Figure 12 follows the PDP variation with respect to voltage variation of proposed 3–2 compressor design 1 and design 2.

Transistor count of proposed 3–2 compressor design 1 is 16, whereas transistor count of proposed 3–2 compressor design 2 is 14. Proposed compressor design 2 displays better performance over all voltage variation, hence it displays superior performance 18%-89% as verified from Table 5 and Fig. 12. Therefore, the proposed circuit is implemented using CNTFET as well and the results obtained are tabulated in Table 6.

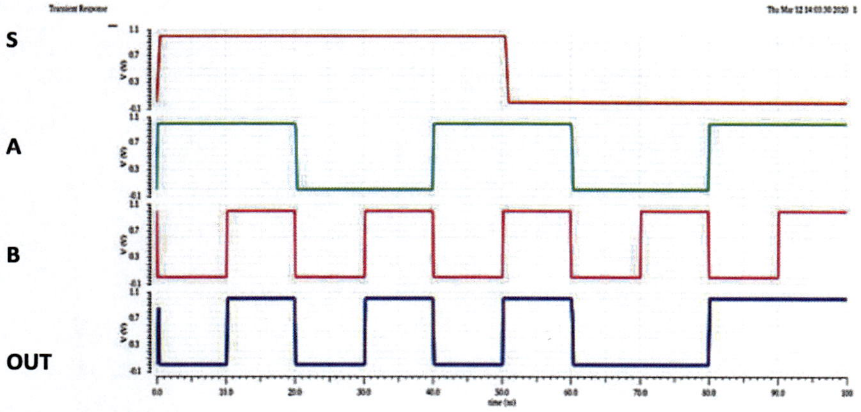

Fig. 10 Simulation waveform of 2:1 multiplexer

Table 4 Performance analysis multiplexer topologies

	CMOS-based 2:1 multiplexer			CNTFET-based 2:1 multiplexer		
2:1 mux type	Power (uW)	Delay (fs)	PDP (10^{-21} J)	Power (uW)	Delay (fs)	PDP (10^{-21} J)
PTL-based mux	0.031	828.4	25.85	0.0043	3900	16.85
TG-based mux	10.53	31.48	331.48	0.0076	4500	34.2

Table 5 3-2 compressor performance analysis

	Proposed 3–2 compressor design 1				Proposed 3–2 compressor design 2			
Supply voltage (V)	Power (uW)	Delay (fs)	PDP (10^{-21} J)	EDP (10^{-18}Js)	Power (uW)	Delay (fs)	PDP (10^{-21} J)	EDP (10^{-18}Js)
0.6	0.36	13.997	4.97	0.069	0.044	13.23	0.59	0.0078
0.8	5.26	46.752	245.91	11.496	3.89	50	194.5	9.725
1	27.2	88.401	2404.50	212.56	28.14	70.27	1977.39	138.951
1.2	64.56	114.75	7408.26	850.097	76.51	85.73	6559.05	562.307
1.4	119.74	146	17,481.6	2552.31	145.4	85.5	12,431.7	1062.910

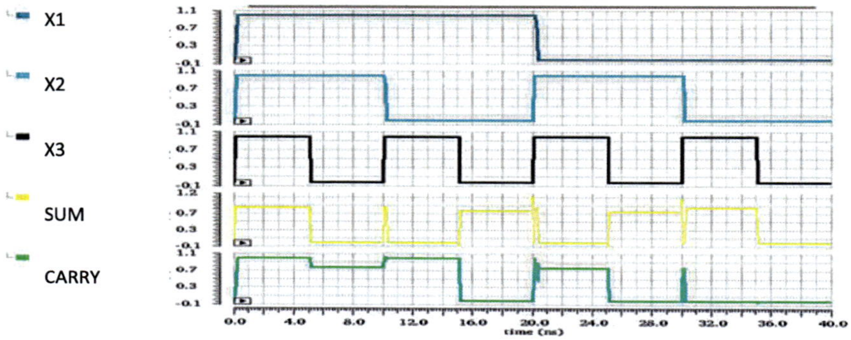

Fig. 11 Simulation waveform of 3–2 compressor

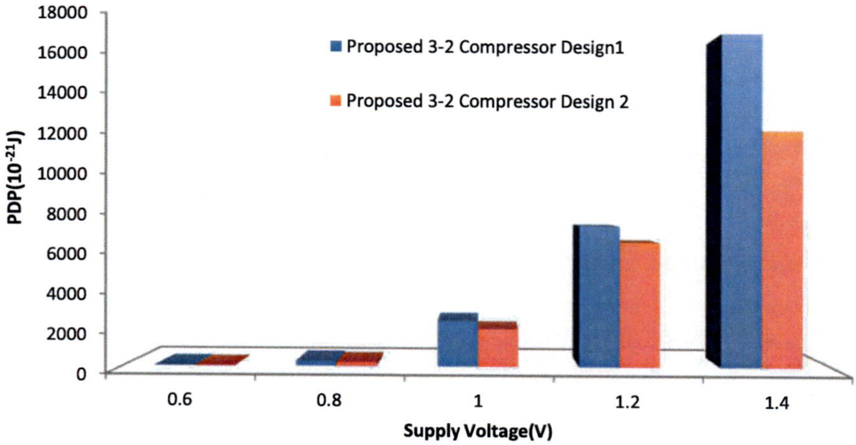

Fig. 12 PDP comparison of implemented and proposed 3–2 compressors

Table 6 Proposed 3–2 compressor design 2 using CNTFET

Supply voltage (V)	Power (uW)	Delay (fs)	PDP(10^{-21} J)	EDP(10^{-21}Js)
0.6	0.022	0.252	0.005	0.001
0.8	0.038	0.227	0.008	0.002
1	0.694	0.899	0.623	0.560
1.2	5.592	1.26	7.046	8.878
1.4	13.958	1.43	19.96	28.514

The proposed CNTFET-based 3-2 compressor design 2 gives PDP improvement of about ~98%.

5 Conclusion

In this paper, two new architecture topologies of 3–2 compressors have been proposed using CNTFET and MOSFET. Apart from that efficient designs of XOR-XNOR gate have been implemented, and their performance has been evaluated in order to determine the most efficient structures for digital applications. The performances of these circuits are compared in terms of power dissipation, delay, and PDP. All parameters are observed at a supply voltage variation of 0.6–1.4 V. All the simulations are done using Cadence Virtuoso at 45 nm gpdk technology and 10 nm Stanford CNTFET model. 10 T XOR-XNOR circuit is depicting the best performance, hence it is deemed as the most efficient compared to others as it shows improvement up to 93%. Its CNTFET implementation provides PDP improvement of 83%. Out of two multiplexer designs, pass transistor logic-based 2:1 mux has been found to be more efficient by 65%. Finally, the proposed 3–2 compressor provides superior PDP up to 89%, and its CNTFET circuit shows up to 98%, thus establishing CNTFET as promising alternative to traditional MOSFET.

References

1. Sridharan K, Srinivasu B, Pudi V (2020) Low-complexity arithmetic circuit design in carbon nanotube field effect transistor technology. Springer Nature
2. Momeni A et al (2015) Design and analysis of approximate compressors for multiplication. IEEE Trans Comput 64:984–994
3. Gavaber MD, Poorhosseini M, Pourmozafari S (2019) Novel architecture for low-power CNTFET-based compressors. J Circuits Syst Comput 28(12)
4. Qi H (2012) A high speed low power modulo 2n þ 1 multiplier design using carbon-nanotube technology, Master's thesis, Department of Electrical and Computer Engineering, Northeastern University, Boston, MA

5. Veeramachaneni S et al. (2007) Novel architectures for high-speed and low-power 3–2, 4–2 and 5–2 compressors. In: Proceedings 20th international conference VLSI design held jointly with 6th international conference embedded systems (VLSID'07)
6. Tonfat J, Reis R (2012) Low power 3–2 and 4–2 adder compressors implemented using ASTRAN. In: Proceedings 2012 IEEE third Latin American symposium circuits and systems (LASCAS)
7. Debashis D (2010) In: Basic electronics, Dorling Kindersley, pp 557. ISBN 9788131710685
8. Radhakrishnan D (2001) Low-voltage low-power CMOS full adder. Circuits Devices Syst IEE Proceedings 148:19–24
9. Vesterbacka M (1999) A 14-transistor CMOS full adder with full voltage- swing nodes. In: Proceedings IEEE workshop signal processing systems (SiPS), October 1999, pp 713–722
10. Hung Tien B, Yuke W, Yingtao J (2002) Design and analysis of low-power 10-transistor full adders using novel XOR- XNOR gates. Circuits and Syst II: Analog and Digital Signal Process IEEE Trans 49:25–30
11. Shams AM, Darwish TK, Bayoumi MA (2002) Performance analysis of low-power 1-bit CMOS full adder cells. Very Large Scale Integration (VLSI) Syst IEEE Trans 10:20–29
12. Hanho L, Sobelman GE (1998) New XOR/XNOR and full adder circuits for low voltage, low power applications. Microelectron J 29:509–517
13. Hanho L, Sobelman GE (1997) New low-voltage circuits for XOR and XNOR. In: Southeastcon '97. 'Engineering new new century', Proceedings IEEE, pp 225–229
14. Jyh-Ming W, Sung-Chuan F, WuShiung F (1994) New efficient designs for XOR and XNOR functions on the transistor level. Solid-State Circuits IEEE J 29:780–786
15. Hung Tien B, Al-Sheraidah AK, Yuke W (2000) New 4- transistor XOR and XNOR designs. In: ASICs, 2000. AP-ASIC 2000. Proceedings of the second IEEE Asia Pacific conference on, 2000, pp 25–28
16. Hamed N, Timarchi S (2018) Low-power and fast full adder by exploring new XOR and XNOR gates. IEEE Trans Very Large Scale Integration (VLSI) Syst 1481–1493
17. Chowdhury SR, Banerjee A, Roy A, Saha H (2008) A high speed 8 transistor full adder design using novel 3 transistor XOR gates. World academy of science, engineering and technology. Int J Electron Commun Eng 2(10)
18. Subodh W, Nagaria RK, Tiwari S (2012) Comparative performance analysis of XORXNOR function based high-speed CMOS full adder circuits for low voltage VLSI design. Int J VLSI Design Commun Syst 3(2)
19. Dan W, Maofeng Y, Wu C, Xuguang G, Zhangming Z, Yintang Y (2009) Novel low power full adder cells in 180nm CMOS technology. In: Industrial electronics and applications, 2009. ICIEA 2009. 4th IEEE conference on, 2009, pp 430–433
20. Kandpal J, Tomar A, Agarwal M, Sharma KK (2020) High-speed hybrid-logic full adder using high-performance 10-T XOR–XNOR Cell. IEEE Trans Very Large Scale Integration (VLSI) Systems
21. Yeole P, Padole DV (2015) Design of basic logic gates using carbon nano tube field effect transistor and calculation of figure of merit. In: 7th international conference on emerging trends in engineering and technology, pp 28–33
22. Deng J (2007) Device modeling and circuit performance evaluation for nanoscale devices: silicon technology beyond 45 nm node and carbon nanotube field effect transistors. Dissertation, June 2007
23. Asma Torkzadeh M, Keshavarzian P (2017) A novel energy-efficient and high speed full adder using CNTFET. Microelectron J 61:79–88
24. Lee C-S, Philip Wong H-S (2015) Stanford virtual-source carbon nanotube field-effect transistors. Model. nanoHUB, 2015

Area-Efficient and Fast Computing Chebyshev Type-I Filter Design

Ajay Kumar Yadav and Sangeeta Singh

Abstract The main feature of this paper is to design a higher-order Chebyshev type-I Low Pass Filter (LPF) that has all-pole circuit by cascading of lower-order Infinite Impulse Response (IIR) filter circuit. Further, the reported filter requires less number of multiplier and adder for circuit realization, and also, it has less computational requirements and lower circuit complexity. Notice that the value of filter coefficients is adequately chosen which is obtained on Xilinx software in Verilog hardware description language and simulation outcomes verify the mathematical model developed.

Keywords Butterworth filter · Chebyshev filter · Infinite impulse response (IIR) · Finite impulse response (FIR) · Low pass filter (LPF) · Verilog

1 Introduction

Noise is generally present in communication channels so it becomes crucial to extract or enhance the useful information from communication path without degrading the relevant information. For this, the signal filtering is one of the best techniques that remove the noises from mixed signal, and we received our desired output (Fig. 1).

In this paper, we have to discuss the Infinite Impulse Response (IIR) digital filters, which is a very important role in Digital Signal Processing (DSP). The primary advantage of IIR filters is they have lower order of filter over Finite Impulse Response (FIR) digital filters at the particular given set of specification [1–6], while IIR filters have also nonlinear phase and low round off noise sensitivity. The frequencies of signals are allowed by the filters, i.e., called passband frequency (in rad), and those frequencies are blocked by the filters, i.e., called stopband frequency (in rad). At passband frequency, the magnitude of system function is large, and ideally, it is constant. Although at stopband frequency, the magnitude of the system function is very small or ideally, it is zero. The main concentration is on design of Nth-order

A. K. Yadav · S. Singh (✉)
Microelectronics and VLSI Design, National Institute of Technology, Patna, India
e-mail: sangeeta.singh@nitp.ac.in

B. Mishra and M. Tiwari (eds.), *VLSI, Microwave and Wireless Technologies*,
Lecture Notes in Electrical Engineering 877,
https://doi.org/10.1007/978-981-19-0312-0_18

Fig. 1 Description of filter

Table 1 Types of IIR Chebyshev filter

Types	Description
Chebyshev type I	Equiripple characteristics in the passband and a monotonic behavior in stopband
Chebyshev type II	Monotonic properties in the passband and equiripple characteristics in the stopband

digital IIR Chebyshev type-I Low Pass Filter (LPF) by which the less number of multipliers and adders are required for evaluating the optimum filter coefficients. Hence, they have required less Computation and low complexity for analyzing Nth-order Chebyshev filters. There are two types of Chebyshev filter [7], which are mentioned in Table 1. The simulation has been done on Xilinx software in Verilog Hardware Description Language (Verilog HDL).

We can derive all other types (High-Pass, All-Pass, Band-Pass, and Band-Stop) filters from LPF by simple frequency transformation [3, 6].

2 Chebyshev Type-I Low Pass Filter (LPF)

The generalized Chebyshev type-I has all-pole filter with transmission zero behavior in the stopband so they are more preferred. This Chebyshev Filter provides faster roll-off factor as compared to Butterworth Filter [8–13]. The magnitude squared of the frequency response characteristics of a Low Pass Chebyshev Type-I filter [6] is

$$|\mathrm{Ha}(j\Omega)|^2 = \frac{1}{1 + \varepsilon^2 C_N\left(\frac{\Omega}{\Omega_P}\right)} \tag{1}$$

where ε is a parameter of the filter related to Passband ripple (A_p) (in decibels), i.e., maximum allowable ripple in the passband and $C_N\left(\frac{\Omega}{\Omega_P}\right)$ is the Nth-order Chebyshev polynomial.

$$\varepsilon = \sqrt{10^{Ap/10} - 1} \tag{2}$$

Computing order of filter by

Table 2 Some low-order Chebyshev polynomials

N	C_N (Ω)
0	1
1	Ω
2	$2\Omega^2 - 1$
3	$4\Omega^3 - 3\Omega$
4	$8\Omega^4 - 8\Omega^2 + 1$

$$N \geq \frac{\cosh^{-1}\sqrt{\left[\left(10^{\frac{A_s}{10}} - 1\right)/\left(10^{\frac{A_p}{10}} - 1\right)\right]}}{\cosh^{-1}(\Omega_s)}$$

Evaluating Chebyshev polynomial by

$$C_N\left(\frac{\Omega}{\Omega_P}\right) = \begin{cases} \cos\left[N\mathrm{Cos}^{-1}\left(\frac{\Omega}{\Omega_P}\right)\right], |\Omega| \leq \Omega_P \\ \cosh\left[N\mathrm{Cosh}^{-1}\left(\frac{\Omega}{\Omega_P}\right)\right], |\Omega| \geq \Omega_P \end{cases} \tag{3}$$

The Chebyshev polynomial can be calculated by the recursive equation

$$C_{N+1}(\Omega) = 2\Omega C_N(\Omega) - C_{N-1}(\Omega), N = 1, 2, 3. \tag{4}$$

The characteristics of Chebyshev polynomial are (Table 2).

1. $C_N(1) = 1$ For all N
2. $C_N(0) = 0$ For odd N
3. $C_N(0) = \pm 1$ For even N
4. The zero crossing of C_N (Ω) occurs for $-1 \leq \Omega \leq 1$

2.1 Design of Low Pass Filter

Low pass filter has defined as for denoising at lower frequency so that without loss of meaningful information.

The transfer function of all-pole LPF in *S*-domain is [6]

$$H(s) = \frac{Y(s)}{X(s)} = \frac{H_0}{\prod_{i=1}^{N}(S - S_i)} \tag{5}$$

where H_0 is zero frequency gain of the system, and Laplace transforms of output and input are $Y(s)$ and $X(s)$, respectively.

S_i is *i*th pole of transfer function, which can be represented by [6, 7]

$$S_i = \sigma_i + j\omega_i \quad (i = 1, 2, 3 \ldots\ldots N) \tag{6}$$

Here, σ_i is real part and ω_i is imaginary part of pole (P_i).
It is calculated by [6]

$$\sigma_i = -\sin(\theta_i)\sinh(\emptyset) \tag{7}$$

$$\omega_i = \cos(\theta_i)\cosh(\emptyset) \tag{8}$$

where,

$$\theta_i = (2i-1)\frac{\pi}{2N} \tag{9}$$

$$\emptyset = \frac{1}{N}\sinh^{-1}\left(\frac{1}{\varepsilon}\right) \tag{10}$$

2.2 *Mapping of S-plane into Z-plane*

Transfer function of first-order LPF in S-plane is [6]

$$H(s) = \frac{1}{S+\sigma} \tag{11}$$

Transformation of $H(s)$ into $H(z)$ is [6]

$$H(z) = \frac{1}{1-e^{-\sigma T_s}Z^{-1}} \tag{12}$$

where T_s is sampling period.

Transfer function of second-order LPF with conjugate pole in S-plane is [8]

$$H(s) = \frac{1}{(s+\sigma+j\omega)(s+\sigma-j\omega)} \tag{13}$$

$H(s)$ has written in the form of partial fraction

$$H(s) = \frac{\frac{j}{2\omega}}{s+\sigma+j\omega} - \frac{\frac{j}{2\omega}}{s+\sigma-j\omega} \tag{14}$$

Now, transformation of $H(s)$ into $H(z)$ by using Eq. (11) and (13), we get

$$H(z) = \frac{\frac{j}{2\omega}}{1-e^{-(\sigma+j\omega)T_s}Z^{-1}} - \frac{\frac{j}{2\omega}}{1-e^{-(\sigma-j\omega)T_s}Z^{-1}} \tag{15}$$

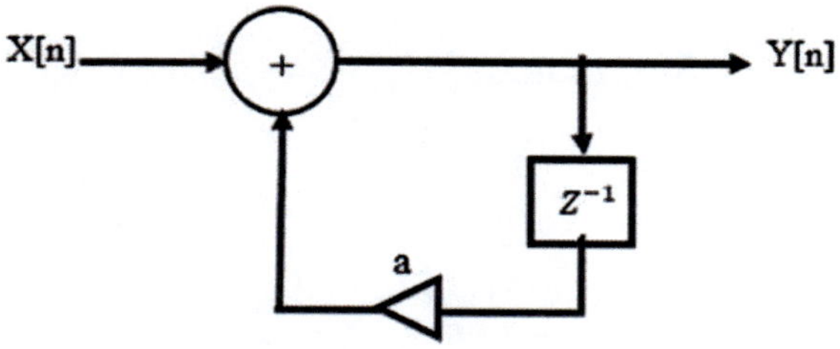

Fig. 2 First-order IIR filter circuit

$$H(z) = \frac{e^{-\sigma T_s}\mathrm{Sin}(\omega T_s)/\omega Z^{-1}}{1 - 2e^{-\sigma T_s}\mathrm{Cos}(\omega T_s)Z^{-1} + e^{-2\sigma T_s}Z^{-2}} \quad (16)$$

The general representations of first- and second-order filters of Eq. (12) and (16) are

$$H(z) = \frac{1}{1 - aZ^{-1}} \quad (17)$$

$$H(z) = \frac{bZ^{-1}}{1 - a_1 Z^{-1} - a_2 Z^{-2}} \quad (18)$$

Comparing the Eq. (12) with (17) and expression (16) with (18), we get

$$a = e^{-\sigma T_s} \quad (19)$$

$$b = \frac{e^{-\sigma T_s}\mathrm{Sin}(\omega T_s)}{\omega} \quad (20)$$

$$a_1 = 2e^{-\sigma T_s}\mathrm{Cos}(\omega T_s) \quad (21)$$

$$a_2 = e^{-2\sigma T_s} \quad (22)$$

Here, a, b, $a1$, and a_2 are the filter coefficients.

Now, the circuit realizations from above Eqs. (17) and (18) are (Figs. 2 and 3).

We can derive the higher order of IIR filter by cascading of second order with first-order filter. Likely, we can implement third-order filter (Figs. 4 and 5).

2.3 Nth-Order Chebyshev Type-I LPF

By the help of second and first-order LPF circuit, we have obtained Nth-order Chebyshev LPF circuit. From Eq. (6), if order of filter is even, then these have only conjugate

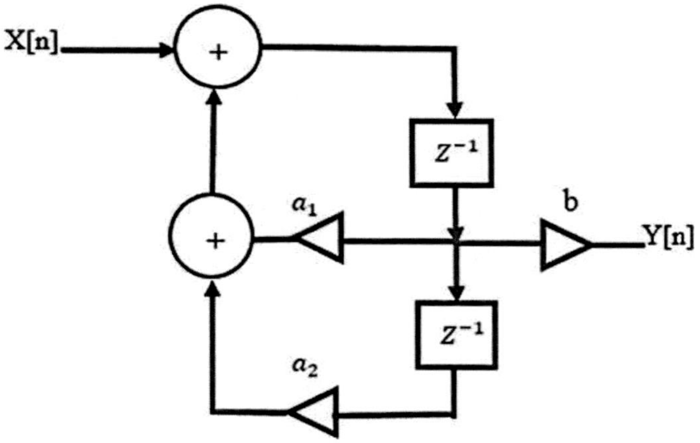

Fig. 3 Second-order IIR filter circuit

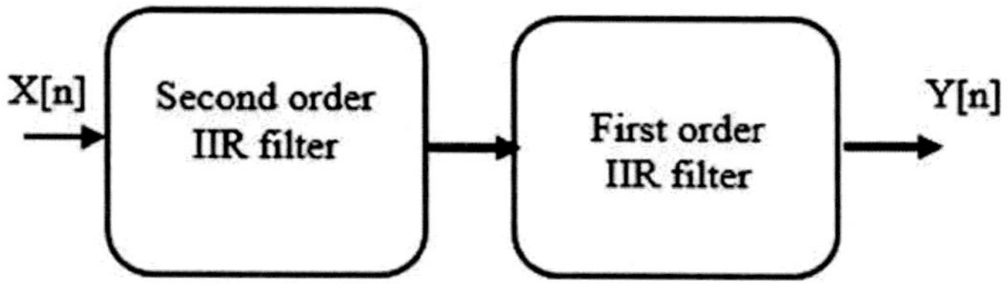

Fig. 4 Block diagram of third filter

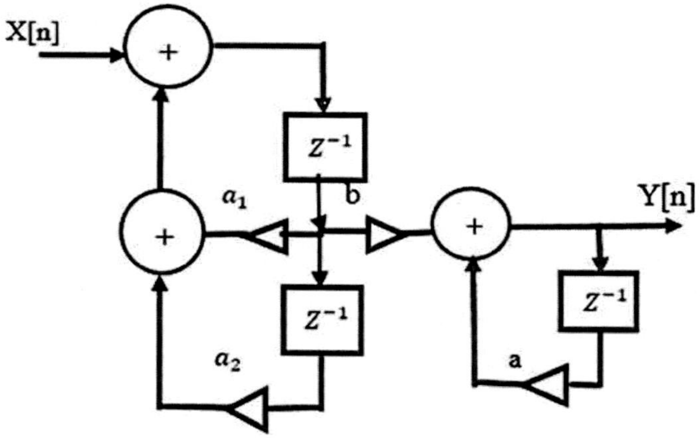

Fig. 5 Third-order filter IIR filter circuit

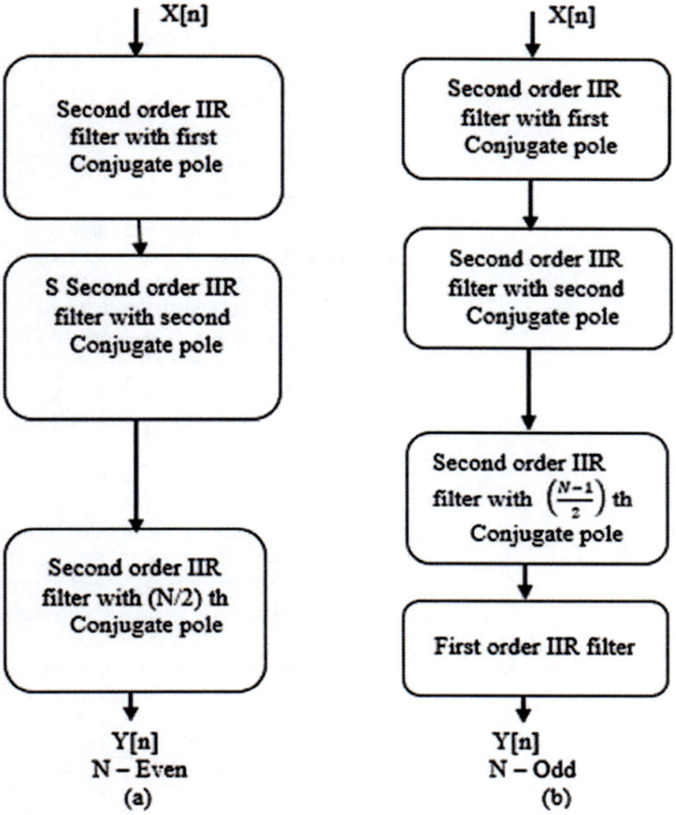

Fig. 6 Block diagram of *N*th-order Chebyshev LPF

pole; otherwise, they have conjugate pole with single real pole for odd order filter (Fig. 6).

3 Simulation and Results

In order to verify the theoretical aspect of *N*th-order Chebyshev type filter, circuits were simulated by using Xilinx software in Verilog hardware description language [4]. The simulation flowchart is given below in Fig. 7.

Here, we are considering the given specification [7].

Passband frequency $(\Omega_p) = 1$ rad.

Stopband frequency $(\Omega_s) = 5$ rad.

Passband ripple $(A_p) = 0.5$ dB.

Stopband attenuation $(A_s) = 30$ dB.

Fig. 7 Flowchart for simulation

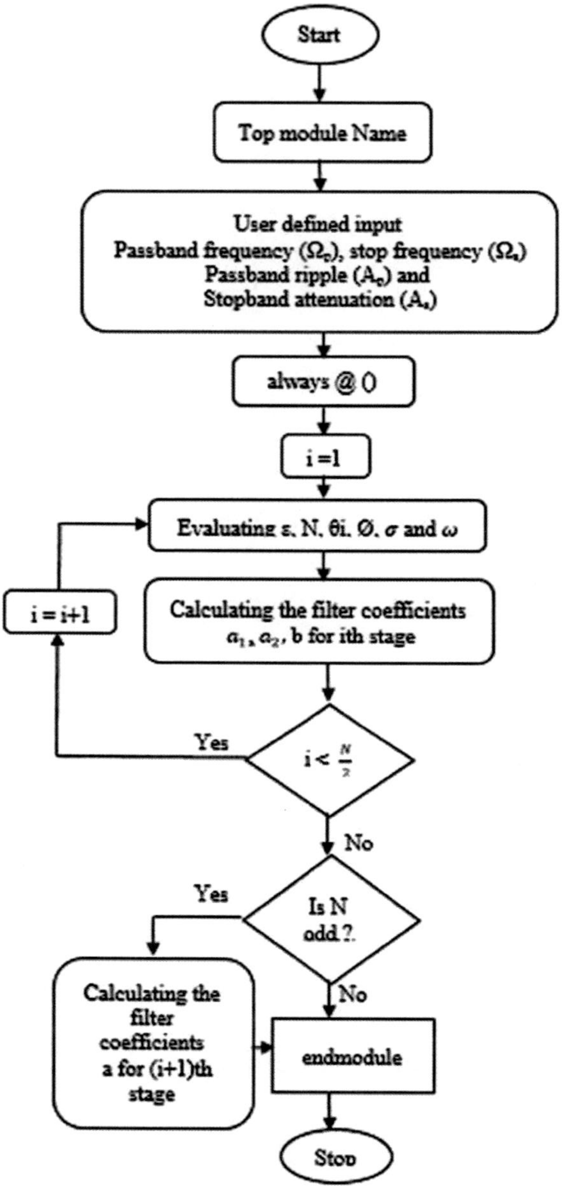

After the simulation, we get result in Fig. 8.

Figure 8 shows the simulation result outcomes of third Chebyshev type LPF circuit in Fig. 5. The simulation outcomes resemble well with the theoretical aspect. For $N = 6$, we have need of three 2nd-order IIR circuit from Fig. 6a. Suppose an example $N = 6, A_p = 1$ dB [9, 14–17]. Their coefficients are at different stages as shown in Fig. 9.

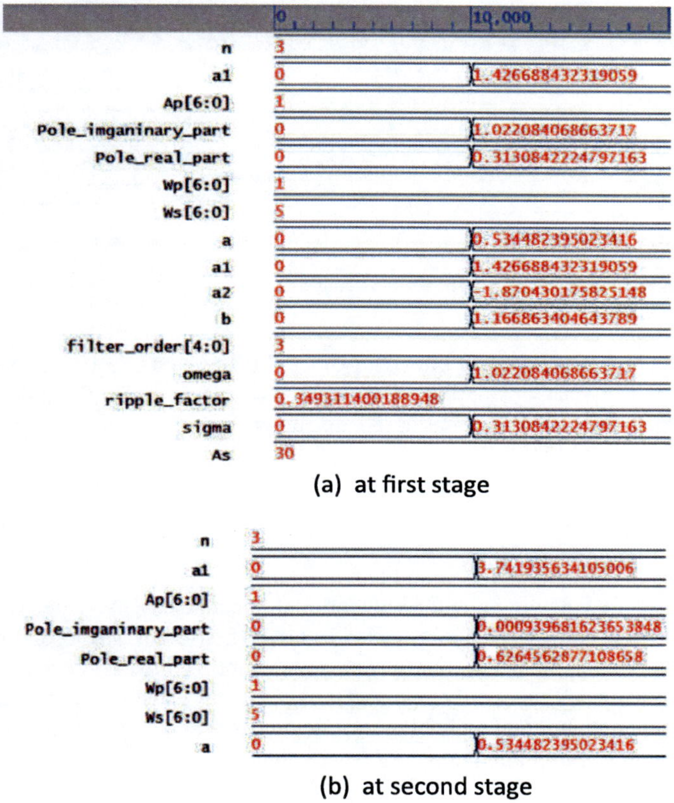

Fig. 8 Simulation result of given above specification

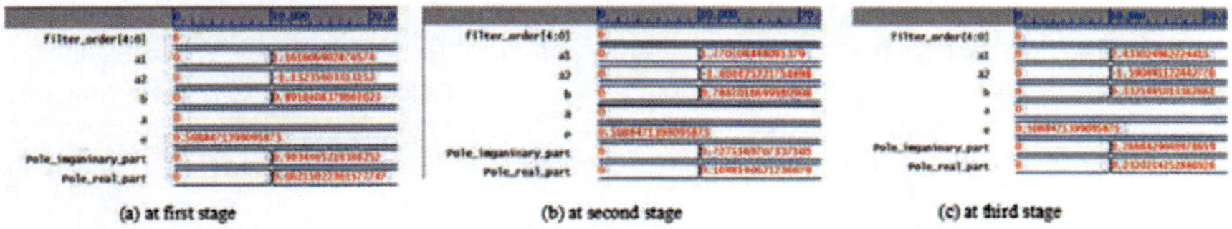

Fig. 9 6th-order filter coefficients at different stages

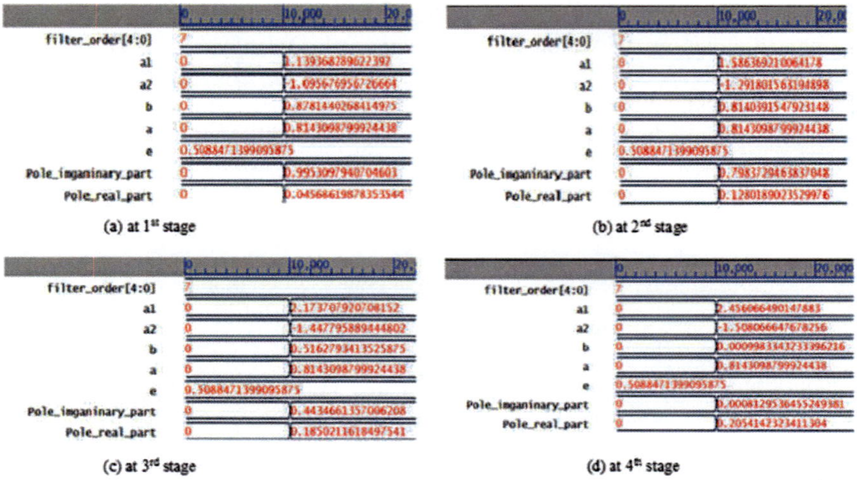

Fig. 10 7th-order filter coefficients at different stages

For $N = 7$, we have need of three 2nd-order IIR filter and one 1st-order IIR filter circuits from Fig. 6b. Considering an example $N = 7$, $A_p = 1$ dB [9], and their filter coefficients at different stages in Fig. 10 are shown.

4 Conclusions

In this paper, the calculation and result for coefficients of higher-order Chebyshev Type-I Low Pass Filter by designing of lower-order (first-order and second-order IIR filter) circuits are shown. They can be simply generalized for Nth order, and their realization has been done easily. We obtained optimum value of coefficients of the circuits for each of the stages in which simulation has been done on Xilinx in Verilog Hardware Description Languages.

References

1. Parks TW, Burrus CS (1987) Digital filter design. John Wiley & Sons, New York Chapter 7
2. Mitra SK (2001) Digital signal processing: a computer-base approach, 2nd Edition.Tata McGraw-Hill
3. Salivahanan S, et al (2010) Digital signal processing, 2nd Edn. Tata McGraw Hill Education Private Limited
4. Palnitkar S (2003) Verilog HDL a guide to digital design and synthesis, 3rd reprint, Pearson EducatonCypress Semiconductor, Addison-Wesley Publishing., Inc.

5. Saini MS, Kaur K (2015) Performance analysis of IIR filter design by using butterworth, chebyshev and cauer. Int J Adv Eng Res Sci (IJAERS). 2(3):29–36, ISSN: 2349-6495
6. Shenoi BA (2006) Introduction to digital signal processing and filter design. John Wiley & Sons, Inc.
7. Proakis JG, Manolakis DG (2002) Digital signal processing principles, algorithms and applications,3rd Edn. PrenticeHall
8. Chauhan RS, Arya SK (2010) Design of IIR digital filter using analog to digital mapping. J Neural Comput Syst 3(Jan. 2010):51–55
9. Tan L, Jiang J (2013) Digital Signal Processing (Second Edition)
10. Zhang S, Zhu L (2012) General synthesis method for symmetrical even-order Chebyshev bandpass filter. In: 2012 Asia Pacific microwave conference proceedings, Kaohsiung, 2012, pp 667–669. https://doi.org/10.1109/APMC.2012.6421697
11. Milosavljevic ZD (2005) Design of generalized Chebyshev filters with asymmetrically located transmission zeros. IEEE Trans Microw Theory Tech 53(7):2411–2415. https://doi.org/10.1109/TMTT.2005.850430
12. Konopacki J (2004) D03: IIR filters design procedures based on digital frequency transformations. In: IFAC proceedings volumes, vol 37, no 20, pp 361–365, ISSN 1474-6670
13. Cameron RJ (1932) Fast generation of Chebyshev filter prototypes with asymmetrically-prescribed transmission zeros. ESAJ. 6:83–95
14. Milic L, Saramäki T (2002) Complementary IIR filter pairs with an adjustable crossover frequency. In: Proceeding IEEE Nordic signal processing symposium. Hurtigruten, Norway
15. Alseyab SA (1982) A novel class of generalized chebyshev low-pass prototype for suspended substrate stripline filters. IEEE Trans Microw Theory Tech 30(9):1341–1347
16. Dutta Roy SC (2010) A new Chebyshev-like low-pass filter approximation. Springer Science+ Business Media, LLC2010, Circuit Syst. Signal process vol 29, pp 629–636
17. Wongprommoon N, Tiamsuphat A, Prommee P (2020) Low-complexity Chebyshev high-pass filter based on OTA-C. In: 2020 43rd international conference on telecommunications and signal processing (TSP). Milan, Italy, pp 373–376. https://doi.org/10.1109/TSP49548.2020.9163437

Analog and Digital Applications of 4-T Based Memristor Emulator

Semba Walli, Jyoti Garg, and Subodh Wairya

Abstract Since the postulation of memristor, numerous memristor emulator networks have been proposed using linear/non-linear elements with active and passive devices to emulate the behavior of memristors for practical applications. Some unique characteristics of memristors are—pinched hysteresis I-V curve, non-volatility, frequency dependent, and dependent on the direction of current flow. Moreover, its ability to overcome the size limitations set by Moore's law makes it an exciting component with many desirable characteristics. In this paper, analog and digital applications of memristor are proposed using a memristive emulator circuit that satisfactorily emulates the behavior of a memristor without employing any complex circuit elements. The emulator circuit referenced is 4-T-based circuit that is used as a fundamental block in the proposed designs. Logic gates, half adder, and full adder are realized in digital applications and Relaxation Oscillator in analog application, respectively. The power dissipation of the circuits are also analyzed. All the proposed digital circuits are simulated in Cadence Virtuoso ADE using gpdk 45 nm technology and analog application using gpdk 180 nm technology.

Keywords MOSFET-based memristor emulators · A/D applications · Logic gates and adders using memristors

1 Introduction

Memristor was first postulated in 1971 by Leon Chua [1], but their popularity increased significantly in the recent decades after the physical implementation of

S. Walli (✉)
Institute of Engineering and Technology, Lucknow 226021, India
e-mail: sembaswami@gmail.com

J. Garg
Dr. A.P.J Abdul Kalam Technical University, Lucknow 226031, India

S. Wairya
Institute of Engineering and Technology, Lucknow 226021, India

B. Mishra and M. Tiwari (eds.), *VLSI, Microwave and Wireless Technologies*, Lecture Notes in Electrical Engineering 877,
https://doi.org/10.1007/978-981-19-0312-0_19

thin film TiO_2 Memristor [2], achieved by H.P. Labs in 2008. Due to its ability to overcome the size limitations set by Moore's Law and high data storage capacities, it is gaining immense popularity.

As memristors were commercially unavailable, so many device models [4, 5] and emulator circuits [6–10] were proposed to emulate its characteristics. The application domain [7–10] of memristor has also widened significantly with the memristor's fabrication process becoming viable. So many analog and digital applications are being proposed due to its increasing usage in various fields.

Earlier active elements like CCIIs (Second Current Conveyors), OTA (Operational Transconductance Amplifier), Multiplier circuits [8], CFOAs (Current Feedback Op-Amp), DDCC (Differential Difference Current Conveyor), etc., were used to achieve the non-linear behavior of memristors. CCIIs, CFOAs [9], OTAs [9] generally employed op-amps and passive elements, and DDCC [7] circuits involved large no. of CMOS transistors which resulted major drawbacks like complex circuitry, high no. of components, large size, and fabrication difficulties. In this paper, a MOSFET-based memristor emulator circuit [3] is used as a fundamental component in the designing of logic gates, adders, and oscillator circuits. The most important advantage of this emulator circuit is that it uses only four MOSFETS with no additional active or passive elements in its circuitry.

2 Reference Circuit

The MOSFET-based memristor circuit [3] shown in Fig. 1 has four transistors, 3-PMOS and 1-NMOS, used to emulate the memristive behavior. Transistor M_4 acts

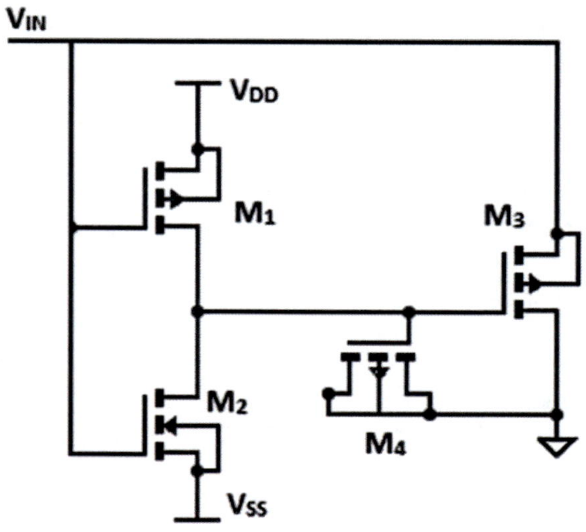

Fig. 1 MOSFET-based memristor emulator circuit

a capacitor when its drain–source and bulk terminals are connected to the ground. Drain currents of M_1 and M_2 transistors create a voltage on the gate terminals of M_4 and drive M_3 transistor. The memductance and memristance at the input port are given by Eqs. 1 and 2, respectively.

$$W(\Phi_{\mathrm{IN}}) = \frac{I_{\mathrm{IN}}}{V_{\mathrm{IN}}} = -\mu_P C_{\mathrm{ox}}\left(\frac{W}{L}\right)_3 V_{\mathrm{TP}} C_{\mathrm{ox}}\left(\frac{W}{L}\right)_3 \frac{g_{m12}\Phi_{\mathrm{IN}}}{C} \tag{1}$$

$$M(q) = \frac{1}{W} = \frac{V_{\mathrm{IN}}}{I_{\mathrm{IN}}} \tag{2}$$

3 Proposed Models

3.1 Logic Gates

NOT, AND, and NAND logic gate configurations are proposed in Figs. 2, 3, 4 and 5. Figure 2a shows the proposed circuit of not gate using single memristor of Fig. 1. As the emulator network uses a simple transconductance stage consisting of PMOS and NMOS at its input side, it acts as an inverter.

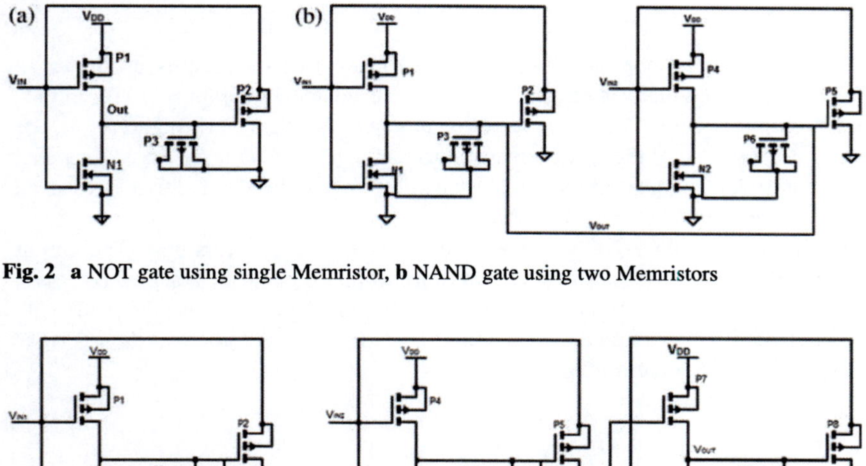

Fig. 2 **a** NOT gate using single Memristor, **b** NAND gate using two Memristors

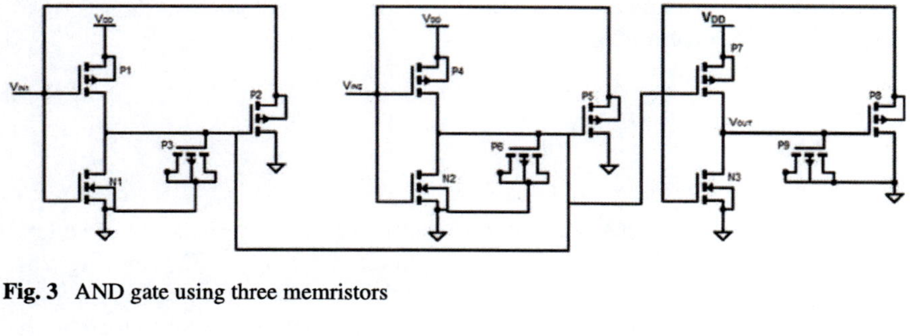

Fig. 3 AND gate using three memristors

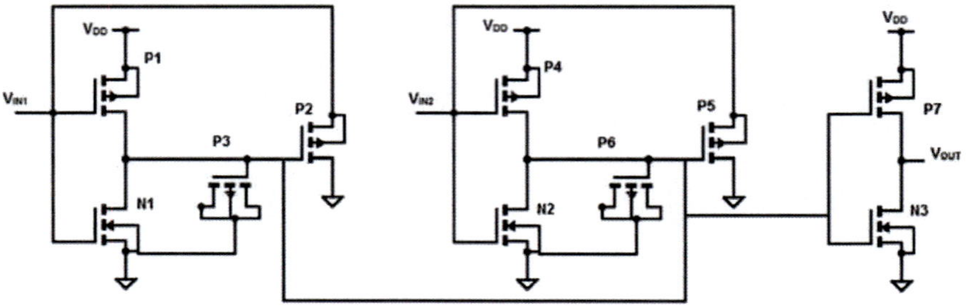

Fig. 4 AND gate using two memristors and CMOS Inverter

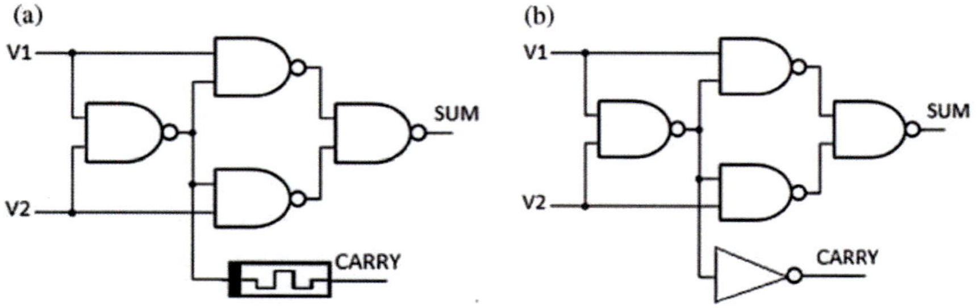

Fig. 5 **a** Half Adder using 4 Memristor-based NAND gates, **b** Half Adder using Memristor-based NAND gates and CMOS Inverter

The NAND gate configuration is realized using two memristors as shown in Fig. 2b. As the emulator circuit itself acts as an inverter, for inputs (0, 0) and (1, 1), the output is 1 and 0, respectively. For input combinations (1, 0) and (0, 1), the output is 1 because of the charge sharing principle. The PMOS, P_3, and P_6 act as a capacitor, so the charge present in either one of the capacitor due to input 1 is shared with the other capacitor that is discharged due to input 0.

Figure 3 shows the AND gate realization based on memristors only. The output of the NAND gate of Fig. 2b acts as an input to the third memristor; hence, AND logic is obtained at the output of the third memristor. Figure 4 shows the AND gate configuration using memristors in combination with a CMOS inverter to show CMOS compatibility of the memristor.

3.2 *Adders*

Memristors-based NAND gates are used in the realization of proposed half adders and full adders as depicted in Figs. 5a, b and 6. In Fig. 5a four NAND gates (Fig. 2b) and a memristor are used to realize the half adder functions. Figure 5b uses four memristor-based NAND gates and an inverter in place of a memristor to realize the

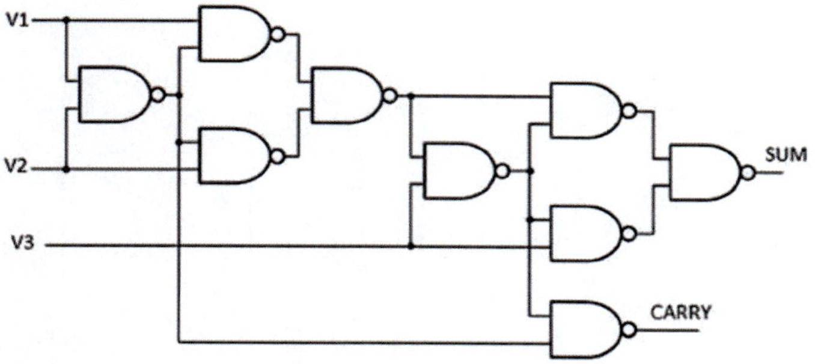

Fig. 6 Full Adder using memristor-based NAND gates

half adder functions. Full adder functions is realized using 9 memristor-based NAND gates as shown in Fig. 6.

3.3 *Relaxation Oscillator*

A relaxation oscillator is a non-sinusoidal oscillator circuit consisting of an energy-storing element (capacitor/an inductor) and a non-linear switching device element in a feedback loop. The switching device periodically charges and discharges the energy stored in the storage element, thus causing abrupt changes in the output waveform. The frequency of a Relaxation Oscillator is given as:

$$f = \frac{1}{2 \times R_1 \times C_1 \times \ln\left(\frac{1+\eta}{1-\eta}\right)}, \quad (3)$$

$$\eta = \frac{R_2}{R_2 + R_3} \quad (4)$$

The memristor-based relaxation oscillator circuit is shown in Fig. 7b that uses a memristor in place of resistor, R_2 of Fig. 7a. Here, the memristor acts as a linear resistor and enhances the output frequency of the relaxation oscillator. R_1 or R_3 can also be replaced instead of R_2 to obtain oscillations of different frequencies.

4 Simulation Results

All the proposed models of Sect. 3 have been modeled and simulated in Cadence Virtuoso Analog Design Environment using gpdk 45 nm technology and gpdk 180 nm technology for digital and analog circuits, respectively. (Figs. 8, 9, 10, 11, 12, 13, 14, 15 and 16)

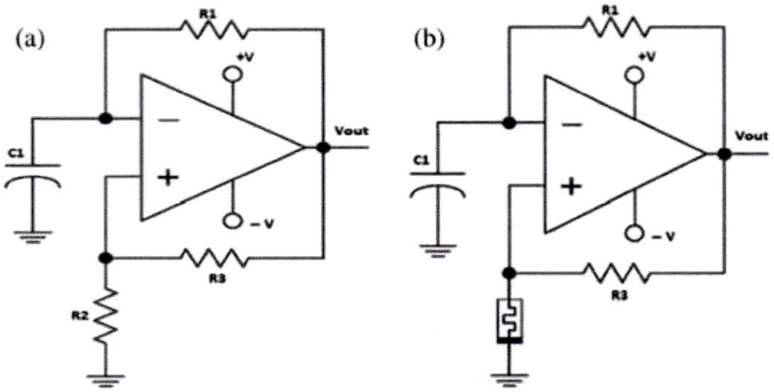

Fig. 7 **a** Conventional RC-based relaxation oscillator, **b** Proposed relaxation oscillator using memristor emulator circuit

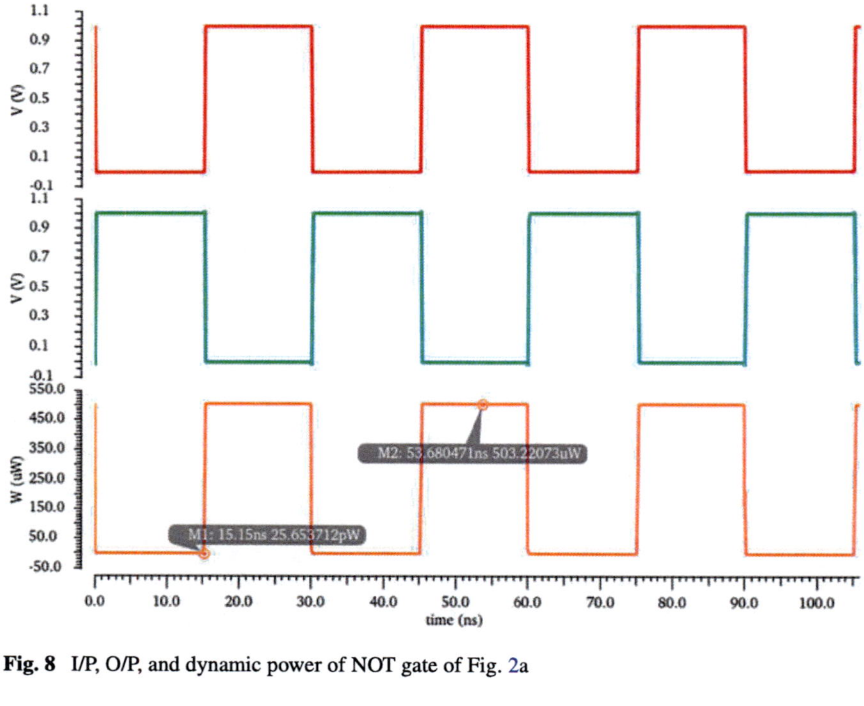

Fig. 8 I/P, O/P, and dynamic power of NOT gate of Fig. 2a

If $R_1 = 5\ \text{k}\Omega$, $R_2 = 11\ \text{k}\Omega$, $R_3 = 10\ \text{k}\Omega$ and $C = 100$ nF and $V_{CC} = \pm 12$ V, the theoretical frequency is 859.733 Hz with a time period of 1.163 ms. However, the output frequency from Fig. 15 is seen to be 846.543 Hz with a time period of almost ≈ 1.18 ms.

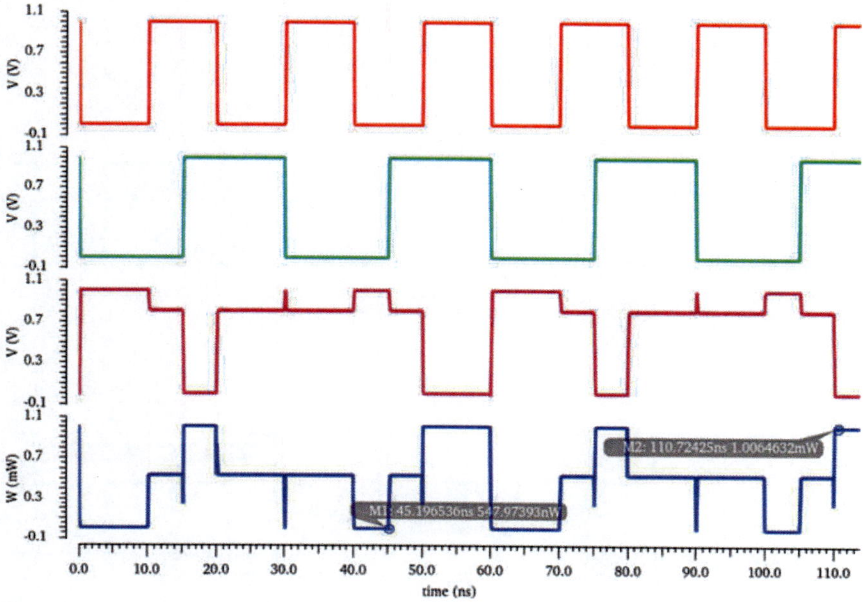

Fig. 9 I/P, O/P, and dynamic power of NAND gate of Fig. 2b

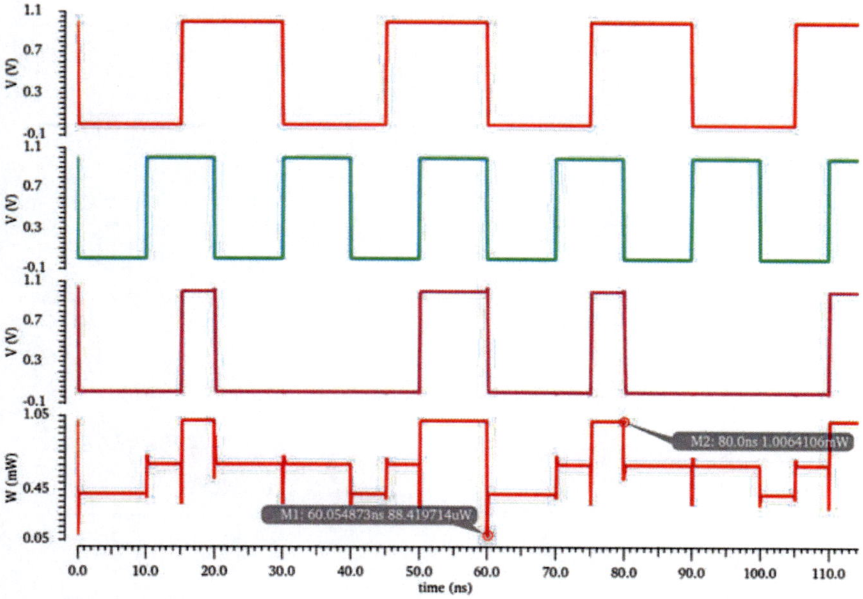

Fig. 10 I/P, O/P, and dynamic power of memristor-based AND gate (Fig. 3)

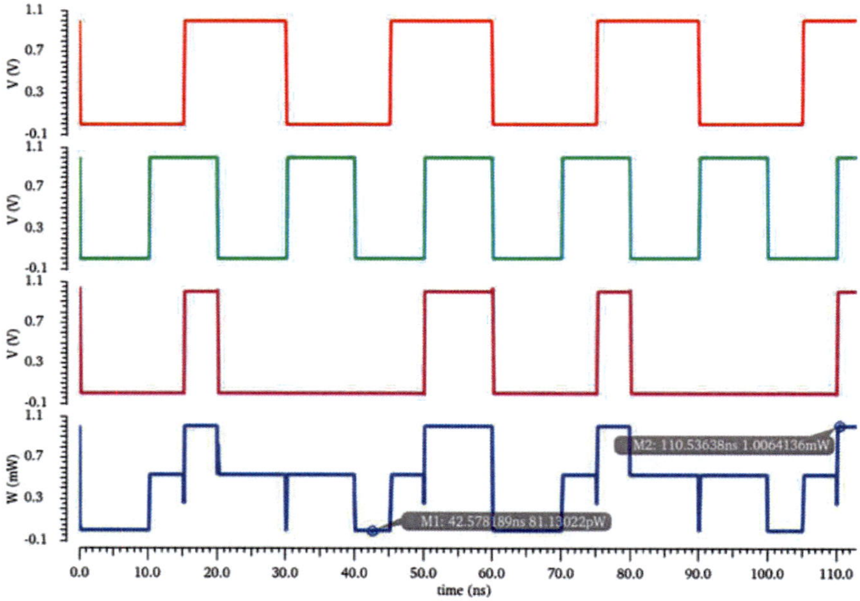

Fig. 11 I/P, O/P, and dynamic power of AND gate using memristors and CMOS inverter (Fig. 4)

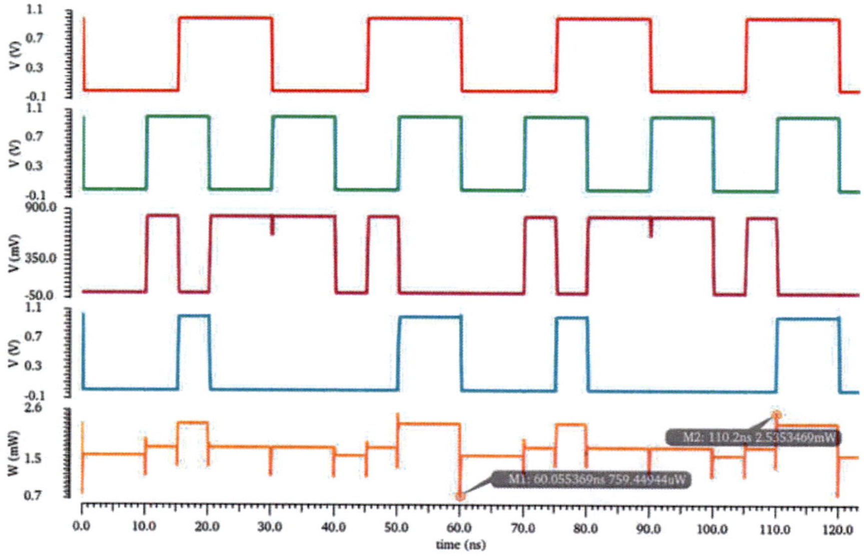

Fig. 12 I/Ps, sum, carry, and dynamic power of memristor-based half adder circuit (Fig. 5a)

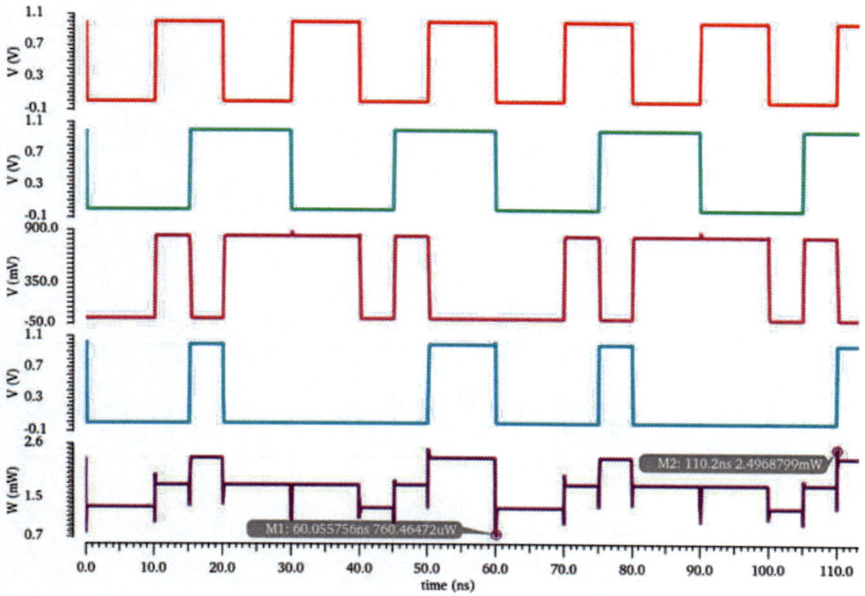

Fig. 13 I/Ps, sum, carry O/P, and dynamic power of half adder using memristors and CMOS Inverter (Fig. 5b)

5 Results and Discussion

The dynamic power, static power, and the average dynamic power dissipated by the digital circuits is shown in Table 1. The power delay product (PDP) that acts as figure of merit (FOM) for any digital circuit is also calculated and depicted in the table below. Calculated PDP is in the range of fJ–pJ for the logic gates and adder circuits proposed which is comparable with other low power digital circuit realization techniques.

In Table 2, it can be seen that there is a difference in the output and input frequencies of conventional RC-based relaxation oscillator, whereas the memristor-based relaxation oscillator has no such errors. Input and output frequencies are perfectly matched. In addition to that, the average dynamic and static power dissipated is 79 and 80% less when compared to the Conventional RC-based Oscillator.

6 Conclusion

The performance of the analog and digital circuits designed using memristor is better than the conventional circuits and has the advantage of low power requirements. The power consumption of the proposed digital circuits is comparable with

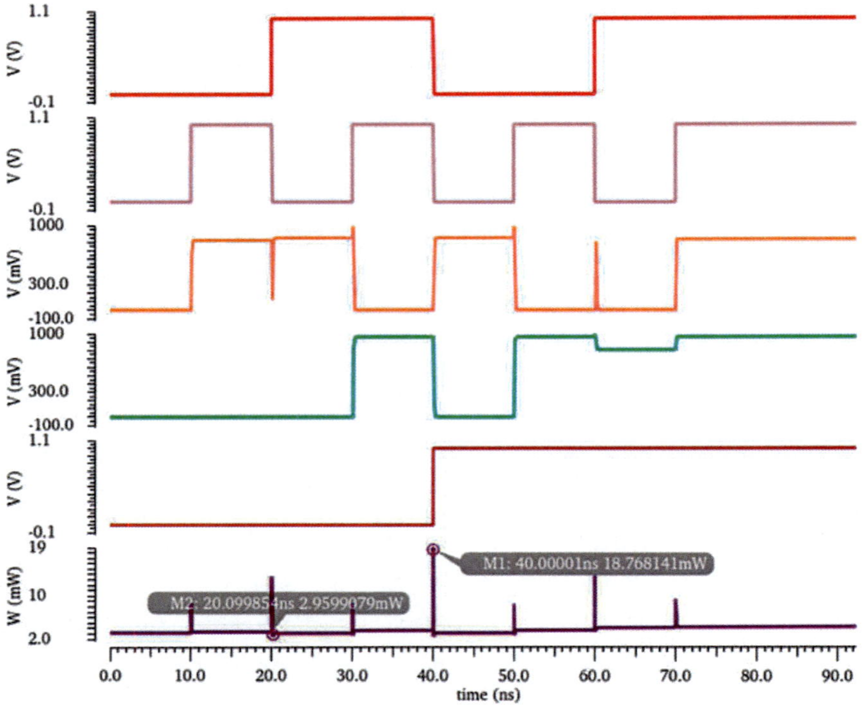

Fig. 14 I/Ps, sum, carry O/P, and dynamic power of full adder circuit using memristors-based NAND gates (Fig. 6)

other low power techniques used for digital circuit realizations. The performance of the memristor-based relaxation oscillator is much better than conventional resistor-based relaxation oscillator as there is no difference between the capacitor's charging discharging frequency and the output frequency. The frequency of the output oscillations is also enhanced by 66%. The power requirement for the memristor-based relaxation oscillator is also very low. The average dynamic power is 79%, and the static power is 80% less than the conventional relaxation oscillator.

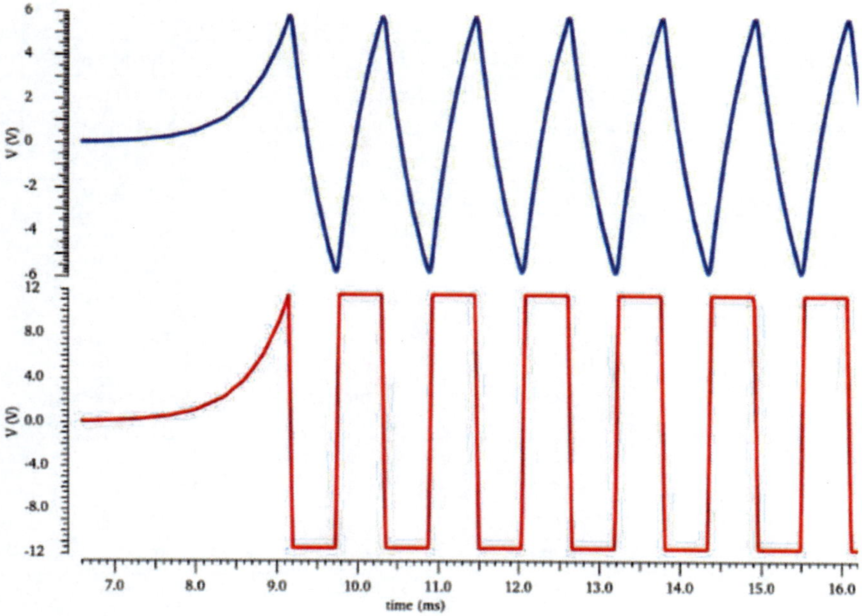

Fig. 15 V across capacitor and V_{out} of conventional RC-based relaxation oscillator shown in Fig. 7a

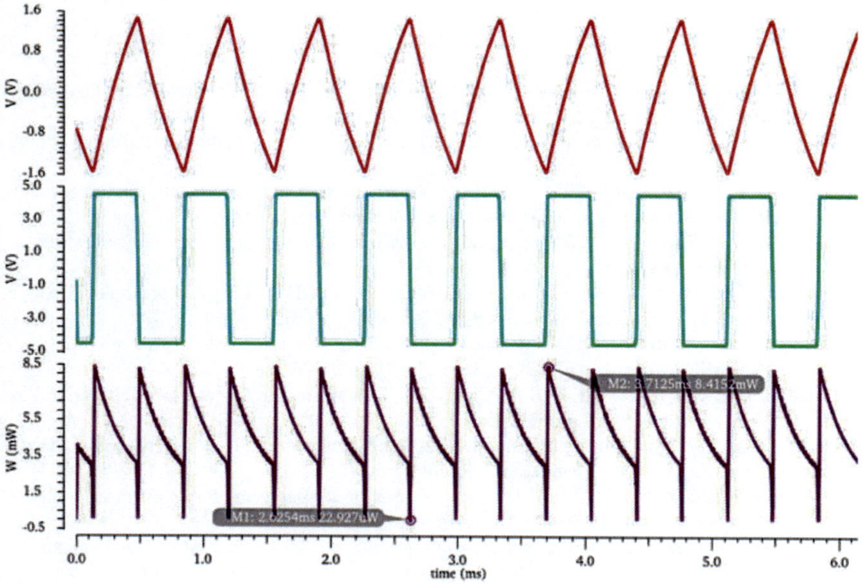

Fig. 16 V_c across capacitor, V_{out}, and dynamic power of memristor-based relaxation oscillator shown in Fig. 7b

Table 1 Average dynamic power, delay, PDP, and static power for digital circuits

Circuits	Avg. Dyn. power	Delay	PDP	Static power
NOT	234.8 uW	44.51 ps	10.45 fJ	503.2 uW
NAND	496.8 uW	45.12 ps	22.42 fJ	1.01 mW
AND	668.5 uW	93.29 ps	62.37 fJ	1.01 mW
AND using Inverter	496.8 uW	58.24 ps	28.93 fJ	1.01 mW
Half Adder	1.864 mW	10.17 ns	18.96 pJ	2.31 mW
H. A. using Inverter	1.766 mW	10.15 ns	17.97 pJ	2.31 mW
Full Adder	3.968 mW	10.08 ns	40 pJ	2.88 mW

Table 2 Comparison between conventional and memristor-based relaxation oscillators

Parameters	Conventional Rel. Osc	Memristor-based Rel. Osc
I/P Time Period	1.192 ms	0.714 ms
O/P Time Period	1.181 ms	0.714 ms
I/P Frequency	839.162 Hz	1.401 kHz
O/P Frequency	846.543 Hz	1.401 kHz
Avg. Dyn. Power	23.984 mW	4.935 mW
Static Power	1.342 mW	273.2 uW

References

1. Chua LO (1971) Memristor—the missing circuit element. IEEE Trans Circuit Theory 5(18):507–519
2. Strukov DB, Snider GS, Stewart DR, Williams RS (2008) The missing memristor found. Nature 453:80–83
3. Babacan Y, Yesil A, Gul F (2018) The Fabrication and MOSFET-only circuit implementation of semiconductor memristor. IEEE Trans Electron Devices 4(65):1625–1632
4. Strukov DB, Williams RS (2009) Exponential ionic drift: fast switching and low volatility of thin-film memristors. Appl Phys A 94:515–519
5. Kvatinsky S, Friedman EG, Kolodny A, Weiser UC (2013) TEAM: threshold adaptive memristor model. IEEE Trans Circuits Syst I Regul Pap 1(60):211–221
6. Pershin Y, Di Ventra M (2009) Practical approach to programmable analog circuits with memristors. IEEE Trans Circuits Syst I Regul Pap 8(57):1857–1864
7. Yesil A, Babacan Y, Kacar F (2014) A new DDCC based memristor emulator circuit and its application. Microelectron J 3(45):282–287
8. Kim H, Sah MP, Yang C, Cho S, Chua LO (2012) Memristor emulator for memristor circuit applications. IEEE Trans Circuits Syst I Regul Pap 10(59):2422–2431
9. Abuelma'atti MT, Khalifa ZJ (2014) A new memristor emulator and its application in digital modulation. Analog Integr Circuits Signal Process 80:577–584
10. Yu D, Iu HC, Fitch AL, Liang Y (2014) A floating memristor emulator based relaxation oscillator. IEEE Trans Circuits Syst Part I: Regular Papers 10(61):2888–2896

Perfomance Analysis of Text Extraction from Complex Degraded Image Using Fusion of DNN, Steganography, and AGSO

Digvijay Pandey and Subodh Wairya

Abstract In present time, any complex degraded image consists of very important and confidential information details which is recognized as a non-textual image and textual information. Due to diversity of text style in image, complicated background, and various interference factors makes detection of text (DOT) from complex degraded image as a field of research. The secure and accurate text in a complex degraded image is found useful for the audience to understand the complete situation. So, a fusion of DNN, adaptive galactic swarm optimization (AGSO), and steganography are applied in this proposed technique to securely, efficiently identify information in the form of text and thereafter, to recognize each character from degraded complex images. In general, images are affected by different type of noise such as structured noise, Poisson–Gaussian noise, periodic noise, and impulse valued noise, and to discard it in the initial preprocessing phase, the guided filter (GF) is used. A very important task in the text identification and recognition process is feature extraction, performed by using Gabor and stroke width transform. The extracted features of the image are required during the classification process. Thereafter, text identification and recognition is done by WNBA. Subsequently, performance comparison of various performance parameters such as precision, F1-scores, and recall was tested using the IIIT5K database for proposed algorithm along with other existing techniques.

Keywords Detection of text (DOT) · Complex degraded images · Steganography

1 Introduction

Complex degraded image document is recognized as an online and offline accessible and important precious media, which contains vital useful information. This image consists of pixels; after that the important information from complex degraded images

D. Pandey (✉) · S. Wairya
Department of Electronics Engineering, Institute of Engineering and Technology, Dr. A.P.J. Abdul Kalam Technical University, Lucknow, India
e-mail: digit11011989@gmail.com

B. Mishra and M. Tiwari (eds.), *VLSI, Microwave and Wireless Technologies*,
Lecture Notes in Electrical Engineering 877,
https://doi.org/10.1007/978-981-19-0312-0_20

are fetched as per on the requirement of computer vision [1]. Textual messages from complex videos and complex images contain accurate meaningful information, same as the textual messages available in images are used by many complex image learning and indulgent implementation like language translator, digitization of book, and recovery of video or image [2, 3]. In latest trend, the use of pre-program text detection process followed by text recognition has received an enormous demand. In past, a lot of research processes are available on removal of text from complex contemplate scene, while this text acquisition process is recognized as one of the important portions of optical way of character recognition (OCR) [1, 4, 5]. OCR commercial is used after performing the extraction of text and further quantization step to recognize the textual information from complex degraded image [6].

Scene image offers suitable and accurate information for blind direction-finding, scene perceptive, and recovery approaches, respectively [7, 8]. A diversity of fonts and other properties are often inbuilt in this complex image [9]. In a complex degraded image, the main focus may include distinct designed characters, information exposed in digital signpost information displayed on monitor. This is a very typical task for the traditional OCR to identify the textual information with different appearances. In this degraded image, the texts are sprinkled regularly, and the preceding information concerning their position is not presented. The input documents from the camera recognize the line spacing number, character, but the complex image text does not comprise any formatting rules, so it is not possible to directly introduce the segmentation approach for complex image [10]. In the complex image, the process of text retrieval from image is precisely predictable by the features of the image. The vision to accomplish high exactness is vastly improved by the heterogeneous features of the image [11]. The process of text detection from complex degraded images contains two main steps; they are text detection, and another is recognition [12, 13]. The key idea of these aforementioned processes in complex degraded scene is to identify and position the textual information available in complex video or complex image. This text extraction procedure is executed in many applications like fetching of images and videos from the database management system (DBMS), data processing multimedia retrieval, understanding the natural complex degraded image connotation, and monitoring of traffic [14]. The text extraction procedure is assessed into three main techniques. They are based on connected component, texture detection and edge detection [15]. The connected component-based method does the quantization of color in image and region expansion to make clusters of nearest pixels, having the same colors in the connected components. An entire structure of every character is not held in reserve by the connected components because of color bleeding and very low contrast available in the each row of textual data. As a result of this, method which is based on connected components is set up to be not suitable for image frames obtained from video. Therefore, the approach based on edge detection is generally replaced by a texture-based (TB) method. In texture-based method, unique text plays a key role to identify the text region. These methods then apply different transforms such as fast Fourier transform (FFT) and Gabor transform (GT) to obtain the features, and these methods will be given to the classifiers for fast neural networks (FNNs) and support vector machines [16]. In this, [17] presents a new scheme for compression of image in both time and frequency domain, and it includes wavelet transform to identify a sub-band of an image and then further decompose it into certain levels. For encoding, the wavelet along with noise shaping bit allocation method was used that might further consider details of less-visible images available at high resolution.

The vital procedure that performed by the complex degraded image text identification is extraction region of candidate character available in image. The most important purpose of this region classification methodizes to obtain the non-text parts from the extracted CCRs [18]. Both the recognition of text and removal of text from the image are considered to be basic and effective for a variety of text identification-based applications. The two processes discussed here are having a difficult obligation due to the presence of some factors like variations of text in scene, complexity of background, and interference factors (Non-uniform illumination).

Some recently developed methods achieve a high accuracy for text extraction, but attaining satisfactory results, such degraded images, are still considered as a challenging issue [19].

The main work of extraction of text process from image do text extraction is mostly used in diverse applications in real-time application. The texts [20], which are removed from complex degraded images, are usually utilized for obtaining the details for various application such as prediction of the car parking and application for visually impaired individuals. The secure and accurate text in complex degraded images is found useful for the audience to understand the complete situation as result fusion of DNN, adaptive galactic swarm optimization (AGSO), and steganography is main motivation for this paper.

2 Review of Literature

A detection of the multi-oriented text in a fused form from complex degraded images by rising a new region-based CNN. A new advance method for text recognition [21] and text detection was a connected component method-based approach that utilized the maximally stable extremely regions. The multiple blur produced by motion and defocus makes the text detection process a challenging one. In this [22], a method for text identification process in distorted or non-distorted images was discussed, and the contrast variants experienced in nearby pixels were identified in this method to evaluate the blur degree; moreover, the low-pass filter was used for deblurring. Mostly, this approach gave pixels under consideration for the purpose of deblurring images. The process of detecting the scene text from videos attained high value in various content removals-oriented video applications like recovery of video as well as investigation. In this [23], a text tracking and recognition approach for frames of videos was given. The public scene text video was included in this method which outperformed the other existing methods.

In [24] an image compression algorithm, containing the property of embedded code. The embedded code denotes a chain of binary decisions that help to discriminate a meaningful image from the "null" image. Additionally, to give a fully bit stream of embedded code, EZW results that are good enough to perform comparison with known compression algorithms are based on standard test images.

Image steganography conceals the confidential message within the digital image. It mainly aims to secure the hidden message from the attackers. In [25], a new IHED (Image hiding encryption and decryption) was developed for image encryption and decryption. The mid-frequency (MF) values were then identified using the MSABM model. With such values, the encoding process was accomplished. Finally, a few attacks like chi-square attack, visual attack, white floor square attack, and RS analysis of complex images were applied for validating the IHED method.

Data security was achieved using a steganography and cryptography approach. Therefore, security was considered the major role in data hiding, which improves data confidentiality. In [26], a 3-bit least significant bit was developed for secret images within the cover image. Then, visual cryptography was used to transfer the secure scene over the Internet. The usage of ECC makes the data extraction tough for various attackers.

3 Proposed Methodology

The proposed approach works well for those with large sizes and attains better performance for color image (see Fig. 1). Initially, a complex degraded image [27] which is affected by noise is given to filter the image for denoising and contrast improvement [28, 29], and then, image is quantized with the MCWA algorithm. Next step is features extraction using latest techniques such as Gabor transform and SWT [16]. After that these features are given to the weighted Naïve Bayes type of classifier for identifying textual and non-textual images. In which, the error will be optimized with the Emperor Penguin Optimization (EPO) algorithm. After identifying the textual and non-textual portions, finally, in the distinct word provided to the deep neural network with AGSO algorithm and steganography. Steganography is used in the final step for secure transmission of text from one place to another place.

4 Result and Discussion

In result and discussion section, performance evaluation of parameters and comparative analysis is performed.

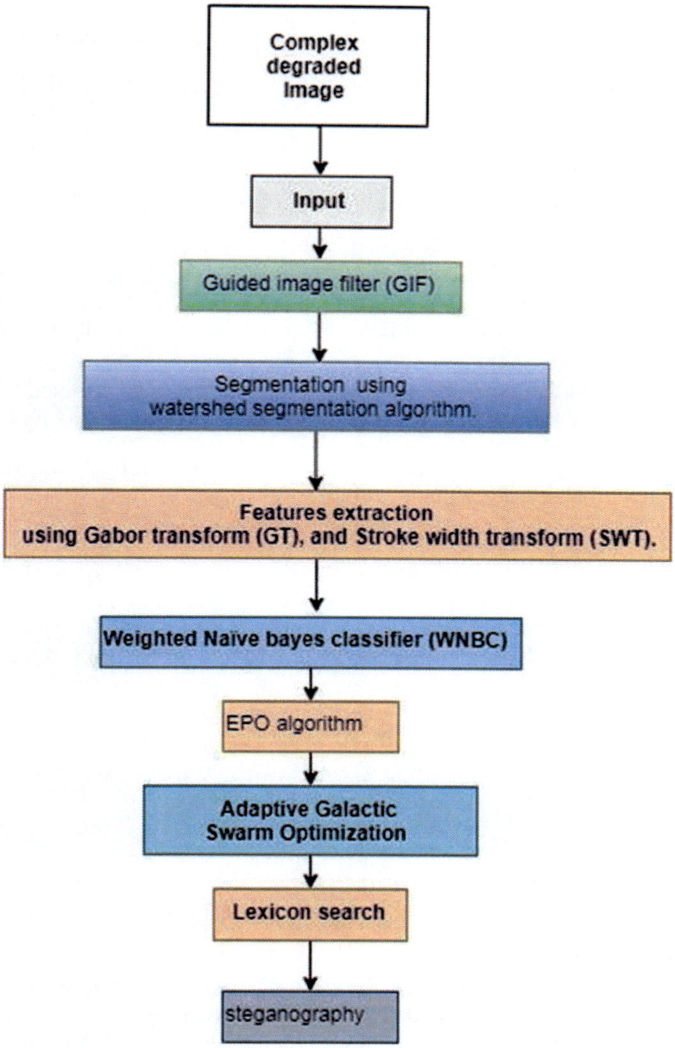

Fig. 1 Flow diagram of proposed algorithm

4.1 Performance Evaluation of Parameters

Different performance evaluation parameters are used for checking validity of proposed algorithms for text detection from complex degraded image. The performance parameters are determined using TP, TN, FP, and FN.

- The text part that is correctly identified as text is determined by TP.

Table 1 Comparative table

Methods	Precision (%)	Recall (%)	F1-score (%)
Proposed	92.89	97.2	95.4
Khlif [17]	89.94	82.28	85.94
Zhu [30]	83	84	84
Zhang [31]	78	88	83
R-FCN [32]	90	76	83
FasterR-CNN [33]	86	75	80

- The text part that is incorrectly identified as the non-text part is determined by FN.
- The non-text part that is correctly identified as non-text is determined by TN.
- The non-text part that is incorrectly identified as text is determined by FP.

Three following performance parameters are used for comparison of existing algorithms and proposed algorithms

$$p(\text{Precision}) = \frac{\text{TP}}{\text{TP} + \text{FP}} \tag{1}$$

$$r(\text{Recall}) = \frac{\text{TP}}{\text{TN} + \text{FP}} \tag{2}$$

$$F_1(\text{F1 - score}) = 2 \times \frac{r \times p}{r + p} \tag{3}$$

4.2 *Comparative Analysis*

The performance parameters of proposed algorithms and previous text detection techniques are used for comparative analysis. The value of various performance parameters like precision, recall, and F1-score of this proposed method is 2.95, 14.92, and 9.46% better than Khilf method, as long as this method is considered as better than other existing one (see Table 1). The proposed algorithm performance parameter shows that this proposed method performs more accurate detection of text from complex degraded scenes than the previous techniques (see Fig. 2).

5 Conclusion

In this work, deep learning-based (DL) optimization technique is deployed for text detection from the complex degraded image. To perform this, a filter such as a guided

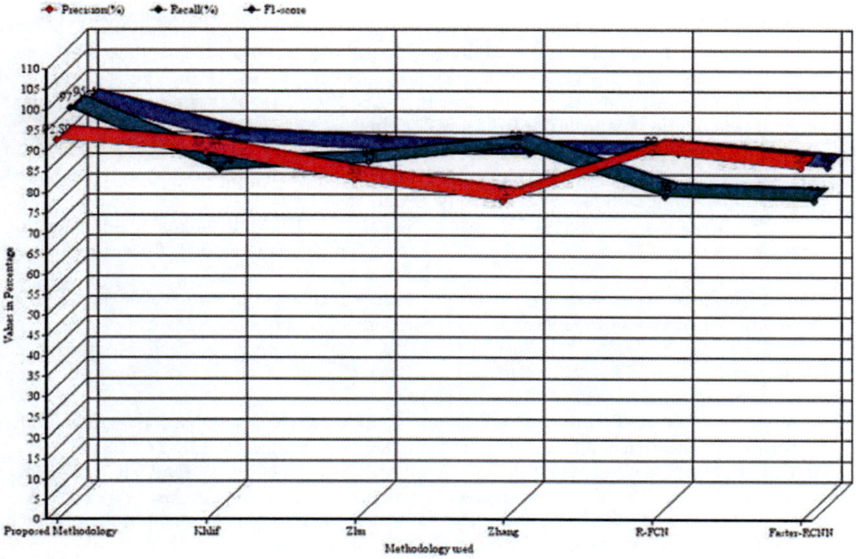

Fig. 2 Comparative analysis precision, recall, and F1-score

image filter is used during an initial preprocessing step to enhance the contrast of the input complex degraded image. Hereinafter, marker-watershed segmentation, SWT, GT, and WNBC techniques are used for segmentation, features extraction, and detection of textual and non-textual part from complex degraded image. In last, a combination of DNN, AGSO, and stenography are used for recognized classified text part. Afterward, performance comparison of various performance parameters such as F1-scores, precision, and recall was performed using the IIIT5K database for proposed algorithms along with previous existing algorithms. The proposed method gives accurate results many times. But, occasionally, the text or characters may be lost, or the analogous character is often extracted at various times which can cause the extraction of incorrect textual data from the natural scene. Therefore, some special techniques must be implemented for secure text detection, recognition, and transmission in future.

References

1. Yin XC, Yin X, Huang K, Hao HW (2013) Robust text detection in natural scene images. IEEE Trans Pattern Anal Mach Intell 36(5):970–983
2. Wang L, Uchida S, Zhu A, Sun J (2018) Human reading knowledge inspired text line extraction. Cognitive Comput 10(1):84–93
3. Wang Y, Shi C, Xiao B, Wang C, Qi C (2018) CRF based text detection for natural scene images using convolutional neural network and context information. Neurocomputing 295:46–58

4. Pandey BK, Mane D, Nassa VKK, Pandey D, Dutta S, Ventayen RJM, Rastogi R (2021) Secure text extraction from complex degraded images by applying steganography and deep learning. In: multidisciplinary approach to modern digital steganography, IGI Global, pp 146–163
5. Ma J, Shao W, Ye H, Wang L, Wang H, Zheng Y, Xue X (2018) Arbitrary-oriented scene text detection via rotation proposals. IEEE Trans Multimedia 20(11):3111–3122
6. Pandey BK, Pandey D, Wairya S, Agarwal G, Dadeech P, Dogiwal SR, Pramanik S (2022) Application of integrated steganography and image compressing techniques for confidential information transmission. Cyber Secur Netw Secur, 169 – 191
7. Dutta IN,Chakraborty N, Mollah AF, Basu S, Sarkar R (2019) Multi-lingual text localization from camera captured images based on foreground homogenity analysis. Recent Dev Mach Learn Data Analytics 149–158
8. Tian C, Xia Y, Zhang X, Gao X (2017) Natural scene text detection with MC–MR candidate extraction and coarse-to-fine filtering. Neurocomputing 260:112–122
9. Ahmed SB, Naz S, Razzak MI, Yusof RB (2019) A novel dataset for English-Arabic scene text recognition (EASTR)-42K and its evaluation using invariant feature extraction on detected extremal regions. IEEE Access 7:19801–19820
10. Khare V, Shivakumara P, Raveendran P, Blumenstein M (2016) A blind deconvolution model for scene text detection and recognition in video. Pattern Recognit 54:128–148
11. Ahmed SB, Naz S, Razzak MI, Rashid SF, Afzal MZ, Breuel TM (2016) Evaluation of cursive and non-cursive scripts using recurrent neural networks. Neural Comput Appl 27(3):603–613
12. Francis LM, Sreenath N (2019) Robust scene text recognition: using manifold regularized twin-support vector machine. J King Saud Univ-Comput Inf Sci
13. Mehmood Z, Mahmood T, Javid MA (2018) Content-based image retrieval and semantic automatic image annotation based on the weighted average of triangular histograms using support vector machine. Appl Intell 48(1):166–181
14. Yin XC, Pei WY, Zhang J, Hao HW (2015) Multi-orientation scene text detection with adaptive clustering. IEEE Trans Pattern Anal Mach Intell 37(9):1930–1937
15. Wu H, Zou B, Zhao YQ, Guo J (2017) Scene text detection using adaptive color reduction, adjacent character model and hybrid verification strategy. Vis Comput 33(1):113–126
16. Pandey D, Pandey BK, Wairya S (2020) Hybrid deep neural network with adaptive galactic swarm optimization for text extraction from scene images. Soft Comput. https://doi.org/10.1007/s00500-020-05245-4
17. Khlif W, Nayef N, Burie JC, Ogier JM, Alimi A (2018) Learning text component features via convolutional neural networks for scene text detection. In: 2018 13th IAPR international workshop on document analysis systems (DAS). IEEE, Apr 2018, pp 79–84
18. Tang Y, Wu X (2018) Scene text detection using superpixel-based stroke feature transform and deep learning based region classification. IEEE Trans Multimedia 20(9):2276–2288
19. Ali A, Pickering M, Shafi K (2018) Urdu natural scene character recognition using convolutional neural networks. In: 2018 IEEE 2nd international workshop on Arabic and derived script analysis and recognition (ASAR). IEEE, Mar 2018, pp 29–34
20. Baliarsingh SK, Vipsita S, Muhammad K, Bakshi S (2019) Analysis of high-dimensional biomedical data using an evolutionary multi-objective emperor penguin optimizer. Swarm Evolut Comput 48:262–273
21. Baran R, Partila P, Wilk R (2018) Automated text detection and character recognition in natural scenes based on local image features and contour processing techniques. In: International conference on intelligent human systems integration. Springer, Cham, Jan 2018, pp 42–48
22. Xue M, Shivakumara P, Zhang C, Lu T, Pal U (2019) Curved text detection in blurred/non-blurred video/scene images. Multimedia Tools Appl 1–25
23. Wang Y, Wang L, Su F (2018) A robust approach for scene text detection and tracking in video. In: Pacific rim conference on multimedia. Springer, Cham, Sep 2018, pp 303–314
24. Antonini M, Barlaud M (1992) Image coding using wavelet transform. IEEE Trans Image Process 1(2)
25. Pandey BK, Pandey D, Wariya S, Agarwal G (2021) A deep neural network-based approach for extracting textual images from deteriorate images. EAI Endorsed Trans on Ind Networks Intell Syst, e3 8(28)

26. Patani K, Rathod D (2020) Advanced 3-bit LSB based on data hiding using steganography. In: Data science and intelligent applications. Springer, Singapore, pp 383–390
27. Xue M, Shivakumara P, Zhang C, Lu T, Pal U (2019) Curved text detection in blurred/non-blurred video/scene images. Multimedia Tools Appl 78(18):25629–25653
28. Pandey D, Pandey BK, Wariya S (2019) Study of various techniques used for video retrieval. J Emerg Technol Innov Res 6:850–853
29. Pandey D, Pandey BK, Wariya S (2019) Study of various types noise and text extraction algorithms for degraded complex image. J Emerg Technol Innov Res 6:234–247
30. Zhu A, Uchida S (2017) Scene text relocation with guidance. In: 2017 14th IAPR international conference on document analysis and recognition (ICDAR), vol 1. IEEE, Nov 2017, pp 1289–1294
31. Zhang Z, Zhang C, Shen W, Yao C, Liu W, Bai X (2016) Multi-oriented text detection with fully convolutional networks. In: Proceedings of the IEEE conference on computer vision and pattern recognition, pp 4159–4167
32. Dai J, Li Y, He K, Sun J (2016) R-FCN: object detection via region-based fully convolutional networks. In: Advances in neural information processing systems, pp 379–387
33. Ren S, He K, Girshick R, Sun J (2015) Faster R-CNN: towards real-time object detection with region proposal networks. In: Advances in neural information processing systems, pp 91–99
34. Pramanik S, Samanta D, Dutta S, Ghosh R, Ghonge M, Pandey D (2020, December) Steganography using improved LSB approach and asymmetric cryptography. In 2020 IEEE international conference on advent trends in multidisciplinary research and innovation (ICATMRI), IEEE pp 1–5
35. Madhumathy P, Pandey D (2022) Deep learning based photo acoustic imaging for non-invasive imaging. Multimedia Tools Appl, 81(5):7501–7518

Simulation-Based Analysis of AlGaN/GaN Gate All Around Field Effect Transistor (AlGaN/GaN GAA-FET)

Yogesh Kumar Verma, Varun Mishra, Rohit Gurjar, Rajeev Kumar Chauhan, and Santosh Kumar Gupta

Abstract The multi-gate field effect transistors (FETs) offer significant merits in comparison with conventional devices such as less switching power, compatibility with current manufacturing processes, and scalability to sub10 nanometer regime. The short channel effects (SCEs) are significantly reduced in the multi-gate FETs along with improved control of gate over the channel. In this brief, the different performance parameters of AlGaN/GaN gate all around field effect transistor (GAA-FET) are computed for different device parameters using TCAD.

Keywords III–V · Gate all around · HFET · AlGaN · GaN · TCAD

Y. K. Verma (✉)
School of Electronics and Electrical Engineering, Lovely Professional University, Jalandhar, Punjab 144411, India
e-mail: yogesh.25263@lpu.co.in

V. Mishra
Department of ECE, Graphic Era (Deemed to be University), Dehradun, Uttarakhand 248002, India
e-mail: varunmishra.ece@geu.ac.in

R. Gurjar
Department of Electronics and Communication Engineering, Netaji Subhas Institute of Technology, New Delhi 110078, India

R. K. Chauhan
Department of Electronics and Communication Engineering, Madan Mohan Malaviya University of Technology, Gorakhpur, UP 273010, India

S. K. Gupta
Department of Electronics and Communication Engineering, Motilal Nehru National Institute of Technology, Allahabad, UP 211004, India
e-mail: skg@mnnit.ac.in

B. Mishra and M. Tiwari (eds.), *VLSI, Microwave and Wireless Technologies*, Lecture Notes in Electrical Engineering 877,
https://doi.org/10.1007/978-981-19-0312-0_21

1 Introduction

GaN-based field effect transistors are utilized as power amplifier for telecommunication applications. Because of the superior material properties of nitrides, AlGaN/GaN FETs inherit great potential for high power microwave devices [1–5]. The focus of the research at present is in nanometer regime, and the devices become more susceptible to SCEs at 20 nm than the preceding technology nodes [6–22]. Accordingly, the traditional planar structures provide poor short channel control, and are unable to cope with this scaling rate; therefore, the improvement in device architecture is essential [23–25]. In multi-gate FETs, the electric field lines from the source and drain terminate on the bottom gate and therefore, cannot reach the channel region; thus, SCEs are reduced [5]. The analog/RF performance of planar bulk FET, double gate (DG) FET, and Fin-FET has been explored extensively by several groups including [26–32], yet there are very few investigations reported on the gate all around FET. Lee et al. [33] fabricated gate all around GaN-based high electron mobility transistor (HEMT) and reported the improvement in breakdown voltage comparing with single gate HEMT. Subramanian et al. [34] compared digital and analog FOMs of Fin-FET and planar bulk MOSFET and revealed that Fin-FET possesses reduced leakage, excellent sub-threshold slope, and better voltage gain in comparison with bulk-MOSFETs at lower frequency. Wambacq et al. [35] revealed that Fin-FETs exhibit less leakage current in comparison with single gate FETs [36]. Therefore, it is essential to replace the planar bulk MOSFET and an intensive investigation is mandatory. The channel materials such as AlGaN and GaN in comparison with Si are being actively explored to deal successfully with the upcoming technology nodes.

2 Structure Description

The gate controllability over the channel can be improved by different methods [2–5]. To reduce the SCEs, dual-metal (or double-metal, DM) and triple-metal (TM) gates have been used with the cylindrical surrounding gate (CSG) structures. These configurations use two or three metals with different work functions [8–12]. By using a triple-metal gate structure (high work function metal in between two smaller work function metals), we can easily create ultra-shallow junctions also known as electrically induced source drain junction (EJ) SOI MOSFETs [6]. Figure 1 represents the structure of MOS-based AlGaN/GaN QG-FET. The channel length and oxide thickness are 20 nm and 2 nm, respectively. The channel height is varied as 18 nm and 10 nm. The doping concentration is varied as 1×10^{19} and 5×10^{18} cm^{-3}.

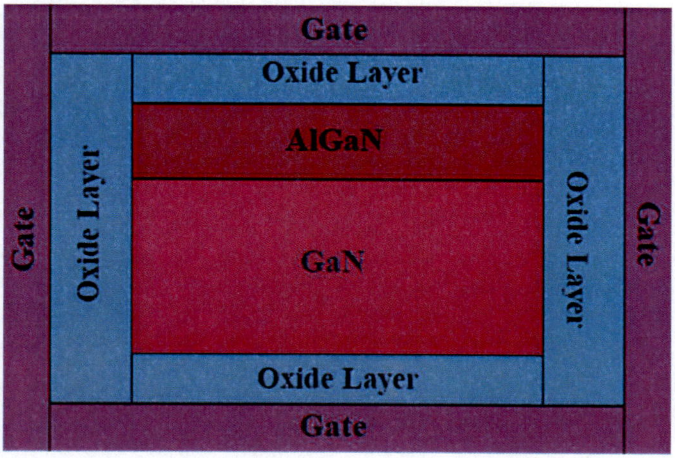

Fig. 1 Structure of QG-AlGaN/GaN FET

3 Results and Discussions

Figure 2a–h represents the analysis of the linearity and distortion performance of AlGaN/GaN GAA-FET. VIP_2 specifies the inferred value of input voltage signifying the equivalence of first and second harmonics ($VIP_2 = 4.\ gm_1/gm_2$). VIP_3 specifies the inferred value of input voltage signifying the equivalence of first and third harmonics ($VIP_3 = \surd 24.\ \surd gm_1/\surd gm_3$). It is obvious from Fig. 2a, b that higher channel height provides higher drain current and transconductance. It can be noticed from Fig. 2c that the lower values of gm_2 occur for lower gate voltage at lower channel height. The lower values of gm_2 for higher value of gate voltage occur at higher values of channel height. It is obvious from Fig. 2d that the lower values of gm_3 occur for lower gate voltage at lower channel height. The lower values of gm_3 for higher value of gate voltage occur at higher values of channel height. Figure 2e represents the calculation of IIP_3. It is worth noting from Fig. 2f that the lower values of IMD_3 occur at lower channel height. Figure 2g, h represents the calculation of VIP_2 and VIP_3, respectively. The higher values of VIP_2 occur at higher channel height. Table 1 represents the percentage improvement in the performance parameters for different variations in channel height. The present device provides better electrostatics and enhanced control over the channel as other semiconductor devices [1–29].

Figure 3a–h represents the analysis of the linearity and distortion performance of QG-AlGaN/GaN FET. The doping concentration is varied as 5×10^{18} and 1×10^{19} cm^{-3}. It can be concluded from Fig. 3a, b that higher doping concentration provides better performance. It is obvious from Fig. 3c that the lower values of gm_2 occur at lower doping concentration. Figure 3d, e represents the calculation of gm_3 and IIP_3, respectively. It is obvious from Fig. 3f that lower channel height provides

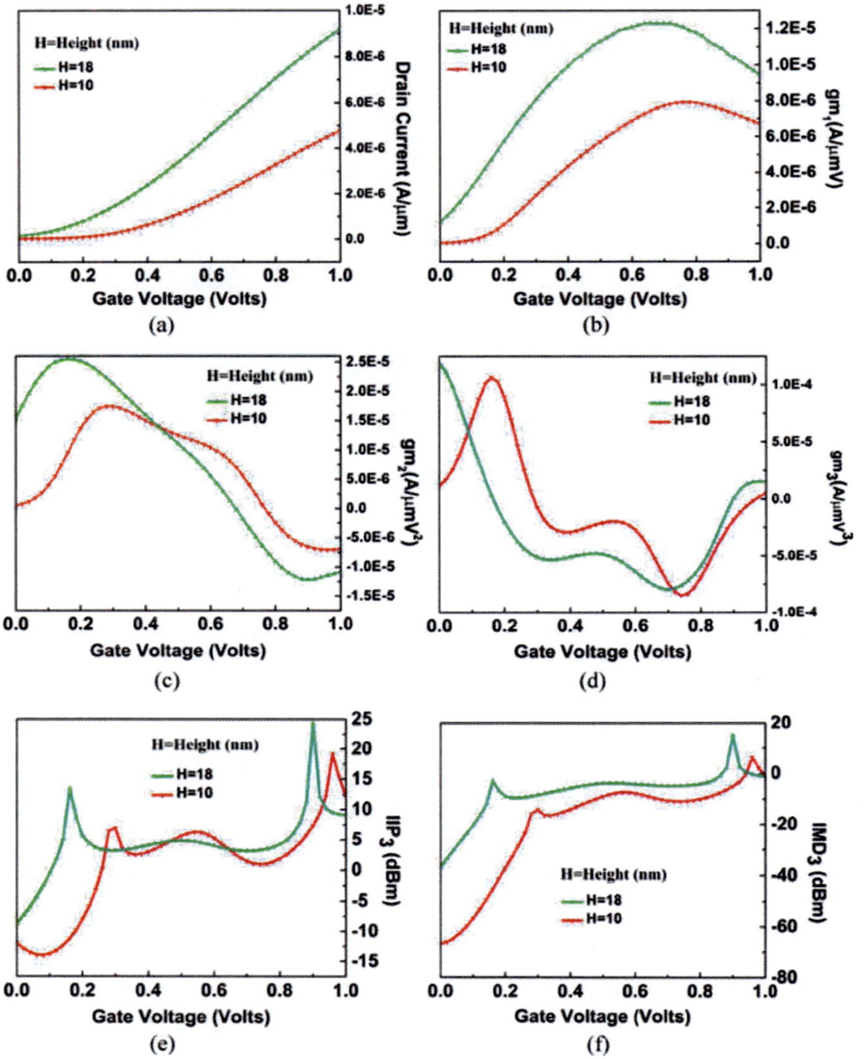

Fig. 2 **a–h**: Calculation of **a** drain current, **b** gm_1, **c** gm_2, **d** gm_3, **e** IIP_3, **f** IMD_3, **g** VIP_2, and **h** VIP_3, respectively for different channel heights

lower IMD_3. Table 2 represents the improvement in the performance parameters for different variations in doping concentration.

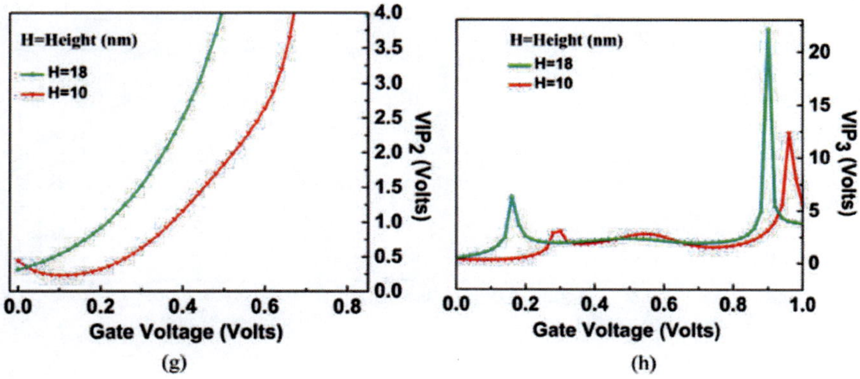

Fig. 2 (continued)

Table 1 Percentage improvement in performance parameters

Variation in channel height (nm)	Variation in peak value
10–18	It increases by 75.90% for VIP_2
10–18	It increases by 44.34% for VIP_3
18–10	It reduces by 56.90% for IMD_3

4 Conclusion

The GAA-FET offers significant merits in comparison with conventional devices. The GAA-FETs are capable to significantly reduce the short channel effects. The different performance parameters of gate all around FET are computed for different values of device parameters. It has been observed that lower channel height, and doping concentration assists in achieving better device performance. The doping concentration significantly increases the ionized impurity scattering and deteriorates the performance of the device, and hence, its concentration should be kept low.

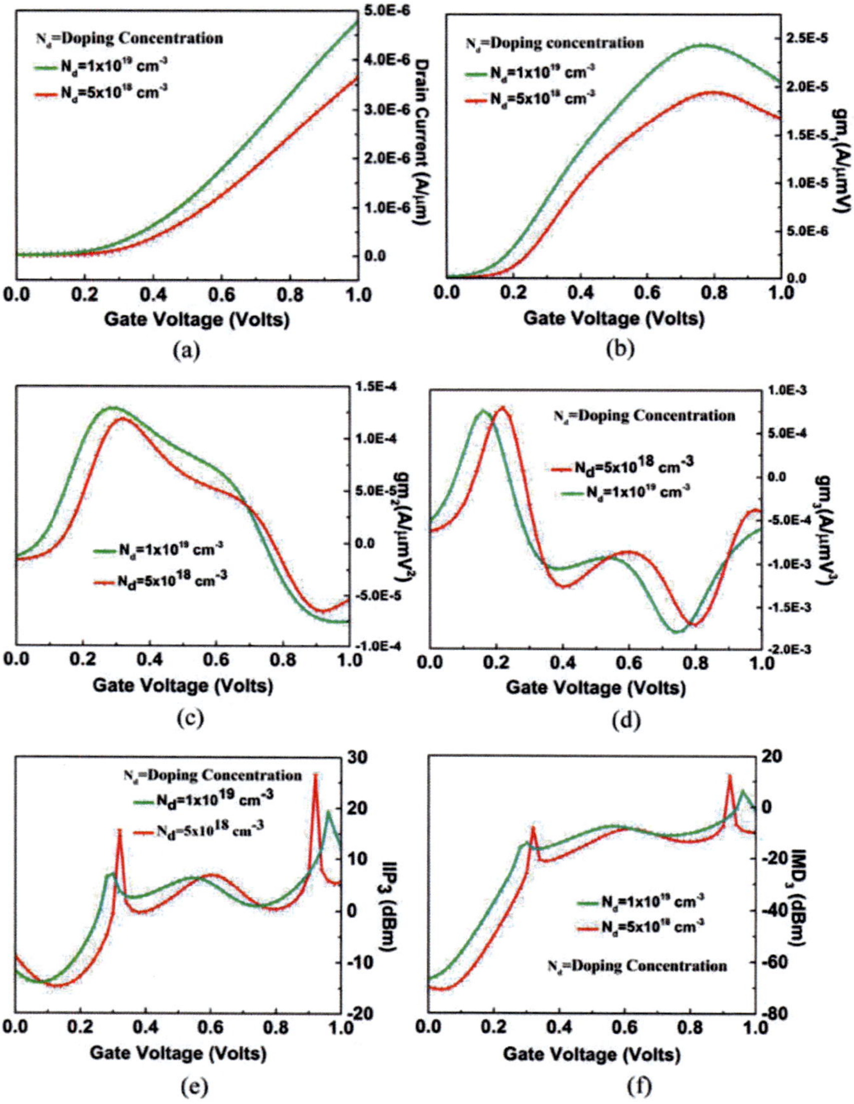

Fig. 3 **a–h**: Calculation of **a** drain current, **b** gm_1, **c** gm_2, **d** gm_3, **e** IIP_3, **f** IMD_3, **g** VIP_2, and **h** VIP_3, respectively for different doping concentrations

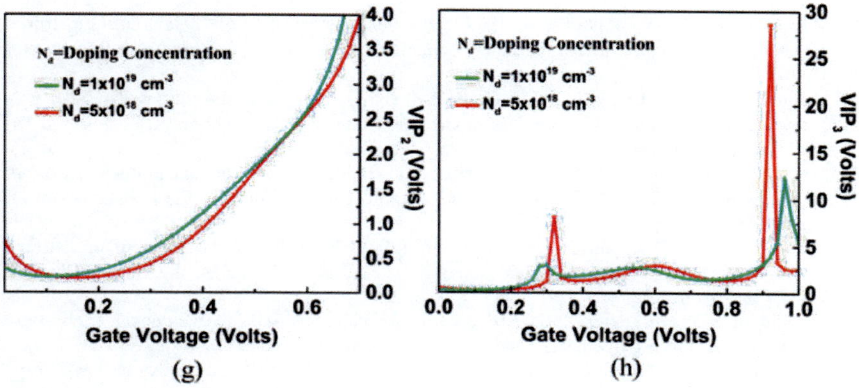

Fig. 3 (continued)

Table 2 Improvement in performance parameters

Variation in doping concentration (cm^{-3})	Variation in peak value
1×10^{19}–5×10^{18}	It decreases by 6.86% for gm_2
5×10^{18}–1×10^{19}	It decreases by 2.8% for gm_3
1×10^{19}–5×10^{18}	It increases by 37.30% for IIP_3
1×10^{19}–5×10^{18}	It increases by 2.32 times for VIP_3
5×10^{18}–1×10^{19}	It reduces by 91.52% for IMD_3

References

1. Yang L et al (2019) High channel conductivity, breakdown field strength, and low current collapse in AlGaN/GaN/Si δ-Doped AlGaN/GaN: C HEMTs. IEEE Trans Electron Devices 66(3):1202–1207
2. Khan AB, Anjum SG, Siddiqui MJ (2018) Effect of barrier layer thickness on AlGaN/GaN double gate MOS-HEMT device performance for high-frequency application. J Nanoelectron Optoelectron 13(1):20–26
3. Kalita S, Mukhopadhyay S (2018) Effect of gate length on the electrical characteristics of nano-electronic AlGaN/GaN high electron mobility transistors to fabricate the biomedical sensors in nanoelectronics. J Nanoelectron Optoelectron 13(8):1123–1127
4. Varghese A, Periasamy C, Bhargava L, Vijayakumar K (2018) Impact of AlN interlayer's in epitaxial and passivation scheme on the DC and microwave performance of doping-less GaN HEMT. J Nanoelectron Optoelectron 13(7):971–979
5. Colinge J-P (2008) FinFETs and other multi-gate transistors. Springer, Heidelberg
6. Yadav C, Agrawal M, Agarwal A, Chauhan YS (2017) Compact modeling of charge, capacitance, and drain current in III-V channel double gate FETs. IEEE Trans Nanotechnol 16(2):347–354
7. Sinha SK, Chaudhury S (2013) Impact of oxide thickness on gate capacitance—a comprehensive analysis on MOSFET, nanowire FET, and CNTFET devices. IEEE Trans Nanotechnol 12(6):958–964
8. Verma YK, Gupta SK (2020) Center potential based analysis of Si and III-V Gate all around Field Effect Transistors (GAA-FETs). SILICON 63:1–17

9. Verma YK, Mishra V, Gupta SK (2019) Electrical characterization of AlGaN/GaN quadruple gate heterostructure field effect transistor for analog applications. J Nanoelectron Optoelectron 14(10):1491–1502
10. Verma PK, Verma YK, Mishra V, Gupta SK (2020) A charge-plasma-based dual-metal-gate recessed-source/drain dopingless junctionless transistor with enhanced analog and RF performance. J Comput Electron 19(3):1085–1099
11. Mishra V, Verma YK, Gupta SK (2019) Investigation of localized charges and temperature effect on device performance of ferroelectric dual material gate all around TFETs. J Nanoelectron Optoelectron 14(2):161–168
12. Mishra V, Verma YK, Verma PK, Gupta SK (2018) EMA-based modeling of the surface potential and drain current of dual-material gate-all-around TFETs. J Comput Electron 17(4):1596–1602
13. Tripathi SL, Sinha SK, Patel GS (2020) Low-power efficient p+ Si 0.7 Ge 0.3 pocket junctionless SGTFET with varying operating conditions. J Electron Mater: 1–9
14. Mendiratta N, Tripathi SL (2020) A review on performance comparison of advanced MOSFET structures below 45 nm technology node. J Semicond 41:0
15. Verma YK, Mishra V, Gupta Sk (2020) Performance enhancement of AlGaN/GaN HEMT by optimization of device parameters considering nanometer barrier layer thickness. Int J Nanosci :2050011
16. Verma YK, Mishra V, Gupta SK (2019) A physics based analytical model for MgZnO/ZnO HEMT. J Circuits Syst Comput 29
17. Subramanian B et al (2020) Switching transient analysis and characterization of an E-mode B-doped GaN-capped AlGaN DH-HEMT with a freewheeling Schottky Barrier Diode (SBD). J Electron Mater :1–9
18. Gupta SK, Rawat AS, Verma YK, Mishra V (2018) Linearity distortion analysis of junctionless quadruple gate MOSFETs for analog applications. SILICON 11:257–265
19. Verma YK, Mishra V, Gupta SK (2020) Analog/RF and linearity distortion analysis of MgZnO/CdZnO quadruple-gate field effect transistor (QG-FET). SILICON 13:91–107
20. Verma YK, Gupta SK, Mishra V, Verma PK (2018) Surface potential based analysis of MgZnO/ZnO high electron mobility transistors. In: 2018 IEEE international students' conference on electrical, electronics and computer science (SCEECS), MANIT Bhopal, pp 1–4
21. Mishra V, Verma YK, Gupta SK (2020) Surface potential-based analysis of ferroelectric dual material gate all around (FE-DMGAA) TFETs. Int J Numer Model Electron Netw Devices Fields 34:1–11
22. Varun M, Yogesh KV, Santosh KG (2019) Investigation of localized charges on linearity and distortion performance of ferroelectric dual material gate all around TFETs, vol 11, p 04014-15
23. Yadav C, Ganeriwala MD, Mohapatra NR, Agarwal A, Chauhan YS (2017) Compact modeling of gate capacitance in III-V channel quadruple-gate FETs. IEEE Trans Nanotechnol 16(4):703–710
24. Ferain I, Colinge CA, Colinge J-P (2011) Multigate transistors as the future of classical metal-oxide-semiconductor field-effect transistors. Nature 479(7373):310
25. He L, Chiang TK, Liou JJ, Zheng W, Liu Z (2014) A new analytical subthreshold potential/current model for quadruple-gate junctionless MOSFETs. IEEE Trans Electron Devices 61(6):1972–1978
26. Verma YK, Mishra V, Verma PK, Gupta SK (2018) Analytical modelling and electrical characterisation of ZnO based HEMTs. Int J Electron 106(5):707–720
27. Korwal M, Haldar S, Gupta M, Gupta RS (2003) Parasitic resistance and polarization-dependent polynomial-based non-linear analytical charge-control model for AlGaN/GaN MODFET for microwave frequency applications. Microw Opt Technol Lett
28. Pradhan KP, Mohapatra SK, Sahu PK, Behera DK (2014) Impact of high-k gate dielectric on analog and RF performance of nanoscale DG-MOSFET. Microelectronics J 45(2):144–151
29. Koley K, Dutta A, Syamal B, Saha SK, Sarkar CK (2013) Subthreshold analog/RF performance enhancement of underlap DG FETs with high-k spacer for low power applications. IEEE Trans Electron Devices 60(1):63–69

30. Mohankumar N, Syamal B, Sarkar CK (2010) Influence of channel and gate engineering on the analog and RF performance of DG MOSFETs. IEEE Trans Electron Devices 57(4):820–826
31. Koley K, Syamal B, Kundu A, Mohankumar N, Sarkar CK (2012) Subthreshold analog/RF performance of underlap DG FETs with asymmetric source/drain extensions. Microelectron Reliab 52(11):2572–2578
32. Kumar SP, Agrawal A, Chaujar R, Gupta RS, Gupta M (2011) Device linearity and intermodulation distortion comparison of dual material gate and conventional AlGaN/GaN high electron mobility transistor. Microelectron Reliab 51(3):587–596
33. Lee C-T, Chen W-S, Lee H-Y (2018) Quadruple gate-embedded T structured GaN-based metal-oxide-semiconductor high-electron mobility transistors. IEEE J Electron Devices Soc 1(6):63–67
34. Subramanian V et al (2006) Planar bulk MOSFETs versus FinFETs: an analog/RF perspective. IEEE Trans Electron Devices 53(12):3071–3079
35. Wambacq P et al (2007) The potential of FinFETs for analog and RF circuit applications. IEEE Trans Circuits Syst I Regul Pap 54(11):2541–2551
36. Sharma D, Vishvakarma SK (2015) Analyses of DC and analog/RF performances for short channel quadruple-gate gate-all-around MOSFET. Microelectronics J 46(8):731–739

Electrical and RF Performance Enhancement of Pocket Doped Junctionless Transistor Using Dual Metal Gate

Priyansh Tripathi, Narendra Yadava, and R. K. Chauhan

Abstract In the pocket doped window-based silicon-on-insulator junctionless transistor (PD-SOIJLT), a dual metal gate (DMG) design is proposed. Using the SILVACO ATLAS-2D device simulator, its electrical and RF performances are analyzed and evaluated against those of single metal gate (SMG) PD-SOIJLT. The simulation outcomes confirm the improvements in the distributions of electric field, electron velocity, and potential, along the channel of the DMG-based structure. For DMG PD-SOIJLT over SMG PD-SOIJLT, drain induced barrier lowering (DIBL) and ON-state current are enhanced by 7.67% and 55.16%, respectively. Moreover, transconductance and cutoff frequency have also improved by 12.39% and 13.72%, respectively in DMG-based structure over SMG-based structure.

Keywords Pocket doped window · Dual metal gate (DMG) · Single metal gate (SMG) · Junctionless transistor · Silicon-on-insulator (SOI) · Electrical performance and radio-frequency (RF) performance

1 Introduction

Bulk MOSFETs are confronting severe complexities with the continual scaling down of the device size, such as the increased leakage of gate current and significant short channel effects (SCEs). As the span of the channel reduces, the gate's controllability is weakened over the depletion region by excessive charge sharing from the drain/source, leading to short channel effects such as threshold voltage roll-off and DIBL. Gate-all-around FETs (GAAFETs) and FinFETs (multigate structures) are perceived, due to their exceptional gate controllability, as viable device structures to broaden the scaling range. The development of steeped channel-source and channel-drain junctions in traditional nano-transistor devices, on the other hand, presents major difficulties to doping and thermal budgeting techniques [1]. With a constant

P. Tripathi (✉) · N. Yadava · R. K. Chauhan
Department of Electronics and Communication Engineering, Madan Mohan Malaviya University of Technology, Gorakhpur, India
e-mail: priyansh.tripathi00@gmail.com

B. Mishra and M. Tiwari (eds.), *VLSI, Microwave and Wireless Technologies*, Lecture Notes in Electrical Engineering 877,
https://doi.org/10.1007/978-981-19-0312-0_22

doping concentration and the same type of dopant all across the extensions of channel, source and drain, junctionless devices are fabricated as the potential substitute to mitigate these issues [1].

Their electrical behavior and temperature reliance have been investigated [1–3], and junctionless transistors (JLTs) have been reported to offer incredibly reduced currents in OFF condition and easy manufacturing processes and are, therefore, less prone to SCEs especially in comparison with conventional devices in inversion mode. Multigate junctionless devices have been also studied which reveal that junctionless transistors give low subthreshold slope [4, 5]. Due to mitigated bulk current leakage, radiation durability, elevated thermal action, and enhanced transconductance, silicon-on-insulator (SOI) technology is better than bulk MOSFET. The dual-material gate (DMG) arrangement in bulk MOSFET lead to improved transconductance and suppressed SCEs as a result of a step function of the potential of the channel [6].

The fabrication [6–8] and theoretical studies [9–14] for DMG devices have already been done. It also enhances the efficiency of transportation of carrier, drain resistance and transfer conductance of conventional MOSFET by modifying channel capacity and distribution of electric field throughout the channel [6, 10, 11]. The incorporation of DMG and SOI technology in junctionless transistor is bring in a novel device configuration with various advantages. The analytical modeling for DMG-SOIJLT is presented in [15]. The DMG-based conventional SOIJLT [16] and DMG-based nanowire transistor [17] have been studied, and their benefits have been also highlighted. Recently, a modified SOIJLT with p-type pocked doped window-based SOIJLT (PD-SOIJLT) with single gate metal has been introduced with the improved electrical and thermal performances [18, 19]. But the impact of DMG structure on its electrical and radio-frequency (RF) performances has been not reported yet.

This paper is thus focused to propose a new structure called DMG PD-SOIJLT. Its electrical and RF behaviors have been replicated by employing SILVACO ATLAS-2D semiconductor device simulator, and are compared with that of SMG PD-SOIJLT. The key parameters chosen for performance evaluation are electric field, potential distribution, electron velocity, current in ON condition (I_{ON}), subthreshold slope (SS), drain induced barrier lowering (DIBL), transfer conductance (g_m), and unity gain frequency (f_T).

2 Device Specifications and Simulation

The schematic view of two-dimensional DMG PD-SOIJLT is shown in Fig. 1a. And the simulated structures of DMG PD-SOIJLT and SMG PD-SOIJLT are represented in Fig. 1b and c, respectively. The DMG PD-SOIJLT has two metals on the gate that are control gate (L_C) and screen gate (L_S) of 11 nm length each while SMG PD-SOIJLT has only single metal on the gate ($L_C = 22$ nm). The work function of L_C is assigned 5.2 eV and L_S is assigned 4.7 eV. L_C is made using gold (Au) as metal, and L_S is made using silver (Ag) [20]. The detailed parameters for both devices are tabulated

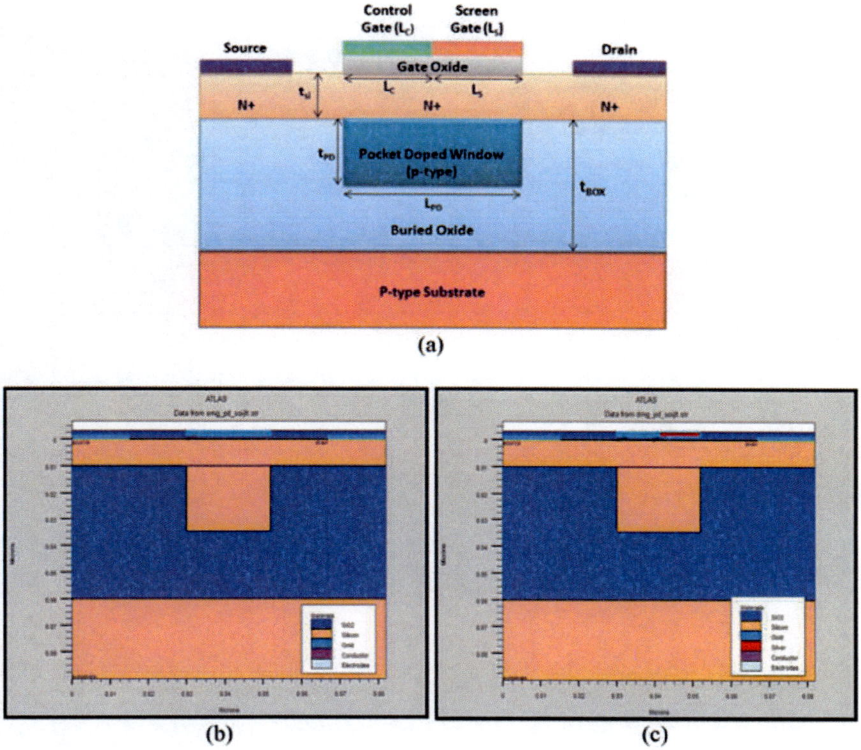

Fig. 1 **a** Schematic representation in two dimensions of DMG PD-SOIJLT, and simulated structures of **b** SMG PD-SOIJLT and **c** DMG PD-SOIJLT

in Table 1. To obtain optimum performance L_C and L_S are assigned equal lengths and work function difference of 0.5 eV is maintained [17]. A p-type pocked window has been opened in the buried oxide layer just below channel region and in vertical alignment of gate [18], and it is doped heavily [19]. Various models considered for simulations are constant voltage and temperature mobility model (CVT), Shockley–Reed–Hall recombination model (SRH), bandgap narrowing model (BGN), Auger recombination model (Auger), and Fermi–Dirac statistics model (fermidirac) [21].

3 Results and Discussion

This section presents the performance features of proposed DMG PD-SOIJLT. Its key parameters are compared with that of SMG PD-SOIJLT. All simulation results and parameters are extracted by using ATLAS-2D simulator [21]. The electrical and

Table 1 Structural specification for device simulation

Parameters	SMG PD-SOIJLT	DMG PD-SOIJLT	
Channel length, L_C (nm)	22	22	
Silicon channel thickness, t_{si} (nm)	10	10	
Buried oxide thickness, t_{BOX} (nm)	50	50	
Gate oxide thickness, t_{ox} (EOT) (nm)	1	1	
Channel doping, N_D (cm^{-3})	1×10^{19}	1×10^{19}	
Gate work function (eV)	5.2	L_C	L_S
		5.2	4.7
Control gate length, L_C (nm)	22	11	
Screen gate length, L_S (nm)	-	11	
Pocket doped window thickness, t_{PD} (nm)	25	25	
Pocket doped window length, L_{PD} (nm)	22	22	
Pocket doped window doping, (*p*-type) (cm^{-3})	5×10^{19}	5×10^{19}	
Substrate doping, N_A (cm^{-3})	5×10^{18}	5×10^{18}	

RF performance parameters are discussed in the following Sections 3.1 and 3.2, respectively.

3.1 *Electrical Behavior of DMG PD-SOIJLT*

Figure 2 shows the potential distribution across the length of channel, obtained at drain-to-source supply $V_{DS} = 1$ V and gate overdrive supply $V_{OV} = 0.2$ V. Gate

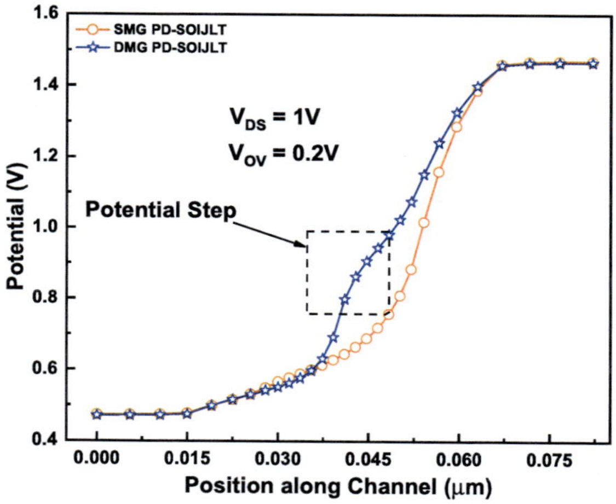

Fig. 2 Potential distribution along the channels of SMG and DMG PD-SOIJLTs

overdrive voltage is defined as $V_{OV} = V_{GS} - V_T$, where V_{GS} is voltage difference between gate and source and V_T is threshold voltage extracted from maximum transconductance method [22].

At the work function transition spot of the dual gates, the potential distribution of DMG PD-SOIJLT changes abruptly (potential step) while it increases monotonically for SMG PD-SOIJLT. This step transition is caused owing to work function distinction of the two metals [6, 23]. Thus, electric field is enhanced in channel region of DMG PD-SOIJLT. The probable drop through the extension of the source/drain in DMG PD-SOIJLT is higher than the SMG PD-SOIJLT, suggesting that the DMG PD-SOIJLT's channel ON-state resistance is less than that of the SMG PD-SOIJLT [24].

Figure 3 shows plot of electric field across the length of channel region. In Fig. 3, there are two peaks in the DMG PD-SOIJLT electric field and just one near the drain in the SMG PD-SOIJLT [6, 23]. Besides, relative to the SMG PD-SOIJLT, the electrical field peak value near the drain of the DMG PD-SOIJLT is decreased by 42.81%. By contrast, for the DMG SOI-FET and DMG bulk MOSFET, the electric field peak value close the drain side is only diminished by 20% relative to those of their complementary SMGs [6, 12]. Thus, in minimizing the influence of short channels and hot carriers, DMG PD-SOIJLT is more efficient by minimizing the drain-channel field [17].

The electric field peak close to the source region induces further electron acceleration in the channel for the ON-state current and boosts the velocity of the carrier pre-saturation next to the source, which is shown in Fig. 4. The speed of the electron (see Fig. 4) in the DMG PD-SOIJLT at the same position can be increased to four times the velocity in the SMG PD-SOIJLT. This controlling of velocity in channel by first peak is due to different work functions of two metals at gate [25]. The second peak of electron velocity in DMG PD-SOIJLT is less as compared to SMG PD-SOIJLT because of minimized drain-channel field.

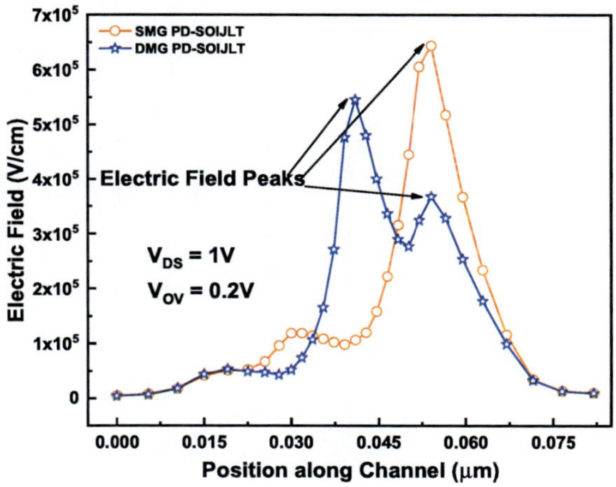

Fig. 3 Electric field distribution along the channels of SMG and DMG PD-SOIJLTs

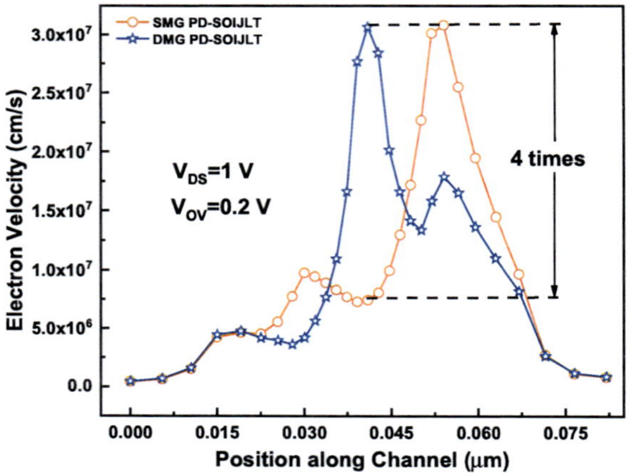

Fig. 4 Distribution of electron velocity along the channels of SMG and DMG PD-SOIJLTs

The transfer characteristics of both JLTs at $V_{DS} = 1$ V are represented in Fig. 5. Due to the high fringing electrical field, the drain-source current (I_{DS}) increases for DMG PD-SOIJLT over SMG PD-SOIJLT on increasing V_{OV}. Figure 6 depicts the output characteristics of the DMG and SMG PD-SOIJLTs at $V_{OV} = 1$ V. DMG PD-SOIJLT produces more output current than corresponding SMG PD-SOIJLT. Table 2 presents the performance comparison of proposed DMG PD-SOIJLT and SMG PD-SOIJLT. Both OFF- and ON-state currents are obtained higher for DMG PD-SOIJLT than corresponding SMG PD-SOIJLT. Higher ON current increases driving capability of the device.

The ratio of the difference in V_T at $V_{DS} = 0.05$ V and $V_{DS} = 1$ V to the difference in drain voltages is determined as drain-induced barrier lowering (DIBL).

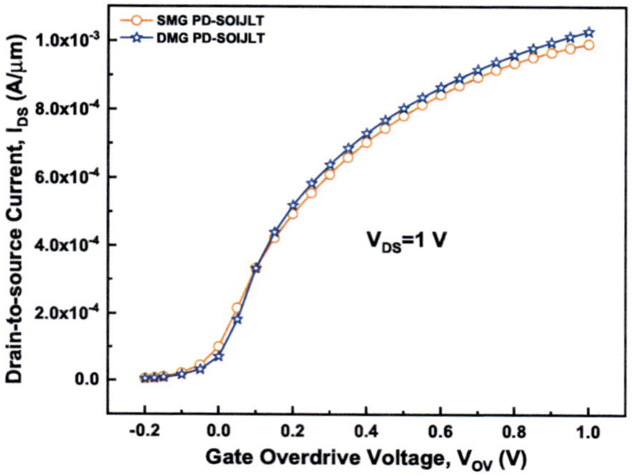

Fig. 5 I_{DS} against V_{OV} plot of SMG and DMG PD-SOIJLTs

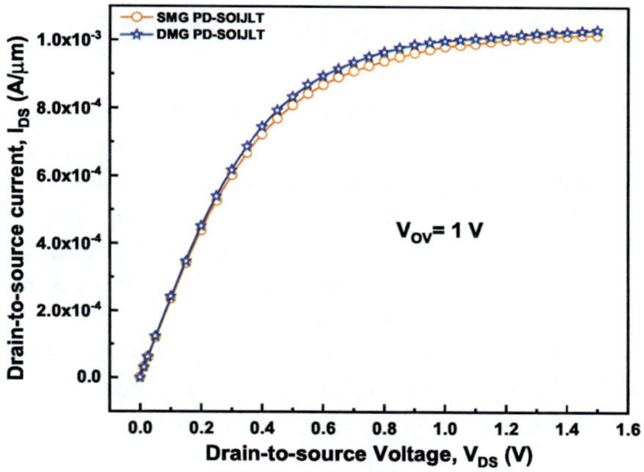

Fig. 6 I_{DS} against V_{DS} plot of SMG and DMG PD-SOIJLTs

Table 2 Performance comparison of proposed DMG PD-SOIJLT and SMG PD-SOIJLT-based on electrical parameters

Device structure	Off state current, I_{OFF} (A)	ON state current, I_{ON} (A)	Subthreshold slope (mV/decade)	DIBL (mV/V)
SMG PD-SOIJLT	1.947×10^{-13}	3.352×10^{-4}	54.0217	236.193
DMG PD-SOIJLT	5.285×10^{-12}	5.201×10^{-4}	67.5659	218.054

$$\text{DIBL} = \frac{V_T(V_{DS} = 50\text{ mV}) - V_T(V_{DS} = 1V)}{V_{DS}(= 1V) - V_{DS}(= 50\text{ mV})} \quad (1)$$

The influence of DIBL is observed to be less in DMG PD-SOIJLT (~218.054 mV/V) as compared to SMG PD-SOIJLT (~236.193 mV/V), with an improvement of 7.67%. The suppressed DIBL in DMG PD-SOIJLT is owned to reorienting the peak electrical field to the drain side due to introduction of screen gate [17].

Subthreshold slope is defined by

$$\text{SS} = \frac{\partial V_{GS}}{\partial \text{Log}_{10} I_{DS}} \quad (2)$$

Subthreshold slope (SS) for DMG PD-SOIJLT is 67.565 mV/decade larger as compared to corresponding SMG PD-SOIJLT (54.021 mV/decade). SS is inversely related to effective length [26], and SMG PD-SOIJLT has a bit larger effective length than channel length. It has also been demonstrated DMG-based JLTs and has effective length only larger than control gate L_C [2, 26]. So SS has increased for DMG PD-SOIJLT.

3.2 RF Behavior of DMG PD-SOIJLT

The small signal ac analysis is performed at 100 MHz after performing dc analysis [21]. The plot for transconductance (g_m) against gate overdrive voltage (V_{OV}) is illustrated in Fig. 7. The highest g_m for DMG PD-SOIJLT is 2.63 mS which is 12.39% more than that of SMG PD-SOIJLT (2.34 mS). In the DMG PD-SOIJLT, higher g_m is induced by the abruptness of potential step closer to the source that contributes to the greater channel potential transition with the same V_{OV}, as already shown in Fig. 2.

As shown in Fig. 8, at higher V_{OV}, the gate capacitance increases for DMG PD-SOIJLT as compared to SMG PD-SOIJLT due to the fact that gate-drain capacitance

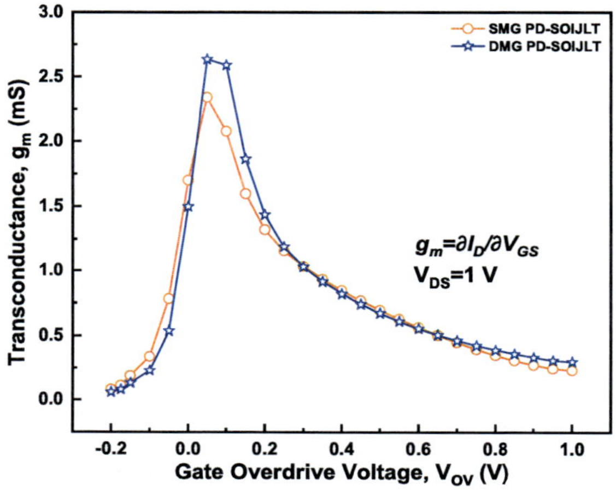

Fig. 7 g_m against V_{OV} plot of SMG and DMG PD-SOIJLTs

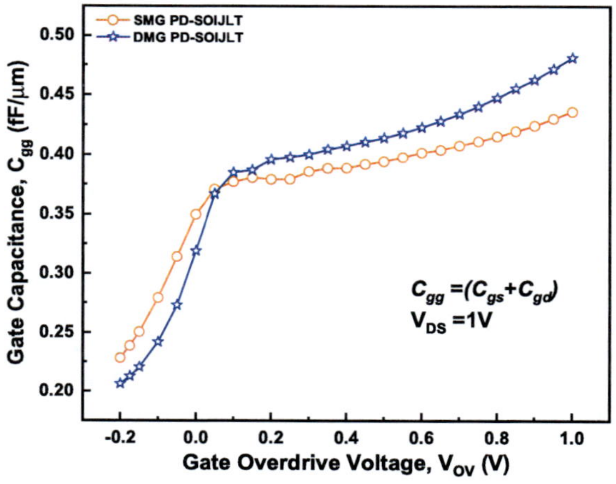

Fig. 8 C_{gg} against V_{OV} plot for SMG and DMG PD-SOIJLTs

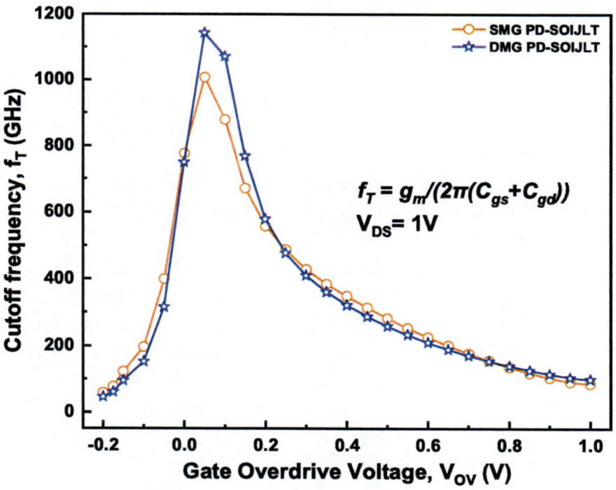

Fig. 9 f_T against V_{OV} plot of SMG and DMG PD-SOIJLTs

is incremented rapidly (than at lower V_{OV}) due to charge piling at drain end because of use of low work function-based screen gate [27]. But the gain of device remains unaffected since the increment in g_m is greater than capacitance, and so the maximum f_T is higher for DMG PD-SOIJLT than SMG PD-SOIJLT.

The unity gain cutoff frequency (f_T) is plotted against the overdrive voltage (V_{OV}) in Fig. 9. The maximum f_T of DMG PD-SOIJLT is 1143.54 GHz which is 13.72% more than maximum f_T of SMG PD-SOIJLT (1005.5 GHz). But f_T reduces for DMG PD-SOIJLT at higher V_{OV} due to increasing parasitic capacitance [17].

4 Conclusion

A new structure has been proposed for PD-SOIJLT using dual metal gate (DMG). Its electrical and RF behavior has been analyzed using SILVACO ATLAS-2D device simulator, and compared with that of single metal gate SMG PD-SOIJLT. The results have shown exceptionally improved performance like improved ON-state current (by 55.6%) and suppressed DIBL (by 7.67%). The distributions of potential, electric field, and electron velocity are also improved in DMG PD-SOIJLT over SMG PD-SOIJLT, but I_{OFF} and SS both are found to be increasing for DMG PD-SOIJLT. Improvements of 12.39% and 13.72% have been obtained in maximum values of g_m and f_T, respectively for DMG PD-SOIJLT as compared to SMG PD-SOIJLT. Hence, it can be a potential substitute for low power RF applications.

References

1. Lee C, Borne A, Ferain I, Afzalian A, Yan R, Akhavan ND, Razavi P, Colinge J (2010) High-temperature performance of silicon junctionless MOSFETs. IEEE Trans Electron Dev 57(3):620–625
2. Colinge J, Lee C, Afzalian A, Akhavan N, Yan R, Ferain I, Razavi P, O'Neill B, Blake A, White M, Kelleher AM, McCarthy B, Murphy R (2010) Nanowire transistors without junctions. Nat Nanotechnol 5(3):225–229
3. Lee C, Ferain I, Afzalian A, Yan R, Akhavan N, Razavi P, Colinge J (2010) Performance estimation of junctionless multigate transistors. Solid State Electron 54(2):97–103
4. Lee CW, Nazarov AN, Ferain I, khavan, ND, Yan R, Razavi P (2010) Low sub threshold slope in junctionless multi gate transistor. Appl Phys Lett 96(10):102106-1–102106-3
5. Lee CW, Afzalian A, Akhavan ND, Yan R, Ferain I, Colinge JP (2009) Junction less multigate field-effect transistor. Appl Phys Lett 94(5):053511-1–053511-2
6. Long W, Ou H, Kuo J, Chin K (1999) Dual-material gate (DMG) field effect transistor. IEEE Trans Electron Dev 46(5):865–870
7. Long W, Chin K (1998) Dual material gate field effect transistor (DMGFET). IEDM Tech Dig 549–552
8. Zhou X, Long W (1998) A novel hetero-material gate (HMG) MOSFET for deep-submicron ULSI technology IEEE Trans. Electron Dev 45(12):2546–2548
9. Shur M (1989) Split-gate field-effect transistor. Appl Phys Lett 54(9):162–164
10. Na K, Kim Y (2006) Silicon complementary metal-oxide-semiconductor field-effect transistors with dual work function gate. Jpn J Appl Phys 45(12):9033–9036
11. Saxena M, Haldar S, Gupta M, Gupta R (2002) Physics-based analytical modeling of potential and electrical field distribution in dual material gate (DMG)-MOSFET for improved hot electron effect and carrier transport efficiency. IEEE Trans Electron Dev 49(11):1928–1938
12. Chaudhry A, Kumar M (2004) Investigation of the novel attributes of a fully depleted dual-material gate SOI MOSFET. IEEE Trans Electron Dev 51(9):1463–1467
13. Tripathi S, Mishra VK, Chauhan RK (2016) Performance analysis of dual metal gate modified source fully depleted SOI MOSFET. i-Manager's J Embedded Syst 5(2)
14. Malviya AK, Chauhan RK (2017) Optimizing performance of dual metal gate modified source FDSOI using symmetric and asymmetric oxide spacers. In: International conference on emerging trends in computing and communication technologies (ICETCCT). Dehradun, pp 1–4
15. Wagaj SC, Patil YS (2016) A two-dimensional analytical model and simulation for dual material gate junctionless transistor. In: International conference on communication and signal processing (ICCASP 2016). Atlantis Press, Paris, France
16. Wagaj SC, Chavan YV (2017) Comparative study of single material gate and dual material gate silicon-on-insulator junction less transistors. In: 11th International conference on intelligent systems and control (ISCO). Coimbatore, pp 272–277
17. Lou HL, Zhang Y, Zhu (2012) A junction less nanowire transistor with a dual material gate. IEEE Trans Electron Dev 59(7)
18. Anvarifard MK (2018) A nanoscale modified junctionless with considerable progress on the electrical and thermal issue. Int J Numer Modeling 32(3):1–15
19. Bashir F, Murshid AM, Khanday FA (2020) Impact of pocket doping on the performance of planar SOI junctionless transistor. Silicon
20. Baruah RK (2015) Modelling and simulation of short channel junctionless transistor from an analog design perspective. Department of Electrical and Electronics Engineering, IIT Guwahati
21. Silvaco A (2016) User's Manual Silvaco, Inc. Santa Clara
22. Wong HS, White MH, Krutsick TJ, Booth RV (1987) Modeling of transconductance degradation and extraction of threshold voltage in thin oxide MOSFET's. Solid State Electron 30:953–968

23. Chakraborty S, Mallik A, Sarkar C (2008) Subthreshold performance of dual-material gate CMOS devices and circuits for ultralow power analog/mixed-signal application. IEEE Trans Electron Dev 55(3):827–832
24. Gnani E, Gnudi A, Reggiani S, Baccarani G, Shen N, Singh N, Lo G, Kwong D (2011) Numerical investigation on the junctionless nanowire FET. In: Proceedings 12th international conference ULIS, pp 1–4
25. Baruah RK, Paily RP (2014) A dual-material gate junction less transistor with High-K spacer for enhanced analog performance. IEEE Trans Electron Dev 61(1):123–128
26. Saxena M, Haldar S, Gupta M, Gupta R (2004) Design considerations for novel device architecture: hetero-material double-gate (HEM-DG) MOSFET with sub-100 nm gate length. Solid State Electron 48(7):1169–1174
27. Bhattacharyya A, Ramesh R (2016) Nanoscale circuit implementation using tri-metal gate engineered nanowire MOSFET with gate stack for analog/RF applications. J Comput Electron 16(1):155–161

Improvement of Leakage Current in Double Pocket FDSOI 22 nm Transistor Using Gate Metal Arrangement

Aryan Kannaujiya, Narendra Yadava, Mangal Deep Gupta, and Rajeev Kumar Chauhan

Abstract This research work shows numerical analysis of double pocket FDSOI MOSFET. The proposed device outperformed at nanoscopic scale and influence of work function on threshold voltage and leakage current have been analyzed thoroughly. Different gate metals are arranged in such a way that it forms stack of three metal layers as gate electrode. This work has been progressed by staking of titanium nitride, silver, and cobalt in 22 nm transistor. TiN/Ag/Co, Ag/Co/TiN, and Co/Ag/TiN are three metal gate arrangement which has been simulated to get improved results in terms of threshold voltage and low leakages. In the proposed device, two identical *p*-type pockets have been introduced in source and drain side of same doping concentration that increase effective channel length to decrease the depletion width (dB) and effects of electric field at junction has been minimized. All simulation work has been performed by using ATLAS TCAD.

Keywords FDSOI · FBE · Leakage · SCE's · Threshold voltage · I_{ON}/I_{OFF} ratio · Work function

1 Introduction

Silicon technologies have grown every year with high speed. The major obstacle that must be concentrate with silicon technologies is impacts of decreasing dimensions of devices [1]. The better device performance and from head to foot integration density can be achieved by scaling down of the devices [2]. Leakage current and short channel effects are major problem that downgrades the device performance which is accounted by decreasing the effective channel length [3]. The attempt of controlling this problem has resulted a new design technique for newer technologies like silicon-on-insulator (SOI) [4]. A very thin layer of silicon has been placed on the top of the insulator in SOI technology. Usually, SiO_2 is referred as buried oxide layer

A. Kannaujiya (✉) · N. Yadava · M. D. Gupta · R. K. Chauhan
Department of ECE, Madan Mohan Malaviya University of Technology, Gorakhpur, U.P. 273010, India
e-mail: aryankannaujiya1@gmail.com

B. Mishra and M. Tiwari (eds.), *VLSI, Microwave and Wireless Technologies*, Lecture Notes in Electrical Engineering 877,
https://doi.org/10.1007/978-981-19-0312-0_23

(BOX) [5]. The fabrication of MOSFET on SOI substrate which involves placing of very thin SOI layer is termed as fully depleted SOI and for thicker layer it is termed as partially depleted SOI [6]. However, bulk depletion width (d_B) is larger than the silicon thickness in FDSOI [7]. Device speed pointedly degraded if power supply is reduced below three times threshold voltage so that power supply scaling must be accompanied by reduction in threshold voltage. BOX layer of FDSOI yields low parasitic junction capacitance, small SCE's, subthreshold swing, and no kink effect [8]. There is a significant decrease in threshold voltage due to reduction in size of FDSOI which account of increasing leakage current and SCE's so that by choosing different metal gate work functions it is possible to set suitable threshold voltage without changing supply voltage resulting reduction in leakage current. The minimum energy in electron volts obligatory to take away an electron from a compact surface to a point in space is termed as work function [9].

Moreover, it has been reported that appropriate threshold voltage of device can be set by the help of thin Si-film method and channel doping but it downgrades the performance of device [10]. Therefore, this points up a need of tri-layer metal gate arrangement, i.e., work function engineering which is better resolution to overcome all short comings arising in nanoscopic devices. Hence, by arranging the metals gate an appropriate threshold voltage. Staking of metals has not been performed in work of Shaik et al. [11], but in the present work, staking of three layers of metals has been studied and the effective work function has been investigated.

2 Device Structure

The proposed device structure is publicized in Fig. 1. By incorporating three metal gate staking one by one a slight modification has been made in the device structure

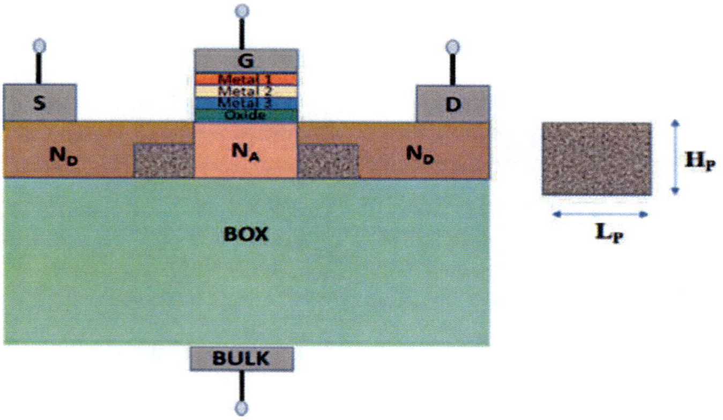

Fig. 1 Proposed structure of double pocket FDSOI

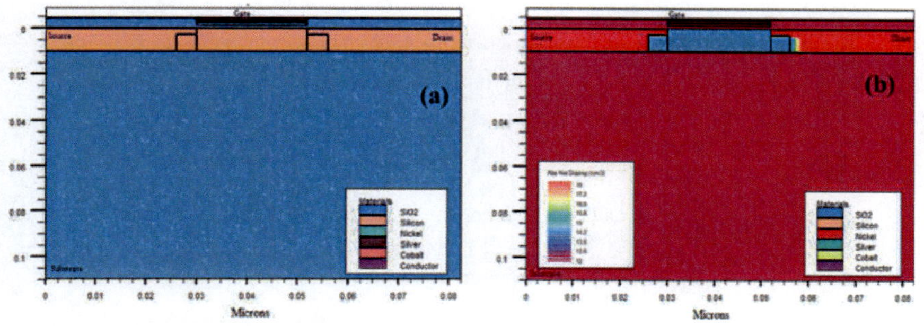

Fig. 2 **(a)** Simulation of proposed FDSOI structure in ATLAS TCAD and **(b)** counter plot of abs net doping profile of FDSOI in ATLAS TCAD

reported by Shaik et al. [11], to attain the proposed device configuration. Both pockets in source and drain side assist to decrease electric field intensity at junction. Figure 2a depicts simulated structure of double pocket FDSOI in Tony plot ATLAS TCAD. Figure 2b shows counter plot of doping profile of FDSOI at different sections in Tony plot. Doping of source and drain has been set 1×10^{18} cm^{-3}, and doping of channel is 1×10^{14} cm^{-3}. Staking of three layers of metal gate arrangement fashion has been configured.

The potential candidate for metal gate materials are TiN, Ag, and Co, respectively. These are stacked layer by layer to reach out appropriate threshold voltage and reduced leakage current. *P*-type material is used to both pockets to create an effective channel. Gate oxide thickness (t_{ox}) of 1.4 nm, channel thickness (t_{si}) of 10 nm, and gate length (L_G) of 22 nm are appropriately designed because electrical parameters of the proposed model depend on this regime. Buried oxide (t_{BOX}) layer of 100 nm thickness, L_p and H_P of 4 nm and 7.5 nm, respectively, has been optimized for better performance. Over all three types of metal gate arrangements have been performed to get optimized results. In the proposed configuration, BOX and gate oxide are made of silicon dioxide which is cheap in cost and source and drain is doped with *n*-type material with doping of 10^{18} cm^{-3} and channel doping of 10^{14} cm^{-3} by *p*-type material. Hence, a comprehensive simulation method is optimized for determining appropriate threshold voltage and low leakage current by arranging metal gate materials. In Table 1, geometrical parameters which have been taken into consideration for simulation and analysis has been illustrated [11].

3 Simulation Methodology

In this work, ATLAS TCAD tool is used for numerical simulation of n-channel FDSOI. Fabrication procedure for the proposed configuration is grounded on convention SOI excluding ion-implantation method to introduce the double pocket beside source and drain as discussed Fig. 1. The contact is declared to be neutral that

Table 1 Geometrical parameters of proposed device configuration [11]

Design parameters	Symbols	Double pocket FDSOI
Gate length (nm)	L_G	22
BOX thickness (nm)	t_{BOX}	100
Gate oxide thickness (nm)	t_{ox}	1.4
Channel thickness (nm)	t_{si}	10
Channel doping (cm^{-3})	N_A	1×10^{14}
Source/drain doping (cm^{-3})	N_D	1×10^{18}
Pocket length (nm)	L_P	4
Pocket height (nm)	H_p	7.5

yields low resistance at contacts [9]. The Selberherr's impact ionization model and Shockley–Read–Hall (SRH) recombination environment have been considered for simulation [12]. Concentration dependent and field dependent models are used for better mobility [13]. To evaluate high current density, Auger recombination model has been reckoned [14]. To evaluate the device under various biasing condition, numerical method in iterative fashion has been used.

4 Results and Analysis

In this section, outcomes of different metal gate arrangement have been discussed. This consists of three sub-sections. In these different metal gate arrangements, threshold voltage and leakage current have been discussed and finally optimized result has been proposed after simulation. Expression for the threshold voltage are as follows:

$$V_{th} = \varnothing_{MS} - \frac{Q_{ss}}{C_{ox}} + 2\varnothing_F + \frac{qN_A x_{d\,max}}{C_{ox}} \tag{1}$$

where V_{th} termed as threshold voltage, $\varnothing_{MS}$ is signified as difference between work function of channel and gate, Q_{ss} is signified as surface state charge, C_{ox} is termed as gate capacitance, $\varnothing_F$ is signified as fermi potential, $x_{(d\,max)}$ and N_A are symbolized as depletion width and doping concentration of channel, respectively.

4.1 Gate Metal Electrode Arrangement TiN/Ag/Co

In this subsection, stacking of metals has been followed by TiN/Ag/Co. Lower work function (WF) metal is incorporated near gate which is recognized as first layer, and higher work function of metal incorporated at third layer, and in middle of them the metal which have smaller value of WF than third layer and higher value of WF than first layer has been incorporated. Titanium nitride (TiN) has been used as the first layer nearest to gate because it has better thermodynamic stability with SiO_2. The work function of TiN, Ag, and Co are 4.5 eV, 4.7 eV, and 5.0 eV, respectively. In this case, it has been observed that the metal which is at the top layer dominate more than the other layer so that titanium nitride have more dominance than silver and cobalt. The rate of change of drain current with respect to gate voltage on linear scale and logarithmic scale has been publicized in Fig. 3a, b.

Since threshold voltage is directly proportional to metal work function, using these arrangements threshold voltage of 0.1 V has been calculated. Switching speed of the device followed by TiN/Ag/Co arrangement is very nice that the arrangement discussed in next sub-section. Neglecting the power consumption but the best switching speed has been achieved. Taking application point of view, device followed by this arrangement can be the best candidate for fast switching circuitry such as orthogonal frequency division multiplexing (OFDM). It is required to modulate carrier signal at high speed in OFDM. High speed digital switching is the primary aspect in this modulation so that TiN/Ag/Co metal gate arrangement make the device potential candidate for meeting the required conditions without any delay. It is observed that in Fig. 3a proposed device start working at 0.1 V with $I_{OFF} = 8.4 \times 10^{-10}$ (low leakage) in comparison with device reported by Shaik et al. as their device start working at 0.16 V and with $I_{OFF} = 3.4 \times 10^{-8}$ (high leakage). The device is cost effective as well and ten times better than the device proposed by Shaik et al. [11]. A comparative analysis between drain current and gate voltage has been depicted in Fig. 3a, b.

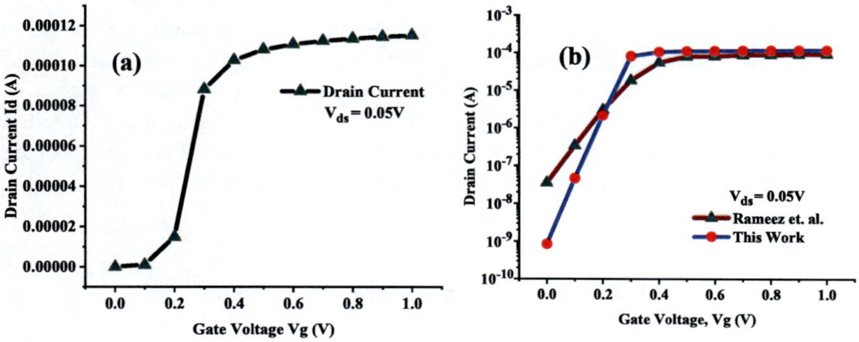

Fig. 3 Rate of change of drain current with respect to gate voltage on **(a)** linear scale and **(b)** logarithmic scale for arrangement of TiN/Ag/Co

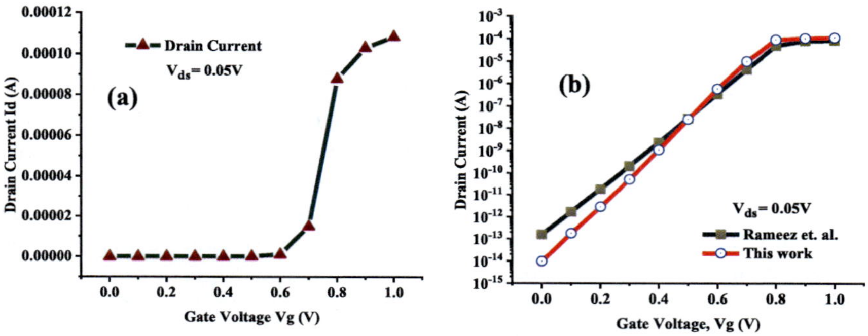

Fig. 4 Rate of change of drain current with respect to gate voltage on (**a**) linear scale and (**b**) logarithmic scale in arrangement of Co/Ag/TiN

4.2 Gate Metal Electrode Arrangement Co/Ag/TiN

This sub-section involves stacking of metals followed by Co/Ag/TiN. Higher work function metal is incorporated near to gate, recognized as first layer and metal of lower work function incorporated at third layer and in middle of them the metal which have smaller value than the first layer and higher value of WF than the third layer is placed. Cobalt has a direct contact with gate resulting higher dominance on the threshold voltage than the other two metals. Silver has been incorporated in the middle of them so that there is less chance of corrosion in electrode. The threshold voltage achieved after investigation in this arrangement is 0.6 V and off current, $I_{OFF} = 9.8 \times 10^{-15}$ with low leakage which is an achievement in comparison with the work reported by Shaik et al. [11] in which he had reported V_{th} of 0.65 V and $I_{OFF} = 6.1 \times 10^{-11}$. The electrode arranged are economical and cheaper than the electrode proposed by Shaik et al., therefore, proposed device is cost effective.

As per the application point of view, since arrangement of Co/Ag/TiN has V_{th} of 0.6 V than that of TiN/Ag/Co, i.e., $V_{th} = 0.1$ V, this arrangement suits for high frequency operation IOT fields. This arrangement has very low power consumption so that the device configured using Co/Ag/TiN can serve its better results in the field of smart irrigation system (IOT) and self-automated IOT area. The rate of change of drain current with respect to gate voltage on linear scale and has been publicized in Fig. 4a. On logarithmic scale, the best comparison between drain current and gate voltage reported by Shaik et al. [11] and proposed work has been shown in Fig. 4b.

4.3 Gate Metal Electrode Arrangement Ag/Co/TiN

This sub-section demonstrates the metal gate arrangement of Ag/Co/TiN. In this arrangement, the metal which has WF value higher than the third layer and lower

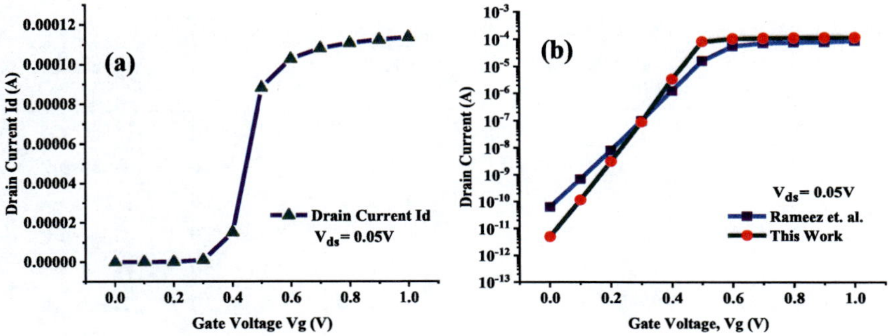

Fig. 5 Rate of change of drain current with respect to gate voltage on **(a)** linear scale and **(b)** logarithmic scale for arrangement of Co/Ag/TiN

than the second layer is incorporated at first layer forming a direct contact with gate. It has high impact on threshold voltage than the other two metals. Silver (Ag), cobalt (Co), and titanium nitride (TiN) have been introduced as first, second, and third layer, respectively.

As silver has better conductivity and chemical stability so that it has been placed at first layer making direct contact with gate. Since threshold voltage obtained by this arrangement is 0.3 V and $I_{\mathrm{OFF}} = 4.9 \times 10^{-12}$ with low leakages which is an attainment than the device reported by Shaik et al. as he demonstrated V_{th} of 0.48 V and $I_{\mathrm{OFF}} = 1.5 \times 10^{-13}$. The metals utilize as electrode are cheaper and easily available than the metal used by Shaik et al. in his device. Therefore, the proposed device is cost effective and ten times better than the device reported by Shaik et al. [11]. This gate metal arrangement has better thermodynamic stability than the other two arrangements as discussed above in Sects. 4.1 and 4.2, therefore, Ag/Co/TiN arrangement is the trade-off between the other arrangement.

As per the application point of view, the device configured with Ag/Co/TiN electrode is potential candidate for digital switching circuit configuration where high rate of switching is needed and for IOT application like smart irrigation system self-automated area like home automation system due to low leakages and fast switching. The proposed device switched on at 0.3 V. Figure 5a, b depicts relation between gate voltage and drain current on linear scale and logarithmic scale, respectively. Performance comparison of proposed model with the reported model by Shaik et al. [11], has been publicized in Table 2.

5 Conclusion

Three different approaches have been considered in present work. It has been well investigated that the different metal gate arrangements (TiN, Co, and Ag) affect threshold voltage significantly. Three metal gate positioning have been examined,

Table 2 Performance comparison with the reported work by Shaik et al. [11]

No.	Gate metal arrangement	Work function (WF) in eV	V_{th}	I_{OFF}	I_{ON}	References
1	Mo	4.6	0.16 V	3.4×10^{-8}	2.3×10^{-5}	[11]
2	TiN/Ag/Co	4.5/4.7/5.0	0.1 V	8.4×10^{-10}	7.9×10^{-5}	This work
3	Ru	4.8	0.38 V	6.1×10^{-11}	5.1×10^{-5}	[11]
4	Ag/Co/TiN	4.7/5.0/4.5	0.3 V	4.9×10^{-12}	8.6×10^{-5}	This work
5	Au	5.1	0.65 V	1.5×10^{-13}	4.6×10^{-5}	[11]
6	Co/Ag/TiN	5.0/4.7/4.5	0.6 V	9.8×10^{-15}	8.7×10^{-5}	This work

and they form tri-layer each other like TiN/Ag/Co, Ag/Co/TiN, and Co/Ag/TiN, respectively. Threshold voltage recorded at on these arrangements are 0.1, 0.3, and 0.6 V. It is found that proposed metal gate staking form better thermal stability in all three cases as addressed and discussed in [15]. There is no lattice mismatch in forming tri-layer as gate electrode with additional benefit cost effectiveness. For electrode, TiN/Ag/Co I_{OFF} is 8.4×10^{-10}, Ag/TiN/Co I_{OFF} is 4.9×10^{-12}, and CO/Ag/TiN I_{OFF} is 9.8×10^{-15}, respectively. Since device is proposed on nanoscopic scale so therefore two *p*-type pockets are made near the source and drain side to overcome the effect of electric field intensity at junction. By the help of ATLAS TCAD, all simulation steps have been performed.

References

1. Vaddi R, Agarwal RP, Dasgupta S (2012) Compact modeling of a generic double-gate MOSFET With gate–S/D underlap for subthreshold operation. IEEE Trans Electron Devices 59(10):2846–2849
2. (2015) International technology roadmap for semiconductors, online at: Itrs.net
3. Kim J, Lee J, Song I, Yun Y, Lee JD (2008) Accurate extraction of effective channel length and source/drain series resistance in ultrashort-channel MOSFETs by iteration method. IEEE Trans Electron Devices 55(10):2779–2784
4. Tsutsui G, Saitoh M, Nagumo T, Hiramoto T (2005) Impact of SOI thickness fluctuation on threshold voltage variation in ultra-thin body SOI MOSFETs. IEEE Trans Nanotechnol 4(3):369–373
5. Ohata A, Bae Y, Beranger CF, Cristoloveanu S (2012) Mobility enhancement by back-gate biasing in ultrathin SOI MOSFETs with thin BOX. IEEE Electron Device Lett 33(3):348–350
6. Peng C, Hu Z, Zhang Z, Huang H, Ning B (2013) A new method for extracting the radiation induced trapped charge density along the STI sidewall in the PDSOI NMOSFETs. IEEE Trans Nucl Sci 60(6):4697–4704
7. Dan SS, Biswas A, Royer CL, Grabinski W, Ionescu AM (2012) A novel extraction method and compact model for the steepness estimation of FDSOI TFET lateral junction. IEEE Electron Device Lett 33(2):140–142
8. Mehrad M, Zareiee M, Orouji AA (2017) Controlled kink effect in a novel high-voltage LDMOS transistor by creating local minimum in energy band diagram. IEEE Trans Electron Devices 64(10):4213–4218

9. Vitale SA, Kedzierski J, Healey P, Wyatt PW, Keast CL (2011) Work-function-tuned TiN metal gate FDSOI transistors for subthreshold operation. IEEE Trans Electron Devices 58(2):419–426
10. Chen Y, Chang S, Hsueh T (2015) Si-based MOSFET and thin film transistor prepared via hot wire implantation doping technique. IEEE Electron Device Lett 36(2):93–95
11. Shaik RR, Arun G, Chandrasekar L, Pradhan KP (2020) A study of work function variation in pocket doped FD-SOI technology towards temperature analysis. SILICON 12(12):3047–3056
12. Subirats A, Garros X, Husseini JE, Royer CL, Reimbold G, Ghibaudo G (2013) Impact of single charge trapping on the variability of ultrascaled planar and trigate FDSOI MOSFETs: experiment versus simulation. IEEE Trans Electron Devices 60(8):2604–2610
13. Silvaco (2016) Atlas user's manual. Silvaco, Inc., CA 95054(408):567–1000
14. Lin Y et al (2019) A study of effects of metal gate composition on performance in advanced n-MOSFETs. IEEE Trans Electron Devices 66(8):3286–3289
15. Deshpande A, Imran A, Paul R, Yuan Z, Peng H, Luo F (2020) High power density1700-V/300-A Si-IGBT and SiC-MOSFET hybrid switch-based half-bridge power module. In: IEEE Energy conversion congress and exposition (ECCE), Detroit, MI, USA, pp 3979–3986

Beat Frequency Detection on Boolean Chaotic Oscillator for True Random Number Generation on FPGA

R. Sivaraman, N. Adithya, Siva Janakiraman, and Amirtharajan Rengarajan

Abstract This work proposes a new True Random Number Generator (TRNG) design on Intel Cyclone II FPGA for cryptographic key generation applications. Boolean Chaotic Oscillator (BCO) has been utilised in this work to generate true randomness where the entropy of TRNG has been harvested through beat frequency detection technique. 50 MHz of the operating clock has been used for true random number generation. Linear Feedback Shift Register (LFSR) has been adopted as a corrector function to enhance the randomness of TRNG. The entropy source requires only 9 Logic Elements (LEs) and the proposed TRNG architecture consumed 783 logic elements. Throughput has been achieved as 27.306666 Mbps. Entropy, hamming distance calculation and NIST 800–22 analyses have been performed to validate the design. Further, true randomness of the proposed TRNG has been verified through restart experiment.

Keywords Key generation · True randomness · FPGA · Sampling · Beat frequency detection · LFSR

1 Introduction

The improvements in the field of communication open a back door for security threats. Cryptography is the best solution to protect the confidentiality of information being shared. The resilience of cryptographic algorithms depends on its key.

R. Sivaraman · N. Adithya · S. Janakiraman · A. Rengarajan (✉)
School of EEE, SASTRA Deemed University, Thanjavur 613401, India
e-mail: amir@ece.sastra.edu

R. Sivaraman
e-mail: sivararaman77uma@sastra.ac.in

N. Adithya
e-mail: 121004010@sastra.ac.in

S. Janakiraman
e-mail: siva@ece.sastra.edu

B. Mishra and M. Tiwari (eds.), *VLSI, Microwave and Wireless Technologies*,
Lecture Notes in Electrical Engineering 877,
https://doi.org/10.1007/978-981-19-0312-0_24

Key is a primary element generated from Random Number Generators (RNGs). In general, RNGs are classified into two: Deterministic Random Number Generators (DRNGs) and True Random Number Generators (TRNGs) [1]. TRNGs are preferred over PRNGs because of its unpredictability property. TRNGs have been utilised in various application fields such as key generation, initialisation vector generation, nonces generation, random simulation and games. TRNG has three basic building blocks on-chip namely, source of entropy, entropy extraction and sampling. Source of entropy is the origin for true randomness. Entropy extraction is used to harvest the true randomness from the source. Finally, the extracted bits are sampled to create n-bit true random numbers. In addition to these blocks, post-processing block has been optionally utilised to increase the randomness of TRNG [2]. Hardware is an eminent source for true randomness where the intrinsic characteristics of an Integrated Circuit (IC) exhibit true random behaviour. Field Programmable Gate Array (FPGA) is reconfigurable hardware. Which is predominantly preferred for TRNGs due to its advantages namely, reconfigurability, reusability, fast turnaround time, easy prototyping, customisable digital design capability, on-chip and off-chip features. Clock jitters and metastability are the two options for entropy source in FPGA [3].

So far, many TRNGs have been proposed with various entropy sources such as oscillator noise [4], on-chip PLLs of Altera Stratix FPGA [5], Ring Oscillator (RO) with two rings [6], RO eth multi-rings [7–10], Galois Field RO [11], Flip flop metastability [12] and chaotic oscillators [13–16]. Three types of entropy sources have been utilised on FPGA to generate true random numbers on analysing the earlier works. Considering metastability, it requires a large number of latches. Chaotic approaches also consume a high volume of logic resources on FPGA. On the other hand, RO is a good option for TRNG, which consumes a relatively low area compared with metastability and chaotic oscillators based implementations. So far, many improvements have been made on RO structure to yield compact TRNG designs. This work aims to provide a one with another solution for TRNG on FPGA by combining Boolean Chaotic Oscillator (BCO) and Beat Frequency Detection. Two rings of BCO serves as an entropy source where the beat frequency detection method has been employed to harvest the randomness. To enhance the statistical properties, LFSR has been used.

2 Proposed TRNG Architecture

The proposed TRNG comprise BCOs as an entropy source, XOR as extraction mechanism and LFSR as post-processing unit. The architectural representation of the proposed TRNG has been displayed in Fig. 1. Entropy source has two rings of BCO where a BCO is constructed using 3 inverters, 6 buffers and an XOR gate. The conventional 3 inverters RO structure has been modified with an XOR gate whose output is delayed with the series of inverters and fed back to one of its inputs. This arrangement produces arbitrary clocks influenced by the jitters produced from the 3 inverters RO structure. Register Transfer Level (RTL) diagram of BCO has been depicted in Fig. 2.

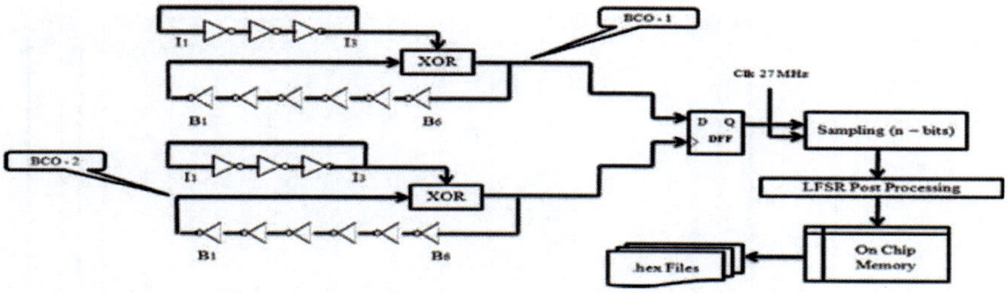

Fig. 1 Architectural representation of TRNG

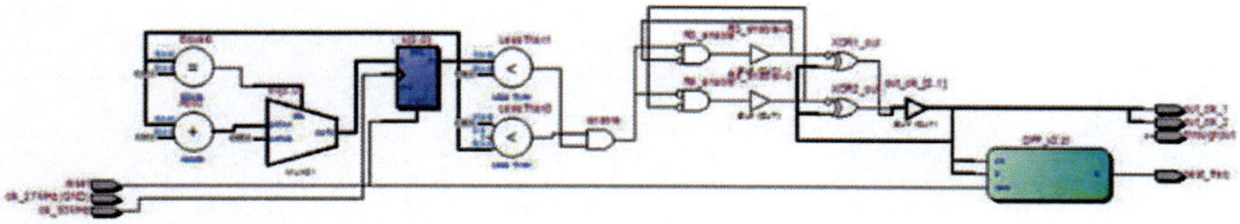

Fig. 2 RTL diagram of BCO

2.1 Linear Feedback Shift Register

LFSR is utilised to intensify the randomness through which the statistical characteristics of TRNG have been improved [17, 18]. Generally, LFSR is being a polynomial governed shift register to generate pseudo-random sequences. Tapings are the key factor in LFSR to bring randomness in its values. In this work, 128-bit LFSR has been designed using d flip flops and XOR gates with respect to the $X127 + X125 + X100 + X98 + 1$ polynomial as shown in Fig. 3. Fibonacci LFSR structure has been adopted in this work where the XORing has been performed after shifting the data. An initial value is required for LFSR to trigger it. Raw true random numbers have been XORed with the LFSR generated random sequences to strengthen the true randomness. Further, true random sequences have been collected in Block Random Access Memory (BRAM) for further analysis. Figures 4 and 5 exhibit the RTL diagram of proposed TRNG and BRAM snapshot true random data, respectively.

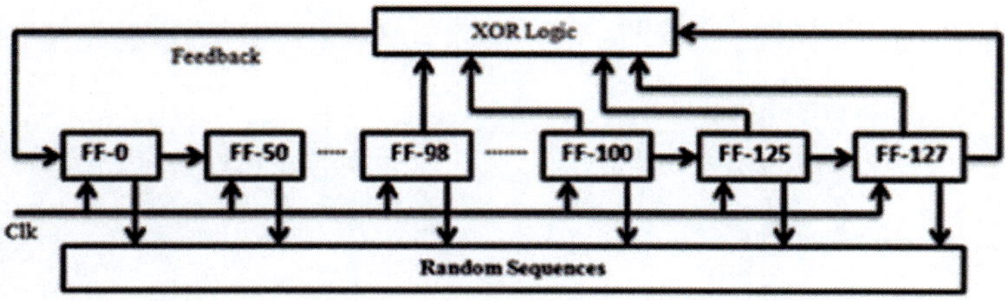

Fig. 3 128 bit LFSR architecture

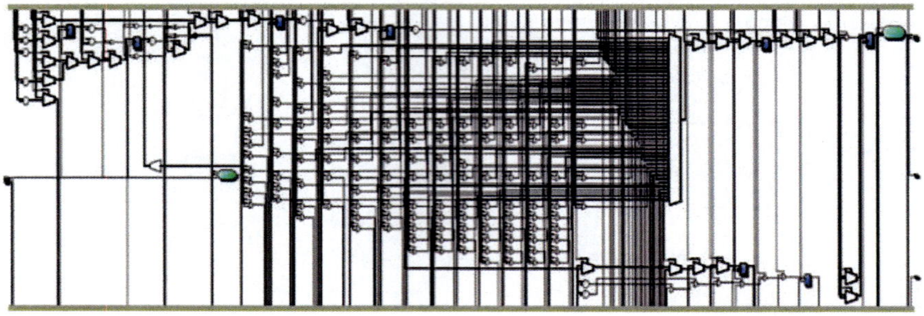

Fig. 4 RTL diagram of the proposed TRNG

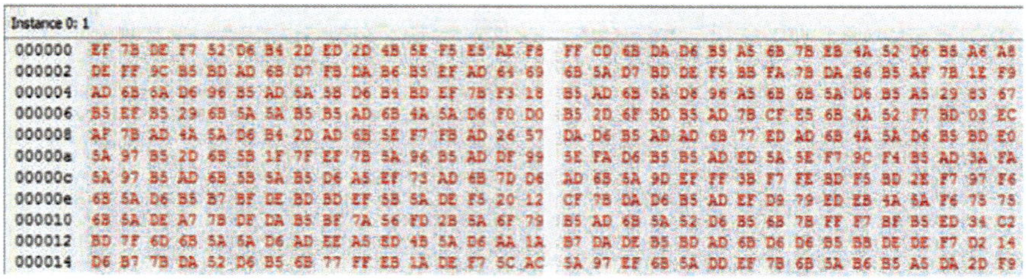

Instance 0: 1

000000	EF	7B	DE	F7	52	D6	B4	2D	ED	2D	4B	5E	F5	E5	AE	F8	FF	CD	6B	DA	D6	B5	A5	6B	7B	EB	4A	52	D6	B5	A6	A8
000002	DE	FF	9C	B5	BD	AD	6B	D7	FB	DA	B6	B5	EF	AD	64	69	6B	5A	D7	BD	DE	F5	BB	FA	7B	DA	B6	B5	AF	7B	1E	F9
000004	AD	6B	6A	D6	96	B5	AD	5A	5B	D6	B4	BD	EF	7B	F3	18	B5	AD	6B	5A	D6	96	A5	6B	6B	5A	D6	B5	A5	29	83	67
000006	B5	EF	B5	29	6B	5A	5A	B5	B5	AD	6B	4A	5A	D6	F0	D0	B5	2D	6F	BD	B5	AD	7B	CF	E5	6B	4A	52	F7	BD	03	EC
000008	AF	7B	AD	4A	5A	D6	B4	2D	AD	6B	5E	F7	FB	AD	26	57	DA	D6	B5	AD	AD	6B	77	ED	AD	6B	4A	5A	D6	B5	BD	E0
00000a	5A	97	B5	2D	6B	5B	1F	7F	EF	7B	5A	96	B5	AD	DF	99	5E	FA	D6	B5	B5	AD	ED	5A	5E	F7	9C	F4	B5	AD	3A	FA
00000c	5A	97	B5	AD	6B	5B	5A	B5	D6	A5	EF	73	AD	6B	7D	D6	AD	6B	5A	9D	EF	FF	3B	F7	FE	BD	F5	BD	2E	F7	97	F6
00000e	6B	5A	D6	B5	B7	BF	DE	BD	BD	EF	5B	5A	DE	F5	20	12	CF	7B	DA	D6	B5	AD	EF	D9	79	ED	EB	4A	5A	F6	75	75
000010	6B	5A	DE	A7	7B	DF	DA	B5	BF	7A	56	FD	2B	5A	6F	79	B5	AD	6B	5A	52	D6	B5	6B	7B	FF	F7	BF	B5	ED	34	C2
000012	BD	7F	6D	6B	5A	5A	D6	AD	EE	A5	ED	4B	5A	D6	AA	1A	B7	DA	DE	B5	BD	AD	6B	D6	D6	B5	BB	DE	DE	F7	D2	14
000014	D6	B7	7B	DA	52	D6	B5	6B	77	FF	EB	1A	DE	F7	5C	AC	5A	97	EF	6B	5A	DD	EF	7B	6B	5A	B6	B5	A5	DA	2D	F9

Fig. 5 BRAM snapshot of 128 bit TRNG

3 Experimentation and Validation

Proposed TRNG architecture is devised using VHDL and realised on Intel Cyclone II FPGA through Quartus 13.0 Electronic Design Automation (EDA) tool. True random numbers have been generated with respect to four different sizes of 128, 64, 32 and 16-bits. To assess the TRNG's statistical characteristics, entropy, switching activity estimation and NIST SP 800–22 batteries of tests were carried out. True randomness has been evidenced through restart experiment. Further, hardware quality analyses have been conducted to validate the hardware efficiency. The following statistical analyse have been carried out by considering the three predominant properties of RNGs such as,

- Uniform distribution of '0' and '1'
- Unpredictability
- No pattern generation

3.1 Entropy Analysis

It is a fundamental metric to test the equidistribution property of any random number generator. It is defined as the probability of occurrences of '1' and '0' in the generated

Table 1 Entropy analysis

TRNG data sets	Untreated TRNG (bits)				Post-processed TRNG (bits)			
	128	64	32	16	128	64	32	16
1	0.983015	0.962718	0.901175	0.892889	1	0.999999	0.999985	0.999976
2	0.981097	0.961699	0.906238	0.893470	0.999986	0.999996	0.999984	0.999983
3	0.988863	0.961442	0.906324	0.896246	0.999995	0.999977	0.999959	1
4	0.981342	0.961349	0.905909	0.895552	0.999997	0.999997	0.999999	0.999992
5	0.982328	0.961761	0.904754	0.894295	0.999991	0.999999	1	0.999999

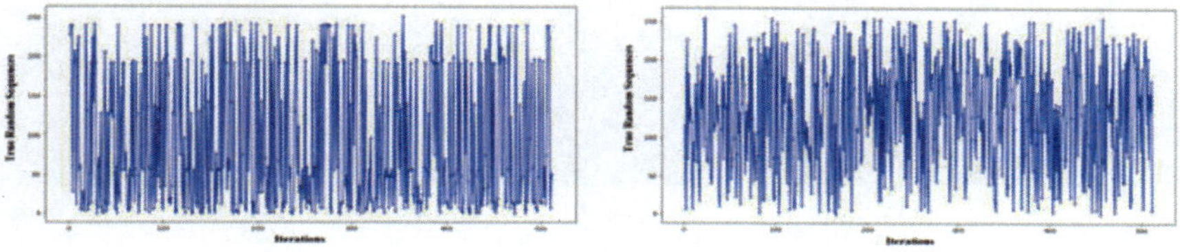
(a) (b)

Fig. 6 Equidistribution plot: **a** untreated TRNG sequences and **b** post-processed TRNG sequences

true random sequences [19]. It is expected from an RNG to have entropy close to 1 to evidence the uniform distribution. In this work, entropy analysis has been performed, and the results have been tabulated. Table 1 presents the entropy for untreated true random bits and post-processed (treated) true random bits. From the results, it has been observed that post-processed true random sequences have attained high equidistribution than raw sequences. Further, Figs. 6a, b shows the pictorial representation of equidistribution for 8-bit TRNGs of raw and post-processed numbers. In Fig. 6a, true random sequences were highly biased to '0' whereas; in Fig. 6b uniform distribution of data between 0 and 255 has been achieved.

3.2 *Switching Activity Estimation Analysis*

Hamming distance is a unique analysis to estimate the switching activity of TRNG. Switching activity is a fundamental criterion of random sequences expected as half of the sequence length to yield high randomness [20]. The transition of '0' to '1' and '1' to '0' influences the randomness. Hence, Hamming distance is defined as the difference between the amounts of change of bits in true random sequences generated from the same TRNG. Table 2 depicts the hamming distance calculation of untreated and post-processed true random sequences. From the table, it has been inferred that post-processed true random numbers have been achieved adequate switching when compared with raw true random numbers.

Table 2 Hamming distance calculation for proposed TRNG

TRNG data sets	Untreated TRNG (bits)				Post-processed TRNG (bits)			
	128	64	32	16	128	64	32	16
1	53.2343	30.5751	13.3554	5.3427	63.6865	32.1152	16.0390	8.0683
2	53.2190	30.1757	13.8125	5.4318	63.8229	32.0930	16.0338	8.0273
3	51.7986	30.4358	13.4670	5.7177	63.8677	32.0735	16.0484	8.0019
4	51.9307	30.4778	13.4380	6.0158	63.9244	32.0478	16.0484	7.9935
5	51.6896	30.5337	13.4834	6.2214	63.9635	32.0253	16.0344	7.9947

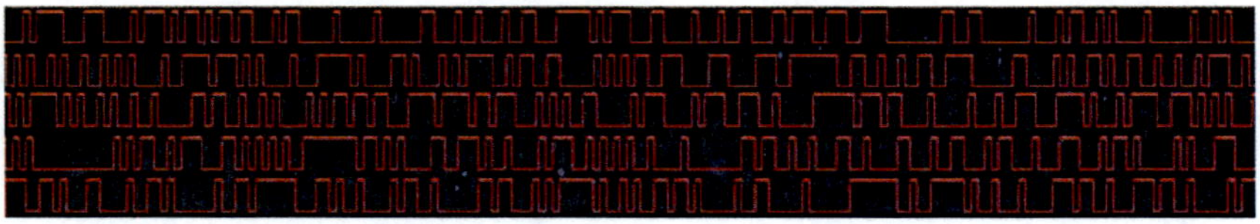

Fig. 7 Restart experiment: untreated true random sequences

3.3 Restart Experiment

Restart experiment distinguishes the TRNG from PRNG in which TRNG generates different true random data for every iteration. In contrast, PRNG produces the same amount of random data for all the '*n*' iterations [21]. True randomness has been evidenced through this analysis. Figure 7 shows the restart experiment of proposed TRNG where each of the sequence exhibits a different bit pattern. TRNG generates a new sequence of bits for every iteration which confirms the true random nature of it.

3.4 NIST SP 800–22 Batteries of Test

National Institute of Standards and Technology (NIST) has published a standard test suite of several tests (T_1 to T_{14}) to grade RNGs [22]. The test batteries (T_1-Frequency, T_2-Block frequency, T_3-Cumulative Sums—I, T_4-Cumulative Sums—II, T_5-Runs, T_6-Longest Run, T_7-Rank, T_8-FFT, T_9-Non-overlapping Template, T_{10}-Overlapping Template, T_{11}-Approximate Entropy, T_{12}-Serial—I, T_{13}-Serial—II, T_{14}-Linear Complexity) will test the given true random sequences in terms of specific parameters such as a total number of bits, block length, block size. In this work, NIST SP 800–22 has been conducted to verify the randomness with 10 different sets of truly random sequences where each set contains 1,32,072 bits. To qualify the NIST test, the *p-value* of each test must be greater than 0.001. All the four-bit widths

Table 3 Results of NIST SP 800–22 for 128 bit TRNG

Test parameters	*p-values* of 5 different test data sets				
	$S1$	$S2$	$S3$	$S4$	$S5$
T_1	0.350485	0.534146	0.534146	0.739918	0.739918
T_2	0.066882	0.350485	0.350485	0.017912	0.534146
T_3	0.739918	0.122325	0.122325	0.739918	0.534146
T_4	0.739918	0.350485	0.213309	0.035174	0.911413
T_5	0.739918	0.350485	0.350485	0.534146	0.534146
T_6	0.739918	0.122325	0.911413	0.213309	0.739918
T_7	0.350485	0.534146	0.350485	0.534146	0.350485
T_8	0.991468	0.122325	0.122325	0.911413	0.534146
T_9	0.911413	0.739918	0.739918	0.911413	0.739918
T_{10}	0.911413	0.739918	0.534146	0.739918	0.534146
T_{11}	0.534146	0.213309	0.739918	0.911413	0.739918
T_{12}	0.534146	0.213309	0.739918	0.350485	0.911413
T_{13}	0.911413	0.739918	0.350485	0.534146	0.350485
T_{14}	0.017912	0.122325	0.350485	0.911413	0.035174

of the proposed TRNGs have been passed the NIST test, and Table 3 presents the NIST test results of 128-bit TRNG.

From the results, it has been evidenced that the proposed TRNG has been cryptographically strong and the NIST SP 800–22 verifies the statistical independency of the TRNG.

3.5 Hardware Analysis

Since the implementation has been done on FPGA, it is necessary to examine the hardware competence of the proposed TRNG in terms of LUTs, logic registers, combinational functions and BRAM bits. Power dissipation (Static—SP, dynamic—DP, Input/Output—I/O) has been determined through Power Play Power Analyzer tool available in the Quartus 13.0 EDA tool where static, dynamic and Input Output (IO) power dissipation has been calculated. Tables 4 and 5 tabulates the hardware analyses of proposed TRNG. Also, timing analysis has been carried out for 128 × 1024 TRNG architecture through interfacing the target FPGA with Zero+ logic analyser. Figure 8 exhibits the timing analysis in which the BCO TRNG has been taken 4.8 ms to generate 1, 32, 072 true random bits which turn into the throughput of 27.306666 Mbps. Further, the proposed TRNG architecture has been compared with the earlier works, and the performance comparison has been presented in Table 6. The comparison confirms that the proposed TRNG architecture consumes low area and yields adequate throughput.

Table 4 Hardware and power dissipation analysis on TRNG: with no post-processing

Parameters (FPGA—Altera cyclone II EP2C20F484C7)		128	64	32	16
LEs consumption	Combinational functions	519	369	296	254
	Logic registers	533	340	243	194
Power dissipation (mW)	SP	47.36	47.36	47.36	47.35
	DP	4.37	2.50	1.44	1.01
	I/O	20.99	20.99	20.99	20.99
BRAM bits (total iterations 'N' = 1024)		131,072	65,536	32,768	16,384

Table 5 Hardware and power dissipation analysis on TRNG: with LFSR post-processing

Parameters (FPGA—Altera cyclone II EP2C20F484C7)		128	64	32	16
LEs consumption	Combinational functions	783	528	403	331
	Logic registers	804	574	418	353
Power dissipation (mW)	SP	47.36	47.36	47.36	47.35
	DP	4.42	2.33	1.42	1.14
	I/O	20.99	20.99	20.99	20.99
BRAM bits (total iterations 'N' = 1024)		131,072	65,536	32,768	16,384

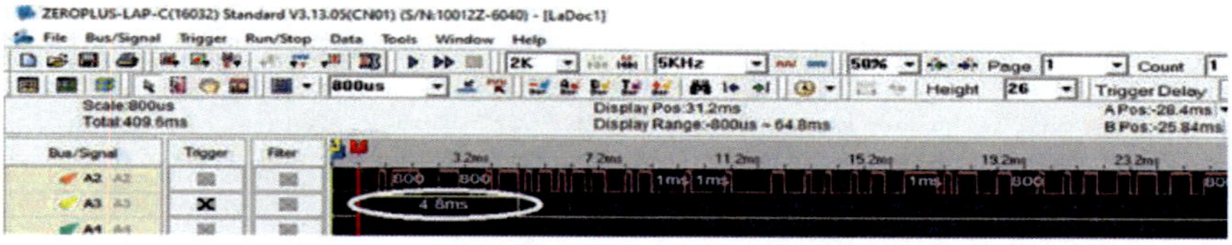

Fig. 8 Timing analysis for 128 × 1024 bits generation

4 Conclusions

Boolean Chaotic Oscillator governed TRNG has been proposed for four different bit length such as 128, 64, 32 and 16 bits. The propose entropy source consumes only 9 logic elements on Cyclone II FPGA where the TRNG architecture requires 783 logic elements and dissipates 72.77 mW of power. Operating clock of 50 MHz (onboard clock) has been used to produce true random numbers in which the TRNG design yields 27.306666 Mbps as throughput. Further, the throughput can be improved by utilising clock synthesisers to increase the clock speed. BCO TRNG have been evaluated using NIST SP 800–22 test suite. Future work is the implementation of non-identical BCO TRNG on FPGA to investigate the effect of true randomness.

Table 6 Performance comparison

Reference	TRNG design	Target device	Hardware utility	Throughput
Ref. [4]	Oscillator phase noise	Xilinx XCV300E–8	310 slices	225 bps
Ref. [5]	On-chip PLLs	Stratix EP1S25F780C5 FPLD	120 logic elements	>1 Mbps
Ref. [6]	RO with beat frequency detection	Xilinx Virtex XCV1000	1 CLB	0.5 Mbps
Ref. [8]	Multi event ROs	Xilinx XC2VP30	973 slices	2.5 Mbps
Ref. [9]	RO with delay lines	Xilinx Virtex 5	973 slices and 334 flip flops	12.5 Mbps
Ref. [10]	Multi event ROs	Intel Cyclone II	167 logic elements	100 Mbps
Ref. [12]	RS latch metastability	Xilinx XC4VFX20	580 slices	12.5 Mbps
Ref. [14]	Memristive chaos	Xilinx Kintex 7	311 LUTs and 133 flip flops	125 Kbps
Proposed TRNG	BCO with beat frequency detection	Intel Cyclone II	783 logic elements(for 128 bit TRNG)	27.306666 Mbps
			9 logic elements (Entropy source)	

Acknowledgements Authors thank Department of Science & Technology, New Delhi for the FIST funding (SR/FST/ET-I/2018/221(C)). Also, authors extend their sincere thanks to Dr. R. Sundararaman, Former Senior Assistant Professor, School of EEE, SASTRA Deemed University for his guidance and technical support.

References

1. Sunar B (2009) True random number generators for cryptography
2. Bagini V, Bucci M (1999) A design of reliable true random number generator for cryptographic applications. Cryptogr Hardw Embed Syst 204–218
3. Elbirt AJ, Yip W, Chetwynd B, Paar C (2001) An FPGA-based performance evaluation of the AES block cipher candidate algorithm finalists. IEEE Trans Very Large Scale Integr Syst 9(4):545–557
4. Tsoi KH, Leung KH, Leong PHW (2003) Compact FPGA-based true and pseudo random number generators. In: IEEE symposium on FPGAs custom computing machines proceedings, vol 2003, Jan 2003, pp 51–61
5. Fischer V, Drutarovský M, Šimka M, Bochard N (2004) High performance true random number generator in altera stratix FPLDs. In: Becker J, Platzner M, Vernalde S (eds) Field programmable logic and application: 14th international conference, FPL 2004, Leuven, Belgium, Aug 30–Sept 1 2004. Proceedings. Springer, Berlin, Heidelberg, pp 555–564

6. Kohlbrenner P, Gaj K (2004) An embedded true random number generator for FPGAs. In: Proceedings of the 2004 ACM/SIGDA 12th international symposium on field programmable gate arrays, pp 71–78
7. Sunar B, Martin WJ, Stinson DR (2007) A provably secure true random number generator with built-in tolerance to active attacks. IEEE Trans Comput 56(1):109–119
8. Schellekens D, Preneel B, Verbauwhede I (2006) FPGA vendor agnostic true random number generator. In: 2006 international conference on field programmable logic and applications FPL, pp 139–144
9. Jessa M, Matuszewski L (2012) The use of delay lines in a ring-oscillator-based combined true random number generator. In: 2012 international conference on signals and electronic systems ICSES 2012—Conference proceedings
10. Wold K, Tan CH (2008) Analysis and enhancement of random number generator in FPGA based on oscillator rings. In: Proceedings—2008 international conference on reconfigurable computing and FPGAs, ReConFig 2008, vol 2009, pp 385–390
11. Jessa M, Matuszewski L (2011) Enhancing the randomness of a combined true random number generator based on the ring oscillator sampling method. In: 2011 international conference on reconfigurable computing and FPGAs, pp 274–279
12. Hata H, Ichikawa S, Member S (2012) FPGA implementation of metastability-based true random. IEICE Trans Inf Syst 2:426–436
13. Bonny T, Al R, Sohaib D, Ahmed M (2019) Hardware optimized FPGA implementations of high-speed true random bit generators based on switching-type chaotic oscillators. Circuits Syst Signal Process 38(3):1342–1359
14. Science N, Phenomena C (2019) A true random bit generator based on a memristive chaotic circuit: analysis, design and FPGA implementation. Chaos Solitons Fractals Interdiscip J Nonlinear Sci Nonequilibrium Complex Phenom 119:143–149
15. ˙Koyuncu I, Tuna M, Bu C (2020) Design, FPGA implementation and statistical analysis of chaos-ring based dual entropy core true random number generator. Analog Integr Circuits Signal Process 0123456789:445–456
16. Murat A, Ebubekir G (2019) Implementation and performance analysis of true random number generator on FPGA environment by using non-periodic chaotic signals obtained from chaotic maps. Arab J Sci Eng 44(11):9427–9441
17. Fischer V, Bernard F (2011) True random number generators in FPGAs. In: Security trends for FPGAS, pp 73–100
18. Schulz RA (1990) Random number generator circuit, Patent No. US 6,807,553 B2
19. Wu Y, Zhou Y, Saveriades G, Agaian S, Noonan JP, Natarajan P (2013) Local Shannon entropy measure with statistical tests for image randomness. Inf Sci (Ny) 222:323–342
20. Pieterse V, Black P (2001) Hamming distance. In: MATH32031: coding theory, pp 706–706
21. Martin H, Peris-Lopez P, Tapiador JE, San Millan E (2016) A new TRNG based on coherent sampling with self-timed rings. IEEE Trans Ind Inform 12(1):91–100
22. Bassham LE et al (2010) A statistical test suite for random and pseudorandom number generators for cryptographic applications. National Institute of Standard Technology, Gaithersburg, MD, USA, Technical Report, no April

Femtocell-Based Interference Reduction Methodology for Self-organized Networks: A Green Network Prospective

Arshita Batra, Simran Wasnik, and Sanjay Kumar Biswash

Abstract Femtocell is a one of the powerful technique to improve network performance in personal communication systems. It provides better network coverage, less network interference, and helps to reduces the load on macrocells and associated base stations (BS) using the self-organizing network (SON), it provides a robust network improvement procedure, automatic configuration, diagnostic action, and healing of networks. In this paper, we analyze technical issues of SON which includes the uplink interference, and we solve it by game theory-based optimization technique. The opponent selects the best strategy subject to the network interference to form the associated Femtocell and access point. If it is less than the threshold then apply a greedy method is applied to improve the performance parameters for enhancing QoS and QoE.

Keywords Femtocell · Self-organizing networks · Uplink interference · Quality of service · Performance

1 Introduction

Day-by-day the mobile users are seeking high data rates and better network services, as it is a fundamental requirement of modern wireless cellular communication [5]. The cellular networks are equipped with the base station (BS) and network-hardware. The BS is unable to cope with these issues of high traffic load and call drop rates of the cellular network with the massive number of end-users. The greed for the better quality of service (QoS) for cellular network leads us to high energy consumption [3, 7]. It was the motivation for green network technologies. To achieve it, many researchers have performed many experiments, i.e., cell breathing or zooming, intelligent network configuration, etc. [2, 10]. These network solutions not only full fill

A. Batra · S. Wasnik · S. K. Biswash (✉)
Department of Computer Science and Engineering, NIIT University, Neemrana, Rajasthan 301705, India
e-mail: skbismu@gmail.com

B. Mishra and M. Tiwari (eds.), *VLSI, Microwave and Wireless Technologies*,
Lecture Notes in Electrical Engineering 877,
https://doi.org/10.1007/978-981-19-0312-0_25

the high demand of network users also focuses on establishing the green cellular communication [1].

The Femtocell technique is one of the best methods to achieve a high data rate and green networking, and our contribution is in line with the same. The Femtocells are capable of resolve the issue of coverage area thereby expanding network connectivity, also providing the fastest and cheapest network source to dynamic mobile users as they are deployed within the cells (smallest unit of network coverage area) without any major configuration change.

In this paper we discuss the technique of self-optimization and configuration of Femtocell, it will lead us to the concept of self-organizing network (SON) in Femtocell-based networks. Its basic schema is shown in Fig. 1. The authors of [9] discuss an algorithm is used and deployed within Femtocell which takes control of manual interventions, performance, and adjusting according to network requirements keeping the cell optimized all the time [8], and it has centralized, distributed, and hybrid model is available in the literature. The Femtocell-based cellular networks help in enhancing the network abilities thereby reducing the manpower as well to achieve green networking. However, Femtocell is unsuccessful to solve the problem of the uplink interference issue. Our next contribution provides a unique solution to overcome the interference problems of SON. Here, we use the dynamic power control mechanism by game theory. It provides an optimal solution to the interference problem [4].

Rest of the paper is organized as follows. The SON and its architecture is discuss in introduction. The proposed solution to achieve the objective is discussed in Sect. 2. To discuss the mathematical model and performance analysis we use Sect. 3. The conclusions are presented in Sect. 4 followed by references.

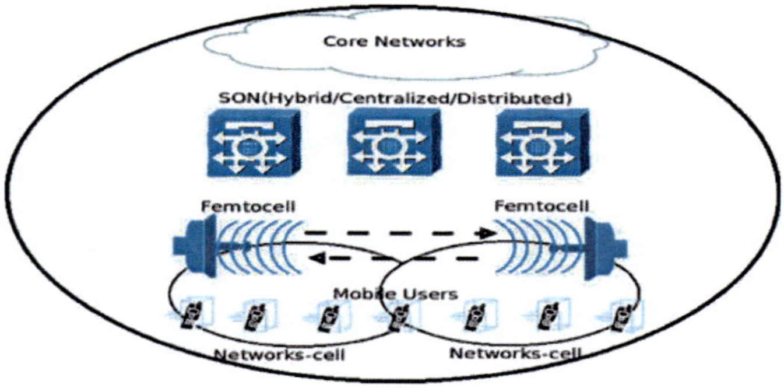

Fig. 1 The SON architecture and control mechanism using Femtocell

2 Proposed Methodology

In this paper, we propose a new method to reduce the interference between Femtocell and mobile access points. The process is illustrated in Figs. 3 and 4. It follows the concept of dynamic power control over the Femtocell network and we apply the game theory technique. Each Femtocell is denoted as a game player that will choose the transmission power level according to their range and keeping other factors in line for improved performance. It provides enhance utility for better throughput and an optimal SINR value. A higher transmission power will act as a first player and having a fair chance for transmission to improve the network performance. The dynamic power control technique shown in Fig. 3. It is used to reduce the transmission power as the other networks are available within proximity area. Associated Femtocell or access points analyzes the traffic and accordingly increases or decreases its output power of transmission as per the game theory model and optimizing it in [6]. Its objective to obtain a satisfactory throughput over the Femtocell and save the transmit power and improve the player's efficiency by a utility function. A simulation-based performance illustration of the deployment scenario is shown in Fig. 4. We focus on the interference between densely populated Femtocell access points thereby improving system performance, efficiency, and node mobility.

3 Analytical Formulation and Performance Analysis

The game theory includes three major components. First player, who have a problem at hand. Here we consider the Femtocell access point as the player and it is denoted by k. Second, the strategies each player have a set of strategies, and it determines what each player can do? Last the utility function, provide the degree of satisfaction as functions of the combination among all player's choices. Is the improved efficiency and satisfaction game known as the payoff function? We map it work with 2 player game. The Femtocell and access points are player and opponent are also same. The player1 and opponent (player2) select Femtocell, then who have the maximum transmission power. We set the transmission power level as p and 0, these are marked as strategies of the game. Therefore we use $K = 2$ as players and strategy sets are $S_k = \{0, p\}$. Here both of the players are supposed to choose their strategy simultaneously so that the game is finite and static (such that no one knows others move). The utility function is the main part of the game, it the degree of satisfaction achieved by both outputs of the transmission and energy spent at power s_k. To describe this situation a mathematical function. Here, we introduce a power gain function $h_{j,k} \in \mathbb{R}^+$, experienced by terminal k's signal when propagating to receiver j. We can define player k's utility as the ratio between throughput and power expenditure, thus accounting for the number of bits correctly delivered per Jouls of energy consumed utility function is calculated as, $u_k(s) = t_k(s) - c_k(s)t_k(s)$ is accounts for the outcome of the transmission, and $c_k(s)$ measures the cost, cost associated to using

s_k. Now, we can define player k's utility as the ratio between throughput and power expenditure, thus according to the number of bits correctly delivered per joule of energy consumed, $u_k(s) = \frac{t_k(s)}{s_k}$. Where $t_k(s)$ accounts for the outcome of transmission. The signal reception depends on SINR at point k. In the proposed game theory model is mathematically represented as follows, $i = 0, 1, 2, \ldots, N$ is the index set of the macro cell and position of the Femtocell is, $p_{i,j} = \sum_{i,j=0}^{\max(BS_i)}$. Where the i, j is the index associated with mobile user and Femtocell, BS_i is the position of a mobile base station within the service area. Now, we talk about the utility function and it can be written as $U_{i,j} = \sum_{u_{i,j}}^{n,M,\max(BS_i)} D\ell c$. It show utility function of base station BS_i, D is the size of data packets, ℓ is the link utilization between mobile node, Femtocell, and BS, c is the constant amount of resource utilization at point $j, i \ldots j$, p_i is the vector of transmission power for all base stations at position i. Here we introduce the term Nash equilibrium function which offers a steady-state transmission power at the Nash equilibrium point. It is the solution for power control to maximize the utility of both the Femtocell, i.e., player1 and player2 with improving throughput performance and reducing the interference as follows, $t_k(s) = t(1 - \exp\gamma_k(s))^L$. In Fig. 5 define the throughput vs utility functions for SINR. L is the number of packets is 20 and SINR is 6.5 dB. The game presents Nash Equilibrium by the fixed point $sk = bk.sk$ for $k = 1, 2$, [1]. Now in Fig. 2b normalized utilities as u_1 and u_2 of player1 and player2, respectively, as (0.269, 0.407). Here the $u_1(s) < u_2(s)$ as player2 experiences better channel conditions achieving optimal SINR with lower power consumption. Inefficiency in utility function mostly occurs due to the selfishness of self-optimization between the players. Now for both players to have better throughput and low interference without one having the upper priority, the function should be efficient and improved by introducing few functions like pricing function, repeating the game, and cooperation of access points. The signal reception depends on SINR k. So for the signal to be received and decoded at the receiver k request, the minimum SINR request should be less and $tk(s) = t$, 't' is the throughput at the destination. We divide our implementation into three parts. Firstly to find Nash Equilibrium for Femtocell. Secondly, improve the efficiency of it, lastly, the difference in throughput via game Theory. Here, calculation of throughput is derived by using $tk(s) = t(1 - \exp -\gamma k(s))^L.D.\ell$. Here the graph is plotted taking (in, L) as parameters where in = SINR, L = the number of information bits per packet, and the output is given by, $(1 - \exp(-in))^L \cdot \ell$. All the methods and improvements made contribute to the achievement of the dynamic power play of Femtocell. Therefore by using game theory in wireless communication we can reduce interference and transmit power with increasing throughput.

4 Conclusion

In this paper, we have successfully found a solution to establish green communication. The use of SON is a cost-effective method to reduce human labor. The use of dynamic transmission power control helps in reducing the interference problem

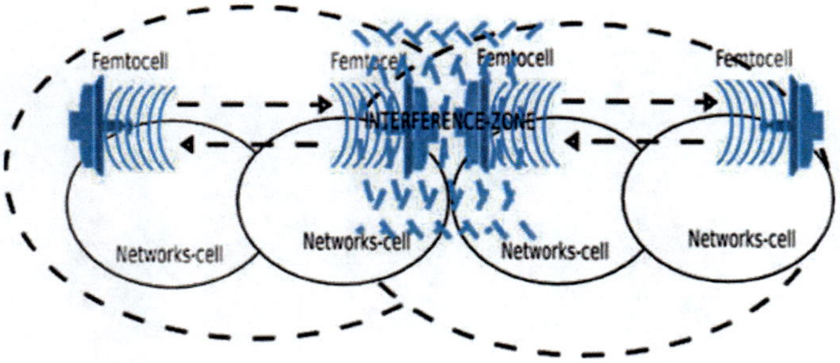

Fig. 2 Interference between Femtocell and access point

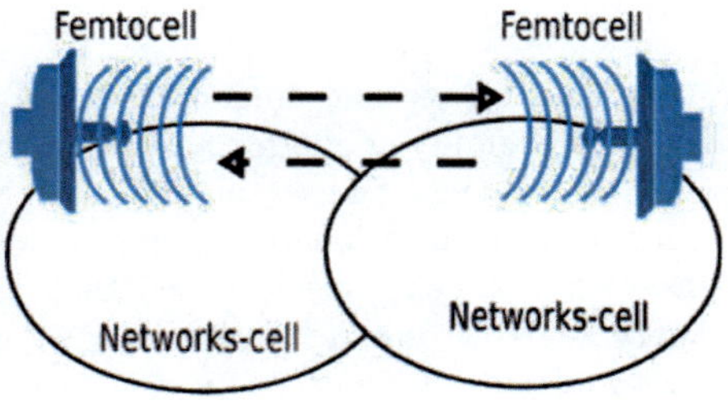

Fig. 3 Power control procedure in proposed work

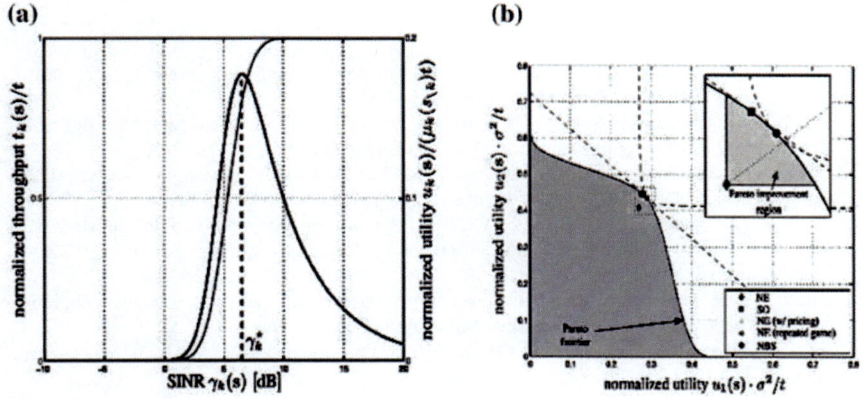

Fig. 4 **a** Throughput (red) and utility (blue) as function of SINR, **b** normalized utility plan

which makes Femtocell the most environment-friendly technology which not only provides excellent quality of service but also helps in improvising the network coverage and reducing the traffic load on macro base stations. The automation of this cell by implementing SON introduced by 3GPP. Our main idea is centralized

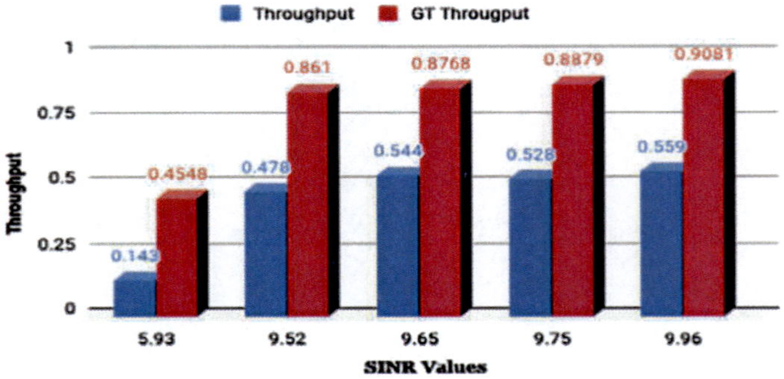

Fig. 5 Throughput versus SINR w.r.t utility function

on the deployment of Femtocell access points around the densely populated system. Therefore there exists a high probability of interference among the Femtocell access point and macro base stations and even the user undergoing uplink and downlinks. Hence, we come up with a proposed solution to use a power controlling method to tackle the issue. We introduce the concept of players, strategies, Nash equilibrium, utilities, etc. in game theory. With our proposed methodology the network condition can become ten times better than the present scenario.

References

1. Bacci G, Sanguinetti L, Luise M (2015) Understanding game theory via wireless power control [lecture notes]. IEEE Sig Process Mag 32(4):132–137
2. Biswash SK, Nagaraj S, Sarkar M, Narra NR (2017) Bee system based base station cooperation technique for mobile cellular networks. Wirel Pers Commun 92(3):1193–1220
3. Chandrasekhar V, Andrews JG, Gatherer A (2008) Femtocell networks: a survey. IEEE Commun Mag 46(9):59–67. https://doi.org/10.1109/MCOM.2008.4623708
4. Chee D, Suk Kang M, Lee H, Jung BC (2011) A study on the green cellular network with femtocells. In: 2011 third international conference on ubiquitous and future networks (ICUFN), June 2011, pp 235–240
5. Ghosh J (2019) Interrelationship between energy efficiency and spectral efficiency in cognitive femtocell networks: a survey. Pervasive Mob Comput 59:101066
6. Hsia KH, Hung CW, Chang HT, Lai YH (2017) Transmission power control for wireless sensor network. J Robot Netw Artif Life 3:279–282
7. Knisely DN, Yoshizawa T, Favichia F (2009) Standardization of femtocells in 3GPP. IEEE Commun Mag 47(9):68–75. https://doi.org/10.1109/MCOM.2009.5277458
8. Ole Grøndalen O (2012) Benefits of self-organizing networks (SON) for mobile operators. J Commun Netw Commun 16. Article ID 862527
9. Xenakis D, Passas N, Merakos L, Verikoukis C (2014) Mobility management for femtocells in lte-advanced: key aspects and survey of handover decision algorithms. IEEE Commun Surv Tutorials 16(1):64–91
10. Yaqoob M, Gemikonakli O, Ever E (2021) Modelling heterogeneous future wireless cellular networks: an analytical study for interaction of 5 g femtocells and macrocells. Futur Gener Comput Syst 114:82–95

Heterodielectric-Based Gate Oxide Stack Engineering in FDSOI Structure with Enhanced Analog Performance

Aditya Kushwaha, Narendra Yadava, Mangal Deep Gupta, and R. K. Chauhan

Abstract This paper presents the examination on the analog execution of a heterodielectric materials gate oxide-based fully depleted silicon-on-insulator (FDSOI) metal oxide semiconductor (MOS) transistor. Heterodielectric structure in gate oxide stack is utilized which helps to improve the control of gate electrode over the channel. In the examination, OFF-state current (I_{OFF}), ON-state current (I_{ON}), I_{ON}/I_{OFF} ratio, transconductance (g_m), subthreshold slope (SS), and drain-induced barrier lowering (DIBL) are improved in stacking of high-k gate dielectric oxide material over the silicon dioxide (SiO_2) layer. The structure is simulated, and parameters are extracted by using the Silvaco ATLAS 2D device simulator.

Keywords FDSOI · MOSFET · High-k · Heterodielectric gate oxide stack (HG) · g_m · I_{ON} · I_{OFF} · DIBL

1 Introduction

Advancements in silicon technology have grown faster year after year. The primary issue that should be concentrated on silicon innovations is the impacts of lessening gadgets components. The downsizing of devices is firmly needed to accomplish high reconciliation thickness and better device execution. The essential role of complementary metal oxide semiconductor (CMOS) scaling is to give more modest and quicker semiconductors starting with one hub, then onto the next, best portrayed by Moore's Law [1]. The short channel impact and leakage current are significant problems that degrade the device's electrical efficiency due to the channel length reduction. Another circuit plan silicon-on-insulator (SOI) method has been present for a fresher innovation to overcome the issues arising due to short channel. SOI alludes to putting a slim layer of silicon on top of an insulator, typically SiO_2, or known as a buried oxide layer (BOX). The SOI MOS has been partitioned into the

A. Kushwaha (✉) · N. Yadava · M. D. Gupta · R. K. Chauhan
Madan Mohan Malaviya University of Technology, Gorakhpur, India
e-mail: kushawahaaditya@gmail.com

B. Mishra and M. Tiwari (eds.), *VLSI, Microwave and Wireless Technologies*, Lecture Notes in Electrical Engineering 877,
https://doi.org/10.1007/978-981-19-0312-0_26

partially depleted (PD)-SOI MOSFET and FDSOI MOSFET. In the current situation, the FDSOI MOS devices are generally utilized in the low power dispersal. A detailed study on fully depleted SOI innovation has been presented in [2].

FDSOI MOSFET is a promising possibility to decrease short channel impacts at 50 nm gate length [3]. In recent years, significant scaling of CMOS technology has decreased the gate dielectric oxide, limiting the dielectric thickness of the silicon dioxide gate to a few nanometers [4]. Due to the scaling down of the conventional gate dielectric oxide layer, direct tunneling of carriers at nanometer dimensions significantly increases the leakage current, which increases the static power dissipation and affects the circuit's operation [5]. Thus, high-k materials are utilized to conquer this issue instead of using SiO_2 gate dielectric material, which has high gate capacitance in contrast to SiO_2. It is also found that directly using high-k material over a silicon wafer debased the gate terminal control over the channel due to longitudinal fringing field. But this issue is settled by utilizing heterodielectric gate oxide stack design [6, 7]. On the MOS gate stack, a high-k dielectric is required to maintain the MOSFET gate capacitance, which controls the tunneling current. High-k material, when stacked over an ultra-thin film of SiO_2 called gate dielectric stacked design, i.e., heterodielectric gate oxide stack (HG) design, improves the performance of the device by suppressing SCEs and leakage current at the gate [8].

In this research work, we examine the DC and analog execution of the 50 nm channel length HG-FDSOI by utilizing distinctive high-k dielectric materials [9–11] stacked over the dainty film of the SiO_2 layer. The Silvaco ATLAS 2D device simulator was used to carry out all the designs and simulations. The system specifications are given in Sect. 2, and the results and discussion have been completed in Sect. 3.

2 Device Design and Simulation

A regular construction of an FDSOI MOSFET is shown in Fig. 1, and HG-FDSOI MOSFET is shown in Fig. 2 and is recreated utilizing a 2D Silvaco ATLAS device simulator [12]. Device specifications are tabulated in Table 1. In Fig. 2, a layer of high-k gate oxide dielectric material is stacked over a dainty film of SiO_2. Various models are in the simulation for mobility conmob and fldmob are used, SRH model and Auger model are used for carrier recombination. Concentration dependent and parallel model and perpendicular field-dependent mobility model are also used in simulations. By utilizing a high-k dielectric oxide layer stacked over the thin SiO_2 layer in the design, the leakage currents and short channel impact diminish while gate capacitance increments.

Equivalent dielectric constant (ε_{eq}) for the fixed gate oxide thickness with variable heterodielectric gate oxide layer stacked over SiO_2 material is given as Eq. 1. Equivalent dielectric constant (ε_{eq}) of heterodielectric gate oxide is tabulated in Table 2. The dielectric constant of different high-k dielectrics is considered from [12].

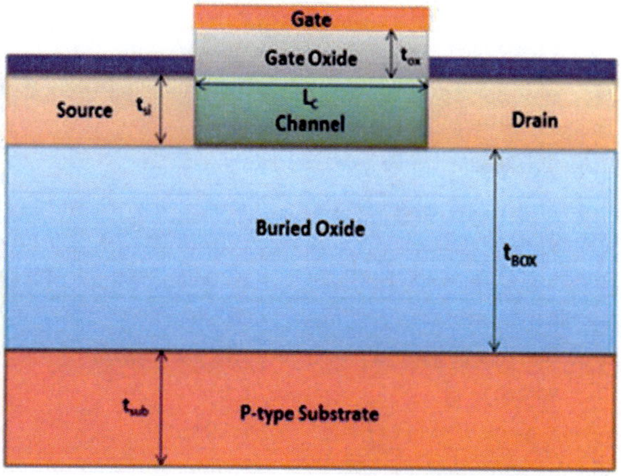

Fig. 1 Conventional design of FDSOI MOSFET

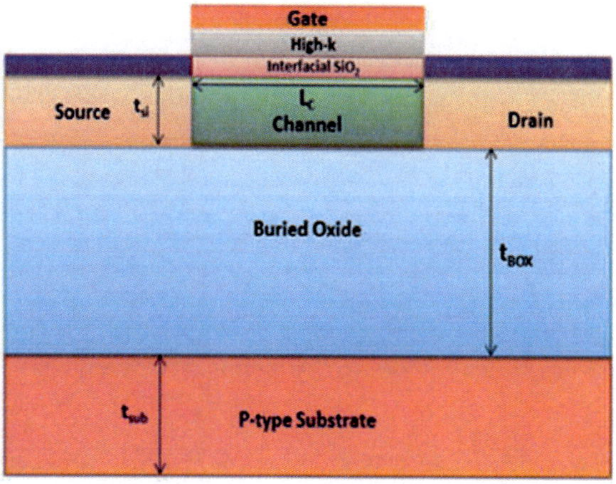

Fig. 2 Design of HG-FDSOI MOSFET

$$\frac{1}{C_{eq}} = \frac{1}{C_1} + \frac{1}{C_2}$$

$$\frac{t_{eq}}{\varepsilon_{eq}} = \frac{t_1}{\varepsilon_1} + \frac{t_2}{\varepsilon_2}$$

$$\frac{1}{\varepsilon_{eq}} = \frac{0.5}{\varepsilon_1} + \frac{0.5}{\varepsilon_2}$$

$$\varepsilon_{eq} = \frac{2\varepsilon_1\varepsilon_2}{(\varepsilon_1 + \varepsilon_2)} \tag{1}$$

Table 1 Device description of different FDSOI structures

Device parameters		Values	
		Conventional FDSOI (C-FDSOI)	HG-FDSOI
Channel length (L_c)		50 nm	50 nm
Oxide thickness	t_{ox}	1 nm	0.5 nm
	$t_{high\text{-}k}$	–	0.5 nm
Channel doping (N_A)		1e16 cm^{-3}	1e16 cm^{-3}
Drain region doping (N_D)		1e20 cm^{-3}	1e20 cm^{-3}
Insulator thickness (t_{BOX})		20 nm	20 nm
Work function of gate (Φ)		4.77 eV (Cu)	4.77 eV (Cu)
Gate voltage (V_g)		1.1 V	1.1 V
Source region doping (N_D)		1e20 cm^{-3}	1e20 cm^{-3}
Silicon film thickness (t_{Si})		10 nm	10 nm
Total width of the device		100 nm	100 nm

Table 2 Equivalent dielectric constant for different heterodielectrics gate oxide materials

S. No.	HG-FDSOI	Equivalent dielectric constant (ε_{eq})
1	SnO_2 ($\varepsilon_2 = 9$)/SiO_2($\varepsilon_1 = 3.9$)	5.44
2	Al_2O_3($\varepsilon_2 = 9.3$)/SiO_2($\varepsilon_1 = 3.9$)	5.49
3	$HfSiO_4$($\varepsilon_2 = 12$)/SiO_2($\varepsilon_1 = 3.9$)	5.88
4	ZrO_2($\varepsilon_2 = 22$)/SiO_2($\varepsilon_1 = 3.9$)	6.62
5	HfO_2($\varepsilon_2 = 24$)/SiO_2($\varepsilon_1 = 3.9$)	6.70
6	Ta_2O_5($\varepsilon_2 = 26$)/SiO_2($\varepsilon_1 = 3.9$)	6.78
7	TiO_2($\varepsilon_2 = 80$)/SiO_2($\varepsilon_1 = 3.9$)	7.43

where ε_1 and ε_2 are a dielectric constant of SiO_2 and high-k material, respectively, and both having an equal thickness of 0.5 nm.

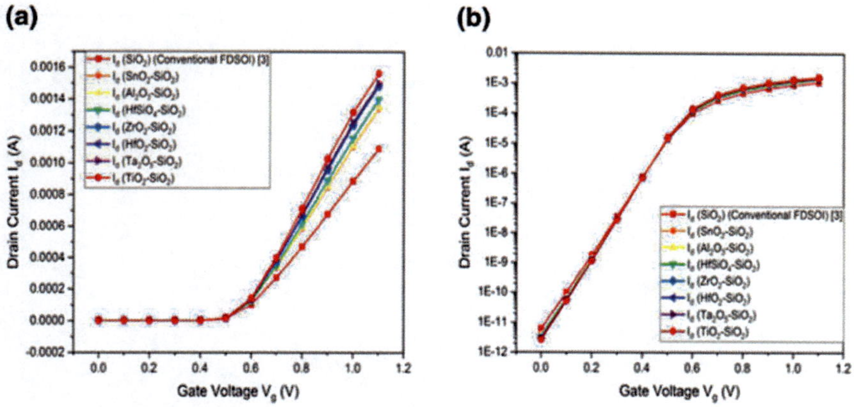

Fig. 3 **a** Shows transfer characteristics at $V_{DS} = 0.5$ V on a linear scale for different HG-FDSOI. **b** Shows transfer characteristics at $V_{DS} = 0.5$ V in the log scale for different HG-FDSOI

3 Results and Discussions

3.1 Transfer Characteristics

Here, a comparative evaluation is performed of HG-FDSOI MOSFET over conventional MOSFET regarding analog and DC examination. Figure 3a shows transfer characteristics in linear, and Fig. 3b represents transfer characteristics in log scales at drain-source voltage, i.e., $V_{DS} = 0.5$ V. One can discover from Fig. 3a, I_{ON} current is increased with different high-k HG-FDSOI MOSFET when contrasted with conventional FDSOI MOSFET. Figure 3b shows the leakage current alteration, and one can see that leakage current is decreased with HG-FDSOI MOSFET. For TiO_2/SiO_2, configuration shows minimum leakage current.

3.2 Output Characteristics

CMOS analog circuits need semiconductors with low output conductance (g_d) to accomplish a high increase in intrinsic gain. High g_d implies low output resistance, which brings about an expansion in drain current (I_{DS}) with V_{DS} in the saturation region. The segments are related to this increment, to be specific channel length modulation (CLM) and DIBL. Figure 4 represents linear scale output characteristics I_{DS}–V_{DS} for different high-k gate dielectric oxide stacked materials.

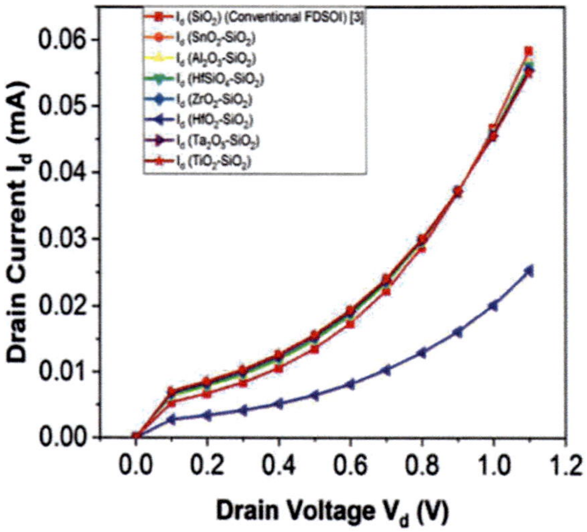

Fig. 4 Output characteristics for different HG-FDSOI devices at $V_{GS} = 0.5$ V

3.3 Output Conductance

Output conductance (g_d) measures the drain current variety with a drain-source voltage variety while keeping the gate-source voltage steady. It is an essential boundary for a device since it chooses the drive current of a device and can be calculated using Eq. 2. Figure 5 shows the drain current as an element of drain voltage for various high-k gate dielectric oxide stacked over the SiO_2 layer.

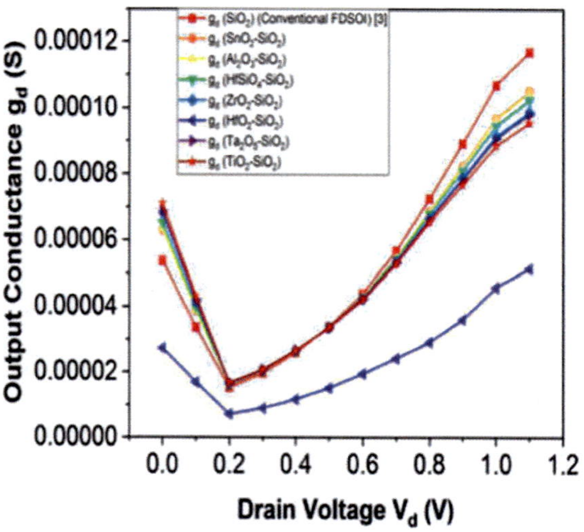

Fig. 5 Output conductance characteristics for different HG-FDSOI devices at $V_{GS} = 0.5$ V

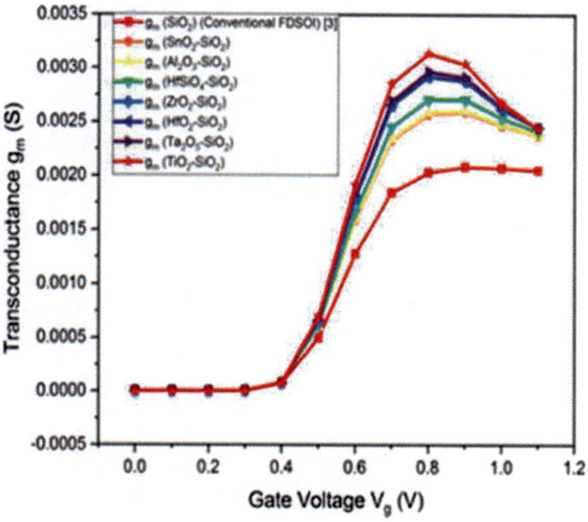

Fig. 6 Transconductance characteristics for different HG-FDSOI devices at $V_{DS} = 0.5$ V

$$g_d = \frac{\partial I_D}{\partial V_{DS}} \tag{2}$$

3.4 *Transconductance*

Transconductance (g_m) measures the channel current variety with a gate-source voltage, i.e., V_{GS} variety, while keeping the drain-source voltage steady and is of pivotal significance since it chooses the capacity of the device to drive a load that can be calculated by using Eq. 3. The g_m has a significant job in deciding the switching speed of a circuit and voltage gain of MOSFET amplifiers. High g_m devices yield circuits able to do a high-speed activity. Figure 6 shows that TiO_2/SiO_2 HG-FDSOI MOSFET displays an enormous transconductance contrasted with the conventional device, which improves the device's gain.

$$g_m = \frac{\partial I_D}{\partial V_{GS}} \tag{3}$$

3.5 *Static Power Dissipation*

There is some static power dissipation ($P_{Diss.}$) due to small reverse bias leakage between drain and source region of FDSOI. Subthreshold conduction can also

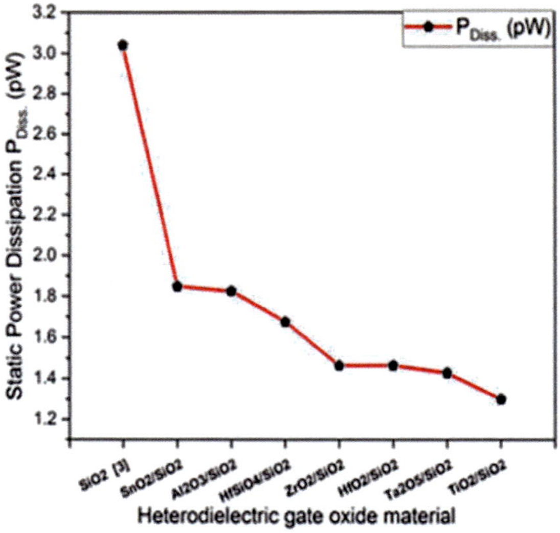

Fig. 7 Static power dissipation variation for different HG-FDSOI devices at $V_{GS} = 0.5$ V and $V_{DS} = 1.0$ V

contribute to the static power dissipation, which is calculated by using Eq. 4. Materials with very large high-k helps in improving static performance by reducing leakage. On the other hand, it increases the intrinsic delay, which will affect the dynamic performance [13]. There is a sort of trade-off between static and dynamic performance. Static power dissipation is a useful construction parameter for determining the digital application of a device. Figure 7 shows that minimum $P_{Diss.}$ occurs in TiO_2/SiO_2 HG-FDSOI configuration in comparison with others.

$$P_{Diss.} = (I_{OFF}.V_{DD}) \tag{4}$$

Figure 8a–c show I_{ON}, I_{OFF}, and I_{ON}/I_{OFF} ratio values respectively for different HG-FDSOI configurations in which TiO_2/SiO_2 shows better performance in comparison with other configurations and on the other hand, it also reduces the SCEs, i.e., DIBL and subthreshold as shown in Figs. 9 and 10, respectively. There is also an improvement in threshold voltage value for TiO_2/SiO_2 configuration from other configurations. The different parameter values are tabulated in Table 3 for different HG-FDSOI structures.

4 Conclusion

From simulated outcomes, heterodielectric-based high-k gate oxide dielectric material creates a higher fringing field, which helps lessen the barrier height between the source and drain in ON state, and increases I_{ON}. ON current is ~49% higher on

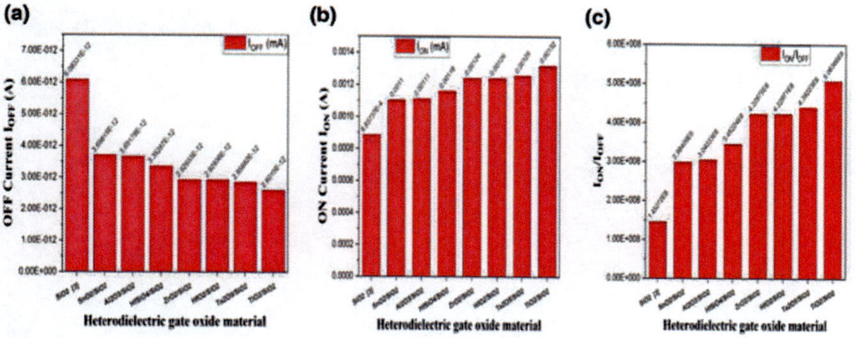

Fig. 8 **a** Shows OFF current values for different HG-FDSOI structures $V_{DS} = 0.5$ V and $V_{GS} = 1$ V. **b** Shows ON current values for different HG-FDSOI structures $V_{DS} = 0.5$ V and $V_{GS} = 1$ V. **c** Shows ON-to-OFF current ratio values for different HG-FDSOI structures $V_{DS} = 0.5$ V and $V_{GS} = 1$ V

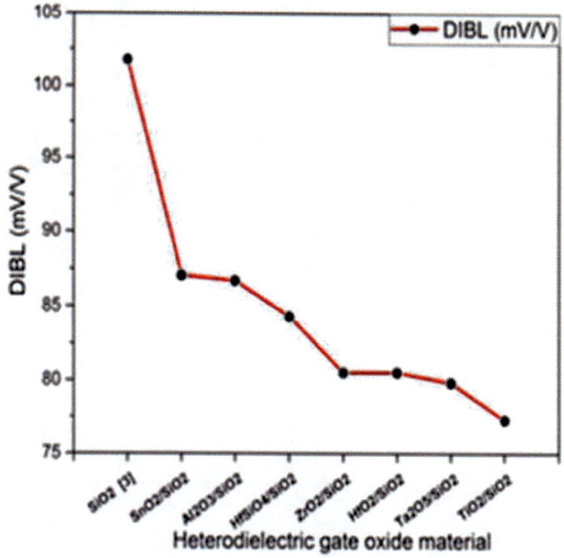

Fig. 9 Shows DIBL values for different HG-FDSOI structures

account of TiO_2/SiO_2 dielectric than the conventional SiO_2 dielectric. In OFF state, barrier height among source and channel is enormous for high-k dielectric stacking, decreasing I_{OFF} flow ~57% than the regular SiO_2 dielectric. High-k stacked dielectric gives higher transconductance than conventional SiO_2 dielectric. It likewise provides higher voltage gain, so we use SOI MOSFET devices with high-k stacked dielectric for enhancement purposes. Additionally, I_{ON}/I_{OFF} is higher on account of a high-k stacked dielectric. There is an improvement of 3.4 times in the I_{ON}/I_{OFF} proportion on the TiO_2/SiO_2 dielectric over SiO_2 dielectric. TiO_2/SiO_2 high-k stacked dielectric

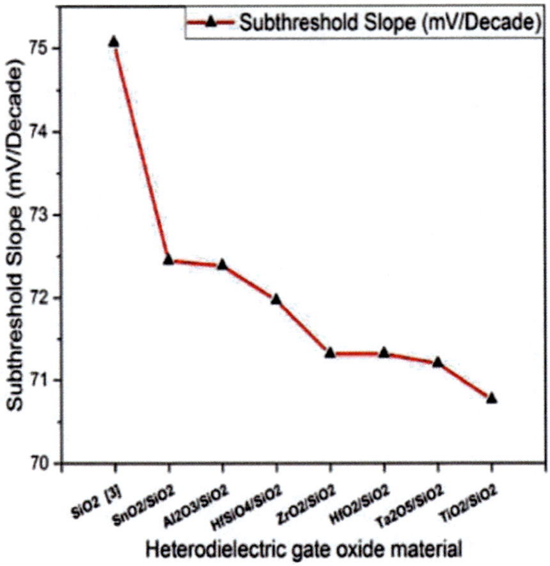

Fig. 10 Shows subthreshold values for different HG-FDSOI structures

improves short channel impacts by improvement in the subthreshold slope by 6% and minimizes DIBL by 25.5% than conventional SiO_2 dielectric so HG-FDSOI is a potential candidate for analog applications.

Table 3 Electrical/analog performance comparison between C-FDSOI and HG-FDSOI on various parameters of the devices

Parameters	C-FDSOI	HG-FDSOI						
	SiO_2	SnO_2/SiO_2	Al_2O_3/SiO_2	$HfSiO_4/SiO_2$	ZrO_2/SiO_2	HfO_2/SiO_2	Ta_2O_5/SiO_2	TiO_2/SiO_2
DIBL (mV/V)	97.4289	85.5244	85.2333	83.2822	80.2467	80.2467	79.6889	77.6244
Subthreshold (mV/decade)	75.0669	72.4524	72.39	71.97	71.3201	71.3201	71.2012	70.7625
I_{OFF} (pA)	6.08331	3.69815	3.65178	3.35287	2.92933	2.92936	2.85682	2.6015
I_{ON} (mA)	0.883737	1.10356	1.11026	1.15749	1.23874	1.23874	1.25478	1.31731
I_{ON}/I_{OFF}	1.45272e+8	2.98409e+8	3.04033e+8	3.45224e+8	4.22875e+8	4.22871e+8	4.39223e+8	5.06366e+8
Threshold voltage (V)	0.326851	0.326163	0.326159	0.326076	0.325704	0.325704	0.325581	0.324859

References

1. Moore GE (1975) Progress in digital integrated electronics. In: International electron devices meeting, IEEE, pp 11–13. http://www.eng.auburn.edu/~agrawvd/COURSE/E7770_Spr07/READ/Gordon_Moore_1975_Speech.pdf
2. Cheng KG, Khakifirooz A (2016) Fully depleted SOI (FDSOI) technology. Sci China Inf Sci 59(6):061402
3. Mishra V, Chauhan RK (2017) Performance analysis of fully depleted SOI tapered body reduced source (FD-SOI TBRS) MOSFET for low power digital applications. Springer Nature Singapore Pte Ltd.
4. Duba A, Jaggi V, Verma A, Mishra A (2013) High K oxides as an alternative gate oxide in CMOS technology and its corresponding effects. Int J Eng Res Technol 02(11)
5. Kim YB (2010) Challenges for nanoscale MOSFETs and emerging nanoelectronics. Trans Electr Electron Mater
6. Kundu A, Dasgupta A, Das R, Chakraborty S, Dutta A, Sarkar CK (2016) Influence of underlap on gate stack DG-MOSFET for analytical study of analog/RF performance. Superlattices Microstruct 94:60–73. ISSN0749-6036
7. Singh J, Yadava N, Chauhan KR (2018) Gate stacked FD-SOI MOS structure for better controllability at high frequencies. In: 5th IEEE Uttar Pradesh section international conference on electrical, electronics and computer engineering (UPCON), Gorakhpur, pp 1–5
8. Basak A, Sarkar A (2020) Drain current modelling of asymmetric junctionless dual material double gate MOSFET with high K gate stack for analog and RF performance. Silicon
9. Pradhan KP, Mohapatra SK, Sahu KP, Behera KD (2014) Impact of high-k gate dielectric on analog and RF performance of nanoscale DG-MOSFET. Microelectron J 45:144–151
10. Belkhiria M, Echouchene F, Jaba N, Bajahzar A, Belmabrouk H (2020) Impact of high-k gate dielectric on self-heating effects in PiFETs structure. IEEE Trans Elect Dev 67(9):3522–3529
11. Singh A, Pandey CK, Chaudhury S, Sarkar CK (2020) Comparative analysis of silicon nanotube FET for switching applications using high K and work function modulation. Int J Nanoparticles 12(3):264–273
12. ATLAS user's manual, SILVACO, Santa Clara, CA, USA (2016)
13. Wagaj CS, Patil CS (2019) Comparison study of dual material gate silicon on insulator junctionless transistor and with junction transistor for analog performance. Int J Mater Mech Manufact 7(3):144–149

Numerical Study of Various ETL Materials for an Efficient Lead-Free Perovskite Solar Cell

Km Shivani Bhardwaj, Shambhavi Rai, Vandita Rao, Pooja Lohia, and D. K. Dwivedi

Abstract Tin(Sn)-based perovskite is a well-known challenger to toxic Lead (Pb)-based perovskite solar cells. Pb-free perovskite ($CH_3NH_3SnI_3$) material attracted attention because of wider absorption and optimum band gap ~1.30 eV. In this work an n-i-p perovskite solar cell has been studied using Solar Cell Capacitance Simulator (SCAPS). The architecture of the device is FTO/ETM/$CH_3NH_3SnI_3$/Cu_2O/ Au. To investigate the efficiency of the proposed device and to optimize its performance various electron transport materials (ETMs) like phenyl-C_{61}-butyric acid methyl ester (PCBM), Indium gallium zinc oxide (IGZO) and WO_3 have been employed. Here, copper oxide (Cu_2O) work as hole transport material (HTM) layer. From the J-V and quantum efficiency curve of various ETL material it has been found that WO_3 is suitable ETL for our device. Further, the input parameters of electron transport layer (WO_3) like as thickness and donor concentration were varied to study their effect on the device performance. The power conversion efficiency of 23.99% was observed for the proposed lead-free PSC.

Keywords SCAPS-1-D · Lead-free · ETL · HTL · WO_3

1 Introduction

The increasing energy demand in the present society has leaded the solar photovoltaic to be one of the important eras to overcome the energy crisis. In recent years, a new candidate in this field has emerged that is the perovskite solar cell (PSC) and has received a lot of attention due to its low fabrication cost and high efficiency [1, 2]. Perovskite solar cell is considered in the third-generation solar cell with an

K. S. Bhardwaj · S. Rai · V. Rao · D. K. Dwivedi (✉)
Photonics and Photovoltaic Research Lab, Department of Physics and Material Science, Madan Mohan Malaviya University of Technology, Gorakhpur 273010, India
e-mail: todkdwivedi@gmail.com

P. Lohia
Department of Electronics and Communication Engineering, Madan Mohan Malaviya University of Technology, Gorakhpur 273010, India

B. Mishra and M. Tiwari (eds.), *VLSI, Microwave and Wireless Technologies*,
Lecture Notes in Electrical Engineering 877,
https://doi.org/10.1007/978-981-19-0312-0_27

experimentally reported efficiency of 3.8% in 2009 which has improved up to 23.8% till date [3, 4] and still on incremental phase with an unexpected upper limit. A compound structure of perovskite is ABX_3 [5] (where A: alkyl group B: metal and X: halogen).Though Perovskite solar cells are good but there are some limitations such as lifetime, instability due to heat and humidity and lack of conformity. Therefore, a lot of effort is needed to deal with these limitations and to take the perovskite solar cell to commercialization stage [6–8]. The basic planar device structure of a perovskite solar cell consists of mainly three layers- (i) perovskites (light harvester), (ii) electron transport layer (ETL) and (iii) hole transport layer (HTL) [9, 10].

The most efficient PSCs till date are mostly lead (Pb)-based PSCs. Though the performance of these PSCs is very good but the presence of Pb in solar cells makes it highly toxic and thus it is the main concern in commercializing these PSCs. Thus, now a days many of the researchers in this era are devoted to make efficient lead-free solar cells which are compatible to those Pb-based PSCs using some substituents. Among such substituents which can replace lead in perovskites, tin (Sn) has drawn most of the attention owing to its similar properties and the most promising output performance. The lead-free perovskite $MASnI_3$ ($CH_3NH_3SnI_3$) has emerged as a most widely used perovskite in the PSCs. The direct band gap of this absorber layer is 1.30 eV. For optoelectrical applications $CH_3NH_3SnI_3$ compound has the majority suitable optical properties [11]. Moreover, copper oxide (Cu_2O) has been used as hole transporting material which is also known as a p-type layer which has a band gap 2.17 eV. It is non-toxic and low-cost material than other hole transporting material [12, 13]. Gold (Au) and fluorine doped tin oxide (FTO) has been used as back and front contact, respectively, which have acceptable work functions to line up themselves according to the band gap of ETL and HTL [14, 15]. An examination of lead-free perovskite solar cell incorporating $CH_3NH_3SnI_3$ is proposed. Device simulation is done by SCAPS-1-D for Glass/FTO/ETM/$CH_3NH_3SnI_3$/Cu_2O/Au structure. We used different electron transport materials and observed their effect on the performance of the device.

2 Device Structure and Simulation Parameters

The lead-free perovskite solar cell Glass/FTO/ETM/$CH_3NH_3SnI_3$/Cu_2O/Au has been simulated by 1-D-solar cell capacitance simulator (SCAPS-1-D). The SCAPS-1-D was developed at Ghent University, Belgium [16]. The proposed device structure and energy band diagram has been illustrated in Fig. 1a–b. The device containing different layer of various thickness, electrical and optical properties. An absorber layer of thickness (350 nm), HTL containing thickness (200 nm) and ETL layer as of (70 nm) thickness which is fixed for various ETL materials. As illustrated in Fig. 1, FTO which is at 500 nm thickness works like a front contact above which a layer of n-type doped (PCBM, IGZO and WO3) works as the electron transport layer [17–19]. Perovskite layer $CH_3NH_3SnI_3$ work as an active layer. Cu_2O work as a hole transport layer which is a p-type doped layer on which gold (Au) contacts are grown. SCAPS-1-D works on Poisson and continuity equation of electron and

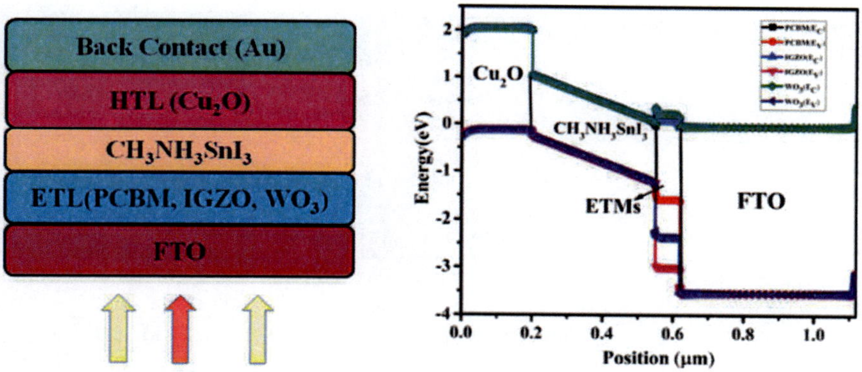

(a) (b)

Fig. 1 **a** Device structure and **b** band diagram of ETMs

hole. Physical parameters utilized in simulation are listed in Table 1 and parameters of the interface are listed in Table 2. The additional parameters and coefficient of those materials were taken from the literature and their references are cited [20–24]. All simulation has been performed to considering the series resistance and shunts resistance 1Ω and 1 × 10^3 Ω, respectively. The usual AM1.5G spectrum as received light source was taken at temperature 300 K for simulation.

Table 1 Substantial parameters of materials used in simulation

Parameters	FTO	PCBM	IGZO	WO_3	$CH_3NH_3SnI_3$	Cu_2O
Thickness (nm)	500	70	70	70	350	200
E_g (eV)	3.5	1.7	3.05	2.6	1.3	2.17
χ (eV)	4.0	3.9	4.16	3.8	4.17	3.2
ϵ_r	9.0	4.2	10	4.8	8.2	7.11
N_c (cm^{-3})	2.2×10^{18}	8×10^{19}	5×10^{18}	2.2×10^{21}	1×10^{18}	2.2×10^{18}
N_v (cm^{-3})	1.8×10^{19}	8×10^{19}	5×10^{18}	2.2×10^{21}	1×10^{18}	2.2×10^{18}
μ_n (cm^2/Vs)	20	0.08	15	30	1.6	200
μ_p (cm^2/Vs)	20	0.035	0.1	30	1.6	80
V_e (cm/s)	1×10^7	1×10^7	1×10^7	1×10^7	1×10^7	
V_h (cm/s)	1×10^7	1×10^7	1×10^7	1×10^7	1×10^7	
N_D (cm^{-3})	2×10^{19}	2.6×10^{18}	1×10^{18}	6.35×10^{17}	–	–
N_A (cm^{-3})	–	–	–	–	3.2×10^{10}	1×10^{17}
N_t (cm^{-3})	1×10^{15}	1×10^{14}	1×10^{15}	1×10^{15}	1×10^{13}	1×10^{14}

Table 2 Interface parameters of the device

Defect type	HTL/Absorber Neutral	Absorber/ETL Neutral
Capture cross-section of electron (cm^2)	1×10^{-18}	1×10^{-15}
Capture cross-section of hole (cm^2)	1×10^{-18}	1×10^{-15}
Energetic distribution	Single	Single
Energy concerning reference (eV)	0.6	0.6
Total density ($1/cm^2$)	1×10^{12}	1×10^{11}

3 Results and Discussion

3.1 J-V and Quantum Efficiency Curve for Various ETL

We have investigated the J-V and quantum efficiency characteristics for different ETLs are revealed in Fig. 2a–b which has been deliberate from Tables 1 and 2. The Output of parameters of PSC using various ETLs are listed in Table 3 using

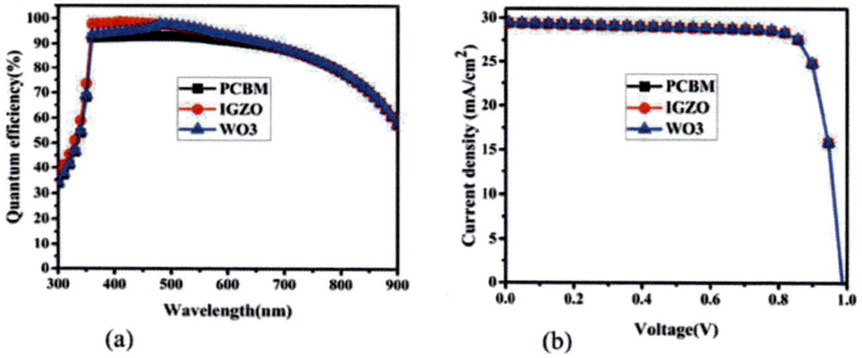

Fig. 2 **a** QE of PSC device **b** J-V characteristic

Table 3 Performance of lead-free PSCs with different ETM layers

Results	PCBM	IGZO	WO_3
Voc (V)	0.9904	0.9848	0.9906
Jsc (mA/cm^2)	29.455	30.073	29.929
FF (%)	80.98	80.88	80.92
PCE (%)	23.62	23.95	23.99

ETL materials PCBM, IGZO and WO_3 examine and found that PCE of IGZO-based mostly device is 23.95% and WO_3-based mostly device is 23.99% that is nearly equal. The FF of IGZO and WO_3 is 80.88 and 80.92%. However we have a tendency to contemplate WO_3 as good ETL material as a result of its thermally and with chemical established semiconductor and having best electron mobility. The quantum efficiency may be operating of wavelength or as energy. Ideally, quantum efficiency has exits in square form. It was reduces suitable to recombination effects. The QE of a PSC is sometimes not measured much below 350 nm because the power from the AM1.5 contained in such low wavelength is low. From Fig. 2a the quantum efficiency initially increases with increment of wavelength due to absorption of light large number of charge carriers increases while after 550–900 nm the QE decreases because recombination of charge carrier. But within the region of 350–450 nm wavelengths, the QE of WO_3 is less than that of the QE of IGZO owing to low diffusion length affect the collection probability and we know that QE applied to the incident photon to converted into electron ratio so the QE of IGZO is higher than WO_3 at the variation of 350 to 450 nm as a result for the IGZO more and more photons are incident so the amount of current is high for IGZO. At open-circuit the voltage is highest and the maximum current is at closed circuit. These conditions do not generate any electrical power however there should be some extent anyplace in amid was the perovskite solar cell generates maximum power. From Fig. 2b as voltage increases up to 0.8 V current density shows constant characteristics and after that value it decreases abruptly because of less recombination of charge carrier. The current density curve is obtained under illumination at constant temperature.

3.2 *Influence of Thickness (WO_3)*

In the previous section, we have analyze that WO_3 is the good ETL material for the reason of large carrier mobility along with various ETL materials. In PSC, the effect of thickness of ETL layer plays an important role as regarded to firmness, efficiency and other photovoltaic parameters of PSC. To observe the effect of thickness of WO_3 electron transport layer, which is done at various thickness 40–300 nm by simulation technique, maintain other parameters same as Tables 1 and 2. From Fig. 3a we have plotted V_{oc} and J_{sc} graph accord with thickness of ETL layer. When we increase the thickness of the ETL layer the open-circuit voltage and short circuit current density decreases gradually because due to the remarkable growth of series resistance. V_{OC} is decreases due to low photocurrent and the generated electron holes may not be effectively collected. Moreover, is that high recombination rate in the perovskite and at the interface of the HTL and the ETL since these are heterofaces with defects. The J_{sc} is affected because of the presence of many defects as like of interface layer that can be act as an effective recombination. In Fig. 3b the fill factor (FF) increases, and efficiency decreases because of high series conductance. As we increment in the thickness of the electron transport layer beyond the diffusion length the PCE was decreases while the fill factor increases because shunt resistance of ETL decreases

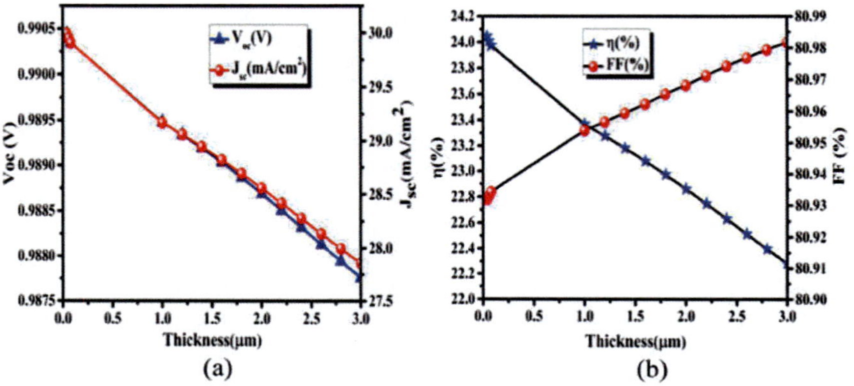

Fig. 3 Influence of changing thickness of ETL layer on cell parameters

and recombination of electron–hole pair takes place. It means if the thickness of ETL (WO_3) is massive then there ought to be less generation carriers.

3.3 *Effect of Doping Concentration of ETL*

The effect of donor density plays a significant role in perovskite solar cell. In this simulation, we have also varied doping concentration of ETL(WO_3) from 1×10^{10} to 1×10^{19} cm^{-3} maintain other parameters same throughout the distinction of donor density. On increasing the doping concentration of WO_3 the efficiency of given device can increase. The doping concentration of WO_3 layer will increases it implies that more photon incident on the surface of device that generate large number of carrier. In Fig. 4a and b with the increase in donor density all parameters such as V_{oc}, J_{sc} FF and PCE increases from particular doping concentration and its gets saturated at doping concentration $10^{18}(cm^{-3})$. Pertaining to donor density because there exhibited generation of the excess of charge carrier which increases conductivity through electron transport layer (ETL) in perovskite solar cell.

4 Conclusion

In this simulation work, we studied the perovskite solar cell with an n-i-p design structure using SCAPS-1-D simulation program. By the help of simulation we carried out the theoretical study. Among various ETL material such as PCBM, IGZO and WO_3 we have observed that WO_3 has higher efficiency. From this simulation study, we have observed that the maximum PCE of device structure $WO_3/CH_3NH_3SnI_3/Cu_2O$ is 23.99%. Furthermore, our results shows that electron transport layer and hole

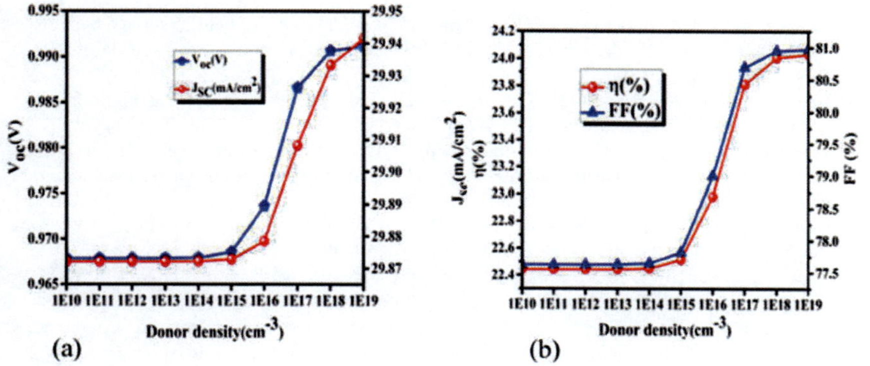

Fig. 4 Effect of donor concentration of ETL on cell parameter

transport layer parameters tremendously affect the cell execution. And on further varying the thickness and donor density of WO_3 we found that at the lower thickness the power conversion efficiency is higher when we increasing the thickness up to 70 nm the efficiency gradually decreases. As on interchanging the donor density of ETL all the parameters of PSC have been increased.

References

1. Yuan S et al (2020) Auger effect assisted perovskite electroluminescence modulated by interfacial minority carriers. Adv Funct Mater 30(12):1909222
2. Snaith HJ (2013) Perovskites: the emergence of a new era for low-cost, high-efficiency solar cells. The J Phys Chem Lett 4(21):3623–3630
3. Yang WS et al (2017) Iodide management in formamidinium-lead-halide–based perovskite layers for efficient solar cells. Science 356(6345):1376–1379
4. Osborne M (2018) Oxford PV takes record perovskite tandem solar cell to 27.3% conversion efficiency. PV-Tech https://www.oxfordpv.com/news/oxford-pv-sets-world-record-perovskite-solar-cell (25 June 2018)
5. Madan J et al (2020) Numerical simulation of charge transport layer free perovskite solar cell using metal work function shifted contacts. Optik 202:163646
6. Liu W-W et al (2019) Recent challenges in perovskite solar cells toward enhanced stability, less toxicity, and large-area mass production. Adv Mater Interfaces 6(9):1801758
7. Baikie T et al (2013) Synthesis and crystal chemistry of the hybrid perovskite (CH 3 NH 3) PbI 3 for solid-state sensitised solar cell applications. J Mater Chem A 1(18):5628–5641
8. Berhe TA et al (2016) Organometal halide perovskite solar cells: degradation and stability. Energy Environ Sci 9(2):323–356
9. Sun X et al (2015) A physics-based analytical model for perovskite solar cells. IEEE J Photovoltaics 5.5:1389–1394
10. Xing G et al (2013) Long-range balanced electron-and hole-transport lengths in organic-inorganic CH3NH3PbI3. Science 342(6156):344–347
11. Hossain MI, Alharbi FH, Tabet N (2015) Copper oxide as inorganic hole transport material for lead halide perovskite based solar cells. Sol Energy 120:370–380

12. Tavakoli MM et al (2018) Surface engineering of TiO_2 ETL for highly efficient and hysteresis-less planar perovskite solar cell (21.4%) with enhanced open-circuit voltage and stability. Adv Energy Mater 8(23):1800794
13. Lee J-W, Kim H-S, Park N-G (2016) Lewis acid–base adduct approach for high efficiency perovskite solar cells. Acc Chem Res 49(2):311–319
14. Ginting RT, Mayaji Ovhal M, Kang J-W (2018) A novel design of hybrid transparent electrodes for high performance and ultra-flexible bifunctional electrochromic-supercapacitors. Nano Energy 53:650–657
15. Chen H, Yang S (2017) Carbon-based perovskite solar cells without hole transport materials: the front runner to the market? Adv Mater 29(24):1603994
16. Guo X et al (2018) A 16.5% efficient perovskite solar cells with inorganic NiO film as hole transport material. IEEE J Photovoltaics 8(4):1039–1043
17. Negi S, Mittal P, Kumar B (2020) In-depth analysis of structures, materials, models, parameters, and applications of organic light-emitting diodes. J Electron Mater 49:4610–4636
18. Kukreti K et al (2017) Comparative analysis of different structures of photovoltaic cell based on organic materials. In: 2017 International conference on emerging trends in computing and communication technologies (ICETCCT). IEEE
19. Negi S, Mittal P, Kumar B (2020) Characteristic performance of OLED based on hole injection, transport and blocking layers. Recent Patents on Eng 14(3):373–383
20. Rai S, Pandey BK, Dwivedi DK (2020) Modeling of highly efficient and low cost CH3NH3Pb (I1-xClx) 3 based perovskite solar cell by numerical simulation. Opt Mater 100:109631
21. Rai N et al (2020) Analysis of various ETL materials for an efficient perovskite solar cell by numerical simulation. J Mater Sci: Mater Electron 31(19):16269–16280
22. Zhang Y, Blom PWM (2011) Electron and hole transport in poly (fluorene-benzothiadiazole). Appl Phys Lett 98(14):143504
23. Abdelkader H et al (2019) An optimized perovskite solar cell designs for high conversion efficiency. Superlattices Microstruct 129:240–246
24. Akimoto K et al (2006) Thin film deposition of Cu_2O and application for solar cells. Sol Energy 80(6):715–722

High Sensitive Surface Refractive Index Sensor Using Multiple Vertical Silicon Nanowire-Based PIC Waveguide

Rukmani Singh and Vishnu Priye

Abstract In this paper, a silicon on insulator-based multiple vertical silicon nanowire comprising waveguide is demonstrated as a surface refractive index sensor with high sensitivity. Waveguide surface sensitivity has been obtained by measuring the variation in effective refractive index due to change in thickness of receptor layer, which occurs because of binding with target materials on receptor layers. Parameters of the device are optimized to provide high sensitivity and compactness. Two-dimensional Finite Element Method (FEM)-based wave optics module of COMSOL software package has been used for rigorous analysis and optimization of the device. The proposed device is highly sensitive, compact and compatible with CMOS technology for manufacturing. The demonstrated device shows waveguide surface sensitivity of 0.0535 nm^{-1}.

Keywords Label-free surface sensor · Refractive index sensor · Silicon–photonics · PIC waveguide

1 Introduction

The development of highly sensitive and efficient sensor has become one of the most demanded research areas from the last few decades. Silicon photonics are extensively used in the field of sensor from last few years because of its high sensitivity, large contrast of refractive index and compatibility with CMOS technology for fabrication

R. Singh (✉) · V. Priye
Department of Electronics Engineering, Indian Institute of Technology (Indian School of Mines), Dhanbad, Jharkhand 826004, India
e-mail: rukmani.singh0@gmail.com

R. Singh
Department of Electronics and Communication Engineering, Madan Mohan Malviya University of Technology, Gorakhpur, Uttar Pradesh 273010, India

V. Priye
Department of Electronics and Communication Engineering, Indian Institute of Information Technology, Ranchi, Jharkhand 834010, India

B. Mishra and M. Tiwari (eds.), *VLSI, Microwave and Wireless Technologies*, Lecture Notes in Electrical Engineering 877,
https://doi.org/10.1007/978-981-19-0312-0_28

[1]. Many silicon on insulator-based devices has been presented as sensors (mainly as a biosensor), such as; Mach–Zehnder Interferometers (MZI) [2], grating assisted waveguides [3], ring resonators [4], grating coupler [5], directional coupler [6] and photonic crystal fibers [7]. Waveguide is a basic building platform of the photonic integrated on-chip sensors [2–7]. The sensitivity of the device and the detection limit of the biosensor can be improved significantly by increasing the surface sensitivity of waveguide. High surface sensitivity of waveguide is achieved by engineering in the waveguide structure to increase the interaction of light and target material. Several types of waveguides such as slot waveguides [8, 9], rib waveguide [10] and silicon nanowire rectangular waveguide [11, 12] have been demonstrated successfully as a refractive index sensor in biomedical fields.

The main working principle of the presented devices is based on the evanescent wave sensing [13]. Evanescent wave is generated for a short time; it spans in nanometers region of cladding and interacts with target molecule [14]. Perturbation in evanescent wave due to target material changes the propagation characteristics of a wave, which is measured to sense the responsible material. The target materials of solution can be detected by two methods: bulk sensing and surface sensing. The change in refractive index due to change in RI of the whole cladding medium is measured in bulk sensing [8] while, in surface sensing: measurement of variation in refractive index is done due to changes in the property of receptor layer on the surface of the device [8]. In biosensing applications, mainly surface biosensing takes place; because, in this application, due to recognition of target/liable biomarkers, only the refractive index near to surface region is varied.

We have demonstrated the application of multiple vertical silicon nanowires-based optical waveguide as a surface refractive index sensor. For the sensing purpose, we have measured change in effective refractive index (n_{eff}) due to variation in thickness of receptor layer on the surface of device. Change in thickness of receptor layer is due to the chemical binding of responsible molecules. This method of sensing is maximally used in biosensing applications. Parameters of waveguide have been optimized to provide very high sensitivity as well as compactness in size. Two-dimensional structure of the waveguide has been optimized and analyzed with the Finite Element Method-based COMSOL software. The silicon nanowires-based waveguide is chosen mainly because of its high surface to volume ratio, which leads to high sensitivity [15]. It is compatible with CMOS fabrication technology [15].

2 Design and Methodology

Here, we have investigated the design of silicon nanowire-based optical rectangular waveguide as a label-free surface sensor. The waveguide structure consisting a SiO_2 substrate contains many vertical silicon nanowires with gap of nanometers in bundle form at the middle part of substrate (outlook like planer-rectangular waveguide as shown by dotted line in Fig. 1a [11]. Silicon nanowires has been functionalized with receptor layer, as shown in Fig. 1b. The thickness of receptor layer is assumed

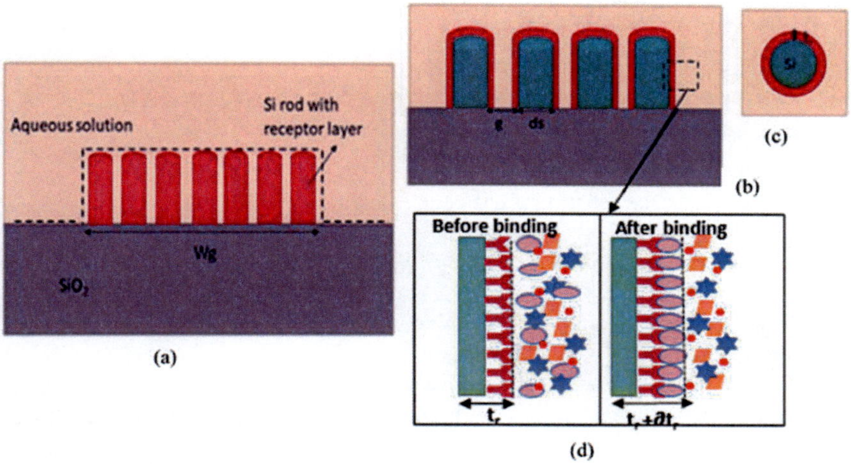

Fig. 1 Schematic structures of device for surface sensor: **a** Front view. **b** Cross-sectional view. **c** Top view of single silicon rod with receptor layer in aqueous solution. **d** Inset showing the thickness variation before and after chemical binding of receptor layer and target biomolecule

of 8 nm [16] on silicon rod. The cladding and slot regions have been filled with aqueous solution of water. The schematic models of the waveguide for sensing of target materials detection have been shown in different views by Fig. 1 where 'w_g' is total guiding width, 'ds' is diameter of silicon rod, 't_r' is thickness of receptor layer and 'g' is gap between two silicon rods. The refractive indices of different layer of the device and materials used for analysis is shown in Table 1 [15–17].

As the target molecule binds with receptor layer, its thickness is increased, as shown in Fig. 1d. The thickness of receptor layer is varied from 8 to 12 nm typically [16] and corresponding effective refractive indices has been measured for analysis of device as a biosensor.

Table 1 The materials and refractive indices of different layers of device [16–18]

Name of layer	Refractive indices
Substrate (SiO_2)	1.45
Silicon nanowire	3.46
Receptor layer	1.45
Aqueous solution of water	$1.324 + (1.38 \times 10^{-5}\, i)$

3 Results and Discussion

3.1 *Calculation of Waveguide Surface Sensitivity and Limit of Detection*

The waveguide surface sensitivity has been calculated by given Eq. (1) [18]

$$S_w = \partial \eta_{\text{eff}} / \partial t_r \tag{1}$$

where ∂n_{eff} is change in effective refractive index and ∂t_r is change in thickness of functionalized layer.

Figure 2 shows the variation in effective refractive index of device with change in thickness of receptor layer, slope of this graph indicates waveguide surface sensitivity of the device structure.

Resonant wavelength shift ($\Delta\lambda_0$) in terms of change in effective refractive (Δn_{eff}) for wavelength shifting scheme (such as ring resonator) is given by Eq. (2) [19].

$$\Delta\lambda_0 = (\Delta n_{\text{eff}} * \lambda_0)/\text{ng} \tag{2}$$

$$n_g = n_{\text{eff}}(\lambda_0) - \lambda_0 \frac{\delta n_{\text{eff}}}{\delta\lambda} \tag{3}$$

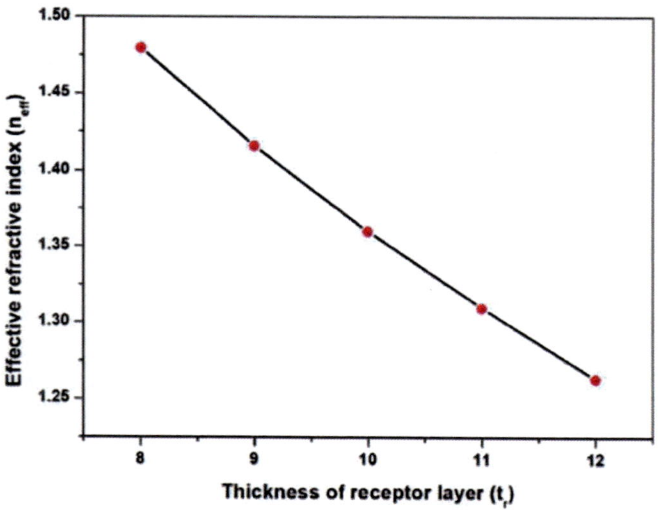

Fig. 2 n_{eff} verses thickness of receptor layer

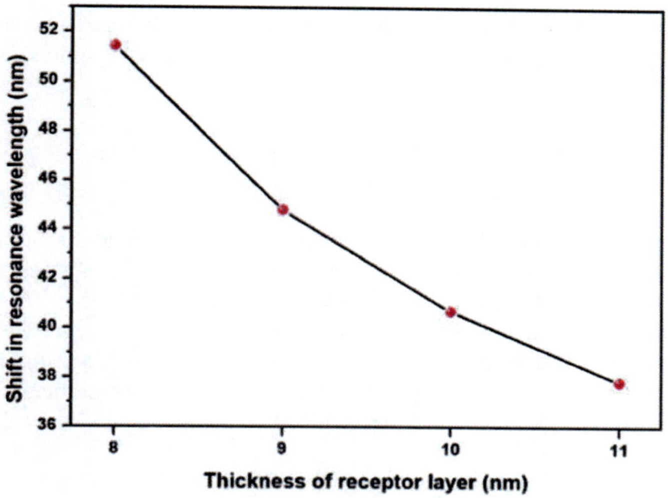

Fig. 3 Shift in resonance wavelength verses thickness of receptor layer

Here, λ_0 is the reference resonance wavelength, which is 1550 nm taken here and 'n_g' is group index, which is calculated to 2.0038 using Eq. (3) [20] by taking air as cladding layer on the optimized structure of SNROW waveguide. Effect on wavelength shift due to different thickness of target analyze is shown in Fig. 3, which has been calculated using Eq. (2).

$$\Gamma = \frac{(n_r - n_{\text{AS}})}{\partial n/\partial c} t_r \tag{4}$$

where n_r is refractive index of receptor layer, and n_{AS} is real part of complex refractive index of the aqueous solution, t_r is thickness of functionalization layer and $\partial n/\partial c$ is variation in index with analyte concentration, which is taken as 187 mm^3/g [18]. The surface mass coverage (Γ) for the given structure is 0.00539 g/m^2. LOD can be given by $\Delta\Gamma_{\min}$ [18] as shown in Eq. 5.

$$\Delta\Gamma_{\min} = \frac{(n_r - n_{\text{AS}})}{\partial n/\partial c} \Delta t_{r_{\min}} \tag{5}$$

where, $\Delta t_{r\min}$ is minimum detectable change in thickness of receptor layer, which could be obtained in terms of resonance wavelength. Usually, the limit of detection of ring resonator in pg/mm^2 is calculated by Eq. (6) [16].

$$\text{LOD} = \Delta\Gamma_{\min} = \frac{(n_r - n_{\text{AS}})}{\partial n/\partial c} \frac{1}{\frac{\partial \lambda_0}{\partial t_r}} \Delta\lambda_{0_{\min}} \tag{6}$$

where λ_0 is the resonance wavelength and $\Delta\lambda_{0\ min}$ is the minimum detectable shift in resonance wavelength. It can also be written in terms of waveguide sensitivity as shown in Eq. (7).

$$\text{LOD} = \frac{(n_r - n_{\text{AS}})}{\partial n/\partial c} \frac{1}{\frac{\lambda_0}{n_{\text{eff}}} \frac{\partial n_{\text{eff}}}{\partial t_r}} \Delta\lambda_{0_{\text{min}}} \tag{7}$$

Thus, the limit of detection (LOD) is inversely proportional to the surface sensitivity, as can be seen from Eq. (7), i.e., improvement in waveguide surface sensitivity leads to improve the LOD, which is an essential metric for the sensor. The proposed waveguide structure can be used as a main basic building block in integrated biosensor to provide improved detection limit because waveguide surface sensitivity is very high. Surface mass coverage Γ (g/m2), which is related to properties of receptor layer as shown Eq. (4) can be used to define LOD [18].

3.2 Optimization of Waveguide's Parameters

In this section, parametric optimization has been performed to provide high waveguide surface sensitivity. The width of guided region of the waveguide can be represented by

$$w_g = n \times d_s + (n - 1) \times g \tag{8}$$

where 'w_g' is width of guided region, 'n' is number of silicon rods, 'd_s' is diameter of silicon rod and 'g' is gap between silicon rods. From the Eq. (8), it could be perceived that the numbers of rods, diameters of Si rods, and gap between silicon rods play important role in defining width of waveguide.

In order to optimize the different parameter, at first, effect on waveguide surface sensitivity with variation in number of silicon rods (n) has been analyzed by keeping $d_s = 15$ nm and $g = 30$ nm initially. Height of silicon wire is taken 220 nm (standard) in whole optimization process. It could be observed by Fig. 4a, the surface sensitivity is increased with number of rods and became almost constant after $n = 13$. Number of rods is chosen 13 to provide high sensitivity as well as compactness.

Further influence of d_s on waveguide surface sensitivity has been investigated by keeping the $n = 13$ and $g = 30$ nm. Figure 2b shows, when the value of d is increasing greater than 35 nm, the increment in surface sensitivity is very low while below this value, it is increasing parabolically. Increment in value of d_s beyond 35 nm to 40 nm increases waveguide surface sensitivity 0.0002 only with the cost of 65 nm increment in guiding width of the waveguide, hence, $d_s = 35$ nm is selected to offer compact size with high sensitivity.

Value of 'g' is taken 30 nm initially by considering the fact that; initial thickness of receptor layer is taken 8 nm, i.e., $8 \times 2 = 16$ nm region of gap is covered when

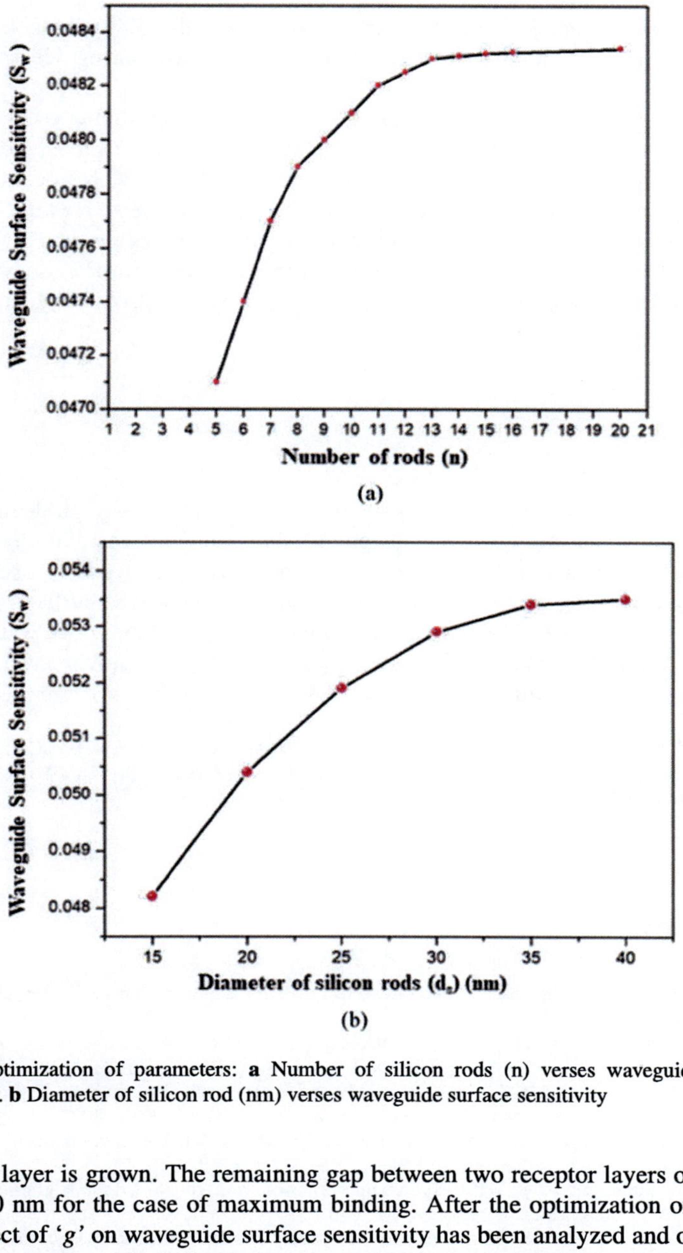

Fig. 4 Optimization of parameters: **a** Number of silicon rods (n) verses waveguide surface sensitivity. **b** Diameter of silicon rod (nm) verses waveguide surface sensitivity

receptor layer is grown. The remaining gap between two receptor layers on silicon rod is 10 nm for the case of maximum binding. After the optimization of '*n*' and 'd_s', effect of '*g*' on waveguide surface sensitivity has been analyzed and observed. Waveguide surface sensitivity is decreased with increment in gap because it reduces the surface to volume ratio of device. Theoretically, the waveguide sensitivity is varying with '*n*', 'd_s' and '*g*' because these (*n*', 'd_s' and '*g*') change the surface to volume ratio of device. Increment in surface to volume ratio leads to enlargement of interaction region of bio-analyte and light and consequently, waveguide sensitivity increases.

Thus for the optimized dimensions $n = 13$, $ds = 35$ nm and $g = 25$ nm, the waveguide surface sensitivity in terms of change in effective refractive index by per unit change in thickness of receptor layer is obtained 0.0535, which is near about 48 and 97 times of surface sensitivity of waveguide structure presented in refs. [16] and [8], respectively, where a specific designed horizontal slot waveguide and SOI rib slot waveguide have demonstrated for label-free surface sensing. The proposed structure of waveguide is showing very high surface sensitivity because of high surface to volume ratio of the device.

4 Conclusion

We have demonstrated application of silicon nanowires-based waveguide in surface refractive index sensing efficiently to provide high surface sensitivity. Impact of waveguide's parameter on waveguide surface sensitivity is analyzed and discussed in contexts of increasing sensitivity. Parametric analysis of device as a surface refractive index sensor is performed by full vectorial finite element method-based wave optics module of COMSOL. The demonstrated device shows high sensitivity and low limit of detection with maintaining the compactness.

Acknowledgements This work was supported by National Project implementation Unit -Ministry of Human Resource Development (MHRD), Government of India via TEQIP Collaborative Research Scheme (ID- 1-5764084542).

References

1. Zinoviev K et al (2008) Silicon photonic biosensors for lab-on-a-chip applications. Adv Opt Technol 2008:1–6
2. Yuan D, Dong Y, Liu Y, Li T (2015) Mach-Zehnder interferometer biochemical sensor based on silicon-on-insulator rib waveguide with large cross section. Sensors 15:21500–21517
3. Liu Q, Kee JS, Park MK (2013) A refractive index sensor design based on grating-assisted coupling between a strip waveguide and a slot waveguide. Opt Exp 21:5897
4. Sarkaleh AK, Lahijani BV, Saberkari H, Esmaeeli A (2017) Optical ring resonators: a platform for biological sensing applications. J Med Sig Sens 7:185–191
5. Singh R, Singh RR, Priye V (2018) Optical biosensor based on guided mode grating coupler. In: Optics infobase conference papers vol. Part F114
6. Luo L, Feng G, Zhou S, Tang T (2016) A refractive-index biosensor based on directional coupling of silica waveguides. Optik 127:4019–4023
7. Hameed MFO, Alrayk YKA, Obayya SSA (2016) Self-calibration highly sensitive photonic crystal fiber biosensor. IEEE Photonics J 8:1–12
8. Dell'Olio F, Passaro VM (2007) Optical sensing by optimized silicon slot waveguides. Opt Exp 15:4977

9. Singh R, Priye V (2021) Si3N4–SiO2 based curve slot waveguide for high confinement factor and low mode effective area along with biosensing application. SILICON. https://doi.org/10.1007/s12633-020-00927-y
10. Mukundan H et al (2009) Waveguide-based biosensors for pathogen detection. Sensors 9:5783–5809
11. Singh RR, Malviya N, Priye V (2016) Parametric analysis of silicon nanowire optical rectangular waveguide sensor. IEEE Photonics Technol Lett 28:2889–2892
12. Singh R, Priye V (2019) Ultrahigh-sensitivity label-free optical biosensor based on a silicon nanowire rectangular optical waveguide for lung cancer detection. J Opt Soc Am B 36:2402
13. Mendez-Astudillo M, Takahisa H, Okayama H, Nakajima (2016) Optical refractive index biosensor using evanescently coupled lateral Bragg gratings on silicon-on-insulator. Japanese J Appl Phys 55:08RE09
14. Brandenburg A, Polzius R, Bier F, Bilitewski U, Wagner E (1996) Direct observation of affinity reactions by reflected-mode operation of integrated optical grating coupler. Sens Actuat B: Chem 30:55–59
15. Khorasaninejad M, Saini SS (2010) Silicon nanowire optical waveguide (SNOW). Opt Exp 18:23442
16. Dell'Olio F, Conteduca D, Ciminelli C, Armenise MN (2015) New ultrasensitive resonant photonic platform for label-free biosensing. Opt Exp 23:28593
17. Palmer KF, Williams D (1974) Optical properties of water in the near infrared*. J Opt Soc Am 64:1107
18. Berini P (2008) Bulk and surface sensitivities of surface plasmon waveguides. New J Phys 10:105010
19. Dar T, Homola J, Rahman BMA, Rajarajan M (2012) Label-free slot-waveguide biosensor for the detection of DNA hybridization. Appl Opt 51:8195
20. Flueckiger J, Schmidt S, Donzella V et al (2016) Sub-wavelength grating for enhanced ring resonator biosensor. Opt Exp 24:15672

A Compact Slot and Notch Loaded Microstrip Antenna for Wireless Applications

Ramesh Kumar Verma and Akhilesh Kumar

Abstract In this article, a compact modified wideband microstrip antenna (MSA) has been designed by cutting slots inside the patch and notches at the upper edge of the patch which provides a wide range of frequency band of 1259 MHz between frequency 1.653 GHz and 2.912 GHz. The proposed antenna design provides a wide bandwidth about 56.73% with −27.63 dB return loss resonating at 2.025 GHz frequency. It has 4.86 dB high gain and antenna efficiency of 95.77% at frequency 2.025 GHz. The antenna is designed with glass epoxy (FR-4) substrate and simulated by IE3D software. The antenna is excited via 50 Ω microstrip line feed. The resonating frequency band can be used for different applications in wireless communication.

Keywords Compact · Modified · Bandwidth · Wireless · FR-4 · IE3D · Microstrip line feed

1 Introduction

The advancement in wireless communication system has increased the interest of low profile as well as compact antennas with high gain and wideband working frequency bandwidth. The microstrip antennas have many advantages such as light weight, low profile, and compactness but the major disadvantages of MSA are its narrow bandwidth, low efficiency, spurious feed radiation, and smaller gain. There are numerous substrate materials whose dielectric constant that lies between 2.2 and 12 can be utilized for designing of an antenna [1]. The substrate having lower dielectric constant provides good efficiency and large bandwidth. The bandwidth of microstrip antenna

R. K. Verma (✉)
Department of Electronics and Communication Engineering, Bundelkhand Institute of Engineering and Technology, Jhansi, UP, India
e-mail: ramesh85.ec@gmail.com

A. Kumar
Department of Electronics and Communication Engineering, Madan Mohan Malaviya University of Technology, Gorakhpur, UP, India

B. Mishra and M. Tiwari (eds.), *VLSI, Microwave and Wireless Technologies*, Lecture Notes in Electrical Engineering 877,
https://doi.org/10.1007/978-981-19-0312-0_29

can be raised by using various strategies, for example, taking different shapes of antenna patch and loading antenna patch with various kinds of notches and slots [2].

Neyestanak et al. [3] presented a W-shape antenna having 36.7% (800 MHz) bandwidth with large size of 70 × 50 mm. Ansari et al. [4] designed a disk shape single as well as multi-layer antenna of size 88 × 62 mm resonating at 3 GHz with fractional bandwidth 36.5% (985 MHz) for WLAN and WiMAX application. Verma and Srivastava [5] designed an H-shape patch antenna of size 39.04 × 47.64 mm showing bandwidth of 35.61% (710 MHz) loaded with three square slots. Wu and Wong [6] proposed direct coupled antenna of size 52.87 × 40 mm resonating at 2.879 GHz with 12.7% (365 MHz) bandwidth. Kamakshi et al. [7] designed an antenna with large size of 120 × 80 mm having three notches and one slot resonating at 1.84 GHz. Rajpoot et al. [8] designed an antenna of I-shape patch with overall size of 40.06 × 48.72 mm resonating at 2.41 GHz with 45.72% (970 MHz) bandwidth. Verma and Srivastava [9] proposed an antenna with inverted T-shape slot of size 38.43 × 46.86 mm having bandwidth of 48.25% (1179 MHz). Zhang et al. [10] designed an antenna of size 88 × 88 mm resonating at 2.45 GHz with bandwidth of 24.8% (650 MHz). Bala et al. [11] presented a metamaterial-based antenna of size 40 × 48 mm with 41% (1118 MHz) bandwidth resonating at 2.73 GHz. Sze and Wong [12] proposed an antenna of size 60 × 50 mm with two slots of right angle shape and a U-shape slot with narrow bandwidth of 53 MHz bandwidth. Mishra et al. [13, 14] designed a slot loaded for both dual and wideband stacked antenna and petal shape dual band gap coupled antenna. Verma and Srivastava [15] presented an inverted Y-shape patch antenna of size 33 × 40 mm with fractional bandwidth of 36.30% (933 MHz). Singh et al. [16] presented a circular patch antenna of hexa-band characteristics with inverted L-shape notch. Gupta et al. [17] presented a wideband gap and direct coupled antenna resonating at 2.399 GHz.

In this paper, bandwidth of MSA of compact size 38.42 × 46.86 mm (ground) has been enhanced by cutting three notches (6 × 6 mm, 6 × 6 mm, and 24 × 8 mm) and three slots (5 × 5 mm, 5 × 5 mm, and 15 × 5 mm) in antenna patch which is excited by microstrip line feed of 50 Ω. The designed antenna operated at 2.025 GHz frequency with 56.73% (1259 MHz) bandwidth between frequency 1.653 GHz and 2.912 GHz which is appropriate for WLAN and WiMAX.

2 Antenna Design

For designing of rectangular shape patch antenna, width and length of patch are calculated by using Eqs. (1)–(6) as given below [18].

$$W = \frac{c}{2f_r}\sqrt{\frac{2}{\varepsilon_r + 1}} \tag{1}$$

where c is speed of light (3×10^8 m/s) in air, f_r is design frequency, and ε_r is dielectric constant of material.

Effective dielectric constant $\varepsilon_{\text{reff}}$ is given as [18]

$$\varepsilon_{\text{reff}} = \frac{\varepsilon_r + 1}{2} + \frac{\varepsilon_r - 1}{2}\left(1 + \frac{12\text{h}}{\text{W}}\right)^{-\frac{1}{2}} \tag{2}$$

ΔL (extension length) of patch is calculated by [18]

$$\frac{\Delta L}{h} = 0.412\frac{(\varepsilon_{\text{reff}} + 0.3)\left(\frac{W}{h} + 0.264\right)}{(\varepsilon_{\text{reff}} - 0.258)\left(\frac{W}{h} + 0.8\right)} \tag{3}$$

Actual length (L) can be calculated by utilizing the value of (ΔL) as [18]

$$L = \frac{c}{2f_{r\sqrt{\varepsilon_{\text{reff}}}}} - 2\Delta L \tag{4}$$

Ground layer length and width of can be calculated as [18]

$$L_g = L + 6h \tag{5}$$

$$W_g = W + 6h \tag{6}$$

3 Design Specifications of Antenna

The antenna specifications that are essentially required for designing of antenna are illustrated in Table 1. FR-4 substrate that has dielectric constant of 4.4 is used for proposed antenna design. The proposed antenna is designed for design frequency 2.45 GHz. The altitude of dielectric substrate (h) is 1.6 mm, and its loss tangent (tan δ) is 0.02. Microstrip line feed of 50 Ω is utilized for the excitation of radiating patch.

Table 1 Antenna design specifications

S. No	Parameter	Value	S. No	Parameter	Value (mm)
1	Dielectric constant ε_r	4.4	5	Patch length (Lp)	28.82
2	Design frequency (f_r)	2.45 GHz	6	Patch width (Wp)	37.26
3	Substrate height (h)	1.6 mm	7	Ground length (Lg)	38.42
4	Speed of light (c)	3×10^8 m/s	8	Ground width (Wg)	46.86

4 Method of Antenna Design

By using above equations and specification given in Table 1, the dimensions of the antenna geometry are calculated at 2.45 GHz frequency. The size of patch has been calculated which are 28.82 mm (length) and 37.26 mm (width), respectively. The size of ground plane is calculated by simply adding 6 h (9.6 mm) in both length (Lp) and width (Wp) of patch. The length and width of ground are considered as 38.42 mm and 46.86 mm, respectively. In designing of proposed antenna geometry, initially a rectangular ground was formed with calculated size and a patch at 1.6 mm above ground. The ground and patch loaded antenna geometry is shown in Fig. 1a. The patch of antenna was modified by slots and notches of appropriate dimension for performance improvement.

The proposed antenna is designed by loading a pair of square notches of size 6 × 6 mm at top corner of patch and slots of different sizes in conventional patch antenna. A vertical notch of dimension 24 × 8 mm is inserted at top middle of patch along with two square slots of size 5 × 5 mm which is y-axis symmetrically on both side of rectangular patch and a horizontal rectangular slot of size 15 × 5 mm at bottom side of patch. The structure of proposed slotted antenna is illustrated in Fig. 1b. The dimensional specifications of slots and notches are shown in Table 2. After the

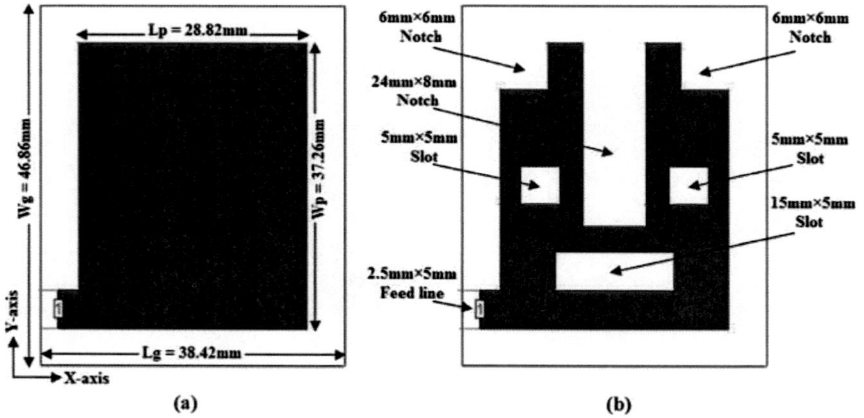

Fig. 1 Design of **a** Ground and patch loaded antenna **b** Notch and slot loaded geometry (proposed)

Table 2 Slot and notch specifications

S. No	Parameter	Value	S. No	Parameter	Value
1	Square notches	6 × 6 mm	4	Horizontal slot	15 × 5 mm
2	Square slots	5 × 5 mm	5	Feed length	2.5 mm
3	Vertical notch	24 × 8 mm	6	Feed width	5 mm

designing of ground and its slotted patch, antenna is excited by a 50 Ω microstrip line feed with the help of strip 2.5 × 5 mm connected at lower left corner of patch.

5 Results and Discussion

The design and analysis of proposed antenna are performed by IE3D [19] simulator at frequency 2.025 GHz between 1 and 3 GHz frequency. The proposed antenna covers 1259 MHz bandwidth between 1.653 GHz and 2.912 GHz frequency. It displays maximum return loss of −27.63 dB at frequency 2.025 GHz, as shown in Fig. 2. According to return loss plot, antenna provides bandwidth about 56.73% at 2.025 GHz resonant frequency. While antenna design without any slots and notches (shown in Fig. 1a) covers range of frequency 2.209–2.401 GHz (192 MHz) resonating at 2.305 GHz having bandwidth 8.33% with −10.66 dB return loss also illustrated in same Fig. 2. Large bandwidth of proposed antenna geometry is achieved after making different modifications inside rectangular patch like slotting as well as notching. The comparative analysis and size comparison of designed antenna with references [3–11] are shown in Table 3. The size and bandwidth comparison with reference antennas [3–11] is shown in Figs. 3 and 4, respectively.

The proposed antenna has VSWR of 1.102 at frequency 2.025 GHz as shown in Fig. 5. The gain of 4.86 dB and directivity of 4.87 dB have been obtained at frequency 2.025 GHz which are simulated by IE3D software, and its plots are presented in Figs. 6 and 7, respectively. The proposed antenna has high antenna efficiency about 95.77% at frequency 2.025 GHz which is shown in Fig. 8. Smith chart and Z parameter of suggested antenna are represented in Fig. 9a, b. The $Z = 47.94\text{-}j4.26\Omega$ ($|Z| = 48.13\ \Omega$) is obtained at frequency 2.025 GHz. Simulated radiation pattern (2D, 3D)

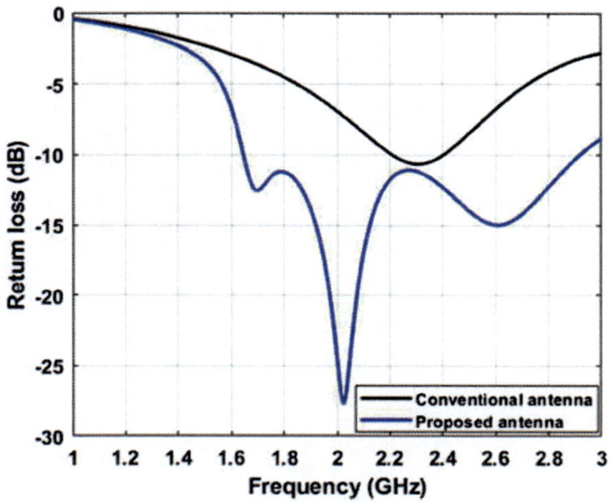

Fig. 2 Bandwidth plot of suggested antenna

Table 3 Comparison of proposed antenna design with references [3–11]

References	Dimensions (mm^2)	Band (GHz)	Resonance Freq.(GHz)	Fractional bandwidth (%)	Uses
[3]	70 × 50 = 3500	1.8–2.6	1.88/2.37	36.7% (800 MHz)	RFID/WLAN
[4]	88 × 62 = 5456	2.382–3.367	3.0	36.5% (985 MHz)	WLAN/WiMAX
[5]	39.04 × 47.64 = 1860	1.639–2.349	1.729/2.223	35.61% (710 MHz)	S-Band
[6]	52.87 × 40 = 2115	2.696–3.061	2.879	12.7% (365 MHz)	Broadband
[7]	120 × 80 = 9600	1.56–2.12	1.84	30.5% (560 MHz)	Broadband
[8]	40.06 × 48.72 = 1951	1.65–2.62	2.41	45.72% (970 MHz)	WLAN
[9]	38.43 × 46.86 = 1800	1.854–3.033	2.477	48.25% (1179 MHz)	WLAN
[10]	88 × 88 = 7744	2.29–2.94	2.45	24.8% (650 MHz)	WLAN/WiMAX
[11]	40 × 48 = 1920	2.233–3.351	2.73	41% (1118 MHz)	WLAN/WiMAX
Proposed	38.42 × 46.86 = 1800	1.653–2.912	2.025	56.73% (1259 MHz)	WLAN/WiMAX

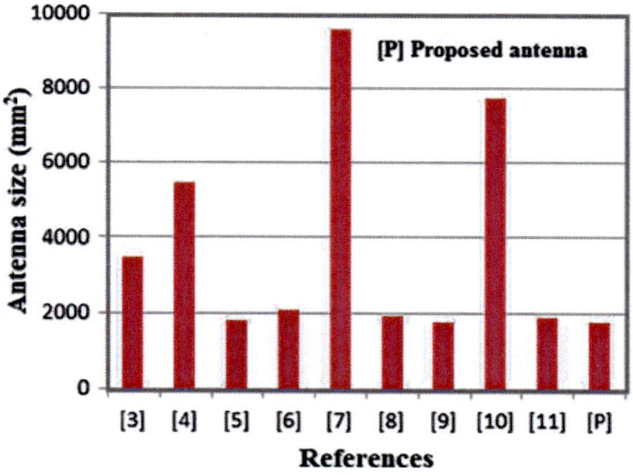

Fig. 3 Antenna size comparisons

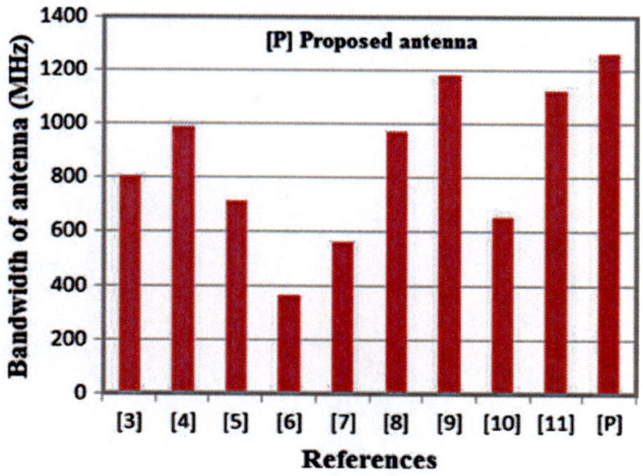

Fig. 4 Bandwidth comparison of antenna

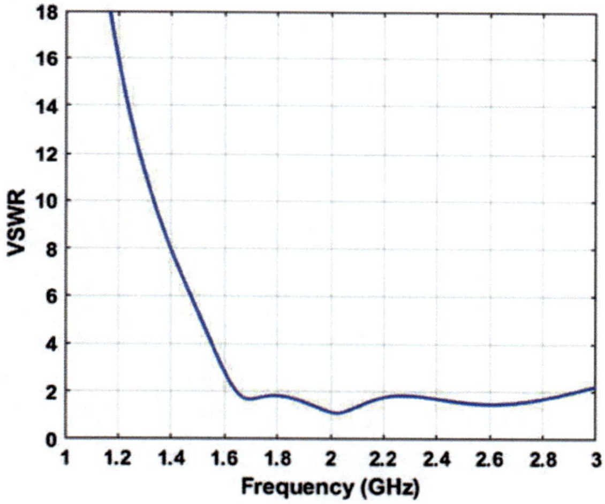

Fig. 5 VSWR of suggested antenna

and current distribution at frequency 2.025 GHz are displayed in Figs. 10, 11, and 12, respectively. 2D radiation pattern is shown at Phi = 0° and Phi = 90° in *E*-plane while at Theta = 0° and Theta = 90° in *H*-plane. 3 dB beamwidth of suggested antenna is 56.54° (72.33°, 128.87°) at frequency 2.025 GHz.

6 Experimental Results

Hardware design (front view and back view) of proposed antenna is displayed in Fig. 13. The measured return loss image measured by vector analyzer is shown in Fig. 14. The measured impedance bandwidth of proposed antenna is achieved

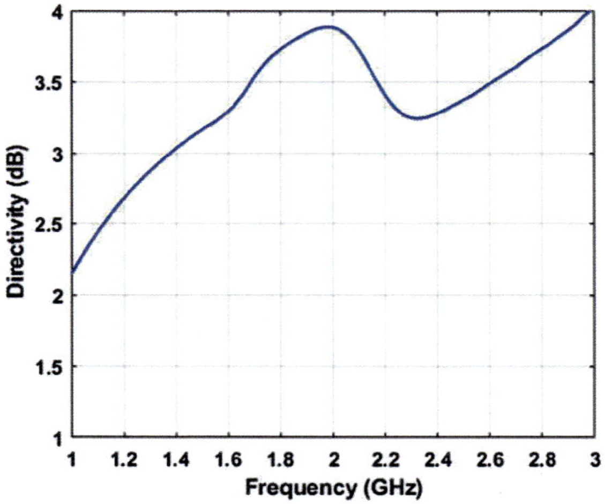

Fig. 6 Directivity of suggested antenna

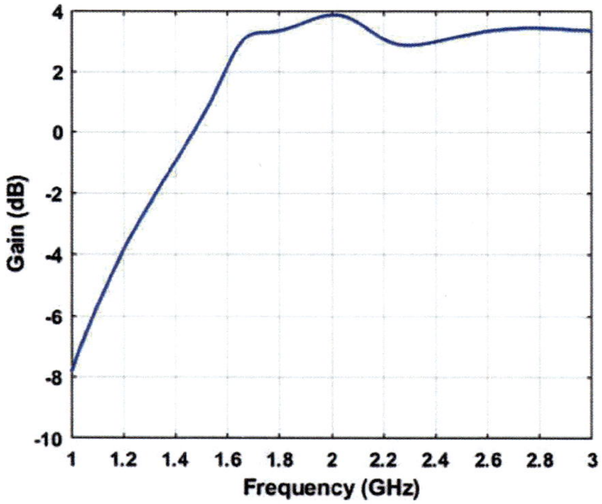

Fig. 7 Gain of suggested antenna

41.38% in the frequency range 1.61–2.45 GHz (840 MHz). Measured antenna design is resonating at frequency 2.04 GHz with −27.28 dB return loss. Bandwidth comparisons of simulated and measured antennas have been displayed in Fig. 15. The small mismatch is occurring in both return losses due to fabrication defect of hardware antenna. Measurement setup image is also shown in Fig. 16.

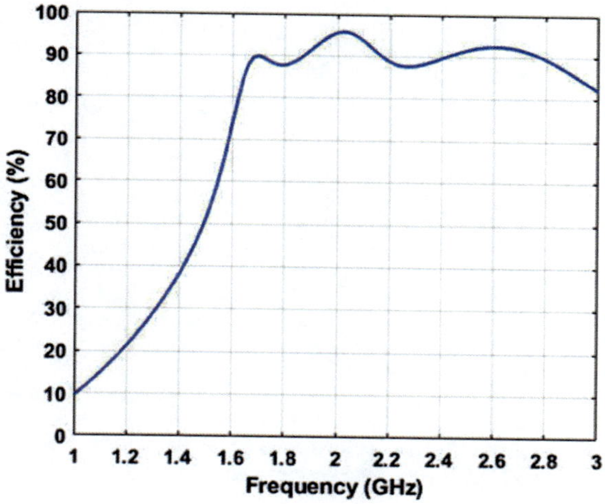

Fig. 8 Antenna efficiency of suggested antenna

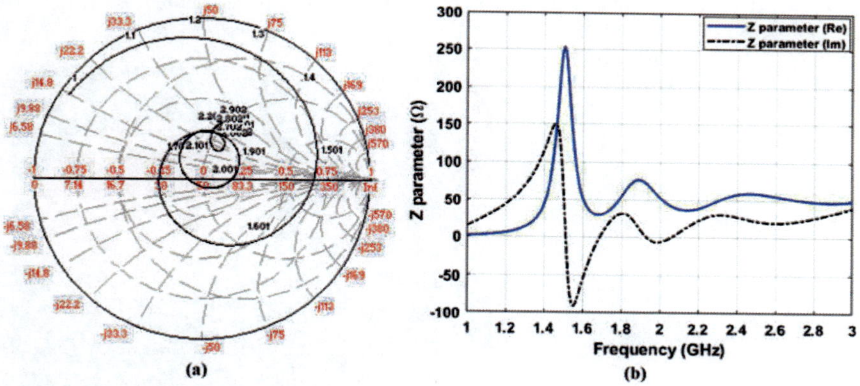

Fig. 9 **a** Smith chart **b** *Z* parameter of suggested antenna

7 Conclusion

Slot and notch loaded rectangular MSA fed by 50 Ω microstrip line fed has been designed and simulated by IE3D on FR-4 (glass epoxy) substrate with wide bandwidth of 56.73% (1259 MHz) resonating at 2.025 GHz between 1.653 and 2.912 GHz. It displays maximum return loss of −27.63 dB and VSWR of 1.102 at frequency 2.025 GHz. The maximum gain of 4.86 dB and antenna efficiency of 95.77% are obtained at resonant frequency. 3 dB beamwidth of suggested antenna is obtained 56.54° (72.33°, 128.87°) at frequency 2.025 GHz. The designed antenna covers frequency band 1.653–2.912 GHz which is suitable for multiple wireless applications.

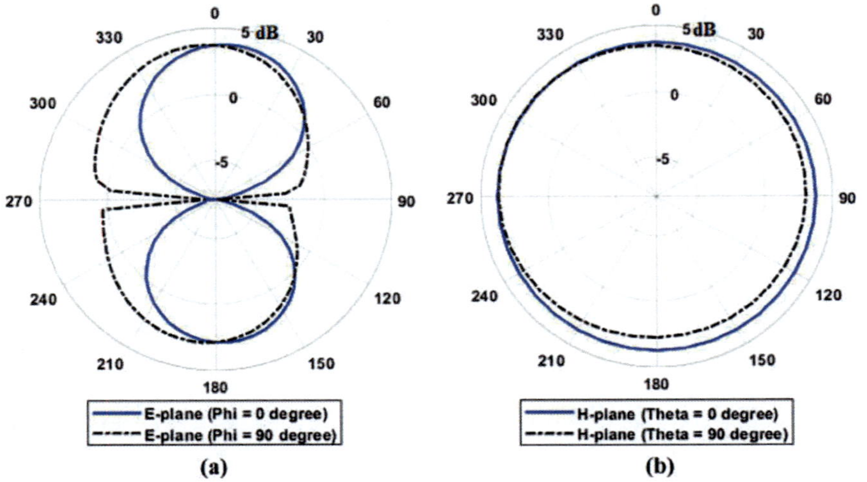

Fig. 10 2D Radiation pattern of suggested antenna at 2.025 GHz for **a** elevation, **b** azimuth

Fig. 11 3D radiation pattern of antenna

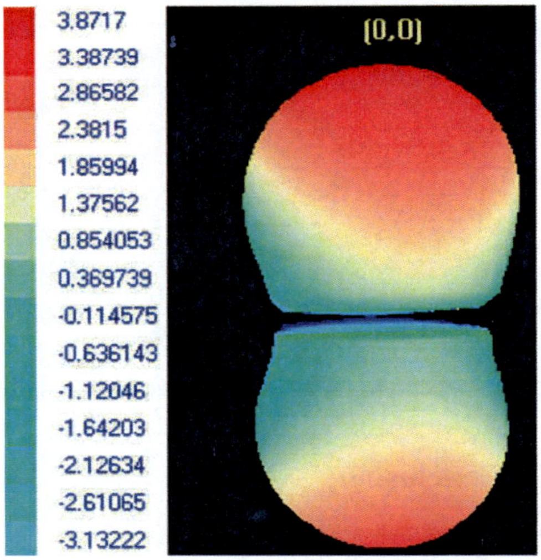

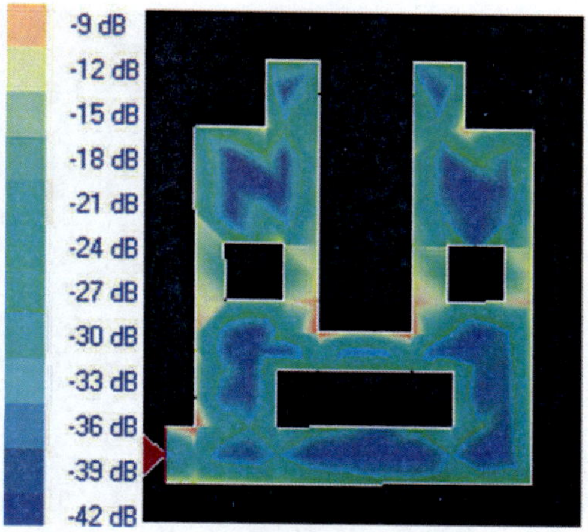

Fig. 12 Current distribution of antenna at 2.025 GHz

Fig. 13 Hardware design of suggested antenna

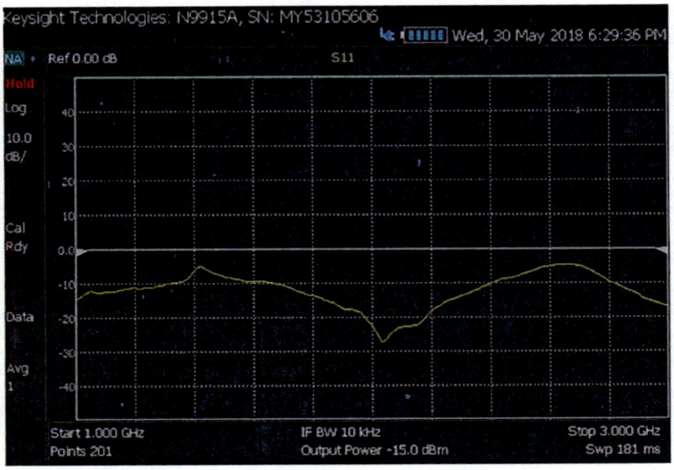

Fig. 14 Measured return loss of suggested antenna

Fig. 15 Return loss of hardware design of antenna

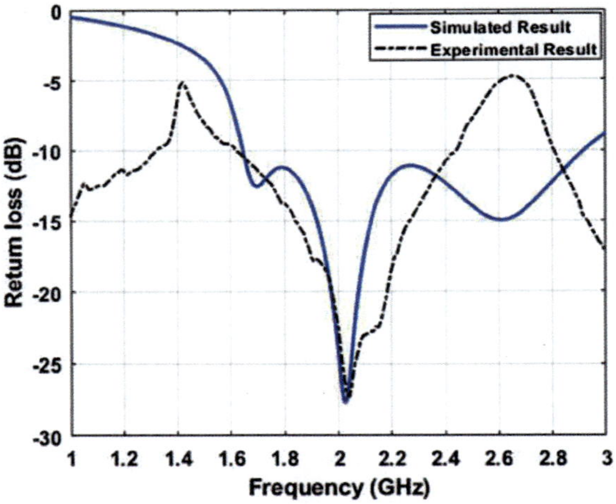

Fig. 16 Setup for return loss measured of suggested antenna

References

1. Balanis CA (2005) Antenna theory, analysis and design. John, New York
2. Pozar DM (1992) Microstrip antennas. Proc IEEE 80(1):79–91
3. Neyestanak AAL, Kashani FH, Barkeshli K (2007) W-shaped enhanced-bandwidth patch antenna for wireless communication. Wireless Pers Commun 43:1257–1265
4. Ansari JA, Yadav NP, Singh P, Mishra A (2010) Analysis of broadband operation of disk patch antenna with parasitic elements in single and two layer structures. Int J Microwave Optical Technol 5(3):140–147
5. Verma RK, Srivastava DK (2019) A compact rectangular microstrip antenna of H-shape patch loaded with notch and slots for S-band applications. Int J Microwave Optical Technol 14(2):100–108
6. Wu CK, Wong KL (1999) Broadband microstrip antenna with directly coupled and parasitic patches. Microwave Opt Technol Lett 22:348–349
7. Kamakshi K, Singh A, Aneesh M, Ansari JA (2014) Novel design of microstrip antenna with improved bandwidth. Int J Microwave Sci Technol 659592:1–7
8. Rajpoot V, Srivastava DK, Saurabh AK (2014) Optimization of I-shape microstrip patch antenna using PSO and curve fitting. J Comput Electron 13(4):1010–1013
9. Verma RK, Srivastava DK (2019) Design, optimization and comparative analysis of T-shape slot loaded microstrip patch antenna using PSO. Photon Netw Commun 38(3):343–355
10. Zhang J, Lu WJ, Li L, Zhu L, Zhu HB (2016) Wideband dual-mode planar endfire antenna with circular polarisation. Electron Lett 52(12):1000–1001
11. Bala BD, Rahim MKA, Murad NA (2015) Bandwidth enhancement metamaterial antenna based on transmission line approach. Microw Opt Technol Lett 57(1):252–256
12. Sze JY, Wong KL (2000) Slotted rectangular microstrip antenna for bandwidth enhancement. IEEE Trans Antennas Propag 48(8):1149–1152
13. Mishra B, Singh V, Singh R (2017) Dual and wide-band slot loaded stacked microstrip patch antenna for WLAN/WiMAX applications. Microsyst Technol 23(8):3467–3475
14. Mishra B, Singh V, Singh R (2018) Gap coupled dual-band petal shape patch antenna for WLAN/WiMAX applications. Adv Electr Electron Eng 16(2):185–198
15. Verma RK, Srivastava DK (2018) A compact inverted Y slot rectangular microstrip patch antenna for bluetooth applications. Indonesian J Electr Eng Comput Sci 11(2):413–418
16. Singh V, Mishra B, Dwivedi AK, Singh R (2018) Inverted L-notch loaded hexa band circular patch antenna for X, ku/K band applications. Microw Opt Technol Lett 60(8):2081–2088

17. Gupta A, Srivastava DK, Saini JP, Verma RK (2020) Comparative analysis of microstrip-line-fed gap-coupled and direct-coupled microstrip patch antennas for wideband applications. J Comput Electron 19(1):457–468
18. Verma RK, Srivastava DK (2020) Design and analysis of triple-band rectangular microstrip antenna loaded with notches and slots for wireless applications. Wireless Pers Commun 114:1847–1864
19. IE3D Electromagnetic Simulation and Optimization Package, Version 9.0

Real-time Face mask Detection Using Deep Learning and MobileNet V2

Akshay Verma and Manish Kumar Srivastava

Abstract COVID-19 has affected the whole world severely. Lockdowns and quarantines are imposed all over the world to prevent its spread. Hand sanitizers and face masks were made compulsory for individuals to apply for safety of their own and their society. This project will check the presence or the absence of masks on the face of a person. There could be more than a single person in the input provided, and the input could vary from images to GIFs to Livestreams.

Keywords Machine learning · Deep learning · Computer vision · TensorFlow · Keras · MobileNet V2

1 Introduction

With its outbreak from a laboratory in Wuhan, COVID-19 spread across the whole world. This deadly virus has caused more than 2 million deaths, and the numbers are not in control even now [1, 2]. People are being vaccinated; but to vaccinate 7 billion people is a tedious task. Utilization of face masks decreases the odds of spread of this sickness by 85% [3, 4].With the help of deep learning and its subfields, a lot of mammoth looking tasks could be solved much easier. This classification project is one such example which detects the presence of masks on a person's face. It could be used for monitoring purposes.

A. Verma (✉)
Master of Technology, Department of Computer Science and Engineering, MMM University of Technology, Gorakhpur, India
e-mail: akshyvrma8@gmail.com

M. K. Srivastava
Assistant Professor, Department of Computer Science and Engineering, MMM University of Technology, Gorakhpur, India

B. Mishra and M. Tiwari (eds.), *VLSI, Microwave and Wireless Technologies*,
Lecture Notes in Electrical Engineering 877,
https://doi.org/10.1007/978-981-19-0312-0_30

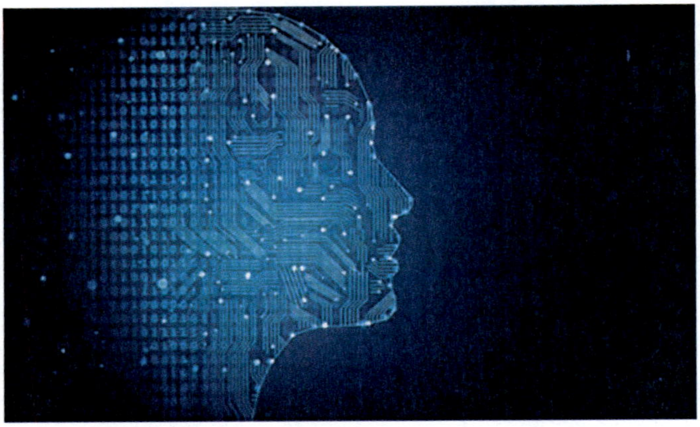

Fig. 1 Machine learning outlook

1.1 Machine Learning

Machine learning is process of training a machine using various algorithms so that it learns to perform tasks better on its own. It is the ability of a machine to find out. It is the facility of giving deciding capabilities to a machine. With given task "U" and time "T," its performance "P" gets more accurate with number of iterations or experience "E." Machine learning may be a quite man-made consciousness (AI) that allows programming applications to be more precise at foreseeing results without being expressly programmed. ML calculations utilize genuine data as commitment to predict new yields whereas old style textual style AI is regularly orchestrated by how an estimation sorts out the best approach to advance to be more exact in its assumptions [5].

There are four major approaches and the sort of count a data scientist chooses to utilize relies upon what very data they need to anticipate [6] (Fig. 1).

1.2 Deep Learning

Deep learning is a subcategory of machine learning. It emphasizes on mimicking tasks of human brain. Deep learning calculations try to use the obscure design in the info dissemination to find great portrayals, regularly at different levels, with more elevated level learned highlights characterized regarding lower-level highlights [7].

Deep learning is an essential advancement behind unpiloted vehicles, empowering them to observe traffic signals, or to see a person by strolling from a light post. Deep learning is getting stacks considered as late and thinking about current conditions.

In deep learning, a PC model sorts out some way to execute portrayal tasks clearly from images, audio files, or texts. Deep learning models can achieve forefront

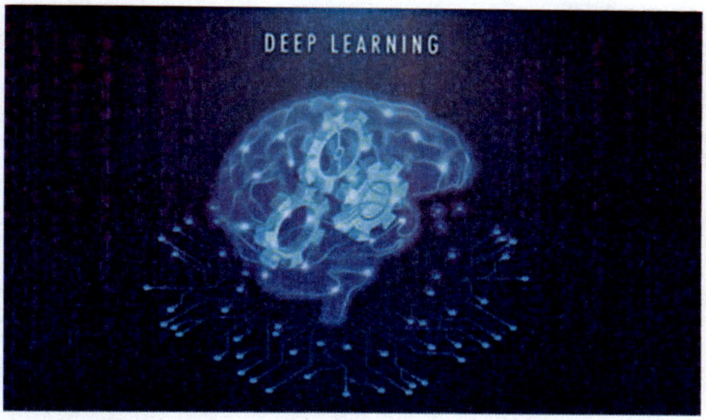

Fig. 2 Deep learning outlook

exactness, to a great extent astounding natural-level execution. Models are built by utilizing an enormous arrangement of marked information and neural affiliation structures comprising different layers [8] (Fig. 2).

Deep learning (DL) is an AI method for man-made consciousness that dissects the information in detail by expanding neuron sizes and number of the concealed layers. DL has a ubiquity with the normal enhancements for the graphical preparing unit abilities. Expanding the number of the neuron sizes at each layer and shrouded layers is straightforwardly identified with the calculation time and training speed of the classifier models. The greater part of the mainstream DL calculations requires long preparing times for enhancement of the boundaries with highlighted learning advances and back-propagation methods. Automated Driving, Clinical Explorations, and Industrial automation are some major fields in which DL is heavily explored in order to take out most from it [7, 9, 10].

1.3 Computer Vision

Computer vision, regularly contracted as CV, is characterized as an area of study that tries to create methods to help PCs "see" and comprehend the substance of pictures, for example, photos and recordings. The issue of computer vision seems basic since it is inconsequentially addressed by individuals, even small kids. By and by, it generally stays an unsolved issue put together both with respect to the restricted comprehension of natural vision and on account of the intricacy of vision discernment in a dynamic and almost vastly fluctuating actual world. The objective of computer vision is to extricate helpful data from pictures. This has demonstrated a gigantic testing task; it has involved large number of insightful and innovative personalities in the course of

Fig. 3 Computer vision outlook

the most recent forty years, and notwithstanding this, we are still a long way from having the option to fabricate a broadly useful "seeing machine" (Fig. 3).

Any famous PC vision applications include attempting to perceive things in photos like object location, check, order, and division [11, 12].

1.4 Keras

"Keras was initially developed as a part of the research effort of the project ONEIROS (Open-Ended Neuro-Electronic Intelligent Robot Operating System)." Keras can be used for the preprocessing of datasets quickly. It can be considered as the framework layer for differentiable programming.

Keras is a high level API of Tensorflow2. It is easy to use with a powerful solution for different AI issues. Related to current scenarios of deep learning, it facilitates transportation of deep learning models with great cycle speeds.

Keras engages architects and scientists to exploit the versatility and cross-stage capacities of TensorFlow2. We can deploy our Keras model to run on mobile devices or in browser. Keras can also be made to run on "Tensor Processing Unit" or on enormous bunches of "Graphical Processing Units" [13].

1.5 OpenCV

"OpenCV also known as 'Open Source Computer Vision Library' is an open source computer vision and AI programming library." OpenCV was attempted to provide a commonplace structure to computer vision practices. It hastens the usage of machine understanding in business applications. OpenCV is authorized under Berkeley Software Distribution.

The library has an assortment of more than 2500 enhanced algorithms, which consolidates a comprehensive plan of both masterpiece and best in computer vision, and AI estimations. These computations can be used to recognize and see faces, recognize objects, organize human exercises in accounts, and track camera advancements. This library is used extensively in associations, research social events, and authoritative bodies.

Alongside grounded organizations like "Google," "Microsoft," and the list goes on, that utilize the library, there are numerous new companies, for example, "Aspiring Minds," "VideoSurf," and "Zeitera" that utilize OpenCV. OpenCV's passed on uses length the compass from perceiving breaks in surveillance video in Israel, noticing mine equipment in China, helping robots, evaluating names on things in plants around the globe on to quick face acknowledgment in Japan [14–16].

1.6 TensorFlow

TensorFlow offers various levels of consultation so you can pick the right one for your necessities. Build and train models by using the huge level Keras API, which makes starting with TensorFlow and ML straightforward. If you need more prominent flexibility, enthusiastic execution considers brief cycle and intuitive investigating. For immense ML planning tasks, use the "Distribution Strategy API" for spread getting ready on different hardware courses of action without changing the model definition. TensorFlow gives you the versatility and control with features like the "Keras Functional API" and "Model Subclassing API" for creation of complex topographies. For straightforward prototyping and snappy examining, use eager execution [17].

1.7 MobileNet V2

MobileNet V2 is a convolutional neural organization engineering that is 53 layers deep. It depends on an altered lingering structure where the leftover associations are between the bottleneck layers. To beat all, the planning of MobileNet V2 contains the underlying completely convolution layer with 32 channels, trailed by 19 remaining bottleneck layers. MobileNet V2 utilizes lightweight depth-wise convolutions to channel highlights within the moderate extension layer. Moreover, we discover that it is imperative to eliminate non-linearities within the thin layers to stay up illustrative power. MobileNet V2 may be a successful element extractor for object location and division. MobileNet V2 gives a particularly proficient versatile arranged model which will be utilized as a base for a few, visual acknowledgments records [18] (Fig. 4).

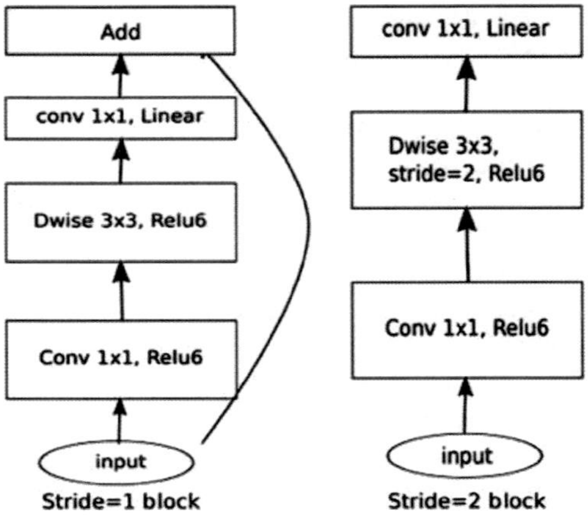

Fig. 4 MobileNet V2 architecture

2 Proposed System

Our model is to load the face mask dataset which consists of approximately 3000 images of different resolutions gathered from different web sources already delegated into two classes, namely "with mask" and "without mask." Images were then resized into (224 × 224) resolution. Next step is to train the classifier with Keras, MobileNet V2, and TensorFlow. The model will learn encodings for facial regions for both classes. Preprocessing for steps in detecting masks in videos and livestreams consists of detecting face using OpenCV and Face Recognition API. This helps in faster recognition of facial region than other methods.

The next step was validation and calculation of accuracy of our model. The train/test split was set at 80:20. We could have used Neural Net in this model, but MobileNet V2 proves itself faster in this case.

3 Conclusion and Future Scope

With the growing number of patients and deaths, it is necessary to wear face masks at public places. Rules must be strictly followed, and actions should be taken against those who are not following rules. This model has shown promising results as the accuracy is 95% and above. There is a further scope for research in this field as we have several types of masks too. This model can be improved further by classifying the masks into various types (Fig. 5).

Fig. 5 A screenshot of precision and related terms from output of my project

	precision	recall	f1-score	support
With_Mask	0.97	0.99	0.98	301
Without_Mask	0.99	0.97	0.98	301
accuracy			0.98	602
macro avg	0.98	0.98	0.98	602
weighted avg	0.98	0.98	0.98	602

Results and Screenshots:

The accuracy for with mask comes out to be 99% and for without mask coming 97%.

It tends to be finished up from the outcomes that an appropriate accuracy is accomplished (Figs. 6–8) and (Table 1).

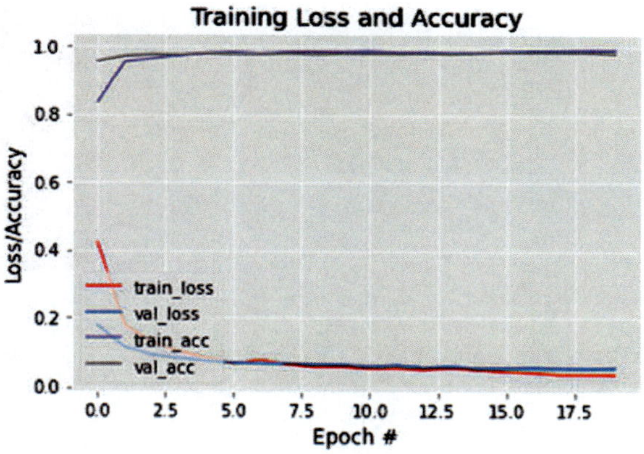

Fig. 6 Graph of testing accuracy versus number of epochs

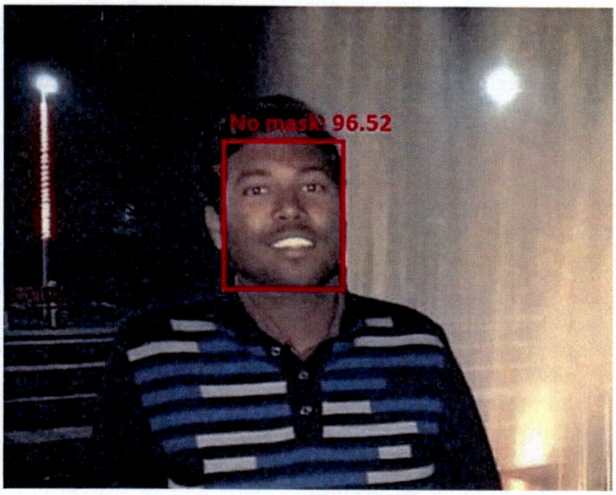

Fig. 7 Result when person is not wearing mask

Table 1 Confusion matrix made from the above output

Predicted/Actual (n = 602)	P ("With mask")	N ("Without mask")
P ("With mask")	299 (True positive)	2 (False positive)
N ("Without mask")	10 (False negative)	291 (True negative)

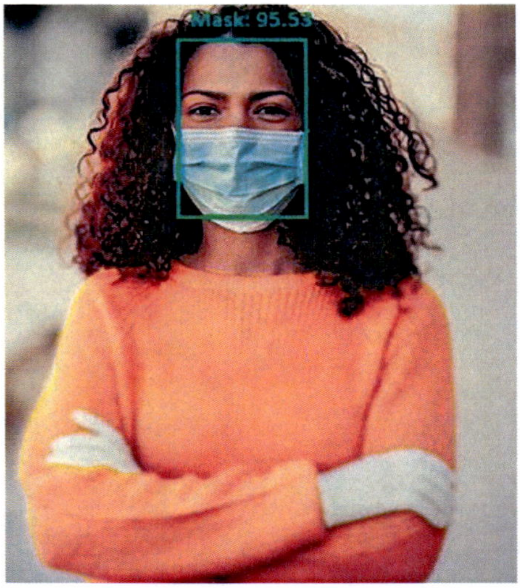

Fig. 8 Result when person is wearing mask

Acknowledgements I am deeply thankful to my supervisor without whose support this project could not have been completed. I am also thankful to Yash for providing me with resources so that I could complete this project in time.

References

1. Rota PA, Oberste MS, Monroe SS, Nix WA, Campagnoli R, Icenogle JP, Penaranda S, Bankamp B, Maher K, Chen M-H et al (2003) Characterization of a novel coronavirus associated with severe acute respiratory syndrome. Science 300(5624):1394–1399
2. Liu Y, Gayle AA, Wilder-Smith A, Rocklöv J (2020) The reproductive number of COVID-19 is higher compared to SARS coronavirus. J Travel Med
3. Feng S, Shen C, Xia N, Song W, Fan M, Cowling BJ (2020) Rational use of face masks in the Covid-19 pandemic. Lancet Respir Med
4. (2020) The economic times website (Online). Available. https://economictimes.indiatimes.com/magazines/panache/does-a-face-mask-protect-me-or-just-the-people-around-studies-show-masks-reduce-infection-risk/articleshow/77782169.cms
5. Burns E (2020) In-depth guide to machine learning in the enterprise (Online). Available. https://searchenterpriseai.techtarget.com/In-depth-guide-to-machine-learning-in-the-enterprise
6. Enterprise IT World (2019) (Online). Available. https://www.enterpriseitworld.com/ai-and-ml-game-changers-for-the-enterprises/

7. Shatnawi A, Al-Bdour G, Al-Qurran R, Al-Ayyoub M (2018) A comparative study of open source deep learning frameworks. In: 2018 9th international conference on information and communication systems (ICICS)
8. Brownlee J (2019) What is deep learning? (Online). Available. https://machinelearningmastery.com/what-is-deep-learning
9. Altan G, Kutlu Y (2021) Superiorities of deep extreme learning machines against convolutional neural networks. (Online). Available. arxiv.org
10. Lee G, Manning CD, Ng AY (2010) Unsupervised feature learning and deep learning. In: Deep learning and unsupervised feature learning workshop
11. Dickson B (2020) What is computer vision. (Online). Available. https://www.pcmag.com/news/what-is-computer-vision
12. (2020) The academia website. (Online). Available. https://www.academia.edu/Documents/in/Computer_Vision
13. (2020) Keras official website. (Online) Available. https://keras.io/about/
14. OpenCV Website (Online). Available. https://opencv.org/about/
15. Srivastava S (2019) Computer vision solving real-world problems with digital visuals. (Online). Available. https://www.analyticsinsight.net/computer-vision-solving-real-world-problems-with-digital-visuals
16. Rosebrock A (2020) COVID-19: face mask detector with OpenCV, Keras/TensorFlow, and deep learning. (Online). Available. https://www.pyimagesearch.com/2020/05/04/covid-19-face-mask-detector-with-opencv-keras-tensorflow-and-deep-learning/
17. Tensorflow website. (Online). Available. https://tensorflow.org/about/
18. Sandler M, Howard A, Zhu M, Zhmoginov A, Chen LC (2018) MobileNetV2: inverted residuals and linear bottlenecks. IEEE/CVF conference on computer vision and pattern recognition

A Novel TD-UTD Coefficients for Evaluation of Diffraction and Transmission from Dielectric Wedges

Vinod Kumar, Sanjay Soni, and N. S. Raghava

Abstract In this paper, a new time-domain six-term heuristic diffraction coefficient is proposed to consider the effect of diffraction and transmission from the dielectric wedge with an arbitrary low wedge angle for UWB applications. The different scenarios are considered to test the overall performance of the TD solution. The results of TD-UTD are confirmed by the inverse fast Fourier transform (IFFT) of frequency-based results in hard polarization. The computational efficiency of the proposed TD-UTD solutions is demonstrated by comparing the time, and they have taken with their IFFT-FD solutions.

Keywords TD-UTD · UWB · IFFT-FD

1 Introduction

The technology for wireless communication is changing every ten years. Each generation of wireless technology has considerable performance improvements. This technology change is mainly due to the demand for high data rates [1]. Fifth-generation (5G) wireless technology is being implemented globally with more features than fourth-generation (4G) communication systems to fulfill the present requirements. Some of the most challenging issues of 5G technology are to increase the bandwidth, capacity, and data rates several times to support industrial automation and other requirements. The ultra-wideband (UWB) communication (3.1–10.6 GHz)

V. Kumar (✉)
Dr. B. R. Ambedkar Institute of Technology, Port Blair, Andaman and Nicobar Islands 744103, India
e-mail: vinodkumar2721@gmail.com

S. Soni
Madan Mohan Malaviya University of Technology, Gorakhpur, Uttar Pradesh 273016, India
e-mail: sksoniec@mmmut.ac.in

V. Kumar · N. S. Raghava
Delhi Technological University, Delhi 110042, India
e-mail: nsraghava@dtu.ac.in

B. Mishra and M. Tiwari (eds.), *VLSI, Microwave and Wireless Technologies*, Lecture Notes in Electrical Engineering 877,
https://doi.org/10.1007/978-981-19-0312-0_31

is the most emerging wireless technology as a potential solution for 5G picocells and femtocells. It is capable to operate in an unlicensed band with high data rates for short-range using very low power parallelly with current radio communications without creating interference [2]. The propagated pulse through the UWB channel is distorted due to its frequency selective nature resulting in performance degradation [3]. Therefore, accurate UWB channel modeling is required to determine the performance of the communication system and network arrangement [4]. It is since the UWB channel has a large bandwidth and thus, working for the entire range of UWB signal in frequency-domain (FD) is difficult. Therefore, many works of literature have a time-domain solution of various propagation scenarios with UWB applications [5–8]. The time-domain analysis of the UWB channel can be done in two ways. (1) Taking the IFFT of the conventional FD solution in the entire UWB frequency range, and (2) directly converting the FD formula into the TD formula using inverse Laplace transform. The second method is more convenient and less time-consuming than its IFFT counterpart. There are three important phenomena to characterize the behavior of UWB channels: (i) reflection, (ii) diffraction, and (iii) transmission. The diffraction phenomenon is considered by the most popular theory known as the uniform theory of diffraction (UTD) [9]. In [10], a six-term UTD formula has been proposed in FD to consider the transmission through the dielectric wedge. However, the TD solution has not been presented. There are various papers where an emphasis on considering transmission through lossy electromagnetic materials has been given [11–13]. In this work, the analytical TD-UTD formulations are obtained by taking the inverse Laplace transform of the FD solutions in [10]. The TD results are verified with the IFFT for the FD solutions, and the results are found to be very consistent with IFFT solutions. Section 2 of the paper explains the propagation scenario with single-order diffraction by a dielectric wedge. In Sect. 3, the formulations for the TD-UTD diffraction coefficient is derived. Finally, Sect. 4 presents a detailed discussion of the proposed outcomes.

2 Propagation Scenario with a Dielectric Wedge

Figure 1 shows the propagation path of diffraction and transmission with single side illumination (SSI) from a dielectric wedge. The edge of the wedge is representing the z-axis of polar coordinates. The incident angle ϕ' and diffracted angle ϕ are measured with the 0-face of the wedge. The interior angle φ is the wedge angle. A line source is used to illuminate the face of the wedge at a distance r_1. The transmitting and receiving points are represented by (r_2, ϕ) (r_1, ϕ') and, respectively. There are four regions of observation with 0-face illumination $(\phi' < \pi - \varphi)$. Region-I $(0 < \phi < \pi - \phi')$ has an incident, reflected, and diffracted fields. The limiting boundary $(\phi = \pi - \phi')$ of the reflected wave is the reflection shadow boundary (RSB). Only incident and diffracted fields exist in Region-II $(\pi - \phi' < \phi < \pi + \phi')$. The limiting boundary $(\phi = \pi + \phi')$ of the incident wave is the incident shadow boundary (ISB). Region-III $(\pi + \phi' < \phi < \pi + g_0(\phi'))$ has diffracted fields only.

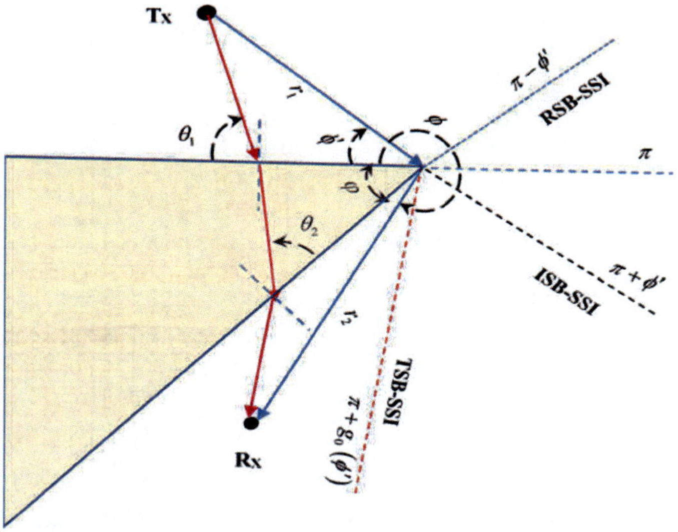

Fig. 1 Propagation scenario by a dielectric wedge

The limiting boundary $\left(\phi = \pi + g_0(\phi')\right)$ of the transmitted ray is the transmission shadow boundary (TSB). Region-IV $\left(\pi + g_0(\phi') < \phi < 3\pi - \varphi - \phi'\right)$ has the diffracted and the transmitted fields. This is the shadow region of the scenario. The $g_0(\phi')$ is defined in [10].

3 Formulation of the Proposed TD-UTD Coefficient

The diffraction and transmission phenomena show a vital role in the modeling of the wireless channel. The diffracted and transmitted fields compensate for the discontinuities of each geometrical optics (GO) field at ISB, RSB, and TSB by adding antisymmetric discontinuities, so that the entire field is continuous around the structure.

The impulse response of the received field in the shadow region of Fig. 1 can be obtained by taking the inverse Laplace transform of (1) in [10] with impulse input. Hence, the impulse response

$$u_d(t) = A(r_2) \cdot \left[d_{s,h}(t) * \delta\left(t - (r_1 + r_2)/c\right)\right] \tag{1}$$

where $A(r_2)$ is the amplitude spreading factor, and the delta function $\delta\left(t - (r_1 + r_2)/c\right)$ is the time delay that occurs between transmitter and receiver. The $d_{s,h}(t)$ is the diffraction coefficient in TD for a dielectric wedge which is obtained by taking the inverse Laplace transform of (3) in [10]

$$d_{s,h}(t) = d_1 + d_2 + r_0 * d_3 + r_n * d_4 + t_0 * t_0' * d_5 + t_n * t_n' * d_6 \tag{2}$$

where $d_i(t)$ with $i = 1 - 4$ is given as in [8]

$$d_i(t) = -\frac{Ln}{2\pi\sqrt{2c}} \times \frac{\sin(2a_i)}{\sqrt{t}\left(t + 2Ln^2\sin^2(a_i)/c\right)} \cdot u(t) \tag{3}$$

$r_{s,h}(t)$ is the Fresnel reflection coefficient in TD. This is obtained from [14]

$$r_{s,h}(t) = \mp\left[P\,\delta(t) + \frac{4p}{1-p^2}\frac{e^{-at}}{t}\sum_{q=1}^{\infty}(-1)^{q+1}qP^qI_q(at)\right] \tag{4}$$

where $P_{s,h} = (1 - p_{s,h})/(p_{s,h} + 1)$, $p_s = \sin\theta_1 / \sqrt{(\varepsilon - \cos^2\theta_1)}$, $p_h = \sqrt{(\varepsilon - \cos^2\theta_1)} / (\varepsilon \cdot \sin\theta_1)$, the modified Bessel function of order q is $I_q(t)$, $a = \sigma / (2\varepsilon_r\varepsilon_0)$, and soft (hard) polarizations represented by the leading $-(+)$ sign.

Taking inverse Laplace transform of (7) and (14) in [10], d_5 and d_6 of (2) can be given as

$$d_5(t) = -\frac{L^i n}{2\pi\sqrt{2v}} \times \frac{\sin(2a_5)}{\sqrt{t}\left(t + 2L^i n^2\sin^2(a_5)/v\right)} \cdot u(t) \tag{5}$$

and

$$d_6(t) = -\frac{L^i n}{2\pi\sqrt{2v}} \times \frac{\sin(2a_6)}{\sqrt{t}\left(t + 2L^i n^2\sin^2(a_6)/v\right)} \cdot u(t) \tag{6}$$

The $t_{s,h}(t) = \delta(t) + r_{s,h}(t)$ and $t'_{s,h}(t) = \delta(t) + r'_{s,h}(t)$ in (2) are the transmission coefficients in TD on the first interface (air/dielectric), and the second interface (dielectric/air) with $r'_{s,h}(t)$ as in (4)

$$r'_{s,h}(t) = \mp\left[P''\,\delta(t) + \frac{4p''}{1-p''^2}\frac{e^{-at}}{t}\sum_{q=1}^{\infty}(-1)^{q+1}qP''^qI_q(at)\right] \tag{7}$$

where $P''_{s,h} = (1 - p''_{s,h})/(p''_{s,h} + 1)$, $p''_s = \cos\theta_2 / \sqrt{(1/\varepsilon - \sin^2\theta_2)}$, $p''_h = \varepsilon\sqrt{(1/\varepsilon - \sin^2\theta_2)} / (\cos\theta_2)$, and $v = c/\sqrt{\varepsilon_r}$. All the other terms are defined as in [10].

4 Results and Discussions

The second order Gaussian pulse [6] shown in Fig. 2 is used to test the TD solutions proposed in Sect. 3 using MATLAB tool. This pulse is crossing the zero line two times with a Gaussian shape. Hence, it is called second order Gaussian pulse. It has the time-scaling factor $\tau = 0.1$ ns.

The second order Gaussian pulse spectrum shown in Fig. 3 is used in IFFT-FD solutions as reported in literatures [5–8]. This spectrum has a frequency range of UWB signals. Therefore, the pulse is used as a transmitted pulse of UWB signals with a short duration in ns.

In Fig. 4, the TD response is compared with its corresponding IFFT-FD response and both have very good matching. Therefore, the proposed method is verified. Here, incidence angle is $\phi' = 15°$ and wedge angle is $\varphi = 17°$ which is selected as in [10] to have transmitted fields. The received pulse is attenuated and distorted in comparison with the input pulse due to the lossy behavior of the dielectric wedge and frequency dependence nature of the UWB signal in the shadow region of the structure.

In Fig. 5, the TD received pulse is also matching with its corresponding IFFT-FD solution for a different scenarios where incidence angle is $\phi' = 75°$. In this case, the received pulse is strongly attenuated and distorted due to the only transmitted and diffracted fields available in the shadow region. The received pulse is also subject to the frequency dependence of UWB signals.

In Fig. 6, a different scenario is selected where the incidence angle is $\phi' = 115°$. Here, the received pulse is less attenuated and distorted than in the earlier cases. This

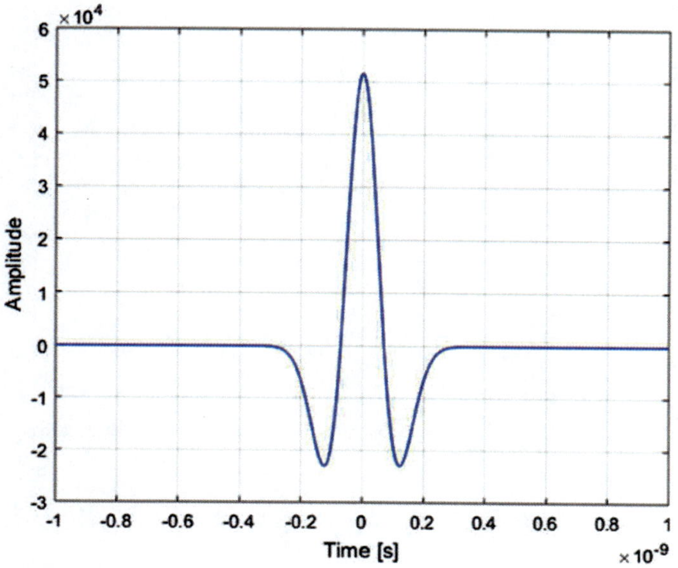

Fig. 2 Second-order Gaussian pulse [6]

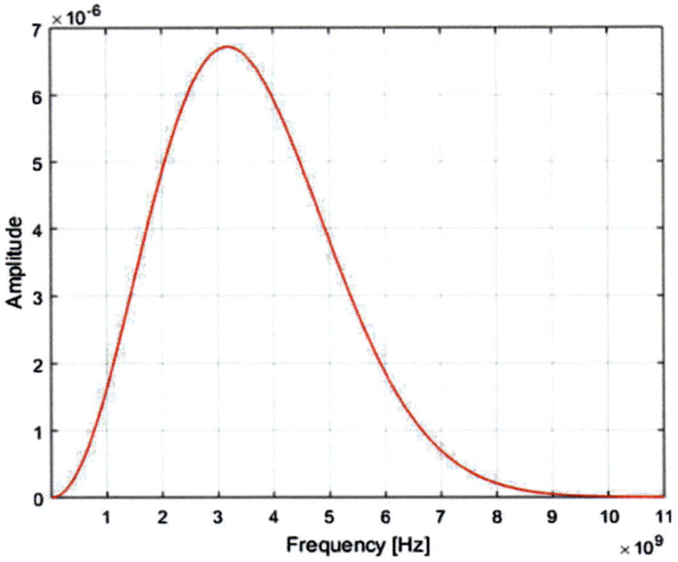

Fig. 3 Second-order Gaussian pulse spectrum

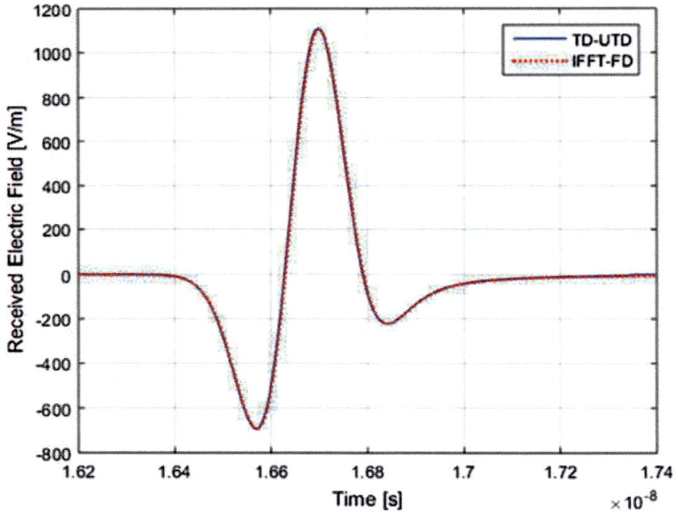

Fig. 4 TD received fields. $\phi' = 15°$, $\phi = 335°$, $r_1 = 3$ m, $r_2 = 2$ m, $\varphi = 17°$, relative permittivity $\varepsilon_r = 10$, conductivity $\sigma = 0.001$ S/m

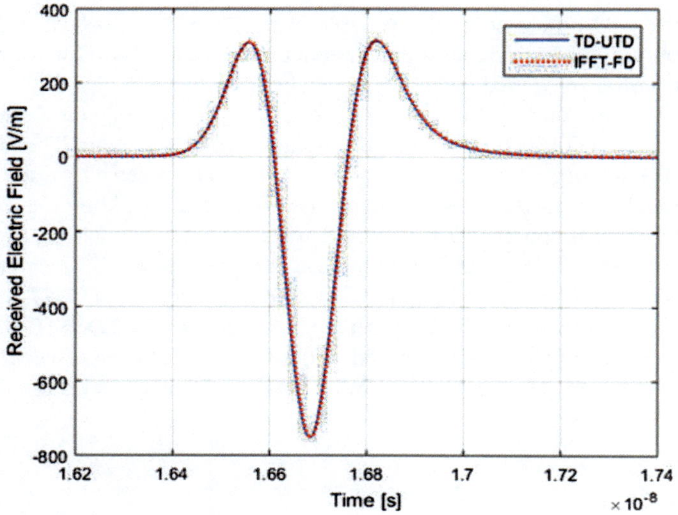

Fig. 5 TD received fields. $\phi' = 75°$, $r_1 = 3$ m, $r_2 = 2$ m, $\varphi = 17°$, relative permittivity $\varepsilon_r = 10$, conductivity $\sigma = 0.001$ S/m

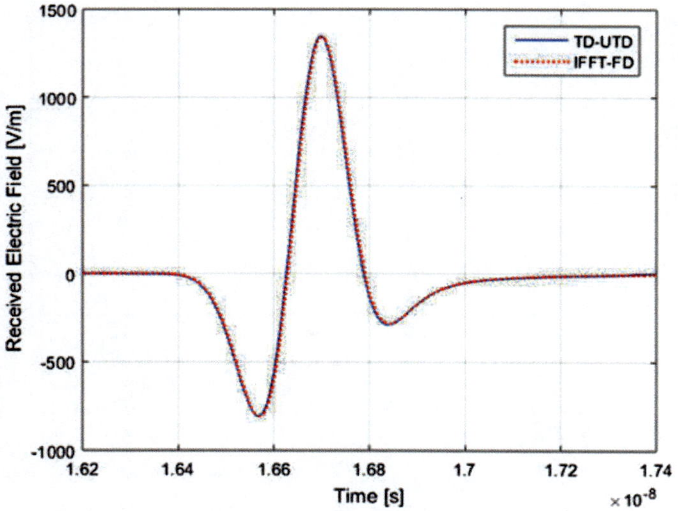

Fig. 6 TD received fields. $\phi' = 115°$, $\phi = 335°$, $r_1 = 3$ m, $r_2 = 2$ m, $\varphi = 17°$, relative permittivity $\varepsilon_r = 10$, conductivity $\sigma = 0.001$ S/m

is due to the only diffracted fields available in the shadow region. Here, both the TD-UTD and IFFT-FD solutions are matching to each other. Therefore, it confirms the accuracy of the proposed method.

In Fig. 7, a very thin wedge structure is selected with a wedge angle $\varphi = 13°$. In this case, the TD-UTD and IFFT-FD solutions are also similar. Therefore, the proposed TD solution is guaranteed to be accurate. The received pulse is strong due to the high transmission fields. Minor distortions are caused by the frequency dependence of the proposed TD-UTD coefficients.

Table 1 presents the computational efficiency of the proposed TD technique with its corresponding IFFT-FD technique. It is quite clear that the TD-UTD solution works much better than the IFFT-FD method. It is because the TD-UTD method is based on the effective convolution method, while the IFFT-FD method takes longer to convert each frequency component of UWB signals to TD due to a shorter Gaussian pulse.

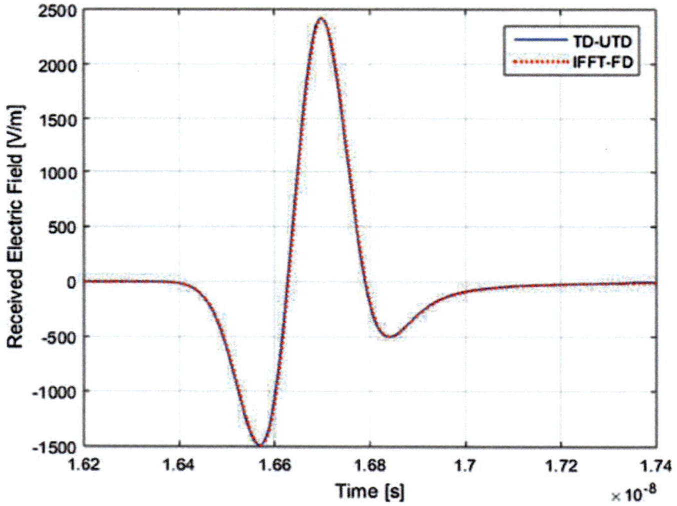

Fig. 7 TD received fields. $\phi' = 75°$, $\phi = 335°$, $r_1 = 3$ m, $r_2 = 2$ m, relative permittivity $\varepsilon_r = 10$, conductivity $\sigma = 0.001$ S/m

Table 1 Details of the time taken in simulation by the TD-UTD and IFFT-FD techniques

Dielectric wedge structure	T(TD-UTD)	T(IFFT-FD)	T(IFFT-FD)/T(TD-UTD)
$\phi' = 15°, \varphi = 17°$	0.670283	28.794093	42.95
$\phi' = 75°, \varphi = 17°$	0.562310	16.318998	29.02
$\phi' = 115°, \varphi = 17°$	2.512664	58.093214	23.12
$\phi' = 75°, \varphi = 13°$	0.217421	26.213103	120.56

5 Conclusion

A novel six-terms UTD diffraction coefficients in TD has been presented to consider transmission and diffraction phenomena from a thin dielectric wedge. The proposed coefficients are based on direct convolution technique resulting from inverse Laplace transform of FD solutions. Using MATLAB tool, the TD-UTD results are compared with its corresponding IFFT-FD results in hard polarization showing excellent matching. Various scenarios have been selected to test the proposed TD solutions. Finally, the TD-UTD process is more effective than the IFFT-FD method.

References

1. Shafi M, Molisch AF, Smith PJ, Haustein T, Zhu P, Silva PD, Tufvesson F, Benjebbour A, Wund G (2017) 5G: a tutorial overview of standards, trials, challenges, deployment, and practice. IEEE J Sel Areas Commun 35(6):1201–1221
2. Puerta R, Rommel S, Ramirez DJ, Monroy IT (2020) Ultra-wideband technology: prospective solution for 5G ultra-small cell networks. Int J Commun Syst 33(16)
3. Qiu RC, Zhou C, Liu Q (2005) Physics-based pulse distortion for ultra-wideband signals IEEE Trans Veh Technol 54(5):1546–1555
4. Rappaport TS, MacCartney GR, Sun S, Yan H, Deng S (2017) Small-scale, local area, and transitional millimeter-wave propagation for 5G communications. IEEE Trans Antennas Propag 65(12):6474–6490
5. Veruttipong TW (1990) Time domain version of the uniform GTD. IEEE Trans Antennas Propag 38(11):1757–1764
6. Bansal B, Soni S, Mishra V, Gupta A, Agrawal A (2019) A Novel heuristic time-domain diffraction model for UWB diffraction by lossy wedges and buildings. Phys Commun 34:80–89
7. Tewari P, Soni S, Bansal B (2015) Time-domain solution for transmitted field through low-loss dielectric obstacles in a microcellular and indoor scenario for UWB signals. IEEE Trans Veh Technol 64(2):541–552
8. Karousos A, Tzaras C (2008) Multiple time-domain diffractions for UWB signals. IEEE Trans Antennas Propag 56(5):1420–1427
9. Tajvidy A (2018) Diffraction loss model at 0.3–6 GHz for 5G cellular system in microcell urban areas. Electromagnetics 39(3):168–185
10. Rouviere JF, Douchin N, Combes PF (1999) Diffraction by lossy dielectric wedges using both heuristic UTD formulations and FDTD. IEEE Trans Antennas Propag 47(11):1702–1708
11. De Jong YLC, Koelen MHJL, Herben MHAJ (2004) A building-transmission model for improved propagation prediction in urban microcells. IEEE Trans Veh Technol 53(2):490–502
12. Soni S, Bhattacharya A (2011) An analytical characterization of transmission through a building for deterministic propagation modeling. Microw Opt Technol Lett 53(8):1875–1879
13. Umul YZ (2014) Diffraction of waves by a resistive half-plane. Optics Commun 323:6–12
14. Holm PD (2000) A new heuristic UTD diffraction coefficient for non-perfectly conducting wedges. IEEE Trans Antennas Propag 48(8):1211–1219

Quad-Band MIMO Antenna for Radar and Satellite Applications

Amrees Pandey, Aditya Kumar Singh, Sweta Singh, Rajeev Singh, and Brijesh Mishra

Abstract In this paper, a compact ($32 \times 32 \times 1.6$ mm^3) two-port guitar-shaped micro-strip patch antenna for radar and satellite applications is presented. Three antenna designs (Ant-1, Ant-2 and Ant-3) are systematically investigated, and the specifications of Ant-3 have been optimized for desired antenna parameters and operation. Two parallel slots are introduced on the radiating patch to achieve multiple resonances. A multiband behaviour at 15.64–17.45, 17.87–18.99, 19.31–21.13 and 22.14–23.79 GHz with impedance bandwidth of 10.95, 6.07, 9 and 7.18%, respectively, for port 1 and 15.55–17.39, 17.92–19.44, 20.27–21.02 and 22.14–23.54 GHz with impedance bandwidth of 11.17, 8.13, 3.63 and 6.12%, respectively, at port 2 is observed. Isolation less than −18 dB, envelope correlation coefficient (ECC) less than 0.05 and diversity gain (DG) between 9.979 and 9.999 are observed.

Keywords Patch antenna · MIMO · Multiband · Isolation · DG · ECC and TARC

1 Introduction

Multiple-input multiple-output (MIMO) antenna system [1, 2] is one of the key technologies in wireless communication systems which provides high data transmission rate, diversity gain and overall gain, and reduces effect of mutual coupling. Such MIMO antennas have high isolation between the ports [1–3] and are capable of enhancing channel capacity which in turn increases transmission and reception data rate [3].

Isolation and suppression of mutual coupling between two antenna elements are two important parameters in the design of a MIMO antenna. In MIMO technology due to the small space between antenna elements in portable devices, the mutual coupling

A. Pandey · A. K. Singh · S. Singh · R. Singh
Department of Electronics and Communication, University of Allahabad, Prayagraj, India

B. Mishra (✉)
Shambhunath Institute of Engineering and Technology, Prayagraj, Uttar Pradesh, India
e-mail: brijesh.mishra0933@gmail.com

B. Mishra and M. Tiwari (eds.), *VLSI, Microwave and Wireless Technologies*, Lecture Notes in Electrical Engineering 877,
https://doi.org/10.1007/978-981-19-0312-0_32

is introduced, which degrades the system performance [4]. Method of current cancelling branches [5] and using defected ground structures and meta-materials [6, 7] are reported for suppression of mutual coupling.

Mutual coupling of perpendicularly placed radiators is reported to have been reduced using stubs on the ground plane [8], in portable devices. By inserting a rectangular stub diagonally between the radiating elements [8, 9], isolation between the MIMO antenna elements is achieved and further stubs and slots with a co-shared radiator and parasitic structures on radiators are reported [10] to achieve high and desired isolation.

Presently multiple-input multiple-output antenna systems have drawn attention due to their potential to enhance the capacity and reliability of the wireless communication system by reducing the effect of mutual coupling, increasing envelope correlation coefficient, decreasing diversity gain and distorting radiation pattern [10, 11].

In this article, a parasitic guitar-shaped MIMO antenna with multiband and high isolation is proposed. Guitar-shaped MIMO antenna is designed and simulated using electromagnetic tool HFSS v 13 from Ansoft. This tool utilizes the mathematical formulation of the finite element method. The antenna design and evolution of the proposed antenna structure, results and discussion and conclusions are presented in Sects. 2, 3 & 4 respectively.

2 Antenna Design and Evolution of the Proposed Antenna Structure

The proposed guitar-shaped antenna (Ant-3) geometry wherein patch side (red colour—top view) and ground side (orange colour—bottom view) of the proposed antenna is shown in Fig. 1 and its specifications are enumerated in Table 1.

Guitar-shaped structure (Ant-2) is achieved by introducing two parallel notches ($W_{P3} \times L_{P3}$: 1.6 × 1.6 mm^2) in Ant-1. Antenna 3 is obtained by introducing two parallel slots ($W_{P4} \times L_{P4}$: 0.5 × 6 mm^2) in Antenna 2. Two symmetrical designs are separated by a distance, W_2 to create a MIMO structure. The antenna structure is designed using FR4 epoxy material ($\varepsilon = 4.4$, $h = 1.6$ mm). The performance of the antenna in terms of return loss, current distribution, gain, radiation patterns, ECC (envelope correlation coefficient), DG (diversity gain) and TARC (total active reflection coefficient) is investigated employing Ansoft HFSS version 13.

The systematic growth of Ant-1, Ant-2 and Ant-3 is presented in the inset of Fig. 2, and their return loss, gain and isolation characteristics are presented in Fig. 2, 3 and 4, respectively.

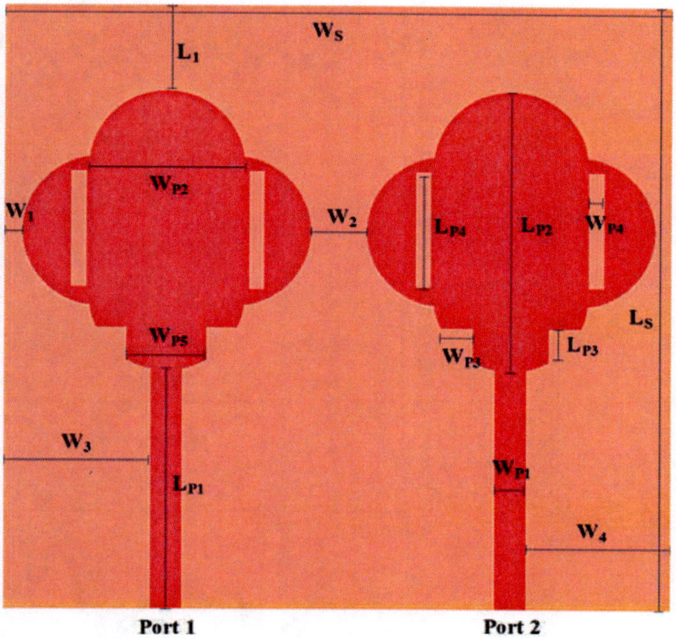

Fig. 1 Top (red colour) and bottom (orange colour) view layout of the proposed antenna

Table 1 Specifications of the proposed antenna (Ant-3)

Parameters	Values
$L_S \times W_S$ (Substrate length × Substrate width)	32 mm × 32 mm
$L_{P1} \times W_{P1}$ (Patch length$_1$ × Patch width$_1$)	13 mm × 1.6 mm
$L_{P2} \times W_{P2}$ (Patch length$_2$ × Patch width$_2$)	14 mm × 8 mm
$L_{P3} \times W_{P3}$ (Patch length$_3$ × Patch width$_3$)	1.6 mm × 1.6 mm
$L_{P4} \times W_{P4}$ (Patch length$_4$ × Patch width$_4$)	6 mm × 0.5 mm
W_{P5} (Patch width$_5$)	4 mm
L_1	5 mm
W_1	1 mm
W_2	2 mm
W_3	8.8 mm
W_4	8.8 mm

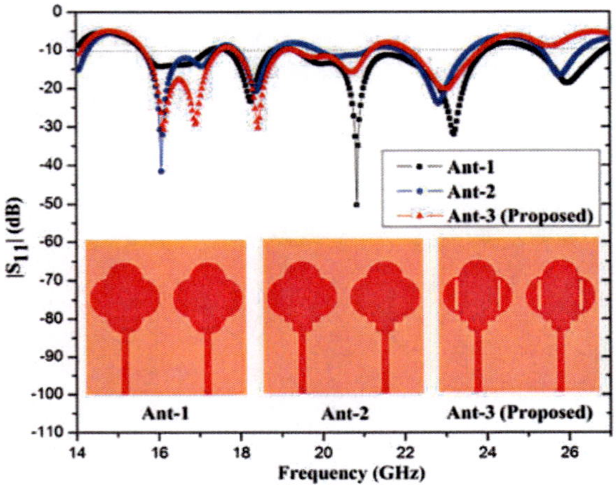

Fig. 2 $|S_{11}|$ versus frequency of Ant-1, Ant-2 and Ant-3

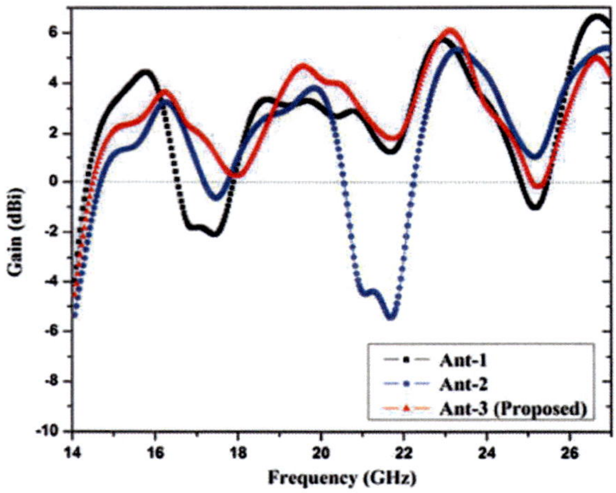

Fig. 3 Simulated gain versus frequency of Ant-1, Ant-2 and Ant-3

3 Results and Discussion

The performance of the proposed antenna is investigated in terms of return loss, current distribution, envelope correlation coefficient (ECC), diversity gain (DG), radiation pattern and total active reflection coefficient (TARC).

Three-band (Ant-1) and four-band (Ant-2, Ant-3) operations have been observed. Table 2 represents the data of Ant-1, Ant-2 and Ant-3 at both ports, i.e. port 1 and 2 for ready reference. Table 2 depicts the number of operating bands and their operating band frequency, their percentage impedance bandwidth, isolation, the resonant frequency of the concerned band, maximum emission point and peak gain. Figures 2,

Fig. 4 Simulated isolation (between ports) of Ant-1, Ant-2 and Ant-3

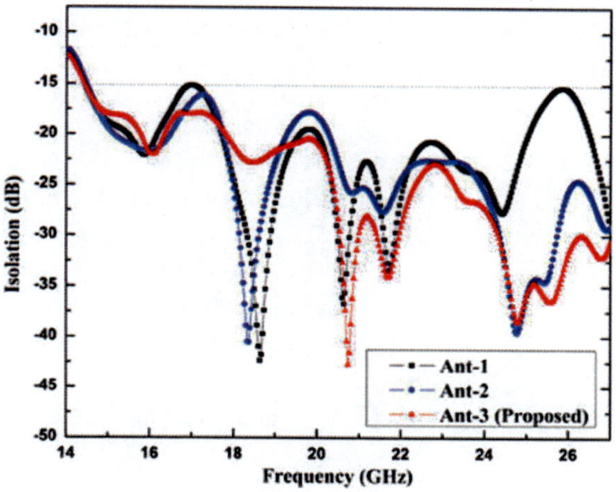

3 and 4 represent the variation $|S_{11}|$, isolation (both in dB) and gain (in dBi) with frequency as tabulated in Table 2.

From perusal of Table 2 and Figs. 2, 3 and 4 it is clearly show that the proposed band and gain characteristics of Ant-3 are matching at both the ports with marginal variation, whereas for Antenna 2 and 3 the variation in band and gain are larger as compared to Antenna 3.

The surface current density of the proposed antenna at 16.1, 16.87, 18.4, 20.7 and 22.96 GHz is observed in Fig. 5. It is depicted from Fig. 5 that port 1 has a large current length as compared to port 2. The maximum surface current density of 96.4 A/m at 20.7 GHz is observed.

The scattering parameters (S_{11}, S_{12}, S_{21} and S_{22}) and the antenna gain of the proposed antenna are portrayed in Fig. 6. The proposed antenna radiates in four different bands (15.64–17.45, 17.87–18.99, 19.31–21.13 and 22.14–23.79 GHz) having impedance bandwidth of 10.95%, 6.07%, 9% and 7.18%, respectively. Isolation is in the acceptable range (<-15 dB), and maximum isolation of < -22 dB is observed at 22.96 GHz. The antenna gain has a maximum peak gain of 5.94 dBi at 22.96 GHz as compared to gain at 16.1, 16.87, 18.4 and 20.7 GHz. From the perusal of Fig. 6 and Table 2, a marginal difference between port 1 and port 2 is observed due to the small gap (2 mm) between ports and compact size ($32 \times 32 \times 1.6$ mm^3) of the proposed antenna.

ECC, DG and TARC as a function of frequency are shown in Figs. 7 and 8, respectively. ECC and DG of the antenna are given in Eqs. 1 and 2. TARC of the antenna represents the apparent return loss of the overall MIMO system as given in Eq. (3).

$$\text{Diversity Gain} = 10 \times \sqrt{1 - (\text{ECC})^2} \tag{1}$$

Table 2 Port characteristics of the Ant-1, Ant-2 and Ant-3 (Proposed)

Antenna	Port No	Number of bands	Operating band (GHz)/impedance BW (in %)	Isolation (dB)	Resonant frequency (GHz)	Reflection coefficients (dB)	Peak gain (dBi)
Ant-1	Port 1	1	17.66–18.78/6.14	≥ −20	18.23	−22.86	2.2
		2	19.32–23.95/21.40	≥ −20	20.81	−49.86	2.8
					23.16	−31.23	5.4
		3	25.12–26.83/6.5	≥ −15	25.98	−18.50	4.2
	Port 2	1	18.12–19.06/5.05	≥ −22	18.57	−40.19	3.3
		2	20.54–21.67/5.35	≥ −22	21.20	−19.09	2.1
		3	22.81–24.51/7.1	≥ −21	23.73	−22.75	3.7
Ant-2	Port 1	1	15.58–17.39/10.98	≥ −16	16.05	−41.54	2.8
					17.05	−14.11	0.63
		2	17.90–18.83/4.9	≥ −25	18.36	−20.48	2.36
		3	21.85–23.58/5.1	≥ −22	22.16	−23.38	-0.51
		4	25.31–26.33/3.9	≥ −24	25.79	−16.63	3.3
	Port 2	1	15.56–18.84/ 19.06	≥ −16	16.04	−24.41	3
					17.04	−26.61	0.5
					18.26	−27.78	1.9
		2	19.30–21/8.4	≥ −17	19.99	−19.38	3.5
		3	21.91–23.81/8.3	≥ −22	22.98	−24.48	4.8
		4	25.34–26.28/3.64	≥ −24	25.77	−12.07	3.5
Ant-3 (Proposed)	Port 1	1	15.64–17.45/10.95	≥ −18	16.1	−30.3	3.5
					16.87	−29.16	2.2
		2	17.87–18.99/6.07	≥ −20	18.4	−29.88	1.56
		3	19.31–21.13/9	≥ −20	20.7	−15.69	3.6
		4	22.14–23.79/7.18	≥ −22	22.96	−20.01	5.94
	Port 2	1	15.55–17.39/11.17	≥ −18	15.99	−34.10	3.5
					16.63	−29.06	2.96
		2	17.92–19.44/8.13	≥ −20	18.84	−26.28	3.03
		3	20.27–21.02/3.63	≥ −20	20.73	−13.22	3.4
		4	22.14–23.54/6.12	≥ −22	22.85	−23.50	5.8

$$\mathrm{ECC} = \frac{|S_{11} * S_{12} + S_{21} * S_{11}|}{\left(1 - |S_{11}^2| - |S_{21}^2|\right)\left(1 - |S_{22}^2| - |S_{12}^2|\right)} \tag{2}$$

$$\mathrm{TARC} = \frac{\sqrt{S_{11} + S_{12})^2 + (S_{22} + S_{21})^2}}{\sqrt{2}} \tag{3}$$

ECC variation of the proposed antenna is in the acceptable limit (0–0.05) which conforms to the minimum mutual coupling effect between antenna elements. The

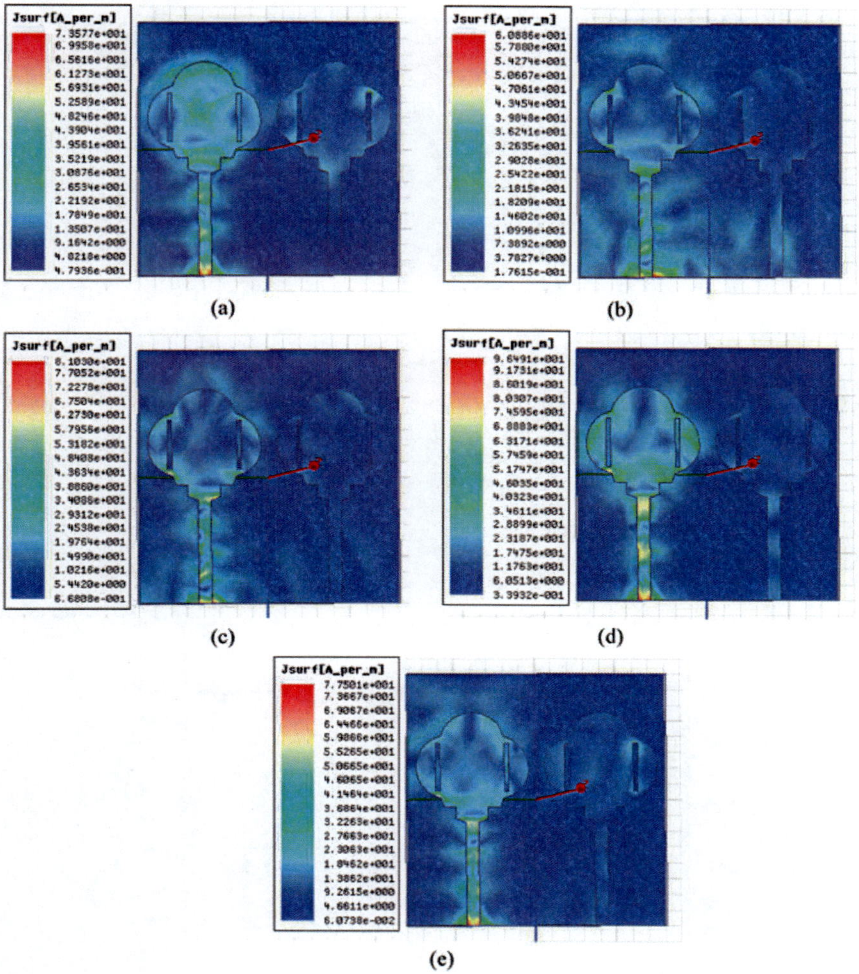

Fig. 5 Surface current distribution at **a** 16.1 GHz **b** 16.87 GHz **c** 18.4 GHz **d** 20.7 GHz **e** 22.96 GHz

minimum mutual coupling of the proposed antenna leads to a high data rate and better diversity gain (9.979 –9.999 dB). For the efficient transmission of MIMO antenna less than 0 dB, TARC value is required. Figure 8 depicts a better TARC value suitable for satellite applications.

The simulated E and H-plane far-field radiation patterns of the proposed antenna are shown in Fig. 9a–e at 16.1 GHz, 16.87 GHz, 18.4 GHz, 20.7 GHz and 22.96 GHz, respectively. The proposed antenna for all resonating frequencies shows omnidirectional broad radiation pattern characteristics.

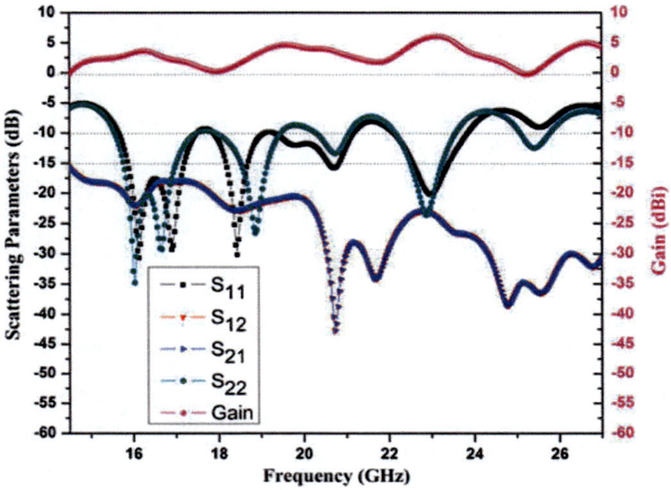

Fig. 6 Simulated scattering parameters S_{11}, S_{12}, S_{21}, S_{22} and gain versus frequency for the proposed MIMO antenna (Ant-3)

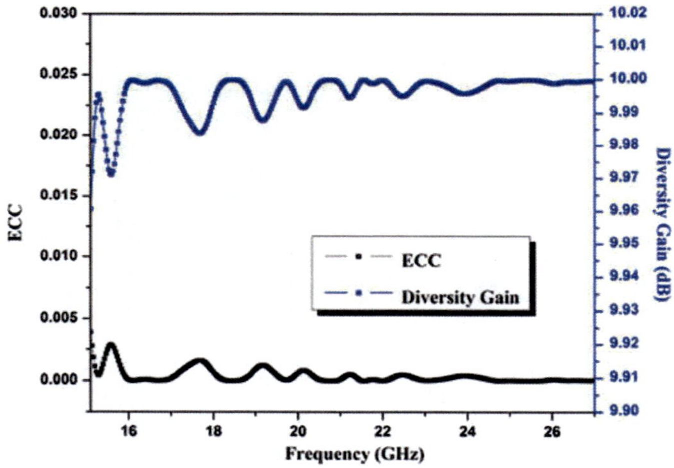

Fig. 7 Simulated envelope correlation coefficient (ECC) and diversity gain (DG)

A comparative overview of the two-port guitar-shaped MIMO antenna in terms of area of the antenna, number of ports, number of bands, ECC, peak gain, isolation and TARC is presented in Table 3. The proposed antenna occupies a lesser area as compared to antennas reported in [7, 8, 11]. However, the antennas reported in [2, 4, 6] occupy a lesser area as compared to the proposed one, but the proposed antenna as tabulated in Table 3 exhibits quad bands which the rest of the antennas do not

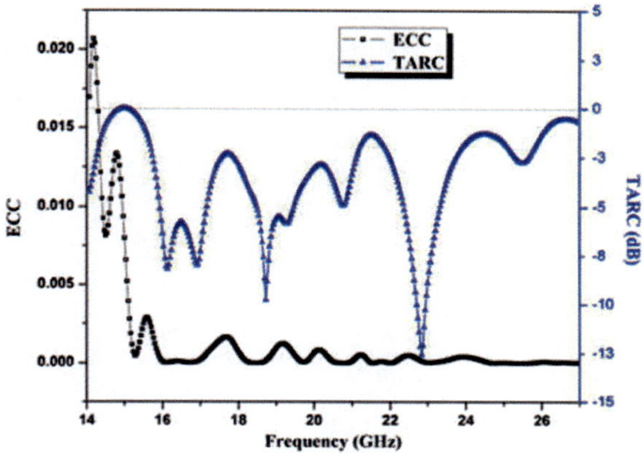

Fig. 8 Simulated ECC and radiation TARC of the proposed antenna (Ant-3)

exhibit. As observed in Table 3, the operating bands of the reported antennas do not fit in with the operating bands of the proposed antenna; therefore, it is not possible to compare the other antenna parameters. However, antenna parameters as reported are tabulated for ready reference.

4 Conclusions

The proposed MIMO antenna design is suitable for quad-band operation with applications for radar and satellite communications. The band and gain characteristics of the proposed antenna are matching at ports 1 and 2 with marginal deviation which conforms to good mutual coupling and better isolation (less than −22 dB) between antenna elements (port 1 and port 2) and improved ECC (0–0.025). Omnidirectional radiation pattern of the proposed antenna is beneficial even for vehicular applications. The proposed antenna structure can be modified for energy harvesting applications as reported in [12].

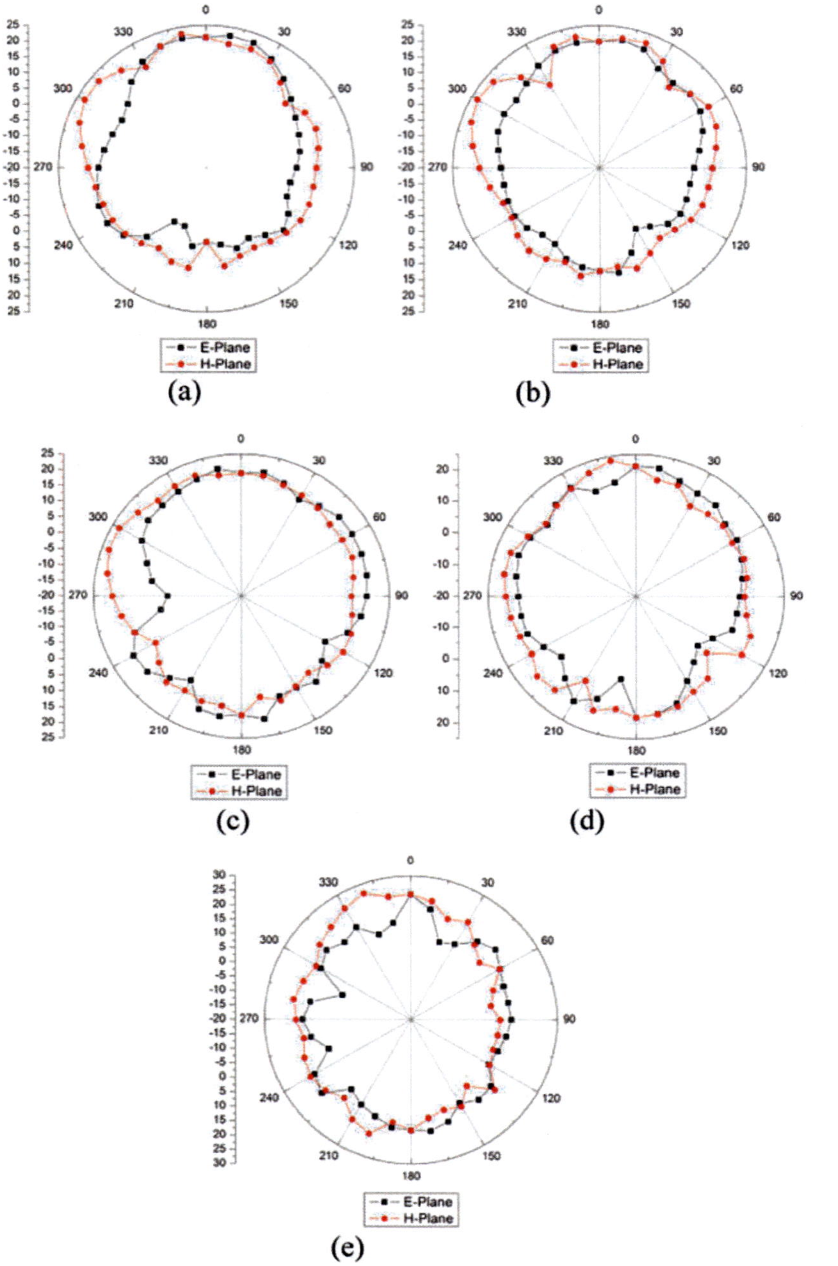

Fig. 9 Radiation pattern in E and H-plane at port 1 for **a** 16.1 GHz **b** 16.87 GHz **C** 18.4 GHz **d** 20.7 GHz **e** 22.96 GHz

Table 3 A comparative overview of the proposed antenna (Ant-3)

References	Area of the antenna (mm^2)	Ports	No. of bands	Operating band (GHz)	Peak gain (dBi)	ECC	Isolation (dB)	TARC (dB)
[2]	728	2	2	5.05–5.86, 6.68–7.43	1.6, 4	0.08	≥ -15	≤ 0
[4]	405	2	1	2.95–15.65	4.8	0.04	≥ -19	≤ 0
[6]	381.51	2	1	3–10.6	4	0.42	≥ -15	–
[7]	2500	2	2	2.3–2.7, 5.4–6.2	4.6, 2.7	0.20	≥ -15	–
[8]	1400	2	1	3.9–10.9	–	0.12	≥ -18	–
[11]	1628	2	1	5.61–5.93	3.4	0.10	≥ -15	–
Proposed work	1024	2	4	15.64–17.45, 17.87–18.99, 19.31–21.13, 22.14–23.79	3.5, 1.56, 3.6, 5.94	0.0025	≥ -18	≤ 0

References

1. Malviya L, Machavaram K (2017) MIMO antennas with diversity and mutual coupling reduction techniques: a review. Int J Microw Wirel Technol 9(8):1763–1780
2. Zhao Y, Zhang FS, Cao LX, Li DH (2019) A compact dual band-notched MIMO diversity antenna for UWB wireless applications. Progress in Electromagnetics Res C 89:161–169
3. Jehangir SS, Sharawi et al (2017) A miniaturized UWB biplanar yagi-like MIMO antenna system. IEEE Antennas Wireless Propagation Lett 16:2320–2323
4. Addepalli T, Anitha R (2020) Compact two-port MIMO antenna with high isolation using parasitic reflectors for UWB, X and Ku band applications. Progress Electromagnet Res C 102:63–77
5. Nadeem I, Choi DY (2019) Study on mutual coupling reduction technique for MIMO antennas. IEEE Access. 7:563–586
6. Khan MS, Capobianco AD, Iftikhar A et al (2017) Ultra-compact dual polarized UWB MIMO antenna with meandered feeding lines. IET Microw Antennas Propag 11:997–1002
7. Alnaiemy Y, Elwi TA, Nagy L (2019) Mutual coupling reduction in patch antenna array based on EBG structure for MIMO applications. Periodica Polytechn Electr Eng Comput Sci 63(4):332–342
8. Ibrahim A, Machac A, Shubair J, Raed M (2017) Compact UWB MIMO antenna with pattern diversity and band rejection characteristics. Microw Opt Technol Lett 59(6):1460–1464
9. Isaac A, Hussain Al-Rizzo A, Yahya S, Al-Wahhamy A, Abushamleh S (2018) Decoupling of two closely-spaced planar monopole antennas using two novel printed-circuit structures. Microw Opt Technol Lett 60(12):2954–2963
10. Radhi AH, Nilavalan R, Wang Y, Al-Raweshidy HS, Eltokhy AA, Ab Aziz N (2018) Mutual coupling reduction with a wideband planar decoupling structure for UWB-MIMO antennas. Int J Microw Wirel Technol 10(10):1143–1154
11. Iqbal A, Saraereh OA, Bouazizi A, Basir A (2018) Metamaterial-based highly isolated MIMO antenna for portable wireless applications. Electronics 7(10):267 (1–8)
12. Kumar A, Kallollu Narayaswamy N, Venkatesh Kumar H, Mishra B, Azam Siddique S, Kumar Dwivedi A (2021) High-isolated WiFi-2.4 GHz/LTE MIMO antenna for RF-energy harvesting applications. AEU - Int J Electron Commun 141:153964. https://doi.org/10.1016/j.aeue.2021.153964

Comparative Study Between Silver and Gold Metal Film-Based Surface Plasmon Resonance Biosensor with Platinum Diselenide

Rashmi Tiwari, Sachin Singh, Pooja Lohia, and D. K. Dwivedi

Abstract Surface plasmon resonance-based (SPR) biosensors serve us an efficient real-time and label-free detection device; this type of sensor offers stability, short response time, ease of use, and excellent sensitivity. In this simulation work, we have proposed two SPR-based biosensors, both consist of a prism, metal, silicon, and a transition metal dichalcogenide (TMDC) material $PtSe_2$. Platinum diselenide ($PtSe_2$) is a new emerging two-dimensional material which has attracted the attention of worldwide researchers due to its new physical and chemical properties. The paper reports that the SPR biosensor consists of Ag film having thickness 50 nm over the prism has higher sensitivity than the SPR biosensor consisting of Au film having thickness 50 nm, both optimized for 2 nm silicon, 2 nm $PtSe_2$, and CaF_2 prism having refractive index value 1.4329. To analyze the performance of the sensors, parameters calculated for both the biosensors are sensitivity, full width half maximum (FWHM), detection accuracy (DA), and figure of merit (FoM). The sensitivity of sensor having Ag film is 220° RIU^{-1}, and sensor having Au film is 206° RIU^{-1}. Sensitivity enhancement shown by the proposed sensor with Ag over proposed sensor with Au is of 6.79%. Such remarkable sensitivity enhancement of the proposed biosensors will make them aspiring candidate for detection different liquid analytes in biosensing application field.

Keywords Resonance · Biosensor · Silicon · $PtSe_2$ · Sensitivity · Refractive index

R. Tiwari · S. Singh · D. K. Dwivedi (✉)
Amorphous Semiconductor Research Lab, Department of Physics and Material Science, Madan Mohan Malaviya University of Technology, Gorakhpur 273010, India
e-mail: todkdwivedi@gmail.com

P. Lohia
Department of Electronics and Communication Engineering, Madan Mohan Malaviya University of Technology, Gorakhpur 273010, India

B. Mishra and M. Tiwari (eds.), *VLSI, Microwave and Wireless Technologies*, Lecture Notes in Electrical Engineering 877,
https://doi.org/10.1007/978-981-19-0312-0_33

1 Introduction

A technique named surface plasmon resonance has emerged as an excellent sensitive technique for identifying a very insignificant change in refractive index of sensing medium while interacting with the metal film of sensor [1–3]. It is a proven technique for the development of label-free and real-time biosensor; real-time measurements provide details regarding bimolecular interaction. Biosensing application of SPR-based biosensors includes enzyme detection, drug diagnostic, medical diagnostic, and food safety. The basic SPR biosensors consist of a thin metal film coated over a prism [4]. In surface plasmon resonance sensor, there is excitation of surface plasmons (SPs), i.e., oscillating free electrons, at the interface of the metal and constant dielectric material [4–6]. Commonly used metal films in SPR sensor are gold, silver, aluminum, copper, indium, and sodium. The durability, great stability, and biocompatibility of gold make it more favorable for SPR biosensor [7]. The accuracy given by the SPR sensor consisting of Ag film is better than that consisting of Au film. Due to Ag having poor stability, the problem of corrosion and oxidation arises [4]. As gold is expensive, to overcome this problem, we can use silver for high-sensitivity sensor if its oxidation can be reduced by coating it with another layer [7]. Silicon can be used to coat the silver layer to prevent oxidation of Ag although for enhancing the sensitivity of sensor by adding silicon layer over metal film is a proven method [8]. For improving the efficiency of sensing materials, two-dimensional nanomaterials can be used such as graphene, black phosphorous, and TMDCs due to their excellent and distinct electrical and optical properties [9]. The unique properties of transition metal dichalcogenides (TMDCs) are electrical, optical, and chemical which make them as promising materials for future [10–12]. A group ten TMDC material $PtSe_2$ has highly large tunable band gap which allows $PtSe_2$ to easily modulate different types of stress that are applied on it [13].

Final layer in proposed design of SPR biosensors is sensing medium which can also be called as sensing analyte. For SPR geometry, there are two commonly used configuration one Kretschmann configuration and other Otto's configuration; each configuration is based on the mechanism of attenuated total internal reflection (ATR). Advantages of Kretschmann configuration over Otto's configuration have made it useful on wider scale [14]. Biosensing application of SPR-based sensor explains us the fundamentals of detecting the concentration of biological objects, such as bacteria, viruses, DNA, and proteins in very small scale. Apart from the biosensing application and bimolecular analysis, SPR sensor is also very operative for the detection of nanostructured film deposition, for the displacement and angular position measurement [15].

2 Theory

The proposed structure of two SPR-based biosensors is shown in Fig. 1 where both sensors contain CaF_2 prism of 1.4329 refractive index at the wavelength of 633 nm, silicon of 2 nm thickness, and monolayer of $PtSe_2$ having 2 nm thickness. One SPR biosensor contains silver (Ag) metal film, and the other contains gold (Au) metal film, over the CaF_2 prism. Sensing medium is the last layer of sensor consisting of biomolecules that interacts with the $PtSe_2$ layer. 1.330 and 1.335 are assumed refractive index of sensing medium. RI change of sensing medium is $\Delta n = 0.005$; adsorption occurs when biomolecules in sensing medium interact with $PtSe_2$ layer which results in the modification of RI of the sensing medium. The refractive index of CaF_2 prism is taken from the reference [15].

The refractive index of silicon (Si) coated over the metal film to reduce the oxidation and to enhance the sensitivity is calculated by the formula given below [16]:

$$n^2(\lambda) = 1 + \frac{10.668493\lambda^2}{\lambda^2 - (0.301516485)^2} + \frac{0.00304347\lambda^2}{\lambda^2 - (1.13475115)^2} + \frac{1.54133408\lambda^2}{\lambda^2 - (1104.0)^2} \quad (1)$$

Ag and Au contain complex RI which is calculated by using Drude–Lorentz model [17]:

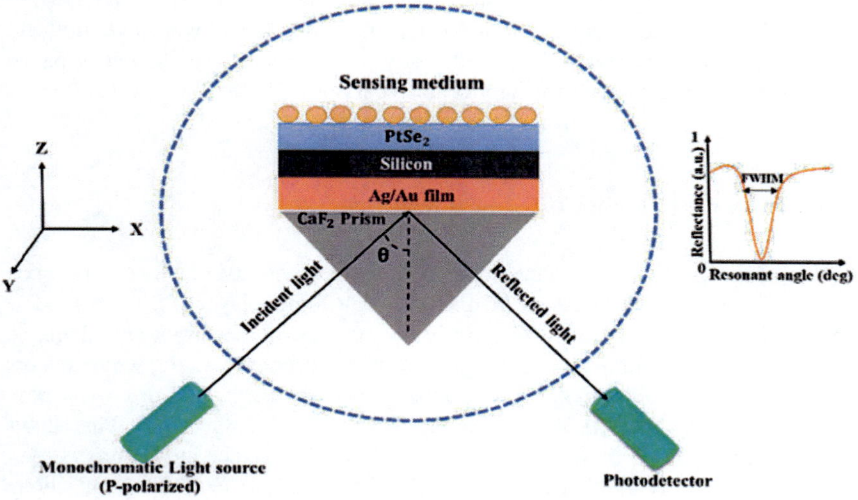

Fig. 1 Sketch diagram of proposed SPR biosensors

Table 1 Dispersion coefficients of gold and silver metal

S. No.	Metal name	Plasma wavelength (λ_p) in nm	Collision wavelength (λ_c) in nm	Reference
1.	Gold (Au)	168.26	8934.2	[18]
2.	Silver (Ag)	145.41	17,614	[18]

Table 2 Tabulated details of each layer of proposed biosensor at 633 nm wavelength

No. of layers	List of materials used	Thickness (nm)	Refractive index	Reference
1	CaF_2 prism	–	1.4329	[15]
2	Gold/Silver film	50 nm	0.1726 + 3.4220i/0.2184 + 3.5113i	[17]
3	Silicon	2 nm	3.420	[16]
4	$PtSe_2$	2 nm	2.9189 + 0.9593i	[4]
5	Sensing medium	–	1.330–1.335	–

$$n_{\text{metal}}(\lambda) = \left(1 - \frac{\lambda^2 * \lambda_c}{\lambda_p^2(\lambda_c - \lambda * i)}\right)^{\frac{1}{2}} \tag{2}$$

where λ_p and λ_c are dispersion coefficients and λ is 633 nm, the values of dispersion coefficients for gold and silver are given in Table 1.

Refractive index of $PtSe_2$ consists of real and imaginary part at 633 nm wavelength [19, 20]. The numerical analysis, transfer matrix method for *n*-layer modeling, and Fresnel equations are applied. MATLAB software is used to evaluate performance parameters of SPR biosensor from reflectance curve [4] (Table 2).

3 Results and Discussion

In this paper, we have done comparison between two biosensors, i.e., CaF_2/Ag/Si/monolayer of $PtSe_2$/S.M. and CaF_2/Au/Si/monolayer of $PtSe_2$/S.M. The comparison is made on the basis of sensitivity of the sensor. Sensitivity is calculated from the reflectance curve by observing shift in the resonance angle, which occurs after the interaction of biomolecules in sensing medium [8]. In case of silver metal, resonance angle at RI 1.330 and 1.335 is 81.68° and 82.78°, respectively. This shows small change in the RI of sensing medium results in resonance angle shift $\Delta\theta_{\text{SPR}}$. Therefore, sensitivity in this case is 220° RIU^{-1} with $\Delta\theta_{\text{SPR}} = 1.10°$. Similarly, in case of gold, small change in RI of sensing medium results in resonance angle shift; the sensitivity calculated for $\Delta\theta_{\text{SPR}} = 1.03°$ is 206° RIU^{-1}. Sensitivity enhancement of CaF_2/Ag/Si/$PtSe_2$ biosensor is 6.79% more than the CaF_2/Au/Si/$PtSe_2$ biosensor. Figure 2a, b illustrates reflectance curve for the corresponding structure shown in it.

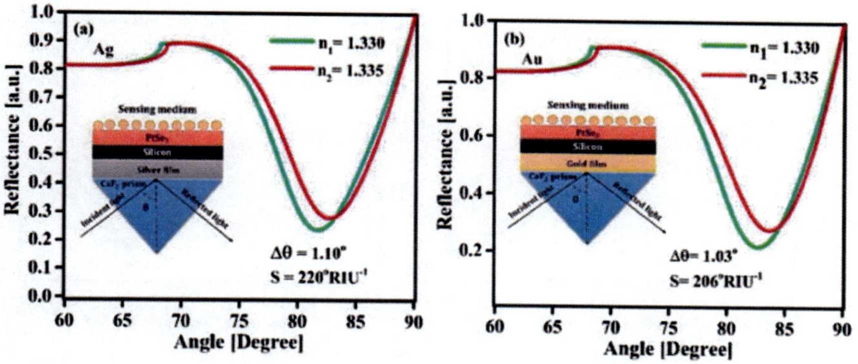

Fig. 2 Plot between incident angle and reflectance **a** proposed SPR biosensor with Ag, **b** proposed SPR biosensor with Au

From Fig. 2, the increment of refractive index of sensing medium has reduced the reflectance of biosensors, while the SPR curve is shifting from left to right [4].

Figure 3a depicts variation of sensitivity with the variation in thickness of silicon for the two SPR biosensors consist of Ag and Au separately, where sensitivity is maximum at 2 nm for both the cases and decreases for the further value on *x*-axis. Variation of sensitivity with number of $PtSe_2$ layers is shown in Fig. 3b which demonstrates that the sensitivity is higher for the monolayer of $PtSe_2$ in both Ag and Au-based sensor, sensitivity decreases for 2 and 3 layers of $PtSe_2$, and then, on further increasing of number of layer, sensitivity increases but not as much for the monolayer of $PtSe_2$. Figure 3c is a graph for quality factor and FWHM versus thickness of $PtSe_2$ layer. Quality factor is a dimensionless parameter; in Fig. 3c, quality factor is observed to be decreasing on further increase in thickness of $PtSe_2$ layer from 0 to 7 nm. It can be clearly seen in Fig. 3c that for lower value of FWHM the corresponding value of quality factor is high. Table 3 gives the information regarding performance parameters of the two proposed SPR biosensor, where we can observe that not only sensitivity, but FWHM and FoM of Au-based sensor are lesser than that of Ag-based sensor.

4 Conclusion

This work reports a comparative analysis of SPR biosensors consisting TMDC material $PtSe_2$ layer to enhance the sensitivity of sensor for biosensing applications, Si layer to prevent oxidation, CaF_2 prism due to its lower refractive index and different metal films Ag and Au. Sensitivity of CaF_2/Au/Si/monolayer of $PtSe_2$ biosensor is 206° RIU^{-1} which is less than the sensitivity 220° RIU^{-1} of CaF_2/Ag/Si/monolayer of $PtSe_2$ biosensor. This analysis can resolve the problem of gold being expensive because we can get higher sensitivity by using silver as plasmonic material in

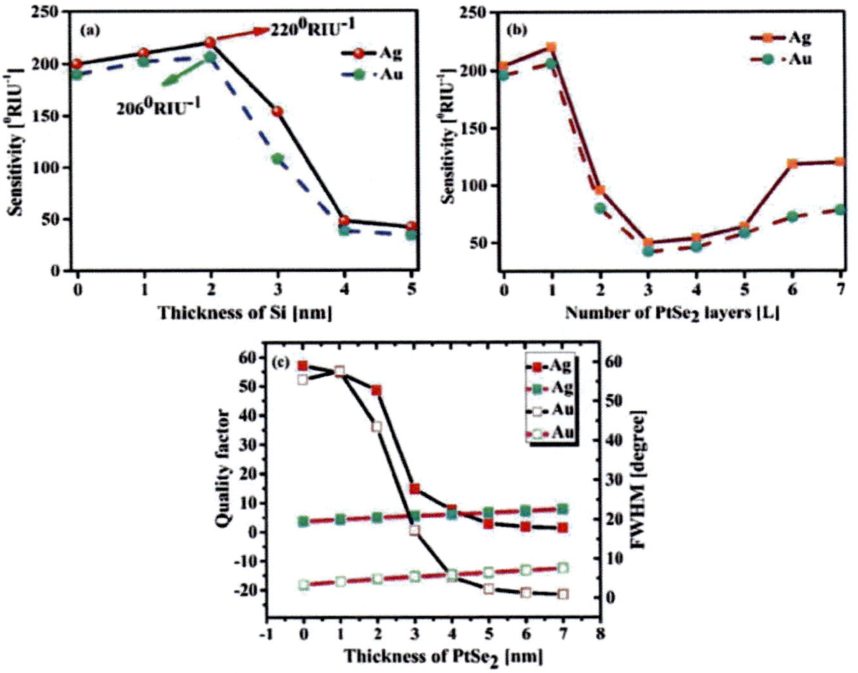

Fig. 3 **a** Variation of sensitivity versus thickness of silicon **b** sensitivity variation versus number of layers of $PtSe_2$ **c** quality factor and FWHM versus thickness of $PtSe_2$ layer

Table 3 Performance parameters of two proposed SPR biosensor

Proposed SPR biosensor	Δn	$\Delta\theta_{SPR}$ (°)	S (°RIU^{-1})	FWHM (°)	DA (deg^{-1})	FoM (RIU^{-1})
CaF_2 + Au + Si + $PtSe_2$	0.005	1.03	206	4.8527	0.2060	42.4504
CaF_2 + Ag + Si + $PtSe_2$	0.005	1.10	220	4.9755	0.2009	44.2161

biosensor, which is cheaper than gold. Our proposed high-sensitivity SPR sensor may a suitable candidate for medical diagnosis and biological applications.

References

1. Schuller JA, Barnard ES, Cai W, Jun YC, White JS, Brongersma ML (2010) Plasmonics for extreme light concentration and manipulation. Nat Mater 9:193–204
2. Gramotnev DK, Bozhevolnyi SI (2010) Plasmonics beyond the diffraction limit. Nat Photonics 4:83–91

3. Bochenkov VE, Frederiksen M, Sutherland DS (2013) Enhanced refractive index sensitivity of elevated short-range ordered nanohole arrays in optically thin plasmonic Au films. Opt Express 21(12):14763–14770
4. Rahman MdM, Rana MdM, Rahman MdS, Anower MS, Mollah MdA, Paul AK (2020) Sensitivity enhancement of SPR biosensors employing heterostructure of $PtSe_2$ and 2D materials. Opt Mater 107, 110123
5. Johansen K, Arwin H, Lundström I, Liedberg B (2000) Imaging surface plasmon resonance sensor based on multiple wavelengths: sensitivity considerations. Rev Sci Instrum 71:3530
6. Ouyang Q, Zeng S, Jiang L, Qu J, Dinh XQ, Qian J, He S, Coquet P, Yong K-T (2017) Two-dimensional transition metal dichalcogenide enhanced phase-sensitive plasmonic biosensors. J Phys Chem C 121(11):6282–6289
7. Prajapati YK, Pal S, Saini JP (2018) Effect of a metamaterial and silicon layers on performance of surface plasmon resonance biosensor in infrared range. Silicon 10:1460
8. Kumar R, Pal S, Prajapati YK, Saini JP (2020) Sensitivity enhancement of MXene based SPR sensor using silicon: theoretical analysis. Silicon
9. Xia G, Zhou C, Jin S, Huang C, Xing J, Liu Z (2019) Sensitivity enhancement of two-dimensional materials based on genetic optimization in surface Plasmon resonance. Sensors 19(5):1198
10. Wang QH, Kalantar-Zadeh K, Kis A, Coleman JN, Strano MS (2012) Electronics and optoelectronics of two-dimensional transition metal dichalcogenides. Nat Nanotechnol 7:712
11. Zhang H, Liu C-X, Qi X-L, Dai X, Fang Z, Zhang S-C (2009) Topological insulators in Bi_2Se_3, Bi_2Te_3 and Sb_2Te_3 with a single Dirac cone on the surface. Nat Phys 5:442
12. Castro Neto AH, Guinea F, Peres NMR, Novoselov KS, Geim AK (2009) The electronic properties of graphene. Rev Mod Phys 81:109
13. Jia Y, Li Z, Wang H, Saeed M, Cai H (2020) Sensitivity of a surface plasmon resonance sensor with platinum diselenide. Sensors 20(1):131
14. Alagdar M, Yousif B, Areed NF, Elzalabani M (2020) Highly sensitive fiber optic surface plasmon resonance sensor employing 2D nanomaterials. Appl Phys A 126, 522
15. Singh S, Singh PK, Umar A, Lohia P, Albargi H, Castaneda L, Dwivedi DK (2020) 2D Nanomaterial based surface plasmon resonance sensors for biosensing applications. Micromachines 11(8):779
16. Ghosh S, Ray M (2015) Analysis of silicon-based surface plasmon resonance sensors with different amino acids. SILICON 7:322
17. Pal S, Prajapati YK, Saini JP, Singh V (2016) Sensitivity enhancement of metamaterial based surface Plasmon resonance biosensor for near infrared. Opt Appl 46(1):131–143
18. Mishra AK, Mishra SK, Verma RK (2016) Graphene and beyond graphene MoS_2: a new window in surface–plasmon–resonance-based fiber optic sensing. J Phys Chem C 120(5):2893–2900
19. Bruna M, Borini S (2009) Optical constants of graphene layers in the visible range. Appl Phys Lett 94, 031901
20. Sharma AK, Pandey AK (2018) Blue phosphorene/MoS_2 heterostructure based SPR sensor with enhanced sensitivity. IEEE Photonics Technol Lett 30(7)

Mobile Application on Drowsiness Detection When Driving Car

Akshat Singhal and Sunil Kumar

Abstract With the increase in the number of vehicles on our roads, the number of road accidents is also increased drastically. One of the major reasons found for this mishappening is driver fatigue. By this proposed system, the author tries to measure a driver's fatigue by detecting drowsiness as this is the only way to calculate fatigue. This paper presents the module for advanced driver assistance system (ADAS) which monitors and calculates driver's drowsiness and warned driver through an alert system which will subsequently lead to decrease in the no. of road accidents. The designing of this system is done in such a way that one need not install any external hardware device. The rear camera of the mobile is used to monitor the eyes and mouth region of the driver. The system will alert you through an alarm if it recognizes the blinking of eyes or yawning or both. If the driver's eye remains closed or mouth remains open for a certain period which can be done through the algorithm discussed below, then it is considered that the driver is drowsy.

Keywords Fatigue · Drowsiness · Dlib · OpenCV · Haar cascade · ADAS · POEMC

1 Introduction

Driver fatigue is one of the main causes of accidents in the world. In fact, "you're more likely to die from drowsy driving than from texting while driving, distracted driving or drunk driving combined". The effects of drowsiness are similar to alcohol—it will make your driving inputs (steering, acceleration, and braking) poorer, destroy your reaction times, and blur your thought processes. About 20% of all fatal accidents in the USA are due to drowsiness! We can only imagine what the stats are like for India

A. Singhal (✉) · S. Kumar
Department of Computer Science and Engineering, Meerut Institute of Engineering and Technology, Meerut, India
e-mail: akshat.singhal.cs.2016@miet.ac.in

S. Kumar
e-mail: sunil.kumar@miet.ac.in

B. Mishra and M. Tiwari (eds.), *VLSI, Microwave and Wireless Technologies*,
Lecture Notes in Electrical Engineering 877,
https://doi.org/10.1007/978-981-19-0312-0_34

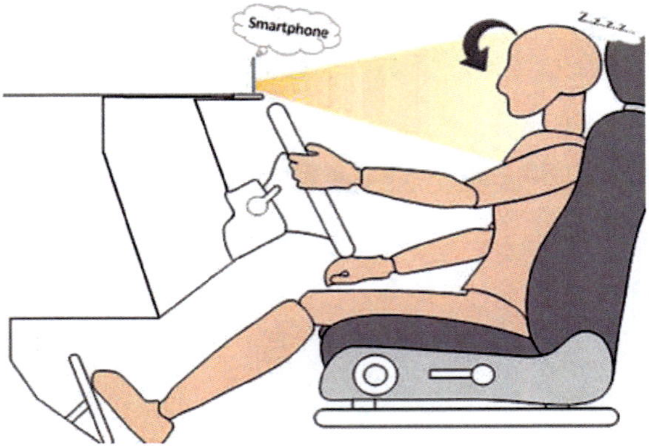

Fig. 1 Drowsiness detection using smartphone

which has a higher road accident rate. Problem is, no one researches this out here. Headlines only scream "speeding car hits biker". The Times of India has merely 800 results for "sleepy driver", but a whopping 2.23 lakh for "speeding car". It is a major issue that is happening in our world. The car manufacturers are adding automation features and accidental safety like airbags and all stuff, but no one wants to have an eye on this major issue. Accidents are happening and innocent people are dying every day. The people sleeping on the footpath and the families sitting inside the car all have the risk of their life.

It is difficult to implement the drowsiness detection system in all the existing cars or vehicles. Therefore, in this project, we are developing an advanced driver assistance system (ADAS) that is presented to decrease the sum of accidents by developing a mobile application for detecting the drowsiness of the driver driving the automobile. Nowadays, a smartphone is an essential part of our life, and the maximum population has a smartphone. Therefore, it is easy to spread this technology in the form of a mobile application.

To address this problem, we have designed and developed a system that can detect and alert the driver drowsiness. The targets of this system are human eyes and mouth by which driver fatigue can be calculated. This fatigue can be calculated in two ways: One is sleepiness, and the other is yawning. Authors use the OpenCV and Dlib libraries (Fig. 1).

2 Existing System

In the existing system, the implementation of driver drowsiness detection is done by using some hardware components like Raspberry Pi 3, camera, gas sensor, vibration sensor, buzzer, IoT module, Wi-Fi module, relay, and motor [1]. The implementation is done by using some hardware components like target board, IR camera, display monitor, windows virtual machine, and power source [2]. The implementation is done using electroencephalogram (EEG), electrocardiogram (ECG), and heart rate variability) (HRV) [3, 4]. Here also, a watch like an external device is used [5]. The implementation is done using the EMG which measures the muscle electrical activity on the skin [6]. But in the following proposed system, the detection is done using software for the smartphone which can easily be distributed to the users as the current market condition in which smartphones are available to all the users which can be used as a medium without any extra cost.

An Android-based application is proposed which is implemented using Haar cascade [7]. The main factor which is missing in all the existing systems is the detection of the face of the driver by eliminating all the other faces which are accidentally coming behind the driver by which the system can get confused and predict the false result which has been resolved in the proposed system.

The implementation is done by extracting the white area of the eyes which can struggle in low lightning conditions [8]. The implementation is done by extracting the face area using skin tone which can result in the wrong prediction due to the different colors of the cloth [9, 10]. The accuracy of the proposed system is also improved as compared to all the existing systems as there is the use of pre-learned libraries which predict the more accurate results and give the quite accurate results in different lighting conditions.

3 Proposed System

The system measures the drowsiness of the driver by capturing each frame from the camera, detecting all the faces present in that particular frame, and extracting the face of the driver. Then, observing the period for which the driver's eye remains closed or mouth being opened. For this, the total no. of frames for which the above action is being performed is calculated. And if this value continuously crosses the given threshold value, then a popup stating the warning is generated on the screen and a warning alarm is also activated to make the driver aware and save him from any mishappening (Fig. 2).

To determine the drowsing state of the driver, the percentage of eye and mouth closure (POEMC) calculation is performed which is generally passed by the following stages.

3.1 Face Detection

Dlib libraries are used to detect the faces present in the given frame and give the coordinates of upper-left corner and lower-right corner of the face as $(x1, y1)$ and $(x2, y2)$.

Then, the task is to extract the driver's face on which the operation is to be performed because without this, the system might get confused by finding more than

Fig. 2 Flowchart

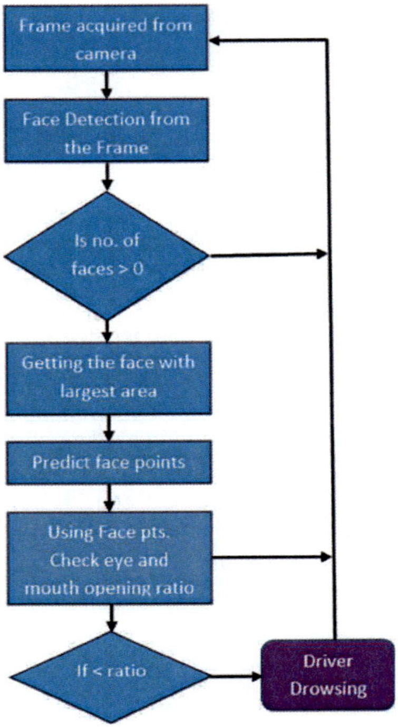

one face in a frame which can be most probably the person sitting behind the driver. In our system, this is achieved by the logic that in the particular frame, the face of the driver is the largest as the driver is sitting nearest to the camera, so we calculate the area of all the frames by the upper-left corner and lower-right corner as:

$$\text{Area} = (x2 - x1) * (y2 - y1)$$

And hence, we can conclude that the face with the largest area belongs to the driver (Fig. 3).

3.2 Eyes and Mouth Tracking

Dlib library is used, with the help of which a pre-trained facial landmark predictor is called for estimating the location of 68 (x, y) coordinates that map to facial structures on the face. These facial points help for calculating the various facial expressions like eye blinking, mouth opening, etc.

The below image can help to visualize the indexing of the 68 coordinates. The landmark points from 37 to 42, 43 to 48, and 49 to 68 give us the exact position of

Fig. 3 Face detection

Fig. 4 Facial dots representation

the left eye, right eye, and mouth, respectively. These all landmark points given by the library contain the corresponding (*x*, *y*) coordinate (Fig. 4).

3.3 *Identification of the State*

From the above-given coordinates, we calculate the opening ratio of both the eyes and then average them. We will do the same for the mouth, but here it is the inner side and outer side for average.

3.3.1 For Eyes

$$T.x = ([38].x + [39].x)/2$$

$$T.y = ([38].y + [39].y)/2$$

$$B.x = ([41].x + [42].x)/2$$

$$B.y = ([41].y + [42].y)/2$$

$$\text{VerticalLength} = \sqrt{\left((T.x - B.x)^2 + (T.y - B.y)^2\right)}$$

$$\text{HorizontalLength} = \sqrt{\left(([37].x - [40].x)^2 + ([37].y - [40].y)^2\right)}$$

$$\text{LeftEyeRatio} = \text{VerticalLength}/\text{HorizontalLength}$$

$$T.x = ([44].x + [45].x)/2$$

$$T.y = ([44].y + [45].y)/2$$

$$B.x = ([47].x + [48].x)/2$$

$$B.y = ([47].y + [48].y)/2$$

$$\text{VerticalLength} = \sqrt{\left((T.x - B.x)^2 + (T.y - B.y)^2\right)}$$

$$\text{HorizontalLength} = \sqrt{\left(([43].x - [46].x)^2 + ([43].y - [46].y)^2\right)}$$

$$\text{RightEyeRatio} = \text{VerticalLength}/\text{HorizontalLength}$$

$$\text{EyeAvgRatio} = (\text{LeftEyeRatio} + \text{RightEyeRatio})/2$$

3.3.2 For Mouth

$$\text{VerticalInner} = \sqrt{\left(([63].x - [67].x)^2 + ([63].y - [67].y)^2\right)}$$

$$\text{HorizontalInner} = \sqrt{\left(([61].x - [65].x)^2 + ([61].y - [65].y)^2\right)}$$

$$\text{VerticalOuter} = \sqrt{\left(([52].x - [58].x)^2 + ([52].y - [58].y)^2\right)}$$

$$\text{HorizontalOuter} = \sqrt{\left(([49].x - [55].y)^2 + ([49].y - [55].y)^2\right)}$$

$$\text{InnerRatio} = \text{VerticalInner}/\text{HorizontalInner}$$

$$\text{OuterRatio} = \text{VerticalOuter}/\text{HorizontalOuter}$$

$$\text{MouthAvgRatio} = (\text{InnerRatio} + \text{OuterRatio})/2$$

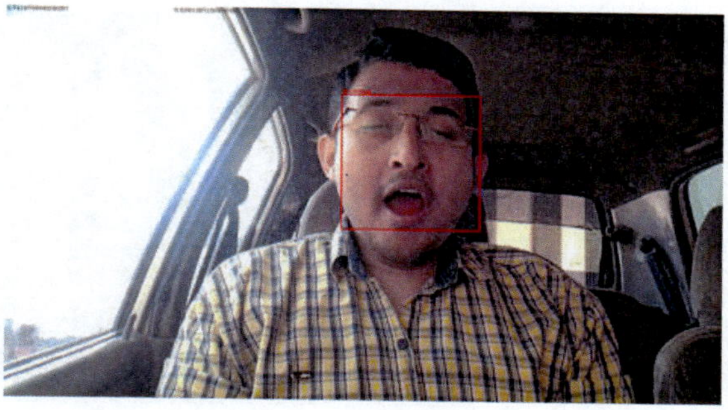

Fig. 5 Yawning detection

3.4 Calculating the POEMC and Identification of the Driver State

Now, the above ratio is checked with the tested threshold and finds the POEMC. If the calculated result is continuous for eight or more frames, then the alarm is activated.

$$\text{Eye}_{\text{AvgRatio}} < \text{Threshold}_{\text{sleep}}$$

OR

$$\text{Eye}_{\text{AvgRatio}} < \text{Threshold}_{\text{Yawn(eye)}} \text{ AND } \text{Mouth}_{\text{AvgRatio}} < \text{Threshold}_{\text{Yawn(mouth)}}$$

The threshold for sleep is different from yawning as at the time of yawning, our mouth opens and eyes got closed (Fig. 5).

4 Experimental Results

To validate our system, authors have tested it on some drivers in different conditions like light, the number of people in the car, and skin tone. The problem mostly arises with the number of people as if someone sitting behind the driver comes in the camera frame the system gets confused by getting multiple faces and gives the inadequate result which can be harmful. Therefore, this is resolved by calculating the area of all the faces, as we know the person who is nearest to the camera gets the biggest face in the photograph; therefore, by using this phenomenon, the face with the highest area gets selected as the driver is the only person to sit nearest to the camera. As per the test performed and their outcomes, our system can determine the state of drowsing and yawn with a high rate of accuracy.

Driver face detection is shown in Fig. 6 and day-by-day face detection in Table 1.

Drowsiness detection is shown in Fig. 7 and day-by-day drowsiness detection in Table 2.

Fig. 6 Driver face detection

Table 1 Driver face detection

Driver	No. of faces present	No. of faces detected
Day 1	2	1
Day 2	3	1
Day 3	2	1

Fig. 7 Drowsiness detection

Table 2 Drowsiness detection

Driver	No. of frames	Face detected	Actual drowsing	Drowsiness detected
Day 1	2045	2032	147	140
Day 2	1986	1965	100	80
Day 3	2489	2457	119	108

5 Conclusions

In this paper, the authors have presented the implementation of an ADAS for driver drowsiness detection intending to warn the driver by calculating driver fatigue. This system can determine the driver state in decent lighting conditions. The results are satisfied with the opportunity to find the driver's face among all faces. This system is developed for the purpose to get in the hand of all the users in the form of a smartphone application.

References

1. Ratna Kaavya M, Ramya V, Franklin RG (2019) Alert system for driver's drowsiness using image processing. In: 2019 international conference on vision towards emerging trends in communication and networking (ViTECoN)
2. Baek JW, Han B-G, Kim K-J, Chung Y-S, Lee S-I (2018) Real-time drowsiness detection algorithm for driver state monitoring systems. In: 2018 tenth international conference on ubiquitous and future networks (ICUFN)
3. Binti Amirudin NA, Saad N, Azhar Ali SS, Hasan Adil S (2018) Detection and analysis of driver drowsiness. In: 2018 3rd international conference on emerging trends in engineering, sciences, and technology (ICEEST)
4. Kim W, Choi H-K, Jang B-T (2018) Study on training convolutional neural network to detect distraction and drowsiness. In: 2018 IEEE region ten symposium (Tensymp)
5. Artanto D, Prayadi Sulistyanto M, Deradjad Pranowo I, Erry Pramesta E (2017) Drowsiness detection system based on eye-closure using a low-cost EMG and ESP8266, ICITISEE 2017
6. Riztiane A, Hareva DH, Stefani D, Lukas S (2017) Driver drowsiness detection using visual information on android device. In: 2017 international conference on soft computing, intelligent system, and information technology (ICSIIT)
7. Tombeng MT, Kandow H, Adam SI, Silitonga A, Korompis J (2019) Android-based application to detect drowsiness when driving vehicle. In: 2019 1st international conference on cybernetics and intelligent system (ICORIS)
8. Kusuma Kumari BM (2015) A real-time driver drowsiness detection system. In: IJCA proceedings on international conference on innovations in computing techniques (ICICT 2015), 17–21 July 2015
9. Alshaqaqi B, Baquhaizel AS, Ouis MEA, Boumehed M, Ouamri A, Keche M (2013) Vision-based system for driver drowsiness detection. In: 2013 11th international symposium on programming and systems (ISPS)
10. Devi MS, Choudhari MV, Bajaj P (2011) Driver drowsiness detection using skin color algorithm and circular Hough transform. In: 2011 fourth international conference on emerging trends in engineering & technology

Recent Advancement in Long-Period Fiber Grating (LPFG)

Pragya Mishra, Sachin Singh, Pooja Lohia, and D. K. Dwivedi

Abstract Inspiring quick, reliable, and real-time measurements, biomass detection, chemical detection for long-period fiber grating (LPFG), for scientific, commercial, and defense applications. In this article, we address the recent development of manufacturing techniques for LPFG in biological and chemical sensor applications. Because of this, the optical fiber field draws great attention. Detailed study of the modulation of LPFG sensors is also explained to reflect on a more realistic approach to their use in sensing in various fields. The possible principles, structure, and parameters are also discussed along with analysis and progressive comparison in this work.

Keywords Long-period fiber grating · Sensor · Sensitivity · Refractive index

1 Introduction

The demand for sensors has reached an anomalous high point for which long-period fiber gratings (LPFGs) could be considered as great platforms for optical/chemical/biosensor due to their capability to record instant changes (light mass, insignificant size, immune electromagnetic interference, high sensitivity, and specificity in the optical properties). These devices exhibit periodic variation in refractive index in the core of the single-mode optical fiber which tends from 100 to 1000 mm; whereas, FBGs are also sensitive to the surrounding catalyst. LPFGs are outstanding as pointer heads because of the capability of minute sensing in the refractive index (10–4). By calculating the optical loss at a fixed wavelength or by recording the complete attenuation spectrum, the resonant shift in LPFG diminishing

P. Mishra · S. Singh · D. K. Dwivedi (✉)
Photonics and Photovoltaic Research Lab, Department of Physics and Material Science, Madan Mohan Malaviya University of Technology, 273010 Gorakhpur, India
e-mail: todkdwivedi@gmail.com

P. Lohia
Department of Electronics and Communication Engineering, Madan Mohan Malaviya University of Technology, Gorakhpur 273010, India

B. Mishra and M. Tiwari (eds.), *VLSI, Microwave and Wireless Technologies*,
Lecture Notes in Electrical Engineering 877,
https://doi.org/10.1007/978-981-19-0312-0_35

spectrum can be detected. Originally, LPFGs were formed as band rejection filters in the telecommunications industry applicability.

An optical fiber-based LPFG sensor is devoted to the latest achievements in research and development with a vast area of physical, chemical, and biochemical sensors, [1] including remote sensing. LPFG sensor was first reported by Vengsarkar et al. [2] in 1996. The propagating light coupled to the cladding modes attenuates rapidly due to scattering losses, thus leaving lossy bands in the guided core mode observed at the output end of the LPFG [3].

Therefore, the transmission spectra of an LPFG show more than one attenuation peak and each one occurring for a different resonant wavelength (λ_{res}^{m}) which corresponds to coupling of light from the fundamental propagating core mode to the co-propagating *m*th cladding mode, which obtained through the phase matching condition as [4]

$$\lambda_{res}^{m} = [n_{eff,co}^{01}(\lambda) - n_{eff,cl}^{m}(\lambda)]\wedge$$

Here, $n_{eff,co}^{01}$ denotes efficient refractive index of core, $n_{eff,cl}^{m}$ denotes the efficient refractive index of the *m*th cladding mode ($m = 1, 2, 3, 4$), and $\wedge$ is the period of the LPFG.

As the increasing demand for LPFGs in the fields of visible communications including sensor technologies, LPFG will be the sensor platform of choice for a vast area of technical employment [5–7] due to its lightweight, small size, economical, superior durability, geometrical versatility, high sensitivity, and electromagnetic interference immunity, and fiber optic sensors have achieved extensive attention. Additionally, it allows immense sensibility, large selectivity, and high security and provides simple operation and rapid online exposure in arduous operating surroundings [8] (Fig. 1).

Because of its benefits and superior flexibility, in the thousands of fields, long-period fiber gratings have been examined and detected in recent years. In this conference paper, we are summarizing the feature, advantages, and advances in the development of LPFG's along with their different fabrication techniques. A comparison between LPFGs and other optical fiber sensors has also been mentioned.

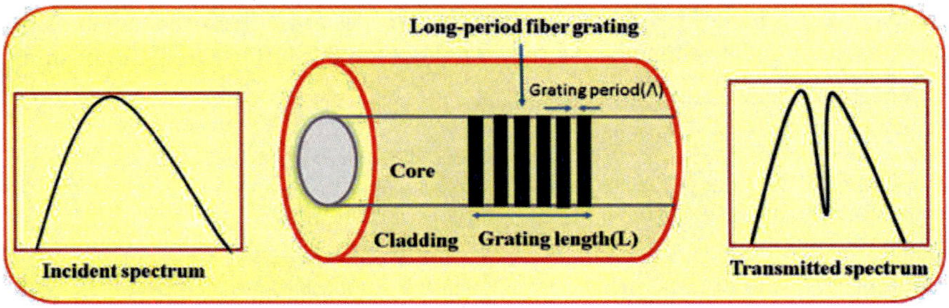

Fig. 1 Diagrammatic representation of LPFG

2 Theory

2.1 *Feature of LPFG*

The basis of LPFG is coupled mode theory which helps to analyze the transmission phenomena in LPFGs [2, 2]. The coupling took place within the fundamental implemented core mode and a specific cladding mode concerning resonance wavelength. It helps to resonant dip at the resonance wavelength in the spectrum of the fiber. Each resonant dip matched with a coupling of light to various cladding order modes.

The modification in various parameters: Temperature, strain, bend radius, waveguide dimensions, and the refractive index of the medium vary for LPFG's fiber convenient and sensitively, which is observed in a shift of the wavelength of a resonant dip in which sensitivity depends on the dispersion properties [9, 10]. With the help of these features, better understanding of these devices and sensors can be achieved.

The variation in coupling coefficient and period along with grating or phase shift along with the grating can provide complex grating profiles and transmission characteristics [11, 12].For the sensitivity calculation of LPFGs, using the transfer matrix method and perturbation theory [13] which show the specialty of the spectrum, at the resonant dip [14]. During changing physical parameters, we get growth in the period of LPFG and various refractive index of core, which involves the variation of resonant dip and its phase-matching conditions from coupling to the cladding modes. This scope of responses makes them attractive for various applications and in the research field [14].

2.2 *Fabrication Technique*

Fabrication of the LPFGs take place by various methods such as UV radiation, CO_2 laser irradiation, ion irradiation, ion implementation, femtosecond laser irradiation, mechanical stress, electric arc discharge, and diffusion of dopant into the core to obtain a periodical modulation of the refractive index within the fiber core [15–22]. The electric arc discharge method is found to be the simplest, flexible, and low-priced technique with several unique properties that differentiate from other fabrication techniques. The electric arc discharge technique for grating fabrication was firstly reported in 1994 in a two-step point fabricating process [23]. The transformation on the surface of fiber into sinusoidal deformation was the main reason to use electric arc in the fabrication process along with surface tension of the silica. A focused CO_2 laser beam is used to create a periodic cut on the fiber surface and individual cut by point-to-point annealing via electric arc discharge.

Dianov et al. reported a new fabrication technique to heat the fiber to the temperature in presence of nitrogen-doped silica fiber and fusion splicer as a result more than 200 μm grating period was produced [24] (Fig. 2).

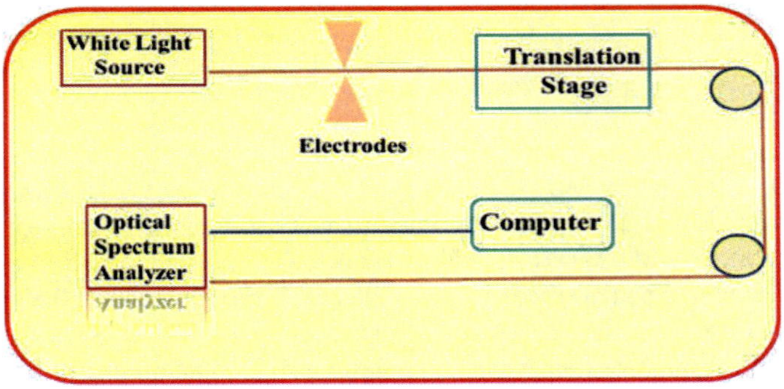

Fig. 2 Diagrammatic representation of fabrication setup

The setup of LPFG fabrication suggested by Kosinski and Vengsarkar with help of a commercial splicer [25] exhibits a couple of electrodes to introduce an electric arc for modification of periodically physical prototype of fiber. This mechanism was encouraged by a motorized stage, in both the electrodes for translation of the fiber.

A LPFGs produced by an electric arc and providing periodic micro bend [26, 27] was reported by Hwang et al. which was later improvised by Kim et al. [28]. In earlier 2014, a periodically tapered, standard single-mode fiber was reported. The fabrication of LPFG setup has of a typical fusion splicer with ZL and ZR motors fixed by fiber holders, and SWEEP motors were focused on a translation stage. The fabrication process took place by fiber heating and was followed by electric arc discharge and synchronously stretching both ends ZR and ZL with the help of fibers in the taper formation along with fibers. SWEEP motor helps to move further for the next point concerning the fiber axis. This fabrication setup solves the issue of alignment along with the effect of mass on the tapers.

2.3 Applications of LPFG

Small optical fiber sensors and their sensitivity to external influences made long-period fiber grating (LPFG) for the great potential of vast applications. Among which a few are given below.

2.3.1 Refractive Index Sensor

Without the need for etched claddings and adaptable nature of refractive index, sensors of LPFGs creates a huge shift in the resonant dip.

Such sensors might help in the determination of the sugar level in the water to assess its feasibility. The standard ellipsometry technique used to find out the refractive index, and the shift in the resonant dip of a grating fabricated was measured for calculation of transition in the refractive index along with great applications in the field of sensitive nanosensors because of the film of the fiber cladding for detection of viruses or bacteria [29, 30]. For the execution of this application, we need to immobilize the antibody on the surface of the cladding and antigen introduction in solution form.

A refractive index change was observed from a reaction between antibody and antigen, which help in epitomizes itself toward a resonant shift of the LPFG. The observed resonant shift provides a determination of antibody and antigen reaction and becomes the cause of the formation of the basis for an adaptable sensor. The long-period fiber gratings can reduce human involvement in the process and help for improvement in performance and operating costs. Various optical techniques like ellipsometry and surface plasmon resonance have been exercised for the detection of antibody–antigen reactions. However, with great advancement, these processes still have limitations especially for complex sensor design, limited sensitivity to refraction index, and polarization sensitivity.

2.3.2 Axial Strain and Temperature Sensor

The effect of axial strain and temperature on LPFG assists to shift in the resonant wavelength of a particular band of spectrum. It corresponds to play a major role in cladding order modes and host fibers. Two or more shifts in the resonant wavelength of the spectrum assist in multiparameter sensing [31].

2.3.3 Highly Effective Pressure Sensor

LPFG is used as a hydrostatic pressure sensor with help of the photo-elastic effect in optical fibers. A thin layer deposition of silicon having a large value of photo-elastic coefficient on the cladding surface and external pressure on the photo-elastic material to refractive index changes with a value which will be proportional to the applied pressure. A resonant change in the spectrum is caused by the reaction of the efficient cladding mode refractive index. Thus, the magnitude of pressure could be determined for a calibrated grating once the photo-elastic coefficient is known [32].

2.3.4 Electric and Magnetic Field Sensor

Thin layer deposition on an outer surface of the cladding made up of electro and magneto-optic materials was used in this technique. The demonstration for monitoring corrosion in metals has been already reported in long-period grating [33], and its feasibility depends on exciting frontiers used for applications in devices of health monitoring. In the evaporation technique, the metal is deposited on the bare cladding region surrounding the grating and placed in a corrosive solution of acids.

As a result, corrosion on metal was observed along with a reduction in thickness of the outer layer, and a corresponding resonant shift was also observed. After the calibration of the sensor, it could be used to monitor the corrosion upon the surface and inside configuration.

2.3.5 Luminescence Spectra and Gain Flattening

The work of LPFGs in the field of the telecommunication system is that it flattens the gain spectrum of fiber amplifiers used in the WDM/DWDM system and also flattens spectra of wideband luminescence sources. The gain spectrum flattening (~0.2 dB) of erbium-doped and neodymium-doped fiber amplifiers was demonstrated by using LPFGs [34–36]. The benefits of these methods are the simplicity of grating fabrication, a high gain value of flattened spectrum and a wide spectral range, low insertion at the pump wavelength, and nonexistence of back-reflected light.

3 Result and Discussion

Comparison with other fiber optic sensor

Moreover, the fabrications of these sensors are expensive and time taking due to a limitation of these sensors, we need a reference signal for extraction and phase modulation sensing observation which shows inherent immunity to transverse strain and temperature fluctuations [37–40].

The intensity-based sensors also needed [41] a reference signal to bypass power alterations in the optical source and bend in the lead-out fibers. These sensors are simple and able to demodulate which made them economically important. The multiplexing process is difficult and is not fit for applications due to the concern of several points along with the fiber length. In fiber Bragg grating sensor [42], multiplexing process is very simple and insensitive to source in fiber Bragg grating sensor intensity fluctuations and bends in the fiber which made it useful in various health monitoring applications with high performance.

Whereas, LPFG sensors are simpler to fabricate and more at a time, however, LPFGs show smaller sensitivity as compared to interferometric sensors but the sensing process is uncomplicated in LPFG. The sensing of the refractive index is done easily by LPFG [43]. To attain the access of evanescent waves, the interferometric sensors need etching of cladding which lowers the strength of the fiber.

Comparison with other Fiber grating structures

Previously, scientists have already explored the various type of gratings which are given in following Table 1.

Table 1 Comparison of LPFG with existing fiber grating structures

Types of fiber Bragg grating (FBG)	Structures of grating	Spectra characteristics	Description
Long-period fiber grating		Co-propagating mode cause resonance at transmission spectra	This fiber Bragg grating exhibits a larger spectral shift This grating structure has a long grating period that couples light to the cladding
Chirped FBG		Broadly reflected spectra are caused by their position	This fiber Bragg grating is a dispersion compensation with a varying grating period It has high dispersion with diffuse bandwidth
Tilted FBG	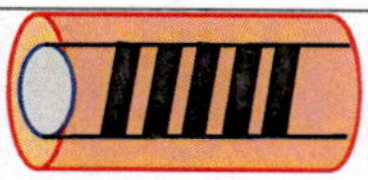	Fine-comb resonance spectra in the transmission spectrum	The grating angle allows light to be coupled with cladding modes and light couples in a backward direction to radiation modes
Uniform FBG		It is characterized by regular perturbation of RI	It is based on the principle of Bragg's law with a constant grating period and low dispersion and has narrow bandwidth
Phase-shifted FBG	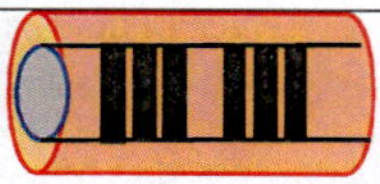	It results in a narrow notch in reflection spectra	By varying, the amount of π-shift the transmission wavelength can be altered
Super-structure FBG	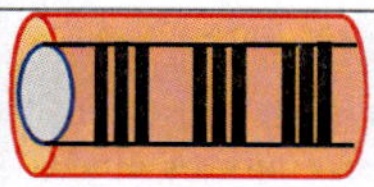	Individual fiber Bragg grating characteristics are found in transmission spectra	It is a combination of the FBG and an LPFG

4 Conclusion

In this study, we sum up the features and advantages of LPFG. Its advantage makes them employable for an appropriate multitude of sensing applications. Several fabrication techniques to write grating on the fiber core have been mention. Along with that, we compiled its various uses as sensing applications and resulted in a comparison of LPFG with other fiber optic sensors and also with existing grating structures such as uniform, chirped, tilted, phase, and superstructure fiber Bragg grating. However, we already know about the great scope of LPFG after enhancing sensitivity and provide specificity, but we need to work in a specialized manner to refractive index matched with cladding material. And for improvements, we must demonstrate a higher mode of the LPFG either by varying the interrogation wavelength as well as the grating periodicity of the LPFG or by modifying its width also. The design of visible features of LPFG will assist further enhancement in near future.

References

1. Garg R, Tripathi SM, Thyagarajan K, Bock WJ (2013) Long period fiber grating based temperature-compensated high-performance sensor for bio-chemical sensing applications. Sens Actuators B Chem 176
2. Vengsarkar AM, Lemaire PJ, Judkins JB, Bhatia V, Erdogan T, Sipe JE (1996) Long-period fiber gratings as band-rejection filters. J Lightwave Technol 14(1):58–65
3. Bhatia V, Vengsarkar AM (1996) Optical fiber long-period grating sensors. Opt Lett 21:692–694
4. Biswas P, Basumallick N, Bandyopadhyay S, Dasgupta K, Ghosh A, Bandyopadhyay S (2015) Sensitivity enhancement of turn-around-point long period gratings by tuning initial coupling condition. IEEE Sens J 15(2):1240–1245
5. Wang S-F, Chiang C-C (2016) A notched long-period fiber grating magnetic field sensor based on nanoparticle magnetic fluid. Appl Sci
6. Coelho TVN et al (2015) A numerical and experimental study of the remote long-period grating fiber sensor with Raman amplification. In: 2015 SBMO/IEEE MTT-S international microwave and optoelectronics conference (IMOC), Porto de Galinhas
7. Askarov D, Kahn JM (2015) Long-period fiber gratings for mode coupling in mode-division-multiplexing systems. J Lightwave Technol 33(19):4032–4038
8. Wang Q, Zhao W-M (2018) A comprehensive review of lossy mode resonance-based fiber optic sensors. Opt Lasers Eng 100
9. MacDougall TW, Pilevar S, Haggans CW, Jackson MA (1998) Generalized expression for the growth of long period gratings. IEEE Photonics Technol Lett 10(10):1449–1451
10. Shu X, Zhang L, Bennion I (2002) Sensitivity characteristics of long-period fiber gratings. J Lightwave Technol 20(2):255–266
11. Chan FYM, Chiang KS (2005) Analysis of apodized phase-shifted long-period fiber gratings. Opt Commun 244(1–6)
12. Ke H, Chiang KS, Peng JH Analysis of phase-shifted long-period fiber gratings. IEEE Photonics Technol Lett 10(11):1596–1598
13. Ng MN, Chiang KS Thermal effects on the transmission spectra of long-period fiber gratings. Opt Commun 208(4–6)
14. Bhatia V (1999) Applications of long-period gratings to single and multi-parameter sensing. Opt Exp 4:457–466

15. Davis DD et al (1998) Long-period fibre grating fabrication with focused CO_2 laser pulses. Electron Lett 34(3):302–303
16. Drozin L, Fonjallaz P-Y, Stensland L (2000) Long-period fibre gratings written by CO_2 exposure of H_2-loaded, standard fibres. Electron Lett 36(8):742–744
17. Fujimaki M et al (2000) Fabrication of long-period optical fiber gratings by use of ion implantation. Opt Lett 25(2):88–89
18. Kondo Y et al (1999) Fabrication of long-period fiber gratings by focused irradiation of infrared femtosecond laser pulses. Opt Lett 24(10):646–648
19. Kim C-S et al (2000) Induction of the refractive index change in B-doped optical fibers through relaxation of the mechanical stress. Opt Commun 185(4–6):337–342
20. Rego G et al (2001) High-temperature stability of long-period fiber gratings produced using an electric arc. J Lightwave Technol 19(10):1574
21. Palai P et al (2001) Characterization and simulation of long period gratings fabricated using electric discharge. Opt Commun 193(1):181–185
22. Dianov EM et al (1998) Long-period fiber gratings and mode-field converters fabricated by thermodiffusion in phosphosilicate fibers. In: 24th European conference on optical communication. ECOC'98 (IEEE Cat. No. 98TH8398), vol. 1. IEEE
23. Poole CD, Presby HM, Meester JP (1994) Two-mode fibre spatial-mode converter using periodic core deformation. Electron Lett 30(17):1437–1438
24. Enomoto T et al (1998) Long-period fiber grating in a pure-silica-core fiber written by residual stress relaxation. In: OFC'98. Optical fiber communication conference and exhibit. Technical digest. Conference edition. 1998 OSA technical digest series, vol 2 (IEEE Cat. No. 98CH36177). IEEE
25. Kosinski SG, Vengsarkar AM (1998) Splicer-based long-period fiber gratings. In: OFC'98. Optical fiber communication conference and exhibit. Technical digest. Conference edition. 1998 OSA technical digest series, vol 2 (IEEE Cat. No. 98CH36177). IEEE
26. Hwang IK, Yun SH, Kim BY (1999) Long-period fiber gratings based on periodic microbends. Opt Lett 24(18):1263–1265
27. Ugale SP, Mishra V (2014) Optimization of reversible LPFG for sensing application. Optik 125(1):111–114
28. Kim MW et al (2002) Performance characteristics of long-period fiber-gratings made from periodic tapers induced by electric-arc discharge. J Korean Phys Soc 40(2):369–373
29. Janczuk-Richter M, Dominik M, Roźniecka E, Koba M, Mikulic P, Bock WJ, Łoś M, Śmietana M, Niedziółka-Jönsson J (2017) Long-period fiber grating sensor for detection of viruses. Sens Actuators B Chem 250
30. Consales M et al (2014) Long period fiber grating biosensor for the detection of drug resistant bacteria: the "OPTObacteria" project. In: 2014 third mediterranean photonics conference, Trani
31. Bhatia V, Campbell D, Claus R, Vengsarkar A (1997) Simultaneous strain and temperature measurement with long-period gratings. Opt Lett 22:648–650
32. Bhatia V (1996) Properties and sensing applications of long-period gratings. Dissertation, Virginia Tech
33. Greene JA, Jones ME, Duncan PG, Kozikowski CL, Bailey TA, May RG, Murphy KA, Claus RO (1997) Grating-based optical fiber corrosion sensors. In: Proceedings of SPIE 3042, smart structures and materials 1997: smart sensing, processing, and instrumentation, 6 June 1997
34. Belov A, Dianov E, Karpov V, Kurkov A, Protopopov V, Tsvetkov M, Guryanov A, Khopin V, Choe W, Kim S (1995) Gain spectrum flattening of erbium-doped fiber amplifier using tapered fiber filter. In: Optical amplifiers and their applications. Vol. 18 of 1995 OSA Technical Digest Series (Optical Society of America, 1995), paper ThD1
35. Singh S, Singh PK, Umar A, Lohia P, Albargi H, Castaneda L, Dwivedi DK (2020) 2D nanomaterial based surface plasmon resonance sensors for biosensing applications. Micromachines 11(8):779
36. Kim HS et al (1997) Single-mode-fiber acousto-optic tunable notch filter with variable spectral profile. In: Optical fiber communication conference. Optical Society of America

37. Giallorenzi TG et al (1982) Optical fiber sensor technology. IEEE Trans Microwave Theor Tech 30(4):472–511
38. Lee CE, Taylor HF (1991) Fiber-optic Fabry-Perot temperature sensor using a low-coherence light source. J Lightwave Technol 9(1):129–134. https://doi.org/10.1109/50.64932
39. Murphy KA et al (1991) Quadrature phase-shifted, extrinsic Fabry–Perot optical fiber sensors. Opt Lett 16(4):273–275
40. Sirkis JS (1991) Phase-strain-temperature model for structurally embedded interferometric optical fiber strain sensors with applications. In: Fiber optic smart structures and skins IV, vol 1588. International Society for Optics and Photonics
41. Dakin J, Culshaw B (1988) Optical fiber sensors: principles and components, vol 1. Boston
42. Meltz G et al (1987) Fiber optic temperature and strain sensors. In: Fiber optic sensors II, vol 798. International Society for Optics and Photonics
43. Mishra AC, Lohia P, Dwivedi DK (2020) Long period grating based optical fiber sensors: fabrication techniques and characteristics. In: 2020 international conference on electrical and electronics engineering (ICE3). IEEE
44. Snyder AW, Love JD (1983) Optical waveguide theory
45. Rastogi V, Chiang K (2002) Long-period gratings in planar optical waveguides. Appl Opt 41:6351–6355
46. Dianov EM et al (1997) Thermo-induced long-period fibre gratings. In: Integrated optics and optical fibre communications. 11th international conference on, and 23rd European conference on optical communications (Conf. Publ. No.: 448), vol 2. IET
47. Yin G et al (2014) Long period fiber gratings inscribed by periodically tapering a fiber. IEEE Photonics Technol Lett 26(7):698–701
48. Nygren H, Stenberg M Calibration by ellipsometry of the enzyme-linked immunosorbent assay. J Immunol Methods 80(1):198524
49. Cullen DC, Brown RGW, Lowe CR (1988) Detection of immuno-complex formation via surface plasmon resonance on gold-coated diffraction gratings. Biosensors 3:211–25

Novel Tri-band Microstrip Bandpass Filter with Stub Loaded in Circular Ring Resonator

Prashant Pandey, Arvind Kumar Pandey, and R. K. Chauhan

Abstract This paper represents a novel and compact design for a tri-band bandpass filter by using short-circuited and open-circuited stub. Circular ring-based geometry with annular feed is used and resulted in miniature of size and satisfactory performance in pass band. Circuit structure is analysed using current distribution at resonant frequency and even odd mode theory. After optimization of different design parameters, three pass bands are obtained at resonant frequencies 1.27 GHz, 4.38 GHz and 6.92 GHz. These bands are used in application of transmitter and receiver system for wireless application, application of satellite communication and recent 5G applications in H band, respectively. Proposed tri-band filter is simple, novel in design, and circuit area has been significantly reduced when compared to latest tri-band filters.

Keywords SLR · Circular ring resonator · Annular feed · Even–odd

1 Introduction

As world is looking to be fully wireless, demand of wireless application is increasing day by day. Same wireless device needs to be operated in multiple frequency bands. Recent 5G applications and emergence of IoT need to be integrated with previous wireless standard. So, in next few years, we will need devices supporting both previous and future wireless standard. In presence of highly dense wireless application, it becomes very important to select different desired bands for operation simultaneously. Frequency band 1.2 GHz is most widely used in transmitter and receiver

P. Pandey (✉)
Department of Electronics Engineering, Rajkiya Engineering College, Sonbhadra, UP 231206, India
e-mail: prashanteceg@gmail.com

A. K. Pandey
Department of Electronics and Communication Engineering, BIT, Gorakhpur, UP 273212, India

R. K. Chauhan
Department of Electronics and Communication Engineering, MMMUT, Gorakhpur, UP 273208, India

B. Mishra and M. Tiwari (eds.), *VLSI, Microwave and Wireless Technologies*,
Lecture Notes in Electrical Engineering 877,
https://doi.org/10.1007/978-981-19-0312-0_36

system such as CCTV surveillance and general purpose application. 4.2 GHz band is widely used for satellite communication, so it is less dense; FCC is looking to open it for general wireless application with certain limitation. Advancement towards 5G application in sub-6 GHz band is now leading new application in H band around frequency band of 6.8 GHz.

Numerous methods have been proposed by researcher to achieve tri-band. In [1], pair of ring resonator resulted in degenerate mode. In [2, 3], stepped impedance-based resonator with tri-section proved to work as good filter. Tri-band filter is also obtained by defected ground structure integrated with SIR and stub loaded, in [4, 5]. Stub-loaded resonator with short-circuit end will also result good performance tri-band filter, in [6]. One-half wavelength resonator loaded with two identical quarter wavelength stub resulted tri-band operation in [7]. Quintuple-mode resonator having symmetrical structure and resulting in tri-band filter is analysed using even and odd method in [8]. Quadruple-mode square ring resonator having stepped impedance is analysed by even/odd mode analysis in [9]. Quad-mode stub-loaded circular resonator having four short-circuited line is analysed in [10]. T-shaped non-symmetrical stub-loaded resonator has given better results in tri-band bandpass filter in [10]. Four open-loop uniform impedance resonators resulting tri-band bandpass filter is explained with different parameters are explained in [11]. Frequency-selective surface based on coupled resonators having stub loading resulting improvement in performance of filter is analysed in [12].

In this work, a novel and compact tri-band microstrip bandpass filter having circular ring loaded with uniform impedance stub is proposed. By use of surface current distribution and even–odd mode theory, design is explained and verified.

2 Filter Design and Analysis

Circular ring-based resonator results in less size and multiple options for generation of various modes. Here, annular feed line is used so that better coupling is obtained, and both ends are feed with coupling angle of 90°. Symmetrical open-circuited stub and long-length short-circuited stub are loaded at circular ring. Novel filter design structure is shown in Fig. 1.

Even and odd-mode analysis will result three equivalent structures one for odd mode and two for even mode. This will result three resonating frequency. Odd mode about line of symmetry will result short circuit about that plane of symmetry. Even mode will result open circuit about line of symmetry.

Characteristics input admittance of resonator is given by Pozar [13]:

$$Y_{\text{in}} = Y_0 \frac{Y_L + jY_0 \tan\theta}{Y_0 + jY_L \tan\theta} \tag{1}$$

Fig. 1 Layout of novel SLCRR

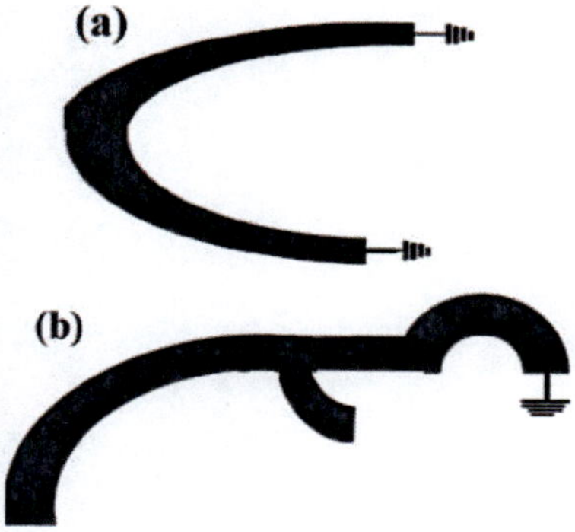

Fig. 2 a Odd-mode equivalent circuit. b Even-mode equivalent circuit

where Y_L is load admittance and Y_0 is characteristics admittance of strip line. We can put $\theta = \beta L$ and propagation constant $\beta = 2\pi/\lambda_g$ and λ_g = guided wavelength. L = electrical length of stub.

For odd-mode analysis, load is shorted as shown in Fig. 2a. So, $Y_L = \infty$ and $Y_0 = Y_1$. Put it in Eq. 1.

Input admittance in odd mode become

$$Y_{\text{in-odd}} = -j\frac{Y_1}{2}\text{Cot}\theta_1 \tag{2}$$

where $\theta_1 = \beta L_1 = \beta(\pi R_1)/2$.

At resonance, $Y_{\text{in-odd}} = 0$, and we will get odd-mode frequency.

In even-mode analysis of circuit, we are having two different paths available one short-circuited and other open-circuited, as shown in Fig. 2b. Now, for $Y_L = 0$ and $Y_L = \infty$, we will get two different admittance, and it will lead to two different resonant frequencies in even mode.

$$Y_{\text{in-even1}} = j\frac{Y_1}{2}\text{Tan}\theta_1 \tag{3}$$

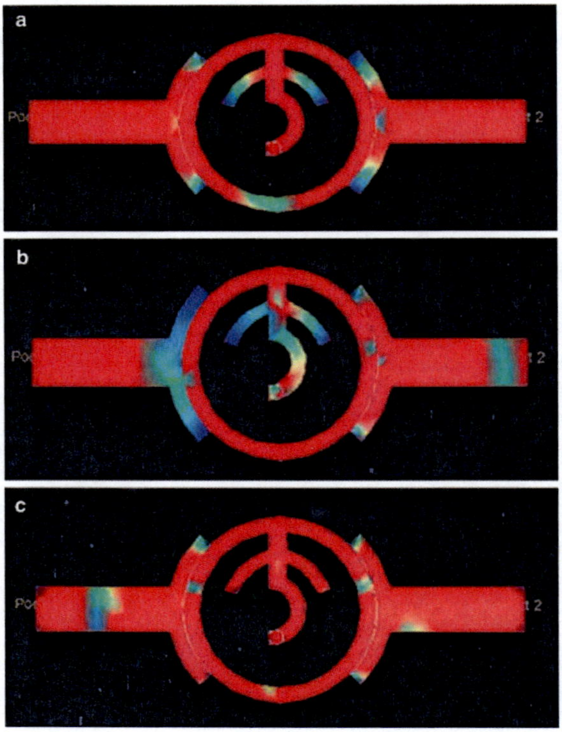

Fig. 3 **a** Current distribution at 1.27 GHz. **b** Current distribution at 4.25 GHz. **c** Current distribution at 6.9 GHz

$$Y_{\text{in-even2}} = -j\frac{Y_1}{2}\text{Cot}\theta_2 \tag{4}$$

where $\theta_2 = \beta L_1 + \beta L_2 = \beta(\pi R_1)/2 + \beta L_2$.

At resonance, $Y_{\text{in-even}} = \infty$, and we will get even-mode frequency.

Surface current distribution at all three resonating frequencies band is studied, and its behaviour is as expected. It is obtained at different frequencies, and resonant frequency behaviour is shown in Fig. 3a–c for frequency 1.27 GHz, 4.38 GHz and 6.92 GHz, respectively.

2.1 Dual-Band Behaviour

Circular ring resonator may work as dual bandpass filter by loading it with stub at plane of symmetry. Here, in two ways, stub is loaded, and result is obtained. With short-circuited stub as shown in Fig. 4 will result additional band at very low frequency around 1.2 GHz. By variation in its length, new pass band resonant frequency can be controlled. Loaded stub chosen shape will allow its length to be

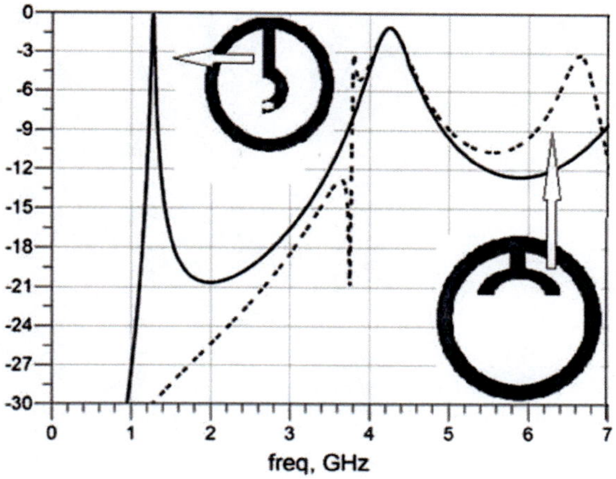

Fig. 4 S21 of circular ring resonator and short-circuit SLCRR

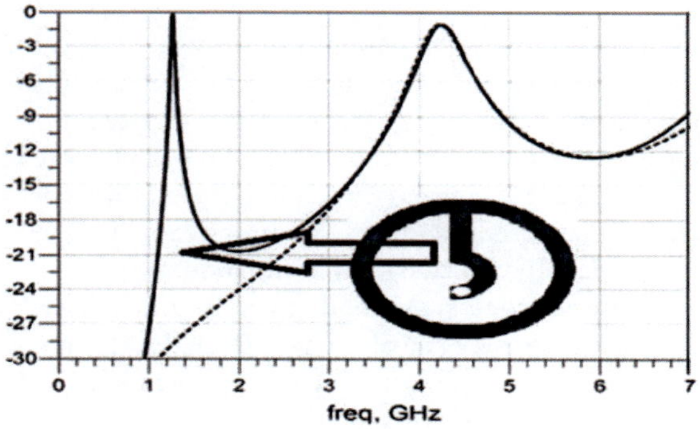

Fig. 5 S21 of short circuited and open-circuited SLCRR

long so that pass band at low frequency is obtained. Figure 5 shows other method of stub loading, and due to this, symmetrical open-circuited stub dual-band filter is obtained which will result new band at higher frequency around 6.5 GHz. Variation in length of this circular open-circuited stub operating frequency may be varied.

2.2 *Tri-band Behaviour*

If in a ring resonator, both open-circuited and short-circuited stubs are loaded; then, it will have band due to effect of both, and tri-band operation obtained is shown in Fig. 6. From Figs. 3, 4 and 5, working of tri-band resonator is very clear.

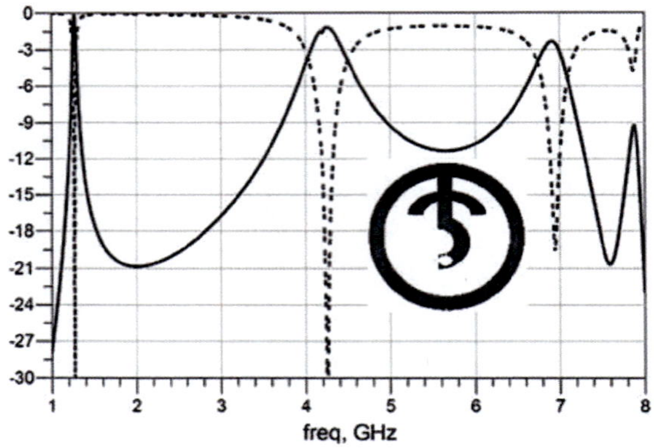

Fig. 6 S21 for tri-band novel resonating structure

3 Optimization of Design Parameter

Filter structure design parameters are varied so that proper matching of impedance can be achieved and band performance can be enhanced. Uniform impedance line is used in overall design of resonator. Proper tuning of loaded stub is performed. Open-circuited stub is in ring form and symmetrical about plane of symmetry. $L1$ is one-sided angle covered by semi-circle. Resonant frequency is obtained for various angle, and result is shown in given by graph in Fig. 7. It clearly indicates that third band is largely affected by this variation.

Short-circuited stub length given by $L2$ is varied, and it will largely affect first resonant frequency and also have affected by third resonant frequency. For variation of $L2$, variation in frequency band is shown in Fig. 7.

Based upon analysis shown in Fig. 7, structure shown in Fig. 8 is having different dimension shown in Table 1.

All above design parameters are calculated for microstrip substrate FR4 (dielectric permittivity 4.4) have both-sided Cu cladding. Effective permittivity is calculated by

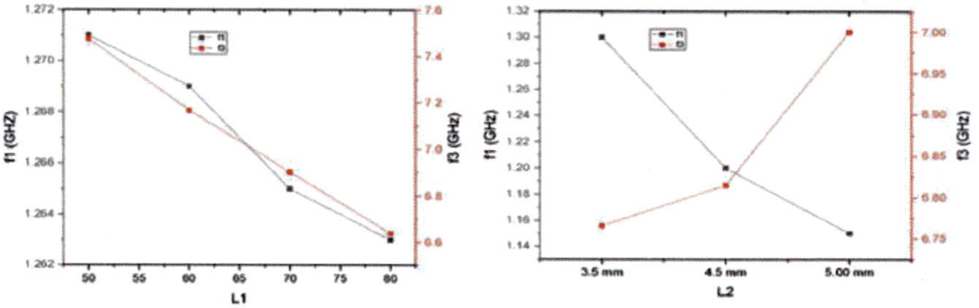

Fig. 7 Effect of variation in $L1$ and $L2$

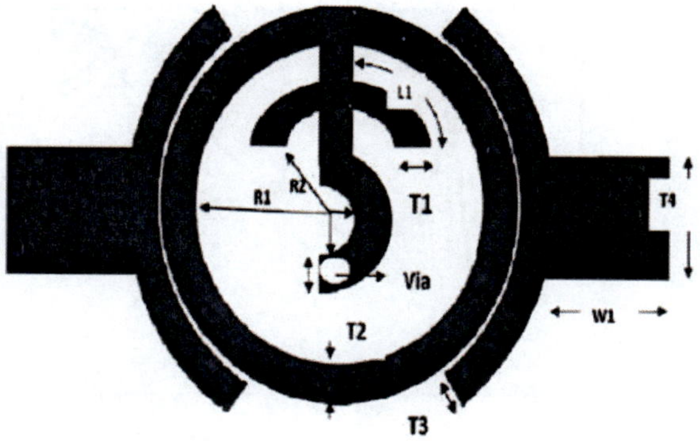

Fig. 8 Filter layout with dimension marking

Table 1 Final dimension of filter

Parameter	Dimension (mm)
Inner radius of outer ring $R1$	5
Inner radius of open stub $R2$	3.5
Inner radius of shorted ring $R3$	2.5
Length of open stub in angle (degree)	70
$T1$, $T2$ and $T3$ width of line	1
Via radius	0.4
Gaping of feed line and resonator	0.2
$L2$ is length from outer ring to inner ring	3.5
$W1$ is length of feed line	9.13
$T4$ is thickness of feed line	3.13

$$\xi_{\text{eff}} = \frac{1+\xi_r}{2} + \frac{\xi_r - 1}{2} \times \frac{1}{\sqrt{1+12\frac{h}{w}}} \tag{5}$$

where $w =$ width of stripline and $h =$ height of FR4.

Proposed filter design area is $15 \times 12\ \text{mm}^2$ or $0.10 \times 0.09{\lambda_g}^2$.

4 Result and Discussion

Proposed structure is fabricated using FR4 shown in Fig. 10 and tested using VNA-measured result, and simulated result is shown in Fig. 9. In proposed tri-band filter, first band for 3-dB range is 1.25 GHz–1.29 GHz (3.14%); second band for 3-dB range is 4.18 GHz–4.48 GHz (5.5%), and third band for 3-dB range is 6.84 GHz–7.00 GHz (2.33%). Insertion loss is 0.32 dB, 1.41 dB and 2.41 dB, whilst return loss

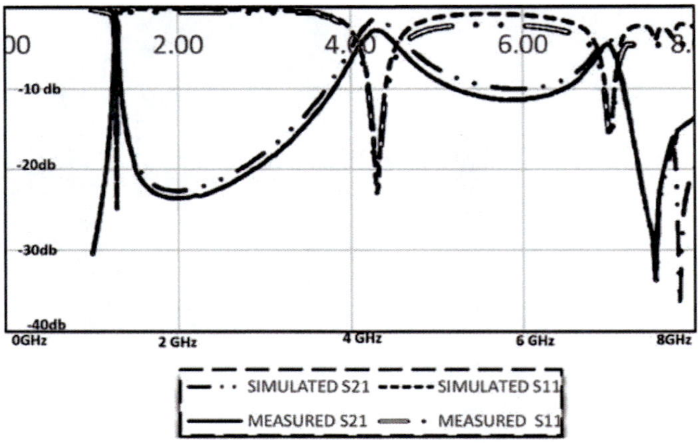

Fig. 9 Measured and simulated result

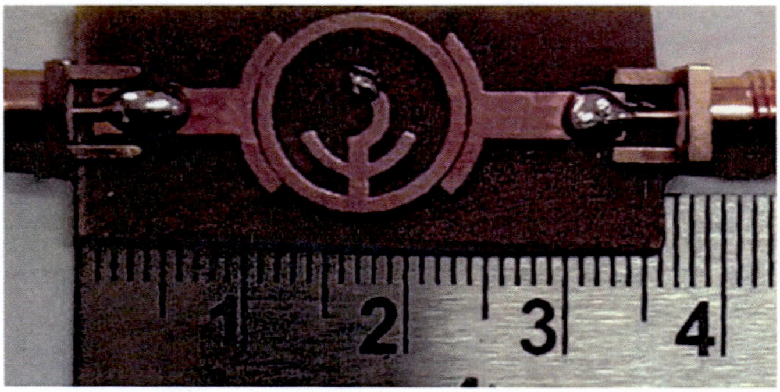

Fig. 10 Fabricate tri-band filter

is better than 22 dB, 20 dB and 15 dB, respectively. Surface current distribution at centre frequency of pass band is shown in Fig. 3.

Proposed novel filter design is compared with recent tri-band filter. We can clearly see that satisfactory filter performance is obtained in comparison with other filters. Design area has been significantly reduced as we can see in Table 2.

5 Future Direction

Uniform stripline in proposed structure can be converted in stepped stripline to obtain more bands. Few more stubs can be added in given design. Tuning diode can make

Table 2 Comparison between the proposed work with other latest tri-band filters

Reference	Freq. band (GHz)	3-dB FBW	IL (dB)	Design size ($\lambda_g \times \lambda_g$)
7	2.50/3.68/5.04	16.7/14.4/13.2	0.24/0.33/0.40	0.31 × 0.32
11	2.48/3.58/4.4	10/12.8/8	0.74/1.14/0.3	0.26 × 0.23
12	1.93/2.6/3.9	5/11/3	1.5/0.6/1.83	0.54 × 0.77
This work	1.27/4.38/6.92	3.14/5.5/2.33	0.32/1.41/2.41	0.10 × 0.09

this filter more effective. Design is simple and small in size, so it can be used in frequency-selective surface.

6 Conclusion

Novel stub-loaded circular ring resonator tri-band bandpass filter was presented in this paper. Resonant characteristics are studied by even–odd-mode theory and surface current distribution at different pass band frequency. Filter shows satisfactory bandpass performance and suitable for multiband wireless communication.

References

1. Luo S, Zhu L, Sun S (2011) Compact dual-mode tri-band bandpass filter using three pair of degenerate modes in a ring resonator. IEEE Trans Microwave Technol 59(5):1222–1229
2. Geschke RH, Jokanovic B, Meyer P (2011) Filter parameter extraction for tri-band composite split-ring resonator and filters. IEEE Trans Microwave Theory Tech 59(6):1500–1508
3. Haiwen L, Lei J, Zhao Y (2013) Tri-band microstrip bandpass filter using dual-mode stepped-impedence resonator. ETRI J 35(2):344–347
4. Ren LY (2010) Tri band bandpass filter based on dual plane microstrip/DGS slot structure. IEEE Microwave Wirel Compon Lett 20(8):429–431
5. Zhang XY, Quan X, Hu BJ (2010) Planner tri bandpass filter with compact size. IEEE Microwave Wirel Compon Lett 9(6):262–264
6. Lai X, Liang CH, Di H, Wu B (2010) Design of tri band filter based on stub loaded resonator and DGS resonator. IEEE Microw Wireless Compon Lett 20(5):265–267
7. Wang ZJ, Wang C, Kim NY (2018) Dual-/triple-wideband microstrip bandpaas filter using independent triple mode-stub-loaded resonator. Microw Opt Technol Letter 60(1):56–64
8. Zhang SF, Wang LT, Zhao SH, Wang ZP, Zhang X, Liang XL, Ji L (2019) Design of dual/tri-band BPF with controllable bandwidth based on a quintuple-mode resonator. PIER Letters 82:129–137
9. Ren B, Liu H, Ma Z, Ohira M, Wen P, Wang X, Guan X (2018) Compact dual band differential bandpass filter using quadruple-mode stepped impedance square ring loaded resonators. IEEE Access 6:21850–21858
10. Liu M, Xiang Z, Ren P, Xu T (2019) Quad-mode dual-band bandpaas filter based on stub-loaded circular resonator. EURASIP J Wireless Com Network 2019, 48
11. Weng M-H, Hsu CW, Lin Y, Tsai C-Y, Yang RY (2020) A simple method to design a tri-bandpass filter using open loop uniform impedence resonators. J Electromagn Waves Appl 34(1):103–115

12. Liao R, Wong SW, Li Y, Lin JY, Liu BY, Chen FC, Quan Z (2020) Quasi-elliptic band-paas frequency selective surface based on coupled stubs loaded ring resonator. IEEE Access 8:113675–11382
13. Pozar DM (2009) Microwave engineering. Wiley, Hoboken, NJ USA

Automatic Detection of Counterfeit and Genuine News Article

Vipula Yadav and Manish Kumar Srivastava

Abstract In the new years, the issue of fake news has developed a lot quicker. Web-based media has incomprehensibly changed its span and effect overall. On the one hand, it is modest and effectively accessible with fast portion of data draws more consideration of individuals to peruse news from it. Then again, it gives wide spread of fake news, which are only off-base data to mislead individuals or per users. Accordingly, Counterfeit News Location assumes an essential part to keep up vigorous on the Web and online media. In this paper, machine learning strategies are utilized to identify the believability of information dependent on the content substance. Furthermore, after that an examination is made to show that calculation utilized here is more solid and compelling as far as distinguishing a wide range of information. The algorithms applied in the work are Logistic Regression, decision tree, Gradient Boosting Classifier, and Random Forest Classifier. The outcomes were discovered to be promising. Here, we present a model which will precisely anticipate the validity of information. Our model outflanks existing model designs.

Keywords Social media · Machine learning · Fake news

1 Introduction

Nowadays, life has become truly pleasant and the people of the world need to thank the wide responsibility of the web advancement for transference and data sharing. There is no uncertainty that Web has made our lives less complex and more comfortable.

There is a development in mankind's set of experiences, and yet it out centers the line between true media and fabricated media. Today anybody can share anything-credible or not-and that can be devoured by the Internet. Tragically, counterfeit news accumulates a lot of consideration over the Web, particularly via online media stages. Subsequent to getting news, individuals do not reconsider prior to circling

V. Yadav (✉) · M. K. Srivastava
Department of CSE, MMMUT, Gorakhpur, India
e-mail: vipulayadav413@gmail.com

B. Mishra and M. Tiwari (eds.), *VLSI, Microwave and Wireless Technologies*,
Lecture Notes in Electrical Engineering 877,
https://doi.org/10.1007/978-981-19-0312-0_37

such mis-data to the most distant apocalypse. This sort of information disappears however not without doing the damage it proposed to cause. The given online media stages that assume a significant part in communicating counterfeit news incorporate Facebook, Twitter, WhatsApp, and so on. Numerous specialists accept that fake news issue might be tended to by the assistance of AI, deep learning, and machine learning. This is conceivable on account of numerous calculations have started to improve work on lots of classification problems (Image Recognition, email detection, Text classification, etc.) on the grounds that equipment is less expensive and greater datasets are unreservedly and effectively available.

There are different models which are utilized to give an accuracy range of 80–95% which includes Naive Bayes Classifier, decision tree, Semantic Model, Logistic Regression, Support Vector Machine, and so on. The considerations that were drawn in thought did not give much precision. The goal is here to enhance the correctness of identifying counterfeit article more than the outcomes that are on hand. A few calculations have been executed that can recognize the distinction between the counterfeit and genuine article with the better correctness. With the help of this calculation which will pass judgment on the fake articles story based on specific rules which are as per the following—jumbled sentences, spelling mistakes, changing into lower case, etc.

2 Related Work

- This paper distinguishes diverse media sources and dissects if the given news story is reliable. This paper furnishes with an instinct on characteristics of news story joined with an alternate substance types available.
- This paper anticipated the fake news using Naive Bayes classifier. This methodology was actualized as a product framework and tried on different dataset of Facebook and so forth, which gives an accuracy of 74%. This paper did not think about the punctuation errors, which prompts a low accuracy.
- This assessed diverse machine learning techniques and broke down their prediction rate. The accuracy of various predictive models which incorporates bounded decision tree, Gradient Boosting Classifier, and Support Vector Machine was recorded. The models are not most dependable which are assessed based on probability threshold.
- This paper examines about the fake news detection and techniques to apply them on different social networking platforms utilizing the classifier Naive Bayes. The information hotspots for news story are from online media locales like Facebook, Twitter, and so forth. The obtained accuracy is a little less as these site's data are not 100% reliable.
- This examines about preventing false information and gossip identification in the continuous. It utilizes novelty-based features and gets its information from Kaggle. The got precision is 74.5%. Misleading content and uncredible sources are not considered here which prompt to low accuracy.

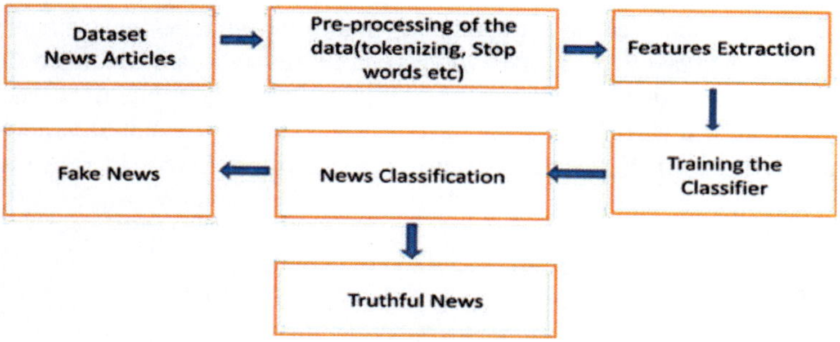

Fig. 1 Workflow of model

- This perceive the fake news through various techniques. The precision achieved is 76% by the utilization of linguistic model. It was conceivable to accomplish the higher precision if predictive models were to be used.
- To distinguish counterfeit news utilizing different machine learning models. The ML models talked about here are Naive Bayes classifier and SVM. No particular precision was noted, and they were utilized for the conversation purpose.
- To perceive whether the given tweets are dependable. Naive Bayes classifier, decision trees, SVMs, and neural networks are the executed machine learning models. The most noteworthy F1 score accomplished is 0.94, with the assistance of both tweet and client highlights. Higher accuracy can be accomplished by bringing non-valid news into account (Fig. 1).

3 Dataset and Preprocessing

3.1 About Dataset

One of the most troublesome issues to unwind in ML has nothing to do with intense estimations. It is the problem of getting the right dataset in the right way. Subsequent to getting right data, it suggests gathering or separating the data that relates with the outcomes which should be anticipated. The datasets ought to be coordinated with the difficulty which is being expected to clarify. In the event that the right data is not introduced by then, the undertakings need to return to the data collection stage.

Choosing the correct dataset for ML is basic to make the model useful with correct methodology. In any event, choosing the correct quality and amount of information is likewise effortful undertaking. There are not many standards which should be followed for ML on enormous data.

On the Internet, there are part of unreservedly and openly accessible datasets for counterfeit news classification. In this undertaking, the utilized dataset has been taken from kaggle.com. Two diverse datasets have been decided for this undertaking, one is for gathering counterfeit information and another for genuine news. Counterfeit news dataset contained 23,481 news while genuine news dataset contained 21,417 news. They have four diverse section alongside these lines. The names of the sections are "title," "text," "subject," and "date" referenced as 0/1; 0 for counterfeit news and 1 for genuine news. On these datasets, the diverse classification models were trained.

3.2 Preprocessing

Preprocessing is a significant advance before the information is prepared for analysis. The noise-free corpus has a diminished size of the example for including extraction which brings about increased accuracy. Data preprocessing is a cycle of delivering the inconsiderate information and making it reasonable for machine learning model. It is the first and key advance while making the AI model. When making machine learning project, it is not generally a case that the information is perfect and arranged. And keeping in mind that doing any activity with information, it is important to utilize clean and arranged information. For this reason, we use information preprocessing task.

Preprocessing is fundamental for improving results from the used model in ML project. The setup of the data should be in lawful manner. Other viewpoint is that the dataset should be orchestrated so more than one ML and deep learning calculations are executed in one instructive list, and out of them best one is chosen. This will assist us with reducing the size of the real information by taking out the immaterial data that exists in the information. The information which is utilized was in CSV format and required preprocessing. All instances were appropriated to prepare: test set in proportion 3:1. Each example relates to another article title and text. NLTK in python was utilized to tokenize the title and text. Eliminating the stop-words helped in Lemmatization.

Text Preparation Social media information is exceptionally chaotic; the greater part of them are casual with errors, slangs and awful sentence structure, and so forth. To accomplish better results, it is important to clean the information before it very well may be utilized for predictive modeling. For this reason, essential preprocessing was done on the dataset. It comprises various advances—

- Convert it into lower case: First step was to change over all the content into lower case, just to keep away from various duplicates of the equivalent words. For example, "Subject" and "subject" is taken as two different words.
- Elimination of Punctuations: Punctuations does not have a lot of significance while treating the content information. By eliminating them, the size of in general content can be reduced.
- Elimination of Stop-words: Stop-words are the most ordinarily seeming utilized words in the corpus. Some stop-words are a, an, for, at, and so forth. They are utilized to characterize the structure of a book, not the specific circumstance.

```
: def wordopt(text):
      text = text.lower()
      text = re.sub('\[.*?\]', '', text)
      text = re.sub("\\W"," ",text)
      text = re.sub('https?://\S+|www\.\S+', '', text)
      text = re.sub('<.*?>+', '', text)
      text = re.sub('[%s]' % re.escape(string.punctuation), '', text)
      text = re.sub('\n', '', text)
      text = re.sub('\w*\d\w*', '', text)
      return text
```

Fig. 2 Text preparation

They may degrade the performance. Thus, they are taken out from the preparation information as a piece of text cleaning.

- Tokenization: It is the way toward changing over delicate information into non-touchy information called "tokens." In straightforward words, we can say that it is a cycle of separating the content into an arrangement of words or gathering of words, similar to bigram, trigram, etc.
- Lemmatization: It is the way toward changing over the words into its promise root. With the assistance of a vocabulary, it performs analysis to get the root word (Fig. 2).

3.3 *Train and Test Set Splitting*

To make an important preparing set, the issues should be comprehend for which it is being agreed to. For instance: what will the ML model do and what sort of yield is normal. ML works with the two data assortments, Training and Testing. Every one of the two need to randomly test a greater variety of data. The principle subset which is being utilized is the Training set, which contains the vast majority of the information, and it is greatest in these two. Also, the other set, i.e., test set has less number of information which is utilized subsequent to preparing the model. Also, if the exact outcomes are not met, re-visitation of the preparation set investigates the misstep made. Taking the right dataset would not make any sort of issues, and the model will work easily. In this work, the dataset is divided into the train set and test set which is in the proportion of 3:1. The dataset is isolated into two subsets (Fig. 3)

- Training Set: a set that used to teach a model
- Test Set: a set that used to test the trained model

```
: x_train, x_test, y_train, y_test = train_test_split(x, y, test_size=0.25)
```

Fig. 3 Splitting of training and testing dataset

4 Evaluation of Models

4.1 Logistic Regression

Logistic relapse is a grouping calculation which is utilized to appoint any occasions to a distinct ordering of classes. It is utilized to foresee the likelihood of event of an occasion in discrete structure, for example, 0/1, valid/invalid, and Yes/No. Dissimilar to linear regression which yields persistent number qualities, logistic regression changes its yield using the determined Sigmoid ability to store a probability regard which would then have the option to be planned to in any event two distinct classes. The linear regression model uses gradient descent to converge into the ideal arrangement of weights for the training set [1]. The hypothesis which is utilized is the Sigmoid function to construct the model:

$$h(x) = \text{sigmoid}(wx + b)$$

$h(x) = \text{sigmoid}\ (wx + \text{b})$.
"Here, $w =$ weight vector,
$x =$ feature vector,
$b =$ bias."
$\text{sigmoid}(\ z\ ) = 1/(1 + e^{(-z)}\ )$.

4.2 Decision Trees

A typical tree incorporates root, branches, and leaves. The same formation is continued in decision tree. It has root hub, branches, and leaf nodes. Testing an aspect is on each inner hub, the result of the test is on branch, and class name therefore is on leaf hub [1, 2]. A root hub is parent, everything being equal, and as the name recommends, it is the highest hub in tree. It is a tree where each node denotes a component (characteristic), each connection (branch) denotes a choice (rule) and each leaf denotes a result (absolute or proceeds with esteem) [2]. As decision trees mirror the human level reasoning translations, the entire thought is to make a tree like this for the whole information and cycle a solitary result at each leaf (Fig. 4).

These algorithms are very powerful [3] in that they give rules of grouping. Near to this, it has two or three inadequacies, one of which is the arranging of all mathematical credits when the tree chooses to part any hub. Such split on sorting all numerical attributes becomes expensive, i.e., efficiency or running time and memory size, especially if decision trees are set on data and the size of which is huge, for example, it has increased number of occurrences.

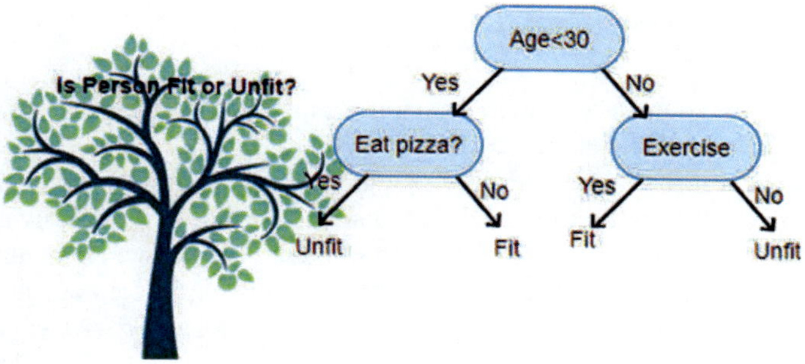

Fig.4 Decision tree

4.3 Gradient Boosting Classifier

This classifier comes under the classification of boosting techniques, which recursively gains from every one of the frail learner to assemble a strong model. This could improve Regression, categorization, and Positioning. The expression "Gradient" in Gradient Boosting classifier to the way that you have at least two subsidiaries of a similar capacity. Gradient Boosting is an iterative functional gradient calculation, for example a calculation which limits a loss function by iteratively picking a capacity that focuses to the negate gradient, a frail hypothesis. It has three primary parts

- Loss Function—"The job of the loss function is to estimate how good the model is at making predictions with the given data. This could vary depending on the problem at hand."
- Weak Learner— "A weak learner is one that classifies our data but does so poorly, perhaps no better than random guessing. In other words, it has a high error rate. These are typically decision trees." [4]
- Additive Model—"This is the iterative and sequential approach of adding the trees (weak learners) one step at a time. After each iteration, we need to be closer to our final model. In other words, each iteration should reduce the value of our loss function" [4] (Fig. 5).

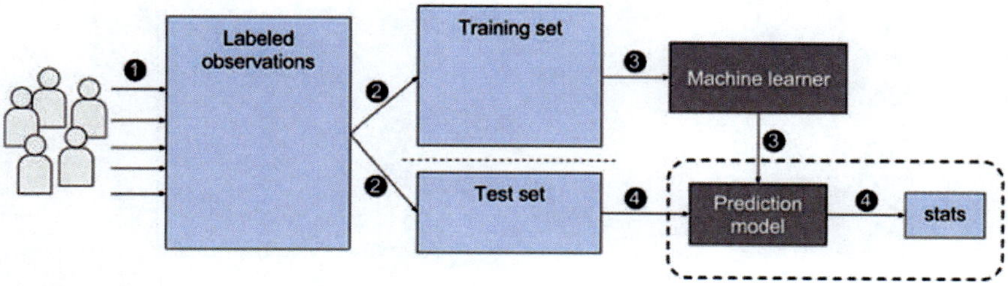

Fig. 5 Steps for gradient boosting

4.4 Random Forest Classifier

Random Forest as characterized in [5] is a non-exclusive guideline of classifier blend that utilizes L tree-organized base classifiers $\{h\ (X, \Theta_n), N = 1, 2, 3, \ldots L\}$, where X denotes the input information and $\{\Theta_n\}$ is a group of indistinguishable and dependent distributed arbitrary vectors. Each decision tree is made by arbitrarily choosing information from the accessible arrangement of data. For instance, Irregular Forest for each decision tree can be worked by haphazardly testing an element subset, and additionally by the arbitrary inspecting of a training data subset for each decision tree.

In a Random Forest, the features are arbitrarily chosen in each decision split. The relationship between trees is less by haphazardly choosing the highlights which improve the predictive ability and outcomes in greater efficiency. As such, the benefits of Random Forests are [5]:

- Overcoming the issue of over-fitting
- In preparing information, they are less delicate to anomaly data
- Parameters can be set effectively and along these lines, dispenses with the need of pruning the trees
- Variable significance and precision will produced automatically (Fig. 6)

It works in four steps:

1. Select irregular examples from a given dataset.
2. Construct a choice tree for each example and get a forecast result from every choice tree.
3. Perform a decision in favor of each anticipated result.

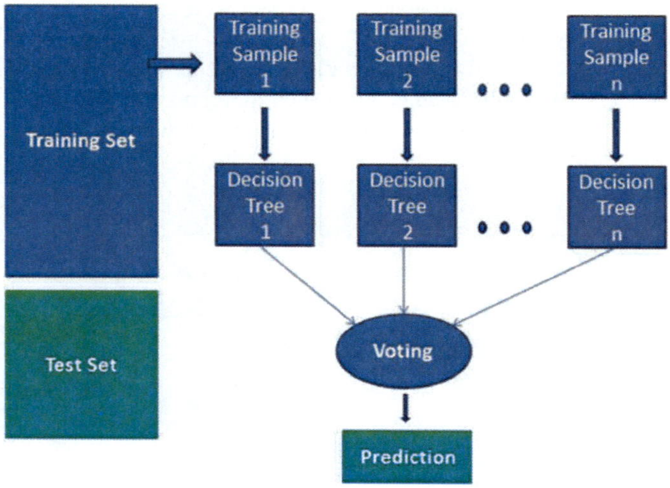

Fig. 6 Steps of random forest

4. Select the expectation result with the most votes as the last prediction.

5 Result and Conclusion

Using the previously mentioned calculations, Strategic Relapse, Choice Tree, Slope Boosting Classifier, and Arbitrary Timberland, the accompanying precision has been accomplished (Table 1).

The maximum accuracy of 99.55% on the given training set was attained by Gradient Boosting classifier, whereas the previous models, which consist of only decision tree, attained low accuracy than the accuracy achieved here (Figs. 7, 8, 9 and 10).

The first algorithm used for classification was Logistic Regression which gives 98.85% accuracy while decision tree has 99.45% and random forest achieves 99.15% accuracy. Gradient Boosting Classifier performs better among all of the algorithms on the given training set (Tables 2, 3, 4, and 5).

After performing testing on the given dataset, using the given different ML techniques, we check the result whether the classifier predicts the news correctly or not. Hence, we perform manual testing on the data and the given classifiers categorize the news accurately (Fig. 11).

The predictions obtained by various ML models on the randomly selected news article are given (Fig. 12).

Table 1 Result for various used models

	Logistic regression	Decision tree	Gradient boosting	Random forest
Accuracy	98.85%	99.45%	99.55%	99.15%

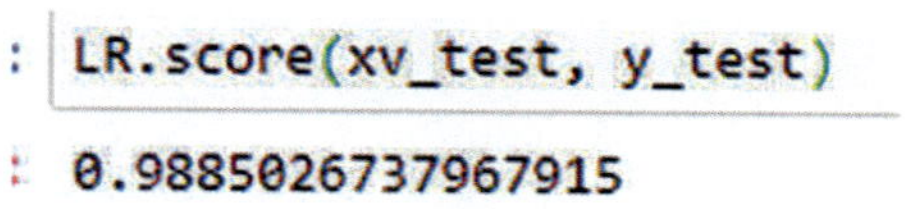

Fig. 7 Accuracy of logistic regression

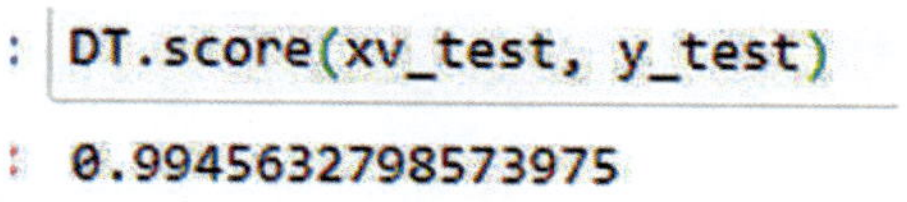

Fig. 8 Accuracy of decision tree

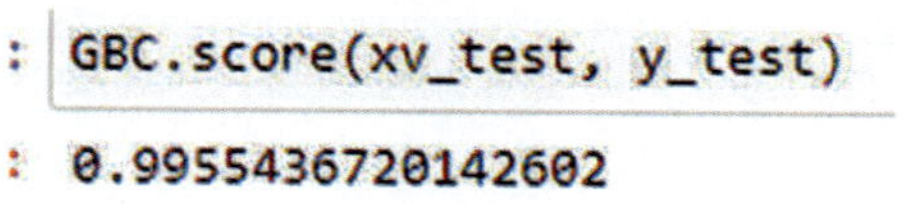

Fig. 9 Accuracy of gradient boosting

```
: RFC.score(xv_test, y_test)

: 0.9915329768270945
```

Fig. 10 Accuracy of random forest

Table 2 Confusion matrix for logistic regression

Predicted/Actual	P("Fake")	N("Real")
P("Fake")	5558 (True positive)	70 (False positive)
N("Real")	60(False negative)	5537(True negative)

Table 3 Confusion matrix for decision tree

Predicted/Actual	P("Fake")	N("Real")
P("Fake")	5592 (True positive)	29 (False positive)
N("Real")	33(False negative)	5571(True negative)

Table 4 Confusion matrix for gradient boosting classifier

Predicted/Actual	P("Fake")	N("Real")
P("Fake")	5557 (True positive)	28 (False positive)
N("Real")	23(False negative)	5616(True negative)

Table 5 Confusion matrix for random forest

Predicted/Actual	P("Fake")	N("Real")
P("Fake")	5575 (True positive)	47 (False positive)
N("Real")	49(False negative)	5554(True negative)

```
: news = str(input())
  manual_testing(news)

  Paul Craig RobertsIn the last years of the 20th century
```

Fig. 11 Manual testing

Fig. 12 Result of testing

```
LR Prediction: Fake News
DT Prediction: Fake News
GBC Prediction: Fake News
RFC Prediction: Fake News
```

References

1. Gershman A, Meisels A, Lüke KH, Rokach L, Schclar A, Sturm A (2010) A decision tree based recommender system. In: IICS 2010 Jun 3, pp 170–179
2. SD Jadhav HP Channe 2016 Efficient recommendation system using decision tree classifier and collaborative filtering Int Res J Eng Technol 3 2113 2118
3. Ben-Haim Y (2010) A streaming parallel decision tree algorithm" Elad Tom-Tov
4. Kurama V (2022) Series: Ensemble Method [Online]. "https://blog.paperspace.com/gradient-boosting-for-classification/"
5. "Introduction to Decision Trees and Random Forests, Ned Horning; American Museum of Natural History's"

Twitter Sentiment Analysis Using Machine Learning

Ayesha Quazi and Manish Kumar Srivastava

Abstract Millions of tweets are created every day on diverse issues. Topical variety in substance demands space autonomous responses for analyzing Twitter sentiments. Twitter sentiment analysis offers associations capacity to screen public feelings toward the occasions and product identified with them progressively. The initial step of the sentiment analysis is the pre-handling of Twitter data. Scalability is another issue while overseeing enormous measure of tweets. This paper presents a supervised method for looking at tweet sentiments. Sentiment analysis by means of electronic media, for instance, Twitter has gotten a crucial and testing task. Because of the attributes of such information such as tweet length, spelling mistakes, abbreviations, and special characters, the sentiment analysis task in such an environment requires a non-conventional approach. Twitter posts are generally short and generated constantly by public and very well-suited for opinion mining. These messages can be elegant as containing either positive or negative sentiment on the basis of specific viewpoints with respect to a term-based query. The previous investigations of sentiment classification are not very conclusive about which features and supervised classification algorithms are good for designing precise and efficient sentiment classification system. In this paper, I mostly centered around existing development of Twitter dataset with approaches like sentiment analysis algorithm using machine learning. The challenging arrangement would be analyzing abbreviations, emoticons, hashtags, thus, so on.

Keywords Sentiment analysis · Twitter · Supervised machine learning

1 Introduction

Twitter, with more than 319 million monthly dynamic clients, has now become a goldmine for associations and affiliations who have a solid political, social, and

A. Quazi (✉) · M. K. Srivastava
Department of Computer Science and Engineering, MMMUT, Gorakhpur, India
e-mail: ayeshaqazi50@gmail.com

B. Mishra and M. Tiwari (eds.), *VLSI, Microwave and Wireless Technologies*,
Lecture Notes in Electrical Engineering 877,
https://doi.org/10.1007/978-981-19-0312-0_38

financial interest in keeping up and upgrading their advantages and notoriety. Sentiment analysis furnishes these associations with the capacity to looking over different web-based media destinations in genuine time. Sentiment analysis is a programmed cycle to decide if a content portions contains objective or obstinate substance, and it can besides decide the content's sentiment polarity. The objective of Twitter sentiment arrangement is to naturally decide if a tweet's estimation extremity is negative or positive. The fundamental thought of sentiment analysis is to detect the polarity of text documents or short sentences and classify them on their assumption. A point becomes moving assuming an ever-increasing number of peoples are contributing their assessment and decisions, along these lines making it an important wellspring of online perception. Social media is a vital source of information where large amount of textual data are found. Among them, Twitter is one of the web-based media webpages which empower clients to tweet, retweet, remark, and adjust the content.

Basically, Twitter is an online journal where user tweets instant reaction and thought about that present circumstances and different challenges. For the sentiment analysis, I have used Twitter instant messages comprising a limit of 140 characters, delivering not just the instant message but hashtags, usernames, and URLs. Twitter messages regularly join abbreviations, contractions, and shortenings and may contain shortened forms, truncated messages, and slang. "Sentiment analysis is a natural language processing techniques which is handling strategies to evaluate a communicated assessment or feeling inside a choice of tweets" [1].

2 Literature Survey

This section summarizes some of the scholarly and research works in the field of artificial intelligence and information mining to analyze sentiments on the Twitter and preparing prediction model for various applications. As the available social platforms are shooting up, the information is becoming vast and can be extracted to turn into business objectives, social campaigns, marketing, and other promotional strategies as explained by Alexander Pak and Patrick Paroubek [2].

The benefit of social media to know public opinions and extract their emotions is considered by Jose et al. who explained how Twitter gives advantage politically during elections [3]. The benefit of social media to know public opinions and extract their emotions is considered by authors Ramteke et al. [4].

Ranganathan and Tzacheva proposed a way to deal with naturally recognizing feelings on Twitter messages that explore attributes of the tweets and the author's feeling using SVM LibLinear model. A total of 520K tweets was collected as database. Tweet To Sparse Feature Vector channel in Weka Emotional tweets bundle was used to extract highlights. Each tweet in the repository was commented on with the relating emotional based on the weightage computed using the extracted features. The results showed that the precision of the SVM classifier was 98% [5].

Barbosa and Feng [6] planned a two-stage programmed sentiment analysis strategy for arranging tweets. They ordered tweets as evenhanded or abstract, and

Fig. 1 Recent five tweets

	Tweets
0	It's great to see President Biden elevate scie...
1	And while COVID-19 will rightfully continue to...
2	The President's commitment to reengage with th...
3	With Americans across the country working toge...
4	I look forward to working with President @JoeB...

afterward in second stage, the emotional tweets were named as positive or negative. The component space utilized included retweets, hashtags, connection, accentuation, and shout marks related to highlights like earlier extremity of words and POS [6].

Xia et al. [7] utilized a gathering structure for sentiment grouping which is gotten by joining different capabilities and order methods. In their work, they utilized two sorts of capabilities and three base classifiers. They applied gathering approaches like fixed combination, weighted combination, and sentiment classifier blend for assumption characterization and acquired better accuracy.

Akilandeswari and Jothi proposed a scoring model joining language and non-language highlights to discover the slant polarity of Twitter messages. The language highlights involve the content which depicts a subject either in a "positive", "negative", or "neutral" way. The non-language highlights comprise the images utilized by the clients of Twitter like emojis and abbreviated words. A sum of 750 tweets was gathered and physically delegated positive, negative, and neutral [8].

3 Methodology

3.1 Twitter Tweets

Figure 1 shows recent tweets from "Bill Gates" Twitter account which is publicly available.

3.2 Preprocessing of Twitter Dataset

Preprocessing of data is done by following different steps like tokenization, cleaning the data, removing stop words, noise removal, and classification.

1. **Tokenization**

   Tokenization is a cycle of parting a paragraph or a sentence into tokens prior to changing it.
2. **Cleaning the data**

In this technique, cleaning of the data is done by eliminating most commonly occurring words, targets, symbols, URLs, and other non-English words, numbers, and all the unnecessary raw data used in the tweet.

3. **Removing stop word**

   In this progression, stop words are taken out. Stop words are generally the regular words like is, the, are, you, etc.

4. **Classification**

   In this step, classification of subjectivity and polarity is done by using supervised-based learning approach. Subjectivity is the subjective part of tweet which does not include any good or bad words. Polarity is the sign of the score which lies between the -1 and 1.

4 Classification Techniques

Sentiment analysis alludes to the overall strategy to separate extremity and subjectivity from interpretation direction, which alludes to the toughness of terms and extremity text and expressions [8]. There are two primary methodologies for separating assessment consequently which are lexicon-based methodology and ML-based methodology [8–12].

4.1 Lexicon-Based Approach

Lexicon-based strategies utilize predefined rundown of terms where each term is related with a particular assumption [11]. The lexicon-based techniques shift as per the setting where they were made and include computing direction for an archive from the semantic direction of writings or expressions in the documents [10]. Aydogan and Akcayol [13] additionally express that a vocabulary supposition is to recognize word-conveying assessment in the corpus and afterward to foresee assessment communicated in the content [13]. Annett and Kondrak have indicated the dictionary strategies which have an essential worldview which are [12]:

i. Preprocess each phrase, tweet, and retweet by eliminating the marks, such as full stop, comma, and brackets.
ii. Start off an all-out extremity result (r) equivalent "$0 \rightarrow r = 0$".
iii. Examine if token is available in a word reference, then

    If token is "positive", r will be "positive".
    If token is "negative", r will be "negative".

iv. Look at the all-out polarity result of tweet

    If "$r >$ threshold", tweet post as "positive"

If "$r <$ threshold", tweet post as "negative"

However, Taboada et al. featured one preferred position of learning-based technique, which is that it can adjust and make prepared models for explicit purposes and settings [10]. Conversely, an accessibility of named information and consequently the low appropriateness of the strategy for new information which causes marking information may be exorbitant or even restrictive for certain assignments [12].

4.2 Machine Learning-Based Approach

Machine learning techniques in the classification of sentiment depend on the use of well-known machine learning technology on text data. The classification of the sentiment based on machine learning can be categorized primarily into supervised and unsupervised methods of learning [13]. The methodology requires named information to prepare classifiers [11]. This methodology becomes obvious that parts of the neighborhood setting of a phrase should be considered. Annett and Kondrak demonstrated an essential worldview for making a component vector [9]:

i. Appeal a grammatical feature tagger to each phrases, tweet.
ii. Assemble all the descriptive word for whole phrase, tweet.
iii. Put together a mainstream term set made out of the top N adjectives
iv. Negotiate the entirety of the tweets in the test set to make the following:

- Number of positive words
- Number of negative words
- Existence, non-existence, or recurrence of each word

However, Goncalves et al. have referenced the limit of machine learning-based approach way to deal with more reasonable for Twitter than the lexical-based strategy [11]. In addition, Annett and Kondrak expressed that the machine learning strategies can produce a fixed number of the most routinely occurring famous terms which appointed a number an incentive for the recurrence of the word in the Twitter [9] (Fig. 2).

4.3 Natural Language Processing (NLP)

NLP methods depend on machine learning and particularly factual realizing which utilizes an overall learning calculation joined with an enormous example, a corpus, of information to gain proficiency with the guidelines [14]. Sentiment analysis has been taken care of as a natural language processing indicated NLP, at numerous degrees of granularity. Beginning from being an archive level order task [15], it has been taken care at the sentence level [16] and all the more as of late at the expression level

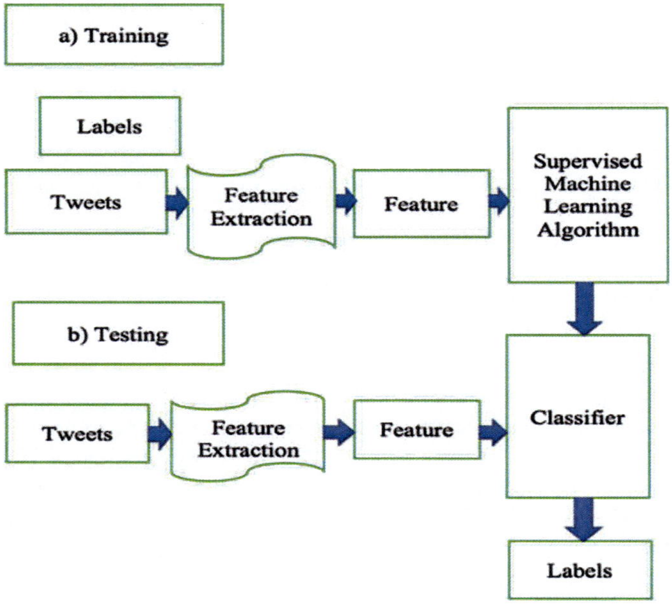

Fig. 2 Sentiment analysis using supervised machine learning algorithm

[17]. However, PCs do not perceive the human words. NLP is utilized to decipher confounded human dialects and cause PCs to comprehend humans.

4.4 Artificial Neural Network (ANN)

"Artificial neural network (ANN) or known as neural network is a mathematical technique that interconnects group of artificial neurons" [17]. It will handle data utilizing the associations way to deal with calculation. ANN is utilized in detecting the connection among information and yield or to discover designs in data [18].

4.5 Support Vector Machine (SVM)

Support vector machine is fundamentally utilized for text arrangement. Support vector machine is to distinguish the assumptions of tweets [19]. Pak and Paroubek along with Saif et al. expressed SVM can concentrate and break down to acquire up to 70–81.3% of precision on the test set [20, 21]. Xia et al. gathered preparing information from three diverse Twitter feeling discovery sites which predominantly utilize

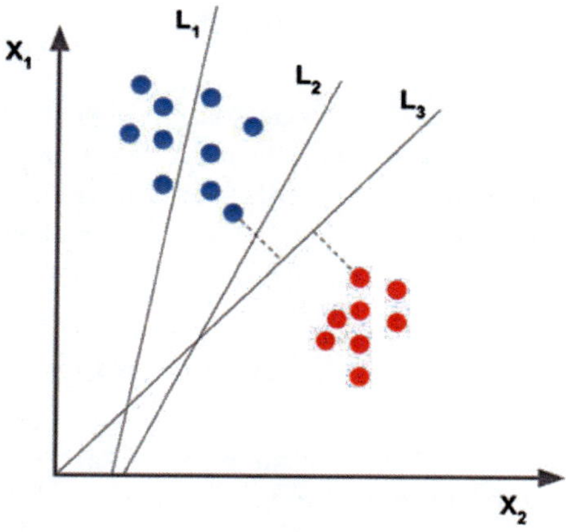

Fig. 3 Support vector machine

some pre-assembled notion vocabularies to mark each tweet as good or negative [22] (Fig. 3).

5 Challenges in Sentiment Analysis

Analysis is a difficult undertaking. Following are the portion of the challenges looked in the sentiment analysis of Twitter [7].

5.1 Recognizing Subjective Pieces of Text

Subjective parts address sentiment-bearing substance. A similar text can be treated as emotional in first case and a target in some other case. This draws up hard to recognize the emotional bits of words. For example, the word crude can be utilized as an assessment in one model, while it can be totally unbiased in the other model.

5.2 Sarcasm Detection

Sarcastic sentences express negative assessment on an objective utilizing positive words in extraordinary manner. For example, "Nice scent, You should shower in it". The statement consist of just positive words; however, it communicates a negative sentiment.

5.3 Entity Recognition

There is a need to isolate out the content about a particular element and afterward investigate estimation toward it. For example: "I hate Microsoft, but I like Linux".

A basic sack-of-words approach will name it as impartial. In any case, it conveys a particular estimation for both the elements present in the statement.

5.4 Applying Sentiment Analysis to Facebook Messages

There has been not so much effort on assumption examination on Facebook information fundamentally because of different limitations by Facebook diagram programming interface and security strategies in getting to data.

6 Results and Discussion

I have used Twitter dataset of "Bill Gates", the founder of Microsoft which is publicly available on Twitter website. Sentiment analysis of 1000 tweets which was first labeled using various feature extraction technique. Then, framework is applied on those with different columns to make it more relevant to understand. Those frames are tweets, subjectivity, polarity, and analysis. Further, the unmistakable machine learning strategies prepare the set of information with highlight vectors which makes accessible the extremity of the content. And on the basis of polarity which lies between 0 and 1, analysis is determined between positive, negative, and neutral (Fig. 4).

There are a considerable number of tweets with "Bill Gates" emotions which are attributed to various governmental and political issues that have been shared by the authorized user. Those tweets are organized in a bigger and brighter manner which basically means for visualization purpose. Visualization is done using WordCloud or TextCloud. As it can be seen from Fig. 5 that words like COVID-19, will, pandemic, people, health, work, vaccine, world, and need are so frequently used in the tweet of "Bill Gates".

The words like COVID-19, vaccine, work, and live are somehow correlated with emotions of joy and relief as we all are aware about the 9 months of lockdown period and other some of the words like pandemic, will, and health are correlated with emotions of fear and sadness. However, all the tweets were analyzed, and there were more number of positive tweets, few are negative tweets, and some of them are neutral tweets.

As the graph shows, after preparing analysis is done on these datasets, where 72% of positive tweets and 11.5% of negative tweets are finished by "Bill Gates" account.

	Tweets	Subjectivity	Polarity	Analysis
0	It's great to see President Biden elevate scie...	0.750000	0.800000	Positive
1	And while COVID-19 will rightfully continue to...	0.541667	0.525000	Positive
2	The President's commitment to reengage with th...	0.600000	-0.400000	Negative
3	With Americans across the country working toge...	0.512500	0.212500	Positive
4	I look forward to working with President and ...	0.000000	0.000000	Neutral
...	...	...	...	...
195	: The world must work together to slow the cor...	0.466667	-0.100000	Negative
196	Roger's foundation partners with local NGOs...	0.279167	0.225000	Positive
197	I'm excited to team up with again for the Mat...	0.750000	0.375000	Positive
198	As we conclude our foundation's second decade ...	0.166667	0.166667	Positive
199	Thank you, Sue, for all of your contributions ...	1.000000	0.400000	Positive

Fig. 4 Dataset description using framework

Fig. 5 WordCloud of tweets

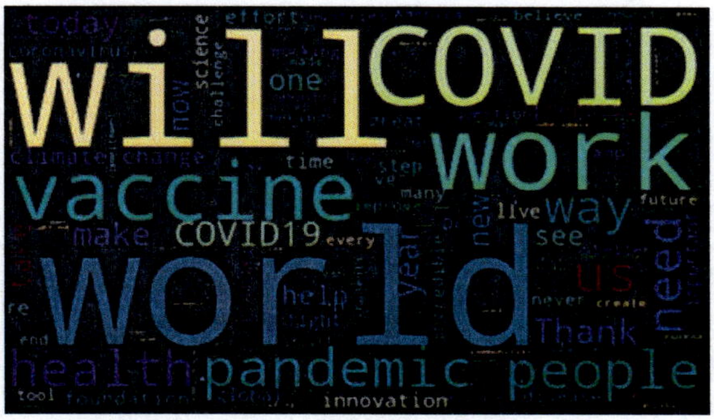

Those blue dots in the diagram beneath proposes the assortment of positive, negative, and unbiased information. Along these lines, the right outcome was discovered utilizing supervised machine learning approach (Fig. 6).

7 Conclusion

Twitter is a demandable micropublishing content to a blog administration which has been worked to find what is going on at any snapshot of time and any place in the world. The proposed framework closes the assessments of tweets which are evoked from Twitter. The trouble increments with the variation and entanglement of sentiments communicated. Product surveys and so forth are generally simple. Books, pictures, workmanship, and music are more troublesome. We can likewise actualize highlights like emojis, balance, and internationalization as they have lately

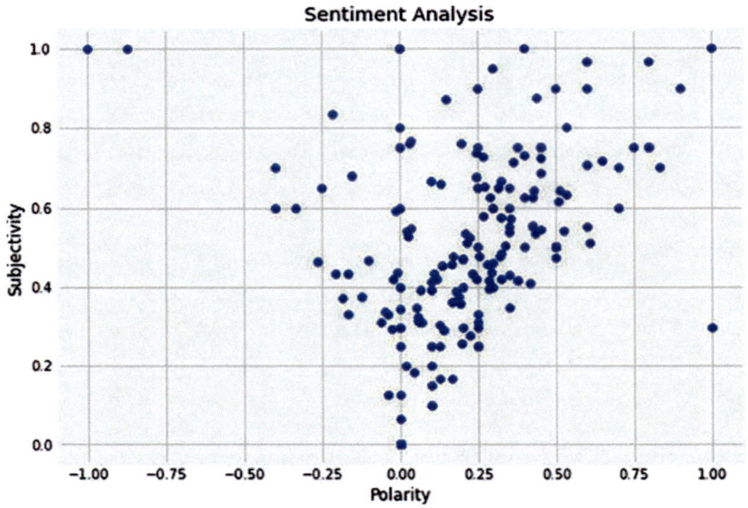

Fig. 6 Graph representing results obtained for supervised machine learning algorithm

become an enormous piece of the Internet. The existing procedures incorporate a machine-based learning approach and lexicon-based methodology together for some assessment measurements which give more precise and more productive outcome for breaking down a notion with natural language preparing technique.

In this paper, diverse techniques for Twitter sentiment analysis methods were discussed, including machine learning and lexicon-based approaches. Research results exhibited that machine learning procedures, for instance, the SVM and ANN created the best exactness, especially when multiple features were included. As an outcome, application will class supposition into positive, negative, and unbiased opinion.

References

1. Carpenter T, Way T (2010) Tracking sentiment analysis through Twitter. In: ACM computer survey. Villanova University, Villanova
2. Pak A, Paroubek P (2010) Twitter as a corpus for sentiment analysis and opinion mining. In: Proceedings of the seventh international conference on language resources and evaluation (LREC'10)
3. Jose AK, Bhatia N, Krishna S (2010) Twitter sentiment analysis. National Institute of Technology Calicut
4. Ramteke J, Shah S, Godhia D, Shaikh A (2016) Election result prediction using Twitter sentiment analysis. Inventive Comput Technol (ICICT) Int Conf 1:1–5
5. Ranganathan J, Tzacheva A (2019) Emotion mining in social media data. Procedia Comput Sci 159:58–66

6. Barbosa L, Feng J (2010) Robust sentiment detection on twitter from biased and Noisy data. COLING 2010: Poster, pp 36–44
7. Xia R, Zong C, Li S (2011) Ensemble of feature sets and classification algorithms for sentiment classification. Inf Sci Int J 181(6):1138–1152
8. Akilandeswari J, Jothi G (2018) Sentiment classification of tweets with non-language features. Procedia Comput Sci 143:426–433
9. Annett M, Kondrak G (2009) A comparison of sentiment analysis techniques: polarizing movie blogs. In: Conference on web search and web data mining (WSDM). University of Alberia, Department of Computing Science
10. Taboada M, Brooke J, Tofiloski M, Voll K, Stede M (2011) Lexicon based methods for sentiment analysis. Assoc Comput Linguist
11. Goncalves P, Benevenuto F, Araujo M, Cha M (2013) Comparing and combining sentiment analysis methods
12. Annett M, Kondrak G (2009) A comparison of sentiment analysis techniques: polarizing movie blogs. In: Conference on web search and web data mining (WSDM). University of Alberia, Department of Computing Science
13. Aydogan E, Akcayol MA (2016) A comprehensive survey for sentiment analysis tasks using machine learning techniques. In: 2016 international symposium on innovations in intelligent systems and applications (INISTA), vol 1, no 1, pp 1–7
14. Blom A, Thorsen S (2012) Automatic Twitter replies with python. In: International conference "Dialog 2012"
15. Pang B, Lee L (2008) Opinion mining and sentiment analysis. In: 2nd workshop on making sense of Microposts, vol 2, no 1. Cornell University, Ithaca
16. Hu M, Liu B (2004) Mining and summarizing customer reviews
17. Agarwal A, Xie B, Vovsha I, Rambow O, Passonneau R (2012) Sentiment analysis of twitter data. In: Annual international conferences. Columbia University, NewYork
18. Spencer J, Uchyigit G (2008) Sentiment or: sentiment analysis of Twitter data. In: Second joint conference on Lexicon and computational semantics. University of Brighton, Brighton
19. Sharma S (2008) Application of support vector machines for damage detection in structure. J Mach Learn Res
20. Pak A, Paroubek P (2010) Twitter as a corpus for sentiment analysis and opinion mining. In: Special issue of international journal of computer application. Universitede Paris-Sud, France
21. Saif H, He Y, Alani H (2012) Alleviating data scarcity for twitter sentiment analysis. Assoc Comput Lingist
22. Wu J, Wang J, Liu L (2006) Kernel-based method for automated walking patterns recognition using klnematics data. In: 5th workshop on natural language processing. Xi'an Jiaotong University, China

Specific Pose Detection Through Efficient Neural Network

Harsh Malik, Mansi Malik, Siddhant Malik, Vimal Kumar, and Mukesh Rawat

Abstract Human pose detection is a process of detecting human body points in images or videos. For the past few years, it has gained the attention of the computer vision community. Some of the achievements such as single pose detection and top-down network were efficient and reliable but suffered some problems. This technology has a very vast scope and is applicable in areas such as medical industry, defense industry, public safety, personal monitoring, and mentoring but due to the limitation of technology and algorithm, implementation has become a big hassle. In this work, we bring a real-time network-based pose estimation system that utilizes CPU processing and generates the normalized body points concerning the frame. It demonstrates how these points can be used to calculate the different body part angles as well as their lengths and by using these calculations, classifying the pose into a category to detect and output the result in the desired format. Our project outperformed the current system in terms of efficiency, cost, and processing speed.

Keywords Real-time pose detection · Open pose deep learning module · TensorFlow · Person counter

---

H. Malik (✉) · M. Malik · S. Malik · V. Kumar · M. Rawat
Meerut Institute of Engineering and Technology, Meerut 250001, India
e-mail: harsh.malik.cs.2016@MIET.ac.in

M. Malik
e-mail: mansi.malik.cs.2016@MIET.ac.in

S. Malik
e-mail: siddhant.malik.cs.2016@MIET.ac.in

V. Kumar
e-mail: vimal.kumar@MIET.ac.in

M. Rawat
e-mail: mukesh.rawat@MIET.ac.in

B. Mishra and M. Tiwari (eds.), *VLSI, Microwave and Wireless Technologies*,
Lecture Notes in Electrical Engineering 877,
https://doi.org/10.1007/978-981-19-0312-0_39

1 Introduction

Real-time pose estimation acts as a constituent for enabling machines to have a comprehension of people in images and videos. As we know, a lot of innovations have been done in various fields, where security is concerned but that requires a lot of investment and equipment. Also, the current monitoring system is human-dependent, where a person has to constantly monitor the screen which makes it prone to errors. The most common approach is to start a single human pose estimation for each detected human. This top-down approach grasps existing techniques for single human pose estimation but faces a commitment in the initial phase. The person detector fails when people are too close. There is no method to recover from such a situation. Further, the runtime is in proportion to the number of people in the image. The bottom-up approach is proved to be more efficient as they reduce runtime complexity from the number of people in the image. In this paper, we present an efficient method for multi-person pose estimation with better performance. We show the bottom-up approach through association via part affinity fields, 2D vector fields that merges the location and orientation of body parts in the image. Also, the system computational performance on body key point estimation is invariant to the number of detected people in the image. This paper proposes a method for detecting multiple human poses using CNN. The rest of the paper is organized into four sections. Section 2 gives the literature review about the human pose detection. Section 3 defines the approach of building the specific pose detection module using CNN and HMS algorithm to identify the specific pose. Section 4 gives the experimental analysis of the developed work for the detection of specific poses. The results and conclusion are given in Sect. 5.

2 Literature Survey

Specific pose detection is a collection of several processes such as single human body point detection using CNN, calculation of different angles and lengths, classification of poses using algorithms, etc. Cao [1] proposed the bottom-up approach that can predict 2D human pose using PAF's, his approach provides results limited to body points [1]. G. Huang, Z. Liu, K. Q. Weinberger, and L. van der Maaten proposed DenseNet, a dense convolutional network, that provides shorter paths between different layers of the network to reduce the parameters and the eliminate vanishing gradient problem [2]. K. Simonyan and A. Zisserman proposed another deep CNN where they used the 3×3 convolution filters that improves accuracy by pushing till 16–19 weight layers [3]. Yang et al. [4] proposed a pyramid residual module which enhanced the invariance in scales of DCNN and also a derivative of the initialization scheme used in multi-branch networks [4]. He et al. [5] proposed the Mask R-CNN which is just the updated version of the faster R-CNN and introduces an overhead of human pose estimation along with the object detection [5]. Fang et al.

[6] proposed a regional multi-person pose estimation, they used SSTN, NMS, and PGPG for estimation and handling inaccurate bounding boxes [6]. Felzenszwalb and Huttenlocher [7] proposed a part-based modeling and object recognition which used the concept of deformable configuration of objects and allows qualitative analysis of appearance [7]. Ramanan et al. [8] developed an algorithm for people tracking and assumed that people take certain canonical poses and used it to build a model using limbs [8]. Andriluka et al. [9] proposed a way to recover a 3D model from real-world scenarios where firstly they generate the 2D model and convert it to 3D model using tracklet-based estimates [9]. Papandreou et al. [10] proposed a top-down approach for multi-person detection using heatmaps from ResNet and NMS for key point confidence scoring [10].

3 Approach

This section talks about architecture, a few modules, and the algorithm.

3.1 Architecture

The complete process starts by capturing the frames one by one using a camera or pre-recorded, which is fed to a convolutional neural network frame by frame. These are then broken into smaller arrays for producing confidence maps to locate body points and affinity fields to determine the orientation and association. Both of these are then combined to form the final pose, and this process is carried out for each person in the frame. Once the pose points are retrieved, they are stored in a different variable that helps us to determine which body part is associated with which point and by using these points, we calculate the different body part lengths and distance between different points. This is based on the purpose the module has to serve and using these distances, we calculate certain body angles giving us the pose details in mathematical form. This enables us to match them with the predefined threshold set for the desired pose selected. This is done by sending these values to the formulation defined for the pose and then matching them to the specified range. If the values lie within the range, the module sends the message to the user which can trigger actions like a display message or an app notification based on the purpose served. If the calculated values do not lie within the range, the process restarts from the initial point, fetches the frame, and evaluates each of them, and all this process goes on in real time (Fig. 1).

Our specialized architecture as shown in Fig. 2 starts with the camera unit which works as an input unit. It intakes real-time images and passes it to the convolutional neural network and produces points and associates them with each other in the parsing phase and finally merge them to obtain the resultant image with pose [5–7]. This resultant image with points travels to the cloud where our detection module finds

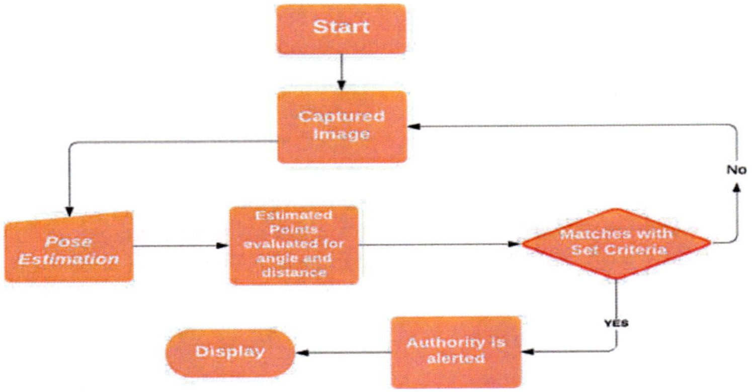

Fig. 1 Generalized architecture

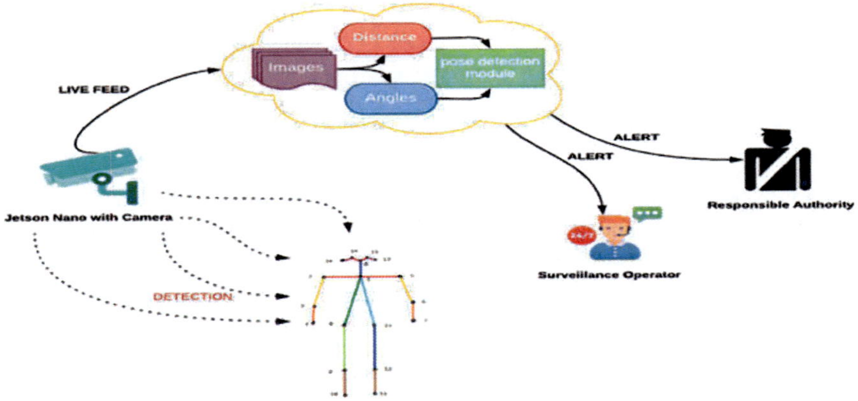

Fig. 2 Specialized architecture

out the distance between the different body parts and the angle between them. These values are matched with the specified angles and distance conditions defined for the pose to determine whether the pose is desired or not, if matches to the minimum threshold, then it generates an alert which can be used to alert the user in any way such as by displaying on the screen or an app according to what purpose it is serving.

3.2 Description of Various Modules of the System

The few modules are taken into consideration:

Fall Detection

(a) The basic logic behind detecting a fall is when a person falls, its head goes down and we get a drastic change in the x and y coordinates of the head with the maximum change in the y-axis.
(b) The first condition is that the difference between the present and the previous coordinates will be positive [8].
(c) The coordinates are in the normalized form, and we convert them to relative points as per the image size [9, 10].
(d) We append every new coordinate into the buffer or array and match it with the previous two coordinates if there is a change of value greater than 27 distances between the current y coordinate of the head point and the previous head coordinate stored in the array [11].
(e) We have to use try and catch method as many times the head is being not detected by the algorithm so we take the highest point possible and perform the same procedure with it.

Yoga pose corrector

(a) For the detection of the yoga pose, we need to constantly measure the distance and angle orientations. For the mountain pose, the hands are stretched upwards so we calculate the angle between the hands relative to the neck and shoulder along with the closeness between the right wrist and left wrist.

$$\sqrt{(x1-x2)^2+(y1-y2)^2}\,\sqrt{(\mathrm{x}1-\mathrm{x}2)^2+(\mathrm{y}1-\mathrm{y}2)^2} \qquad (1)$$

(b) The above Euclidean distance formula Eq. (1) lets you to calculate the distance between two points in the plane with coordinates $(x1, y1)$ and $(x2, y2)$ as shown in Fig. 3, where $x1$, $x2$, $y1$, and $y2$ are the respective x and y coordinates of the two points plotted on the human body present in the image such as head point, hand point or leg points. Here, 'x' represents the pixel location on the x-axis in the frame, and 'y' represents the pixel location on the y-axis in the frame.

(c) For calculating angle, first, we form a triangle using the three points as shown in Fig. 4, where P_0 and P_2 are the points between which we want to find the angle and $P1$ is the central point. Then, we find the distance between the points, i.e., the distance between P_0 and P_1, P_1 and P_2, and P_2, and P_0 which is calculated through the Euclidean distance formula denoted by a, b, and c, respectively.

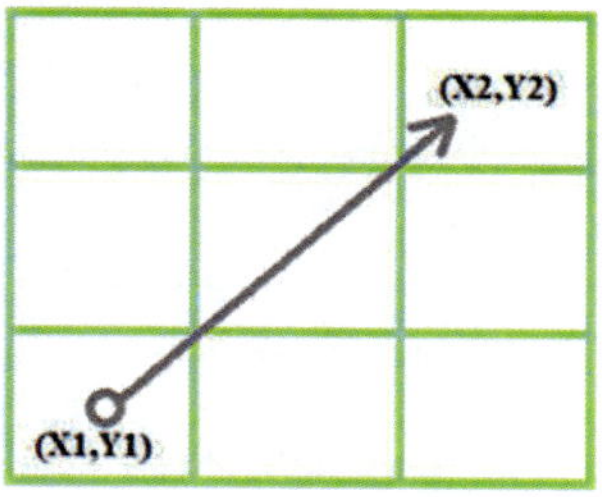

Fig. 3 Euclidean distance

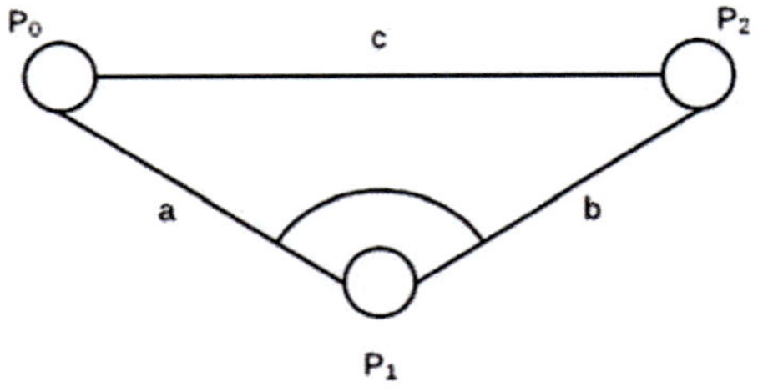

Fig. 4 Determining angle at P_1

When we have received all the lengths, we use the inverse cosine formula for determining the angle at $p1$ as shown in below Eq. (2)

$$\cos^{-1}\frac{\left(a^2+b^2-c^2\right)}{2ab}*\frac{180}{\pi} \tag{2}$$

(d) Also, we match the length from the tip of the hand to the bottom which should be maximum than any other body length.

Plank Correction

Particularly, for the plank position, we used four different body angles and two distances to detect the correctness of the pose.

(a) Head point to left wrist point distance (used points: 0, 7).
(b) Head point to right wrist point distance (used points: 0, 4).
(c) Angle between left shoulder and left wrist (used points: 7, 6, and 5).
(d) Angle between right shoulder and right wrist (used points: 4, 3, and 2) (e). Angle between the left hip and the left ankle (used points: 11, 12, and 13)
(f) Angle between right hip point and right ankle point (used points: 8, 9, and 10)

We take all the angles as input and checks if they are in the range of our predefined thresholds that represent a plank in real life. Here, body points are depicted in Fig. 5

Body Ratio Calculation

Calculation of body ratio is through a simple equation (3), defined as the ratio of the total leg length to the total height. Leg-to-body ratio (LBR)

$$\mathrm{LBR}=\frac{\text{leg length}}{\text{total height}}\mathrm{LBR}=\frac{\text{leg length}}{\text{total height}} \tag{3}$$

Algorithm-1

```
Start
 human.detect_points()
 module = input()
 If(module):
  formulate(dis,ang)
```

```
Display(detected)
defdetect_points():
 for w in maps:
  for v in fields:
   point = merge(m,o)
   return point
defcalc_distance():
 × 1, × 2 = fetch_points(0,1)
 len = Euclidian(×1, × 2)
 return(len)
defcalc_angle():
 × 1, × 2, × 3 = fetch_points(0,1,2)
 angle = calci(×1, × 2, × 3)
 return angle
```

4 Experimental Result Analysis

For analyzing the performance of the developed system, various parameters were kept in mind such as lighting, angles of the image, size of frames, and visibility, and a custom test set was fed in the network whose results are compared below [12].

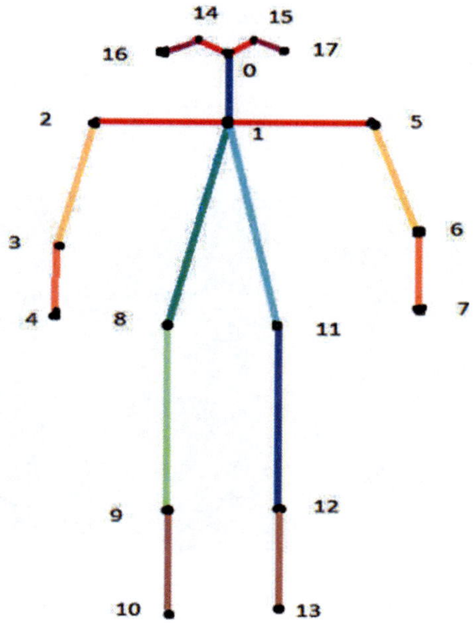

Fig. 5 Human body points

Table 1 Comparison table

Parameter in consideration	Previous version	Improved version
Efficiency	57.7	77.3
Accuracy (detected poses)	62.2%	78%
Storage	High	Low or limited
Cost	Heavy hardware cost	Less
Processing speed	2409 ms	2317 ms
Analysis and interpretation	Top-down approach	Bottom-up approach

Considering most of the situations, the module was able to achieve an accuracy of 78% during the test run.

The above comparison Table 1 compares the previous open pose-based detection approach and the current version with 18 body points and trained through the coco dataset. The test was performed on the Nvidia Jetsonnano device, where the improved version was able to throw a maximum of 23 fps, and the previous version was able to throw 17 fps thus giving an efficiency of 77.3, and the tests conducted on a certain set of samples showed that the improved version was able to detect poses in 78% of the total samples where the previous was able to detect only the 62.2%. Some of the samples are shown below.

Figure 6 shows an output example where a plank position is detected with a message displayed on the left top of the screen and in the console with the time stamp.

Figure 7 shows a fall detection demo in which as soon as the person falls, the system generates an alert in the console.

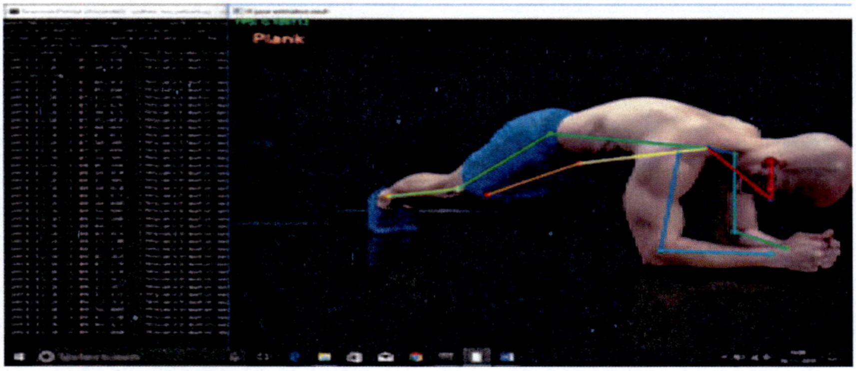

Fig. 6 Plank detection

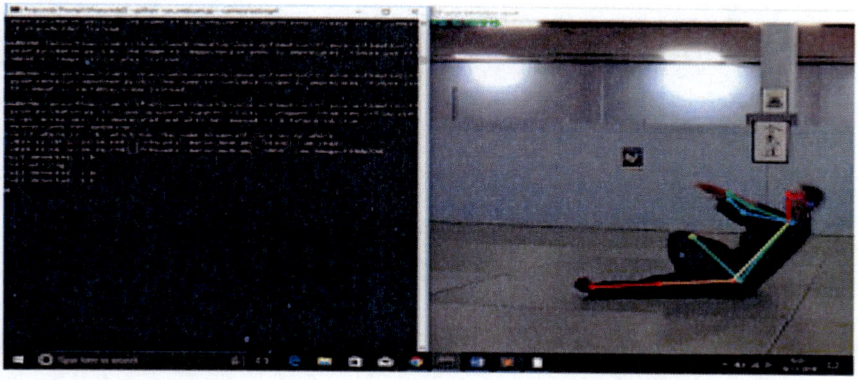

Fig. 7 Fall detection

5 Conclusion and Future Works

Specific pose detection using an efficient neural network is a way to enable machines to visually understand and interpret humans. In this paper, we present a way of representing the key point association that encodes both position and orientation of human limbs. We have considered a few basic modules, viz yoga pose correction, planking/push up corrector, fall detection, and body ratio. We have shown how the length and the different angles between the body parts can be used to determine the specific pose, and how these angles and lengths can be calculated relatively for every human detected in the frame. All the demo modules developed to show how the determination of pose can be done using the pose estimation algorithm and serves their purpose with overall 78% accuracy taking all the natural as well as coding anomalies in mind. This approach is much more convenient than parsing the points in another neural network for estimating whether the pose is desired one or not and increases system performance, thus enabling it to run on multiple devices and soon can be optimized to run on mobile devices making it more portable and easy to use and also it can be extended for object detection.

References

1. Cao Z, Simon T, Wei S-E, Sheikh Y (2017) Realtime multi-person 2d pose estimation using part affinity fields. In: CVPR
2. Huang G, Liu Z, Weinberger KQ, van der Maaten L (2017) Densely connected convolutional networks. In: CVPR
3. Simonyan K, Zisserman A (2015) Very deep convolutional networks for large-scale image recognition. In: ICLR
4. Yang W, Li S, Ouyang W, Li H, Wang X (2017) Learning feature pyramids for human pose estimation. In: ICCV
5. He K, Gkioxari G, Doll´ar P, Girshick R (2017) Mask r-cnn. In: ICCV

6. Fang H-S, Xie S, Tai Y-W, Lu C (2017) RMPE: regional multiperson pose estimation. In: ICCV
7. Felzenszwalb PF, Huttenlocher DP (2005) Pictorial structures for object recognition. In: IJCV
8. Ramanan D, Forsyth DA, Zisserman A (2005) Strike a pose: tracking people by finding stylized poses. In: CVPR
9. Andriluka M, Roth S, Schiele B (2010) Monocular 3D pose estimation and tracking by detection. In: CVPR
10. Papandreou G, Zhu T, Kanazawa N, Toshev A, Tompson J, Bregler C, Murphy K (2017) Towards accurate multi-person pose estimation in the wild. In: CVPR
11. Chen Y, Wang Z, Peng Y, Zhang Z, Yu G, Sun J (2018) Cascaded pyramid network for multi-person pose estimation. In: CVPR
12. Xiao B, Wu H, Wei Y (2018) Simple baselines for human pose estimation and tracking. In: ECCV

Tri-forked (Trishul Shape) Triple-band MIMO Antenna for Satellite Applications

Aditya Kumar Singh, Amrees Pandey, Sweta Singh, and Rajeev Singh

Abstract A tri-forked (Trishul shaped) MIMO antenna ($30 \times 25 \times 1.6$ mm^3) is presented for X, Ku, and K band applications. The proposed structure shows two ultra-wideband (UWB), first at 7–9.2 GHz with impedance bandwidth of 27.2% and second at 10.9–14 GHz with impedance bandwidth of 20.7%.The third band is broadband and is observed at 23.4–26.2 GHz (K band) with an impedance bandwidth of 11.2% at port-1. Isolation (less than −15 dB) between the two radiating antenna elements is observed by using defected ground structure (DGS) at the ground plane. Ansys HFSS v13 is used to obtain simulated reflection coefficient, gain, current distribution, and radiation efficiency. Envelope correlation coefficient (ECC), diversity gain, and isolation parameters are calculated to characterize MIMO.

Keywords MIMO · X; Ku; and K bands · Port isolation · Diversity gain · Envelope correlation coefficient

1 Introduction

With the rapid evolution in wireless communication technology, high-data transmission rate, large channel capacity, and good reliability are of paramount importance. Due to these features, MIMO antennas have garnered the attention of engineers working in the field of wireless communication [1]. MIMO antenna systems are capable of providing a good quality of mobile communication if the designed antenna has desired isolation and envelope correlation coefficient (ECC) between its antenna elements [2]. In MIMO systems, it is also desirable to achieve low-mutual coupling between antenna elements, reduce the effect of multi-fading, and easy to integrate and embed multiple antennas together resulting in improved performance [3]. The antenna presented herein covers X, Ku, and K bands, and X band finds applications in the fields of the satellite, mobile, radio-location, etc., [4] whereas the Ku/K band is used for radar and broadcasting satellite television [4]. Ultra-wideband (UWB)

A. K. Singh · A. Pandey · S. Singh · R. Singh (✉)
Department of Electronics and Communication, University of Allahabad, Prayagraj, India
e-mail: rsingh68@gmail.com

B. Mishra and M. Tiwari (eds.), *VLSI, Microwave and Wireless Technologies*, Lecture Notes in Electrical Engineering 877,
https://doi.org/10.1007/978-981-19-0312-0_40

antenna systems were proposed due to their attractive features such as excellent immunity to multipath interference and high-data rates distortion and also are easy to fabricate and preferred in MMICs and small devices [5, 6]. Wideband and ultra-wideband with single and dual-band operation in either of the X/Ku/K bands [7] are reported, but triple-band antenna covering the X/Ku/K bands is rare, therefore, the proposed antenna is novel in this sense.

In this paper, two-element multiband broadband, wideband, and ultra-wideband (UWB) MIMO patch antenna on FR-4 substrate for X, Ku, and K band applications are presented. Two-port tri-forked (Trishul shaped) MIMO antenna is designed and simulated by using an electromagnetic tool high-frequency structure simulator (HFSS) Ansys v13.Tri-forked (Trishul shaped) patch is introduced between the centre positions of elements to provide good isolation between the ports. The proposed antenna exhibits bands in the frequency range 7–9.2 GHz, 10.9–14 GHz, and 23.4–26.2 GHz which covers X band (8–12 GHz), Ku (12–18 GHz), and K band (18–27 GHz) applications. DGS has been introduced to increase the gain and other antenna parameters. The overall geometry, results, and conclusion have been discussed in the forthcoming sections.

2 Evolution of Antenna Design

Figure 1a shows the systematic growth of the proposed tri-forked (Trishul shaped) MIMO patch antenna. The specifications of the proposed antenna are shown in Table 1. The evolution of the proposed antenna starts with Ant-1.The patch and ground structure of Ant-1 with two-port MIMO micro-strip line feeding are presented in Fig. 1a and b, respectively. Ant-2 has been obtained by introducing defected ground structure in Ant-1, resulting in the geometry as shown in Fig. 2a and b. A swastika shape slot (a, b, c: 2.5, 1.5, 0.5 mm; cf-Fig. 3a) is introduced in the patch, and

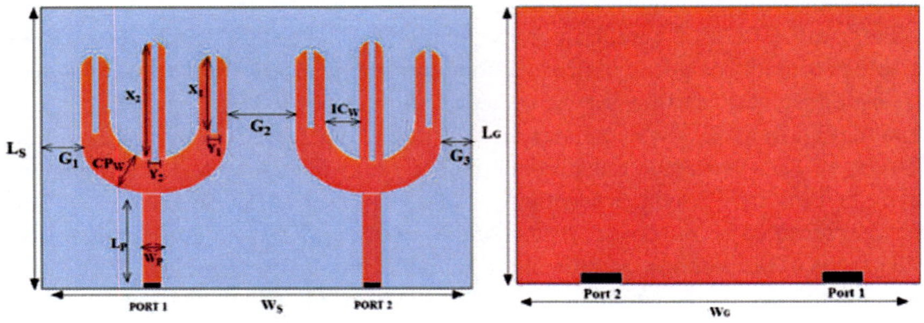

(a) (b)

Fig. 1 **a** and **b** front (top) and back (bottom) view of the Ant-1, respectively

Table 1 Specifications of the proposed antenna (Ant-3)

Parameters	Values
$L_S \times W_S$ (substrate length × substrate width)	25 mm × 30 mm
$L_P \times W_P$ (patch length1 × patch width1)	10 mm × 2 mm
$X_2 \times Y_2$ (inner slot)	10.8 mm × 0.5 mm
$X_1 \times Y_1$(inner slot)	8 mm × 0.5 mm
CP_W (circular width)	2 mm
IC_W (inner circular width)	2.5 mm
GP_1, GP_2, GP_3 (gap patch)	1, 4, 1 mm
a, b, c (swastika patch)	2.5, 1.5, 0.5 mm
Lg × Wg (ground length × ground width)	25 mm × 30 mm
g, h, i (swastika ground)	3.5 mm × 8 mm × 1 mm
d, e, f, m, j (defected ground)	4, 1, 6, 10, 12 mm

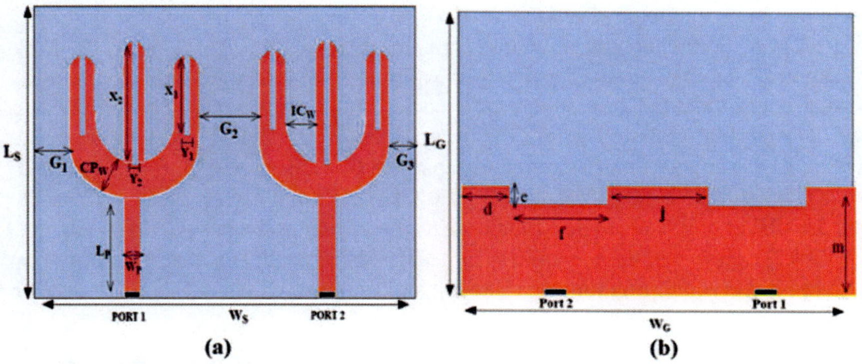

Fig. 2 **a** and **b** front (top) and back (bottom) view of the Ant-2, respectively

also swastika shape slot is introduced in the ground of Ant-2 (g, h, i: 3.5, 8, 1 mm; cf-Fig. 3b) resulting into Ant-3, which is the proposed antenna.

3 Results and Discussion

The antenna parameters such as return loss, gain, surface current distribution, diversity gain, envelope correlation coefficient (ECC), and radiation pattern are presented to evaluate its performance.

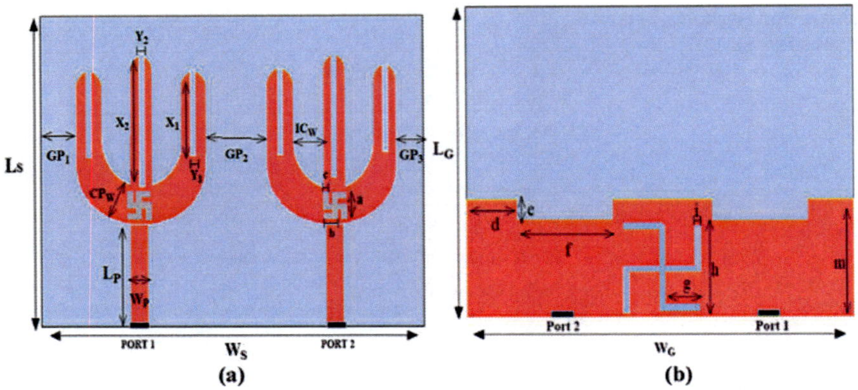

Fig. 3 **a** and **b** front (top) and back (bottom) view of the Ant-3 (proposed), respectively

Port 1 demonstrates three band sat 7–9.2 GHz, 10.9–14 GHz, 23.4–26.2 GHz, respectively, and port 2 exhibits bands at 7–9.2 GHz, 10.8–14.1 GHz, 23.2–26.2 GHz, respectively, as shown in Fig. 4.

Port 1 and port 2 antenna parameters are presented in Table 2 and Figs. 4–8. Table 2 demonstrates that both the ports of the proposed antenna are exhibiting nearly similar operational bands, resonant frequencies, impedance bandwidths, return loss barring marginal variation within the permissible limits. A large impedance bandwidth (27.2% at port 1 and 2 both) and peak gain (4.2 dBi at both ports-cf. Figure 5) for X band and 25% (at port 1) and 26.5% (at port 2) and peak gain (5.8 dBi-port1 and 3.6 dBi at port 2-cf. Figure 5) is observed partially at the lower end of Ku band (cf Table 2). However, the variation in impedance bandwidth and peak gain from port 1 and port 2 is more pronounced at the K band as observed from Table 2. However,

Fig. 4 $|S_{11}|$ versus frequency of Ant-1, Ant-2, and Ant-3

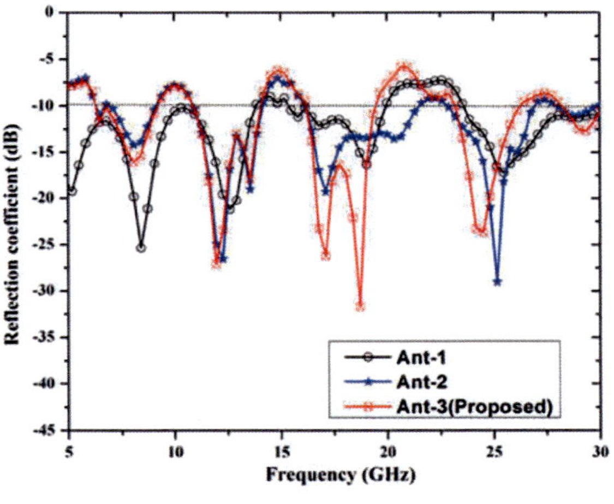

Table 2 Port characteristics of the proposed (Ant-3) MIMO antenna

Port1	Number of bands	Band frequency (GHz) (cf Fig. 4)	Resonant frequency (GHz) (cf Fig. 4)	Reflection coefficient (dB) (cf Fig. 4)	Impedance BW (%) (cf Fig. 4)	Peak gain (dBi) (cf Fig. 5)
	1	7–9.2	8.1	−16	27.2 (UWB)	4.2
	2	10.9–14	12	−26.7	25 (UWB)	5.8
	3	23.4–26.2	24.5	−23.5	11.2 (WB)	2.4
Port 2	1	7–9.2	8	−17.1	27.2 (UWB)	4.2
	2	10.8–14	12	−26.5	26.5 (UWB)	3.6
	3	23.2–26.2	24.5	−32.2	12.14 (WB)	1.6

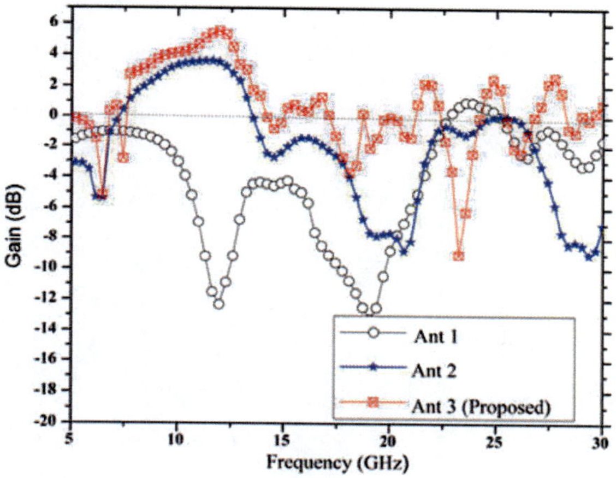

Fig. 5 Simulated gain versus frequency of Ant-1, Ant-2, and Ant-3

the variation in impedance bandwidth and gain at higher frequencies could be due to poor and varying isolation (cf Fig. 6) at frequencies above 10 GHz.

Also, isolation less than −15 dB is observed (cf Figs. 6 and 7). The diversity gain response of the proposed MIMO antenna Ant-3 varies between 9.4 and 10 dB as presented in Fig. 7. Radiation efficiency is changing from 95 to 65% in the operating frequency range 5–30 GHz as shown in Fig. 8.

The current distribution at three different resonant frequencies, i.e. 8.1, 12, and 24.5 GHz is presented in Fig. 9a–c. The maximum surface current density at resonating frequencies 8.1, 12, and 24.5 GHz is 70, 66, 62 A/m, respectively, which is observed in Fig. 9a–c. We observe the highest current density at 8.1 GHz whereas the lowest current density at 24.5 GHz is observed. At port 2, matching surface current distributions are observed at resonating frequencies 8, 12, 24.5 GHz confirming the MIMO antenna system. However, the same has not been shown here to avoid duplicity.

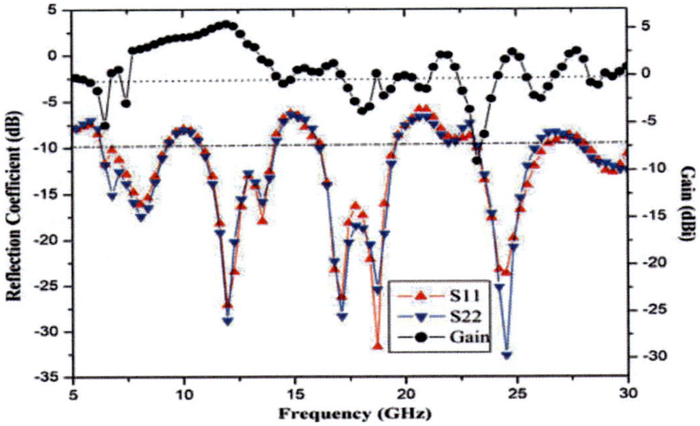

Fig. 6 Simulated S_{11}, S_{22} and gain versus frequency of the proposed MIMO antenna (Ant-3)

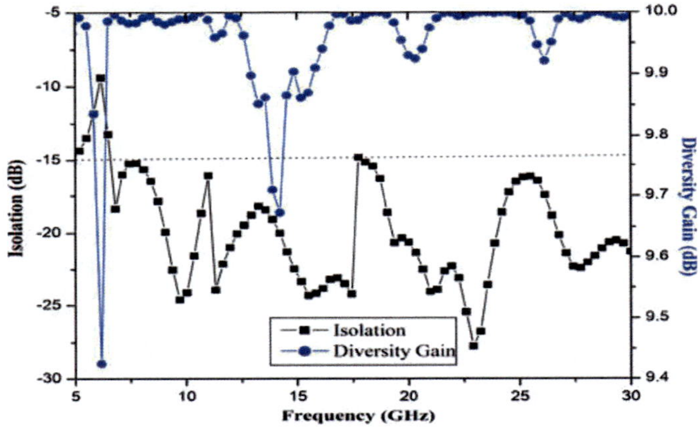

Fig. 7 Simulated isolation (between ports) and diversity gain (DG) of the Ant-3 (proposed)

Fig. 8 Simulated radiation efficiency of the proposed antenna (Ant-3)

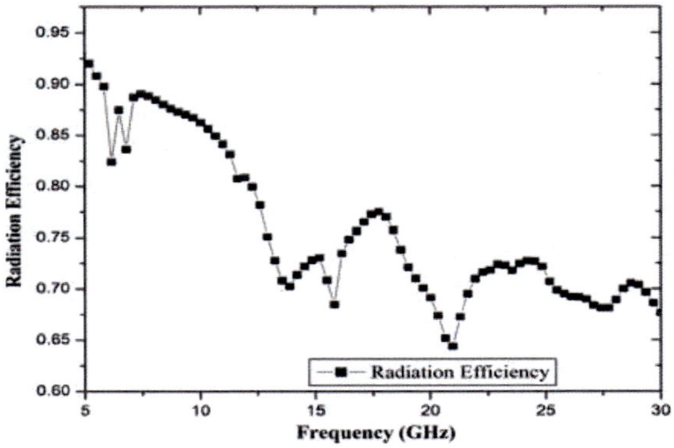

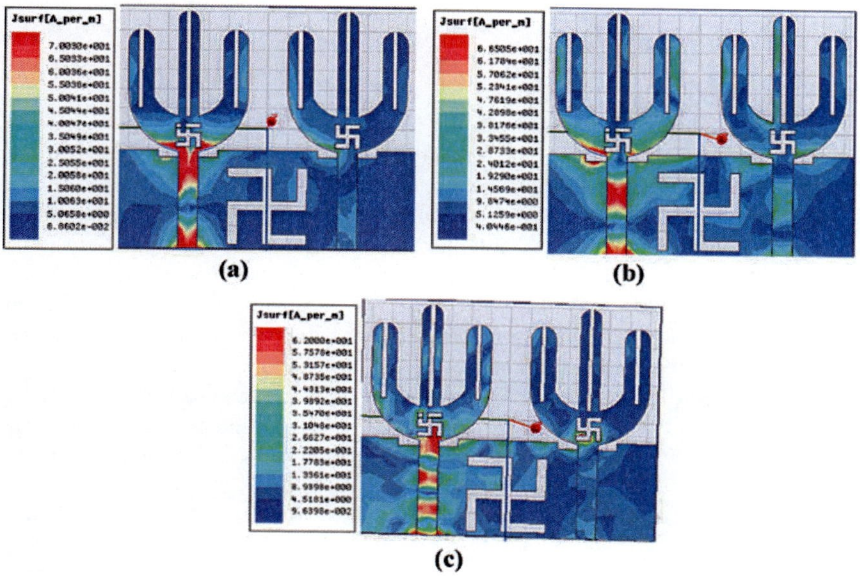

Fig. 9 Surface current distribution at port 1 for **a** 8.1 GHz, **b** 12 GHz, and **c** 24.5 GHz

Envelope correlation coefficient and diversity gain as a function based on frequency are presented in Figs. 10 and 6. Both the antenna parameters are related as given by Eqs. (1) and (2) [8].

$$\text{Diversity Gain} = 10\sqrt{1 - \text{ECC}^2} \tag{1}$$

Fig. 10 Simulated envelope correlation coefficient (ECC)

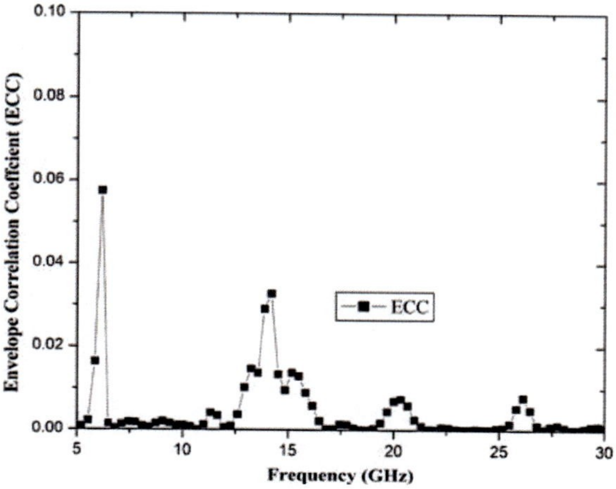

$$\mathrm{ECC} = \frac{|S_{11}{*}S_{12} + S_{21}{*}S_{22}|}{\left(1 - |S_{11}^2| - |S_{21}^2|\right)\left(1 - |S_{22}^2| - |S_{12}^2|\right)} \tag{2}$$

From Eqs. 1 and 2, it is clear that the lower the value of ECC, the higher the diversity gain. A lower value of ECC in the range of 0–0.06 is considered to be appropriate for a MIMO antenna to operate efficiently. When the ECC value is close to zero, the efficiency of the MIMO antenna is maximum as the mutual coupling, and signal interruptions between antenna elements are minimized. The envelope correlation coefficient (ECC) for the proposed MIMO Trishul shaped antenna Ant-3 shows a maximum value of 0.06, which is near to zero and indicates the minimum mutual coupling effect between antenna elements leading to higher data rates and diversity gain for the proposed antenna.

The simulated E-plane and H-plane for the far-field radiation pattern of Ant-3 are presented in Fig. 11a–c (at port 1 at 8.1, 12, and 24.5 GHz) and at port 2 is the same resonant frequency. The antenna for all resonating frequencies shows omnidirectional radiation pattern.

4 Conclusions

The proposed MIMO antenna simulated characteristics reveal the wideband at higher frequency (23.2–26.2 GHz-K band) and ultra-widebands at (10.9–14 GHz-X and partially Ku band) and from (7–9.2 GHz-X band) with impedance bandwidths of 27.2, 25 and 11.2% at port 1 and 27.2, 26.5 and 12.14% at port 2. A maximum peak gain of 5.8 dBi is observed amongst both the ports. As the MIMO antenna presented here has good isolation and omnidirectional radiation pattern, therefore it is suitable for 4G/5G mobile applications and satellite communications.

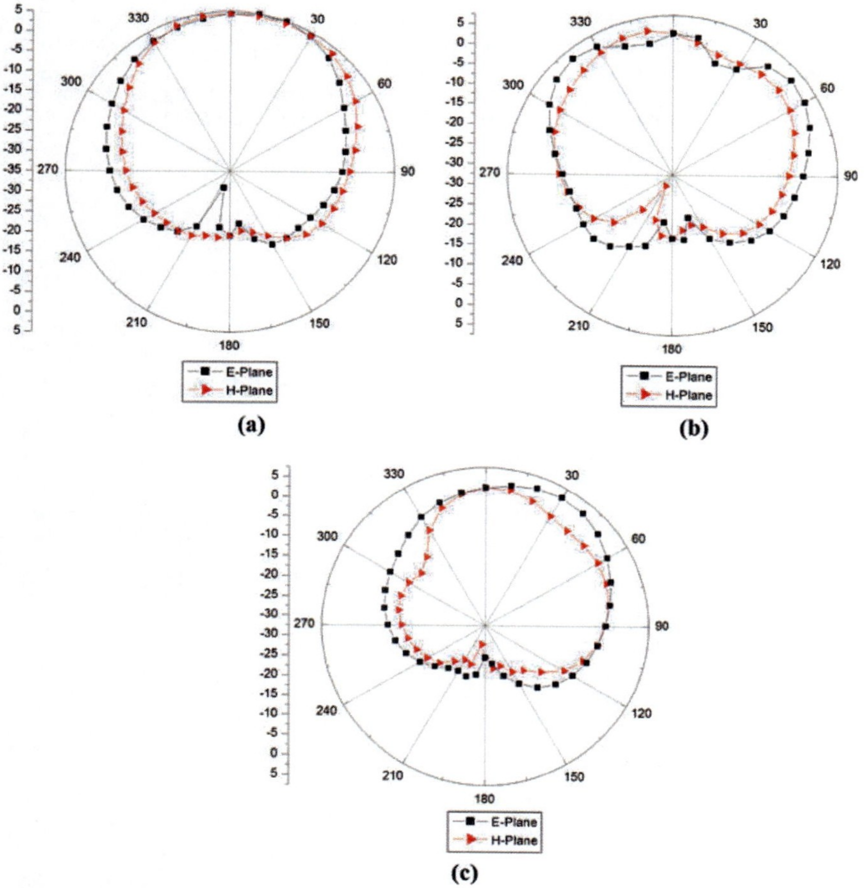

Fig. 11 Radiation pattern in E- and H-plane at port 1 at **a** 8.1 GHz, **b** 12 GHz, and **c** 24.5 GHz

References

1. Choukiker YK, Sharma SK, Behera SK (2014) Hybrid fractal shape planar monopole antenna covering multiband wireless communications with MIMO implementation for handheld mobile devices. IEEE Trans Antennas Propag 62(3):1483–1488
2. Chen Z, Geyi W, Zhang M, Wang J (2016) A study of antenna system for high order MIMO device. Int J Antennas Propag 1–14, Article ID 1936797
3. Nirdosh, Kakkar A, Sah S (2018) A two-element wideband MIMO antenna for X-band, Ku-band, K-band applications. In: 5th international conference on signal processing and integrated networks (SPIN), Noida, India, pp 94–97
4. Mishra B (2019) An ultra compact triple band antenna for X/Ku/K band applications. Microw Opt Technol Lett 61(7):1857–1862
5. Dwivedi AK, Mishra B, Singh V, Tripathi PN, Singh AK (2020) Design of a compact CPW-Fed monopole antenna with asymmetrical hexagonal slot loaded ground structure for C/X/Ku band

applications. Electr Control Commun Eng 16(1):15–22
6. Arabi O, See CH et al (2020) Compact wideband MIMO diversity antenna for mobile application using multi-layered structure. Electronics 9(8):1307
7. Mishra B, Singh V, Singh RK, Singh N, Singh R (2017) A compact UWB patch antenna with defected ground for Ku/K band applications. Microw Opt Technol Lett 60(1):1–6

Effect of Packet Size on the Performance of Wireless Ad Hoc Network with Various Data Rates

Alok Singh, Saurabh Sharma, and Pramod Narayan Tripathi

Abstract Wireless ad hoc networks are self-organizing, self-healing, adaptive, and dynamic networks that might be configured anyplace without needing any pre-existing infrastructure or remote access point. All this is due to the virtue of wireless medium. With so many advantages, the wireless medium has its limitations, viz., limited bandwidth, propagation losses, interference, security, and all this can lower the throughput of the network. In order to boost and achieve the maximum throughput, one must strategically select each parameter. Among all parameters, application packet size is also an important parameter that must be chosen wisely to achieve maximum overall throughput of the network. In this paper, we have examined the impact of packet size choice on wireless ad hoc networks with the help of packet delivery ratio (PDR), lost packets, average end to end delay, and throughput. This simulation-based study has been done using the ns-3 network simulator, and the result shows that the wireless ad hoc network can perform well when the application packet size is taken 512 or 1024 bytes.

Keywords Data rate · NS-3 · Packet size · Traffic generator · Wireless ad hoc network

1 Introduction

Today, the wireless ad hoc network has become a significant niche in the communication industry due to its outstanding characteristic of being deployed anywhere, any time without pre-existing infrastructure. In this type of network, devices can communicate with each other hop-by-hop, increasing its transmission range compared to single-hop networks. The selection of the right packet size in wireless ad hoc network

A. Singh (✉) · S. Sharma
Department of Electronics and Communication, University of Allahabad, Prayagraj, India
e-mail: aloksingh.eic@gmail.com

P. N. Tripathi
Department of Radio Engineering and Cybernetics, MIPT, National Research University, Moscow, Russia

B. Mishra and M. Tiwari (eds.), *VLSI, Microwave and Wireless Technologies*, Lecture Notes in Electrical Engineering 877,
https://doi.org/10.1007/978-981-19-0312-0_41

has always been the topic of extensive research in the literature. There is a trade-off between selecting a large packet size to get the enhanced payload to overhead ratio or a small packet for low end to end delay [1].

Selecting the appropriate packet size in a point-to-point link is done to ensure successful transmission [2–4]. However, this approach does not capture the multi-hop nature of wireless ad hoc networks. In the ad hoc network, packet size has a direct effect on the error rate. A large-size packet is more prone to error as compared to small-size packet. Another aspect in selecting the packet size of the ad hoc network is the nature of the wireless medium. The wireless medium is very much unpredictable.

It is possible to improve the ad hoc network performance by selecting an appropriate packet size of the application layer data. Selection of the packet size in wireless ad hoc network has been made in the literature for optimizing various network metrics. It was found that the congestion of the network reduced with the reduction in the packet size from 1024 to 64 bytes [5].

Average end to end delay (A-EED) of the wireless network has a linear relationship with packet size. This is due to the large packet size results in longer processing, transmission, and propagation delay. It was found in a 1-hop scenario with increasing packet size from 64 bytes to 48,856 bytes; the change in A_EED was found to be 0.69 ms increase as measured per 100 bytes. For 2-hop scenario, with increasing packet size from 64 to 1448 bytes, the change in A_EED was found to be 1.44 ms increase as measured per 100 bytes which is double compared to the 1-hop scenario. And for the 3-hop scenario, with increasing packet size from 64 to 1448 bytes, the change in A-EED was found to be 2.13 ms increase as measured per 100 bytes which is three times larger 1-hop scenario. The throughput of the wireless network is high for the large packet size, but throughput and packet size are not linear. A specific scenario with the change in packet size from 64 to 1008 bytes reflects the variation of throughput from 157.5 kbps to 775.4 kbps, which is approximately 392% increases. As packet size increased up to 5056 bytes that results in the throughput attain 1.05 Mbps. Finally, throughput saturated at 1.16 Mbps when the size of the packet reaches 10,056 bytes [6].

Lee [7] reported that if there is no movement, increased throughput is proportional to the maximum transmission unit (MTU). Even in the case of small movement, throughput is not heavily influenced. Under no BER circumstances, the MTU between 250 and 750 bytes has higher throughput. In contrast, the best throughput is for 500 bytes MTU in 10^{-4} BER. A report [8] shows that throughput of the external transmitter is proportional to the packet size, whereas that in the central emitter remains constant. Ideally, the throughput of the IEEE 802.11b network is proportional to MAC packet size, but due to the wireless network's instability, it is impossible [9].

With theoretical and experimental studies on the ad hoc network, it has been proved that packet size optimization in wireless networks is a major research problem. Overhead can be reduced by using a larger transmitting packet, but the loss rate may increase due to the wireless link. On the other hand, using the smaller packet size may have more overhead [10]. In a specific work [11], packet delivery ratio (PDR), average delay, and throughput of the wireless network using AODV as the routing protocol were found to decrease, with an increase in packet size.

The result obtained using network simulator OMNET++ shows that throughput and PDR increase in parallel with packet size. It also shows that packet size variation has better performance in homogeneous MANET compared to heterogeneous MANET [12].

In the literature, lots of work have been done on the impact of packet size on the wireless network using various network simulators such as OMNET++, QualNet, and NS-2 with the various mobility model [13, 14]. Perhaps, no such work has been reported to date on the impact of packet size on the wireless ad hoc network using steady-state random waypoint mobility model (SSRWPM) with the latest version of NS-3.

The effect of application packet size on wireless ad hoc networks is presented in this paper. The remaining portion of the paper is systematized as follows: Overview of different traffic generators that has been provided by ns-3 is given in Sect. 2; Sect. 3 provides the detail of simulation parameters used in this work; result and discussion were provided in Sect. 4. Finally, the conclusion and future work is provided in Sect. 5.

2 Traffic Generator

There are many traffic generators in the NS-3 simulator. All the traffic generators are also known as applications, and it is the base class used for all NS-3 applications, as shown in Fig. 1. Each node is associated with applications. This applications class has the following module associated with it, viz., BulkSendApplication, OnOffApplication, PacketSink, ThreeGppHttpClientServer, UdpEcho, UdpClientServer, and application module tests [15]. The OnOff application is the most popular traffic generator in NS-3 used by researchers.

2.1 On–Off Application

As the NS-3 calls a function and goes to this application class, it starts generating the traffic with the on and off states in an alternate manner. A random variable controls the duration of both states and this duration measured in seconds. No traffic will generate during the off state, while CBR-type traffic will generate in another state. The characteristic of this CBR traffic depends upon data rate and packet size [15].

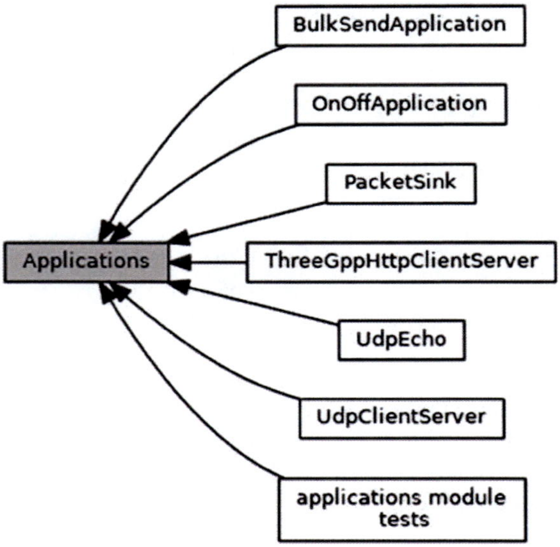

Fig. 1 Collaboration diagram of NS-3 applications

3 Simulation Parameter

The resources used for the execution of this work comprise Dell Vostro 3471 (Intel Core i3-9100 CPU-3.60 GHz, 8 GB RAM), and operating system is Ubuntu 20.04. All the experiments have been carried out using network simulator-3 (ns-3). We have selected 1 Mbps data rate of IEEE 802.11b standard in the wireless ad hoc network. Table 1 exhibits all the essential simulation parameters. Figure 2 shows the time window related to simulation, and it depicts the start and stops time of various applications as well as total simulation duration. All the simulations are averaged of 10 runs, and the confidence level is 95% [16].

4 Result and Discussion

The number of the lost packet of the network increases with increasing the data rate from 10 to 100 kbps, as shown in Fig. 3. This may be due to at high data rate of on–off Application and limited queue size, thus causing the queue overflow at the forwarding nodes, and packets will be drop.

The traffic source with on–off application will generate more packets for small packet size compared to large packet size at the same data rate. This means the traffic source will generate more packets in the case of 64 bytes as compared to 512 bytes and 1024 bytes. Due to this, overflow of the queue will also occur. Therefore, the

Table 1 Simulation environment

Parameters	Value
NS-3 simulator version	NS-3.32
Run ID	1, 2, 3, 4, 5, 6, 7, 8, 9, 10
Seed	1
Total nodes	50
Source destination pairs	10
Mobility model	Steady-state random waypoint mobility model
	Speed (m/s)—[Min = 10.0, Max = 20.0]
	Pause (s)—[Min = 1.0, Max = 3.0]
Area (m^2)	100 × 100
Application traffic	CBR
Traffic generator	On–Off Application
	On State—1 s, Off State—0 s
Payload size	64 bytes, 512 bytes, 1024 bytes
Application data rate (kbps)	10, 20, 30, 40, 50, 60, 70, 80, 90, 100
Ad hoc routing protocol	AODV
Transport layer protocol	User datagram protocol
MAC mode	Ad hoc mode
Physical	IEEE 802.11b standard
Propagation loss model	Log distance propagation loss model
Reference loss at 2.4 GHz	40.0459 dBm
Propagation delay model	Constant speed propagation delay model
Simulation time (s)	200 s
Confidence interval	95%

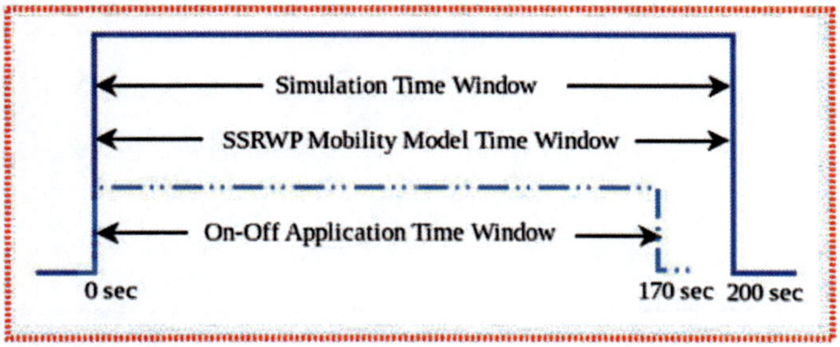

Fig. 2 Time windows for the simulation environment

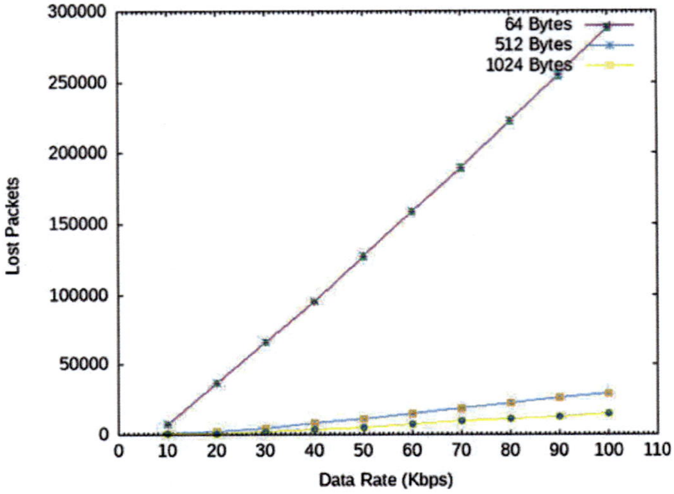

Fig. 3 Lost packet versus traffic data rate using different packet sizes

number of lost packets due to 64 bytes packet size is more than 1024 bytes and 512 bytes.

Figure 4 shows the PDR of the wireless ad hoc network to the traffic data rate that changes from 10 to 100 kbps. As shown in the figure, there is decreasing PDR, which may be due to an increasing number of packets that will cause contention in the channel, due to which packet may be lost. PDR of the network using 1024 bytes

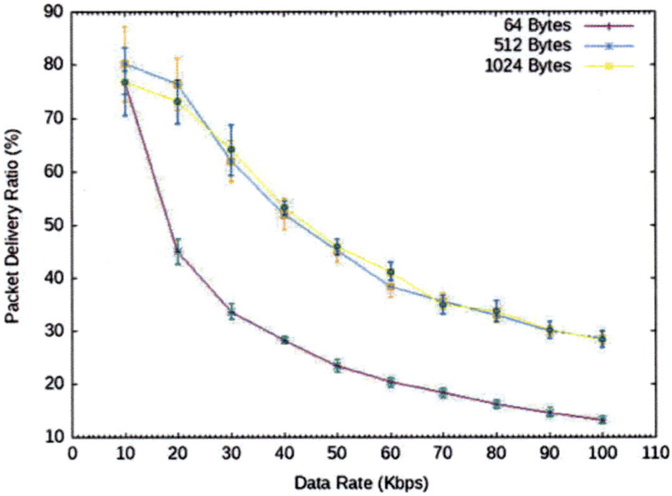

Fig. 4 Packet delivery ratio versus traffic data rate using different packet sizes

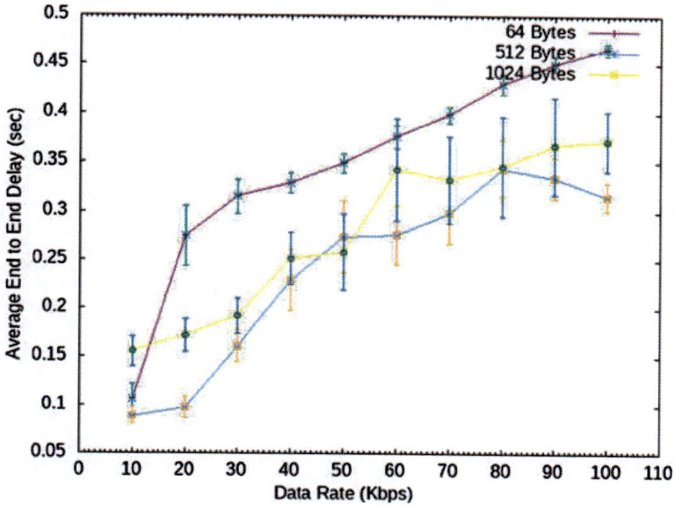

Fig. 5 Average end to end delay versus traffic data rate using different packet sizes

is more as compared to 64 bytes because the number of lost packets is less in the case of 1024 bytes.

Figure 5 shows that as traffic data rates are increasing from 10 to 100 kbps; the average end to end delays of the network are also rising. This may due to the channel is so busy resulting in excess back off time. As shown in Fig. 5, the average end to end delay of the network using 64 bytes is much more than 512 bytes and 1024 bytes. This may be due to more number of packets generate that will result in more RTS/CTS handshake in the case of 64 bytes.

Figure 6 shows that as traffic data rates are increasing from 10 to 100 kbps; the throughput of the network is also rising. It may be due to more packets are injecting into the network. Throughput of the network using 64 bytes is increasing but less as compared to packet sizes of 512 bytes and 1024 bytes, and after 80 kbps data rate, throughput is showing saturation; this is due to the channel capacity of the network is limited. Throughput of the network using 512 bytes and 1024 bytes packet size may look like also saturation after 100 kbps data rate. Throughput of the network using packet size 1024 byte and 512 bytes is high and approximately close to each other as compared to 64 bytes; this is due to the number of the lost packet is less when using large packet size.

5 Conclusion and Future Work

In this paper, we have analyzed the consequence of the packet size on the performance of wireless ad hoc networks. This performance analysis has been done with the help

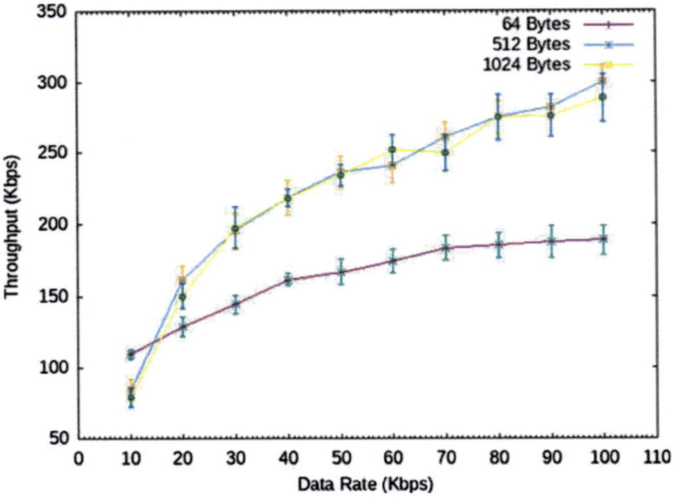

Fig. 6 Throughput versus traffic data rate using different packet sizes

of such metrics, viz., lost packets, PDR, average end to end delay, and throughput. From the obtained result, we can conclude that when the packet size is 512 bytes and 1024 bytes, the packet delivery ratio is much better as compared to that of 64 bytes. The lost packet and average end to end delay are low when using 512 bytes and 1024 bytes. The throughput of the network is high with 512 bytes and 1024 bytes as compared to 64 bytes. Throughput of the network is saturating beyond 80 kbps data rate in the case of 64 bytes, while in the case of 512 bytes and 1024 bytes, it may be saturated after 100 kbps, which needs further investigation in the future.

Acknowledgements The authors want to say thanks to "Prof. Rajneesh Kumar Srivastava, Department of Electronics and Communication, University of Allahabad, Prayagraj," for his valuable support and guidance.

References

1. Kurt S, Yildiz HU, Yigit M, Tavli B, Gungor VC (2017) Packet size optimization in wireless sensor networks for smart grid applications. IEEE Trans Ind Electron 64:2392–2401. https://doi.org/10.1109/TIE.2016.2619319
2. Lettieri P, Srivastava MB (1998) Adaptive frame length control for improving wireless link throughput, range, and energy efficiency. In: Proceedings—IEEE INFOCOM. IEEE, pp 564–571. https://doi.org/10.1109/infcom.1998.665076
3. Poppe F, De Vleeschauwer D, Petit GH (2001) Choosing the UMTS air interface parameters, the voice packet size and the dejittering delay for a voice-over-IP call between a UMTS and a PSTN party. In: Proceedings—IEEE INFOCOM, pp 805–814. https://doi.org/10.1109/infcom.2001.916271

4. Sankarasubramaniam Y, Akyildiz IF, McLaughlin SW (2003) Energy efficiency based packet size optimization in wireless sensor networks. In: Proceedings of the 1st IEEE international workshop on sensor network protocols and applications, SNPA 2003. Institute of Electrical and Electronics Engineers Inc, pp 1–8. https://doi.org/10.1109/SNPA.2003.1203351
5. Broch J, Maltz DA, Johnson DB, Hu Y-C, Jetcheva J (1998) A performance comparison of multi-hop wireless ad hoc network routing protocols. In: Proceedings of the 4th annual ACM/IEEE international conference on Mobile computing and networking—MobiCom '98. Association for Computing Machinery (ACM), New York, USA, pp 85–97. https://doi.org/10.1145/288235.288256
6. Toh CK, Delwar M, Allen D (2002) Evaluating the communication performance of an ad hoc wireless network. IEEE Trans Wirel Commun 1:402–414. https://doi.org/10.1109/TWC.2002.800539
7. Lee J, Kim G, Park S (2002) Optimum UDP packet sizes in ad hoc networks. In: IEEE international conference on high performance switching and routing, HPSR. IEEE Computer Society, pp 214–218. https://doi.org/10.1109/HPSR.2002.1024238
8. Chaudet C, Dhoutaut D, Guérin Lassous I (2005) Performance issues with IEEE 802.11 in ad hoc networking. IEEE Commun Mag 43:110–116. https://doi.org/10.1109/MCOM.2005.1470836
9. Ilić Ž, Bažant A, Čolak I, Jakšić D (2021) Optimal MAC packet size in wireless LAN. https://www.researchgate.net/publication/228984807_Optimal_MAC_Packet_Size_in_Wireless_LAN. Last accessed 2021/02/03
10. Korhonen J, Wang Y (2005) Effect of packet size on loss rate and delay in wireless links. In: IEEE wireless communications and networking conference, vol 3, pp 1608–1613. https://doi.org/10.1109/WCNC.2005.1424754
11. Tuteja A, Gujral R, Thalia S (2010) Comparative performance analysis of DSDV, AODV and DSR routing protocols in MANET using NS2. In: ACE 2010—2010 international conference on advances in computer engineering, pp 330–333. https://doi.org/10.1109/ACE.2010.16
12. Ismail Z, Hassan R (2011) Effects of packet size on AODV routing protocol implementation in homogeneous and heterogeneous MANET. In: Proceedings—CIMSim 2011: 3rd international conference on computational intelligence, modelling and simulation, pp 351–356. https://doi.org/10.1109/CIMSim.2011.70
13. Yigit M, Yildiz HU, Kurt S, Tavli B, Gungor VC (2018) A survey on packet size optimization for terrestrial, underwater, underground, and body area sensor networks. Int J Commun Syst 31:e3572. https://doi.org/10.1002/dac.3572
14. Freschi V, Lattanzi E (2019) A study on the impact of packet length on communication in low power wireless sensor networks under interference. IEEE Internet Things J 6:3820–3830. https://doi.org/10.1109/JIOT.2019.2891841
15. ns-3: Applications. https://www.nsnam.org/doxygen/group__applications.html. Last accessed 2021/02/04
16. Singh A, Sharma S, Srivastava RK (2020) Investigation of random waypoint and steady state random waypoint mobility models in NS-3 using AODV. J High Speed Netw 26:267–274. https://doi.org/10.3233/JHS-200643

Performance of Wireless Ad Hoc Network with All the Data Rate of IEEE 802.11b Using NS-3

Saurabh Sharma, Alok Singh, and Pramod Narayan Tripathi

Abstract Throughout the past couple of decades, wireless devices have obtained a huge place in the marketplace as a consequence of appealing and promising attributes such as easy setup, mobility, dynamic character, and low price. This industry got the main kick-start after the deployment of IEEE 802.11 wireless LAN (Wi-Fi). Now, the IEEE 802.11 standards have become a landmark in the wireless sector on account of their various amazing capabilities. The IEEE 802.11b is the most well-known standard of the IEEE 802.11 family, which is broadly utilized in many wireless devices. IEEE 802.11b standards are very much used in wireless ad hoc network (WANET). WANET is the autonomous network that is formed by wireless devices without the need for infrastructure. IEEE 802.11b offers four different data rates, viz., 1, 2, 5.5, and 11 Mbps, and all of these work in the 2.4 GHz ISM band. In this paper, we have analyzed the wireless ad hoc networks performance using different data rates offered by IEEE 802.11b, and we have also explained why the network is outperforming at the higher data rate. The performance metrics used within this work are packet delivery ratio, lost packet ratio, average end-to-end delay, and throughput. Performance analysis about ad hoc network reveals different aspects those not focused on in the previous studies. The analysis has been completed with the assistance of network simulation 3 (NS-3).

Keywords Ad hoc network · Data rate · IEEE 802.11b · NS-3 · Packet size

1 Introduction

Today, the entire world is moving from static devices toward mobile devices; these devices communicate with each other either with the aid of infrastructure or without

S. Sharma (✉) · A. Singh
Department of Electronics and Communication, University of Allahabad, Prayagraj, India
e-mail: saurabh5771@gmail.com

P. N. Tripathi
Department of Radio Engineering and Cybernetics, MIPT, National Research University, Moscow, Russia

B. Mishra and M. Tiwari (eds.), *VLSI, Microwave and Wireless Technologies*,
Lecture Notes in Electrical Engineering 877,
https://doi.org/10.1007/978-981-19-0312-0_42

infrastructure. Those devices that can communicate with each other without pre-existing infrastructure are referred to as ad hoc networks.

In the ad hoc network, every node plays the routers' role, forwarding data packets to other nodes. These types of networks are very much use full in disaster management, battel field, etc. This entire paradigm shift could not even be imagined without the wireless networks. The most common and most popular wireless network is still IEEE 802.11 wireless LAN (Wi-Fi).

The limitation of wireless mediums, such as unpredictable channel conditions due to mobility, fading of the channel, and interference, impacts wireless networks' performance. The packet data rate is among the most significant parameters that control how quickly a node can send data onto the medium/channel.

In contrast to the wired network, the channel condition is very much unpredictable in the wireless network. To combat these channel dynamics, IEEE 802.11b is designed to work with multiple data rates to maximize network performance. Different data rates have different fading characteristics, bandwidth, and range. High-transmission rates might not work well in low-channel quality. For this reason, it is critical to gage the present channel requirements right and pick a reasonable data rate for wireless channel requirements to strengthen the throughput in a wireless network significantly.

There are many network simulators such as "QualNet, NS-2, NS-3, OPNET, OMneT++ , and GloMoSim" for wired and wireless networks. According to the research community, for wireless ad hoc networks, network simulator 3 (NS-3) is the best open-source network simulator. This simulator forces the user to learn the basic concept of the respected field due to the virtue of mimicking the real-world scenario. We have selected NS-3 as a simulator because it supports various protocols, mobility models, and communication systems. Moreover, it supports dynamic communication systems and discrete event simulation. NS-3 is based on object-oriented programming (OOPs). It is platform-independent open-source software that supports large simulation environments resulting in a more scalable and realistic simulation. NS-3 supports hierarchical and modular components. It includes advanced debugging functionality. NS-3 supports multithreading environments. It has the system visualization and user interface (UI) tool and provides a library support framework for integration with third-party components [1].

This paper targets a simulation-based approach to evaluate the wireless ad hoc networks performance using different data rates offered by IEEE 802.11b. The remaining paper is organized as follows: Sect. 2 provides a summary of IEEE 802.11b, Sect. 3 provides the literature review and the motivation behind the work, Sect. 4 details the simulation parameters used in this work, Sect. 5 provides the results and discussion, and Sect. 6 concludes the work.

2 IEEE 802.11b Overview

In 1997, the "Institute Of Electrical and Electronics Engineers (IEEE)" presented the very initial wireless LAN (WLAN) standard referred to as IEEE 802.11 that offered the maximum data rate of 2 Mbps only. There were some applications for which this data rate was significantly less. Due to this, IEEE 802.11 product could not make the grip into the market. Subsequently, many variants of IEEE 802.11 have been launched. A succinct description of each is provided in Table 1 [2, 3]. It is clearly depicted in Table 1 that some other variants of the IEEE 802.11 family have additional features compared to IEEE 802.11b. However, IEEE 802.11b standard will not disappear from the market since it is widely deployed worldwide.

The exact first variant of IEEE 802.11 WLAN delivers only two data rates of 1 and 2 Mbps with all the support of "direct sequence spread spectrum (DSSS) or frequency hopping spread spectrum (FHSS)." It is essential to mention that the two techniques, such as FHSS and DSSS, are essentially distinct and not interoperable with each other.

Over the initial IEEE 802.11 WLAN standard, the most crucial contribution of 802.11b was to support the physical layer (Phy) standard service by two fresh data

Table 1 Summary of IEEE 802.11

802.11 variant	Explanation	Maximum data rate
IEEE 802.11	Modulation technique uses DSSS/FHSS that works in 2.4 GHz band	2 Mbps at 22 MHz bandwidth
IEEE 802.11a	Modulation technique uses OFDM that works in the 5 GHz band	54 Mbps at 5/10/20 MHz bandwidth
IEEE 802.11b	Modulation technique uses DSSS that works in the 2.4 GHz band	11 Mbps at 22 MHz bandwidth
IEEE 802.11 g	Modulation technique uses ERP-OFDM that works in the 2.4 GHz band	54 Mbps at 5/10/20 MHz bandwidth
IEEE 802.11p	Modulation technique uses OFDM that works in the 5.9 GHz band	54 Mbps at 5/10/20 MHz bandwidth
IEEE 802.11n	Modulation technique uses HT-OFDM that works in two ISM bands, viz., 2.4 GHz and 5 GHz	288.8 Mbps at 20 MHz bandwidth 600 Mbps at 40 MHz bandwidth
IEEE 802.11ac	Modulation technique uses VHT-OFDM that works in the 5 GHz band	346.8 Mbps at 20 MHz bandwidth 800 Mbps at 40 MHz bandwidth 1733.2 Mbps at 80 MHz bandwidth 3466.8 Mbps at 160 MHz bandwidth
IEEE 802.11ax	Modulation technique uses HE-OFDMA that works in three ISM bands, viz., 2.4, 5, and 6 GHz	1147 Mbps at 20 MHz bandwidth 2294 Mbps at 40 MHz bandwidth 4804 Mbps at 80 MHz bandwidth 9608 Mbps at 160 MHz bandwidth

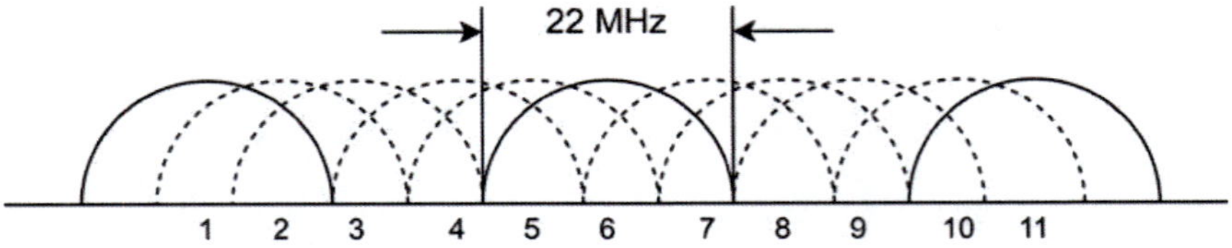

Fig. 1 Non-overlapping channels of IEEE 802.11b

rates viz 5.5 and 11 Mbps. The modulation technique chosen at the physical layer was DSSS since frequency hopping could not support the higher data rate without breaking up "Federal Communications Commission (FCC)" regulation.

Figure 1 displays IEEE 802.11b channel arrangement in the 2.4 GHz band, in which channels 11, 6, and 1 are the three non-overlapping channels exhibited by solid lines. It has been reported that in case the space between the antennas is less, then there will be no non-overlapping channels. Under such conditions, no channel will be completely free of interference [4].

3 Related Work and Motivation

Over the last couple of years, substantial research has been conducted in wireless ad hoc networks to be more specific on the method to increase ad hoc networks' performance under different channel conditions. The advice given by some of this literature [5–7] about controlling the data rate suggests that the rate adaptation algorithm needs to lower the present data rate on a certain number of packet loss. The theory behind the fact is that if the channel condition deteriorates between the sender and receiver node, causing packet loss at the current data rate, the sender node switches into a decrease data rate to accommodate such a worst situation. This strategy is readily broken up in practice whenever there is a hidden node issue. There is a substantial loss at the receiver node regardless of the channel condition due to the hidden node issue. This, then, activates a data rate adaptation algorithm to lower the sender's data rate. This measure does not solve contention at the channel. Instead, a decline in the data rate creates the channel's contention more pitiful since this advances the channel occupancy period, which subsequently causes the channel crashes, which further reduces the data rate.

Through the proper examination, it has been found that if a single node tries to send the data to the access point (AP), the auto rate fallback (ARF) algorithm works well even if there is noise present in the signal due to the location of the node. However, the network's performance degrades using ARF when multiple nodes try to use the channel, which causes packet loss due to collision [8].

Anastasi et al. [9] revealed several aspects of the IEEE 802.11b standard neglected in the simulation-based studies. Firstly, they have said that since different data rates were used for control and data frames, there will be two different ranges, first for the transmission and second for the carrier sensing in the network. Secondly, through an experimental study, they have calculated the transmission range of the node and found that the practical range is 2–3 times less than that assumed in the simulation. Moreover, this range is not constant. It is highly variable even in the same scenario.

The above work in the literature converges that it is essential to select an appropriate data rate to improve the wireless ad hoc network's performance, which is the driving force behind this investigation. This work investigates the wireless ad hoc network's performance by varying the data rate that was offered by IEEE 802.11b.

4 Simulation Parameter

This work's computational facility includes Dell Vostro 3471 (Intel Core i3-9100 CPU-3.60 GHz, 8 GB RAM) and Ubuntu 20.04 Operating System. All the simulations have been carried out using network simulator 3 (NS-3). We have used IEEE 802.11b standard for the wireless network in which 1 Mbps data rate has been selected. The speed of nodes movement has been chosen up to 50 m/s since adequately published works in ad hoc networks have also taken the same speed [10–12].

Table 2 shows all the important simulation parameters. Figure 2 shows the time window related to simulation, and it depicts the start and stop time of various application as well as total simulation duration. All the simulations are averaged for 10 runs with a 95% confidence interval [13].

5 Result and Discussion

Figure 3 shows the packet delivery ratio (PDR) of the ad hoc network while varying data rates offered by IEEE 802.11b. PDR of the network has been measured with respect to varying constant speed of nodes using steady-state random waypoint mobility. As the speed of nodes increases, the chance of link breakage is comparatively high compared to the low speed of nodes. Therefore, the PDR at 10 m/s is comparatively high as compared to other higher speeds of nodes. PDR of the network using 11 Mbps data rate is significantly higher as compared to 1, 2, and 5.5 Mbps data rates. This may be due to the channel occupancy time being more for low-data rate than a high-data rate of 802.11b [14].

Figure 4 shows the lost packets ratio with respect to the varying speed of nodes. As the node speed increases from 10 to 30 m/s, the lost packet ratio increases gradually, but beyond the 30 m/s speed of nodes, the lost packet ratio changes abruptly. This may be due to the routing information changes frequently at high mobility. At low mobility, the impact of change in routing information is not significant. As shown in Fig. 4, the lost packet ratio of the network using 1 Mbps data rate is relatively high as compared to other data rates of IEEE 802.11b. This may be attributed due to the

Table 2 Simulation attributes

Attributes	Value
Network simulator	NS-3.32
Run number	1, 2, 3, 4, 5, 6, 7, 8, 9, 10
Seed	1
Source destination pairs	10
Total number of nodes	50
Mobility model	Steady-state random waypoint mobility model
	Speed (m/s)—10, 20, 30, 40, 50
	Pause (s)—[Min = 1.0, Max = 3.0]
Area (m^2)	100 × 100
Application layer traffic	CBR
Application	On–off application
	On state—1 s, off state – 0 s
Application data rate	16 kbps
Packet size (payload)	500 bytes
Transport layer protocol	UDP
Routing protocol	AODV
Wi-Fi	IEEE 802.11b
MAC mode	Ad hoc
Data rate (Mbps)	1, 2, 5.5, 11
Propagation loss model	Log distance propagation loss model
Reference loss	40.0459 dBm
Propagation delay model	Constant speed propagation delay model
Simulation time (s)	200
Confidence interval (%)	95

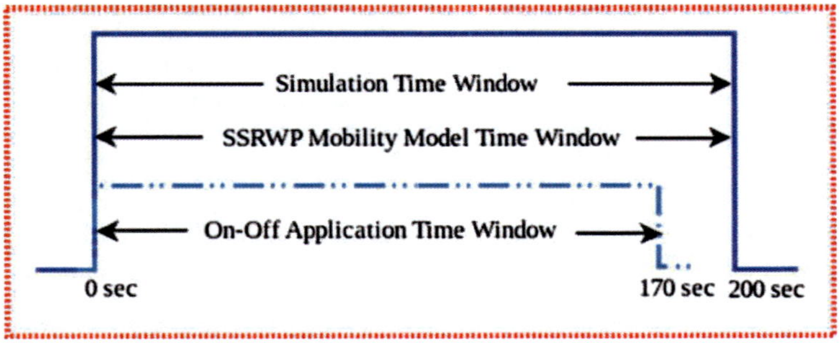

Fig. 2 Different time windows

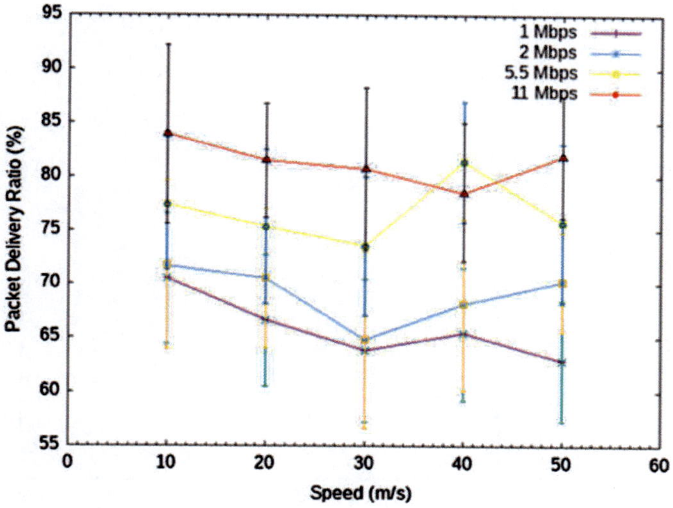

Fig. 3 Packet delivery ratio with the varying speed of nodes

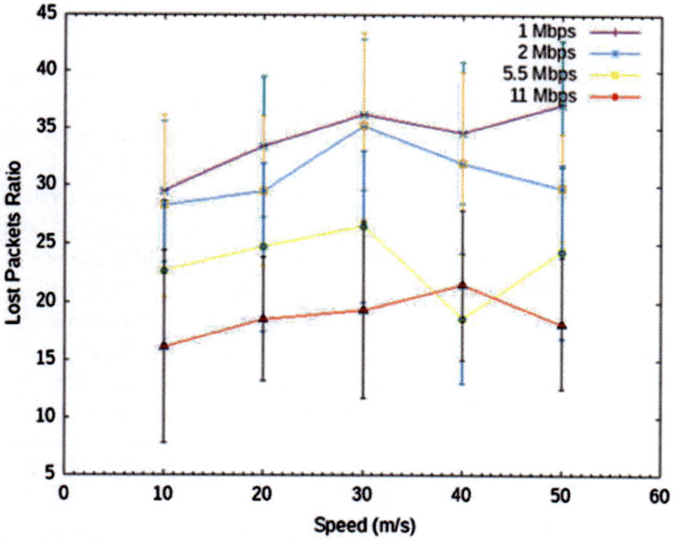

Fig. 4 Lost packets ratio with the varying speed of nodes

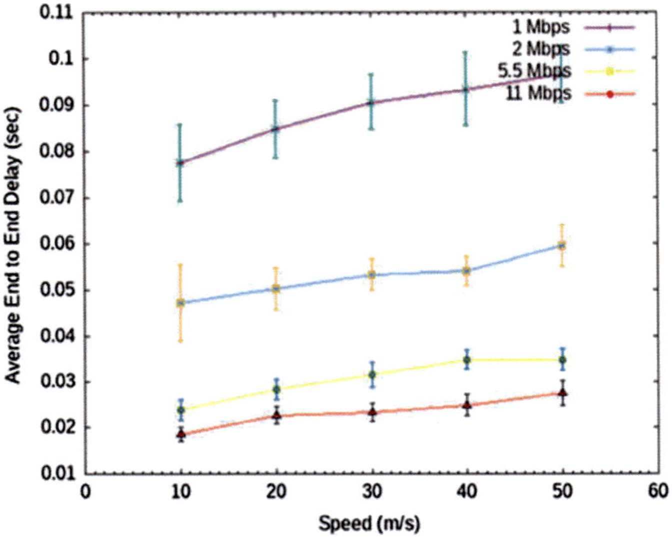

Fig. 5 Average end-to-end delay with the varying speed of nodes

channel will be occupied by the transmitting nodes, and the backoff time of other waiting nodes for the packet transmission will exceed the limit, and hence, the packet will be dropped. The lost packet ratio of higher data rates will be comparatively less due to lesser channel occupancy time.

Average end-to-end delay on varying the speed of nodes is depicted in Fig. 5. It is exhibited in the figure as the speed of nodes increases, the network delay gradually increases. This could possibly be caused by the more intermediate nodes increase between destination and source, which in turn are going to result within the most quantity of hops, and thus, the distance between destination and source may increase because of that propagation delay may be increased. As shown in Fig. 5, the average end-to-end delay of the network using 11 Mbps data rate is less than others lower data rate of IEEE 802.11b. This may be due to the modulation technique used in 11 Mbps, QPSK with 8 bits/symbol. While 5.5, 2, and 1 Mbps data rates are using QPSK with 4 bits/symbol, QPSK with 2 bits/symbol, and BPSK with 1 bits/symbol, respectively.

Figure 6 shows the throughput for varying the speed of nodes. Throughput decreases with increasing the speed of nodes. This may be due to network topology changes frequently at high speed, resulting in fewer bits received by the destination. As shown in Fig. 6, the throughput of the network using 11 Mbps data rate is high compared to other IEEE 802.11b data rates. This may be attributed to channel occupancy time, which is less using 11 Mbps than 1, 2, and 5.5 Mbps data rates. Due to this, an 11 Mbps data rate channel will be ideal for the maximum duration of time, resulting in the maximum number of received bits.

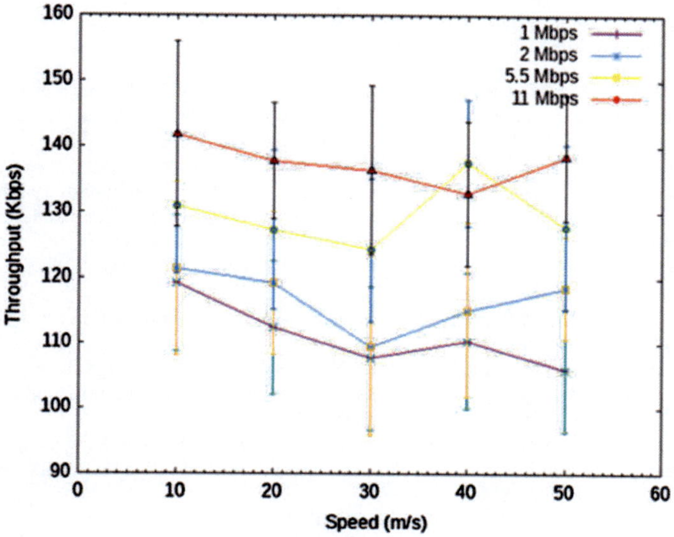

Fig. 6 Throughput with the varying speed of nodes

6 Conclusion

Unlike wired networks, wireless channel conditions change over time and space. The WLAN specification IEEE 802.11b offers multiple data rates that can maximize the system's performance against the diversity of radio signals in order to meet the variable channel conditions. Data rates should be selected according to the wireless channel conditions.

With the simulation results in this paper, we have shown the performance of wireless ad hoc network using data rates, viz., 1, 2, 5.5, and 11 Mbps that were offered by IEEE 802.11b. The performance of the network has been discussed with the help of metrics such as packet delivery ratio, lost packet ratio, average end-to-end delay, and throughput.

All the data rates of IEEE 802.11b affect the network performance in wireless ad hoc networks. As the results of our research and performance evaluations, we have discovered that the total number of packets received using 11 is more than 5.5, 2, and 1 Mbps. Further, we have observed that the PDR and throughput of the network using an 11 Mbps data rate are high compared to other data rates of IEEE 802.11b in the same type of network. We have planned to explore other aspects of wireless ad hoc networks on varying the number of nodes, source–destination pair, etc., in future.

Acknowledgements The authors would like to thank "Prof. Rajneesh Kumar Srivastava, Department of Electronics and Communication, University of Allahabad, Prayagraj," for his valuable support and guidance.

References

1. Aldalbahi A, Rahaim M, Khreishah A, Ayyash M, Little TDC (2017) Visible light communication module: an open source extension to the ns3 network simulator with real system validation. IEEE Access 5:22144–22158. https://doi.org/10.1109/ACCESS.2017.2759779
2. Athanasopoulos A, Topalis E, Antonopoulos C, Koubias S (2006) Evaluation analysis of the performance of IEEE 802.11b and IEEE 802.11g standards. In: Proceedings of the international conference on networking, international conference on systems and international conference on mobile communications and learning technologies, ICN/ICONS/MCL'06. p 141. IEEE Computer Society (2006). https://doi.org/10.1109/ICNICONSMCL.2006.92
3. Ohrtman F (2004) Voice over 802.11
4. Holt A, Huang C-Y (2010) 802.11 wireless networks. Springer London, London. https://doi.org/10.1007/978-1-84996-275-9
5. Kamerman A, Monteban L (2002) WaveLAN®-II: a high-performance wireless LAN for the unlicensed band. Bell Labs Tech. J. 2:118–133. https://doi.org/10.1002/bltj.2069
6. Bicket JC (2005) Bit-rate selection in wireless networks
7. Tseng S-M (2016) Study of mobility effect on Ad-Hoc 802.11b networks with automatic rate fallback in impulse noise channels. J Commun Technol Electron Comput Sci 4:25. https://doi.org/10.22385/jctecs.v4i0.51
8. Kim N, Choi S, Yoon H (2005) IEEE 802.11b WLAN Performance with variable transmission rates: in view of high level throughput. In: Lecture notes in computer science. Springer Verlag, pp 1080–1087. https://doi.org/10.1007/978-3-540-31957-3_123
9. Anastasi G, Borgia E, Conti M, Gregori E (2003) IEEE 802.11 ad hoc networks: performance measurements. In: 23rd international conference on distributed computing systems workshops, Proceedings, IEEE, pp 758–763. https://doi.org/10.1109/ICDCSW.2003.1203643
10. Chen Z, Zhou W, Wu S, Cheng L (2020) An Adaptive on-demand multipath routing protocol with QoS support for high-speed MANET. IEEE Access. 8:44760–44773. https://doi.org/10.1109/ACCESS.2020.2978582
11. Theerthagiri P (2020) FUCEM: futuristic cooperation evaluation model using Markov process for evaluating node reliability and link stability in mobile ad hoc network. Wirel Netw 26:4173–4188. https://doi.org/10.1007/s11276-020-02326-y
12. Yang S, Yeo CK, Lee BS (2012) Toward reliable data delivery for highly dynamic mobile ad hoc networks. IEEE Trans Mob Comput 11:111–124. https://doi.org/10.1109/TMC.2011.55
13. Singh A, Sharma S, Srivastava RK (2020) Investigation of random waypoint and steady state random waypoint mobility models in NS-3 using AODV. J High Speed Netw 26:267–274. https://doi.org/10.3233/JHS-200643
14. Yin W, Hu P, Indulska J, Portmann M, Mao Y (2020) MAC-layer rate control for 802.11 networks: a survey. Wirel Netw 26:3793–3830. https://doi.org/10.1007/s11276-020-02295-2

Impact of Parasitic Component on Bandwidth of Printed Dipole Antenna

Harshit Kumar Dhameniya, Ramesh Kumar Verma, Dheeraj Tripathi, and Shadab Azam Siddique

Abstract Impact of parasitic component on bandwidth of printed dipole antenna has been presented in this paper. A simulation using HFSS software from ANSYS has been performed. Bandwidth of the proposed antenna has been enhanced by utilizing the parasitic component. A colossal change in bandwidth from 17.69 to 54.93% demonstrates the effect of parasitic component. The proposed antenna has −18.49 dB return loss and maximum gain of 4.73 dB at 3.31 GHz frequency. A comparison between two antennas without parasitic and with parasitic component is also appeared to clarify the impact of parasitic component. FR-4 epoxy substrate of 1.6 mm thickness has been used to fabricate the proposed antenna having permittivity 4.4. Because of its wide bandwidth, the proposed antenna can be used in extensive number of applications of wireless communications.

Keywords Printed dipole antenna · Parasitic component · HFSS · Bandwidth · Gain · Wireless

1 Introduction

We are living in the age of wireless. The technology grows day by day as the small antennas take place of larger and heavier antennas. Small antennas which are made on little size PCBs are known as microstrip antennas. Compact size of such antennas is the real advantage over the other regular antennas. Distinctive shapes of the patches can be designed on the PCB. Dipole antennas are the revolutionary antennas as Yagi-Uda antenna took revolution in the field of wireless communication. Dipole antennas

H. K. Dhameniya · R. K. Verma (✉) · D. Tripathi
Department of Electronics and Communication Engineering, Bundelkhand Institute of Engineering and Technology, Jhansi, UP, India
e-mail: ramesh85.ec@gmail.com

S. A. Siddique
Department of Electronics and Communication Engineering, National Institute of Technology Patna, Patna, India

B. Mishra and M. Tiwari (eds.), *VLSI, Microwave and Wireless Technologies*, Lecture Notes in Electrical Engineering 877,
https://doi.org/10.1007/978-981-19-0312-0_43

are notable for their directivity, high bandwidth, and efficiency [1]. Presently, printed dipole antennas are in trend because of their high bandwidth and directivity.

The drawback of the microstrip patch antennas is its narrow bandwidth. Printed dipole antennas may provide the solution for such drawbacks. Printed dipole antennas may have several shapes and sizes, and they may likewise vary in their applications. In [2], parasitic components are utilized for the pattern reconfiguration of printed dipole antenna. The introduced antenna is dealt with 2.40 GHz and could radiate signals in two directions. In [3], a printed structure of a cross-dipole antenna is recommended which is useful in the WLAN applications. A novel antenna with H-shaped structures is displayed in [4] utilizing numerous PIN diodes are executed on H-formed structure for pattern reconfiguration. In [5], a compact size end fire printed dipole antenna has been designed at 45° edge, and a balun-type structure is used at the feed to get wide frequency band. A double-band printed dipole antenna has been acquainted in [6] with diminishing the space attenuation difference between the frequencies of 2.45 and 5.80 GHz. This antenna contains a dipole and a reflector. It is valuable in the uses of Bluetooth and other wireless applications. In [7], a printed dipole antenna is depicted for wireless communication. Verma and Srivastava [8] proposed an inverted Y-shape patch antenna of size 33 $\times$ 40 mm^2 with fractional bandwidth of 36.30% (933 MHz). Fan et al. [9] proposed a wideband printed dipole antenna for enhancing the bandwidth. In [10], Li et al. display an aperture-coupled double-polarized filtering antenna. Bandwidth enhancement of an "inverted SHA" shape of microstrip patch antenna with defected ground has been presented in [11]. The bandwidth improvement [12] has been done using double-layered substrate and metamaterial layer structure. Mbinack and Tonye [13] proposed a printed dipole antenna having BW 22% using strip line feed for WLAN application. Rahman et al. [14] designed a wideband printed dipole antenna using CST Microwave Studio for 5G application. Jain et al. [15] presented a wideband dipole antenna of circularly polarized with I-shape slot for C-band application. Bala et al. [16] presented a metamaterial-based antenna of size 40 $\times$ 48 mm^2 with 41% (1118 MHz) bandwidth resonating at 2.73 GHz. Neyestanak et al. [17] proposed a W-shape antenna having 36.7% (800 MHz) bandwidth with large size of 70 $\times$ 50 mm^2. Wu and Wong [18] proposed direct coupled antenna of size 52.87 $\times$ 40 mm^2 resonating at 2.879 GHz with 12.7% (365 MHz) bandwidth. Verma and Srivastava [19] proposed an H-shape patch antenna of size 39.04 $\times$ 47.64 mm^2 showing bandwidth of 35.61% (710 MHz) loaded with three square slots. Deshmukh and Ray [20] proposed a broadband antenna of size 51 $\times$ 110 mm^2 having L-shape slots with 28% (290 MHz). Rajpoot et al. [21] designed an antenna of I-shape patch with overall size of 40.06 $\times$ 48.72 mm^2 resonating at 2.41 GHz with 45.72% (970 MHz) bandwidth. Verma and Srivastava [22] presented an antenna with inverted T-shape slot of size 38.43 $\times$ 46.86 mm^2 having bandwidth of 48.25% (1179 MHz). Kamakshi et al. [23] proposed an antenna with large size of 120 $\times$ 80 mm^2 having three notches and one slot resonating at 1.84 GHz. Deng et al. [24] designed a compact antenna excited by tuned loop of size 56 $\times$ 56 mm^2 with bandwidth of 460 MHz. Verma and Srivastava [25] designed an antenna with slotted geometry of size 39 $\times$ 47.6 mm^2 having bandwidth about 48.68% (1155 MHz). Sze and Wong [26] proposed an antenna of size 60 $\times$ 50 mm^2 with two slots of right angle shape and a U-shape slot with narrow bandwidth of 53 MHz. Mishra et al. [27, 28] proposed a petal shape dual-band gap coupled antenna and slot loaded for both dual-band and wideband stacked antenna. Gupta et al. [29] designed a wideband gap and direct coupled antenna resonating at 2.399 GHz. Singh et al. [30] presented a circular patch antenna of hexa-band characteristics with inverted L-shape notch.

The design and the consequences of the portrayed antenna are talked about. The parasitic component has been used to improve the bandwidth of the antenna. Designing process, simulation, and results have been discussed. In order to design the printed dipole antenna with parasitic component, a conventional antenna is designed without parasitic component. After that, a parasitic component is executed in the design. Then simulation results of both the designs are compared to demonstrate the changes in bandwidth. HFSS [31] software from ANSYS has been used for simulation and analysis. To assess the antenna execution, some fundamental parameters such as return loss, operating frequency, working bandwidth, VSWR, gain, and radiation pattern are explored and analyzed.

2 Printed Dipole Antenna with No Parasitic Component

Remembering the ultimate objective to setup printed dipole antenna with parasitic component, an antenna has been designed with no parasitic component at resonant frequency utilizing 1.6 mm thick substrate of FR-4 epoxy [32]. The horizontal length of substrate is 47 mm, and vertical length of substrate is 39 mm. The permittivity of the FR-4 epoxy is 4.4. All the design specifications and dimensional parameters are shown in Table 1. Design and the dimensional measurements of the depicted antenna with no parasitic component are shown in Fig. 1.

To outline substrate in HFSS, the measurements are given to the chosen material of the substrate. After designing the substrate, radiating dipole has been designed. To design the radiating structure, we design reflector, base, and dipole at the top of substrate as mentioned in Fig. 1. Measurements of the reflector, base, and dipole are additionally appeared in Fig. 1. At the base of the substrate, the ground plane of

Table 1 Design and dimensional parameters

S. No.	Parameters	Value
1	Substrate thickness	1.6 mm
2	Dielectric constant	4.4
3	Design frequency	3.2 GHz
4	Loss tangent	0.02
5	Horizontal length of substrate	47 mm
6	Vertical length of substrate	39 mm
7	Feed length	7 mm
8	Feed width	3 mm

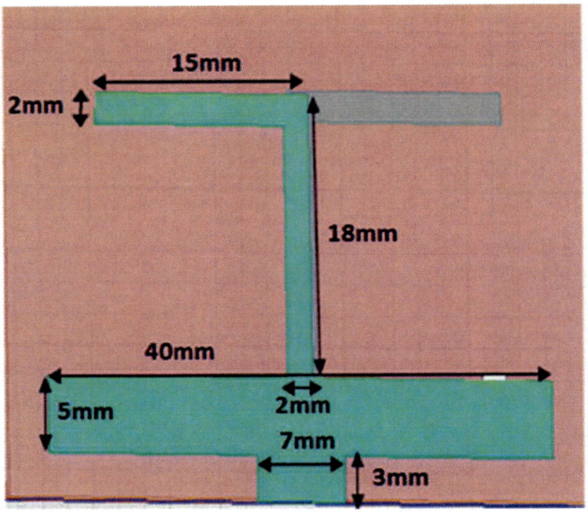

Fig. 1 Geometry of printed dipole antenna with no parasitic component

the dipole is outlined. In the wake of planning the all-radiating segment, a perfect E boundary is allocated for each radiating segment. After assigning the boundary, a strip of length 7 mm and width 3 mm has been associated at lower edge of proposed antenna to provide feed. After that, a setup is prepared for the analysis of the antenna design for 1–6 GHz. First, the simulation consequences of the portrayed antenna with no parasitic component have been inspected. The return loss, bandwidth, and resonant frequency of the depicted antenna have been gotten from Fig. 2. The antenna is radiating at frequency 3.23 GHz with return loss of −18.63 dB. Bandwidth of proposed antenna accomplished 17.69% between lower frequency 3 GHz and higher frequency 3.59 GHz.

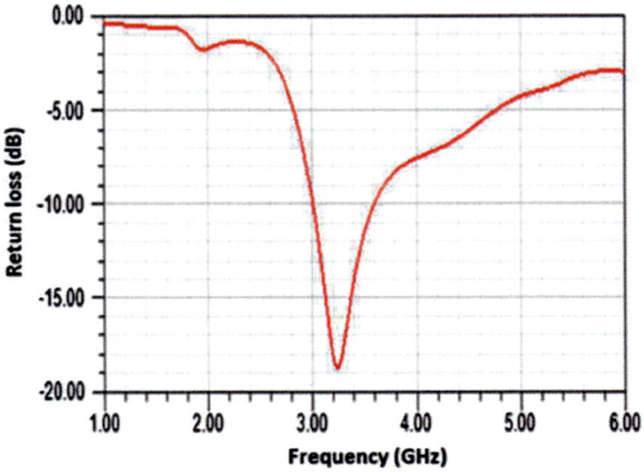

Fig. 2 Return loss plot of printed dipole antenna with no parasitic component

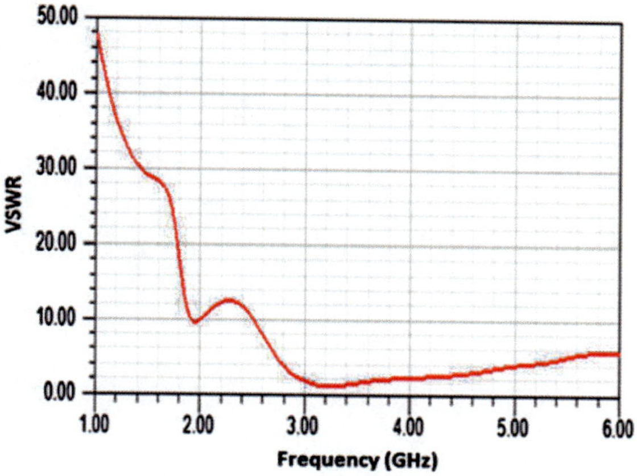

Fig. 3 VSWR plot of printed dipole antenna with no parasitic component

Estimation of VSWR is 1.265 at radiating frequency 3.23 GHz. The VSWR plot of printed dipole antenna with no parasitic component has been shown in Fig. 3.

3 Printed Dipole Antenna with Parasitic Component

To design the printed dipole antenna, a parasitic component of length 15 mm and width 2 mm has been composed on the front side of substrate at 2 mm above from the dipole. Entire procedure of designing is same as the printed dipole antenna with no parasitic component. Figure 4 shows the HFSS design of depicted antenna with a parasitic component.

By introducing parasitic component in the traditional dipole antenna, the bandwidth of antenna has been expended. The bandwidth of the antenna is 54.93% which

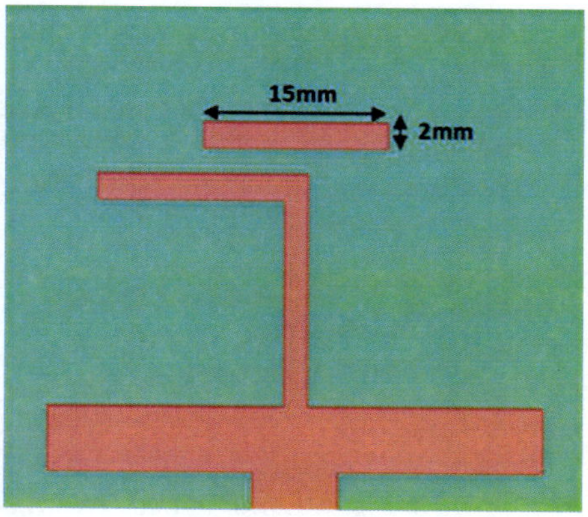

Fig. 4 Printed dipole antenna with parasitic component

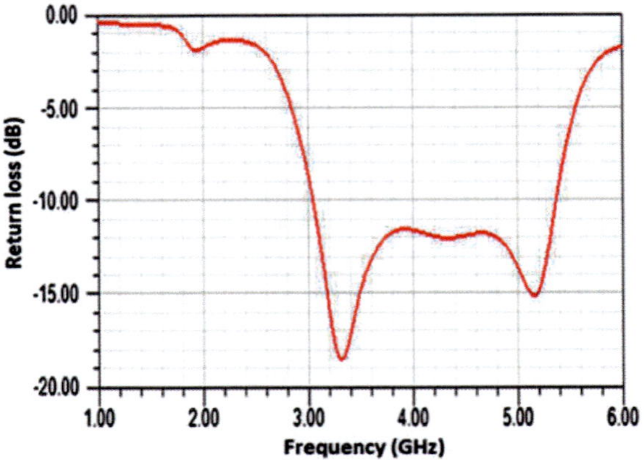

Fig. 5 Return loss plot of printed dipole antenna with a parasitic component

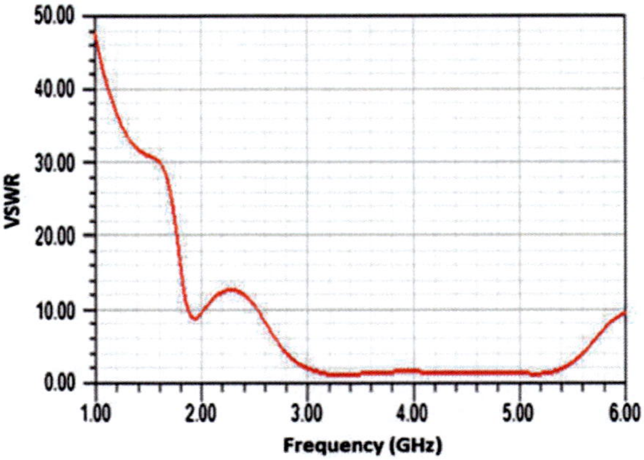

Fig. 6 VSWR plot of printed dipole antenna with a parasitic component

lies in the vicinity of 3.05 and 5.36 GHz. The return loss versus frequency plot of printed dipole antenna with a parasitic component has been appeared in Fig. 5.

The return loss of proposed antenna is −18.49 dB at the resonant frequency of 3.31 GHz. The bandwidth of printed dipole antenna with a parasitic component is expended from 17.69 to 54.93%. The VSWR of the antenna is 1.270 at resonant frequency 3.31 GHz. The VSWR plot of printed dipole antenna with a parasitic component is appeared in Fig. 6.

4 Results and Discussion

A comparison of return loss between two antennas which are mentioned above has been shown in Fig. 7. From figure, we can see an increment in bandwidth in the wake

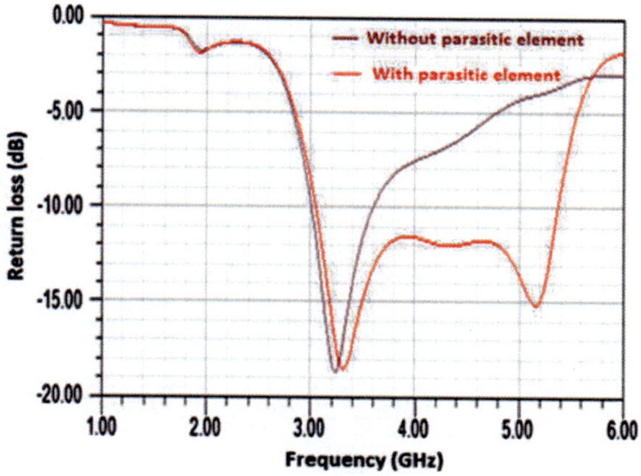

Fig. 7 Comparison between printed dipole antenna with parasitic and without parasitic component

of utilizing the parasitic component. When no parasitic component is exhibited, the bandwidth of the proposed antenna is 17.69%. After using the parasitic component, it becomes 54.99%. Hence, the bandwidth has been expended by using parasitic component. The red color graph indicates the return loss of the antenna with parasitic component, while blue color graph shows the return loss of the antenna with no parasitic component. The red color graph shows the more bandwidth in comparison of blue color graph. Thus, this comparison graph shows the impact of the parasitic component on the antenna's bandwidth. The comparative analysis of presented antenna with references [16–26] is shown in Table 2. The comparison of antenna size and bandwidth with references [16–26] is shown in Figs. 8 and 9, respectively.

From Fig. 7, we conclude that both proposed antennas are resonating at almost same resonant frequency and return loss but bandwidth is enhanced. The 2D radiation pattern of the proposed printed dipole antenna with parasitic element at $\varphi = 0°$ and $\varphi = 90°$ is shown in Fig. 10. The gain of the antenna is 4.73 dB at resonant frequency.

5 Hardware Design and Experimental Result

Hardware of the proposed printed dipole antenna with parasitic component has been planned on FR-4 glass epoxy PCB. The proposed hardware design of the antenna is appeared in Figs. 11 and 12, respectively. The experimental return loss of printed dipole antenna is compared with its simulated results. This comparison has been shown in Fig. 13. The experimental bandwidth and return loss of printed dipole antenna are 32.42% and −16.78 dB, respectively, at radiating frequency 3.28 GHz. The simulated bandwidth and return loss of described antenna are 54.93% and −18.49 dB, respectively, at 3.31 GHz. Experimental plot of return loss from vector analyzer is also shown in Fig. 14.

Table 2 Comparison of presented antenna with references [16–26]

References	Dimensions (mm^2)	Frequency range (GHz)	Resonating freq. (GHz)	Bandwidth (%)	Application
[16]	40 × 48 = 1920	2.233–3.351	2.73	41% (1118 MHz)	WLAN/WiMAX
[17]	70 × 50 = 3500	1.8–2.6	1.88/2.37	36.7% (800 MHz)	RFID/WLAN
[18]	52.87 × 40 = 2115	2.696–3.061	2.879	12.7% (365 MHz)	Broadband
[19]	39.04 × 47.64 = 1860	1.639–2.349	1.729/2.223	35.61% (710 MHz)	S-Band
[20]	55 × 110 = 6050	0.88–1.17	0.89/1.16	28% (290 MHz)	GSM
[21]	40.06 × 48.72 = 1951	1.65–2.62	2.41	45.72% (970 MHz)	WLAN
[22]	38.43 × 46.86 = 1800	1.854–3.033	2.477	48.25% (1179 MHz)	WLAN
[23]	120 × 80 = 9600	1.56–2.12	1.84	30.5% (560 MHz)	Broadband
[24]	56 × 56 = 3136	1.71–2.17	NA	(460 MHz)	DCS/PCS/UMTS
[25]	39 × 47.6 = 1854	1.795–2.950	2.391	48.68% (1155 MHz)	WLAN/WiMAX
[26]	60 × 50 = 3000	2.197–2.25	2.224	2.4% (53 MHz)	S-Band
Proposed	47 × 39 = 1833	3.05–5.36	3.31	54.93% (2310 MHz)	S- and C-Band

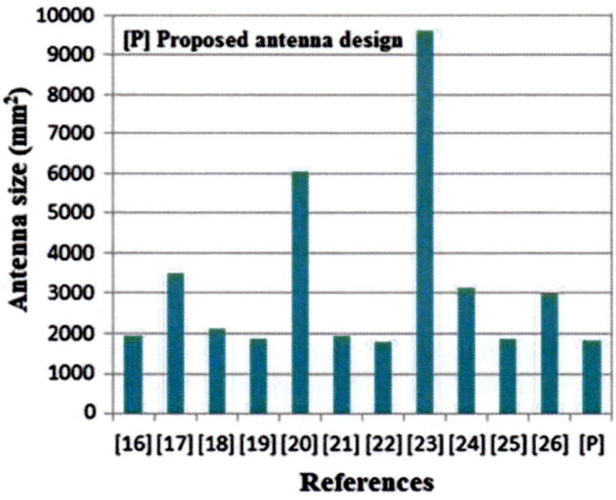

Fig. 8 Comparison of antenna size

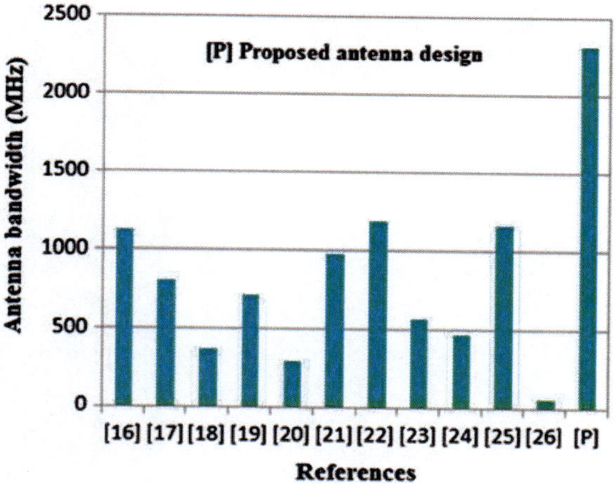

Fig. 9 Comparison of antenna bandwidth

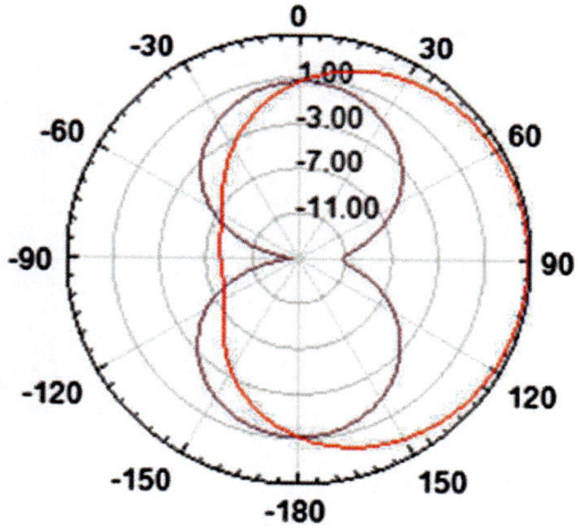

Fig. 10 Radiation pattern of printed dipole antenna with parasitic component at $\varphi = 0°$ (—) and $\varphi = 90°$ (—)

6 Conclusion

A printed dipole antenna with a parasitic component has been outlined, manufactured, and analyzed. The depicted antenna looks like a printed Yagi-Uda antenna. It additionally demonstrates a maximum radiation gain of 4.73 dB at radiating frequency 3.31 GHz. The radiation efficiency of the antenna is also very high (89.29%). The return loss of the printed dipole antenna is −18.49 dB at transmitting frequency, and this return loss helps for the impedance coordinating between the antenna and the microstrip feed line. The VSWR of this antenna is 1.270 at the radiating frequency 3.31 GHz. The estimation of VSWR is near to the 1, and we can state that printed dipole antenna with 50 Ω impedance is perfectly matched with the microstrip feed line. The described antenna is also overcome from the issue of narrow bandwidth

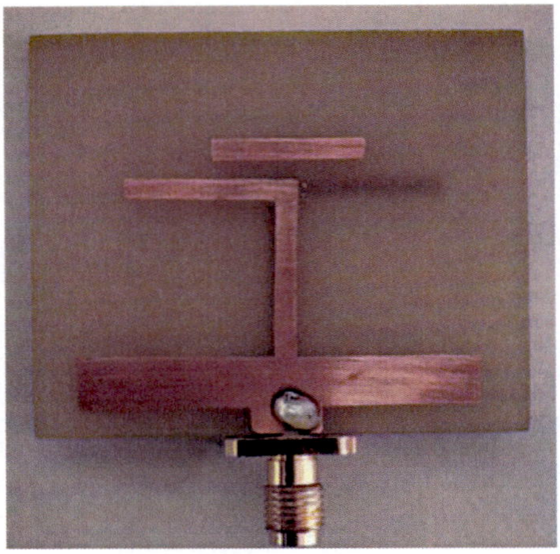

Fig. 11 Front view of fabricated antenna

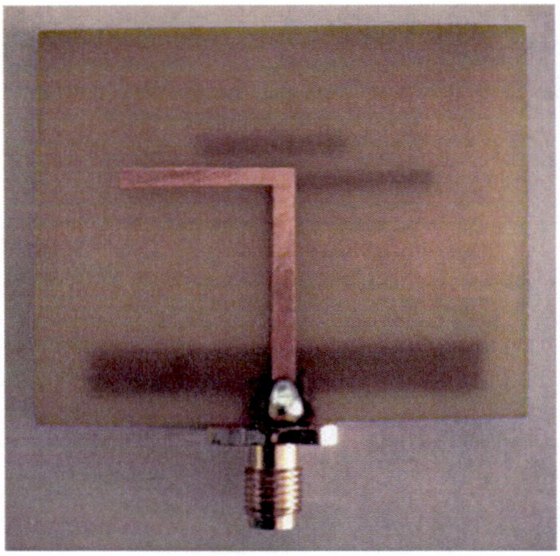

Fig. 12 Base view of fabricated antenna

by achieving 54.93% bandwidth, which is generally not occurred in printed dipole antennas.

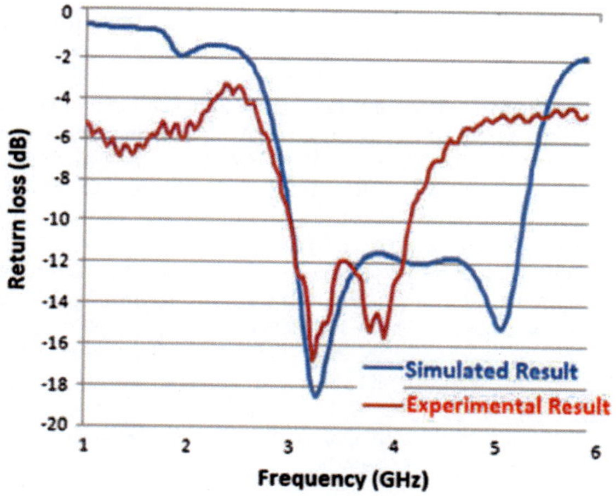

Fig. 13 Comparison between simulated and experimental return loss plots

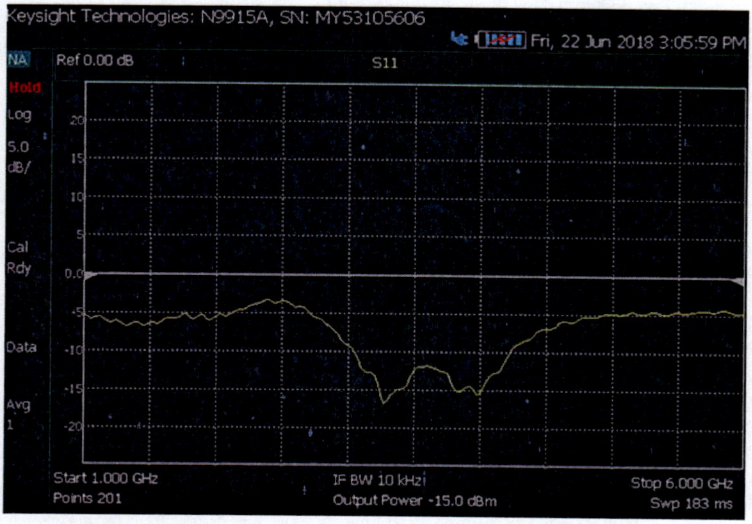

Fig. 14 Experimental result of printed dipole antenna with a parasitic component

Conflict of Interest The author declares no potential conflict of interest.

References

1. Yagi H, Uda S (1926) Projector of the sharpest beam of electric waves. Proc Imperial Acad 2(2):49–52

2. Smida B, Farzami F, Khaledian S, Erricolo D (2016) Pattern reconfigurable printed dipole antenna using loaded parasitic elements. IEEE Antennas Wirel Propag Lett 16:1151–1154
3. Park SB, Eom SY (2017) Printed cross-dipole antenna with slow wave structure for highly efficient and directive WLAN applications. Electron Lett 53(13):826–828
4. Ren J, Yang X, Yin J, Yin Y (2015) A novel antenna with reconfigurable patterns using H-shaped structures. IEEE Antennas Wirel Propag Lett 14:915–918
5. Ta SX, Park I (2017) Broadband printed-dipole antennas for millimeter-wave applications. In: IEEE Conferences, pp 65–67
6. Floc JM, Ahmad AES (2012) Dual-band printed dipole antenna with parasitic element for compensation of frequency space attenuation. Int J Electromagn Appl 2(5):122–128
7. Munir A, Aviolanda S, Bharata E (2016) Wideband printed dipole antenna fed using modified planar marchand balun. In: IEEE Conferences, pp 295–298
8. Verma RK, Srivastava DK (2018) A compact inverted Y slot rectangular microstrip patch antenna for Bluetooth applications. Indonesian J Electr Eng Comput Sci 11(2):413–418
9. Fan ST, Yin YZ, Hu W, Li B, Yang JH (2012) Bandwidth enhancement of a printed dipole antenna for wideband applications. Microw Opt Technol Lett 54(7):1585–1590
10. Li R, Hua C, Lu Y, Wu Z (2017) Dual-polarized aperture-coupled filtering antenna. In: IEEE Conferences, pp 152–153
11. Fistum D, Mali D, Ismail M (2016) Bandwidth enhancement of rectangular microstrip patch antenna using defected ground structure. Indonesian J Electr Eng Comput Sci 3(2):428–434
12. Rao VS, Reddy KVVS, Prasad AM (2017) Bandwidth enhancement of metamaterial loaded microstrip antenna using double layered substrate. Indonesian J Electr Eng Comput Sci 5(3):661–665
13. Mbinack C, Tonye E (2018) Microstrip center-fed Bi-plane printed dipole antenna design and the excitation gap contribution for impedance control. Wireless Pers Commun 103(2):1713–1724
14. Rahman DA, Mohamad SY, Malek NA, Rahman DA, Zabri SN (2018) A wideband mm-wave printed dipole antenna for 5G applications. Indonesian J Electr Eng Comput Sci 10(3):943–950
15. Jain AK, Kanaujia BK, Dwari S, Panday GP, Singh DK (2018) Wideband circularly polarized magnetoelectric dipole antenna with I-slot for C-band applications. J Comput Electron 18:660–670
16. Bala BD, Rahim MKA, Murad NA (2015) Bandwidth enhancement metamaterial antenna based on transmission line approach. Microw Opt Technol Lett 57(1):252–256
17. Neyestanak AAL, Kashani FH, Barkeshli K (2007) W-shaped enhanced-bandwidth patch antenna for wireless communication. Wireless Pers Commun 43:1257–1265
18. Wu CK, Wong KL (1999) Broadband microstrip antenna with directly coupled and parasitic patches. Microw Opt Technol Lett 22:348–349
19. Verma RK, Srivastava DK (2019) A compact rectangular microstrip antenna of H-shape patch loaded with notch and slots for S-band applications. Int J Microwave Opt Technol 14(2):100–108
20. Deshmukh AA, Ray KP (2013) Analysis of L-shaped slot cut broadband rectangular microstrip antenna. Int J Electron 100(8):1108–1117
21. Rajpoot V, Srivastava DK, Saurabh AK (2014) Optimization of I-shape microstrip patch antenna using PSO and curve fitting. J Comput Electron 13(4):1010–1013
22. Verma RK, Srivastava DK (2019) Design, optimization and comparative analysis of T-shape slot loaded microstrip patch antenna using PSO. Photon Netw Commun 38(3):343–355
23. Kamakshi K, Singh A, Aneesh M, Ansari JA (2014) Novel design of microstrip antenna with improved bandwidth. Int J Microwave Sci Technol 1–7, 659592
24. Deng J, Guo L, Yin Y, Qiu J, Wu Z (2013) Broadband patch antennas fed by novel tuned loop. IEEE Trans Antennas Propag 61(4):2290–2293
25. Verma RK, Srivastava DK (2021) Optimization and parametric analysis of slotted microstrip antenna using particle swarm optimization and curve fitting. Int J Circ Theory Appl 1–16
26. Sze JY, Wong KL (2000) Slotted rectangular microstrip antenna for bandwidth enhancement. IEEE Trans Antennas Propag 48(8):1149–1152

27. Mishra B, Singh V, Singh R (2018) Gap coupled dual-band petal shape patch antenna for WLAN/WiMAX applications. Adv Electr Electron Eng 16(2):185–198
28. Mishra B, Singh V, Singh R (2017) Dual and wide-band slot loaded stacked microstrip patch antenna for WLAN/WiMAX applications. Microsyst Technol 23(8):3467–3475
29. Gupta A, Srivastava DK, Saini JP, Verma RK (2020) Comparative analysis of microstrip-line-fed gap-coupled and direct-coupled microstrip patch antennas for wideband applications. J Comput Electron 19(1):457–468
30. Singh V, Mishra B, Dwivedi AK, Singh R (2018) Inverted L-notch loaded hexa band circular patch antenna for X, ku/K band applications. Microw Opt Technol Lett 60(8):2081–2088
31. ANSYS HFSS (High frequency electromagnetic field simulation software)
32. Balanis CA (2005) Antenna theory, analysis and design. Wiley, Hoboken

Charge Selective Layer Optimization of a Double Perovskite Solar Cell by Numerical Simulation

Shambhavi Rai, B. K. Pandey, and D. K. Dwivedi

Abstract The use of lead in conventional metal halide perovskite leads to the toxicity and hindrance in commercialization of these PSCs. The present work is a study about the Cesium Titanium (IV) Halide-based perovskite solar cell which is lead-free and eco-friendly. The planar device structure FTO/ZnO (ETL)/Cs_2TiX_6 (X = I, Br)/P3HT (HTL)/Au has been initially studied and simulated using SCAPS-1D solar cell simulation software. Further the device FTO/ETL/Cs_2TiBr_6/HTL/Au structure was optimized using diverse charge selective layers as ETL and HTL to find the most suitable ETL and HTL for our device. It was found that WS_2 as ETL and Cu_2O and CuSCN as HTL gives the optimum solar cell performance parameters. Moreover, the thickness of absorber was optimized and a good PCE of 15.9% was achieved at a moderate thickness of 400 nm Cs_2TiBr_6 for FTO/WS_2/Cs_2TiBr_6/Cu_2O/Au device.

Keywords Lead-free · Perovskite · ETL · HTL · SCAPS-1D

1 Introduction

In the past decade, a material, with superior optoelectronic properties for application in solar cells, called metal halide perovskite got a lot of attention amongst the researchers. These metal halide perovskite material possesses excellent electron-to-photon conversion efficiency, large tolerance to defects, good light capturing ability, long diffusion length and easy fabrication technique which makes it suitable for optoelectronic application. The intensive research in this era have raised the power conversion efficiency (PCE) of Perovskite solar cells (PSCs) from 3.5 to 25.2% in last few years [1].

The organic–inorganic hybrid perovskite structure ABX_3, where A is Cs^+, FA^+ or MA^+, B is most commonly lead (Pb) and X is I, Br or their mixture, have been widely studied due to their suitable bandgap and attractive optoelectronic properties

S. Rai · B. K. Pandey · D. K. Dwivedi (✉)
Department of Physics and Material Science, Photonics and Photovoltaic Research Lab, Madan Mohan Malaviya University of Technology, Gorakhpur 273010, India
e-mail: todkdwivedi@gmail.com

B. Mishra and M. Tiwari (eds.), *VLSI, Microwave and Wireless Technologies*,
Lecture Notes in Electrical Engineering 877,
https://doi.org/10.1007/978-981-19-0312-0_44

[2–4]. However, the organic cation in the perovskite causes instability in the device structure whilst toxicity and disposal of lead-based perovskite is one of the biggest obstacle in the path of commercialization of PSCs. To overcome the issues rising due to lead various attempts have been made to replace Pb by other inorganic metal substituents like tin (Sn^{2+}), silver (Ag^{+}), antimony (Sb^{3+}), bismuth (Bi^{3+}) and copper (Cu^{2+}) but the resulting PSC shows less stability and poor device output parameters. Moreover, most of these substituents are still less environment friendly [5, 6].

Recently, titanium (Ti) have been proposed as an alternative to bypass the problems arising due to lead. Chen et al. introduced Cesium titanium (IV) bromide (Cs_2TiBr_6) thin film perovskites which showed a PCE of 3.28%. The tunable bandgap of cesium titanium halide from 1.4 to 1.8 eV facilitates its application in photovoltaic devices [7].

In the present work, initially a comparative study amongst Cs_2TiI_6 and Cs_2TiBr_6 has been done and the device structure FTO/ZnO/Cs_2TiBr_6/P3HT/Au has been optimized by employing various charge transport layers (CTLs). Since a large number of CTLs are proposed in the literature, thus the objective of this work is to find the most appropriate CTL for our proposed device so that we get optimal device output parameters. Further the impact of absorber layer thickness variation on the solar cell device output parameters has been studied.

2 Device Structure and Simulation Parameters

The simulation in this work has been done using SCAPS-1-D, i.e. solar cell capacitance simulation software which was developed by University of Gent [8]. On the basis of three basic equations of semiconductor which are electron continuity equation, hole continuity equation and Poisson's equation the SCAPS calculates and extracts the device output parameters like open circuit voltage (V_{OC}), short circuit current density (J_{SC}), fill factor (FF), PCE, quantum efficiency (QE), band alignment, etc.

The lead-free device structures used for simulation are shown in Fig. 1. In the structure shown in Fig. 1, zinc oxide (ZnO) acts as electron transport layer and [poly(3-hexylthiophene-2,5-diyl)] P3HT acts as hole transport layer. The Cs_2TiI_6 and Cs_2TiBr_6 with bandgap 1.8 and 1.6 eV were employed as the light harvesting material in the proposed planar device structure. Fluorine doped tin oxide (FTO) with work function 4.4 eV and gold (Au) with work function 5.1 eV have been employed as front and back contact, respectively. The physical parameters used in our simulation are given in Table 1 which are taken from the literature and their references are cited [9, 10]. All the simulation was done under AM1.5G solar spectrum with an incident power density of 100 mW/cm^2 at 300 K.

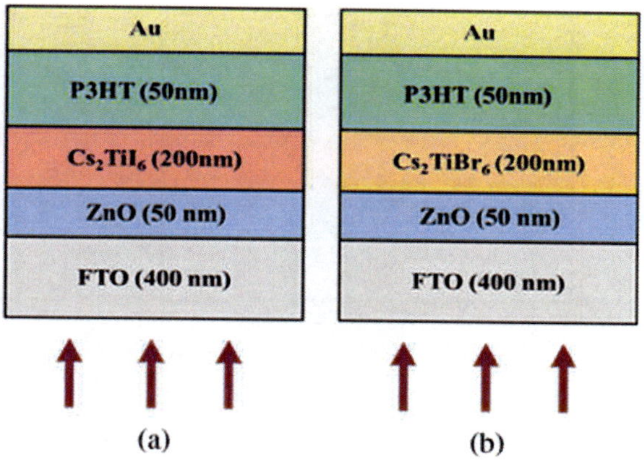

Fig. 1 Device structure **a** with Cs_2TiI_6 absorber and **b** with Cs_2TiBr_6 absorber

Table 1 Physical parameters of the materials used in simulation

Parameters	FTO	ZnO	Cs_2TiI_6	Cs_2TiBr_6	P3HT
Thickness (nm)	400	50	200	200	50
E_g (eV)	3.5	3.3	1.55	1.8	1.2
χ (eV)	4	4	3.9	4.2	3.88
ϵ_r	9.0	9	6.5	18	6.5
N_c (cm^{-3})	2.2×10^{18}	4×10^{18}	2.2×10^{18}	1×10^{19}	2.2×10^{18}
N_v (cm^{-3})	1.8×10^{19}	1×10^{19}	1.8×10^{19}	1×10^{18}	1.8×10^{19}
μ_n (cm^2/Vs)	20	100	2	150	2
μ_p (cm^2/Vs)	20	25	2	585	2
V_e (cm/s)	1×10^7	1×10^7	1×10^7	1×10^7	1×10^7
V_h (cm/s)	1×10^7	1×10^7	1×10^7	1×10^7	1×10^7
N_D (cm^{-3})	2×10^{19}	1×10^{18}	–	–	1×10^{13}
N_A (cm^{-3})	–	–	–	1×10^{14}	1×10^{17}
N_t (cm^{-3})	1×10^{15}	2×10^{17}	2.5×10^{14}	1×10^{18}	1×10^{15}
		(Donor, uniform 1.7 eV above E_v)			

3 Results and Discussion

The resulting *J*-*V* curve and QE curve of both the device structures have been shown in Fig. 2. The extracted output parameters have been given in Table 2. From Fig. 2 we observe that the device based on Cs_2TiBr_6 outperforms the Cs_2TiI_6-based device. The higher performance parameters like open circuit voltage (V_{OC}), short circuit current density (J_{SC}), fill factor (FF) and PCE of the Br-based device (as shown in Table 2) is attributed to the lower bandgap of Cs_2TiBr_6 which results in absorption of more light in visible region than that of Cs_2TiI_6 (as illustrated in Fig. 2b). Also this

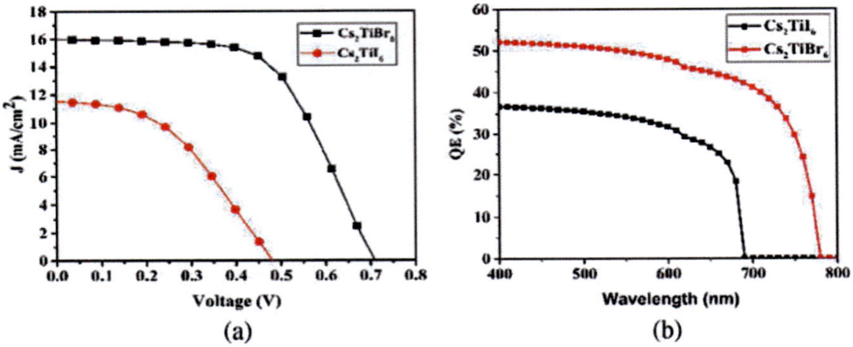

Fig. 2 **a** *J-V* curve of Cs_2TiI_6 and Cs_2TiBr_6-based devices **b** QE curve of Cs_2TiI_6 and Cs_2TiBr_6-based devices

Table 2 The solar cell output parameters of Cs_2TiI_6 and Cs_2TiBr_6-based devices

Parameters	FTO/ZnO/Cs_2TiI_6/P3HT/Au	FTO/ZnO/Cs_2TiBr_6/P3HT/Au
V_{oc} (V)	0.5077	0.7078
J_{sc} (mA/cm^2)	14.6957	15.9981
FF (%)	62.42	59.56
PCE (%)	4.66	6.74

can be attributed to the combined effect of electron affinity which results in proper band alignment and high carrier mobility of Cs_2TiBr_6-based device. Thus, as per our simulation results we opted Cs_2TiBr_6-based device for further optimization as it was showing better performance.

3.1 *Impact of Different Electron Transport Layers (ETLs)*

The performance of PSCs extensively depend upon the ETL layer parameters like electron affinity, charge conductivity, band alignment, carrier mobility, relative permittivity and its interfaces with other layers. Moreover, an ETL with higher conductivity is beneficial to lower the ohmic losses whereas large carrier mobility of ETLs facilitates fast carrier extraction thus there is less charge accumulation at the interface [11]. Initially we employed ZnO as ETL in all the devices. Though ZnO is an excellent electron transporter with high mobility but in practical a serious recombination rises at the perovskite/ZnO interface due to lack of crystal symmetry which results in defects at the interface and limits the practicability of the device. That is why a variety of ETL materials has been proposed in the literature to overcome such problems. In order to get better performance of the proposed device a number of ETLs were employed. As an ETL 50 nm of tungsten disulphide (WS_2),

Table 3 Input parameters of proposed ETL materials

Parameters	TiO_2	WO_3	ZnO	WS_2	SnO_2	ZnSe	PCBM
Thickness (nm)	50	50	50	50	50	50	50
E_g (eV)	3.2	2.6	3.3	1.8	3.6	2.9	2.0
χ (eV)	3.9	3.8	4	4.2	4.0	4.02	3.9
ϵ_r	9.0	4.8	9	13.6	9	10	4.0
N_c (cm^{-3})	1×10^{21}	2.2×10^{21}	4×10^{18}	2.2×10^{17}	2.2×10^{18}	2.2×10^{18}	1×10^{21}
N_v (cm^{-3})	2×10^{20}	2.2×10^{21}	1×10^{19}	2.2×10^{16}	1.8×10^{19}	1.8×10^{19}	2×10^{20}
μ_n (cm^2/Vs)	20	30	100	100	20	25	0.01
μ_p (cm^2/Vs)	10	30	25	100	10	100	0.01
V_e (cm/s)	1×10^7	1×10^7	1×10^7	1×10^7	1×10^7	1×10^7	1×10^7
V_h (cm/s)	1×10^7	1×10^7	1×10^7	1×10^7	1×10^7	1×10^7	1×10^7
N_D (cm^{-3})	1×10^{19}	6.5×10^{17}	1×10^{18}	1×10^{18}	1×10^{17}	1×10^{17}	1×10^{15}
N_A (cm^{-3})	1	–	–	–	–	–	–
N_t (cm^{-3})	1×10^{15}	1×10^{15}	2×10^{17}	1×10^{15}	1×10^{15}	1×10^{15}	1×10^{15}
			(Donor, uniform 1.7 eV above E_v)				

tungsten trioxide (WO_3), organic [6]-phenyl-C60-butyric acid methyl ester (PCBM), zinc selenide (ZnSe), titanium dioxide (TiO_2) and tin oxide (SnO_2) have been studied and compared to ZnO. All the relevant references from which the ETL parameters have been taken are cited [10, 12–17]. The parameters of these ETLs used in simulation are summarized in Table 3. The *J-V* curve and the FF, PCE variation for distinct ETLs used in the PSC are shown in Fig. 3a, b, respectively, and the PSC output parameters thus extracted using distinct ETLs has been given in Table 4.

From the simulation results we observe that all the parameters for WS_2 ETL-based device are better amongst all the ETLs except the FF (58.6%) which may be due to the high series resistance. The carriers mobility in WS_2 is maximum (100 cm^2/Vs) which is one of the governing factor in the improving the performance of WS_2-based PSC. PCBM also showed the parameters comparable to WS_2 with maximum FF of 59.6%. TiO_2-based device showed an efficiency of 6.92% which is close to one that is based on PCBM. All other ETLs showed the comparable performance parameters with efficiency nearly 6.7% as summarized in Table 4. Thus, WS_2 is found to be the most efficient ETL for the proposed device structure and further the simulation was proceeded by taking WS_2 as the ETL.

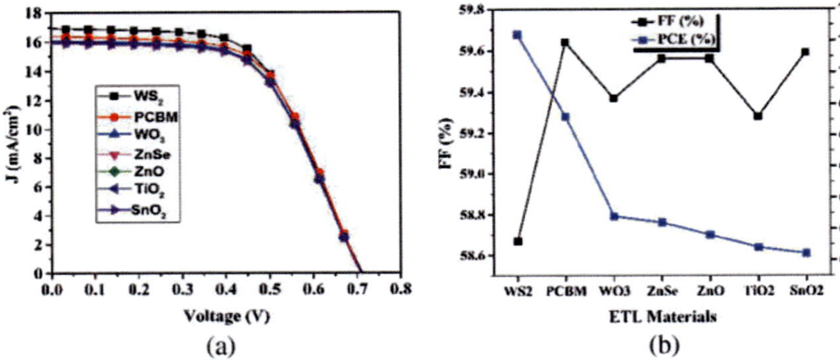

Fig. 3 **a** The *J-V* curve for different ETLs and **b** the PCE, FF variation for distinct ETLs used in the PSC

Table 4 The output parameters obtained using different ETL materials

S. No.	ETL materials	V_{oc} (V)	J_{sc} (mA/cm^2)	FF (%)	PCE (%)
1	WS_2	0.7110	16.9244	58.67	7.06
2	PCBM	0.7086	16.4074	59.64	6.93
3	WO_3	0.7083	16.0943	59.37	6.77
4	ZnSe	0.7079	16.0286	59.56	6.76
5	ZnO	0.7078	15.9980	59.56	6.74
6	TiO_2	0.7081	16.0016	59.28	6.92
7	SnO_2	0.7076	15.9146	59.59	6.71

3.2 *Impact of Different Hole Transport Layers (HTLs)*

In the previous section we got the most suitable ETL for our device. In the similar way in this section a number of organic and inorganic HTLs have been employed to the simulated structure having WS_2 as the ETL material. Thus the various materials which were employed are mainly of two types—(i) inorganic HTL like copper iodide (CuI), copper oxide (Cu_2O) and copper thiocyanate (CuSCN) and (ii) organic HTLs like Spiro-OMeTAD (2,2′,7,7′-tetrakis-(*N*,*N*-di-p-methoxyphenyl-amine)-9,9′-spirobifluorene), PEDOT:PSS [Poly(3,4-ethylenedioxythiophene)-poly(styrenesulfonate)] and P3HT [poly (3-hexylthiophene-2,5-diyl)]. The input parameters used in the simulation are summarized in Table 5 and their references have been cited [10, 18, 19]. The *J-V* curve corresponding to each of these HTLs and their FF and PCE variation have been shown in Fig. 4 and the resulting values of the device output parameters have been listed in Table 6.

From the *J-V* curve for different ETLs shown in Fig. 4a it is observed that the P3HT HTL is showing a larger series resistance amongst all other HTLs except NiO_x (as can be observed from *J-V* near V_{OC} in Fig. 4a). NiO_x-based device is

Table 5 Input parameters of proposed HTL materials

Parameters	CuI	Spiro-OMeTAD	Cu_2O	PEDOT:PSS	NiO_x	CuSCN	P3HT
Thickness (nm)	50	50	50	50	50	50	50
E_g (eV)	2.9	2.9	2.17	1.5	3.6	3.6	1.2
χ (eV)	2.1	2.2	3.2	3.6	2.1	1.7	3.88
ϵ_r	6.5	3.0	7.11	10.0	11.75	10.0	6.5
N_c (cm^{-3})	2.8×10^{19}	2.5×10^{20}	2.02×10^{17}	1.0×10^{21}	2.5×10^{20}	2.2×10^{19}	2.2×10^{18}
N_v (cm^{-3})	1×10^{19}	2.5×10^{20}	1.1×10^{19}	1.0×10^{21}	2.5×10^{20}	1.8×10^{18}	1.8×10^{19}
μ_n (cm^2/Vs)	1.7×10^{-4}	2.1×10^{-3}	200	1	1×10^{-3}	100	2
μ_p (cm^2/Vs)	2.0×10^{-4}	2.6×10^{-3}	80	40	1×10^{-3}	25	2
V_e (cm/s)	1×10^7	1×10^7	1×10^7	1×10^7	1×10^7	1×10^7	1×10^7
V_h (cm/s)	1×10^7	1×10^7	1×10^7	1×10^7	1×10^7	1×10^7	1×10^7
N_D (cm^{-3})	–	–	–	–	–	–	1×10^{15}
N_A (cm^{-3})	1×10^{18}	1×10^{18}	1×10^{18}	1×10^{18}	1×10^{16}	1×10^{18}	1×10^{17}
N_t (cm^{-3})	1×10^{15}	1×10^{15}	1×10^{15}	1×10^{15}	1×10^{14}	1×10^{14}	1×10^{15}

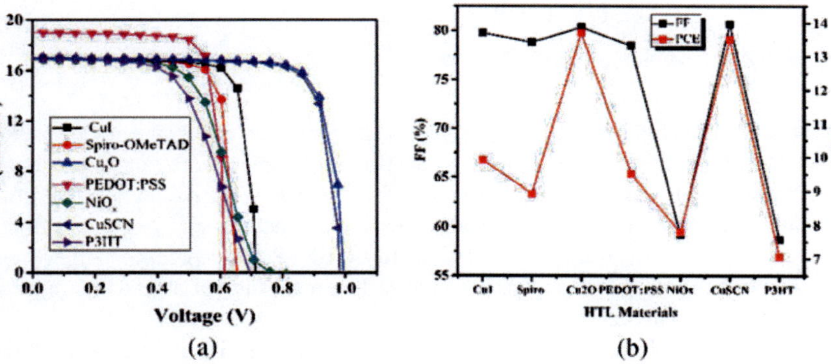

Fig. 4 **a** The *J-V* curve for different HTLs and **b** the PCE, FF variation for different HTLs used in the PSC

Table 6 The output parameters obtained using different HTL materials

S. No.	ETL materials	V_{oc} (V)	J_{sc} (mA/cm^2)	FF (%)	PCE (%)
1	Cu_2O	1.0019	17.0122	80.33	13.69
2	CuSCN	0.9907	16.8949	80.65	13.50
3	CuI	0.7356	16.8855	79.76	9.91
4	PEDOT:PSS	0.6395	18.9674	78.43	9.51
5	Spiro-OMeTAD	0.6713	16.8883	78.76	8.93
6	NiO_x	0.7788	16.8785	59.24	7.79
7	P3HT	0.7110	16.9244	58.67	7.06

showing similar curve as that of P3HT. It can also be seen from Table 6 that the parameters obtained from these two HTLs are similar. The device based on most widely applicable HTL Spiro-OMeTAD showed a PCE of 8.93% which is better than that of P3HT and NiO_x-based devices. The efficiencies of CuI and PEDOT:PSS were 9.91% and 9.51%, respectively. It can be observed from Fig. 4 that CuSCN and Cu_2O are the showing comparable and quite higher performance parameters than all other HTLs. Both CuSCN and Cu_2O-based devices showed PCE of 13.50 and 13.69% with improved open circuit voltage and FF. CuSCN and Cu_2O outperforms all other HTL because of the higher mobility of electron and holes which facilitates in fast collection of charge carriers in the device. Also it may be attributed to the proper band alignment which will create low hindrance in the hole flow. Thus, we conclude that both CuSCN and Cu_2O are suitable for our proposed device as they showed optimal output parameters as shown in Table 6.

3.3 Impact of Absorber Thickness

Since the absorber thickness is a governing parameter which affects the device performance at large extent as it determines the amount of light absorbed and the length of path travelled by charge carriers to reach their charge selective layers. Therefore the optimization of the thickness of absorber is very necessary. Owing to this the thickness of Cs_2TiBr_6 was varied in a range of 100–800 nm keeping other layer thickness to be the same as shown in Table 1. The resulting variations in V_{OC}, J_{SC}, FF and PCE are illustrated in Fig. 5.

From Fig. 5 we observe that both J_{SC} and PCE increased with increasing thickness till 450–500 nm. This increase in these parameters is due to the increase in light absorption and more charge carrier creation with the increase in perovskite absorber's thickness. Further in the vicinity of 500 nm and at higher thicknesses the J_{SC} becomes almost constant whilst the PCE falls down in a small amount. This fall in PCE in PSC with extensively thicker perovskite is because of increase in probability of carrier recombination in such PSCs with thickness of absorber much larger to the carrier diffusion length. From Fig. 5a we observe that V_{OC} is not much affected due to

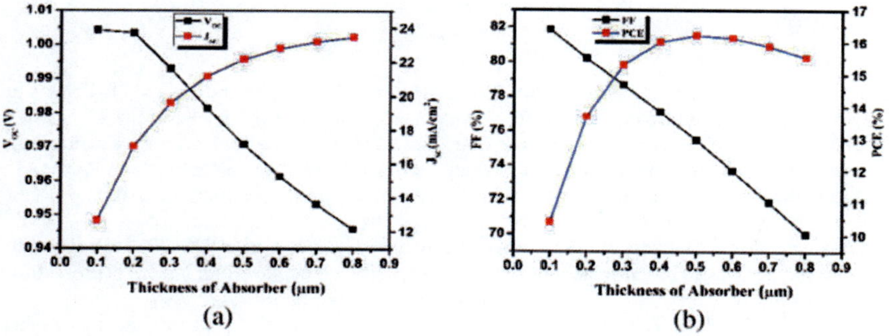

Fig. 5 **a** Variation of V_{OC} and J_{SC} with absorber thickness and **b** variation of FF and PCE with absorber thickness

increase in absorber thickness as it showed a very small decrement of 0.06 (V) as the thickness was varied from 100 to 800 nm. Moreover, the FF decreased from 82% (at 100 nm) to nearly 70% (at 800 nm) as the thickness was increased which is due to increase in series resistance in the thicker absorber. Thus, according to our simulation results and keeping in harmony in all the PSC performance parameters we found that the range of absorber thickness should be between 300 and 500 nm to get an optimum device performance. For 400 nm thickness of Cs_2TiBr_6 the device FTO/WS_2/Cs_2TiBr_6/Cu_2O showed promising solar cell parameters given as—$V_{OC} =$ 0.98 V, $J_{SC} = 21.15$ mA/cm^2, FF = 76.17% and PCE = 15.90% (as can be observed in Fig. 5).

4 Conclusion

This research work studies the optimization of structure FTO/ZnO/Cs_2TiX_6 (X = I, Br)/P3HT/Au for optimizing the solar cell output parameters using SCAPS-1D. First the Cs_2TiX_6-based device with two different halide contents I and Br was comparatively studied and it was found the device with Br content outperforms to that with I content. Next the FTO/ZnO/Cs_2TiBr_6/P3HT/Au was studied extensively by applying various charge transport layers to get the most suitable ETL and HTL for an efficient device structure. From the simulation results we found that WS_2 as ETL and Cu_2O and CuSCN as HTL gave the optimum solar cell performance parameters. Moreover, the thickness of absorber was varied and 300–500 nm range of absorber thickness was concluded be most appropriate for the proposed device. Finally, a favourable PCE of 15.90% was achieved at a moderate thickness of 400 nm Cs_2TiBr_6 for FTO/WS_2/Cs_2TiBr_6/Cu_2O/Au device. These simulation results will give an idea about suitability of charge transport layers and will help in the development of such lead-free PSCs.

References

1. Ansari MIH, Qurashi A, Nazeeruddin MK (2018) Frontiers, opportunities, and challenges in perovskite solar cells: a critical review. J Photochem Photobiol C Photochem Rev 35:1–24
2. Jung K, Lee JH, Oh K, Im C, Do J, Kim J, Chae WS, Lee MJ (2018) Efficient composition tuning via cation exchange and improved reproducibility of photovoltaic performance in FAxMA1-xPbI3 planar heterojunction solar cells fabricated by a two-step dynamic spin-coating process. Nano Energy 54:251–263
3. Prakash J, Singh A, Sathiyan G, Ranjan R, Singh A, Garg A, Gupta RK (2018) Progress in tailoring perovskite based solar cells through compositional engineering: materials properties, photovoltaic performance and critical issues. Mater Today Energy 9:440–486
4. Hamed MSG, Mola GT (2020) Mixed halide perovskite solar cells: progress and challenges. Crit Rev Solid State Mater Sci 45:85–112
5. Kung P, Li M, Lin P, Jhang J, Pantaler M, Lupascu DC, Grancini G, Chen P (2020) Lead-free double perovskites for perovskite solar cells 1900306:1–32
6. Ke W, Kanatzidis MG (2019) Prospects for low-toxicity lead-free perovskite solar cells. Nat Commun 1–4
7. Chen M, Ju M, Carl AD, Zong Y, Grimm RL, Gu J (2018) Cesium titanium (IV) bromide thin films based stable lead-free perovskite solar cells. Joule 1–13
8. Burgelman M, Nollet P, Degrave S (2000) Modelling polycrystalline semiconductor solar cells. Thin Solid Films 361:527–532
9. Chakraborty K, Gupta M, Paul S (2019) Numerical study of Cs_2TiX_6 ($X = Br$, I, F and Cl) based perovskite solar cell using SCAPS-1D device simulation. Sol Energy 194:886–892
10. Azri F, Meftah A, Sengouga N, Meftah A (2019) Electron and hole transport layers optimization by numerical simulation of a perovskite solar cell. Sol Energy 181:372–378
11. Yang D, Yang R, Zhang J, Yang Z, Liu S, Li C (2015) High efficiency flexible perovskite solar cells using superior low temperature TiO_2. Energy Environ Sci 8:3208–3214
12. Madan J, Garg S, Gupta K, Rana S, Manocha A, Pandey R (2020) Numerical simulation of charge transport layer free perovskite solar cell using metal work function shifted contacts. Optik (Stuttg) 202:163646
13. Zhao P, Liu Z, Lin Z, Chen D, Su J, Zhang C, Zhang J, Chang J, Hao Y (2018) Device simulation of inverted CH3NH3PbI3–xClx perovskite solar cells based on PCBM electron transport layer and NiO hole transport layer. Sol Energy 169:11–18
14. Sobayel K, Akhtaruzzaman M, Rahman KS, Ferdaous MT, Al-Mutairi ZA, Alharbi HF, Alharthi NH, Karim MR, Hasmady S, Amin N (2019) A comprehensive defect study of tungsten disulfide (WS_2) as electron transport layer in perovskite solar cells by numerical simulation. Results Phys 12:1097–1103
15. Ovchinnikov D, Allain A, Huang Y, Dumcenco D, Kis A (2014) Engineering, electrical transport properties of single-layer WS_2. ACS Nano 8:8174–8181
16. J. Feng, Z. Yang, D. Yang, X. Ren, X. Zhu, Z. Jin, W. Zi, Q. Wei, S. (Frank) Liu, E-beam evaporated Nb2O5 as an effective electron transport layer for large flexible perovskite solar cells, Nano Energy. 36 (2017) 1–8
17. Macco B, Black LE, Melskens J, van de Loo BWH, Berghuis WJH, Verheijen MA, Kessels WMM (2018) Atomic-layer deposited Nb_2O_5 as transparent passivating electron contact for c-Si solar cells. Sol Energy Mater Sol Cells 184:98–104
18. Samanta M, Ijaj S, Chattopadhyay KK, Bose C (2020) Optik role of various transport layer and electrode materials in enhancing performance of stable environment-friendly Cs_2TiBr_6 solar cell. Opt Int J Light Electron Opt 217:164805
19. Rai S, Pandey BK, Dwivedi DK (2020) Modeling of highly efficient and low cost CH3NH3Pb(I1-xClx)3 based perovskite solar cell by numerical simulation. Opt Mater (Amst) 100:109631

An Improved Pseudo-Domino Technique for Low-Power Applications

Shivangi Pandey and Manish Kumar

Abstract In the era of nanometer technology, power consumption is the most important factor in digital circuit systems. In this paper, power consumption in a pseudo-domino logic-based buffer circuit has been minimized using dual-threshold MOS technique. The average power consumption of the proposed pseudo-domino-based buffer is reduced by 72.28, 64.13, 68.20, 72.11, 69.97, 67.10, 66.71, and 53.86% in comparison with the conventional domino, pseudo-domino, DOIND logic, controlled current and comparison-based domino circuit (C-3D logic), modified true single-phase clock pulse (M-TSPC) logic, foot driven with stacked transistor-based domino logic (FDSTDL), CDDK logic, and stacked pseudo-domino, respectively. Monte Carlo simulation is also carried out for 1000 samples. The simulation has been performed using Cadence Virtuoso EDA tool in gpdk 90 nm CMOS technology at 1 V supply voltage, 1 GHz operating frequency, and at a temperature of 27 and 110 °C.

Keywords Pseudo-domino · Dual-threshold MOS technique · Average power · Process variations · Monte Carlo

1 Introduction

Average power consumption has become a primary concern in VLSI circuits with scaling down in technology into the nanometer regime [1]. Domino circuits are widely used in many applications such as arithmetic circuits and microprocessors, and this logic provides advantages in terms of power consumption and chip area [2, 2]. Domino logic operates in two stages: The first one is a precharge stage, and the second one is the evaluation stage, where both stages are controlled by clock [4].

S. Pandey (✉) · M. Kumar
Electronics and Communication Engineering Department, Madan Mohan Malaviya University of Technology, Gorakhpur, India
e-mail: shivangipandeydt.126@gmail.com

M. Kumar
e-mail: mkece@mmmut.ac.in

B. Mishra and M. Tiwari (eds.), *VLSI, Microwave and Wireless Technologies*, Lecture Notes in Electrical Engineering 877,
https://doi.org/10.1007/978-981-19-0312-0_45

When the clock is low, then the standard domino logic goes into the precharge stage, and the node is charged by supply voltage (V_{DD}) using PMOS precharge transistor, whereas if clock is high, then the domino logic goes into evaluation stage in which NMOS evaluation transistor becomes OFF, and the output is evaluated by the inputs of pull-down network; the node discharges. Otherwise, it will remain in the precharge stage V_{DD} [5].

Consequently, the average power consumption of domino logic circuit is calculated and can be written by the equation which is given [6]:

$$P_{avg} = P_{switching} + P_{shortckt} + P_{leakage} \tag{1}$$

Here, $P_{switching}$ is switching power consumption because of the charging and discharging the circuit capacitance present at the output. $P_{shortckt}$ is the short-circuit power due to the short path between the V_{DD} to ground, and $P_{leakage}$ is leakage power present in the MOS transistors.

Switching power is due to the charging and discharging of the capacitor and can be expressed as:

$$P_{switching} = \alpha C_L V_{DD}^2 f \tag{2}$$

Here, α is the activity factor which shows the transition of the circuit node; C_L is the load capacitance; f is the operating clock frequency.

When the PMOS and NMOS transistors are turned ON simultaneously, a short-circuit current occurs and forms a path between the supply voltage (V_{DD}) and ground. Here, the short-circuit power can be represented by [7].

$$P_{shortckt} = V_{DD} * I_{SC} \tag{3}$$

where V_{DD} is DC power supply voltage and I_{SC} is a short-circuit current.

The power consumption due to leakage current that flows when all the MOS transistors are in OFF condition and can be written as [7]:

$$P_{leakage} = V_{DD} * I_{leakage} \tag{4}$$

$$I_{leakage} = I_{GT} + I_{GIDL} + I_{REV} + I_{SUB} \tag{5}$$

where $I_{GT,}I_{GIDL,}I_{REV}$, and I_{SUB} are the gate tunneling leakage current, gate-induced drain leakage current, reverse bias junction leakage current, and the sub-threshold leakage current, respectively [7].

2 Background

Figure 1a shows the conventional clock-controlled footed domino buffer circuit [8]. It operates in mainly two stages: the precharge stage and the evaluation stage. When clock (CLK) = 0, the precharge transistor (M_{pre}) turns ON, making the dynamic node high. If input (IN) = 0, then the OUT node becomes low, and if input (IN) = 1, then output (OUT) of the precharge stage is low as dynamic node is still charged. But, in the evaluation stage as CLK goes from 0 to 1 transition, then the dynamic node starts discharging and becomes 0 leading to high OUT voltage. Here, the precharge pulse propagates by using static inverter that results in unbalanced output and high-power consumption [8].

Figure 1b shows the pseudo-domino buffer with conventional-footed domino [9] the source of NMOS which is present at pull-down network of inverter is connected to the drain of the footer transistor. When IN = 0, the operation is same as the conventional-footed domino buffer [8]. This approach eliminates the problem of propagation of precharge pulses by using static inverter, which degrades the performance.

Figure 1c shows the DOIND logic [10] that comprises of two transistors: M_{P4} and M_{N3}, which depend on input and clock. The clock pulse denoted by V_0 is dependent on the gate terminal of M_{P4}, while the gate terminal of M_{N3} is dependent on the input, denoted by V_1. In the precharge stage, the clock = 0 and $V_0 = 0$, due which M_{P1} and M_{P4} become activated, and starts charging dynamic node to the supply voltage V_{DD} with precharge transistor M_{p4}. And in evaluation stage, when the clock = 1, and V_0 = 1, then dynamic node becomes low based on applied inputs, it means when IN = 1; subsequently, $V_1 = 1$ that pulls the dynamic node at "low," and if IN = 0 and V_1 = 0, it maintains Node_dynamic to supply voltage V_{DD}. This circuit improves the elimination of leakage by reason of the formation of stacking effect in DOIND-based approach [10].

Figure 1d shows the controlled current and comparison-based domino circuit (C-3D) [11]. There is a comparison of the currents, I_x and I_y, at the varying junction voltages, "X" as well as "Y," respectively. Dynamic node starts charging or discharging based on the difference between currents I_x and I_y and the node potential applied to the inputs. This technique can reduce the power by lowering the swing of voltages at PDN (pull-down network) in an efficient manner. The stacked effect with the NMOS M_{diode} transistor, by using this approach, improves the area overhead at the cost of decrease in the speed.

Figure 1e shows the modified true single phase clock pulse" (M-TSPC) [12] circuit. There is use of two ultralow power transistors ML_P and ML_N. Here, ML_p and ML_N are connected with each other's source. In this circuit, M_{S1} and M_{S2} are NMOS transistors that are used for the formation of dynamic node with charging and discharging process. ML_P and ML_N transistors are used based on applied value of inputs, which helps to achieve the zero resistance path at the time of discharging of the dynamic node. This shows the reduction in power dissipation, but this technique has disadvantages in the improvement of noise and reliability of the circuit.

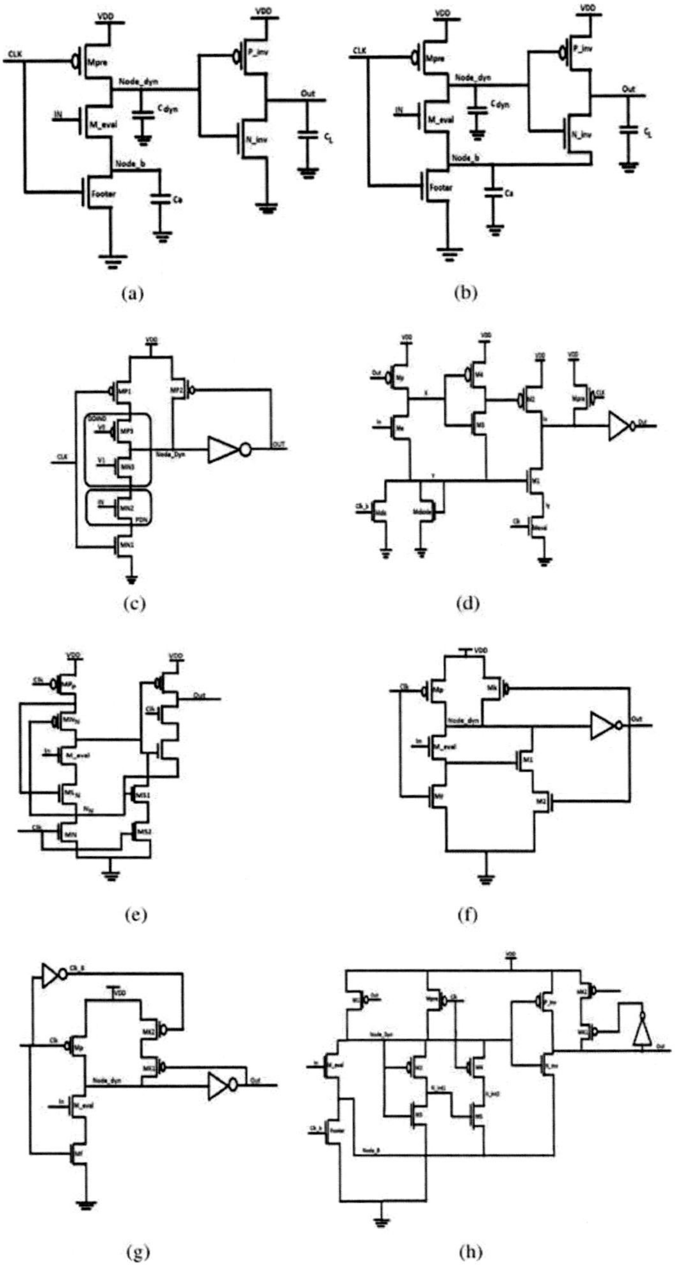

Fig. 1 **a** Conventional-footed domino buffer. **b** Pseudo-domino-based buffer. **c** DOIND logic-based buffer. **d** C-3D logic-based buffer. **e** M-TSPC technique domino-based buffer. **f** FDSTDL logic-based buffer. **g** CDDK-based logic buffer. **h** Pseudo-stacked domino logic-based buffer

Figure 1f shows the foot driven with stacked transistor-based domino logic (FDSTDL) [13]. This logic includes dynamic node discharges with the help of M_1 and M_2. This way, it improves the speed and can be driven by node of NMOS footer transistor and output (out).

Figure 1g shows the clock pulse-delayed domino logic with the help of two keeper transistors (CDDK) [14]. This logic consists of dual keeper transistors which can be driven by output (out) and delayed operating clock. The delayed operating clock is used at the gate terminal of the keeper to enhance the circuit's performance. It reduces delay of the circuit and improves noise immunity. The figures which are given below are the existing domino-based buffer.

Figure 1h shows the pseudo-stacked domino logic [15]. This circuit minimizes the leakage, less power dissipation, and better margin in terms of noise. The stacking effect is used in the precharge stage with the aid of transistors and a pseudo-domino buffer to control the VGS of the PDN for low power consumption and faster operation. The keeper circuit includes dual PMOS transistors M_{K1} and M_{K2}, respectively. The transistor, M_{K1}, is connected with the inverted out, and the dynamic node is connected with M_{K2}. The complemented output-controlled keeper M_{K1} assists in preserving dynamic node voltage, while keeper M_{K2} supports the enhancement of noise immunity and the speed of the circuit. In this, the length of channel for these two transistors M_{K1} and M_{K2} is same, whereas the transistors width is as $W_{K1} + W_{K2} = W_K$ where W_K is width of the standard keeper. We have used the modified keeper for the elimination of charge sharing problem.

3 Proposed Technique

The proposed method employs the dual-threshold MOS technique to design the pseudo-domino logic-based buffer logic. In this, we have used the "High-threshold voltage transistors" (HVT) in the non-critical path of the circuit (in the precharge stage) to reduce the leakage current due to which the average power dissipation is minimized. On the other hand, low-threshold voltage transistors are used in the critical path. High-threshold voltage transistors are M_1, M_{pre}, M_4, and P_inverter, and these all are the only active precharge stage of the circuit, while the low threshold voltage transistors are M_eval, M_footer, M_2, M_3, M_5, and N_inverter active in evaluation stage and are used for the performance enhancement of the circuit.

In this method, we have used the value of threshold voltage for high V_{th} PMOS as −0.546 V and high V_{th} NMOS as 0.446 V, respectively. Besides this, the value of threshold voltage of low V_{th} PMOS and NMOS are −0.201 V and 0.086 V, respectively. The PMOS transistors M_{K1} and M_{K2}, which are used as a keeper, are also low threshold transistors (Fig. 2).

The proposed pseudo-domino logic-based buffer circuit operates in two stages: precharge stage and evaluation stage.

In the precharge stage, input In = 1, clock CLK = 0, and complemented clock CLKB = 1, then, both the high threshold transistors M_1 and M_{pre} are ON and make the

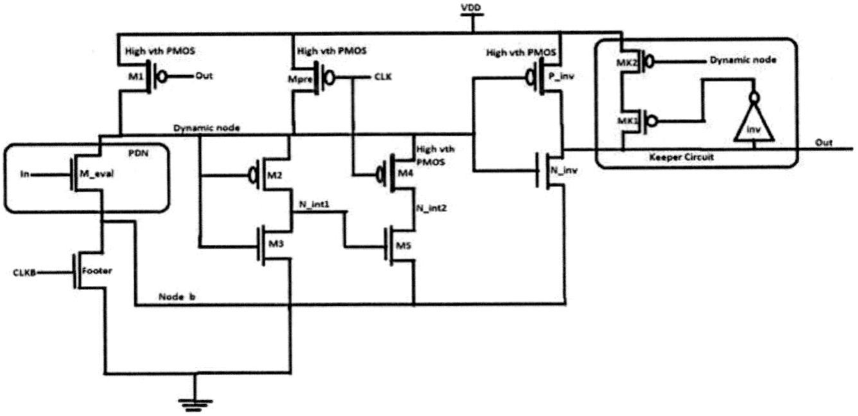

Fig. 2 Proposed pseudo-domino-based buffer

dynamic node as logic 1. Since CLKB = 1 is applied to the footer NMOS transistor, so the footer transistor also turns ON. During this stage, the NMOS transistor M_3 and PMOS transistor M_4 are in a high state due to the connection of gate terminal with the dynamic node = 1 and CLK = 0, respectively. Low V_{th} NMOS is used to make the N_int1 node low because it acts as an inverter with the PMOS transistor M_2. Since N_int1 becomes low, the NMOS transistor M_5 becomes OFF. It forms the stacking effect, which lowers the voltage at node B and increases the voltage between source-to-gate of the NMOS, which is the pull-down network of the static inverter, present at the output. It shows the dominating nature at the previous evaluation stage output as high.

In evaluation stage, when the CLK = 1 and CLKB = 0, the high V_{th} PMOS M_1 and M_{pre} are available between the evaluation network, which makes the M_5 transistor ON, due to which the discharging of voltage of dynamic node starts from one to zero. At the same time, there is a reduction in the voltage at node B, which guides to switch ON the NMOS transistor of the inverter which makes the output to logic 1. The proposed technique provides lesser power consumption at the cost of speed.

4 Simulation and Results

The proposed pseudo-domino logic-based buffer circuit is designed using the Cadence Virtuoso tool in the gpdk 90 nm technology at 1 V supply. In this, we used operating clock frequency of 1 GHz, at 110 and 27 °C.

The average power dissipation in the case of the proposed pseudo-domino logic-based buffer by dual-threshold MOS technique is 2.273 uW at 110 °C and clock frequency of 1 GHz. The delay of the proposed circuit is 391.25 ps.

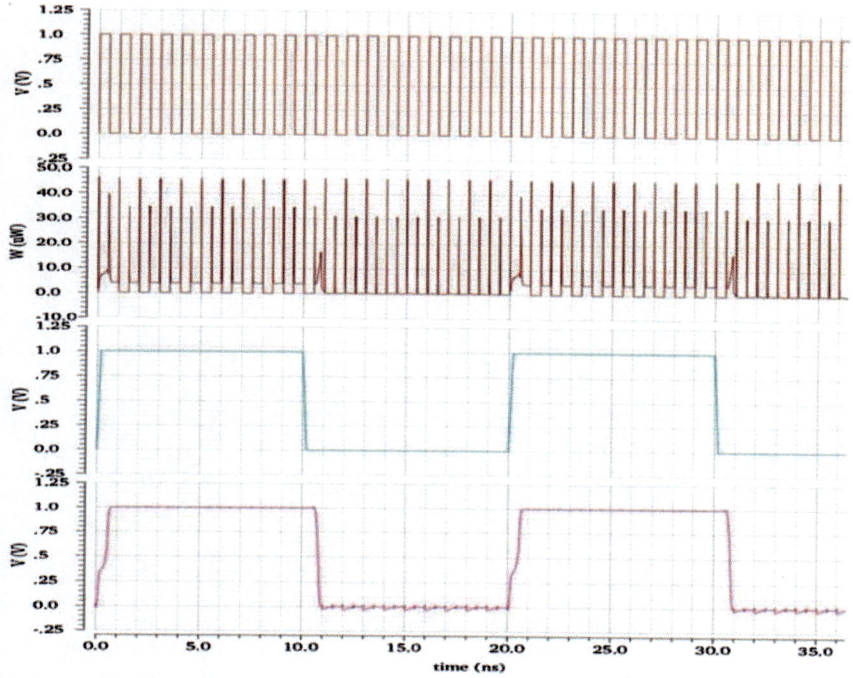

Fig. 3 Transient analysis of proposed pseudo-domino-based buffer logic

The transient analysis of output waveform of the proposed circuit is illustrated in Fig. 3.

Table 1 shows the performance analysis of the proposed technique and other existing techniques at 1 V supply voltage and temperature of 110 °C.

The average power dissipation for the proposed circuit is also analyzed by varying the supply voltage V_{DD} from 0.8 to 1.2 V which is shown in Fig. 4. Here, we have observed that the average power dissipation of the proposed circuit is 4.398 uW at 1.2 V which is also lower in comparison with the earlier work.

The proposed circuit is also simulated for different process corner analysis and observed the best scenario and worst scenario conditions such as fast NMOS and fast PMOS Transistors (FF), slow NMOS and Slow PMOS transistors (SS), fast NMOS and slow PMOS Transistors (FS), slow NMOS and fast PMOS (SF) and finally nominal NMOS and nominal PMOS (NN). Here, the SS corner's best scenario condition is 1.5857 uW, whereas the worst case at the FF corner is 2.8455 uW.

By the simulation, we compared the power dissipation for all the process corners at 1 V supply. The operating clock frequency for all the simulation is 1 GHz. Table 2 shows the effect of process variation of the proposed circuit:

The statistical analysis of the process parameters using Monte Carlo analysis of the proposed circuit is validated with 1000 samples which is illustrated in Figs. 5a–c. We

Table 1 Performance analysis of the proposed work and existing techniques

Parameters	Conventional domino [8]	Pseudo-domino [9]	DOIND [10]	C-3D [11]	M-TSPC [12]	FDSTDL [13]	CDDK [14]	Stacked pseudo-domino [15]	Proposed domino
Average power (μW)	8.20	6.33	7.14	8.14	7.56	6.90	6.82	4.92	2.273
Normalized power	1	0.77	0.87	0.98	0.92	0.84	0.83	0.6	0.27
Delay (ps)	32.22	31.29	32.16	32.16	32.83	32.63	31.22	29.47	391.25
Power delay product (*10^{-16} J)	2.64	1.98	2.24	2.61	2.48	2.25	2.12	1.44	8.88

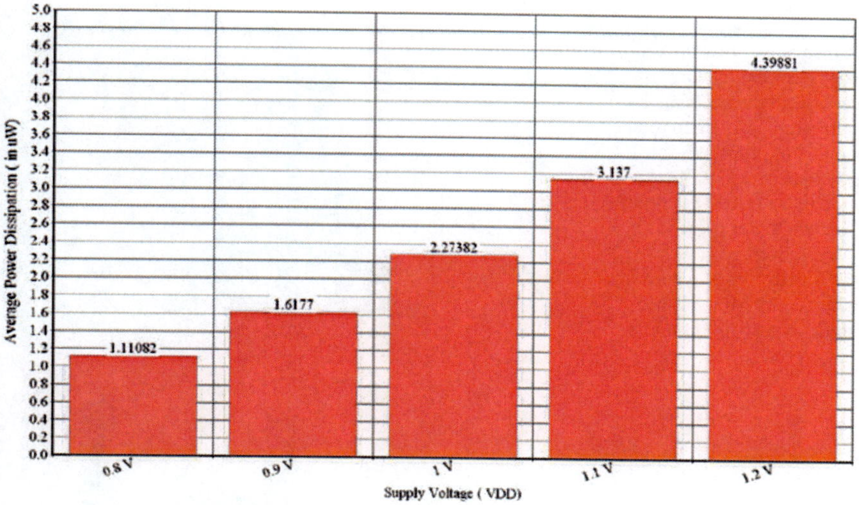

Fig. 4 Average power in buffer circuit using proposed technique (change in V_{DD})

Table 2 Process corners analysis of the proposed circuit for average power (in uW)

Technique	FF	FS	NN	SF	SS
Conventional [8]	10.4	9.28	8.10	7.44	6.23
Pseudo-domino [9]	8.12	7.33	6.11	5.43	4.90
DOIND [10]	8.43	7.68	6.98	5.88	5.03
C-3D [11]	9.83	9.02	7.99	7.12	6.44
M-TSPC [12]	9.44	8.63	7.14	6.83	6.08
FDSTDL [13]	7.43	6.91	6.53	6.04	5.48
CDDK [14]	7.12	6.63	6.42	6.00	5.40
Stacked pseudo-domino [15]	5.58	5.35	4.60	3.75	3.325
Proposed domino	2.8455	1.7191	2.273	2.610	1.585

obtained the mean (μ) and standard deviation (σ) of the power dissipation and delay of the proposed pseudo-domino-based buffer logic by using dual V_{th}. This simulation is having a mean $\mu = 2.2918$ uW and standard deviation $(\sigma) = 125.421$ nW in terms of average power dissipation. The proposed pseudo-domino-based buffer design's mean (μ) with standard deviation (σ) for delay is 406.609 and 95.69 ps at 1000 samples, with 1 V supply at 110 °C.

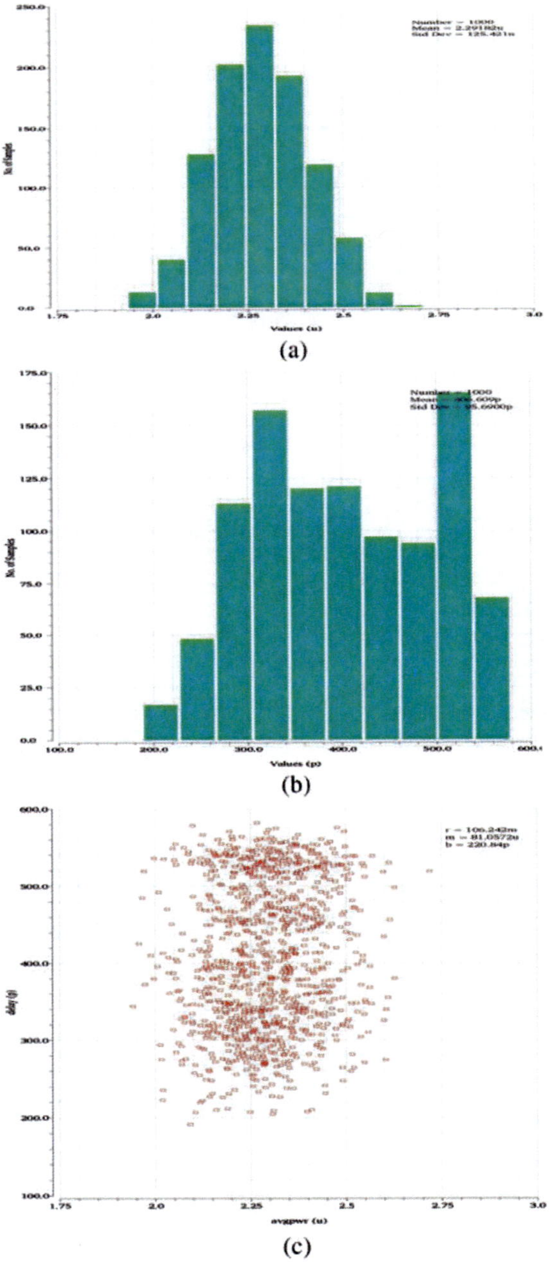

Fig. 5 **a** Histogram of average power dissipation of proposed technique by Monte Carlo simulation at 1000 samples. **b** Histogram of delay of proposed technique by Monte Carlo simulations at 1000 samples. **c** Scattered plot graph between average power versus delay for proposed technique by Monte Carlo simulations at 1000 sample

5 Conclusion

The dual-threshold MOS technique is used to design the proposed pseudo-domino logic-based buffer. The proposed buffer shows less power dissipation and applicable for low-power applications. Here, we have calculated different parameters to analyze the circuit in terms of power, delay, and power delay product. A significant reduction in the average power dissipation is observed in the proposed circuit by using the dual V_{th}.

References

1. Dadoria AK, Khare K, Gupta TK, Singh RP (2017) Ultra low power FinFET based domino circuits. Int J Electron. https://doi.org/10.1080/00207217.2017.1279227
2. Pandey AK, Verma PK, Verma R (2018) Analysis of noise immunity for wide OR footless domino circuit using keeper controlling network. Circuits Syst Sig Process. https://doi.org/10.1007/s00034-018-0781-0
3. Garg S, Gupta TK (2020) A 4:1 multiplexer using low-power high-speed domino technique for large fan-in gates using FinFET. Circuit World. https://doi.org/10.1108/CW-09-2019-0128
4. Asyaei M (2015) A new leakage-tolerant domino circuit using voltage-comparison for wide fan-in gates in deep sub-micron technology. Integr VLSI J. https://doi.org/10.1016/j.vlsi.2015.06.003
5. Singhal S, Mehra A, Tripathi U (2019) Power reduction in domino logic using clock gating in 16 nm CMOS technology. Sig Process Integr Netw. https://doi.org/10.1109/SPIN.2019.8711713
6. Kumar A, Nagaria RK (2019) A new process variation and leakage-tolerant domino circuit for wide fan-in OR gates. Analog Integr Circuits Sig Process. https://doi.org/10.1007/s10470-019-01473-3
7. Langpoklakpam C, Ghimiray SR, Dutta PK (2019) A novel high-performance leakage-tolerant keeper domino circuit for wide fan-in gates. Adv Commun Dev Netw. https://doi.org/10.1007/978-981-13-3450-4_2
8. Kumar S, Singhal S, Pandey AK, Nagaria RK (2013) Design and simulation of low power dynamic logic circuit using footed diode domino logic. In: Students conference on engineering and systems. https://doi.org/10.1109/SCES.2013.6547509
9. Tang F, Bermak A, Gu Z (2012) Low power dynamic logic circuit design using a pseudo dynamic buffer. Integr VLSI J. https://doi.org/10.1016/j.vlsi.2011.08.003
10. Shah AP, Neema V, Daulatabad S (2019) Dual threshold voltage and sleep switch dual threshold voltage DOIND approach for leakage reduction in domino logic circuits. Microsyst Technol. https://doi.org/10.1007/s00542-017-3437-2
11. Asyaei M, Moradi F (2018) A domino circuit technique for noise-immune high fan-in gates. J Circuits Syst Comput. https://doi.org/10.1142/s0218126618501517
12. Verma P, Sharma AK, Noor A, Mishra AK, Pandey VS (2019) A novel approach for noise tolerant energy efficient TSPC dynamic circuit design. Analog Integr Circ Sig Process. https://doi.org/10.1007/s10470-019-01444-8
13. Garg S, Gupta TK (2019) FDSTDL: low-power technique for FinFET domino circuits. Int J Circuit Theor Appl. https://doi.org/10.1002/cta.2627
14. Angeline AA, Bhaaskaran VSK (2019) Design impacts of delay invariant high speed clock delayed dual keeper domino circuit. IET Circuits Dev Syst. https://doi.org/10.1049/iet-cds.2018.5410

15. Ghimiray SR, Meher P, Dutta PK (2020) An improved charge-sharing elimination pseudo domino logic. Int J Circ Theor Appl. https://doi.org/10.1002/cta.2798

Memristor Emulator Circuits an Emerging Technology with Applications

Jyoti Garg, Aishita Verma, and Subodh Wairya

Abstract In this paper, the fourth fundamental passive circuit element called memristor has been discussed along with models, emulator circuits, characteristics, and applications in different engineering areas. Memristors are able to hold the information about net amount of charge gone through it. Memristors are also compatible with the CMOS fabrication process and enable hybrid CMOS/Nanodevices integration, which imports that this device can be used in various applications. Memristor has a wide application in memory logic design, where memristor acts as a building block. Other applications of memristor include Nanocomputing, Neuromorphic circuits, and digital computation, which includes implication logic. This review imitates a compendious elucidation of memristor, in addition to future scope in a proper manner.

Keywords Memristor model · Characteristics · Emulator circuits · Engineering application

1 Introduction

In 1971, Prof. Leon Chua perceived that existed no entity relating magnetic flux and electric charge [1]. The resistor capacitor and inductor are passive elements as they need no internal power source to operate, and as their behavior cannot be mimicked by other basic passive elements. Prof. Leon Chua realized that out of six relations only five had been recognized. Based on these simple symmetry arguments, the Prof. proposed the memristor to complete the missing link. Therefore, a prediction was made based on criteria of symmetry that there should be another element of circuit and named this hypothetical element, "memristor." Till 1971, circuit theory was ruled

J. Garg (✉)
Department of Electronics and Communication Engineering, Dr. A.P.J. Abdul Kalam Technical University, Lucknow, Uttar Pradesh, India
e-mail: jyotigarg16@gmail.com

A. Verma · S. Wairya
Department of Electronics and Communication Engineering, Institute of Engineering and Technology, Lucknow, India

B. Mishra and M. Tiwari (eds.), *VLSI, Microwave and Wireless Technologies*,
Lecture Notes in Electrical Engineering 877,
https://doi.org/10.1007/978-981-19-0312-0_46

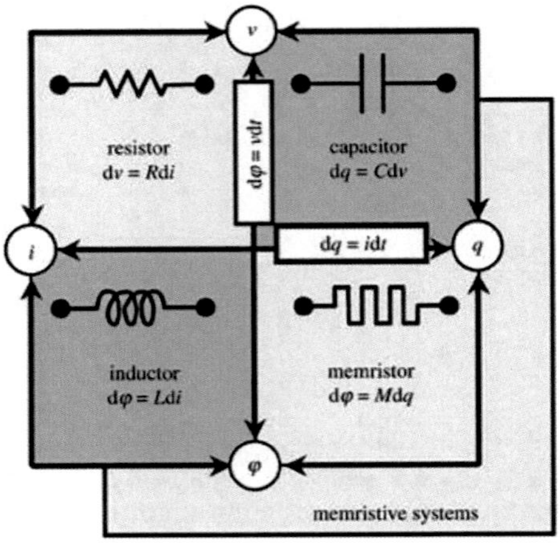

Fig. 1 The four fundamental passive components with the relations in between them and memristor [1]

by three fundamental circuit elements, resistor which relates Voltage and Current ($\mathrm{d}v = R.\mathrm{d}i$) where R is the Resistance, Capacitor-Relates Voltage and Charge ($\mathrm{d}q = C.\mathrm{d}v$) where C is the Capacitance, and Inductor-Relates Current and Magnetic Flux ($\mathrm{d}\Phi = L.\mathrm{d}i$) where L is the Inductance. Memristor relates Charge and Magnetic flux ($\mathrm{d}\Phi = M\ \mathrm{d}Q$) where M is the memristance. Figure 1 shows the fourth fundamental passive element with resistor, inductor, and capacitor. Memristors have important characteristics, which are current voltage, flux charge, etc. A memristor represents an increasing or decreasing function of monotonic nature as shown in Fig. 3a–c. The tangent of the $\Phi - Q$ curve grants memristance of a memristor. On increasing the frequency pinched hysteresis loop shrinks and at very large input frequencies, the linear resistance behavior of the device is observed. If the hysteresis curve is pinched, then it is a memristor; if the curve is not pinched, then it is not a memristor. Characteristic curve is shown in Fig. 2. If flux and charge is defined in terms of charge, then term is coined as charge-controlled memristor and when in expressed as function of function flux it is a flux-controlled memristor. It is known as charge-controlled flux, and φ is expressed as a function of the charge,

$$\varphi = f(q) \tag{1}$$

The voltage across the charge-controlled memristor is expressed as

$$v(t) = M(q(t)).i(t) \tag{2}$$

M is called as the memristance/resistance of the memristor.

It is called as flux or voltage-controlled if the charge (q) can be expressed as a single valued function of the flux (φ), i.e.,

$$q = f(\varphi) \tag{3}$$

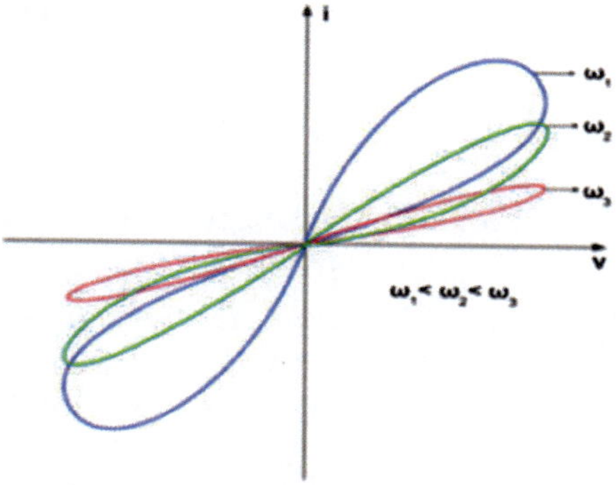

Fig. 2 Characteristic of the memristor with respect to frequency [1]

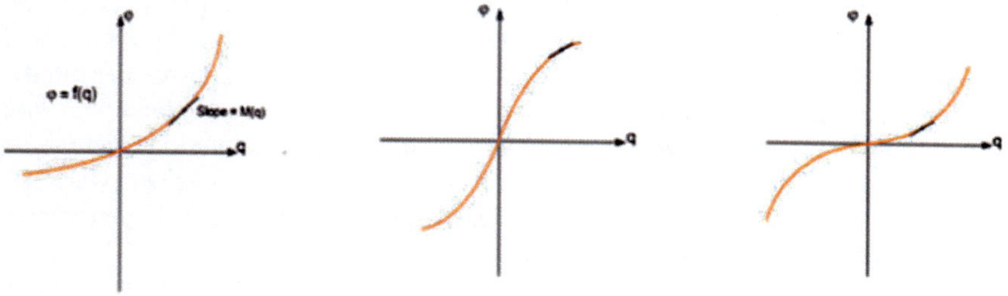

(a) (b) (c)

Fig. 3 **a–c** are monotonically increasing charge-flux characteristics of memristor [2]

Similarly, current across the flux or voltage-controlled is given as:

$$i(t) = W(\varphi(t)).v(t) \tag{4}$$

$$W(\varphi(t)) = \frac{dq(\varphi)}{d\varphi} \tag{5}$$

The unit of memristance of the memristor is the same as the resistance unit, and memconductance unit is the same as the conductance unit. In particular, a description of the first memristor, manufactured at HP Laboratories, is presented in memristor model. This paper is explained in detail followed by the respective sections. Section 2 gives a brief description of memristor emulator circuit and memristor models. Section 3 gives the research review of memristor emulator circuits, Section 4 gives the brief of Memristor Applications, and Conclusion and Future scope are discussed in Sect. 5, respectively.

The dissipation of power by the memristor is given by Eq. 6, and since $M(q) \geq 0$, the power has a positive value (Fig. 3).

$$P(i) = M(q)i((t))^2 \tag{6}$$

2 Memristor Emulator Circuit and Memristor Models

The memristors are commercially inaccessible, few circuit which perform as memristors are obligatory to develop application circuits and these must have some important characteristics like memristors. The memristance should be capable of being programmed, devoid of volatility, and connected to different circuit elements [3]. The memristor emulator circuit response is always limited to first and third quadrants, which indicate that the device is passive. The response always crosses the origin, which means that once the input signal is removed from the device, the output always falls to zero. In order to obtain, analyze, and simulate memristor-based circuits, appropriate model is mandatory. The HP Lab for the first time presented a brief on memristor implementation, and also models for various applications were been discussed. Figure 4 shows the HP memristor and its implementation where there is a doped and undoped side and a TiO_2 layer between two platinum electrodes. The voltage has been applied at the platinum contacts, respectively, and the characteristics are obtained. Other models of memristors are discussed in detail.

2.1 *Linear Ion Drift Model*

This model was presented by HP, based on the physical dimensions of the device. It is presumed that the physical device of width D had a couple of regions where one side is doped with oxygen ions having a positive charge and the other side is not doped. The modeling of the regions has been done using a resistor. The region which is doped has width w resulting in lower resistance and is more conductive, and the region that is not doped retains resistance of higher value. Figure 5 shows the HP memristor model. This model's advantages are its simplicity and easy to analyze, but the demerit of this model is the lowest accuracy.

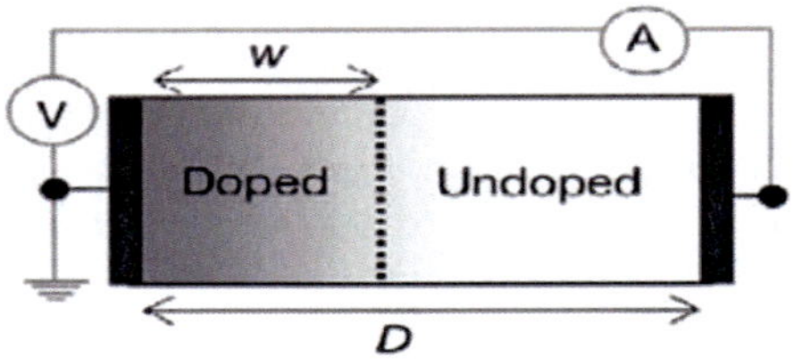

Fig. 4 HP memristor [2]

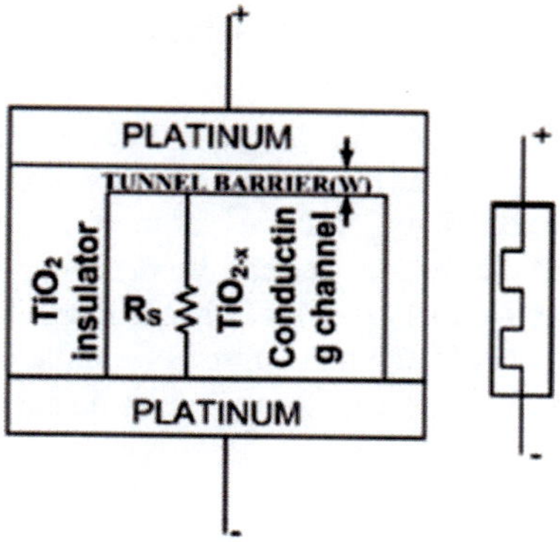

Fig. 5 Tunnel barrier model with symbol [4]

2.2 *Nonlinear Ion Drift Model*

A memristor that is voltage controlled is presumed in the nonlinear ion drift model, which has nonlinear dependency interjacent the voltage and internal state derivative [4]. The linear drift model has the lowest accuracy; hence, another model called nonlinear ion drift model was developed. It was noted that the comportment of memristors were nonlinear, and the linear ion drift model had lower accuracy, so there was a need to develop this model. For some of the significant applications, nonlinear characteristics are needed, which can be achieved by this model.

2.3 *Simmons Tunnel Barrier Model*

Pickett et al. presented an exact model of the memristor. This model worked as a substitute of two resistors in series, and a resistor is in series with an electron tunnel barrier, as shown in Fig. 5. It has been insisted that it is the more definite model of memristor; however, in terms of complexity, this model is more intricate as compared to other existing models, and this model also consists of complex equations to study the memristor characteristics. This model is not pertinent to different types of memristors and only fits a specific type of memristor [5]. Figure 5 is the schematic of the memristor model with its symbol. The titanium dioxide TiO_2 is in between two platinum (Pt) electrodes, and a tunnel barrier is placed along withTiO_{2-X}, which is a conductant.

2.4 *Threshold Adaptive Memristor Model (TEAM)*

The TEAM model, proposed by Kvatinsky et al., is the elementary model. This model has simple expressions to study the memristor characteristics. The model could be fitted to any of the actual memristor models, such as the linear ion drift model [6]. The above model can be pertained to different memristor models and also to the above-discussed memristor model. The TEAM model is precise enough with a mean error of 0.2% and improved simulation by approximately 47.5%. It justifies sufficient efficiency needed for simulation engines and also essentials for the memristor-based applications.

3 Research Review

The Concept of Memristor by L. O. Chua, "Memristor—The Missing Circuit Element" hypothesized based on relations among the four prominent circuit elements and symmetry, there should be another circuit element to complete missing link, whose resistance was not constant it can depend on one or more basic differential equations. In fact, a symbol was presented along with three fundamental circuit realizations [1]. In 2012, memristor emulator for memristor circuit applications was proposed which shows the behavior of a TiO memristor and also this emulator can be connected in serial, in parallel, or in hybrid. The most significant drawback is the memristive behavior [2] was only seen at some particular frequencies 100 Hz and also more number of component were used which increased chip area. In [3], another memristor emulator circuit was proposed with simple dimensionless expression models double-loop behavior, and this memristor emulator circuit was proposed couple of current conveyor (CCII) devices, voltage multiplier, and non-inverting or inverting buffer (using Op-amp). Simultaneously, they also proposed two emulator circuits, for the current-controlled memristor and for the voltage-controlled memristor, but more number of active elements were used, which resulted in increased power dissipation and the device area [7]. A memristor emulator circuit comprising DDCC (differential difference current conveyor) based on CMOS was presented, this emulator was quite less complex as comparable to the other designed emulator circuits and it consisted 1-DDCC, 2-resistors, 1-capacitor, and 1-multiplier circuit, but the single DDCC circuit consists of a high number of transistors and also this emulator cannot be controlled tuned electronically. The authors in research article [8] have addressed an emulator circuit consisting of 2-CCII+s, 1-multiplier, 2-resistors, 1-capacitor, and a switch. A behavioral model of the emulator circuit has been analyzed in this work. It was concluded that this emulator was much more improved over the existing emulator, but this circuit requires more passive elements [9]. The emulator circuit comprises three operational transconductance amplifiers (OTA) and four second-generation current conveyors (CCII), six resistors, and one capacitor element for the emulator circuit. Since this circuit consists of OTA, which has low output impedance, so the gain was less [10]. In this, the emulator circuit was designed with CCII+, a four-quadrant analog multiplier, capacitor, resistor, and two DC voltage sources, and also the derivation of memductance of the proposed memristor emulator circuit was done with the equations but since this design consisted of more active elements, there was much power dissipation. Hence, an emulator circuit with less transistors counts should be designed to reduce its chip area and power dissipation [11]. Memristor emulator consists of only four MOS transistors. This circuit is convenient both for emulating the fabricated memristor and also for some applications of memristor. This circuit did not consist of a circuit block as multiplier or an active element to obtain memristive characteristics. The analysis was done at a different temperature and also on different voltages. The chip area was also reduced since this design consisted of a significantly less number of MOS transistors, the layout area was also reduced as compared to other existing emulator circuits, also this design did not consist of active elements and the power dissipation was also less as compared to other circuits of memristor emulator. However, [12] proposed meminductor models that can be used in wide applications consist of only two active voltage differencing transconductance amplifier (VDTA) and grounded capacitors. The meminductor emulator circuit depicts broad operating frequency range clarity, and also, this circuit was congruent for memristor applications [13]. This memristor emulator circuit consists of a second-generation current conveyor CCII and OTA as active elements besides few passive elements. This circuit is compatible with both CMOS realizations and breadboard implementations. The circuit resembles a comprehensible design of an ideal memristor and also functional verification of circuit is realized with 0.18-μm CMOS technology and voltage of $\pm$1.2 V. In [4], emulator circuits presented could be used for the implementation of several analog circuits, as it can operate to the 26.3-MHz frequency in which designing was done with CMOS technology and consisted of a single resistor and capacitor as passive elements. In addition to this, the designing of the emulator circuit [14] was done from BJT-based ICs and performed up to 1.7 MHz and can be used in many applications efficiently.

4 Applications of Memristor

Memristor has wide applications in digital and analog designing [15]. Their division is based on how the memristance of the device is utilized. In digital applications, only a discrete number of resistance states is used, while in analog applications, the complete resistance changes between the minimum and the maximum values. Memristor has several applications in different discipline of engineering, including neuromorphic application, chaos circuits, image processing, neural networks, memories, etc. Apart this, hybrid memory circuit can be designed as can see in Fig. 6. Some significant memristor applications are discussed in detail in following section.

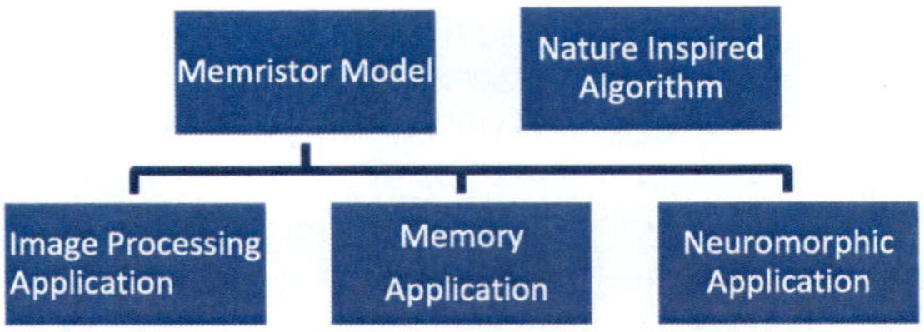

Fig. 6 Hybrid memristor circuit applications in different areas

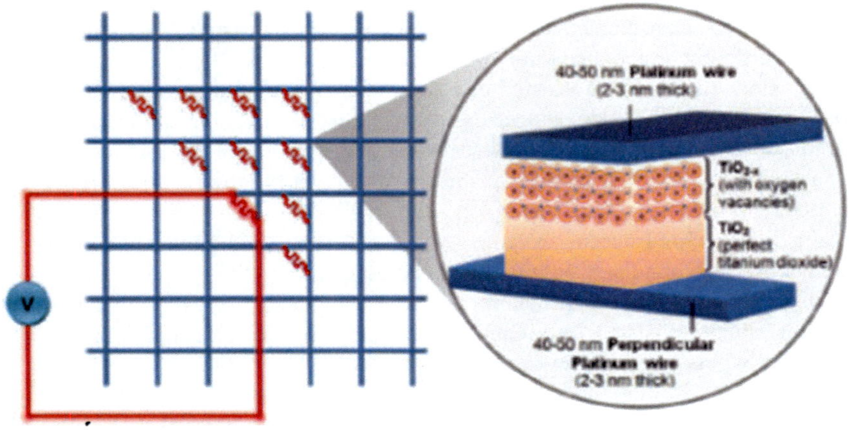

Fig. 7 Crossbar structure [17]

4.1 Crossbar

Crossbar structures are versatile nanostructures that consist of many switches in nanoscale. These are scalable as well as flexible. Crossbar grids can be used in many applications such as digital circuits and memories [16]. These switches have a grid of nanowires consisting of vertical wires which are intersected, which is a memristor switch. Junction size is about 2–3 nm. Nanoimprint lithography (NIL) is a method for fabricating nanowires. The fabrication method used for crossbar nanowires is simpler, is cheaper, and has a higher resolution. Figure 7 shows the memristor's crossbar structure, and Fig. 8 shows the mapping of the NOR gate.

4.2 Hybrid Chip

The hybrid chip consists of an amalgamation of transistors and memristors [19]. It has a single layer of transistors to operate memristor layers effectively. This entire structure is power-efficient and intakes less energy, resulting in less heat generation.

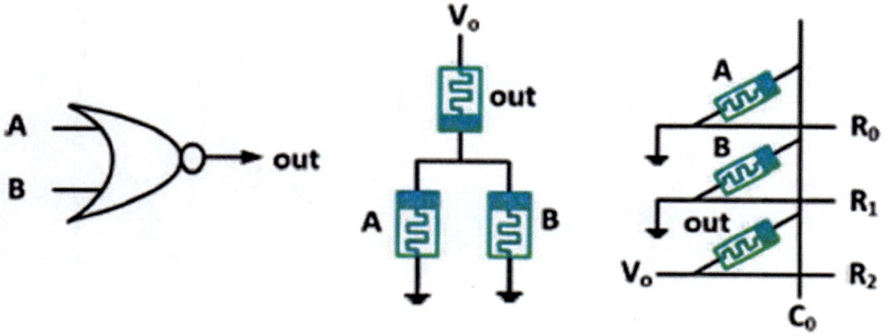

Fig. 8 Mapping NOR gate to crossbar column [18]

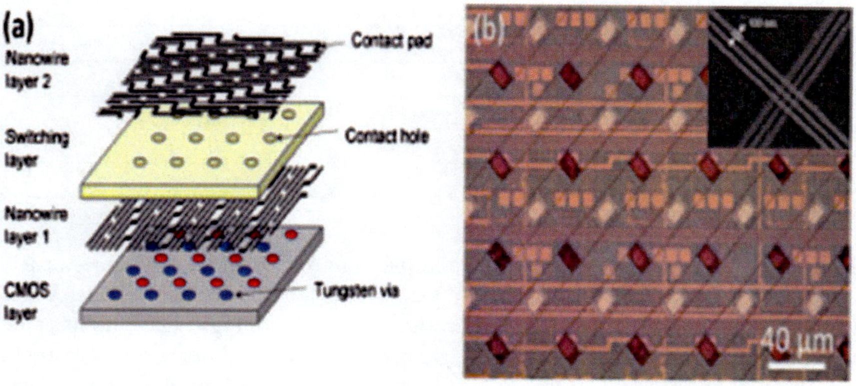

Fig. 9 Memristor reconfigurable hybrid chip

There is a CMOS Layer at the bottom, and above the CMOS layer, there is nanolayer1, and the switching layer exists in between the nanolayer1 and nanolayer2 as shown in Fig. 9.

4.3 IMPLY Logic Gate and Memristor-Aided Logic (MAGIC)

A cluster of memristors can be used to realize Boolean logic because memristors are passive elements. Some of these approaches are Wired AND logic and implication logic [20]. The best approach is implication logic which uses the implication operator because it can implement all the Boolean functions [21]. IMPLY function is one of the elementary 2 input Boolean functions. The schematic of the memristor-based IMPLY is shown in Fig. 10. The IMPLY logic gate is based on one resistor R_G connected to two memristors from memory crossbar array p and q. The resistance of

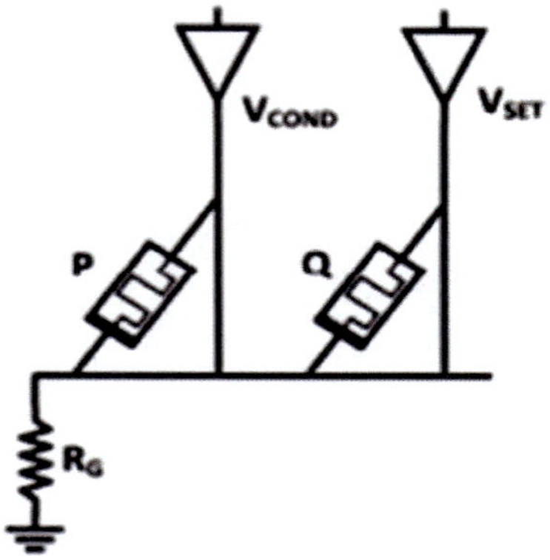

Fig. 10 Memristor-based IMPLY gate [14]

Table 1 Truth table of IMPLY function [18]

p	q	pIMPq
0	0	1
0	1	1
1	0	0
1	1	1

R_G has a value between the minimum and maximum resistances of memristor R_{on} and R_{off}, and the memristor p and q are the input of logic gate, the memristor q is also the output of logic gate as its initial value is the input of logic gate, and its value after computation is the output of the logic gate. The essential operation is to apply two different voltages, V_{SET} and V_{COND}, to the memristors where the magnitude of voltage V_{SET} is higher than the magnitude of voltage V_{COND}. Table 1 illustrates the IMPLY function, and a schematic of memristor-based IMPLY is shown in Fig. 10.

Implementing NAND gate with the IMPLY logic requires three computational steps, so three clock cycles will be required but if a more complex function e.g., a 1-bit full adder is to be implemented with IMPLY logic, then with Naive approach it will require 89 computation steps but with the parallel approach it will require only 5 computational steps. There is a need for refresh because of drift in IMPLY logic, whereas MAGIC-Memristor-Aided LOGIC consists of one applied voltage V_G in the gateway of logic gate and the separate input and output memristors, unlike IMPLY logic gate. The NOR schematics are shown in Fig. 11. It contains two inputs with separate outputs. In the first phase, there is a need to commence the output memristors to logical 1, and applied voltage causes current to flow from left to right. MAGIC NOR is good for logic inside memory and is easy and intuitive.

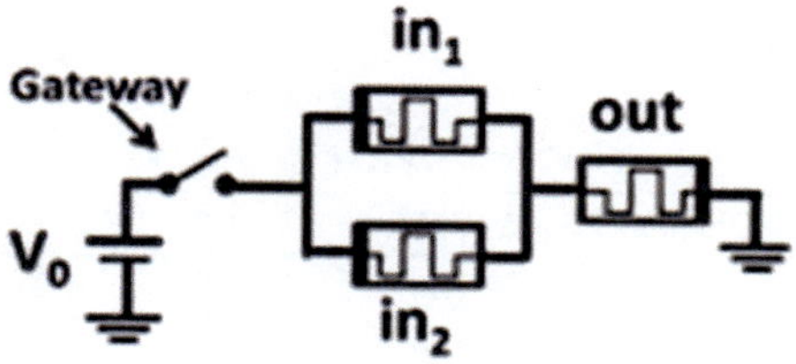

Fig. 11 Schematic of a two-input NOR gate [18]

4.4 Neuromorphic Computing

In the analog domain, memristors can be used to enable new unconventional applications, and the most exciting applications are large-scale Neuromophic circuits [22]. Artificial neural networks are implemented in the hardware mimicking functionalities of the human brain. The synapse in our brain initiates the learning, and it dictates how strong the connection in the middle of two neurons is and how a signal propagates from neuron one to two. Depending how frequently neuron1 sends signal to neuron2 to readjust its weight is according to nonlinear function. Memristor is placed in the middle of the pre-neuron and post-neuron, and synaptic weight crossbar of the memristor is created. Figure 12 depicts an illustration of two neurons and their connection.

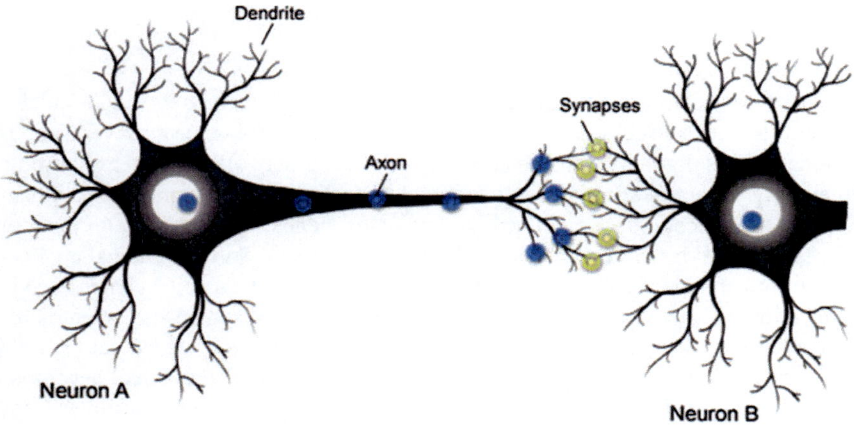

Fig. 12 An illustration of two neurons and their connection [22]

Table 2 Comparison of memristor with other memory

	Traditional memory			Emerging technology		
	DRAM	SRAM	Flash	FeRAM	MRAM	Memristor
Cell element	ITIC	6 T	1 T	1T1C	1T1R	1D1T1R
Feature size (nm)	36–45	45	22	180	65	20
Read time (ns)	2–10	0.2	100	45	35	<50
Write time (ns)	2–10	0.2	10^6	65	35	0.3
Retention time	4–64 ms	NA	10 yr	10 yr	>10 yr	>10 yr

4.5 Non-Volatile Memory

Memristor device has many advantages: compatibility with CMOS, low power consumption, no leakage current, high scalability, nonvolatility, fast access time, high retention time, and high switching speed. Due to these properties, it has a wide application in memory circuit designing, low power designing. The memristor device has a unique feature that its memristance can be varied in a very effective controlled way by only changing biasing. Memristor has been compared with other memories, as can see in Table 2 [23].

5 Conclusion

The literature review of memristors, characteristics, various models, and emulator circuits have been discussed. Different emulator circuits are feasible for several applications such as digital circuits, analog circuits, neural networks, and neuromorphic computing. Future works include memristor modeling and neuromorphic applications, pattern recognition, and ternary logic gates; hence, there is a requirement of a definite model for circuit analysis and simulation to suggest more radical applications of memristors. Memristors can offer promising higher capacities when used in applications of memory or hybrid memory circuits. The expedience of memristors over other memories is the non-volatile property of memristors which can be used effectively in memories; hence, more research is being conducted. Finally, it can be concluded the need to develop memristor emulator circuits in order to discover many more application areas with fast computation and less power dissipation.

References

1. Chua L (1971) Memristor-the missing circuit element. IEEE Trans Circuit Theory 18:507–519
2. Keshmiri V (2014) A study of the memristor models and applications (Dissertation). Retried from https://urn.kb.se/resolve?urn:nbn:se:liu:diva-112930

3. Elwakil AS, Fouda ME, Radwan AG (2013) A simple model of double-loop hysteresis behavior in memristive elements. IEEE Trans Circuits Syst II Express Briefs 60(8):487–491
4. Prodromakis T, Toumazou C (2010) A review on memristive devices and applications. In: 17th IEEE international conference on electronics, circuits and systems, Athens, pp 934–937
5. Kim H, Sah MP, Yang C, Cho S, Chua LO (2012) Memristor emulator for memristor circuit applications. IEEE Trans Circuits Syst I Regul Pap 59:2422–2431
6. Kvatinsky S, Friedman EG, Kolodny A, Weiser UC (2013) TEAM: threshold adaptive memristor model. IEEE Trans Circuits Syst I Regul Pap 60:211–221
7. Yeşil A, Babacan Y, Kaçar F (2014) A new DDCC based memristor emulator circuit and its applications. Microelectron J 45:282–287
8. Sánchez-López C, Carrasco-Aguilar MA, Muñiz-Montero C (2015) A 16 Hz–160 kHz memristor emulator circuit. AEU-Int J Electron Commun 69:1208–1219
9. Sözen H, Çam U (2016) Electronically tunable memristor emulator circuit. Analog Integr Circuits Signal Process 89(3):655–663
10. Sánchez-López C, Aguila-Cuapio LE (2017) A 860 kHz grounded memristor emulator circuit. AEU-Int J Electron Commun 73:23–33
11. Babacan Y, Yesil A, Gul F (2018) The fabrication and MOSFET-only circuit implementation of semiconductor memristor. IEEE Trans Electron Devices 65(4):1625–1632
12. Vista J, Ranjan A (2020) High frequency meminductor emulator employing VDTA and its application. IEEE Trans Comput Aided Des Integr Circuits Syst 39(10):2020–2028
13. Raj N, Ranjan RK, Khateb F (2020) Flux-controlled memristor emulator and its experimental results. IEEE Trans Very Large Scale Integr (VLSI) Syst 28:1050–1061
14. Kvatinsky S, Satat G, Wald N, Friedman EG, Kolodny A, Weiser UC (2014) Memristor-based material implication (IMPLY) logic: design principles and methodologies. IEEE Trans Very Large Scale Integr (VLSI) Syst 22:2054–2066
15. Gregory MD, Werner DH (2015) Application of the memristor in reconfigurable electromagnetic devices. IEEE Antennas Propag Mag 57:239–248
16. Gharpinde R, Thangkhiew PL, Datta K, Sengupta I (2018) A scalable in-memory logic synthesis approach using memristor crossbar. IEEE Trans Very Large Scale Integr (VLSI) Syst 26:355–366
17. Truong SN, Van Pham K, Yang W, Min K (2016) Sequential memristor crossbar for neuromorphic pattern recognition. IEEE Trans Nanotechnol 15:922–930
18. Kvatinsky S et al (2014) MAGIC—memristor-aided logic. IEEE Trans Circuits Syst II Express Briefs 61:895–899
19. Pershin YV, Di Ventra M (2010) Practical approach to programmable analog circuits with memristors. IEEE Trans Circuits Syst I Regul Pap 57:1857–1864
20. Liang Y, Lu Z, Wang G, Yu D, Iu HH (2019) Threshold-type binary memristor emulator circuit. IEEE Access 7:180181–180193. https://doi.org/10.1109/ACCESS.2019.2957371
21. Guckert L, Swartzlander EE (2017) MAD gates—memristor logic design using driver circuitry. IEEE Trans Circuits Syst II Express Briefs 64:171–175
22. Xu L, Li C, Chen L (2015) Analog memristor based neuromorphic crossbar circuit for image recognition. In: Sixth international conference on intelligent control and information processing (ICICIP), Wuhan, pp 155–160
23. Zidan MA, Fahmy HAH, Hussain MMH, Salama KN (2013) Memristor-based memory: the sneak paths problem and solutions. Microelectron J 44(2):176–183

Sensitivity Enhancement of Graphene and Blue Phosphorene/Mos_2 Heterostructure-Based SPR Biosensor Using Gold (Au) Metal Layer: Theoretical Insight

Sachin Singh, Pravin Kumar Singh, A. K. Sharma, Pooja Lohia, and D. K. Dwivedi

Abstract In this study, an angular interrogation technique has been used for modeling a highly sensitive surface plasmon resonance (SPR)-based biosensor. The large surface area of graphene layer facilitates the biomolecules absorption. A five-layer Kretschmann configuration of the SPR biosensor containing the BP/MoS_2 heterostructure with a gold layer is proposed. Compared to the traditional gold film-based SPR biosensors, the sensitivity of the proposed SPR biosensor has been significantly improved. By optimizing the proposed structure with a 50-nm-thick gold layer and a monolayer of graphene and heterostructure BP/MoS_2 with a thickness of 0.34 and 0.75 nm, respectively, enhanced sensitivity 229.12°/RIU has been achieved. Moreover, the proposed SPR sensor design offers extremely small FWHM, high detection accuracy (DA), and high-quality factor (QF) parameters. The highest sensitivity of 251.6°/RIU was found with two layers of graphene with fixed monolayer of heterostructure BP/MoS_2 configuration. It is also noted that the proposed SPR biosensor shows better results as compared to previously recorded SPR sensor parameters.

Keywords Surface plasmon resonance · Heterostructure · Biosensor · Kretschmann configuration · Sensitivity

---

S. Singh · P. K. Singh · D. K. Dwivedi (✉)
Photonics and Photovoltaic Research Lab, Department of Physics and Material Science, Madan Mohan Malaviya University of Technology, Gorakhpur 273010, India
e-mail: todkdwivedi@gmail.com

A. K. Sharma
Department of Applied Sciences, National Institute of Technology Delhi, Narela, New Delhi 110040, India

P. Lohia
Department of Electronics and Communication Engineering, Madan Mohan Malaviya University of Technology, Gorakhpur 273010, India

P. K. Singh
Institute of Advanced Material, Sweden, Sweden

B. Mishra and M. Tiwari (eds.), *VLSI, Microwave and Wireless Technologies*, Lecture Notes in Electrical Engineering 877,
https://doi.org/10.1007/978-981-19-0312-0_47

1 Introduction

A technique of surface plasmon resonance-based sensing has been emerging as an excellent technique to identify very minute change in refractive index of sensing medium while interacting with the metal film of sensor [1–3]. It is an established technique for the development of label-free and real-time biosensor, real-time measurements provide details regarding bimolecular interaction. Biosensing application of SPR-based biosensors includes enzyme detection, drug diagnostic, medical diagnostic, and food safety. The basic SPR biosensors consist of a thin metal film coated over a prism [4]. In surface plasmon resonance sensor, there is excitation of surface plasmons (SPs), i.e., oscillating free electrons, at the interface of the metal and constant dielectric material [5, 6]. Commonly used metal films in SPR sensor are gold, silver, aluminum, copper, indium, and sodium. The durability, great stability, and biocompatibility of gold make it more favorable for SPR biosensor [7, 8]. The accuracy of gold metal film provides maximum sensitivity corresponding to the other metals. AU metal has good stability, no oxidation issue but cost is high. Two-dimensional nanomaterials, such as graphene, black phosphorous (BP) and transition metal dichalcogenides (TMDCs), heterostructure of TMDCs can be used to enhance SPR biosensor sensing. Since 2D heterostructure nanomaterials have excellent electrical and optical properties [9]. The unique properties of TMDCs are electrical, optical, and chemical which make them as promising materials for future [10–12]. A group-ten TMDC materials graphene and heterostructure BP/MoS_2 have highly large tenable band gap which allows to easily modulate different types of stress that are applied on it [13].

For SPR sensor geometry, there are two commonly used configuration one Kretschmann configuration and other Otto's configuration; each configuration is based on the mechanism of attenuated total internal reflection (ATR). Advantages of Kretschmann configuration over Otto's configuration have made it useful on wider scale [14]. Biosensing application of SPR-based sensor explains us the fundamentals of detecting the concentration of biological objects, such as bacteria, viruses, DNA, and proteins in very small scale. Apart from the biosensing application and bimolecular analysis, SPR sensor is also very operative for the detection of nanostructured film deposition, for the displacement and angular position measurement [15].

2 Theoretical Modeling

The proposed five layers Kretschmann configuration-based [16] surface plasmon resonance-based (SPR) biosensor is shown in Fig. 1, where a monochromatic p-polarized light source is used having wavelength 633 nm at an angle (θ) on a metal-coated glass prism. The angle of incidence (θ) is changed according to the angle interrogation method. The first layer is CaF_2 glass prism with refractive index (RI) 1.4329 [17].

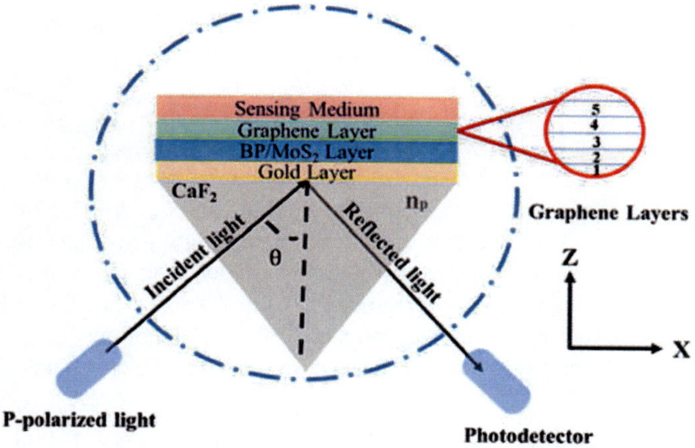

Fig. 1 Schematic design of a proposed SPR biosensor based on the Kretschmann configuration

A gold metal (Au) layer is the second layer deposited on the glass prism. When light is incident on glass-metal interface, then surface plasmons are generated. Graphene and heterostructure BP/MoS_2 are the third and fourth layers used because these materials have immense potential for sensing applications [18]. The fifth layer in proposed design of SPR biosensors is sensing medium which can also be called as sensing analyte.

The design parameters such as RI and thickness of different layers of the proposed SPR biosensor are arranged in Table 1.

When metal-coated prism layer is illuminated by plane polarized light, there is complete reflection through metal dielectrics interface of prism. This generates evanescent waves in the region of interface. The generated evanescent wave thus passes through the layer of gold and then propagates along x-direction with propagation constant $K_x = n_{\text{Prism}} \left(\frac{2\pi}{\lambda}\right) \text{Sin}\theta$. Thus, the plane-polarized charge density oscillates, and resonance condition is obtained at the metal dielectrics interface. The condition of resonant oscillation exists only if the incident wave vector matches with

Table 1 Parameters of proposed SPR sensor design at wavelength 633 nm

No of layers	Material layer	Thickness (nm)	Refractive index (n + ik)		Refs.
			Real part (n)	Imaginary part (k)	
Layer-1	Prism CaF_2	–	1.4329	–	[17]
Layer-2	Gold layer (Au)	50.0	0.19683	3.09505	[17]
Layer-3	BP/MoS_2 layer	0.75	2.7915	0.335	[19]
Layer-4	Graphene (G)	0.34	3.0	1.149106	[19]
Layer-5	Sensing medium (SM)	–	1.33 to 1.35	–	–

the surface plasmon wave vector. A sharp dip is obtained the condition of perfect resonance which is given by the following equation:

$$n_d\left(\frac{2\pi}{\lambda}\right)\text{Sin}\,\theta_{\text{SPR}} = \text{Real}\left(\frac{2\pi}{\lambda}\right)\sqrt{\frac{\in_m \in_s}{\in_m + \in_s}} \tag{1}$$

From Eq. (1), n_d is the RI of the dielectric medium, and $\in_m$ is dielectric constants of the metal layer, and $\in_s$ is dielectric constant of sensing layer, respectively. Equation (2) shows the relation between wavelength and dielectric constant of the metal layer, with the help of these two parameters Drude–Lorentz calculated the RI of metals, such as gold (Au), silver (Ag), copper (Cu), and platinum (Pt).

The Drude–Lorentz expression for RI of gold (Au) metal is given as [17]:

$$n_{\text{Au}} = \sqrt[2]{(\varepsilon_r + i\,\varepsilon_i)} = \left[1 - \lambda^2\lambda_c/\lambda_p^2(\lambda_c + i\,\lambda)\right]^{1/2} \tag{2}$$

where λ_c, λ_p, and λ are collision, plasma, mochoromatic light wavelength having values 8.9342×10^{-6} mt, 1.6826×10^{-7} mt, and 6.33×10^{-7} mt (633 nm), respectively.

Since p-polarized light penetrates the metal prism layer, the radiative properties take place. Basically, we have three methods to calculate the mathematical expression of radiative properties, such as the transmittance and reflectance of a multilayer thin film (i) Transfer matrix method (ii) Resultant wave method (iii) Field tracing method. This paper deals with N-layer model using the transfer matrix method approach. The transfer matrix method offers accuracy, simplicity, and precision. So, there is no need for approximation in this process [20, 21]. This technique also allows to calculate the reflectance curve and various performance parameters of SPR biosensor.

3 Performance Parameters

The performance parameters of SPR sensing depend upon four parameters like sensitivity (S), full width half maximum (FWHM), detection accuracy (DA), and quality factor (QF). The extraction of these parameter values depends on the curve of reflectivity. The values of these parameters for proposed SPR biosensor designs are shown in Table 2.

4 Results and Discussions

Figure 2 shows the reflectance curve of SPR biosensor. It shows that when we change the RI (1.330 to 1.345) of analyte in sensing medium, the dip of SPR angle is also

Table 2 Performance parameter analysis of proposed SPR sensor designs at wavelength 633 nm

Sensor design	Δn	$\Delta\theta_{SPR}$	S (°RIU^{-1})	FWHM (°)	DA	QF (RIU^{-1})
Prism + Au + SM (Conventional)	0.005	1.011	202.20	3.8022	0.2660	53.180
Prism + Au + BP/MoS_2 + SM (Proposed structure)	0.005	1.072	214.40	4.0702	0.2634	52.675
Prism + Au + BP/MoS_2 + G + SM (Proposed structure)	0.005	1.146	229.12	7.2071	0.1590	31.791

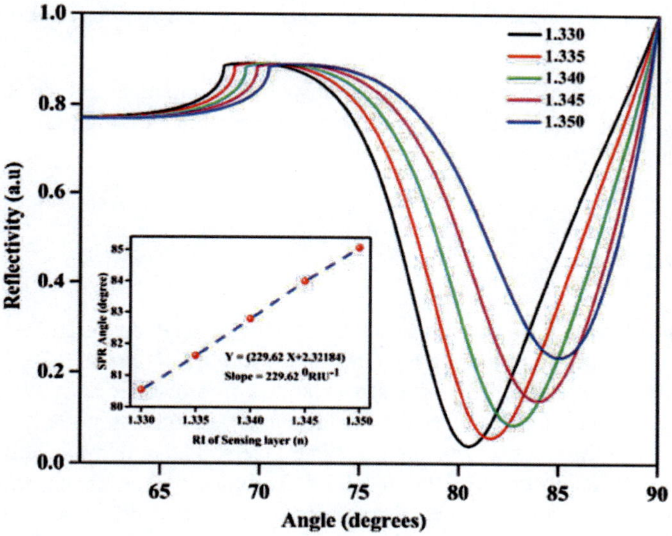

Fig. 2 Reflectance curve at different sensing layer RI for the proposed SPR biosensor

shifted. Therefore, the raise in the SPR angle shift corresponding to the change in RI (n = 0.005) provides an improvement in the performance of SPR biosensor parameters. Another plot inserted in Fig. 2 which describes the changes in RI corresponding to the SPR angle, it gives the slope value or sensitivity of SPR biosensor. The other parameters, such as DA, QF, and FWHM, were also calculated. The sensitivity of proposed SPR biosensor is 229.12 °RIU^{-1}, correspondingly the values of other parameters are 7.2071°, 0.1590, and 31.791°RIU^{-1}, respectively.

Figure 3a–b show the effect of change in number of graphene layer on SPR biosensor parameters. The SPR angle dip varies in Fig. 3a according to the number of layers of graphene (L = 1 to L = 5), which provides the change in values of performance parameters of SPR biosensor. The maximum sensitivity of proposed SPR sensor observed with bilayer (L = 2) of graphene is 251.6 ^{0}RIU^{-1}. Similarly,

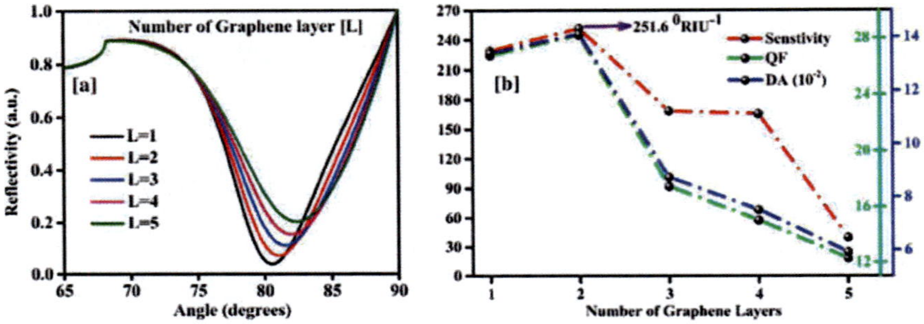

Fig. 3 Effect of rising the number of graphene layers, **a** Reflectance curve for the different number of graphene layers, **b** Combined plot of sensitivity, quality factor and detection accuracy

the QF and DA are also maximum with bilayer of graphene as shown in Fig. 3b. Thus, we can say that in our design graphene layers play an important role for the performance of SPR biosensor.

5 Conclusion

Current paper focuses on the impact of the graphene layer on parameters of SPR sensor. In this paper introduces a new design for the low RI prism, SPR biosensor has been introduced. The proposed sensor shows improved sensitivity compared to the conventional gold film-based SPR biosensor and other reported work. This study optimizes the thickness of the graphene layers. The maximum sensitivity of the proposed SPR biosensor is 251.6°RIU^{-1} for bilayer of graphene; the FWHM increases with varying the number of graphene layers, but the other parameters, such as QF and DA parameters, are slightly decreasing in angular interrogation mode because of the optical property components. Proposed SPR sensor may a suitable candidate for medical diagnosis and biological applications.

References

1. Schuller JA, Barnard ES, Cai W, Jun YC, White JS, Brongersma ML (2010) Plasmonics for extreme light concentration and manipulation. Nature Mater 9:193–204
2. Gramotnev DK, Bozhevolnyi SI (2010) Plasmonics beyond the diffraction limit. Nature Photon 4:83–91
3. Bochenkov VE, Frederiksen M, Sutherland DS (2013) Enhanced refractive index sensitivity of elevated short–range ordered nanohole arrays in optically thin plasmonic Au films. Optic Expr 21(12):14763–14770
4. Rahman MM, Rana MM, Rahman MS, Anower MS, Mollah MA, Paul AK (2020) Sensitivity enhancement of SPR biosensors employing heterostructure of PtSe2 and 2D materials. Opt Mater 107:110123

5. Johansen K, Arwin H, Lundström I, Liedberg B (2000) Imaging surface plasmon resonance sensor based on multiple wavelengths: Sensitivity considerations. Rev Sci Instrum 71:3530
6. Ouyang Q, Zeng S, Jiang L, Qu J, Dinh XQ, Qian J, He S, Coquet P, Yong K-T (2017) Two-dimensional transition metal Dichalcogenide enhanced phase-sensitive plasmonic biosensors. J Phys Chem C 121(11):6282–6289
7. Prajapati YK, Pal S (2018) Saini JP (2018) Effect of a Metamaterial and silicon layers on performance of surface Plasmon resonance biosensor in infrared range. SILICON 10:1460
8. Kumar R, Pal S, Prajapati YK, Saini JP (2020) Sensitivity enhancement of MXene based SPR sensor using silicon: theoretical analysis, Silicon
9. Xia G, Zhou C, Jin S, Huang C, Xing J (2019) Liu Z (2019) Sensitivity enhancement of two-dimensional materials based on genetic optimization in surface Plasmon resonance. Sensors 19(5):1198
10. Prajapati YK, Srivastava A (2019) Effect of BlueP/MoS_2 heterostructure and graphene layer on the performance parameter of SPR sensor: theoretical insight. Superlattices and Microstructures 129:152–162
11. Zhang H, Liu C-X, Qi X-L, Dai X, Fang Z, Zhang S-C (2009) Topological insulators in Bi_2Se_3, Bi_2Te_3 and Sb_2Te_3 with a single Dirac cone on the surface. Nat Phys 5:442
12. Castro Neto AH, Guinea F, Peres NMR, Novoselov KS, Geim AK (2009) The electronic properties of graphene. Rev Mod Phys 81:109–Published 14 January 2009
13. Jia Y, Li Z, Wang H, Saeed M, Cai H (2020) Sensitivity of a surface plasmon resonance sensor with platinum diselenide. Sensors 20(1):131
14. Alagdar M, Yousif B, Areed NF, Elzalabani M (2020) Highly sensitive fiber optic surface plasmon resonance sensor employing 2D nanomaterials. Appl Phys A 126(522)
15. Pal S, Prajapati YK, Saini JP, Singh V (2016) Sensitivity enhancement of Metamaterial based surface Plasmon resonance biosensor for near infrared. Opt Appl 46(1):131–143
16. Kretschmann E (1971) The determination of the optical constants of metals by excitation of surface plasmons. J Phys 241:313–324
17. Singh S, Singh PK, Umar A, Lohia P, Albargi H, Castaneda L, Dwivedi DK (2020) 2D Nano-material based surface plasmon resonance sensors for biosensing applications. Micromachines 11(8):779
18. Zeng S, Hu S, Xia J, Anderson T, Dinh XQ, Meng XM, Coquet P, Yong KT (2015) Graphene–MoS2 hybrid nanostructures enhanced surface plasmon resonance biosensors. Sensor Actuator B Chem 207:801–810
19. Mishra AK, Mishra SK, Verma RK (2016) Graphene and beyond graphene MoS2: a new window in surface–plasmon–resonance–based fiber optic sensing. J Phys Chem C 120(5):2893–2900
20. Bruna M, Borini S (2009) Optical constants of graphene layers in the visible range. Appl Phys Lett 94:031901
21. Sharma AK, Pandey AK (2018) Blue phosphorene/MoS_2 heterostructure based SPR sensor with enhanced sensitivity. IEEE Photon Technol Lett 30(7)
22. Pal S, Verma A, Raikwar S, Prajapati YK, Saini JP (2018) Detection of DNA hybridization using black phosphorus-graphene coated surface plasmon resonance sensor. Appl Phys A 124:394

Recent Development of Hardware-Based Random Number Generators on FPGA for Cryptography

Mangal Deep Gupta and R. K. Chauhan

Abstract The most important component in the cryptographic system is the cryptographic keys generator. These keys are generated by a random number generator (RNG) since the security of the cryptographic system depends entirely on the quality of generated keys. This paper summarizes the recent development of FPGA-based hardware efficient and secure RNGs. The main aim of this study is to summarize the knowledge of hardware performance, security strength, and suitability for the cryptographic system from the different classes of RNGs. It discusses the different classes of RNGs, recalls the basic ideas, and provides the details of several well-known RNGs. This work presents a comprehensive discussion on the hardware implementation of RNGs on FPGAs. A complete list of LCGs-based pseudorandom number generators (PRNGs) is presented with deep technical details on their mathematical formation and implementations. Finally, the performance of RNGs with respect to utilization of FPGA resources, frequency, latency, power consumption, security strength using the national institute of statistical testing (NIST), and weaknesses is presented.

Keywords RNGs · Cryptography · VLSI · FPGA

1 Introduction

With the rapid development of information technology, the cryptosystem is used widely to protect the data or information. A various encryption techniques are used to make ensure of information security. So the random number generator (RNG) used in cryptography determines the system security. The RNGs are widely used in various applications related to cryptography such as key generation, encryption/decryption, masking protocols, Internet gambling, and block ciphers [1–4]. A different RNG is proposed by the researchers that enhance the security of information. In the evolutionary development of smart mobile devices that are connected to the internet, the

M. D. Gupta (✉) · R. K. Chauhan
Department of Electronics and Communication Engineering, MMMUT, Gorakhpur, Uttar Pradesh, India
e-mail: mangaldeepgcet@gmail.com

B. Mishra and M. Tiwari (eds.), *VLSI, Microwave and Wireless Technologies*,
Lecture Notes in Electrical Engineering 877,
https://doi.org/10.1007/978-981-19-0312-0_48

information exchange over an insecure network is a major concern over the years. The physical systems like environment monitoring, health care system, advanced metering in smart grids, etc., which are connected over the Internet of things (IoT) that generate a large amount of data that leads to privacy and security issues. To assure the security of associated information over an Internet network, the RNG is the primary requirement [5]. Due to trade-off between different factors (like hardware performance, security, and cost), most of the cryptosystems are unsuitable for real-time implementation on IoT-based resource constraint devices [6]. To accomplish this request, it is required to implement efficient hardware architectures capable of generating pseudorandom bit sequences to provide the public and private keys for effective data cryptography. Therefore, the hardware-based cryptosystem is compulsory in IoT applications for secure information exchange over smart mobile devices that are connected with IoT. So, the primary requirement of a hardware-based cryptosystem is low-hardware complexity, high speed, low-power consumption, secure key generation, and high randomness. In this regard, different RNG methods were proposed for the generation of the random number to satisfy the randomness behavior as required for cryptography.

The FPGA implementation of the RNG is more useful in real tile applications like cryptography, secure communications, etc. The fully digital circuit or/and embedded systems with high-speed and low-power consumption are suitable for IoT, cybersecurity, and Industry 4.0 security applications. The RNGs based on FPGA open the opportunity to use the large number of combinational blocks that are connected through programmable logic. So, this powerful platform is widely used in digital circuit implementations [7].

This research work surveys the large set of FPGA implementations of RNGs. First, summarize the different classifications of RNGs and provide the advantages and weakness of different well-known RNGs both pseudorandom and truly random, while linear and nonlinear system-based generators are discussed in the PRNG case. The LCGs-based PRBGs are explained in detail and also discuss the choices of the RNGs. Finally, present the performance of FPGA-based PRNGs in terms of FPGA resources, timing performance, security strength, and weakness.

The remainder of the article is as follows. Sect. 2 refers to the classification of the RNGs. Sect. 2.1 presents the FPGA implementations of nonlinear PRNGs, whereas the next Sect. 2.2 focuses on linear system-based PRNGs. This section summarizes the mathematical formation and VLSI architectures corresponding PRBGs with a short comparison regarding area resources and timing performance of the FPGA implementations. The FPGA implementation results (in terms of FPGA resources, frequency, throughput, and power consumptions) and security status of linear and nonlinear PRNGs are detailed in Sect. 3. This article ends with a conclusion section that summarizes the review.

2 Classification of the RNGs

RNGs are categorized mainly into two families: true random number generators (TRNGs) and pseudorandom number generators (PRNGs). In TRNGs, the physical process like jitter or thermal noise is used to generate random numbers. Therefore, the TRNGs cannot be used in the encryption/decryption process because they cannot be able to generate the same sequences corresponding to ciphering and deciphering operations [7]. For this problem, there is only one possible solution is generated sequences from TRNGs stored in memory. Additionally, TRNGs also suffer from a low-throughput rate, therefore they cannot be used in high-speed applications.

In general, there are two types of PRNG: (1) linear and (2) nonlinear PRNG. The linear system-based PRNG, linear feedback shift registers (LFSRs) [8, 9], and linear congruential generators (LCGs) [6, 10–17] are used for generating pseudorandom number sequences. In nonlinear PRNGs, the nonlinear output function or nonlinear transition function is used to convert the linear system into a nonlinear [18–30].

2.1 Nonlinear PRNG

In nonlinear PRNGs, the nonlinear output function or nonlinear transition function is used to convert the linear system into a nonlinear. Various PRNGs based on nonlinear system are used in the cryptography for their good randomness properties. Nonlinear dynamical systems consist of simple mathematical equations that can exhibit chaos behavior. So, the cryptographic properties of generated random sequences from the chaotic map are very crucial for the security of encryption algorithms. Chaotic systems generate a pseudorandom sequence, which can be applied in designing cryptographic keys to get their valuable characteristics like random behavior, sensitivity to the initial conditions, and ergodicity.

Mathematically, a hyperchaotic system can be defined as a chaotic system with two or more than two positive Lyapunov exponents. Its dynamic behavior is expended in more than two directions. So, the hyperchaotic attractor has more complex dynamic behaviors as compared to a chaotic system. The expansion of this dynamic behavior happens at the same time in two or more than two directions that make the hyperchaotic system, which shows better performance in many chaos-based applications including technological applications, than chaotic systems. Nowadays, hyperchaos has attracted attention from various scientific and engineering communities. So, the application of hyperchaos is becoming more popular in the field of chaos-based cryptography. Though, the well-known disadvantage of ordinary chaotic attractors for topological applications possesses only a single positive Lyapunov exponent (LE), hence its degree of disorder is not high as compare to hyperchaotic systems.

The recent literature of FPGA-based PRNG using chaotic and hyperchaotic attractors is discussed. The FPGA implementation of six different multiplierless chaotic PRNGs using Chua, Lorenz, Rössler, and the other three systems has been done in [7]. To increase the randomness as well as prevent the digital chaotic system to fall into short-period orbits of the generated sequences, a PRNG based on the one-dimensional logistic map was implemented on FPGA [23]. In [24], Rezk et al. proposed an FPGA-based PRNG that is using the Lü and Lorenz chaotic attractors. The PRNG based on a hyperchaotic system with a self-shrinking perturbance generator was proposed by Yang Liu et al. in [25]. A new 4D hyperchaotic oscillator was proposed by wu et al. and analyzed its nonlinear dynamic behavior. Furthermore, an analog circuit of this system is implemented on a chip for some relevant engineering applications such as information encryption [26]. A hyperchaotic system and its qualitative properties were discussed by Rajagopal et al. in [27]. This system was also implemented in FPGA to prove that the system is hardware realizable.

2.2 *Linear PRNG*

The linear PRNG, LCGs, and LFSRs are used for generating pseudorandom number sequences. The linear PRNGs are suitable for high-speed and low-power applications in a hardware-based cryptosystem, but there is some limitation, i.e., limitation of state and a short period of generated bit sequences. To mitigate this limitation, many-related literature surveys are presents in detail thereafter.

2.2.1 LCG-Based PRNGs

The most popular random number generation method is linear congruential on modular arithmetic. An LCG is originated on the system of linear recurrence equations, which is defined as $x_{i+1} = [(a_1 \times x_i) + x_i + b_1]\text{mod}2^K$, where a (the "multiplier"), b (the "increment"), where $0 \leq a, b \leq 2^{k-1}$ are parameters of the generator. LCG is convenient for high-speed and low-power constraints, but it is not capable to generate more secure pseudorandom numbers. Because of this, many hardware implementations are proposed to increase the security and period.

The authors of [14] proposed the high-secure dual-CLCG algorithm-based PRBG. The dual-coupled-LCG blocks are used to design this architecture, and these blocks are designed by following recurrence relations:

$$x_{i+1} = [(a_1 \times x_i) + x_i + b_1]\text{mod}2^n \quad (1)$$

$$y_{i+1} = [(a_2 \times y_i) + y_i + b_2]\text{mod}2^n \quad (2)$$

$$p_{i+1} = [(a_3 \times p_i) + p_i + b_3]\text{mod}2^n \quad (3)$$

$$q_{i+1} = [(a_4 \times q_i) + q_i + b_4]\text{mod}2^n \quad (4)$$

$$B_i = \begin{Bmatrix} 1 \text{ if} x_{i+1} > y_{i+1} \\ 0 \text{ if} x_{i+1} < y_{i+1} \end{Bmatrix} \tag{5}$$

$$C_i = \begin{Bmatrix} 1 \text{ if} p_{i+1} > q_{i+1} \\ 0 \text{ if} p_{i+1} < q_{i+1} \end{Bmatrix} \tag{6}$$

$$z_i = B_i, if C_i = 0 \tag{7}$$

Here, constant parameters $(a_1, a_2, a_3, a_4, b_1, b_2, b_3, b_4)$ and initial seeds (x_0, y_0, p_0, q_0) are used in recurrence relation as given in corresponding Eqs. (1)–(4). The comparator output, i.e., B_i and C_i is given by Eqs. (5) and (6). The random bit sequences (z_i) are given by Eq. (7).

Authors of [15] optimized the implementation of the dual-CLCG algorithm [14] that involves arithmetic operations such as multiplication. In this architecture, the author uses the logical left shifting rather than multiplication operation, which reduces the hardware complexity of dual-coupling of LCG [14]. Therefore, Eqs. (1)–(7) can be rewritten as

$$x_{i+1} = \left[\left(2^{r1} \times x_i\right) + x_i + b_1\right] \text{mod} 2^n \tag{8}$$

$$y_{i+1} = \left[\left(2^{r2} \times y_i\right) + y_i + b_2\right] \text{mod} 2^n \tag{9}$$

$$p_{i+1} = \left[\left(2^{r3} \times p_i\right) + p_i + b_3\right] \text{mod} 2^n \tag{10}$$

$$q_{i+1} = \left[\left(2^{r4} \times q_i\right) + q_i + b_4\right] \text{mod} 2^n \tag{11}$$

$$B_i = \begin{Bmatrix} 1 \text{ if} x_{i+1} > y_{i+1} \\ 0 \text{ if} x_{i+1} < y_{i+1} \end{Bmatrix} \tag{12}$$

$$C_i = \begin{Bmatrix} 1 \text{ if} p_{i+1} > q_{i+1} \\ 0 \text{ if} p_{i+1} < q_{i+1} \end{Bmatrix} \tag{13}$$

$$z_i = B_i \oplus C_i \tag{14}$$

Gupta and Chauhan of [6] further optimized the implementation of the dual-CLCG algorithm [15], in which LCG blocks are designed using 2-operands modulo adder instead of 3-operands modulo adder. So, the nth bit of final addition will be calculated by XOR between nth bit of 2-operands modulo adder's output, i.e., $(S_{x1i}[n-1], S_{y1i}i[n-1], S_{p1i}[n-1] \text{and} S_{q1i}[n-1])$ and nth bit of the shifted value of variables $(x_i, y_i, p_i \text{and} q_i)$, i.e., $(x_i[0], y_i[0], p_i[0]$ and$q_i[0])$ corresponding to each LCG block and given by $\left(S_{x2i}, S_{y2i}, S_{p2i} \text{and} S_{q2i}\right)$ as shown in Eqs. (15)–(18). The values $(S_{x2i}, S_{x1i}[n-2:0])$, $\left(S_{y2i}, S_{y1i}[n-2:0]\right)$, $\left(S_{p2i}, S_{p1i}[n-2:0]\right)$, and $\left(S_{q2i}, S_{q1i}[n-2:0]\right)$ are stored in four registers corresponding to each LCG block that hold the value for further processing. These register values are assigned to the next iterative values as given in Eqs. (19)–(22).

$$S_{x1i}[n-1:0] = b_1[n-1:0] + x_i[n-1:0],\ S_{x2i} = S_{x1i}[n-1] \oplus x_i[0] \quad (15)$$

$$S_{y1i}[n-1:0] = b_2[n-1:0] + y_i[n-1:0],\ S_{y2i} = S_{y1i}[n-1] \oplus y_i[0] \quad (16)$$

$$S_{p1i}[n-1:0] = b_3[n-1:0] + p_i[n-1:0],\ S_{p2i} = S_{p1i}[n-1] \oplus p_i[0] \quad (17)$$

$$S_{q1i}[n-1:0] = b_4[n-1:0] + q_i[n-1:0],\ S_{q2i} = S_{q1i}[n-1] \oplus q_i[0] \quad (18)$$

$$x_{i+1}[n-1:0] = \{S_{x2i}, S_{x1i}[n-2:0]\} \quad (19)$$

$$y_{i+1}[n-1:0] = \{S_{y2i}, S_{y1i}[n-2:0]\} \quad (20)$$

$$p_{i+1}[n-1:0] = \{S_{p2i}, S_{p1i}[n-2:0]\} \quad (21)$$

$$q_{i+1}[n-1:0] = \{S_{q2i}, S_{q1i}[n-2:0]\} \quad (22)$$

Authors of [16] proposed the high-secure PRBG architecture based on variable-input coupled-LCG. This architecture is designed using the coupling of LCG and input seeds of these LCG blocks are change by another two LCG blocks in each iteration. Each LCG block is defining by recurrence relations. It is given by Eqs. (23)–(26). The Eqs. (2) and (24) are named as variable-input linear congruential generators, were, the variables p_i and q_i are attained from two different LCGs recurrence relations as mentioned in Eqs. (25) and (26). The random bit sequence Z_i is obtained from the inequality condition, which is given in Eq. (27). Here, x_0, y_0, p_0, and q_0 are the initialization values corresponding to each recurrence equation. The b_1 and b_2 are the constant parameter of LCG blocks, as shown in Eqs. (25) and (26) (Figs. 1, 2, 3, and 4).

$$x_{i+1} = \left[\left(2^{r1} \times x_i\right) + x_i + p_i\right]\text{mod}2^n \equiv f_1(x_i, p_i)\text{mod } 2^n \quad (23)$$

$$y_{i+1} = \left[\left(2^{r2} \times y_i\right) + y_i + q_i\right]\text{mod}2^n \equiv f_2(y_i, q_i)\text{mod } 2^n \quad (24)$$

$$p_{i+1} = \left[\left(2^{r3} \times p_i\right) + p_i + b_1\right]\text{mod}2^n \quad (25)$$

$$q_{i+1} = \left[\left(2^{r4} \times q_i\right) + q_i + b_2\right]\text{mod}2^n \quad (26)$$

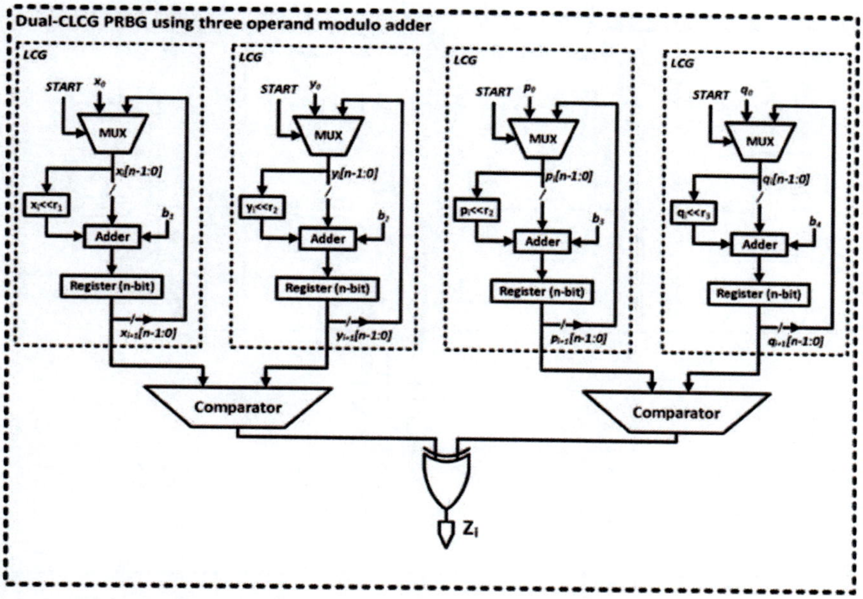

Fig. 1 VLSI architecture of dual-CLCG-based PRBG using 3-operands modulo adder [15]

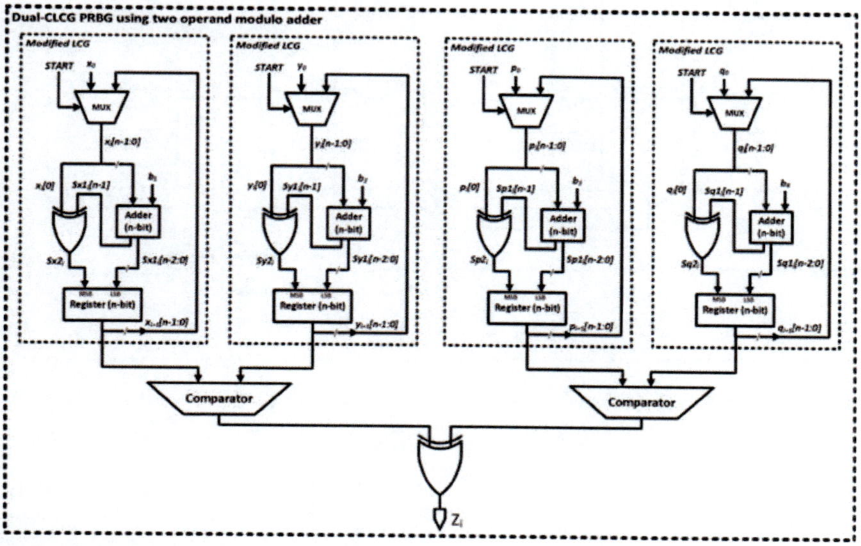

Fig. 2 PRBG architecture of dual-CLCG using 2-operands modulo adder [6]

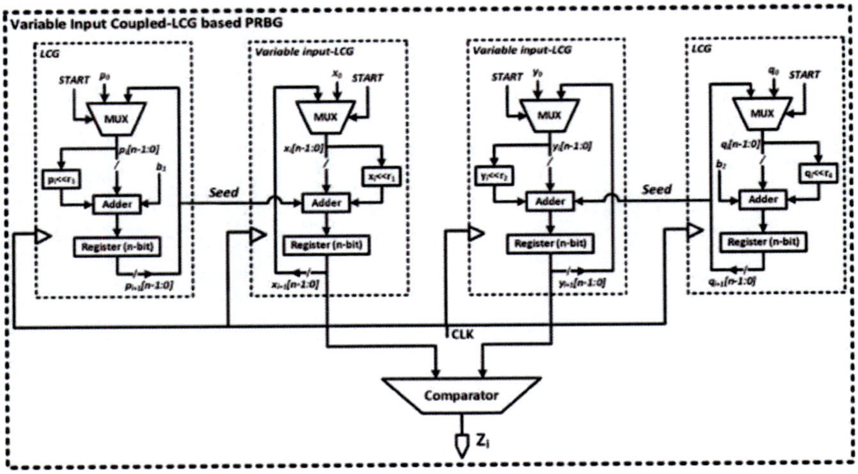

Fig. 3 PRBG architecture is based on the variable-input coupled-LCG [16]

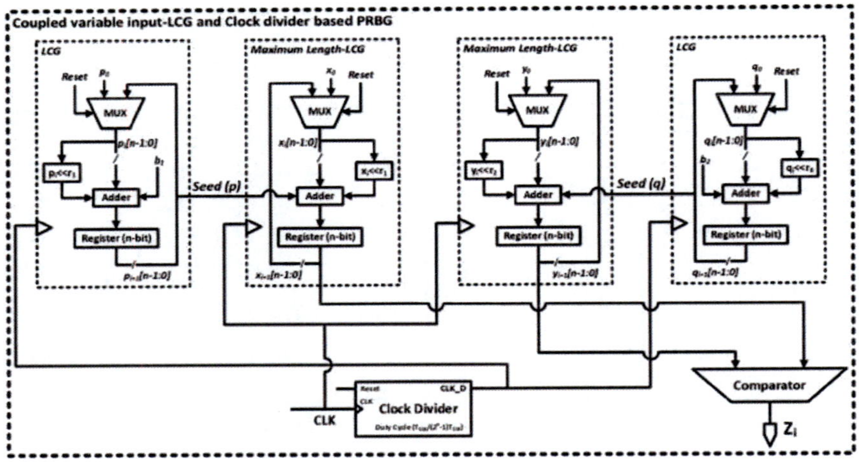

Fig. 4 PRBG architecture is based on the variable-input coupled-LCG and clock divider [17]

$$Z_i = \begin{Bmatrix} 1, \text{ if } x_{i+1} > y_{i+1} \\ 0, \text{ otherwise} \end{Bmatrix} \tag{27}$$

Gupta and Chauhan of [17] further improve the period length, security, and hardware performance of variable-input coupled-LCG architecture [16]. It is designed using several values of seed, i.e., $p(p_0, p_1, \ldots, p_{2^n-2}, p_{2^n-1})$ and $q(q_0, q_1, \ldots, q_{2^n-2}, q_{2^n-1})$ change periodically instead of changing in every iteration. Benefits of these techniques, the sequence of 2^{2n} maximum elements, i.e.,

$\{x_i\}$ is generated by periodically changing the value of p. In this method, the first 2^n elements are obtained by $x_{i+1} = f_1(x_i, p_0) \bmod m$, the next 2^n elements are obtained as $x_{i+1} = f_1(x_i, p_1) \bmod m$, and so on. After using all values of p, it generates 2^{2n} total elements. All value of p (from initial value) is reused to circularly continuing the process. Similarly, the sequence of 2^{2n} elements, i.e., $\{y_i\}$ is generated by periodically changing the value of q. With this method, the first 2^n elements are obtained by $y_{i+1} = f_1(y_i, q_0) \bmod m$, the next 2^n elements are obtained as $y_{i+1} = f_1(y_i, q_1) \bmod m$, and so on. After using all values of q, it generates 2^{2n} total elements. Now, this sequence of 2^{2n} elements $\{x_i\}$ and $\{y_i\}$ is computed by inequality condition to generate pseudorandom bits sequence of 2^{2n} period, without paying extra resources.

3 Result and Discussion

The performance parameters in terms of FPGA resources (i.e., number of flip-flops (FFs), look-up-table (LUT), slices, and DSP blocks), timing performance (critical path delay, frequency, and bit-rate), power consumption per unit frequency, and security strength are presented in Table 1.

Let us start to conclude the results obtained with different linear and nonlinear PRNGs, as demonstrated in Table 1. It appears demonstrated that the linear system-based PRNGs have the lowest area resources, high throughput, and less power consumable while maintaining the same security strength as compared with the nonlinear system (chaotic and hyperchaotic ordinary differential equations)-based PRNG approaches. These advantages leading the utility of linear system-based PRNG in lightweight IoT enable devices for cryptographic applications, i.e., better and more recommended schemes of PRNG.

The performance parameters of PRBGs belonging to the LCGs category are demonstrated in Table 1. The one that is based on the dual-coupled-LCG using two-operands modulo adder [6] has the lowest area occupation and high throughput. However, some other LCGs PRNGs can be presented as good competitors, namely (1) the variable-input coupled-LCG [16], (2) variable-input coupled-LCG with clock divider [17]-based PRBGs. Regarding the FPGA resources, throughput, the period length of the bit sequence, and security strength, the LCG-PRBG-based on variable-input coupled-LCG with clock divider [17] outperforms other linear PRNGs. In a conclusion, if we compare linear PRNGs to other PRNGs, it can play an important role in high-speed uses due to their parallel and rapidity generation.

4 Conclusion

This manuscript provided a widespread report on recent development in the FPGA implementation of RNGs. We have first recalled the different types of RNGs and

Table 1 Table captions should be placed above the tables

Methods	Size (bits)	FPGA	Total FFs	No. of slices	No. of LUTs	DSP blocks	Maximum frequency (MHz)	Clock period (T_{CLK}) (ns)	Power/freq. (mW/MHz)	NIST status
CLCG [14]	32	Virtex 5	64		311	–	219.78	4.550	0.146	Pass
		Virtex 7	64		294	–	282.64	3.54	0.084	
Dual-CLCG [15]		Virtex 5	128		603	–	219.78	4.550	0.228	Pass
		Virtex 7	128		571	–	282.64	3.54	0.146	
Modified dual-CLCG [6]		Spartan 3E	133	236	453	–	183.051	5.463	–	Pass
VCLCG [16]		Virtex 5	128		480	–	219.35	4.559	0.193	Pass
		Virtex 7	128		440	–	282.64	3.54	0.130	
VCLCG and clock divider [17]	16	Virtex 5	81		304	–	341.384	2.929	7.38	Pass
		Virtex 7	81		243	–	441.014	2.267	2.46	
Reconfigurable chaotic [24]	32	Virtex 5	96	100	276	8	78.149	12.796	–	Pass

discuss their properties in terms of hardware complexity, performance, and security strength. Thereafter, deeply investigate the nonlinear and linear system-based PRNG. Then, a large review of the FPGA-based PRNGs using LCGs systems is presented. For each type of PRNG, a hardware analysis regarding FPGA resources, timing performance, and security strength has been provided. Each RNG technique is discussed in detail. Finally, the performance parameter of FPGA-based PRNGs in terms of FPGA resources, frequency, throughput, power consumption, security strength, and weaknesses is presented.

References

1. Hossain MS, Kong Y, Saeedi E, Vayalil NC (2017) High-performance elliptic curve cryptography processor over NIST prime fields. IET Comput Digit Tech
2. Imran M, Rashid M, Jafri AR, Kashif M (2019) Throughput/area optimised pipelined architecture for elliptic curve crypto processor. IET Comput Digit Tech
3. Canivet G, Maistri P, Leveugle R, Clédière J, Valette F, Renaudin M (2011) Glitch and laser fault attacks onto a secure AES implementation on a SRAM-Based FPGA. J Cryptol 24:247–268
4. Beyne T (2020) Block cipher invariants as eigenvectors of correlation matrices. J Cryptol
5. Alaba FA, Othman M, Hashem IAT, Alotaibi F (2017) Internet of Things security: a survey. J Netw Comput Appl
6. Gupta MD, Chauhan RK (2020) Efficient hardware implementation of pseudo-random bit generator using dual-CLCG method. J Circuits, Syst Comput
7. Rezk AA, Madian AH, Radwan AG, Soliman AM (2020) Multiplierless chaotic Pseudo random number generators. AEU—Int J Electron Commun
8. Babu AS, Anand B (2020) Modified dynamic current mode logic based LFSR for low power applications. Microprocess Microsyst 72:102945
9. Singh SK, Gupta MD, Mani S, Chauhan RK (2020) Design of LFSR circuit based on high performance XOR gate. Int Conf Electr Electron Eng ICE3 2020
10. Gupta MD, Chauhan RK (2020) Design of modified dual-CLCG algorithm for pseudo-random bit generator. Int Conf Electr Electron Eng ICE3 2020
11. Gupta MD, Chauhan RK (2021) Improved VLSI architecture of dual-CLCG for pseudo-random bit generator. Lect Notes Electr Eng
12. Katti RS, Kavasseri RG (2008) Secure pseudo-random bit sequence generation using coupled linear congruential generators. Proc—IEEE Int Symp Circ Syst
13. Katti RS, Srinivasan SK (2009) Efficient hardware implementation of a new pseudo-random bit sequence generator. Proc—IEEE Int Symp Circ Syst
14. Katti RS, Kavasseri RG, Sai V (2010) Pseudorandom bit generation using coupled congruential generators. IEEE Trans Circuits Syst II Exp Briefs
15. Panda AK, Ray KC (2019) Modified dual-CLCG method and its VLSI architecture for pseudorandom bit generation. IEEE Trans Circuits Syst I Regul Pap 66:989–1002
16. Kumar Panda A, Chandra Ray K (2020) A coupled variable input LCG method and its VLSI architecture for pseudorandom bit generation. IEEE Trans Instrum Meas 69:1011–1019
17. Gupta MD, Chauhan RK (2021) Coupled variable input-LCG and clock divider based large period pseudo-random bit generator on FPGA. IET Compute Dig Tech. Accepted on 14 Feb 2021
18. Lambić D, Nikolić M (2019) New pseudo-random number generator based on improved discrete-space chaotic map. Filomat 33:2257–2268
19. Alhadawi HS, Zolkipli MF, Ismail SM, Lambić D (2019) Designing a pseudorandom bit generator based on LFSRs and a discrete chaotic map. Cryptologia 43:190–211

20. Alawida M, Samsudin A, Sen TJ (2020) Enhanced digital chaotic maps based on bit reversal with applications in random bit generators. Inf Sci (Ny) 512:1155–1169
21. Zhao Y, Gao C, Liu J, Dong S (2019) A self-perturbed pseudo-random sequence generator based on hyperchaos. Chaos, Solitons Fractals X 4:100023
22. Tolba MF, Elwakil AS, Orabi H, Elnawawy M, Aloul F, Sagahyroon A, Radwan AG (2020) FPGA implementation of a chaotic oscillator with odd/even symmetry and its application. Integration.
23. Garcia-Bosque M, Perez-Resa A, Sanchez-Azqueta C, Aldea C, Celma S (2019) Chaos-based bitwise dynamical pseudorandom number generator on FPGA. IEEE Trans Instrum Meas 68:291–293
24. Rezk AA, Madian AH, Radwan AG, Soliman AM (2019) Reconfigurable chaotic pseudo random number generator based on FPGA. AEU—Int J Electron Commun 98:174–180
25. Liu Y, Tong X (2016) Hyperchaotic system-based pseudorandom number generator. IET Inf Secur 10:433–441
26. Wu YL, Yang CH, Wu CH (2017) Chip implementation of a new hyperchaotic oscillator. Electron Lett 53:226–228
27. Rajagopal K, Guessas L, Vaidyanathan S, Karthikeyan A, Srinivasan A (2017) Dynamical analysis and FPGA implementation of a novel hyperchaotic system and its synchronization using adaptive sliding mode control and genetically optimized PID control. Math Probl Eng
28. Rajagopal K, Jafari S, Karthikeyan A, Srinivasan A, Ayele B (2018) Hyperchaotic memcapacitor oscillator with infinite equilibria and coexisting attractors. Circuits, Syst Signal Process 37:3702–3724
29. Bonny T (2020) Chaotic or hyper-chaotic oscillator? numerical solution, circuit design, MATLAB HDL-coder implementation, VHDL code, security analysis, and FPGA realization. Circuits, Syst Sig Proc
30. Gupta MD, Chauhan RK (2021) Secure image encryption scheme using 4D-Hyperchaotic systems based reconfigurable pseudo-random number generator and S-Box. Integr the VLSI J, Publisher: Elsevier, 81:137 –159. https://doi.org/10.1016/j.vlsi.2021.07.002
31. Di Patrizio SG, De Marcellis A, Palange E, Faccio M (2019) A true random number generator architecture based on a reduced number of FPGA primitives. AEU—Int J Electron Commun 105:15–23

Ergodic Secrecy Capacity Analysis Over Composite Weibull/Inverse Gamma Fading Channel

Vipin Kumar Upaddhyay, Puspraj Singh Chauhan, and Sanjay Kumar Soni

Abstract In this work, the secrecy performance of traditional Wyner's model over more realistic composite wireless fading channel, i.e., Weibull/Inverse Gamma, is investigated. The closed-form expression of average ergodic secrecy capacity is developed in terms of Fox's H-function. The efficacy of the proposed solution is validated through Monte–Carlo simulation. Moreover, the importance of channel state information of eavesdropper and the multipath parameter in compensating the secrecy concern at the physical layer is comprehensively discussed.

Keywords Secrecy rate · CSI · Transmit power

1 Introduction

Nowadays, the development trends in wireless applications have witnessed different areas, such as wearable communication [1, 2] and Internet of things (IoT) [3], where privacy and security play a vital role to ensure reliable communication among nodes in the wireless network. The intercommunication link among the nodes is generally prone to eavesdropping and conventionally different algorithms like RSA or AES at the data link layer are utilized to enhance security [4]. The aforementioned algorithms are based on the supposition that error-free transmission is provided at the physical layer, but in reality, randomness in wireless propagation nature significantly introduces error in the packet reception. UM, Maurer in [5] suggested that the nature of the physical channel must be taken into consideration to provide the strictest security during communication. Following it, Wyner in [6] considered a realistic wiretap channel and exhibits that a targeted secrecy capacity can be achieved in the eavesdropper's presence over the AWGN channel. Moreover in [7], secrecy capacity is analyzed over Gaussian wiretap channel under the assumption that the achieved capacity of the legitimate user is higher than that of the eavesdropper's capacity. This

V. K. Upaddhyay · P. S. Chauhan · S. K. Soni (✉)
Department of Electronics and Communication Engineering, M. M. M. University of Technology, Gorakhpur, Uttar Pradesh, India
e-mail: sksoniec@mmmut.ac.in

B. Mishra and M. Tiwari (eds.), *VLSI, Microwave and Wireless Technologies*,
Lecture Notes in Electrical Engineering 877,
https://doi.org/10.1007/978-981-19-0312-0_49

work is further extended to incorporate the detrimental impact of a more realistic fading channel in [8]. There are different performance metrics such as strictly positive secrecy capacity (SPSC) and secrecy outage probability (SOP) to investigate the security concern over a physical channel. Considering Nakagami-m fading channel, the SPSC and SOP metrics are analyzed under the presence of multiple eavesdroppers [9], and later, it is extended to incorporate mobile vehicular network application for the same in [10]. The physical layer security considering imperfect channel state information (CSI) over Nakagami-m channel is studied in [11]. The secrecy capacity investigation over other well-known fading distributions such as Weibull and Rice channel is carried out in [12, 13]. Moreover, upper bound analysis of SOP is carried out in [14]. In [15], the novel approximate expressions of SPSC and SOP are developed over a more generalized $\alpha - \eta - \kappa - \mu$ fading channel. Considering a more realistic propagation scenario, where multipath and shadowing occurs simultaneously, numerous work related to the physical layer statistics have been done in [16–18]. Specifically, to study the security aspects in the line of sight (LOS) environment, authors in [17, 18] derived the SOP and SPSC expressions along with its high power solutions. In [19], authors comprehensively discussed the practical utility of Weibull/Inverse Gamma distribution and also carried out the numerical investigation about energy detection in a cognitive radio environment. Additionally, the practical utility of Weibull in mobile to fixed and mobile-to-mobile environment has been discussed in [20]. However, the ergodic secrecy capacity at the physical layer over Weibull/Inverse Gamma fading channel is not presented yet in the open literature.

Henceforth, in this work, we have derived ergodic secrecy capacity over the Weibull/Inverse Gamma fading channel. This work is organized as follows. Section 2 is dealt with the mathematical formulation of ergodic secrecy capacity for the same. The numerical discussion about the proposed solutions is discussed in Sect. 3. The concluding remark is provided in Sect. 4.

2 System Model

The concept of secrecy capacity for single-input and single-output (SISO) channel is well described in [21]. Here, three nodes—source (S), destination (D), and eavesdropper (E)—are considered respectively, where eavesdropper tries to overhear the legitimate link between S and D nodes. Moreover, channel state information (CSI) is considered to be available at the transmitter (Tx). The legitimate and eavesdrop channel are supposed to undergo Weibull/Inverse Gamma fading. Before transmission, the packets at the Tx are encoded to codeword and at the receiver Rx, it is received as

$$Y(j) = h(j)X(j) + n(j) \tag{1}$$

where $h(j)$ and $n(j)$ are channel fading coefficient and AWGN noise coefficient corresponding to the jth symbol, respectively. The SNR PDF related to Weibull/Inverse Gamma fading is given as [19, Eq. (5)]

$$f_Y(\gamma) = \frac{cA}{2\Gamma(k)(\beta\gamma_s)^{\frac{c}{2}}} \sum_{i=1}^{N} w_i t_i^{c/2+k-1} \gamma^{\frac{c}{2}-1} \exp\left(-A\left(\frac{t_i\gamma}{\beta\gamma_s}\right)^{\frac{c}{2}}\right) \tag{2}$$

where γ_s and c denotes average fading power and multipath parameter, respectively, and $A = \Gamma(1 + 2/c)^{c/2}$. The shape parameter k and scale parameter β are Inverse Gamma shadowing parameters. The expression for cumulative distribution function (CDF) can be obtained by plugging 2 in $\int_0^{\gamma} f_Y(\gamma)d\gamma$ as

$$F_Y(\gamma) = \frac{1}{\Gamma(k)} \sum_{i=1}^{N} w_i t_i^{c/2+k-1} \gamma\left(1, A\left(\frac{t_i\gamma}{\beta\gamma_s}\right)^{c/2}\right) \tag{3}$$

where $\Gamma(.)$ is gamma function and $\gamma(.,.)$ is lower incomplete gamma function defined in [22].

3 Secrecy Analysis

The modeling of secrecy capacity is based on the fundamental assumption that the CSI of the eavesdropper channel is available at Tx. This helps to achieve the targeted capacity in the active presence of an eavesdropper. The secrecy capacity is defined as the maximum capacity average over the difference between legitimate channel capacity and eavesdropper capacity. Over fading channel, numerically the instantaneous secrecy capacity can be given as [21, Eq. (5)], $C_{esc} = \max[\ln(1 + \gamma_D) - \ln(1 + \gamma_E)]$. Here, the main and eavesdropper capacity is given by $\ln(1 + \gamma_E)$ and $\ln(1 + \gamma_D)$ with instantaneous SNR γ_E and γ_D respectively.
Under quasi-static fading condition, the average secrecy capacity can be represented as [21], $\bar{C}_{\text{esc}} = I_1^{\text{esc}} + I_2^{\text{esc}} - I_3^{\text{esc}}$. The expression for integrals are provided below

$$I_1^{\text{esc}} = \int_0^{\infty} \log_2(1 + \gamma_M) f_M(\gamma_M) F_E(\gamma_M) d\gamma_M$$

$$I_2^{\text{esc}} = \int_0^{\infty} \log_2(1 + \gamma_E) f_E(\gamma_E) F_M(\gamma_E) d\gamma_E$$

$$I_3^{\text{esc}} = \int_0^{\infty} \log_2(1 + \gamma_E) f_E(\gamma_E) d\gamma_E \tag{4}$$

We now derive I_1^{esc} by substituting (2) and (3), yields

$$I_1^{\text{esc}} = \frac{cA}{2\Gamma(k_M)(\beta_M\gamma_{sM})^{\frac{c}{2}}\log(2)\Gamma(k_E)}\sum_{i=1}^{N} w_i t_i^{\frac{c}{2}+k_M-1}\sum_{l=1}^{N} w_l t_l^{k_E-1}\int_0^\infty \gamma_M^{\frac{c}{2}-1}$$
$$\log(1+\gamma_M)\exp\left(-A\left(\frac{t_i\gamma_M}{\beta\gamma_{sM}}\right)^{\frac{c}{2}}\right)\gamma\left(1, A\left(\frac{t_i\gamma_M}{\beta_E\gamma_{sE}}\right)^{c/2}\right)d\gamma_M \tag{5}$$

The above expression consists lower incomplete gamma function that makes integral computation complex. To overcome it, we employed a substitution as [23], $\gamma(1, x) = 1 - \exp(-x)$ in (5).

$$I_1^{\text{esc}} = \frac{cA}{2\Gamma(k_M)(\beta_M\gamma_{sM})^{\frac{c}{2}}\log(2)\Gamma(k_E)}\sum_{i=1}^{N} w_i t_i^{\frac{c}{2}+k_M-1}\sum_{l=1}^{N} w_l t_l^{k_E-1}$$
$$\Bigg\{\underbrace{\int_0^\infty \gamma_M^{\frac{c}{2}-1}\log(1+\gamma_M)\exp\left(-A\left(\frac{t_i\gamma_M}{\beta_M\gamma_{sM}}\right)^{\frac{c}{2}}\right)d\gamma_M}_{I_{\text{esc}}^{1'}}$$
$$\underbrace{-\int_0^\infty \gamma_M^{\frac{c}{2}-1}\log(1+\gamma_M)\exp\left(-A\left(\left(\frac{t_i}{\beta_M\gamma_{sM}}\right)^{\frac{c}{2}}+\left(\frac{t_l}{\beta_E\gamma_{sE}}\right)^{c/2}\right)\gamma_M^{\frac{c}{2}}\right)d\gamma_M}_{I_{esc}^{1''}}\Bigg\} \tag{6}$$

The solution for integral $I_{\text{esc}}^{1'}$ can be obtained by substituting Fox'H equivalents of Logarithm [24, Eq. (2.9.11)] and exponential [25, Eq. (8.4.3.2)] and atlast identity [24, Eq. (2.8.4)] is utilized which completes its proof.

$$I_{\text{esc}}^{1'} = H_{23}^{31}\left[A\left(\frac{t_i}{\beta_M\gamma_{sM}}\right)^{\frac{c}{2}}\Bigg|_{(0,1)(-\frac{c}{2},\frac{c}{2})(-\frac{c}{2},\frac{c}{2})}^{(-\frac{c}{2},\frac{c}{2})(1-\frac{c}{2},\frac{c}{2})}\right] \tag{7}$$

Similarly, integral's $I_{\text{esc}}^{1''}$ solution involves similar steps as we did for (7) that yields

$$I_{\text{esc}}^{1''} = H_{23}^{31}\left[A\left(\frac{t_i}{\beta_M\gamma_{sM}}\right)^{\frac{c}{2}}+A\left(\frac{t_l}{\beta_E\gamma_{sE}}\right)^{\frac{c}{2}}\Bigg|_{(0,1)(-\frac{c}{2},\frac{c}{2})(-\frac{c}{2},\frac{c}{2})}^{(-\frac{c}{2},\frac{c}{2})(1-\frac{c}{2},\frac{c}{2})}\right] \tag{8}$$

The final solution for integral I_1^{esc} is obtained after plugging (7) and (8) into $I_1^{\text{esc}} = I_{\text{esc}}^{1'} - I_{\text{esc}}^{1''}$. Similarly, the analytical solution of integral I_2^{esc} is very similar to I_1^{esc}, except some change of composite fading parameters; following it, we get

$$I_{\text{esc}}^{2} = \Bigg(H_{23}^{31}\left[A\left(\frac{t_i}{\beta_E\gamma_{sE}}\right)^{\frac{c}{2}}\Bigg|_{(0,1)(-\frac{c}{2},\frac{c}{2})(-\frac{c}{2},\frac{c}{2})}^{(-\frac{c}{2},\frac{c}{2})(1-\frac{c}{2},\frac{c}{2})}\right]$$
$$-H_{23}^{31}\left[A\left(\frac{t_i}{\beta_E\gamma_{sE}}\right)^{\frac{c}{2}}+A\left(\frac{t_l}{\beta_M\gamma_{sM}}\right)^{\frac{c}{2}}\Bigg|_{(0,1)(-\frac{c}{2},\frac{c}{2})(-\frac{c}{2},\frac{c}{2})}^{(-\frac{c}{2},\frac{c}{2})(1-\frac{c}{2},\frac{c}{2})}\right]\Bigg) \tag{9}$$

Atlast, the solution of I^3_{esc} can be be obtained following similar steps as in (9) that concludes the proof

$$I^3_{\text{esc}} = \frac{cA}{2\Gamma(k_E)(\beta_E\gamma_{sE})^{\frac{c}{2}}\log(2)} \sum_{i=1}^{N} w_i t_i^{\frac{c}{2}+k_E-1} H^{31}_{23}\left[A\left(\frac{t_i}{\beta_E\gamma_{sE}}\right)^{\frac{c}{2}} \middle| \begin{matrix}(-\frac{c}{2},\frac{c}{2})(1-\frac{c}{2},\frac{c}{2}) \\ (0,1)(-\frac{c}{2},\frac{c}{2})(-\frac{c}{2},\frac{c}{2})\end{matrix}\right] \tag{10}$$

The final solution for ergodic secrecy capacity can be obtained after utilizing the integral solutions, which complete the proofs.

4 Numerical Analysis

This section consists of numerical discussions of Monte–Carlo validated derived ergodic secrecy capacity expression. For numerical simulation, we have chosen summation term for Laguerre polynomial solution (2) as $N = 45$ and 10^6 samples for Monte–Carlo. Figure 1 shows the importance of eavesdropper CSI at source in enhancing communication secrecy at the physical layer. The composite fading parameters values are provided in the figure. The increase in transmits power helps the user to adapt its coding strategy to increase the SNR difference between the destination node and eavesdrop node. On the contrary, the increase in eavesdropper SNR significantly degraded the secrecy performance rendering to lessen the SNR gap. Additionally, the relative shift in the saturation of the curve occurs at smaller SNR as γ_{sE} increases. Figure 2 depicts the impact of multipath over the secrecy performance

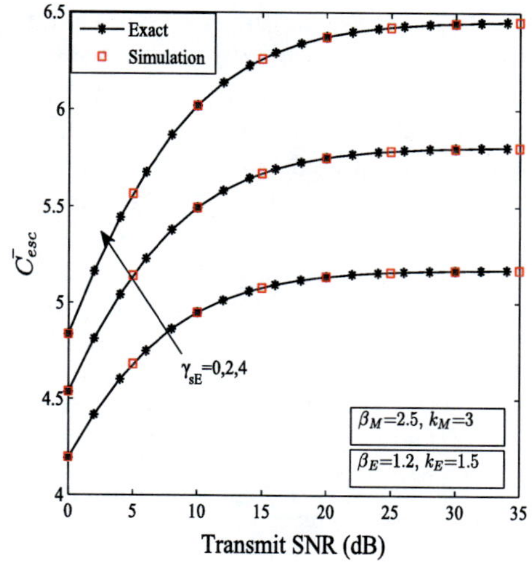

Fig. 1 Average secrecy capacity versus transmit SNR at $\gamma_{sM} = 20$ (dB)

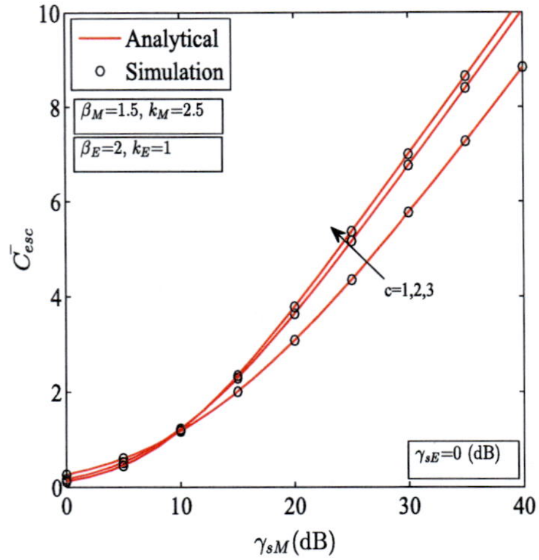

Fig. 2 Average secrecy capacity versus Main channel SNR (γ_{sM}) at $\gamma_{sE} = 0$ (dB)

of the wireless channel. It is evident from the plot that as the multipath parameter c increase from 1 to 3, an improvement in secrecy capacity is observed. This happens due to the improvement in legitimate channel condition occurs over the eavesdropper channel. Moreover, the growth rate in $\bar{C_{\text{esc}}}$ curves at the higher value of γ_{sM} reflects approximately linear behavior.

5 Conclusion

This work contributes derivation of analytical closed-form solution of ergodic secrecy capacity over Weibull/Inverse Gamma fading channel. The practical significance of secrecy capacity in the composite Weibull/Inverse Gamma fading environment lies in indoor, outdoor, and mobile-to-mobile communication scenario. Moreover, this work can also be extended to incorporate secrecy outage analysis along with its high and low power analysis. This work also contains the scope of secrecy analysis in multi-hop networks and decode-and-forward relay.

References

1. Cotton SL, Scanlon WG (2007) Characterization and modeling of the indoor radio channel at 868 MHz for a mobile bodyworn wireless personal area network. IEEE Antennas Wireless Propag Lett 6:51–55
2. Cotton SL, Scanlon WG (2006) Indoor channel characterisation for a wearable antenna array at 868 MHz. IEEE wireless communications and networking conference. WCNC, Las Vegas, NV, USA, pp 1783–1788

3. Vermesan O, Friess P (2004) Internet of things-from research and innovation to market deployment. River Publishers, Aalborg, Denmark
4. Rivest RL, Shamir A, Adleman L (1978) A method for obtaining digital signatures and public-key cryptosystems. Commun ACM 21(2):120–126
5. Maurer UM (1993) Secret key agreement by public discussion from common information. IEEE Trans Inf Theo 39(3):733–742
6. Wyner AD (1975) The wire-tap channel. Bell Syst Tech J 54(8):1355–1387
7. Leung-Yan-Cheong S, Hellman M (1978) The gaussian wire-tap channel. IEEE Trans Inf Theo 24(4):451–456
8. Barros J, Rodrigues MRD (2006) Secrecy capacity of wireless channels. IEEE international symposium on information theory. Seattle, WA, USA, pp 356–360
9. Sarkar M, Ratnarajah T, Sellathurai M (2009) Secrecy capacity of Nakagami-m fading wireless channels in the presence of multiple eavesdroppers. In: Proceedings of 2014 mediterranean microwave symposium (MMS2014). Marrakech, Morocco, pp 1–6
10. Xu L, Yu X, Wang H et al (2019) Physical layer security performance of mobile vehicular networks. Mob Netw Appl 25:1–7
11. Kumar R, Chauhan SS (2018) Secrecy analysis of MRT/RAS system under Nakagami-m fading channels in the presence of imperfect channel state information. AEU—Int J Electron Commun 85:68–73
12. Liu X (2013) Probability of strictly positive secrecy capacity of the Weibull fading channel. 2013 IEEE Global communications conference (GLOBECOM). Atlanta, GA, pp 659–664
13. Liu X (2013) Probability of strictly positive secrecy capacity of the Rician-Rician fading channel. IEEE Wireless Commun Lett 2(1):50–53
14. Kong L, Tran H, Kaddoum G (2016) Performance analysis of physical layer security over $\alpha - \mu$ fading channel. Electron Lett 52(1):45–47
15. Mathur A, Yun A, Bhatnagar MR, Cheffena M, Ohtsuki T (2018) On physical layer security of $\alpha - \eta - \kappa - \mu$ fading channels. IEEE Commun Lett 22(10):2168–2172
16. Alexandropoulos GC, Peppas KP (2018) Secrecy outage analysis over correlated composite Nakagami-m/Gamma fading channels. IEEE Commun Lett 22(1):77–80
17. Sun J, Li X, Huang M, Ding Y, Jin J et al (2018) Performance analysis of physical layer security over $\kappa - \mu$ shadowed fading channels. IET Commun 12(8):970–975
18. Srinivasan M, Kalyani S (2018) Secrecy capacity of $\kappa - \mu$ shadowed fading channels. IEEE Commun Lett 22(8):1728–1731
19. Upaddhyay VK, Chauhan PS, Soni SK, Mishra B (2020) On the performance analysis of cognitive radio based wireless system over Weibull-inverse gamma composite fading channel. International conference on electrical and electronics engineering (ICE3). Gorakhpur, India, pp 484–488
20. Karadimas P, Kotsopoulos SA (2018) The Weibull-lognormal fading channel: analysis, simulation, validation. IEEE Trans Veh Technol 58(7):3308–813
21. Chauhan PS, Kumar S, Soni SK (2020) On the physical layer security over Beaulieu-Xie fading channel. AEU—Int J Electron Commun 113:152940
22. Gradshteyn I, Ryzhik I (2007) Table of Integrals, Series, and Products. 7th: Academic Press
23. Abramowitz M, Stegun I (1988) Handbook of mathematical functions: with formulas, graphs, and mathematical tables
24. Kilbas AA, Saigo M (2004) H-transforms theory and applications
25. Prudnikov AP, Brychkov YA, Marichev OI (1986) Integrals and series more special functions 1st ed. Gordon

IoT-Based Efficient Parking System Using IntelliP

Dheeraj Tiwari and Rajan Mishra

Abstract Idea of smart transportation has gained momentum in smart cities. The increasing numbers of vehicles lead to a overcrowding in the traffic network in urban areas. Therefore, it is becoming a challenging task to seek out parking slots. To solve the above-mentioned problem, in this paper, designed a system named intelliP, In which Android Application will help the people who have an extra space with their house they can use it as a business for profit without spending any extra effort in it. People who need the parking can go through the intelliP to book and use the parking, and also, it provides direction towards the parking slots. This is completely IoT-based which reduces the user's effort to locate the road side existing parking slots in real time. Simulation and testing of intelliP system exhibit that the allocative efficiency is high and provides low average waiting time (AWT) to user.

Keywords Smart parking · IoT · ESP8266 (Node MCU) · Firebase · Real time · intelliP

1 Introduction

The Internet of things (IoT) is universally recognised as one of the main technology enablers for the development of future intelligent environments. It is driving the digital transformation of many different domains like environment, mobility, health care, industry, etc., of our everyday life. This is happening by realising the illustrations of more interconnected, instrumented and well-advised scenarios. New (IoT) Internet of things applications that levitated comprehensive connectivity, system interoperability and analytics are enabling smart city initiatives everywhere the planet. These two technologies, Internet of things (IoT) and big data, are going to play a very important role in smart cities and smart parking because large number of data are generated in machine-to-machine (M2M) communications [1, 2].

D. Tiwari (✉) · R. Mishra
Electronics and Communication Engineering Department, Madan Mohan Malaviya University of Technology, Gorakhpur, India
e-mail: tdheeraj9a@gmail.com

B. Mishra and M. Tiwari (eds.), *VLSI, Microwave and Wireless Technologies*, Lecture Notes in Electrical Engineering 877,
https://doi.org/10.1007/978-981-19-0312-0_50

Per capita vehicles increase rapidly throughout the years, and parking has become a big issue everywhere. According to WHO, the percentage of the world's population living in urban areas is projected to increase from 54% in 2015 to 60% in 2030 and to 72% by 2050. Increasing the number of people in the cities leads to increase the number of vehicle. "On average 30% of traffic is caused while looking for an available parking slots. According to Paidi et al. [3], on an average 4–15 min is wasted to find a parking slots. In addition, it increases traffic congestion, accidents, fuel consumption and air pollution. This leads to parking hazards because of taking the multiple rounds trips to get the parking slot which causes burning of additional fuel and ultimately producing excessive CO2 emission" [4]. "Smart parking can mitigate all these problems. The car parking solution from Smart Parking Ltd. is the automated parking system through which the users can check the occupancy of the parking sites" [5]. "The smart parking system enables the drivers by plotting the vacancy slots of the desired locations. This application can be tracked through a mobile or Web application to identify the free slots and also can reserve the empty slots in prior. It is possible to divide the existing studies for smart parking applications as outdoor (open type) and indoor (close area types). When it is observed carefully, it can be analysed that the most of the literature mainly focus on indoor parking" [3]. Many different types of sensors are used to detect the available parking slots. Selection of the appropriate sensors depends on the requirements of parking slot whose main focus should be to preserve high accuracy while reducing the overall cost [6]. RFID is used to the check the available parking slots in [7]. RFID uses tags and sensors that is the main disadvantage of RFID implementation because it is costlier. Nandyal et al. [8] use the number plate recognition through image processing-based smart parking system in which they automate car parking monitoring called CPMMS. Internet of thing (IoT)-enabled real-time smart parking system (iERS) [9] is proposed. This is the state-of-the-art smart parking system; it replaces the conventional parking system, and it reduces the average waiting time (AWT) and medium response time. Main disadvantage is that they are not using real-time database. Instead of using drone which makes the system more complex as well as costlier, it can use sensors which are more cost effective. Intelligent parking reservation (IPR) systems [10] allow users to select a parking slots according to their own choice. Its prediction does not guarantee slot availability, average waiting time (AWT) is also medium, and the pros is their algorithm. Paper [11] presents the learning automata using Markov model to ascertain the percentage of the parking area for old hat parking.

Based on the above examples, there are a lots of benefit of our proposed (intelliP) system—(i) efficient use of vacant space (ii) reduce traffic (iii) reduce CO2 emission (iv) cost effective.

In the overcrowded metro cities, people cannot find parking space at right place at right time, and the people having the extra parking area or parking area empty at the moment are also unable to use it in any fashion due to lack of time for making deals of these spaces. We have implemented, in this paper, an intelligent parking (intelliP) system in the under-mentioned manner: we have designed an Android application for the people who have extra space with their house so that they can use it as a business for profit without spending any extra effort in it. Apart from this, people who need the parking can easily go through the Android application to book and use the parking near them. The solution is completely IoT-based; i.e. it is fully automated and smart enough that the parking owner need not to indulge in the fuss while he gets the status of his parking area in real time. And he can monitor that remotely from anywhere whether his space is booked or not.

2 Proposed System

Proposed architecture consists primarily of three layers; first is "sensors network" layer, "IoT middleware layer" and end user layer as final "user interface" (Android application) that provide real data update to the user as well as to the Admin. For the real-time data collection, we used ultrasonic sensors. ESP8266 WiFi Module (Node MCU) have been used sensor networks and firebase database (Cloud). Concise details of various components used in the intelliP system architecture are as follows:

Ultrasonic Sensor (HC-SR04): Ultrasonic or HC-SR04 sensor is also known as distance measurement sensor. Travelling speed of ultrasonic waves is faster as compared to speed of audible sound (i.e. the sound that humans can hear). Ultrasonic sensor is a transceiver module (transmitter and receiver). It provides benefits like it intercepts the waves reflected by an obstacle, it works efficiently in darkness, and it is cost effective. It is not affected by environment parameter like dust, dirt and moisture.

Node MCU (ESP8266 Wi-Fi Module): The ESP8266 is low-power, highly integrated Wi-Fi microcontroller. It requires minimum of seven external components to get it in action. Operating temperature range is wide −40 C to +125 C. It consists all the important elements of the state-of-the-art computer. Arduino IDE is used to program it.

MIT App Inventor: It is an open-source Web application used to create the Android app. Using MIT app inventor, even a novice can create an advanced full functional Android Application. intelliP system would be developed using this platform. Google Firebase database is integrated with MIT app inventor, data come from the ultrasonic sensor HC-SR04 to Google Firebase.

Firebase Database: FirebaseUI is an open-source library for Android that allows you to quickly connect common UI elements to Firebase APIs. It supports JSON data that is why it is real-time database After any changes all users will get updated in real time. FireBase is a backend-as-a-service (Baas) very useful for mobile app development. It provides many features like authentication and security, real-time database and file storage, analytics, push notifications, AdMod and many others. It provides the SDK for Android, iOS, Web, NodeJS, C++ and Java Server. Advantage of the firebase database is that it is user friendly. Configurations are not complicated.

Descriptions of the components which are used to find the availability of parking slots are encapsulated in Table 1.

Table 1 Specification of components

Components	Specification
(HC-SR04) Ultrasonic sensor	Distance sensor module—HC-SR04. It operates at +5 V. It can easily integrate with any type of controller. The ultrasonic sensor has high sensitivity, high frequency, and it can even penetrate the transparent object also. Thus, it can easily detect the deep or external objects
Node MCU	ESP8266 is a low-cost open-source IoT platform. Operating voltage: 3.3 V. Current consumption: 10uA–170 mA. Flash memory 16 MB max (512 K normal). Number of GPIO is 17 (multiplexed with other functions)
Cloud (Firebase)	Firebase is a "backend-as-a-service" (BaaS). It is a real-time database. When we deploy our app to Firebase, we are not connecting through normal HTTP. You are connecting through a WebSocket. HTTP is much, much slower than WebSockets
MIT app inventor	It is a Web browser-based Android app development platform. Android phone can be connected to this app, or emulator is also provided through which we can modify the application before the deployment. Tracking of project can also be done because it stores our work on MIT app server

2.1 IntelliP System Architecture

There are two Android application that have been developed; one is Admin side, and another is user side. After installation of intelliP system, Admin will be able to lend his parking space to the users. For that, owner will have to login the Admin side Android application and enter all the details like his location, how much parking space he want to lend and time interval in between he wants to rent his space.

The system architecture of intelliP is depicted in Fig. 1.

2.2 IntelliP Application Flow

The working of intelliP is exhibited in Fig. 2, where the client can open intelliP Android app, and for the new user, sign-up is compulsory. If the existing user forgot the password, then he can reset the password, and then, he can login. The intelliP system asks for location to the user, searches for the nearby available parking lots and shows all the parking slots. After choosing the parking slot, that particular slot will be allocated to the user. When the client reaches to the reserve lots, after this user needed to authenticate, through the OTP. After OTP verification, user is allowed to park. User can make the payment using both methods using traditional methods of payment as well as digital payment methods. If there is no available parking slots in that particular area, then it will show no parking. The database server (FireBase)

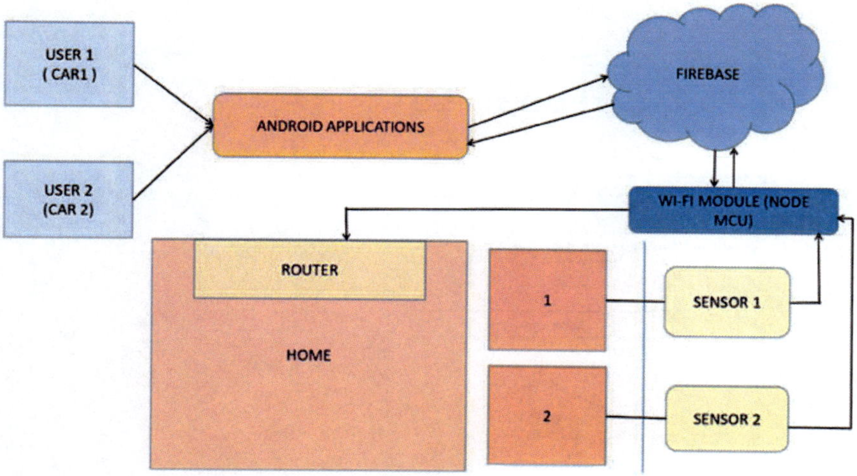

Fig. 1 Intellip system architecture

will be updated according to the status of the parking slots. The parking owner will notify in each and every steps of the booking in real time.

2.3 *IntelliP Working*

The working of the intelliP is demonstrated below with the help of the screenshots. The complete flow of working of the intelliP is shown in Fig. 2. It comprises the following steps.

After installation of our system, the owner will be able to allot his space to the car parking. Firstly, the owner has to login intelliP Android application and enter his location and time interval for which he want to rent his space. In the same fashion client (user) will also login to the intelliP application to booked his parking space, the owner will get notified through text-message and he will get fixed amount of money into his account after completion of booking process, the ultrasonic sensor turned on and waiting for the car status. And when the car will cover the parking space, ultrasonic sensor will detect the car, Node MCU will update the status of the database (FireBase), and that space will be occupied. It will update the owner's page with an app notification. The parking will be engaged as per the time interval inputted by the user at the client's page. Client has to inform the parking owner if he is unable to leave after the interval. If owner is not convinced for extension, he can use help button and call to the owner. The owner will get notified by the text messages as the booking time is over. If the user wants to renew the space, then he can do so, and all the actions are repeated accordingly.

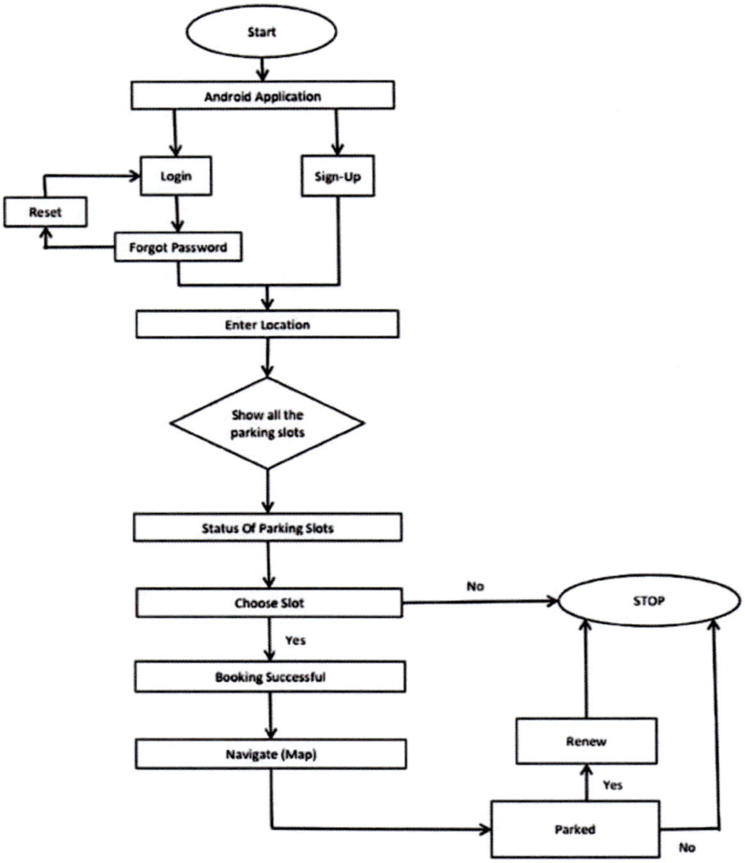

Fig. 2 Flow of intelliP

User login: Login page is shown in Fig. 3a. This page contains the information of the user like user id or e-mail and password. User can login into his/her account using his login credentials such as e-mail and correct password. If the credentials are correct, then the user will be redirected to the parking slots page.

User Sign-Up: The sign-up page requires the clients to enter the credentials like full name, e-mail id, password and vehicle number as well as phone number. The sign-up screen of intelliP is depicted in Fig. 3b.

Reset Password: If the existing user forgot the register credentials, then he need not have to register again; he can reset the credentials using reset which is shown in Fig. 4a. To reset, user has to enter username or e-mail and has to choose new password.

Parking Slots Page: Parking slots page will open after successful login or sign-in. This page shows how much parking slots the owner have. It will display the status of

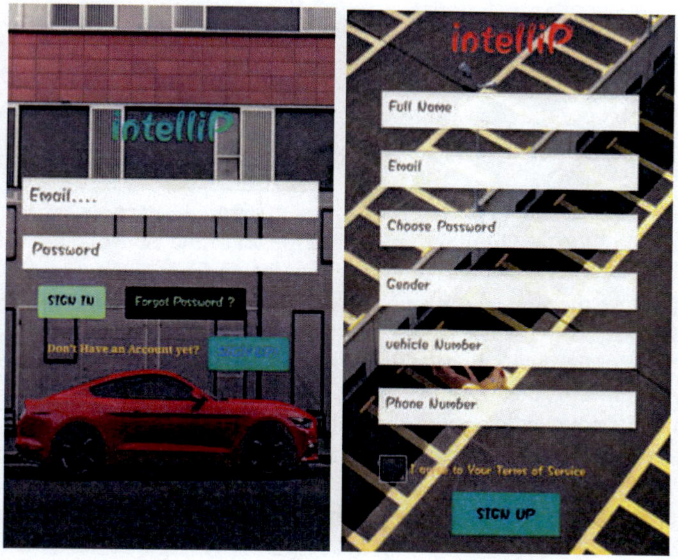

(a) (b)

Fig. 3 **a** User login, **b** User sign-up

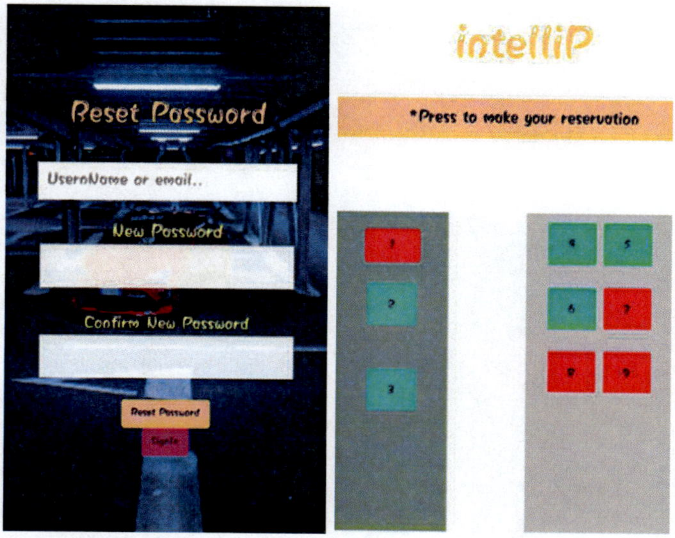

(a) (b)

Fig. 4 **a** Reset password, **b** Parking slots

each parking slot, which one is occupied and which one is available. After clicking on the particular slots, that slot will be allocated to the user.

3 Results

We have checked the proposed design within the real-time environment using HC-SR04. In which we have created a prototype intelliP Android application. This application consists of the several screens, and screenshots are portrayed in Figs. 3 and 4. The page Fig. 3a is for user login and Fig. 3b for registration to the new user. Password reset option is also available for the existing user. Figure 4b displays the real-time parking slots details to the user. The occupied slots will be turned into red colour, while the green colour indicates that the slot is available for booking. intelliP proposed system using a Wi-Fi connection to communicate between Node MCU and Firebase database. The data are sent from the sensor to the firebase and updated in the real time which are sync to the Android application, and then, the application will display the data parking slot in real time. After successful booking, user will be redirected to the navigation page shown in Fig. 5 which will help the user to reach at the parking area.

Fig. 5 GPS navigation

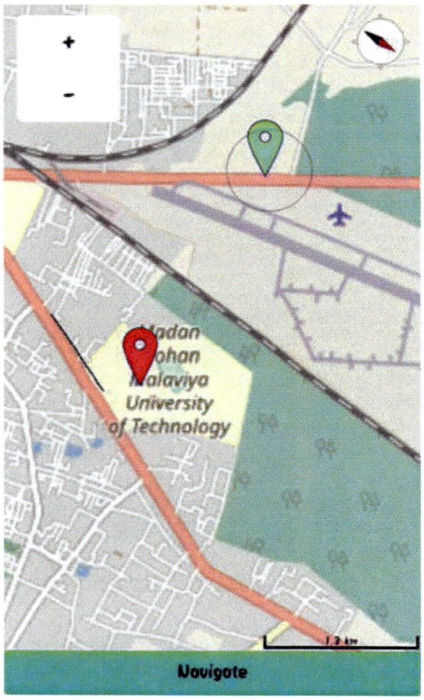

4 Conclusion

The number of cars per capita increases exponentially; this is just because of the cost of automobiles decreases gradually as well as the living standards of people increase. But the number of parking are not increasing, the rate at which vehicles are increasing rapidly. Hence, to alleviate these problems, we propose intelliP in which people having extra space with their house can install intelliP to their area which they want to use as parking. Apart from the houses, churches, bars or malls are closed on fixed days, so they can also use intelliP automated system in off days so that another people can use their parking slots. On the other hand, it gains the profit from these off days/off hours without any man-handling effort. In the future, this can also implemented in corporate organisations.

References

1. Jiang L, Xu LD, Cai H, Jiang Z, Bu F, Xu B (2014) An IoT-oriented data storage framework in cloud computing platform. IEEE Trans Ind Inform 10:1443–1451
2. Mitton N, Papavassiliou S, Puliafito A, Trivedi KS (2012) Combining cloud and sensors in a smart city environment. EURASIP J Wirel Commun Netw 2012:247
3. Paidi V, Fleyeh H, Håkansson J, Nyberg RG (2019) Smart parking sensors, technologies and applications for open parking lots: a review. IET Intell Transp Syst
4. Ramaswamy P (2016) IoT smart parking system for reducing green house gas emission. In: 2016 International conference on recent trends in information technology (ICRTIT), Chennai, pp 1–6
5. Smart Parking Ltd., Smart parking: car parking solution. (Online). Available. https://www.smartparking.com
6. Al-Turjman F, Malekloo A (2019) Smart parking in IoT-enabled cities: a survey Sustain Cities Soc 49:101608
7. Thorat SS, Ashwini M, Kelshikar A, Londhe S, Choudhary M (2017) IoT based smart parking system using RFID. Int J Comput Eng Res Trends (IJCERT) 4(1):9–12
8. Nandyal S, Sultana S, Anjum S (2017) Smart car parking system using Arduino UNO. Int J Comput Appl 169:13–18. https://doi.org/10.5120/ijca2017914425
9. Chauhan V, Patel M, Tanwar S, Tyagi S, Kumar N (2020) IoT Enabled real-time urban transport management system. Comput Electr Eng 86
10. Caicedo F, Blazquez C, Miranda P (2012) Prediction of parking space availability in real time. Expert Syst Appl 39(8):7281–7290
11. Waheed A, Krishna PV, Gitanjali J, Sadoun B, Obaidat M (2021) Learning automata and reservation based secure smart parking system: Methodology and simulation analysis. Simul Model Pract Theory 106:102205

Study of Area-delay and Energy Efficient Multi-operand Binary Tree Adder

Km Jyoti Singh and Manish Kumar

Abstract The study of construction of various multi-operand adders is portrayed in this article. The power dissipation and propagation delay of various multi-operand adders are evaluated. Ripple carry adder (RCA) which is designed using binary tree adder (BTA) is evaluated to reveal the probability of delay minimization. That structure and organizational formulation and also the accompanying configuration of RCA and for BTA centered mostly on results of the study. The outcome of both the comparison reveals that the best RCA design has greater performance in terms of delay, area, and energy than for the current RCA design. Adders are simulated using cadence virtuoso software. The gpdk90 nm CMOS technology is used to implement the adder.

Keywords Ripple carry adder · Carry save adder · Carry increment adder · Carry-select adder · Binary tree adder

1 Introduction

Multi-operand adders (MOA) have been widely used in modern high-speed, low power, lightweight, systems integration on a massive scale for image and data transmission applications in optical filters. This simple MOA is actually a binary tree adder (BTA), which should give more operands binding binary tree configuration of two operands adders.

Both delay and energy consumption of the BTA change the nature of adders used in the device. In this, there are many types of adders used which are as follows: ripple carry adder (RCA), carry-look-ahead adder (CLA), parallel prefix adder (PPA), carry-select adder (CSLA), carry-skip adder (CSKA), these adders are used to build the BTA structure for energy with higher delay.

K. Jyoti Singh (✉) · M. Kumar
Department of Electronics and Communication Engineering, Madan Mohan Malaviya University of Technology, Gorakhpur, Uttar Pradesh, India
e-mail: jyotisingh291@gmail.com

B. Mishra and M. Tiwari (eds.), *VLSI, Microwave and Wireless Technologies*, Lecture Notes in Electrical Engineering 877,
https://doi.org/10.1007/978-981-19-0312-0_51

The fundamental building blocks of several digital devices, such as microprocessors, digital signal processors, and much more specialized signal processing circuits, were arithmetic processor units [1–6]. Owing to their use of energy as well as the delay involved in the processing, adders are among the most critical components for just about any arithmetic processing system. Adders are part of multipliers as well, that is, just another arithmetic circuit resource-intensive part [7]. The selection of a specific adder thus has become an interesting technical factor for the successful application for arithmetic circuits [1–9]. Given the value of the introduction of even an adder, several of the current adder implementations throughout this paper were contrasted in relation to prolonged delay, resource utilization, and power consumption.

A half adder can also be used to add two bits, while the simple adder that is only been was doing the addition is indeed a complete adder. You will be using this 1-bit full adder and add multiple operands utilizing multiple full adder numbers [8]. A multi-operand adder adds additional upwards of 2 operand numbers [10, 11]. During 2005, R. D Kenney, M. 3 techniques were developed and studied by Schulte [1] to conduct rapid operand addition. For 6–12 input operands, multi-operand adder designs were built and synthesized by S. R Singh. A scheme for both the matrix multiplication of so many operands was defined by Waxman [8], trying to apply the technique called bit reformatting.

The propagated adder (CPA) method even with no extra hardware costs throughout particular. Throughout addition, the CPA-based high-resolution hold adder arrays (CSAs) promote flexibility use and the functionality. The multi-operand adder can indeed be clearly expressed by either a compressor shaft architecture [12] that decreases the transcending and propagating bearing [10]. There are many styles with multi-operand adders, but the tree-to-matrix circuit is really the adder mentioned in this study, that adder of the Wallace tree, another balanced delay tree adder as well as the adder of both the reversed scale tree [13]. The operands required to inclusion can also be single bits or multi-bits, so multi-bits could become the input and output of the adder. Currently, 8-bit, 16-bit, and 32-bit multi-operand adders have been used on several circuits, above and the parameters have been used for comparative purposes.

The content of this paper could be described through following points; in Sect. 2, there is the classification of array multi-operand adders is given. In Sect. 3, it is representing the literature reviews. In Sect. 4, we will see comparison of performing parameters, simulation using Cadence software and implemented the circuit using 90 nm CMOS technology, and Sect. 5 is conclusion.

2 Classification of Array Multi-operand Adders

2.1 Array Tree Adder

A multi-operand adder including adding and collecting partial sums is the array tree adder. Figure 1a displays the architecture of the six-operand array adder, each 8-bit.

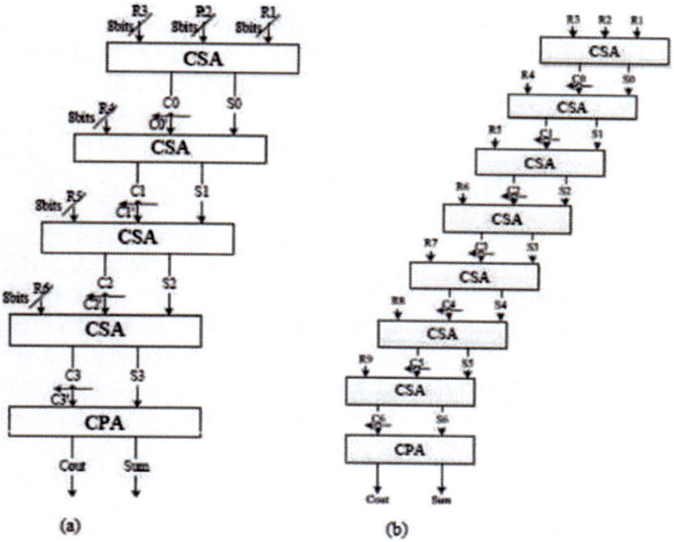

Fig. 1 Array tree adders; **a** six-operand array tree adders, **b** nine-operand array tree adders

R_i is operands, and S_i and C_i are their partial sums and carries. The final sum as well as the outputs was Sum and C_{out}.

2.2 *Wallace Tree Adder*

Figure 2 shows the Wallace tree adder circuits for six and nine operands, respectively. In Fig. 2a, b, both operands are being used in parallel, utilizing multiple carry save

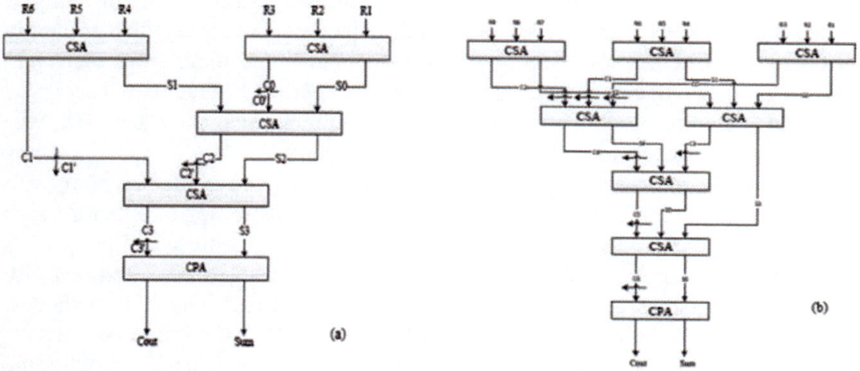

Fig. 2 Wallace tree adder; **a** six-operand, **b** nine-operand

adders during the first level alone (CSAs). Throughout the subsequent levels of both the CSA tree, the partial carries and sums produced from the very 1st level will then be worked to produce that input for both the carry propagate adder.

2.3 *Balanced Delay Tree Adders*

Balanced delay tree adder architectures with six and nine operands are possible. Some of the input operands have been taken in the 1st level of this adder, while others have been taken afterward. In these adders, the amount of delay obtained with each input is approximately equivalent to the amount of delay obtained with the other inputs [15]. Whenever the operation of those first stage partial outputs became complete, some other left outputs were selected toward operation.

3 Literature Review

As listed below "Comminiello et al. [1]," generally, evolutionary sensors have been used, namely applications of image processing, signal prediction/detection, namely echo cancelation. The finite impulse response adaptive filter (FIR), comprising even of a vector coefficient, is indeed one of the commonly utilized proposed approach (or weight) FIR filter and a plug in for weight changes. These coefficients are regulated either by adaptive algorithm. The development about such an adaptive FIR filter hardware is indeed very problematic for both the closed-loop adaptive system as well as its algorithm. Furthermore, its power consumption, wide area either long critical path of both weighted quantity operation of both the linear filter greatly reduces the applicability of this digital signal processing system (DSP).

Meher and Park [2] compared both to the exact application, their integrated interface recognition or saccadic system implementations show what; the planned approximate FIR adaptive filters suffer a small loss of accuracy. The results of both the synthesis show that sometimes the proposed project achieves a decline for energy per operation (EPO) of about 55% as well as a yield of 3.2 times from each field (TPA). The modeling definition needs equivalent to 60% less EPO for better accuracy compared to the process that relies more on deferred monthly mean square (DLMS) of [2] (i.e., lower mean squared error and misalignment).

Wu et al. [3], many previous works even centered also on implementation with FIR filters using specific multipliers then after canonical definition of either the sign digit of a filter coefficient. Similarly, a common method of multiplication also has been traditionally applied and is a variation of adding but converting operations with common numerical results. The major problem with this multiplication method in filling, however, is that when the number of bits had to define their filter coefficients grows exponentially, sharing the calculations would have provided a large additional virtual memory. Subsequently, with in chosen profession, they support multiplication but use the form through addition by removing each form by removing canonical representation of signed digits (CSD).

According to Haykin and Widrow [4], adaptive filters (ADFs) applications in different digital signal processing (DSP) implementations, noise cancelation, echo cancelation, channel equalization, including interface detection are often used. The FIR filter based on either the minimum squares mean (LMS) algorithm was the most used algorithm for simple implementation and satisfying convergence behavior. For both the LMS-based ADF, a simple structure with the long critical path is being used. And the conventional LMS algorithm does not really support piped execution including its recursive behavior. Depending mostly on implementation of the following pipeline LMS, the LMS algorithm becomes interpreted as delayed LMS (DLMS) to ADF in such a new guise. The DLMS algorithm is quite similar to both LMS, except then if the definition for weight increase was calculated while using previously available error value to adjust the weight vector and during the previous frame.

Ramkumar et al. [5], the construction of high-speed, environmentally sustainable data path logic systems continues as one of the most critical research areas of both the VLSI system architecture. That addition of rate frequency through digital adders is limited by both the time needed to spread another drag via the adder. When the previous bit position has already has been connected and a carry is already inserted was its total with each decomposition level sequentially produced to next place while on an elementary adder, propagated with each bit position.

4 Result

4.1 New Logic Formulation

To minimize RCA, carry propagation delay the goal of the entire logical formulation should be measuring intermediate carrier signals employing AOI doors. Therefore, the expression of which leads (1) is expressed as in the type AOI (Fig. 3):

$$\overline{C_i} = \overline{g_i + p_i . C_{i-1}} \quad (1)$$

The calculated ith carry signal is now in compliment form which cannot be used explicitly to measure the $(i + 1)$th carry signal.

Consequently, and use the ith carry c_i, that expression with true $(i + 1)$, the carry signal becomes supplied as (Fig. 4)

$$C_{i+1} = \overline{\overline{g_{i+1}} . \left(\overline{P_{i+1}} + \overline{C_i}\right)} \quad (2)$$

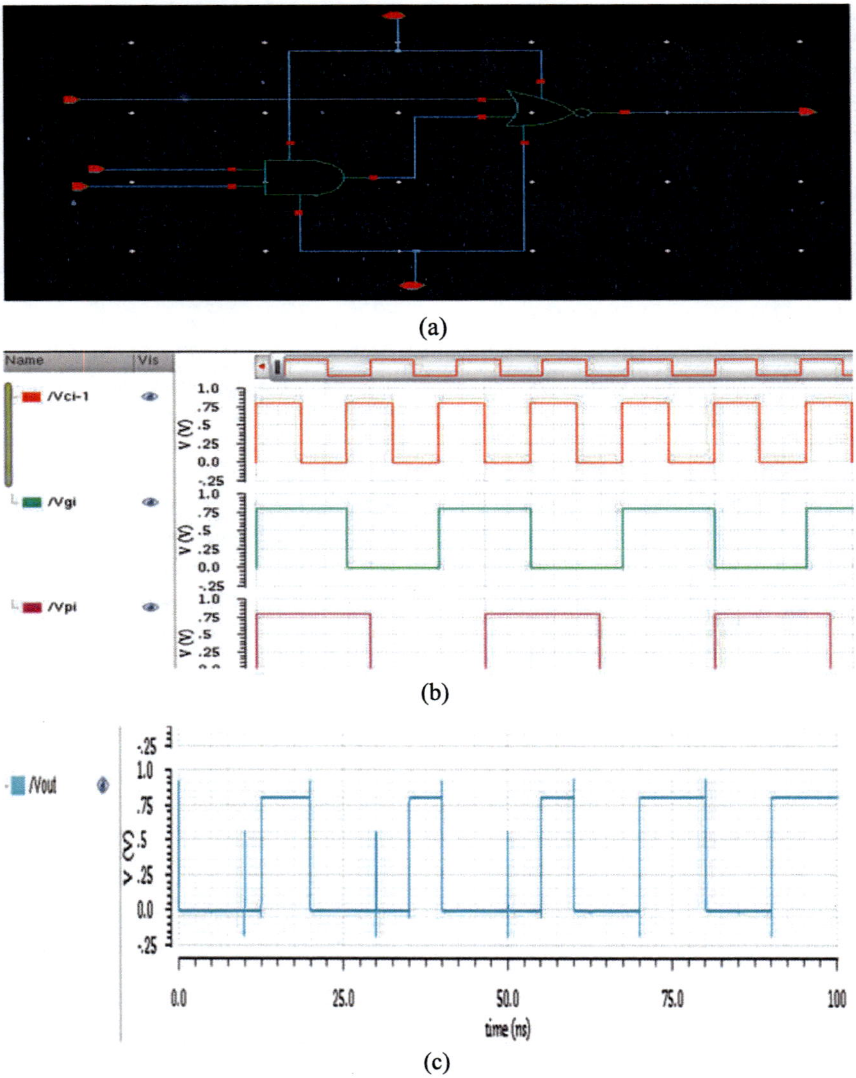

Fig. 3 **a** Schematic diagram of AOI, **b** input waveform, **c** output waveform

The operation defined in (2) could be determined by means of an OAI complementary gate that also generates the normal form of the carrying c_{i+1} signal. It is indeed clear, never the less, that now the OAI-based carry calculation involves p_{i+1}, g_{i+1}, and c_i.

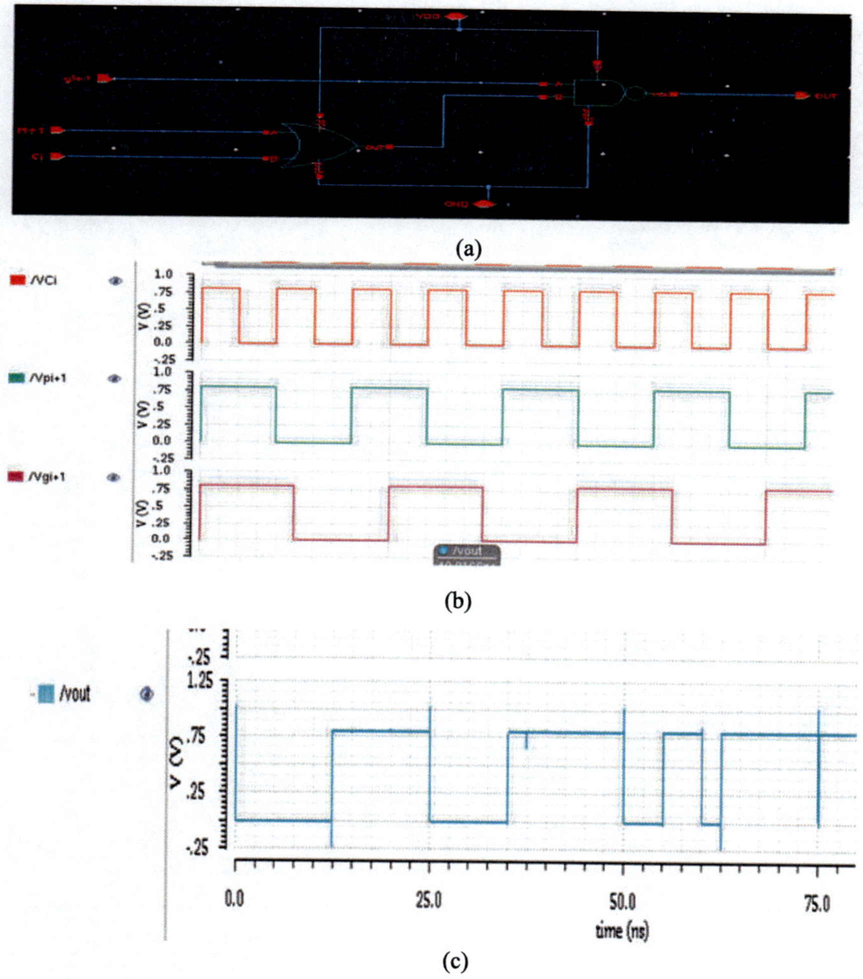

Fig. 4 **a** Schematic diagram of OAI, **b** input waveform, **c** output waveform

4.2 RCA Design

Reported in table for comparison, RCA design is calculated for the different input bit width ($m = 8, 16, 32$). Figure 5a is showing RCA schematic diagram, and 5(b) and 5(c) are input and output waveform of RCA. It can delay saving increase with an increase in input bit width to validates the theoretical results. As compare to CLA, RCA design has better performance (Table 1).

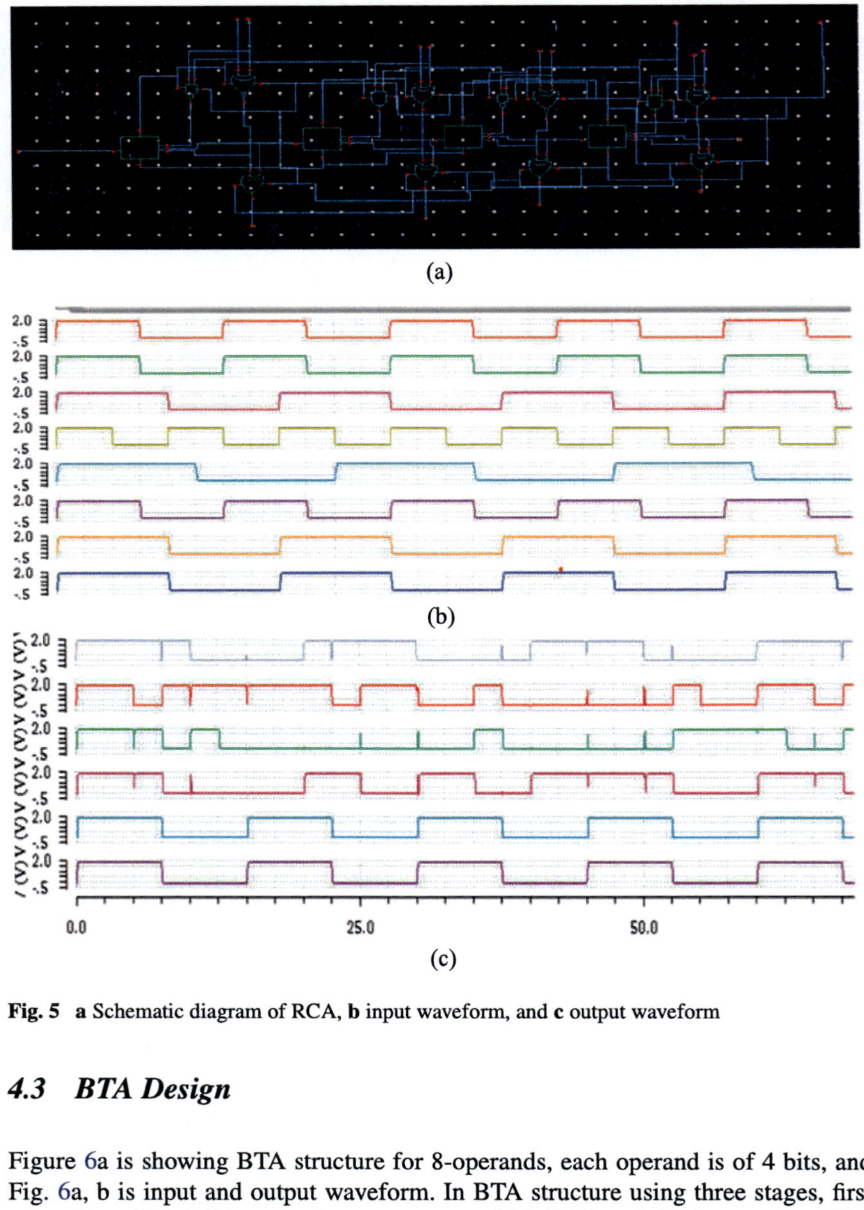

Fig. 5 **a** Schematic diagram of RCA, **b** input waveform, and **c** output waveform

4.3 BTA Design

Figure 6a is showing BTA structure for 8-operands, each operand is of 4 bits, and Fig. 6a, b is input and output waveform. In BTA structure using three stages, first stages four-4-bit RCA, second stage uses two-5-bit RCA and third stages one-6-bit RCA. BTA structure has better performance as compare to RCA.

Table 1 Performance comparison of power and delay for different adder topologies

Adder topology	N	Power (mW)	Delay (ns)
RCA	288	0.206 (mW)	4.208
CSA	576	1.082 (mW)	2.924
CLA	272	0.312 (mW)	3.1
CSLA	–	68.54 (mW)	7.438
CSA(FIFB)	–	0.13368 (mW)	951 (ps)
CSA(FIEB)	–	0.31523 (mW)	1053 (ps)
CSA(FISB)	–	0.68930 (mW)	2290
OAI	–	53.58 (mW)	17.56
AOI	–	68.85 (mW)	18.69
–	Bit width *m*		
	8	153.9 (μW)	0.318
RCA	16	173.0 (μW)	0.356
	32	185.6 (μW)	0.646
CLA	8	374.2 (μW)	–
WTA using CSLA	8	1.473 (mW)	0.839
BTA using CLA	8	2.121 (mW)	1.251
BTA using RCA	8	0.956 (mW)	–
BTA structure	8	0.818 (mW)	–

5 Conclusion

The significant variables that decide the efficiency of any circuit in the VLSI design phase are power, delay, and area. More power usage in a wide region is the downside of standard CSLA. In reducing area of overall control, the decreased number for gates provides the benefit. The RCA which is designed using BTA has outperformed the currently existing adder. CSA (FIFB) has best optimization in terms of delay and power with values 951 psec and 0.133 mW, respectively. BTA using RCA improves the performance by saving 22.5% in ADP and 28.7% in EDP as compared to BTA using CLA. So, BTA design using RCA is most suitable candidate to be used in the field of digital system for signal and image processing application.

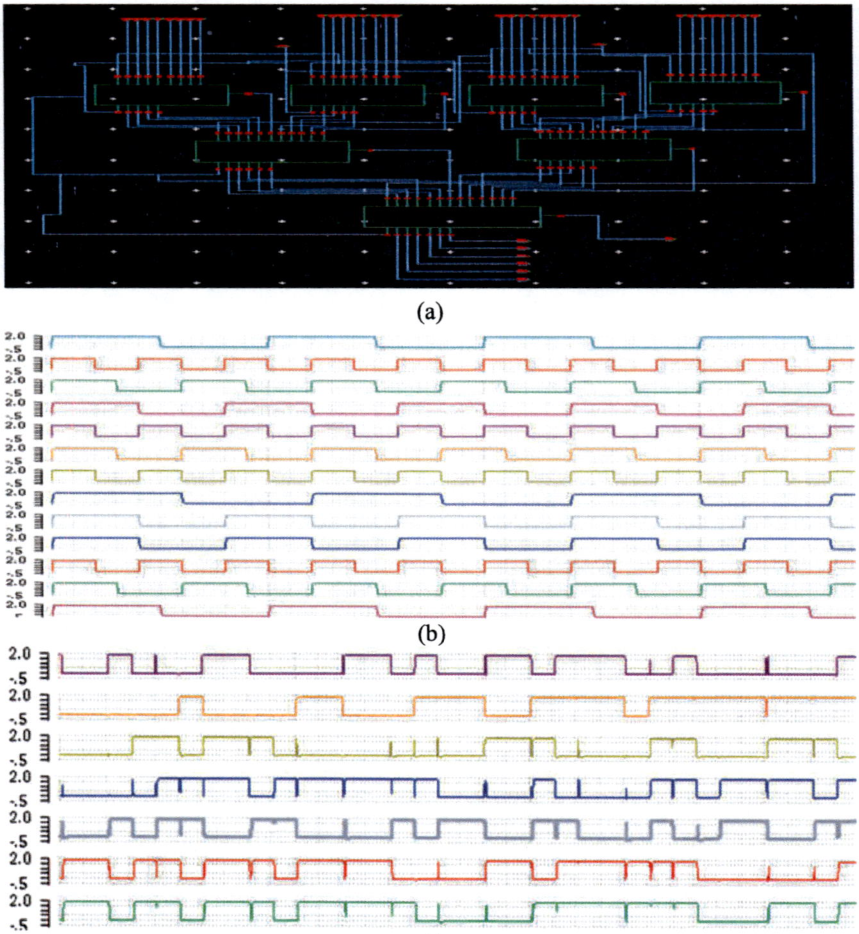

(a) (b) (c)

Fig. 6 **a** Schematic diagram of BTA structure, **b** input waveform, **c** output waveform

References

1. Comminiello D, Scarpiniti M, Azpicueta-Ruiz LA, Arenas-Garcia J, Uncini A (2013) Functional link adaptive filters for nonlinear acoustic echo cancellation. IEEE Trans Audio Speech Lang Process 21(7):1502–1512
2. Meher PK, Park SY (2014) Area-delay-power efficient fixed-point LMS adaptive filter with low adaptation-delay. IEEE Trans VLSI Syst 22(2):362–371
3. Wu L, Cui Y, Huang J (2010) Design and implementation of an optimized FIR filter for IF GPS signal simulator. In: IEEE conference on microelectronics and electronics, September 2010, pp 25–28

4. Haykin S, Widrow B (2003) Least-mean square adaptive filters. Wiley-Interscience, Hoboken, NJ
5. Ramkumar B, Kittur HM (2012) Low-power and area-efficient carry select adder. IEEE Trans Very Large-Scale Integr (VLSI) Syst 20(2)
6. Kenney RD, Schulte MJ (2005) High-speed multioperand decimal adders. IEEE Trans Comput 54(8):953–963
7. Chabini N, Belkouch S (2015) Area and delay aware approaches for realizing multioperand addition on FPGAs using two-operand adders. In: Proceedings of IEEE/ACS international conference of computer systems and applications (AICCSA), pp 1–4
8. Kumar P, Sharma RK (2016) Real-time fault tolerant full adder design for critical applications. Eng Sci Technol Int J 19(9):1465–1472
9. Sushmidha M, Premalatha B (2016) Design of high-performance parallel self-timed adder. In: Proceedings of international conference on communication and signal processing (ICCSP), pp 1400–1404
10. Balzola PI, Schulte MJ, Glossner JRJ, Hokenek E (2001) Design alternatives for parallel saturating multioperand adders. In: Proceedings of IEEE international conference on computer design: VLSI in computers and processors, pp 172–177
11. Singh KJ, Sharan T, Tarunkumar H (2018) High speed and low power basic digital logic gates, half-adder and full-adder using modified gate diffusion input technology. J VLSI Des Tool Technol 8(1):34–42
12. Saini J, Agarwal S, Kansal A (2015) Performance, analysis and comparison of digital adders. In: Proceedings of international conference on advances in computer engineering and applications (ICACEA), pp 80–83
13. Singh S, Waxman D (1973) Multiple operand addition and multiplication. IEEE Trans Comput C-22(2):113–120
14. Albeck A, Wimer S (2016) Energy efficient computing by multi-mode addition. Integr VLSI J 55(9):176–182
15. Hormigo J, Villalba J, Zapata EL (2013) Multioperand redundant adders on FPGAs. IEEE Trans Comput 62(10):2013–2025
16. Cini U, Kurt O (2016) MAC unit for reconfigurable systems using multi-operand adders with double carry-save encoding. In: International conference on design and technology of integrated systems in nanoscale era (DTIS), pp 1–4
17. Villalba J, Hormigo J, Prades JM, Zapata EL (2005) On-line multioperand addition based on on-line full adders. In: Proceedings of international conference on application-specific systems, architecture processors (ASAP'05), pp 322–327
18. Parandeh-Afshar H, Brisk P, Ienne P (2008) Efficient synthesis of compressor trees on FPGAs. In: Proceedings of Asia and south pacific design automation conference, March 2008, pp 138–143

Source/Drain Engineered Silicon-on-Insulator Transistor with Improved Analog Performance

Shailesh Shankar Upadhyay, Narendra Yadava, Mangal Deep Gupta, and R. K. Chauhan

Abstract This paper examines the analog behavior of various modified source and drain structures for fully depleted silicon-on-insulator (FDSOI) technologies. In the design engineering, the source/drain locale is partitioned into two vertical subportions termed as n^+ and n^-. The doping concentration is kept high (n^+) in the upper region and less (n^-) in the lower part of the engineered, modified source (MS), modified drain (MD), and modified source modified drain (MSMD) FDSOI structures. The aim is to compare the conventional FDSOI technology's analog output with the evolving FDSOI technologies involving various source/drain engineered structures. The MD-FDSOI structure shows the best improvement in terms of intrinsic gain (A_V), ON–OFF current ratio, and transconductance (g_m) over other structures and by 1.5, 5, and 2.35 times, respectively, over conventional.

Keywords MS · MSMD · MD · FDSOI · I_{OFF} · I_{ON}

1 Introduction

The universe of integrated circuit innovation sees an interminable and unbridled surge toward scaling down, keeping the famous Moore's Law [1]. The leakage power and circuit power problems with continuous scaling efficiency have gained much attention [2]. To solve the scalability issues with better short-channel functionality, the leading industrial communities with two opposite visions came up with their respective promising solution [3]. As of late, STMicroelectronics, followed by Global Foundries, is continuing with the planar MOS innovation called FDSOI [4]. FDSOI represents fully depleted silicon-on-insulator. The important development lies in presenting a thin silicon film that executes the channel and afterward utilizes a thin insulating layer just beneath the channel called buried oxide (BOX) [5].

Cheng et al. examine the FDSOI technology and describe different doping methods for the last decade, such as source/drain engineering, gate engineering [6].

S. S. Upadhyay (✉) · N. Yadava · M. D. Gupta · R. K. Chauhan
Madan Mohan Malaviya University of Technology, Gorakhpur, India
e-mail: shailesh10195@gmail.com

B. Mishra and M. Tiwari (eds.), *VLSI, Microwave and Wireless Technologies*, Lecture Notes in Electrical Engineering 877,
https://doi.org/10.1007/978-981-19-0312-0_52

The paper concluded that doping in the silicon layer presented under the buried oxide layer would accomplish the multiple thresholds, and the channel doping will be kept undoped [7, 8]. The effect of Drain Engineering on Nano-scaled FDSOI-MOSFET is shown by Narendra et al. [9].

This paper conducts a comparative analysis of various source/drain engineered FDSOI structures at 50 nm regime to examine potential structures with better analog performance over conventional FDSOI (C-FDSOI) designs. All the plans and recreations were performed utilizing the SILVACO ATLAS 2-D device simulator. The device description is given in segment 2, and the outcomes and conclusion have been finished in segments 3 and 4, respectively.

2 Device Design and Simulation

The proposed structure employs engineering in both drains and source regions. The source region has been engineered in the MS-FDSOI structure, while the drain region has been engineered in the MD-FDSOI structure, and both the source and drain region are engineered in the MSMD-FDSOI structure. In the design engineering, the source/drain locale is partitioned into two vertical sub-portions termed as n^+ and n^-. The conventional FDSOI structure has a source and drain regions both are formed only with n^+ sub-portion.

All the plans and recreations were performed utilizing Silvaco Atlas 2-D device simulator [10]. The system structure with ATLAS dopant concentration is shown in Fig. 1. The proposed device's outcome is compared with the previous result [11]. The device specifications are listed in Table 1.

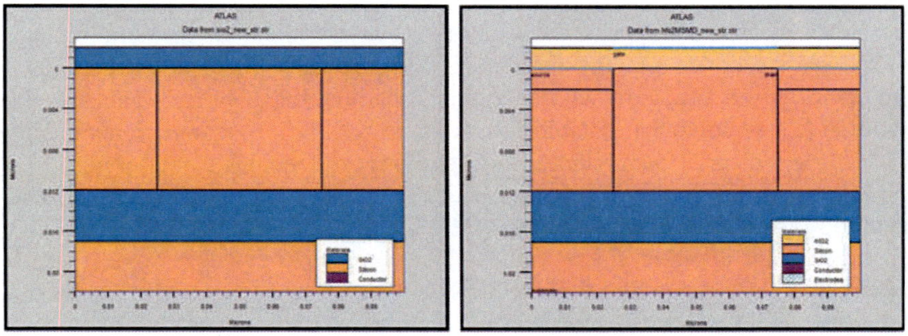

(a) (b)

Fig. 1 **a** Conventional design of FDSOI. **b** MSMD-FDSOI design

Table 1 Device specification for the proposed designs

Device parameters	C-FDSOI	MSMD-FDSOI
Gate length (L_g)	50 nm	50 nm
Oxide thickness (T_{ox})	2 nm (SiO_2)	2 nm (HfO_2) [11]
Channel doping (N_A)	1e17 cm^{-3}	1e17 cm^{-3}
Substrate doping	1e17 cm^{-3}	1e17 cm^{-3}
Source/drain region doping (n^+)	1e19 cm^{-3}	1e19 cm^{-3}
Source/drain region doping (n^-)	–	1e17 cm^{-3}
Work function of gate (Φ)	4.77 eV (Cu)	4.77 eV (Cu)
Insulator thickness (T_{BOX})	5 nm	5 nm
Gate voltage (V_{gs})	1.1 V	1.1 V
Silicon film thickness (T_{Si})	12 nm	12 nm
Silicon Substrate thickness (T_{BOX})	5 nm	5 nm

3 Results and Discussions

3.1 *Transfer Characteristics*

Transfer characteristics ($I_{DS}-V_{GS}$), shows variation of drain current (I_D) with gate-source voltage (V_{GS}), when drain-source voltage (V_{DS}) keep constant. Figure 2 represents the transfer characteristics (I_{DS}–V_{GS}) of various FDSOI structures. From Fig. 2a and b, MD-FDSOI MOSFET reveals a steeper curve of I_{DS}–V_{GS} than conventional FDSOI. So, it gives enhanced amplification than the other conventional and modified structures of FDSOI devices.

3.2 *Transconductance*

A gate-source voltage (V_{GS}) variance tests the drain current variation while holding the drain-source voltage (V_{DS}) constant and critical importance since it determines the device's ability to drive a load. The transconductance is determined by the slope of the curve for $I_{DS}-V_{GS}$ and is equated by Eq. 1. In Fig. 3, MD-FDSOI shows better transconductance in comparison to other FDSOI structures.

$$g_m = \frac{\partial I_D}{\partial V_{GS}} \tag{1}$$

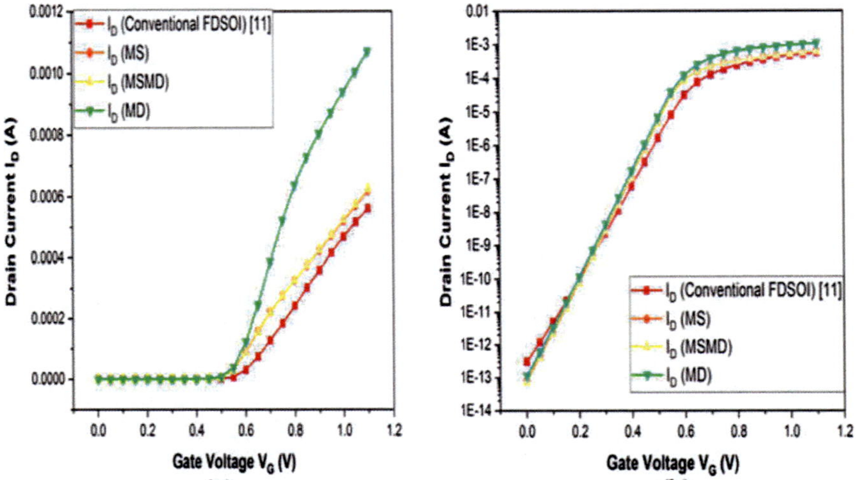

(a) (b)

Fig. 2 **a** shows linear scale transfer characteristics of various modified FDSOI structures at $V_{DS} = 1.0$ V. **b** shows log scale transfer characteristics of various modified FDSOI structures at $V_{DS} = 1.0$ V

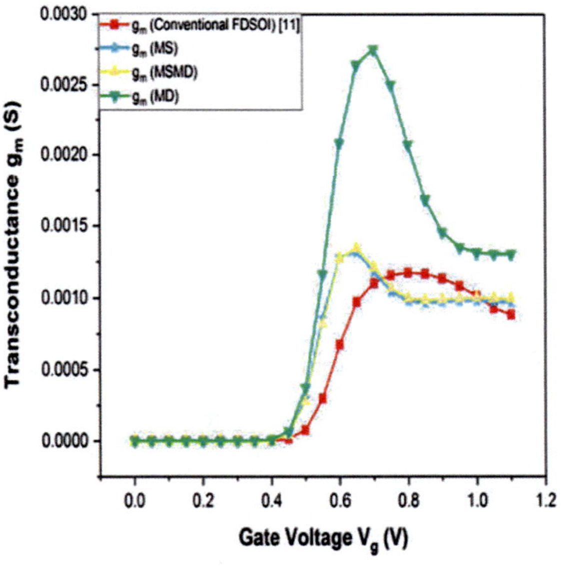

Fig. 3 $g_m - V_{GS}$ characteristics of various modified FDSOI structures at $V_{DS} = 1.0$ V

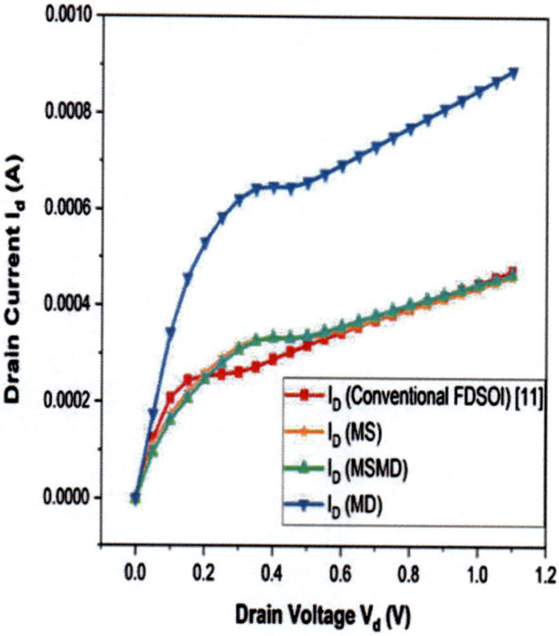

Fig. 4 Output characteristics of various modified FDSOI structures at $V_{GS} = 1.0$ V

3.3 *Output Characteristics*

From Fig. 4, the FDSOI structure of MS and MSMD shows comparable performance with the traditional FDSOI structure. The MD-FDSOI structure provides a better output drain current due to the low parasitic capacitance induced by the modified drain structure [9]. It offers better output characteristics than the MD-FDSOI structure.

3.4 *Output Conductance*

Figure 5 shows the modified FDSOI-MOSFET structures output conductance characteristics for varying gate voltages (refer to Eq. 2). MSMD and MS-FDSOI structures show comparable output conductance in the saturation region, and it is the least among all structures. So, both these structures can provide better amplification. Low parasitic capacitance in the drain region explains a higher drive current in the modified drain structure [9]. When used in amplifier circuits, the device can display large voltage swings [11, 12].

$$g_d = \frac{\partial I_D}{\partial V_{DS}} \tag{2}$$

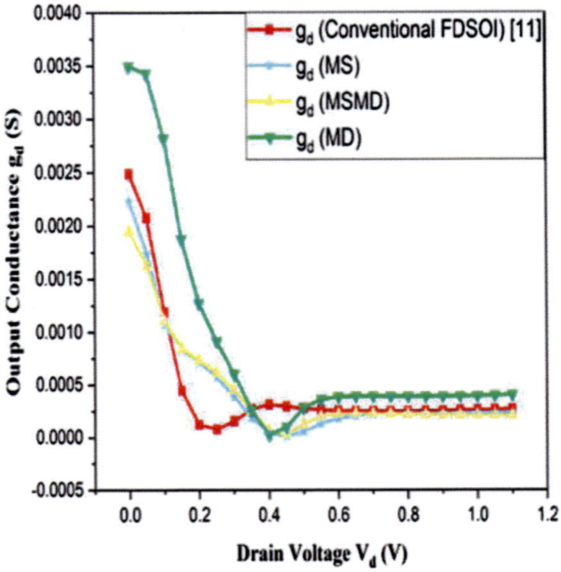

Fig. 5 $g_d - V_{DS}$ characteristics of various modified FDSOI structures at $V_{GS} = 1.0$ V

3.5 *Intrinsic Gain*

An essential statistic of interest for the operational transconductance amplifier is the intrinsic gain (A_V). The intrinsic gain should be as high as possible for better analog efficiency. The intrinsic gain varies as a function of gate voltage is formulated as Eq. 3. As shown in Fig. 6, FDSOI structures for MS and MD demonstrate more significant and comparable intrinsic gain than other FDSOI structures.

$$A_V = \frac{g_m}{g_d} \tag{3}$$

For different modified FDSOI configurations, Figs. 7, 8, and 9 shows I_{OFF}, I_{ON}, and I_{ON}/I_{OFF} ratio values, respectively, where MD-FDSOI structure indicates a higher I_{ON}/I_{OFF} ratio compared to other configurations. On the other hand, the SCEs, i.e., DIBL and subthreshold, as shown in Figs. 10 and 11, are also reduced by modified structures. There is also an improvement in threshold voltage value for modified configurations from conventional configuration. The different parameter values are tabulated in Table 2 for different modified structures.

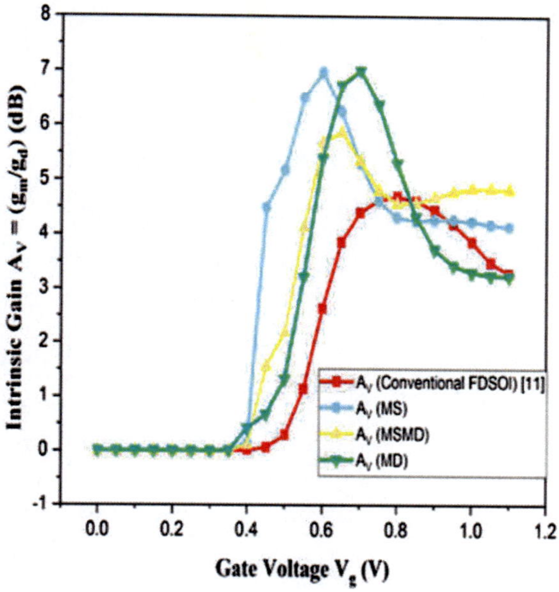

Fig. 6 $A_V - V_{GS}$ characteristics of various modified FDSOI structures

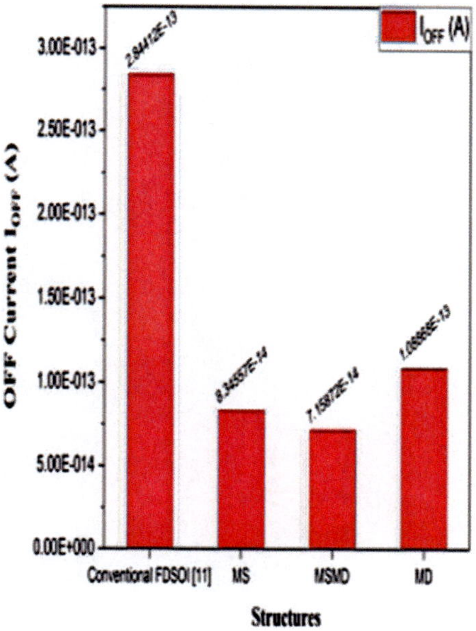

Fig. 7 OFF current value for the modified and conventional structures at $V_{DS} = 1.0$ V and $V_{GS} = 1.1$ V

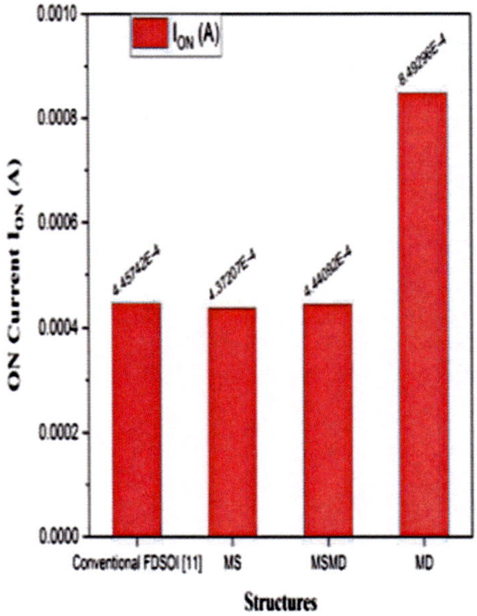

Fig. 8 ON current value for the modified and conventional structures at $V_{DS} = 1.0$ V and $V_{GS} = 1.1$ V

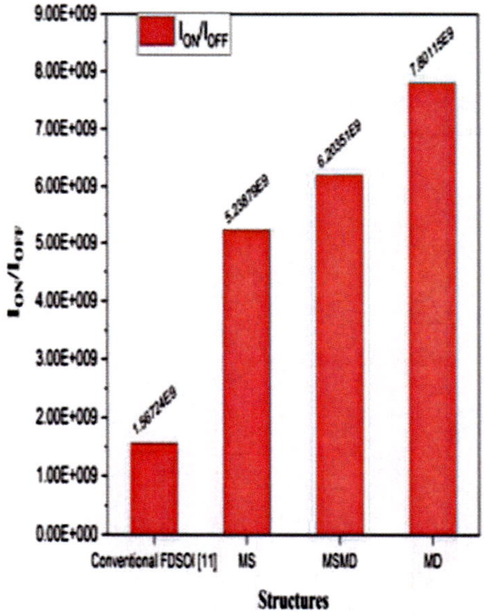

Fig. 9 ON-to-OFF current ratio value for the modified and conventional structures at $V_{DS} = 1.0$ V and $V_{GS} = 1.1$ V

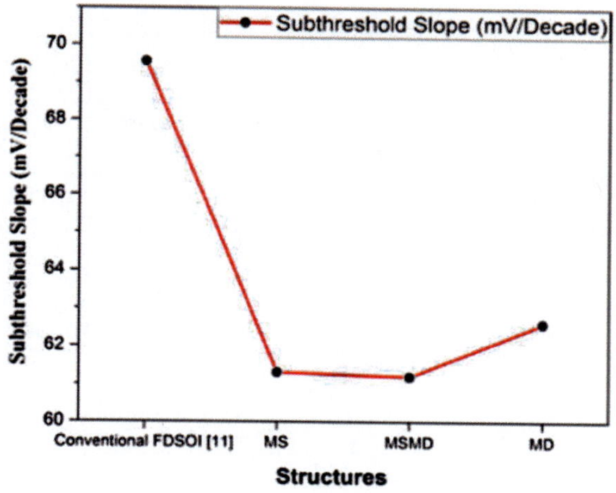

Fig. 10 Subthreshold slope (SS) value for the modified and the conventional structures

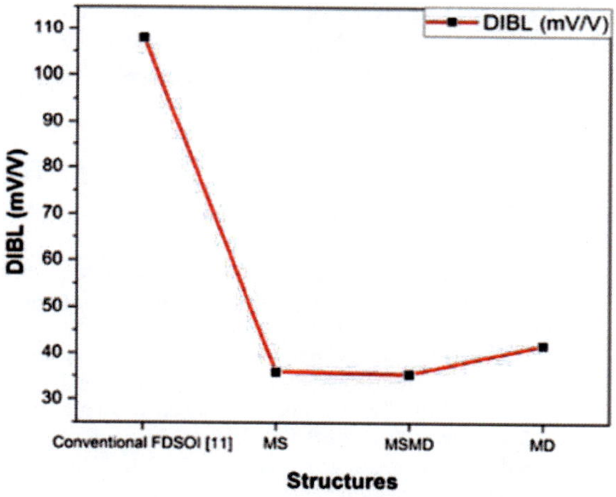

Fig. 11 DIBL value for the modified and the conventional structures

4 Conclusion

For SCEs such as SS, DIBL, and I_{OFF}, comparative analysis of modified and conventional FDSOI structures is also done for significant analog output. Transconductance is found to be maximal for MD-FDSOI and is increased by ~2.35 times over conventional FDSOI. MS and MSMD-FDSOI have almost the same transconductance but are better than conventional FDSOI performance. MD-FDSOI demonstrates improved performance of output features that will display a better analog performance alternative. MS and MSMD-FDSOI both have reduced the DIBL by approx. ~3 times than conventional FDSOI. MSMD-FDSOI has improves the subthreshold slope approx. by ~1.13 times over conventional. MD-FDSOI shows a higher I_{ON}/I_{OFF}

Table 2 Tabulates the extracted values of DIBL, SS, I_{OFF}, I_{ON}, I_{ON}/I_{OFF} ratio, and V_t for various configurations

Parameters	Structures			
	Conventional FDSOI	MS-FDSOI	MSMD-FDSOI	MD-FDSOI
DIBL (mV/V)	108.012	35.8526	35.4347	41.8284
Subthreshold (mV/Decade)	69.5588	61.3059	61.1948	62.5987
I_{OFF} (pA)	0.284412	0.0834557	0.0715872	0.108868
I_{ON} (mA)	0.445742	0.437207	0.444092	0.849296
I_{ON}/I_{OFF}	1.56724e+009	5.23879e+009	6.20351e+009	7.80115e+009
Threshold voltage (V)	0.0973384	0.0308755	0.034958	0.062627

ratio over conventional FDSOI, and intrinsic gain is also higher than the conventional structure in MD-FDSOI structure. Hence, source/drain engineered FDSOI are better alternatives than conventional FDSOI for analog performance, which reduces the SCEs and increases the gain by a more excellent ratio.

References

1. Moore GE (1975) Progress in digital integrated electronics. international electron devices meeting, IEEE 11–13. http://www.eng.auburn.edu/~agrawvd/COURSE/E7770_Spr07/READ/Gordon_Moore_1975_Speech.pdf
2. Iwai H (2008) CMOS technology after reaching the scale limit, extended abstracts—2008 8th international workshop on junction technology (IWJT '08), Shanghai, pp 1–2
3. Nowak JE (2004) Turning silicon on its edge double gate CMOS/FinFET technology. IEEE Circ Dev Magaz 20(1):20–31
4. Magarshack P, Flatresse P, Cesana G (2013) UTBB FD-SOI: a process/design symbiosis for breakthrough energy-efficiency. In: 2013 design, automation and test in Europe conference and exhibition (DATE), Grenoble, France, pp 952–957
5. Karel A, Comte M, Galliere J, Azais F, Renovell M (2016) Comparative study of Bulk, FDSOI and FinFET technologies in presence of a resistive short defect. In: 17th Latin-American test symposium (LATS), Foz do Iguacu, pp 129–134
6. Cheng KG, Khakifirooz A (2016) Fully depleted SOI (FDSOI) technology. Sci China Inf Sci 59(6):061402
7. Kim DS, Park MC, Woo SCJ (2002) Advanced source/drain engineering for box-shaped ultra-shallow junction formation using laser annealing and pre-amorphization im-plantation in sub-100-nm SOI CMOS. IEEE Trans Electron Dev 49(10):1748–1754
8. Li L, Yuan C, Li L (2011) A two-dimensional analytical model of fully depleted asym-metrical dual material gate double-gate strained-Si MOSFETs. J Semiconductors 32
9. Yadava N, Chauhan KR (2020) Impact of drain engineering on RF performance of nano-scale FD-SOI-MOSFET. In: International conference on electrical and electronics engineering (ICE3), Gorakhpur, India, pp 666–669
10. ATLAS User's manual (2016) SILVACO, Santa Clara, CA, USA

11. Mishra V, Chauhan KR (2017) Performance analysis of modified source and tunnel diode body contact based fully-depleted silicon-on-insulator MOSFET for low power digital applications. J Nanoelectron Optoelectron 12:59–66
12. Deepesh R, Rana KA, Yadav KR, Giri D (2011) Performance analysis of FD-SOI MOSFET with different gate spacer dielectric. Int J Comput Appl 18:22–27

Design of Partial Product Generator Circuit for Approximate Radix-8 Booth Multiplier with Lower Delay

Pooja Singh and Manish Kumar

Abstract Booth's algorithm is used to design fast multipliers. An approximate radix-8 booth multiplier reduces the number of steps for which the addition is done in the multiplication process. This algorithm encodes the multiplier bits and that is called Booth Recoding Process. Radix-8 booth recoding is done with grouping of 4 bits of multipliers. The main benefit of using radix-8 recoding is a faster circuit with an acceptable degradation in accuracy as it needs a bit large noise margin for logic 1 and logic 0. The multiplier consists of a left shift circuit to evaluate $2Q$ and $4Q$ and a circuit to determine $3Q$ (where Q is multiplicand). The partial product (PP) generator circuit is proposed to determine the partial product $3Q$. In this proposed partial product (PP) circuit, a precise adder which have lesser propagation delay is used. Its delay is calculated and compared with other previously proposed circuits. The radix-8 approximate booth multiplier with proposed partial product generator having precise adder is faster than the accurate Booth multiplier. The logics circuit are designed and simulated on Cadence Virtuoso Analog Design Environment using 180 nm technology for NMOS and PMOS at supply voltage of 0.8, 1 and 1.8 V.

Keywords Adder · Booth multiplier · Radix-8 · Delay · Partial product · Power consumption

1 Introduction

Computer arithmetic is largely used in digital signal processing. Different types of image processing with other digital signal processing are used in mobile devices these days. These devices use a digital signal processor separately other than the main processor because the processing is done at a large scale in these types of

P. Singh (✉) · M. Kumar
Department of Electronics and Communication Engineering, M.M.M. University of Technology, Gorakhpur, India
e-mail: ps739499@gmail.com

M. Kumar
e-mail: mkece@mmmut.ac.in

B. Mishra and M. Tiwari (eds.), *VLSI, Microwave and Wireless Technologies*,
Lecture Notes in Electrical Engineering 877,
https://doi.org/10.1007/978-981-19-0312-0_53

processors. A very common practice for the signal processing is the convolution of signal. The convolution in digital domain is done using multipliers and adders so a fast multiplier is needed to process large data. A slight change in propagation delay causes significant overall latency in cascaded and loop system. The process of multiplication is the most power hungry and consumes many clock cycles in a digital processor. A common multiplier performs the processes of multiplication in as many number of steps as the number of bits in the multiplier. But the Booth multiplier algorithm encodes the multiplier and the number of steps is reduced. The Booth recoding algorithm is generally done in three steps partial product (PP) generation, partial product accumulation and carry propagate addition. The multiplier is encoded differently for different radix. If a multiplication of two numbers is to be done where the multiplicand is Q has N bits and multiplier R has M bits and radix-2^r algorithm is used. This algorithm encodes the bit of R in $r + 1$ bits the number of partial products is reduced to M/r. The radix-4 reduces the number of partial products from M to $M/2$ and for radix-8 the number is reduced to $M/3$. The number of partial product accumulation required is directly one third or less. So, for fast application it uses less hardware resources and thus the critical time to complete an instruction is less. The only drawback of radix-8 is that the encoding of the multiplier creates 0, ±1, ±2, ±3, ±4 multiples of Q that is 0, $\pm1Q$, $\pm2Q$, $\pm3Q$, $\pm4Q$. The $1Q$, $2Q$ and $4Q$ is done with left shift circuit but $3Q$ is needed to apply some logic to reduce the hardware. So, to remove this complexity approximate adder is used. The proposed work of this paper reduces the delay significantly with a little trade off with total power.

1.1 Related Work

A short review on other available methods to overcome the $3Q$ problem and possible algorithm with error tolerant circuit in application specific requirements is recounted. An error tolerant multiplier is proposed in [1]. In that paper the input operand is split into two parts and for higher order of bit the partial product is generated. Due to discarding the lower order part a manageable error is generated. A DRUM algorithm is proposed in [2] and this uses the k-significant bit from a leading logic 1. These k-significant bit are added k times with one shift (i.e., $k \times k$ multiplier). Mitchell logarithmic multiplier-based system is discussed in [3, 4]. The approximation of logarithmic multiplier is proposed in that literature. Approximate log-based less error multiplier [5] reduces the successive error. Piece wise linear approximation algorithm is proposed in [6] and a worst case error is discussed. Approximate accumulation of partial products is widely mentioned and studied in the literature [7–10]. The seven bit precise adder with approximate bit is proposed in [11]. The error is reduced due to the use of precise adder and approximate adder.

In this paper the approximate circuit is used to process this $3Q$. The organization of this paper is as follows: Sect. 2 includes the previous work on radix-8 and working of radix-8 Booth multiplier with encoding table and calculation of partial product, Sect. 3 includes the proposed circuit model and its simulation on cadence virtuoso,

Sect. 4 includes the results and discussion on the proposed circuit, Sect. 5 concludes the paper and discuss the application where it can be used.

2 Conventional Radix-8 Booth Multiplier

The circuit of conventional radix-2 and radix-4 Booth multiplier can be created easily as it needs only shifting by 2 (i.e., $2Q$) but in radix-8 the complexity occurs due to the $3Q$ multiplier. The $2Q$ and $4Q$ is done easily using 2 times shift and 4 times shift. The 2 bit left shift circuit is shown in Fig. 1. The dotted rectangle shows the single unit of the shift circuit, for n-bit left shift circuit it is repeated $n - 1$ times. If the SHIFT input is one then this circuit will left shift the bits. Zero is inserted at R_0 and Q_0 will appear at R_1 and this will continue for every Q_i until $i = n - 1$. For SHIFT input zero Q_i will be appear at R_i.

A lot of work is done and different algorithms are developed to design the circuit to get $3Q$ multiplier. Let Q and R are multiplicand and multiplier, with M number of sign bit. The 2's complement representation of Q and R is

$$Q = -q_{M-1}2^{M-1} + \sum_{i=0}^{M-2} q_i 2^i \tag{1}$$

$$R = -r_{M-1}2^{M-1} + \sum_{i=0}^{M-2} r_i 2^i \tag{2}$$

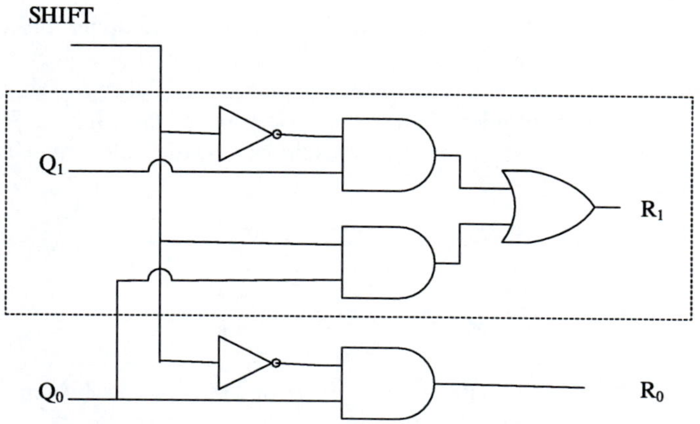

Fig. 1 Two bit left shift circuit

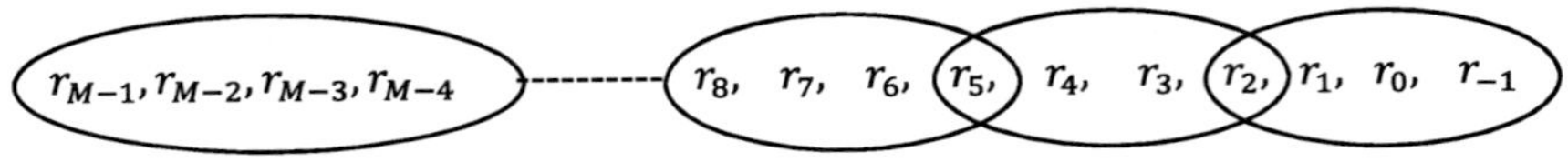

Fig. 2 Radix-8 multiplier bit grouping and overlapping

Table 1 Truth table of recoding of multiplier bit

r_{3j+2}	r_{3j+1}	r_{3j}	r_{3j-1}	R_j	PP_j
0	0	0	0	0	0
0	0	0	1	+1	Q
0	0	1	0	+1	Q
0	0	1	1	+2	$2Q$
0	1	0	0	+2	$2Q$
0	1	0	1	+3	$3Q$
0	1	1	0	+3	$3Q$
0	1	1	1	+4	$4Q$
1	0	0	0	−4	$4Q$
1	0	0	1	−3	$-3Q$
1	0	1	0	−3	$-3Q$
1	0	1	1	−2	$-2Q$
1	1	0	0	−2	$-2Q$
1	1	0	1	−1	$-Q$
1	1	1	0	−1	$-Q$
1	1	1	1	0	0

To obtain the radix-8 Booth recoding the four bits of the multiplier R is taken and recoded. The group of four bits $\{r_{3j+2}, r_{3j+1}, r_{3j}, r_{3j-1}\}$ with one bit overlapping, as shown in Fig. 2, is made and it is encoded in the manner shown in Table 1.

Here R_j is the sign multiplier of jth row and r_k represent the kth bit of R and PP_j represents partial product of jth row and it is obtained by the following equation

$$R_j = -4r_{3j+2} + 2r_{3j+1} + r_{3j} + r_{3j-1} \text{ For } 0 \le j \le \left[\frac{M}{3}\right] - 1 \quad (3)$$

$$\mathrm{PP}_j = Q * R_j \text{ For } 0 \le j \le \left[\frac{M}{3}\right] - 1 \quad (4)$$

The encoded bit having value $2Q$ and $4R$ is obtained by one bit left shift and two left bit shift, respectively. But the term $3Q$ can be obtained as $Q + 2Q$ using recoding. The left shift of one bit and addition with itself is done to obtain $Q + 2Q$. This recoding adder can be implemented as shown in Fig. 3.

The partial product (PP) is given by the following equation

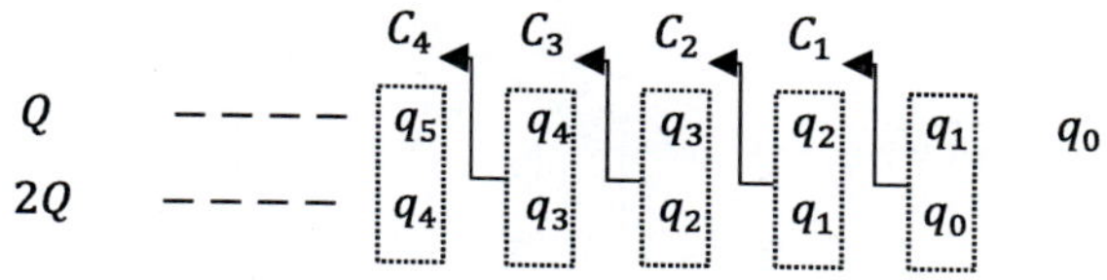

Fig. 3 $Q + 2Q$ adder design

$$\begin{aligned}\mathrm{PP}_{ij} = \big[&q_i(r_{3j+2} \oplus r_{3j+1})(r_{3j} \oplus r_{3j-1}) + q_{i-1}(r_{3j+1}r_{3j}r_{3j-1} + r_{3j+1}r_{3j}r_{3j-1}) \\ &+ s_i(r_{3j+2} \oplus r_{3j+1})(r_{3j} + r_{3j-1}) \\ &+ q_{i-2}(r_{3j+2}r_{3j+1}r_{3j}r_{3j-1} + r_{3j+2}r_{3j+1}r_{3j}r_{3j-1})\big] \oplus r_{3j+2}\end{aligned} \tag{5}$$

Various methods are used to design the circuit for $3Q$ but the most efficient circuit with tolerable error. The design and working of the proposed model is described in the section ahead.

3 Proposed Circuit and Simulation

The partial product for $2Q$ and $4Q$ are done without any difficulty. The circuit proposed in this paper is to reduce the complicacy of calculation of $3Q$. This proposed circuit is combination of 8 bit approximate adder and seven bit precise adder shown in Fig. 5b. The 2 bit approximate adder uses the 3 input XOR (Fig. 5a) as described in Eq. (6).

$$S_i = C_{\text{in}} \oplus r_i \oplus r_{i-1} \tag{6}$$

The probability of error for can be calculated taking the input combination of $C_{\text{in}}, r_{i+1}, r_i, r_{i-1}$ and it comes out to be 1/16 as 1 and 0 are equally likely to occur with probability 1/2. Thus, the error can be estimated for 2 bit approximate adder and it would be 1/4.

The precise adder is designed using two XOR and one multiplexer Fig. 6a which have lower delay [12, 13]. The transistor level circuit of this XOR is shown in Fig. 6b.

The circuit is combination of XOR and multiplexer. The complementary design of XOR (Fig. 6b) with mux as carry generator is implemented in FA. The advantage of this circuit is that it substantially reduces the delay. The circuit shown in Fig. 5b is used at the place of partial product generator shown in Fig. 4 and the adder used in the circuit 8A7PA is the circuit shown in Fig. 6 This combination is used to reduce the overall delay. Figure 7 shows the 1 bit precise full adder with a little but acceptable noise and drop in voltage level. Consideration of high noise margin is recommended to use this circuit. Figure 8 shows the design of PP generator and simulation result in discussed in next section and delay and power is calculated.

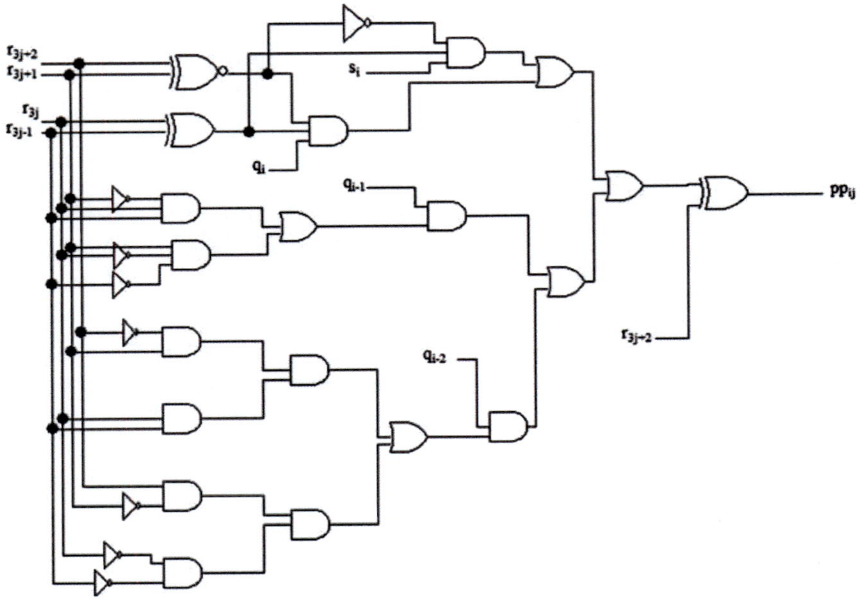

Fig. 4 The conventional partial product generator

4 Result and Discussion

The delay and power are measured and a comparison table is shown in Table 2. The performance of precise adder is compared. Change in adder affects the overall delay for partial product (PP) circuit. The best suitable adder is used in PP circuit and simulated. The simulation is done at different supply voltage to find out the optimum working voltage of different adder and proposed adder. This proposed adder is further used in the proposed PP generator circuit. The cell-2 full adder circuit is taken from [13] and verified at these voltages.

The proposed partial product (PP) generator circuit shown in Fig. 8 is designed using proposed FA with 2 XOR and 1 MUX and the power and delay is calculated and comparison is done with previously proposed work [14] and table is given. The delay in the proposed work is very much less than the previously proposed circuits. The power and delay product (PDP) is calculated and compared with the previously proposed circuit. The PDP is lesser than ARA8 and ARA6 (Table 3).

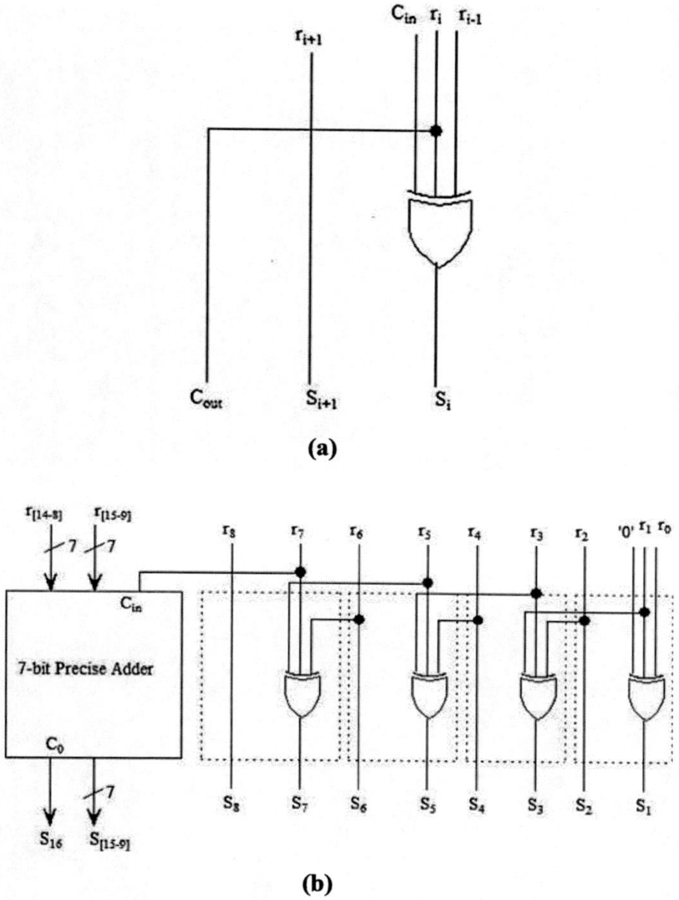

Fig. 5 **a** A 3 input 2 bit approximate adder. **b** 8 bit approximate adder with 3 input XOR with 7 bit precise adder (8A7PA)

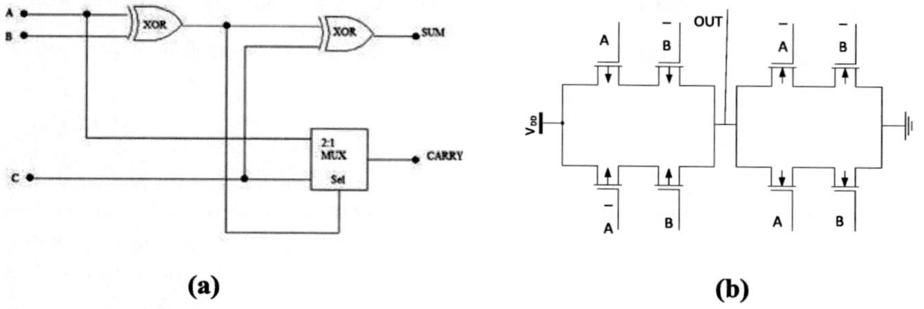

Fig. 6 **a** One bit precise full adder. **b** XOR with less delay

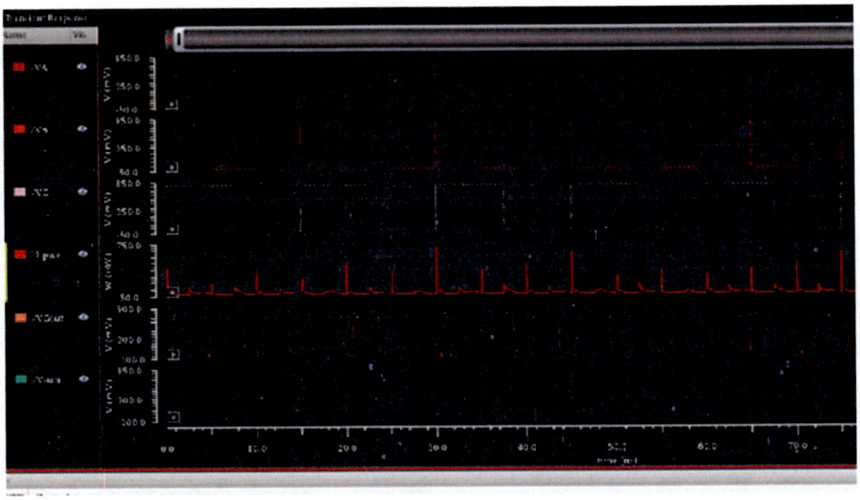

Fig. 7 One bit precise full adder at 0.8 V supply (V_{dd})

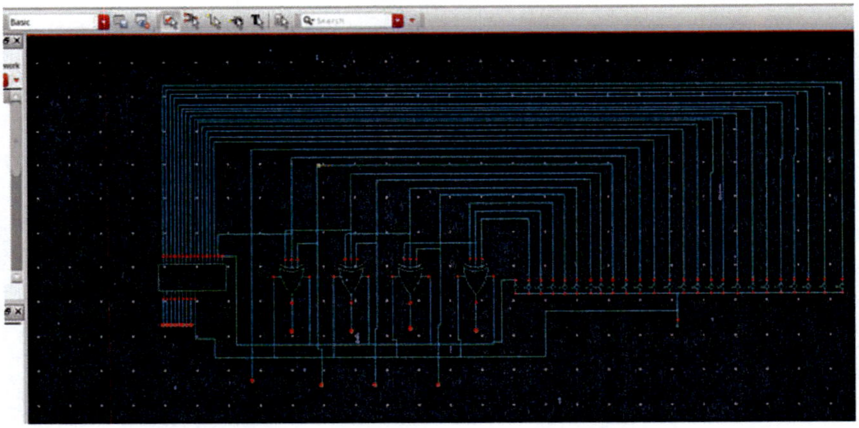

Fig. 8 Design of 8 approximate bit and 7 precise bit PP generator for $3Q$ calculation (8A7PA)

5 Conclusion

The circuit of precise adder which is proposed in this paper is used for the calculation of partial product of factor $3Q$ in Booth multiplier. The circuit is simulated and delay and power are analyzed. The delay of the proposed circuit is significantly reduced with slight increase in power. The result indicates that approximate 8 bit adder and 7 bit precise adder which is proposed, achieved the better optimized performance in terms of minimizing the propagation delay. The design is suitable for the applications

Table 2 The comparison among various 1 bit adder delay and power of precise adder

Types of adder (180 nm)		CMOS complementary FA	Cell-2 FA	Proposed FA with 2 XOR and 1 MUX
1.8 V supply	Delay (s)	Sum 65.86E−12 C_{out} 122.8E−12	Sum 2.7E−9 C_{out} 8.07E−9	Sum 65.81E−12 C_{out} 89.67E−12
	Power (W)	Sum 839.8E−3 C_{out} 943.8E−3	Sum 884.0E−3 C_{out} 854.0E−3	Sum 838.8E−3 C_{out} 949.3E−3
1 V supply	Delay (s)	Sum 105.6E−12 C_{out} 255.7E−12	Sum 2.86E−9 C_{out} 2.55E−9	Sum 105.7E−12 C_{out} 212.4E−12
	Power (W)	Sum 449.3E−3 C_{out} 533.8E−3	Sum 515.8E−3 C_{out} 443.5E−3	Sum 448.4E−3 C_{out} 536.8E−3
0.8 V supply	Delay (s)	Sum 130.7E−12 C_{out} 454.5E−12	Sum 3.10E−9 C_{out} 2.56E−9	Sum 23.0E−12 C_{out} 5.70E−12
	Power (W)	Sum 340.1E−3 C_{out} 442.6E−3	Sum 428.5E−3 C_{out} 357.4E−3	Sum 339.3E−3 C_{out} 442.0E−3

Table 3 Power and delay for the proposed PP generator circuit

SI. No.	PP generator circuit	Power (W)	Delay (s)	Power delay product
1.	ARA8	18.37E−6	730E−12	13.410E−15
2.	ARA6	22.63E−6	940E−12	21.018E−15
3.	8A7PA (proposed in this paper)	142.0E−6	3.909E−12	0.555078E−15

where the delay is main concern and cannot be overlooked. The drawback of this circuit is that it requires a bit more power. So, a trade of exists and the application, where the power is not the main concern but the delay is, it can be used.

References

1. Kyaw KY, Goh WL, Yeo KS (2010) Low-power high-speed multiplier for error-tolerant application. In: IEEE International conference on of electron devices and solid-state circuits (EDSSC), Hong Kong, pp 1–4
2. Hashemi S, Bahar RI, Reda S (2015) DRUM: a dynamic range unbiased multiplier for approximate applications. In: IEEE/ACM International conference on computer-aided design (ICCAD), Austin, TX, pp 418–425
3. Low JYL, Jong CC (2015) Unified Mitchell-based approximation for efficient logarithmic conversion circuit. IEEE Trans Comput 64(6):1783–1797
4. Liu W, Xu J, Wang D, Wang C, Montuschi P, Lombardi F (2018) Design and evaluation of approximate logarithmic multipliers for low power error-tolerant applications. IEEE Trans Circ Syst I Regul Pap 65(9):2856–2868
5. Saadat H, Bokhari H, Parameswaran S (2018) Minimally biased multipliers for approximate integer and floating-point multiplication. IEEE Trans Comput-Aided Design Integr Circ Syst 37(11):2623–2635
6. Loukrakpam M, Choudhury M (2019) Implementation of energy efficient approximate multiplier with guaranteed worst case relative error. Microelectron J 88:1–8

7. Liu W, Qian L, Wang C, Jiang H, Han J, Lombardi F (2017) Design of approximate Radix-4 booth multipliers for error tolerant computing. IEEE Trans Comput 66(8):1435–1441
8. Venkatachalam S, Adams E, Lee HJ, Ko S (2019) Design and analysis of area and power efficient approximate booth multipliers. IEEE Trans Comput 68(11):1697–1703
9. Leon V, Zervakis G, Soudris D, Pekmestzi K (2018) Approximate hybrid high radix encoding for energy-efficient inexact multipliers. IEEE Trans Very Large Scale Integr (VLSI) Syst 26(3):421–430
10. Jiang H, Han J, Qiao F, Lombardi F (2016) Approximate Radix-8 booth multipliers for low-power and high performance operation. IEEE Trans Comput 65(8):2638–2644
11. Jiang H, Han J, Qiao F, Lombardi F (2016) Approximate Radix-8 booth multipliers for low-power and high-performance operation. IEEE Trans Comput 65(8)
12. Shoba M, Nakkeeran R (2016) GDI based full adders for energy efficient arithmetic applications. Eng Sci Technol Int J 19(2016):485–496
13. Dokania V, Verma R, Guduri M, Islam A (2018) Design of 10 T full adder cell for ultralow-power applications. Ain Shams Eng J 9:2363–2372
14. Boroa B, Manikantta Reddya K, Nithin Kumara YB, Vasanthaa MH (2020) Approximate Radix-8 booth multiplier for low power and high speed applications. Microelectron J. https://doi.org/10.1016/j.mejo.2020.104816

Power Optimisation of UAV-Based Relay for Selective Decoding and Forward Protocol Co-operation in Wireless Communication

Deepak Kumar, Satish Chandra, and Seemanti Saha

Abstract In this paper, co-operation in wireless communication with an unmanned aerial vehicle (UAV)-based relay has been analytically studied and simulated. UAV works on selective decoding and forward (SDAF) protocol relay. The transmitted power of source and UAV-based relay have been optimized by minimizing the average end to end bit error rate (e2e BER) of the communication link. A new solution to an old analytical expression of average e2e BER is derived by convex optimisation and simulation results have also been shown. Obtained results show that the diversity order increases with increase in the system performance.

Keywords UAV · Co-operative communication · Power optimisation

1 Introduction

One of the state-of-the-art features of 4G and 5G wireless communication has recently been migrated to co-operative communications. This generally alters the abstraction of a wireless connection and provides wireless communication networks with substantial potential benefits.

Tao et al. [1] explores collaborative strategies such as: transmission, distributed antenna systems (DAS), multicell integration, group cell, multi-point connectivity and reception (COMP), which transforms a traditional mobile system into a cooperative system. They proposed a number of co-operative processing policies to reflect the concept of a cell group (COMP). The authors in [2] proposed how to utilize increasing transmission (IR) for spectral efficiency. Specifically, it is also defined based on the end to end error number (e2e SER) and the discovery uncertainty region. Only if the

D. Kumar (✉) · S. Saha
ECE Department, National Institute of Technology, Patna 800005, India
e-mail: deepak.purnia@gmail.com

D. Kumar · S. Chandra
ECE Department, M. M. M. University of Technology, Gorakhpur 273010, India

B. Mishra and M. Tiwari (eds.), *VLSI, Microwave and Wireless Technologies*, Lecture Notes in Electrical Engineering 877,
https://doi.org/10.1007/978-981-19-0312-0_54

processed signal falls within this region will the amplifier-forward (AF) transfer be caused by the included relay [3, 4].

The optimization of the resource allocation has become a major challenge for the last few decades to maximize the average sum-rate. The authors in [5] recently suggested two effective matching algorithms for many-to-many two-sided source–destination pair-subchannel, in which the source–destination pairs and sub-channels are regarded as two sets of players chasing their own interests. The algorithms in [5] can provide an affordable time to provide a sub-optimal solution to this resource allocation problem. It is also explained that after a limited number of iterations, both the static matching algorithm and the dynamic matching algorithm converge to a pair-wise stable matching.

Transmission nodes are installed on unmanned aerial vehicles (UAVs) and are therefore capable of high-speed travel. Mobile relaying provides a new level of freedom to improve performance with a flexible trajectory, compared to standard static relaying. The study of the problem of maximization of throughput in mobile relaying systems was carried out in [6, 7] by optimizing the source and relay power, subject to practical mobility limitations as well as the limitation of information causality at the relay. By optimizing the UAV trajectory via successive convex optimization, the throughput can be further improved. An iterative algorithm is also derived for the joint optimal allocation of power and the relay trajectory [6, 7]. An analytical approach is discussed in [8] to optimize the altitude of such UAV to provide highest radio coverage on the ground.

It has also been shown that absolute height is the function of the most permissible pathloss and the limits of urban environmental statistics. The analytical method of increasing the height of this UAV is to provide a large radio coverage on the ground. In [8] it shows that the right height is the function of the most permissible pathloss and the boundaries of the urban natural statistics.

In this paper, a half-duplex enabled uplink UAV has been used, which is based on co-operative communication model having source, destination and UAV-based relay. The UAV works on selective decoding and forward (SDAF) protocol-based relay which is having ability to control and optimize the power. The optimization of power has been formulated for minimum probability of end to end bit error rate (e2e BER) at the destination. The system model is shown in (Fig. 1).

2 System Model and Mathematical Formulation

UAV-based co-operation is an important technique for modern wireless communication system, where one or more UAV can work as relay. The relay receives the signal from source, which performs some signal processing and retransmit to the destination for upgrading the reliability of communication system. According to the nature of operation, the relay can have protocol such as amplify-forward protocol and decode-forward protocol. The relay has function to forward the signal only when it decodes the information correctly, so it is termed as selective decode and

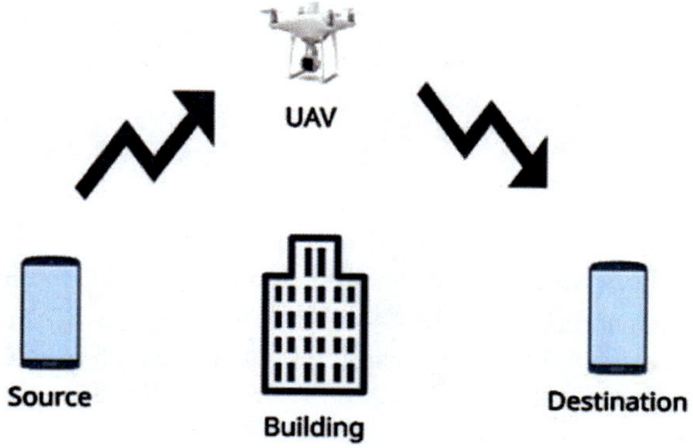

Fig. 1 System model for UAV-based relay in wireless network

forward protocol (SDAF) for co-operation. SDAF processing occurs in two phases. Considering h_{sd}, h_{sr} and h_{rd} are the channel coefficients between source to destination, source to relay and relay to destination, respectively. The flying altitude of UAV is set to a minimum certain height, which is required for proper coverage in mountainous area. Also assumed that the UAV is at equal distance from source and destination.

Co-operative communication occurs in two phases, in phase-1: source broadcast the signal to both destination and relay nodes. At destination and relay, the received signal is given as $y_{\text{sd}} = \sqrt{P_1} h_{\text{sd}} x + n_{\text{sd}}$ and $y_{\text{sr}} = \sqrt{P_1} h_{\text{sr}} x + n_{\text{sr}}$, respectively. Where P_1 is the source power and n_{sd} and n_{sr} the noises in the respective links. Considering average transmitted symbol power, $E\{|x|^2\} = 1$. Also, let the average channel gain be given as $E\{|h_{\text{sd}}|^2\} = \delta_{\text{sd}}^2$ and $E\{|h_{\text{sr}}|^2\} = \delta_{\text{sr}}^2$. In Phase-2: relay retransmit the decoded symbol to the destination, only if it decodes information correctly. The received signal at the destination is presented as $y_{\text{rd}} = \sqrt{\tilde{P}_2} h_{\text{rd}} x + n_{\text{rd}}$. Where the average channel gain as $E\{|h_{\text{rd}}|^2\} = \delta_{\text{rd}}^2$ and $\tilde{P}_2$ is the power employed at the relay which can be written as:

$$\tilde{P}_2 = \begin{cases} P_2; & \text{if symbol decoded correctly} \\ 0; & \text{if symbol decoded incorrectly} \end{cases} \tag{1}$$

The decoding criteria can be achieved by using a suitable threshold condition at the relay. The relay can decode the symbol correctly if the received SNR is greater than threshold value, else decoding fails at the relay with high probability.

2.1 NR Analysis

If decoding error occurs at relay it could not forward the information, so only source to destination link is available and the system model will be. $y_{\text{sd}} = \sqrt{P_1} h_{\text{sd}} x + n_{\text{sd}}$

and therefore SNR $= \rho_1|h_{sd}|^2$. In other case, when no error at relay, the signal from both the source and the relay is received at destination that can be written in the matrix form as below.

$$\begin{bmatrix} y_{sd} \\ y_{rd} \end{bmatrix} = \begin{bmatrix} \sqrt{P_1}h_{sd} \\ \sqrt{P_2}h_{rd} \end{bmatrix} x + \begin{bmatrix} n_{sd} \\ n_{rd} \end{bmatrix} \tag{2}$$

Performing maximum ratio combining (MRC) at the destination and SNR at the destination is $\mathrm{SNR}_d = \frac{\|h\|^2}{\sigma^2} E\{|x|^2\} = \rho_1\beta_{sd} + \rho_2\beta_{rd}$. In [9], the issue of beamforming and combining with transmission path selection in a decode and forward (DF) relay network is addressed. For the selection of the strong transmission path between the source and destination nodes, a maximum value-based criterion is used.

2.2 *Symbol Error Rate (SER)*

The outage probability of a cellular user and the average achievable rate from a transmitter to receiver in analytic form is described in [10]. The long-term average problem of SER minimization for DF co-operative communications with a relay node for energy harvesting is discussed in [11]. Also known as end to end error (e2e) or symbol error rate, the probability of error at the destination is also known. Let Φ denotes the relay error event, SER can be written as,

$$\Pr(e) = \Pr(e \cap \Phi) + \Pr(e \cap \overline{\Phi}) = \Pr(e|\Phi)\Pr(\Phi) + \Pr(e|\overline{\Phi})\Pr(\overline{\Phi}) \tag{3}$$

2.2.1 Calculation of $\mathbf{Pr}(e|\Phi)$ and $\mathbf{Pr}(\Phi)$

When decoding error occurs at relay the information is not retransmitted, so only source to destination link is available and the model of the system will be $y_{sd} = \sqrt{P_1}h_{sd}x + n_{sd}$ therefore the SNR $= \rho_1|h_{sd}|^2$. Let $|h_{sd}|^2 = \beta_{sd}$ and $E\{|h_{sd}|^2\} = \delta_{sd}^2$. Instantaneous probability of error at destination e given Φ will be $\widetilde{\Pr}(e|\Phi) = Q\left(\sqrt{\mathrm{SNR}}\right) = Q\left(\sqrt{\rho_1\beta_{sd}}\right)$. Using Craig's formula: $Q\left(\sqrt{\rho_1\beta_{sd}}\right) = \int_0^{\pi/2} \frac{1}{\pi}\exp\left(-\frac{\rho_1\beta_{sd}}{2\sin^2\theta}\right)d\theta$. We can calculate the average probability of error at destination e given Φ will be $\Pr(e|\Phi)$ is equal to $E\left\{\widetilde{\Pr}(e|\Phi)\right\} = \int_0^\infty Q\left(\sqrt{\rho_1|h_{sd}|^2}\right) f_{|h_{sd}|}(|h_{sd}|)d|h_{sd}|$, Where $f_{|h_{sd}|}(|h_{sd}|)$ is the pdf of $|h_{sd}|$. Here we are considering that all the fading channel coefficients are zero mean symmetric complex Gaussian therefore pdf of magnitude of channel coefficient $|h_{sd}|$ will be Rayleigh pdf. $f_{|h_{sd}|}(|h_{sd}|) = \frac{2|h_{sd}|}{\delta_{sd}^2}\exp\left(-\frac{\beta_{sd}}{\delta_{sd}^2}\right)$, therefore $\Pr(e|\Phi) = \int_0^\infty Q\left(\sqrt{\rho_1|h_{sd}|^2}\right) f_{|h_{sd}|}(|h_{sd}|)d|h_{sd}|$ and solving we get the probability of error,

$\Pr(e|\Phi) = \frac{1}{2\rho_1\delta_{\text{sd}}^2}$. Similarly, the probability of error at relay is obtained as $\Pr(\Phi) = \frac{1}{2\rho_1\delta_{\text{sr}}^2}$.

2.2.2 Calculation of $\Pr\left(e|\overline{\Phi}\right)$

When relay correctly decode the information and retransmit to destination, we have signal from both the source and the relay, hence received signal at destination can be written in the matrix form as below.

$$\begin{bmatrix} y_{\text{sd}} \\ y_{\text{rd}} \end{bmatrix} = \begin{bmatrix} \sqrt{P_1}h_{\text{sd}} \\ \sqrt{P_2}h_{\text{rd}} \end{bmatrix} x + \begin{bmatrix} n_{\text{sd}} \\ n_{\text{rd}} \end{bmatrix} \tag{4}$$

Performing maximum ratio combining (MRC) at the destination. The SNR at the destination is $\text{SNR}_d = \rho_1\beta_{\text{sd}} + \rho_2\beta_{\text{rd}}$. Let $|h_{\text{sd}}|^2 = \beta_{\text{sd}}$, $|h_{\text{rd}}|^2 = \beta_{\text{rd}}$, $E\{|h_{\text{sd}}|^2\} = \delta_{\text{sd}}^2$, $E\{|h_{\text{rd}}|^2\} = \delta_{\text{rd}}^2$. Instantaneous probability of error at destination e given Φ will be $\widetilde{\Pr}(e|\overline{\Phi}) = Q\left(\sqrt{\text{SNR}_d}\right) = Q\left(\sqrt{\rho_1\beta_{\text{sd}} + \rho_2\beta_{\text{rd}}}\right)$. Using Craig's formula, average probability of error at destination e given $\overline{\Phi}$ will be $\Pr\left(e|\overline{\Phi}\right)$ is equal to

$$E\{\widetilde{\Pr}(e|\overline{\Phi})\} = \int_0^\infty \int_0^\infty Q\left(\sqrt{\rho_1\beta_{\text{sd}} + \rho_2\beta_{\text{rd}}}\right) f_{|h_{\text{sd}}|}(|h_{\text{sd}}|)d|h_{\text{sd}}| f_{|h_{\text{rd}}|}(|h_{\text{rd}}|)d|h_{\text{rd}}| \tag{5}$$

where $f_{|h_{\text{sd}}|}(|h_{\text{sd}}|)$ and $f_{|h_{\text{rd}}|}(|h_{\text{rd}}|)$ is the pdf of $|h_{\text{sd}}|$ and $|h_{\text{rd}}|$, respectively. Here we are considering that all the fading channel coefficients are zero mean symmetric complex Gaussian, therefore pdf of magnitude of channel coefficient $|h_{\text{sd}}|$ and $|h_{\text{rd}}|$ will be Rayleigh pdf. $f_{|h_{\text{sd}}|}(|h_{\text{sd}}|) = \frac{2|h_{\text{sd}}|}{\delta_{\text{sd}}^2}\exp\left(-\frac{\beta_{\text{sd}}}{\delta_{\text{sd}}^2}\right)$ and $f_{|h_{\text{rd}}|}(|h_{\text{rd}}|) = \frac{2|h_{\text{rd}}|}{\delta_{\text{rd}}^2}\exp\left(-\frac{\beta_{\text{rd}}}{\delta_{\text{rd}}^2}\right)$. By solving (5), we obtain.

$$\Pr\left(e|\overline{\Phi}\right) = \int_0^\infty \int_0^\infty \int_0^{\pi/2} \frac{1}{\pi}\exp\left(-\frac{\rho_1\beta_{\text{sd}} + \rho_2\beta_{\text{rd}}}{2\sin^2\theta}\right) d\theta f_{|h_{\text{sd}}|}(|h_{\text{sd}}|)d|h_{\text{sd}}| f_{|h_{\text{rd}}|}(|h_{\text{rd}}|)d|h_{\text{rd}}| \tag{6}$$

$$= \int_0^{\pi/2} \int_0^\infty \int_0^\infty \frac{1}{\pi}\exp\left(-\frac{\rho_1\beta_{\text{sd}}}{2\sin^2\theta}\right) f_{|h_{\text{sd}}|}(|h_{\text{sd}}|)d|h_{\text{sd}}| \exp\left(-\frac{\rho_2\beta_{\text{rd}}}{2\sin^2\theta}\right) f_{|h_{\text{rd}}|}(|h_{\text{rd}}|)d|h_{\text{rd}}| d\theta \tag{7}$$

$$= \frac{1}{\pi}\int_0^{\pi/2} \left(\frac{1}{1 + \frac{\rho_1\delta_{\text{sd}}^2}{2\sin^2\theta}} \times \frac{1}{1 + \frac{\rho_2\delta_{\text{rd}}^2}{2\sin^2\theta}}\right) d\theta \tag{8}$$

At very high SNR, we can approximate that $1+\frac{\rho_1\beta_{\text{sd}}}{2\sin^2\theta}\approx\frac{\rho_1\beta_{\text{sd}}}{2\sin^2\theta}$ and $1+\frac{\rho_2\beta_{\text{rd}}}{2\sin^2\theta}\approx\frac{\rho_2\beta_{\text{rd}}}{2\sin^2\theta}$. And therefore, $\Pr(e|\overline{\Phi})=\frac{1}{\pi}\int_0^{\frac{\pi}{2}}\left(\frac{4\sin^4\theta}{\rho_1\delta_{\text{sd}}^2\rho_2\delta_{\text{rd}}^2}\right)d\theta=\frac{3}{4\rho_1\delta_{\text{sd}}^2\rho_2\delta_{\text{rd}}^2}$. End to end symbol error rate (SER) can be obtained by assumption is that $\Pr(\Phi)$ is very small at high SNR and hence $\Pr(\overline{\Phi})$ is approximated to one. Therefore SER $\Pr(e)$ is approximated as

$$\begin{aligned}\Pr(e)=\Pr(e|\Phi)\Pr(\Phi)+\Pr(e|\overline{\Phi})&=\frac{1}{2\rho_1\delta_{\text{sd}}^2}\times\frac{1}{2\rho_1\delta_{\text{sr}}^2}+\frac{3}{4\rho_1\delta_{\text{sd}}^2\rho_2\delta_{\text{rd}}^2}\\&=\frac{1}{4\rho_1\delta_{\text{sd}}^2}\left(\frac{1}{\rho_1\delta_{\text{sr}}^2}+\frac{3}{\rho_2\delta_{\text{rd}}^2}\right)\end{aligned}\tag{9}$$

2.3 Diversity Order

Let, the total SNR be constrained as $\rho_1+\rho_2=\rho$, α is the power allocation factor. So, the total power be distributed between source and relay as $\rho_1=\alpha\rho$ and $\rho_2=(1-\alpha)\rho$, respectively. So (9) is reduces to

$$\Pr(e)=\frac{1}{\rho^2}\left(\frac{1}{4\alpha^2\delta_{\text{sd}}^2\delta_{\text{sr}}^2}+\frac{3}{4\alpha(1-\alpha)\delta_{\text{sd}}^2\delta_{\text{rd}}^2}\right)\tag{10}$$

Therefore, we can conclude $\Pr(e)\propto\frac{1}{\rho^2}$ and diversity order can be seen, $d=2$, this is termed as co-operative diversity. For K relay node system, the diversity order will be $K+1$. Co-operative communication reduces end to end BER of the system when diversity order increases.

2.4 Optimal Power Allocation

To find α which minimizes the probability of error $\Pr(e)$. We solved the optimization problem given by

$$\min_{\alpha}\Pr(\mathrm{e})=\min_{\alpha}\frac{1}{\rho^2}\left(\frac{1}{4\alpha^2\delta_{\text{sd}}^2\delta_{\text{sr}}^2}+\frac{3}{4\alpha(1-\alpha)\delta_{\text{sd}}^2\delta_{\text{rd}}^2}\right)\tag{11}$$

Equivalently equation (11) can be rewritten as,

$$\min_{\alpha}\left(\frac{1}{\alpha^2\delta_{\text{sr}}^2}+\frac{3}{\alpha(1-\alpha)\delta_{\text{rd}}^2}\right)\tag{12}$$

Above optimization problem is solved and we found $\alpha = \frac{\delta_{sr}+\sqrt{\delta_{sr}^2+\frac{8}{3}\delta_{rd}^2}}{3\delta_{sr}+\sqrt{\delta_{sr}^2+\frac{8}{3}\delta_{rd}^2}}$. Therefore, optimal power allocation at source, $P_1 = \alpha P$ and at destination, $P_2 = (1-\alpha)P$. Further we observed that if $\delta_{sr} \gg \delta_{rd}$ then $P_1 = P_2 = \frac{1}{2}P$, i.e., Equal power is distributed among source and relay when strong channel condition is present between source and relay, and if $\delta_{sr} \ll \delta_{rd}$ then $P_1 = P$ and $P_2 = 0$, i.e., for week channel condition between source and relay, all the power is given to source.

In other words, when a co-operative link does not agree enough, the proposed optimal solution may be unable to get the significant operational benefits. It is also apparent that at an appropriately high altitude, the SDAF protocol is slower and tends to operate smoothly as the UAV flies high [12].

3 Simulation Results and Discussion

Simulation results based on Simulink tool of MATLAB software are provided here to justify our proposed UAV-based relay for co-operative communication. A system with the binary bit stream having unity energy of block-length 1000 have been considered. The separation between source and the destination is $L = 500$ m, the noise variance is one. Transmitted power is varied from 5 to 25 dB for 10,000 iteration. A significant decrease in BER in co-operative link have been observed, which is shown in Fig. 2. The power distribution factor for direct and co-operative link is simulated, which is shown in Fig. 3. Further, to obtain optimal power allocation at source and relay, total transmitted power $P = 21$ dB is considered, and optimal power at source and relay are P_1 and P_2 equal to 10.5 dB is found, where the optimized power has been obtained through following flowchart shown in Fig. 4.

A comparison of BER with UAV-based relay and direct link is also shown in Fig. 2 from source to destination and an optimal UAV power splitting ratio profile for the

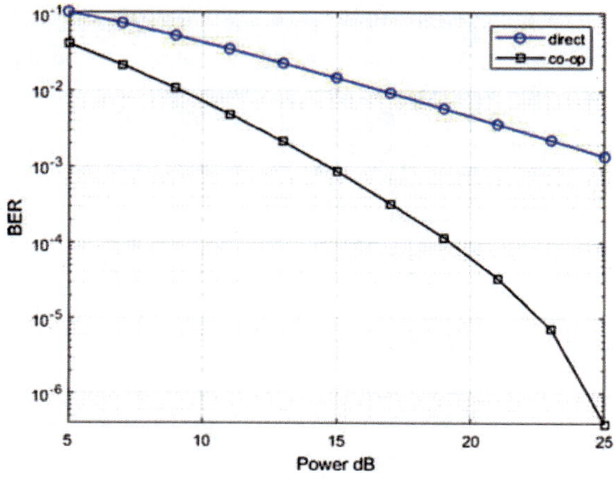

Fig. 2 Power versus BER of direct link and co-operative link

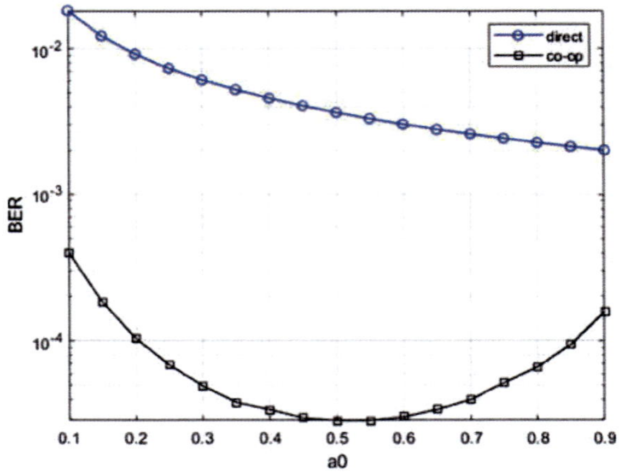

Fig. 3 Power distribution factor (*a*0) versus BER of direct and co-operative link

SDAF protocol is shown in Fig. 3, and the power distributed between source and relay to achieve the minimum BER is illustrated. Where strong channel conditions are present between source and relay, equal power is distributed between source and relay. The optimal power profile and power splitting ratio were taken at a height of 100 m for the appropriate location of UAVs from source to destination. This paper describes the end to end bit error rate and transmission power at the source and relay.

Therefore, only optimization of power for the UAV profile is required for the co-operative throughput optimization issue to solve it in a simpler way. In general, because of the additional power received from the source, the proposed solution beats the foundation [12, 13].

The system can continuously take advantage of a UAV that serves as a portable relay. As indicated in Fig. 1, with higher transmitting power in the source, the source power ratio tends to decrease because the higher transmission power boosts the obtained SNR of direct connection. However, due to the poor state of the co-operative link channel, the amount of resource capacity likely to decrease, when the UAV travels to a greater altitude. The successful operation of the partnership always results in a negligible amount of end to end bit error rate. This is because the status of the co-operative channel is far better than the direct link as the performance of the system can be greatly improved by UAV such as mobile transmission even with low transmission capacity.

4 Conclusion

The study of end to end bit error rate maximization problem for a UAV-based co-operative communication system is shown where power optimization is done among

Fig. 4 Flowchart for power optimization

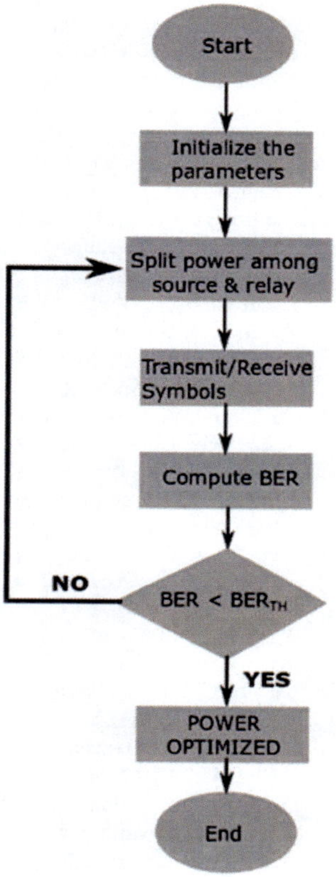

source and relay. The UAV works as an information relay and its transmission capability is driven by the source. It is also seen that the diversity order increases by using relay and hence system performance improves. Simulation shows that the proposed solution results and demonstrate the efficiency of the proposed solution in terms of bite error rate. The main advantage of present models is the simplicity of the formulas, which only uses power optimization.

References

1. Tao X, Xu X, Cui Q (2012) An overview of cooperative communications. IEEE Commun Mag 50(6):65–71. https://doi.org/10.1109/MCOM.2012.6211487
2. Lee Y (2013) End-to-end error-rate based incremental relaying for af cooperative communications. IEEE Commun Lett 17(9):1806–1809. https://doi.org/10.1109/LCOMM.2013.072313.131305

3. Singh SK, Agrawal K, Singh K, Li CP, Huang WJ (2020) On UAV selection and position-based throughput maximization in multi-UAV relaying networks. IEEE Access 8:144039–144050. https://doi.org/10.1109/ACCESS.2020.3014513
4. Zhong X, Xu K, Xu Y (2013) Comments on 'an opportunistic-based protocol for bidirectional cooperative networks.' IEEE Trans Wirel Commun 12(1):412–413. https://doi.org/10.1109/TWC.2012.112312.120495
5. Zhang S, Di B, Song L, Li Y (2017) Sub-channel and power allocation for non-orthogonal multiple access relay networks with amplify-and-forward protocol. IEEE Trans Wirel Commun 16(4):2249–2261. https://doi.org/10.1109/TWC.2017.2661281
6. Zeng Y, Zhang R, Lim TJ (2016) Throughput maximization for UAV-enabled. IEEE Trans Commun 64(12):4983–4996
7. Al-Tous H, Barhumi I (2015) Resource allocation for multiuser improved AF cooperative communication scheme. IEEE Trans Wirel Commun 14(7):3655–3672. https://doi.org/10.1109/TWC.2015.2409258
8. Al-Hourani A, Kandeepan S, Lardner S (2014) Optimal LAP altitude for maximum coverage. IEEE Wirel Commun Lett 3(6):569–572. https://doi.org/10.1109/LWC.2014.2342736
9. Bhatnagar MR, Arti MK (2013) Selection beamforming and combining in decode-and-forward MIMO relay networks. IEEE Commun Lett 17(8):1556–1559. https://doi.org/10.1109/LCOMM.2013.062113.130757
10. Lee J, Lee JH (2019) Performance analysis and resource allocation for cooperative D2D communication in cellular networks with multiple D2D pairs. IEEE Commun Lett 23(5):909–912. https://doi.org/10.1109/LCOMM.2019.2907252
11. Ku ML, Li W, Chen Y, Ray Liu KJ (2015) On energy harvesting gain and diversity analysis in cooperative communications. IEEE J Sel Areas Commun 33(12):2641–2657. https://doi.org/10.1109/JSAC.2015.2481215
12. Yin S, Zhao Y, Li L, Yu FR (2020) UAV-assisted cooperative communications with time-sharing information and power transfer. IEEE Trans Veh Technol 69(2):1554–1567. https://doi.org/10.1109/TVT.2019.2956167
13. Yang Z, Pan C, Shikh-Bahaei M, Xu W, Chen M, Elkashlan M, Nallanathan A (2018) Joint altitude, beamwidth, location, and bandwidth optimization for UAV-enabled communications. IEEE Commun Lett 22(8):1716–1719

AUC Analysis for Generalised Bessel K Fading Model

Manisha Bhatt, Sanjay Kumar Soni, and Sandeep Bhatt

Abstract An important metric for the performance analysis of energy detector is the area under the curve (AUC) pertaining to receiver operating characteristic (ROC). Energy detector is utilised for spectrum sensing in case of applications requiring cognitive radio. So in the present work, focus is laid upon deriving the analytical expression for AUC of Generalised Bessel K (GBK) fading channel. Due to flexibility offered by the GBK model in choice of its parameters, it gets converted into different fading distribution and hence in this context it is generalised in nature. The effect of different parameters on performance of a system undergoing GBK fading is also evaluated by varying them.

Keywords Generalised Bessel K fading channel · Receiver operating characteristic · Energy detector · Area under the receiver operating characteristic curve · Probability of detection

1 Introduction

Spectrum sensing has gained much attention of the researchers due to the wide popularity and advancement in the field of cognitive radio [1]. Although there are various methodology available for signal detection with their pros and cons, energy detector (ED)-based sensing method has gained a lot of attention among research community. The reason behind wide popularity of energy detection method is due to hardware simplicity and cost-effectiveness [2].

In case of an ED, decision about the presence and absence of an unknown signal is based upon the fact that, whether the measured energy level over an observation time is above or below a threshold value. Numerous study has already been done for conducting performance analysis of energy detector based upon its receiver operat-

M. Bhatt · S. Bhatt
G.B. Pant Institute of Engineering and Technology, Pauri Garhwal, Uttarakhand, India

S. K. Soni (✉)
M. M. M. U. T., Gorakhpur, Uttar Pradesh, India
e-mail: sksoni98fece@gbpec.ac.in

B. Mishra and M. Tiwari (eds.), *VLSI, Microwave and Wireless Technologies*, Lecture Notes in Electrical Engineering 877,
https://doi.org/10.1007/978-981-19-0312-0_55

ing characteristic (ROC) curves. In [3], probability of detection (PD) is derived for different fading scenarios such as Rayleigh, Rician, and Nakagami, while in [4] performance analysis of ED has been carried out for a cognitive radio network which is relay based. For the case of composite fading scenario where the effect of multipath and shadowing is taken in account, investigation of performance for ED has already been done in [5]. A rigorous study to evaluate both PD and area under the curve (AUC) of ROC, for the case of Rice-Gamma Fading model, has been conducted in [6]. Apart from ED performance analysis, ROC curve finds its application in various diverse scenarios such as drug testing and diagnostic tests in medical field [7].

In literature [8, 9], much emphasis has already been laid upon, for using ROC as a performance analysis metric. But there are some limitation associated with this metric. It became very difficult to compare the performance of energy detectors on the basis of visual examination of their ROC curve if they are crossing each other. So in such a scenarios, the advantage of AUC become quite significant as it evaluates area inside the ROC curve. And thus it helps in comparing the performance of energy detectors. Apart from this, it is a compact performance analysis metric and overall performance analysis in case of energy detector can be done with the help of AUC. Motivated by this fact, the present work focuses upon carrying out AUC analysis for Generalised Bessel K (GBK) fading channel as it is not available in literature to the best of author's knowledge. First of all, system model elaboration is done followed by mathematical derivation of analytical expression AUC. Finally, validation of the proposed analytical expression of AUC is done with the help of some simulation results.

2 System Model Elaboration

This section deals with system model elaboration for an energy detector undergoing GBK fading distribution. The working of an energy detector is based upon the fact that the received signal is measured over a particular time interval. Then this received signal is compared with a threshold value to find out whether unknown signal is present or absent. Energy detector is assumed to have two-state, one representing presence while other for absence of the observed signal. As already derived in [10], the mathematical formulae for calculating out probability of detection $P_d = Q_u(\sqrt{2\gamma}, \sqrt{\lambda})$ and probability of false alarm are $P_f = \frac{\Gamma(u,\lambda/2)}{\Gamma(u)}$.

As received signal exhibits random nature in case of propagation through a fading channel, average probability of detection is evaluated in such a case, which is mathematically evaluated as

$$\bar{P_d} = \int_0^\infty Q_u(\sqrt{2y}, \sqrt{\lambda}) f(y) dy \tag{1}$$

In present work, we are interested in carrying out the performance analysis of energy detector undergoing GBK fading distribution. As already derived in [11], GBK fading model is the product of two random variable following generalized gamma distribution. The instantaneous SNR, probability distribution function of GBK fading model, is mathematically written as [11]

$$f(y) = 2\lambda_1 \left(\frac{y}{\chi}\right)^{\frac{\lambda_1(m+n)}{2}} \frac{1}{y\Gamma(m)\Gamma(n)} K_{m-n}\left\{2\left(\frac{y}{\chi}\right)^{\lambda_1/2}\right\}, \tag{2}$$

the value of the parameters $\alpha = \frac{\bar{X}\Gamma(m))}{\Gamma(m+1/\lambda_1)}$, $\beta = \frac{\bar{Y}\Gamma(n)}{\Gamma(n+1/\lambda_1)}$, and χ is the product of α and β. Shape parameters are denoted with λ_1, m, and n, while scaling factors are represented as α and β. $K_n(.)$ stands for the modified Bessel function of order n and is of second kind [12]. GBK distribution reduces to various well-known fading model such as Gamma, K, exponential and Rayleigh to name a few, for different values of its shaping and scaling parameters as already mentioned in table in [11].

For simplifying the notation used in Eq. (2), it can be re-expressed as follows

$$f(y) = A_1(y)^{\frac{P\lambda_1}{2}-1} K_{m-n}\left\{a_1(y)^{\lambda_1/2}\right\}, \tag{3}$$

where $P = m + n$, $A_1 = \frac{2\lambda_1}{\chi^{P\lambda_1/2}\Gamma(m)\Gamma(n)}$, and $a_1 = \frac{2}{\chi^{\lambda_1/2}}$.

3 AUC Analysis

In literature, much emphasis has already been laid upon, for using ROC as a performance analysis metric. But there are some limitations associated with this metric. It become very difficult to compare the performance of energy detectors on the basis of visual examination of their ROC curve if they are crossing each other. So in such a scenarios, the advantage of AUC become quite significant as it evaluates area inside the ROC curve. And thus it helps in comparing the performance of energy detectors. Apart from this, it is a compact performance analysis metric, and overall performance analysis in case of energy detector can be done with the help of AUC. The average AUC of energy detector is evaluated as [13]

$$\overline{A} = \frac{1}{2^u \Gamma(u)} \int_0^\infty \lambda^{u-1} e^{-\frac{\lambda}{2}} \overline{P}_d(\lambda)\, d\lambda. \tag{4}$$

As already derived in [14], the average probability of detection in case of GBK fading model is given by

$$\bar{P}_d = \frac{A_1\lambda_1}{2\rho^{\frac{\lambda_1 P}{2}}(2\pi)^{\frac{\lambda_1-1}{2}}}\sum_{k=0}^{\infty}\lambda_1^{\frac{\lambda_1 P}{2}+k-0.5}\frac{\Gamma(u+k,\lambda/2)}{k!\Gamma(u+k)}$$
$$G^{2\ \lambda_1}_{\lambda_1\ 2}\left(\left\{\frac{a_1^2}{4}\left(\frac{\lambda_1}{\rho}\right)^{\lambda_1}\right\}\Bigg|^{\Delta(\lambda_1,1-\lambda_1-P-k)}_{\Delta(1,b_1),\Delta(1,b_2)}\right). \tag{5}$$

Substituting (5) in (4) yields

$$\overline{A} = \frac{1}{2^u\Gamma(u)}\int_0^{\infty}\lambda^{u-1}e^{-\frac{\lambda}{2}}\frac{A_1\lambda_1}{2\rho^{\frac{\lambda_1 P}{2}}(2\pi)^{\frac{\lambda_1-1}{2}}}\sum_{k=0}^{\infty}\lambda_1^{\frac{\lambda_1 P}{2}+k-0.5}\frac{\Gamma(u+k,\lambda/2)}{k!\Gamma(u+k)}$$
$$G^{2\ \lambda_1}_{\lambda_1\ 2}\left(\left\{\frac{a_1^2}{4}\left(\frac{\lambda_1}{\rho}\right)^{\lambda_1}\right\}\Bigg|^{\Delta(\lambda_1,1-\lambda_1-P-k)}_{\Delta(1,b_1),\Delta(1,b_2)}\right)d\lambda, \tag{6}$$

rearranging the terms in above equation results in

$$\overline{A} = \frac{1}{2^u\Gamma(u)}\frac{A_1\lambda_1}{2\rho^{\frac{\lambda_1 P}{2}}(2\pi)^{\frac{\lambda_1-1}{2}}}\sum_{k=0}^{\infty}\frac{\lambda_1^{\frac{\lambda_1 P}{2}+k-0.5}}{k!\Gamma(u+k)}$$
$$G^{2\ \lambda_1}_{\lambda_1\ 2}\left(\left\{\frac{a_1^2}{4}\left(\frac{\lambda_1}{\rho}\right)^{\lambda_1}\right\}\Bigg|^{\Delta(\lambda_1,1-\lambda_1-P-k)}_{\Delta(1,b_1),\Delta(1,b_2)}\right)\int_0^{\infty}\lambda^{u-1}e^{-\frac{\lambda}{2}}\Gamma(u+k,\lambda/2)d\lambda, \tag{7}$$

and with the aid of [15, eq. (6.455/1)] one can obtain

$$\overline{A} = \frac{1}{2^u\Gamma(u)}\frac{A_1\lambda_1}{2\rho^{\frac{\lambda_1 P}{2}}(2\pi)^{\frac{\lambda_1-1}{2}}}\sum_{k=0}^{\infty}\frac{\lambda_1^{\frac{\lambda_1 P}{2}+k-0.5}}{k!\Gamma(u+k)}$$
$$G^{2\ \lambda_1}_{\lambda_1\ 2}\left(\left\{\frac{a_1^2}{4}\left(\frac{\lambda_1}{\rho}\right)^{\lambda_1}\right\}\Bigg|^{\Delta(\lambda_1,1-\lambda_1-P-k)}_{\Delta(1,b_1),\Delta(1,b_2)}\right)\frac{\Gamma(2u+k)}{u!2^{2u+k}}{}_2F_1\left(1, 2u+k; u+1; \frac{1}{2}\right), \tag{8}$$

where ${}_2F_1(.)$ is the Gaussian hypergeometric function [12].

4 Simulation Results

Validation of the analytical expression of AUC, which has already been deduced previously, is done in the present section with the help of some simulation results. These simulated plots are also aided with some exact results which are computed by solving the equation numerically. The influence of varying the shaping parameter and scaling factor on system performance is also analysed. Figure 1 depicts the complementary AUC (CAUC) on y-axis and average SNR in decibel scale on x-axis. This plot portrays that with increase in the value of shape parameter m, CAUC curve shift downward hinting that decrease in CAUC value, which implies that detection capability improves.

Fig. 2 represents average AUC versus number of sample plot for the case of GBK distribution. This plot investigates, how the number of samples impacts the value of average AUC for different value of average SNR. The inconsistency in plot shows

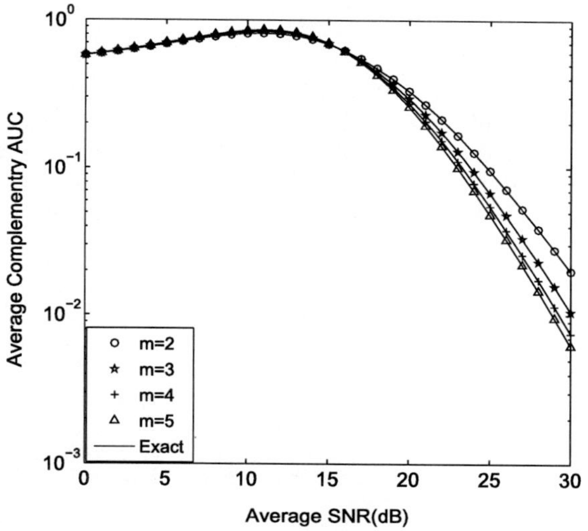

Fig. 1 Average complementary AUC versus average SNR over GBK fading model

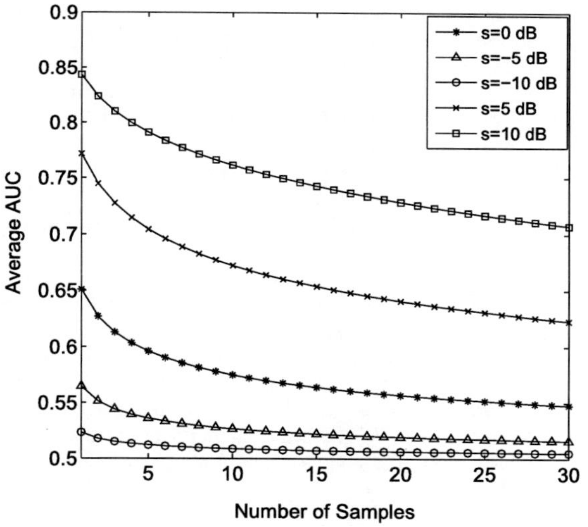

Fig. 2 Average AUC versus number of samples for GBK distribution

that with increase in the number of samples, there is decrease AUC value. It is due to the fact that rate of increase of false alarm probability is greater as compared to detection probability which leads to decrease in overall value of AUC.

5 Conclusion

AUC is an important performance analysis metric in case of energy detector. So in the present work, analytical expression of AUC is derived for GBK fading model. Influence of various parameters on AUC curve is also investigated through different simulation result. Finally, the impact of number of samples on the value of average AUC for different value of average SNR is also deduced.

References

1. Haykin S (2005) Cognitive radio: brain-empowered wireless communications. IEEE J Select Areas Commun 23(2):201–220
2. Cabric D, Mishra S, Brodersen R (n.d.) (2004) Implementation issues in spectrum sensing for cognitive radios. In: Conference record of the thirty-eighth asilomar conference on signals, systems and computers
3. Kostylev VI (2002) Energy detection of a signal with random amplitude. In: Proceedings of IEEE International Conference on Communications (ICC), April 2002, pp 1606–1610
4. Atapattu S, Tellambura C, Jiang H (2009) Relay based cooperative spectrum sensing in cognitive radio networks. In: Proceedings of IEEE Global Communications Conference (GLOBECOM), Nov–Dec 2009
5. Atapattu S, Tellambura C, Jiang H (2010) Performance of an energy detector over channels with both multipath fading and shadowing. IEEE Trans Wireless Commun 9(12):3662–3670
6. Peppas KP, Efthymoglou G, Aalo VA, Alwakeel M, Alwakeel S (2015) Energy detection of unknown signals in Gamma-shadowed Rician fading environments with diversity reception. IET Commun 9(2):196–210
7. Swets JA, Dawes RM, Monahan J (2000) Better decisions through science. Sci Am 283(4):82–87
8. Bhatt M, Soni SK (2018) Performance analysis of energy detection based cognitive radio over GBK fading channel. In: 2018 5th IEEE Uttar Pradesh section international conference on electrical, electronics and computer engineering (UPCON), Gorakhpur, pp 1–4
9. Bhatt M, Soni Sk (2018) Performance analysis of shadowed-Rice fading channel with cooperative spectrum sensing. Int J Electronics 105(11):1816–1830
10. Digham FF, Alouini MS, Simon MK (2007) On the energy detection of unknown signals over fading channels. IEEE Trans Commun 55(1):21–24
11. Shanker PM (2012) Fading and shadowing in wireless systems. Springer, New York, Dordrecht, Heidelberg, London
12. Gradshteyn IS, Ryzhik IM, Jeffrey A (1994) Table of integrals, series and products. Academic Press, San Diego
13. Al-Hmood H, Al-Raweshidy HS (2017) Unified modeling of composite $\kappa - \mu/Gamma$, $\eta - \mu/Gamma$, and $\alpha - \mu/Gamma$ fading channels using a mixture gamma distribution with applications to energy detection. IEEE Antennas Wireless Propag Lett 16:104–108
14. Bhatt M, Soni SK (2018) Performance analysis of energy detection based cognitive radio over GBK fading channel. In: 2018 5th IEEE Uttar Pradesh section international conference on electrical, electronics and computer engineering (UPCON), Gorakhpur, pp 1–4. https://doi.org/10.1109/UPCON.2018.8597095
15. Gradshteyn IS, Ryzhik IM (2009) Table of integrals, series, and products, 7th edn. Academic Press, Amsterdam

Analysis of Erroneous SIC and CSI on the Outage Performance of Cooperative NOMA Systems

P. Bachan, Aasheesh Shukla, and Atul Bansal

Abstract Due to the growing demand of transmission reliability and massive connectivity in 5G network cooperative non-orthogonal multiple access (C-NOMA) has been widely utilized for its superior spectral efficiency. In this work we investigate impact of erroneous successive interference cancelation (SIC) and channel state information (CSI) on the outage performance of cooperative half-duplex amplify and forward (HD-AF) relay-based downlink NOMA system. To evaluate the system performance we derive probability of outage expressions for pair of downlink users considering SIC and CSI errors. Simulation-based numerical analysis is presented to verify the analytical expressions of outage probability. Our results reveal that there is an existence of error floor in outage probability due to the impact of both interference cancelation error and CSI error.

Keywords Successive interference cancelation · Channel state information · Outage probability · Cooperative NOMA · Amplify and forward relay

1 Introduction

Recently with the increasing utilization of Internet of things (IoT)-based smart devices in 5G and beyond wireless networks, power domain non-orthogonal multiple access (PD-NOMA) scheme is becoming popular due to its superior throughput and spectral efficiency [1]. The main principle of PD-NOMA scheme allow the multiple users to share the same frequency channel by employing superposition coding (SC)

P. Bachan (✉) · A. Shukla · A. Bansal
Department of Electronics and Communication Engineering, G. L. A. University, Mathura 281406, India
e-mail: p.bachan@gla.ac.in

A. Shukla
e-mail: aasheesh.shukla@gla.ac.in

A. Bansal
e-mail: atul.bansal@gla.ac.in

B. Mishra and M. Tiwari (eds.), *VLSI, Microwave and Wireless Technologies*, Lecture Notes in Electrical Engineering 877,
https://doi.org/10.1007/978-981-19-0312-0_56

at the transmitter by allocating different power levels, whereas SIC strategy is utilized at the receiver to decode the information of different users [2]. Recently authors have investigated application of NOMA in cooperative relay system where the reliability of the system can be enhanced by exploiting the information of weaker users' message feature by encouraging strong users to relay the decoded message to weaker user [3, 4]. In a two-user cooperative NOMA system base station broadcasts the information to the user experiencing better channel conditions directly whereas the information of user experiencing poor channel conditions will be transmitted with the help of a dedicated relay [5].

Even though cooperative NOMA is very promising and spectrally efficient scheme but still authors have not fully explored the influence of CSI errors on the performance of such system. In many of the previous literatures [6–9] at the receiver authors have considered accurate channel state information. Furthermore successful SIC is very essential in all C-NOMA-based systems. The success of SIC in C-NOMA depends on various factors such as power allocation, CSI error [10, 11]. Error in CSI leads to loss to high SNR signal which can cause errors in interference cancelation.

In [12] authors have studied EH-based C-NOMA system and evaluated outage and sum rate performance while considering the accurate CSI/SIC. Exact and lowed bound analysis of outage probability with CSI error is investigated in [13] by employing AF relay assisted downlink transmission. In [14] authors have analyzed two-way relay NOMA with and without SIC errors whereas joint effect of SIC error and I/Q imbalance in C-NOMA system is investigated in [15]. A C-NOMA assisted cognitive radio network is considered in [16, 17] to evaluate system performance with SIC errors.

Motivated by the challenges presented in previous literature this work studies and analyzes the outage performance of a two-user HD-relay-based NOMA system in which realistic assumptions of erroneous SIC and CSI are considered. In this system we consider a pair of NOMA assisted downlink users where only near user is directly receiving the message from base station whereas the information to far user is transmitted from BS via a dedicated relay by employing amplify and forward strategy. For this model we derive the mathematical expressions of outage probability for both the users and analyze the joint influence of residual interference and channel estimation error on the outage performance.

The organization of this paper is as follows: Mathematical expressions for system model and SINR models are discussed in Sect. 2. In Sect. 3 we have provided the mathematical expressions for probability of outage for the same system and in the subsequent Sect. 4 the simulation results for outage performance analysis of the system model are presented. Finally, we conclude our work in Sect. 5.

2 System Model

We consider a two-user cooperative downlink NOMA system which includes one base station (BS), a dedicated relay (R), a near user (D_n) and a far user $\left(D_f\right)$ as shown in Fig. 1. Near user has direct link from BS but due to severe shadowing there is no direct link between BS and D_f. Nakagami-m fading is considered for all the links between any two nodes of this system with channel fading index (λ) and mean power (d^{-n}), where d is the distance between the nodes is and n is path-loss exponent.

Under practical wireless network scenario it is assumed that there will be channel reporting errors so on the basis of channel estimate $\overline{h}_i$, channel coefficient is modeled as $h_i = \overline{h}_i + e_i$, where error e_i is modeled as additive white Gaussian noise with zero mean and variance σ_e^2.

As shown in Fig. 1 downlink transmission from BS to D_f is divided into two phases. In phase-I the superimposed signal is transmitted from BS to D_n and relay node (R) which is given by,

$$S_B = \sqrt{aP_B}s_n + \sqrt{bP_B}s_f \tag{1}$$

where the BS transmit power is P_B and s_n and s_f are intended message signals for D_n and D_f, respectively. Power distribution coefficients a and b are selected on the basis of NOMA principle such that $a < b$ and $a + b = 1$.

The signals received at nodes D_n and R are given by,

$$r_{D_n} = S_B\left(\overline{h}_n + e_n\right) + n_{D_n} \tag{2}$$

$$r_R = S_B\left(\overline{h}_R + e_R\right) + n_R \tag{3}$$

where n_{D_n} and n_R represents the Additive White Gaussian noise at D_n and R, respectively.

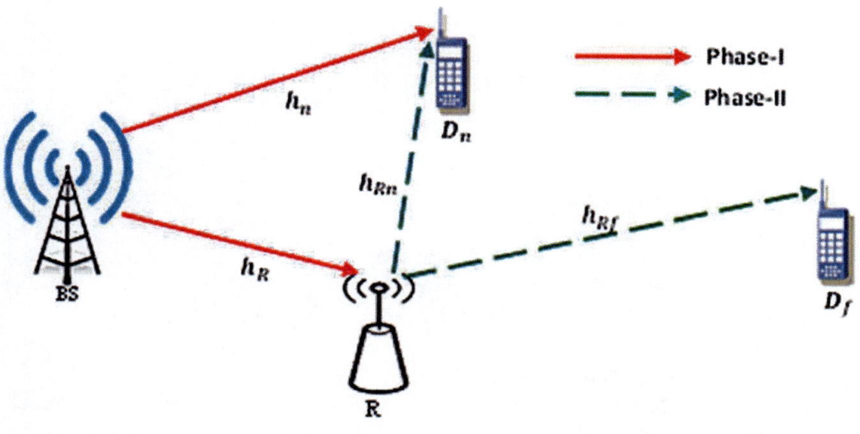

Fig. 1 AF relay assisted two-user downlink NOMA system

In the same phase firstly signal of D_f is decoded by D_n and then decodes its own signal by using successive interference cancelation strategy. The SINR for successful detection of s_f signal at D_n is given by,

$$g_{nf}^{n} = \frac{b\gamma\left|\overline{h_n}\right|^2}{a\gamma\left|\overline{h_n}\right|^2 + \gamma\mu_n d_n^{-n} + 1} \tag{4}$$

where $= \frac{P_B}{N_0}$, μ is CSI error index.

When D_n decodes its own signal then SIC error may be present due to the impact of residual interference signal from far user D_f which results in SINR given by,

$$g_{n}^{n} = \frac{a\gamma\left|\overline{h_n}\right|^2}{\delta_1 b\gamma\left|\overline{h_n}\right|^2 + \gamma\mu_n d_n^{-n} + 1} \tag{5}$$

where δ_1 is the residual interference signal level and $0 \leq \delta_1 \leq 1$.

In phase-II AF relay amplifies the signal received in phase-I and relays it to both D_n and D_f. The received signals at D_n and D_f are given by,

$$r_{RD_n} = \left[S_B\left(\overline{h}_R + e_R\right) + n_R\right]\alpha\left(\overline{h}_{Rn} + e_{Rn}\right) + n_{D_n} \tag{6}$$

$$r_{RD_f} = \left[S_B\left(\overline{h}_R + e_R\right) + n_R\right]\alpha\left(\overline{h}_{Rf} + e_{Rf}\right) + n_{D_f} \tag{7}$$

where n_{D_n} and n_{D_f} denote AWGN at D_n and D_f, respectively, α is AF relay gain coefficient and is denoted as, $\alpha = P_B\sigma_e^2 + \frac{P_R}{P_B\mathbb{E}|\mathrm{h_R}|^2} + N_0$.

For far user SIC cannot be performed so SINR for detection of s_f at D_f is given by,

$$g_{f}^{f} = \frac{b\gamma\left|\overline{h_R}\right|^2\left|\overline{h_{Rf}}\right|^2}{a\gamma\left|\overline{h_R}\right|^2\left|\overline{h_{Rf}}\right|^2 + \Omega_1\left|\overline{h_{Rf}}\right|^2 + \Omega_2\left|\overline{h_R}\right|^2 + \Omega_3} \tag{8}$$

where $\Omega_1 = \gamma\mu_R d_R^{-n} + 1$, $\Omega_2 = \gamma\mu_{Rf} d_{Rf}^{-n}$, and $\Omega_3 = \frac{\Omega_2}{\gamma}[\Omega_1 + 1] + \frac{1}{\alpha^2}$.

By employing SIC, D_n can also decode s_f for which SINR is given by,

$$g_{nf}^{f} = \frac{b\gamma\left|\overline{h_R}\right|^2\left|\overline{h_{Rn}}\right|^2}{a\gamma\left|\overline{h_R}\right|^2\left|\overline{h_{Rn}}\right|^2 + \Omega_1\left|\overline{h_{Rn}}\right|^2 + \Omega_4\left|\overline{h_R}\right|^2 + \Omega_5} \tag{9}$$

where $\Omega_4 = \gamma\mu_{Rn} d_{Rn}^{-n}$, and $\Omega_5 = \frac{\Omega_4}{\gamma}[\Omega_1 + 1] + \frac{1}{\alpha^2}$.

Finally after detection of s_f, near user D_n will decode its own signal for which SINR with the residual interference signal level δ_2 is given by,

$$g_n^f = \frac{a\gamma\left|\overline{h_R}\right|^2\left|\overline{h_{Rn}}\right|^2}{\delta_2 b\gamma\left|\overline{h_R}\right|^2\left|\overline{h_{Rn}}\right|^2 + \Omega_1\left|\overline{h_{Rn}}\right|^2 + \Omega_4\left|\overline{h_R}\right|^2 + \Omega_5} \tag{10}$$

3 Probability of Outage Analysis

In this section we provide the mathematical analysis of Probability of outage in cooperative non-orthogonal multiple access (NOMA) system considering the joint impact of erroneous SIC and CSI. To begin, let us first define the outage events that may take place at nodes D_n and D_f:

O_1: When D_n fails to detect s_f or D_n is not able to detect s_n after SIC process during phase-I and phase-II.
O_2: In phase-II, D_f fails to detect its own signal s_f.

Let R_n, R_f denote the desired data rates for near user D_n and far user D_f, respectively. Then to satisfy the QoS requirement the target SINR thresholds for both the users must be satisfied which are $g_{thi} = 2^{2R_i} - 1$, where $i\epsilon\{n, f\}$.

So the outage probability of near user in case of an event O_1 is given as,

$$\begin{aligned} PO_{D_n} &= \left[1 - P_r\left(g_{nf}^n > g_{thf}, g_n^n > g_{thn}\right)\right] \\ &\times \left[1 - P_r\left(g_{nf}^f > g_{thf}, g_n^f > g_{thn}\right)\right] \end{aligned} \tag{11}$$

By combining and substituting (4), (5), (9) and (10) in (11) and with mathematical manipulations PO_{D_n} is expressed as,

$$\begin{aligned} PO_{D_n} = C.\Bigg\{1 - \frac{2q_1^{\lambda}}{\Gamma(\lambda)}\sum_{l_1=0}^{\lambda-1}\sum_{l_2=0}^{\lambda-1}\sum_{l_3=0}^{l_2}\frac{1}{l_2!}\binom{\lambda-1}{l_1}\binom{l_2}{l_3} \\ \times\, \varepsilon_2^{\lambda-1-l_1} e^{-q_1\varepsilon_2-u_2} u_1^{\frac{l_1+l_3+1}{2}} u_2^{l_2-l_3} q_1^{-\frac{l_1-l_3+1}{2}}\Bigg\} \end{aligned} \tag{12}$$

where $C = \left\{1 - e^{-\frac{\varepsilon_1\lambda d_n^n}{1-\mu_n}}\sum_{i=0}^{\lambda-1}\frac{1}{i!}\left(\frac{\varepsilon_1\lambda d_n^n}{1-\mu_n}\right)^i\right\}$, with $\varepsilon_1 = \max(\rho_1, \xi_1)$, $\varepsilon_2 = \max(\rho_2, \xi_2)$ $\rho_1 = \frac{(\gamma\mu_n d_n^{-n}+1)g_{thn}}{\gamma(b-ag_{thf})}$, $\xi_1 = \frac{(\gamma\mu_n d_n^{-n}+1)g_{thn}}{\gamma(a-\delta_1 bg_{thn})}$, $u_1 = \frac{q_2 g_{thf}(\Omega_1\rho_2+\Omega_5)}{\gamma(b-ag_{thf})}$, $u_2 = \frac{q_2 g_{thf}\Omega_1}{\gamma(b-ag_{thf})}$ $q_1 = \frac{\lambda d_n^n}{1-\mu_n}$, $q_2 = \frac{\lambda d_R^n}{1-\mu_R}$, $\rho_2 = \frac{g_{thf}\Omega_4}{\gamma(b-ag_{thf})}$, $\xi_2 = \frac{g_{thn}\Omega_4}{\gamma(a-\delta_2 bg_{thn})}$.

Similarly, the outage probability of far user in case of an event O_2 is given as,

$$PO_{D_f} = \left[1 - P_r\left(g_f^f > g_{thf}\right)\right] \tag{13}$$

By substituting (8) into (13) and with mathematical manipulations PO_{D_f} is expressed as,

$$PO_{D_f} = \left\{1 - \frac{2q_3^{\lambda}}{\Gamma(\lambda)} \sum_{m_1=0}^{\lambda-1} \sum_{m_2=0}^{\lambda-1} \sum_{m_3=0}^{m_2} \frac{1}{m_2!} \binom{\lambda-1}{m_1} \binom{m_2}{m_3} \right. \\ \left. \times \rho_3^{\lambda-1-m_1} e^{-q_3\rho_3 - u_2} u_3^{\frac{m_1+m_3+1}{2}} u_2^{m_2-m_3} q_3^{-\frac{m_1-m_3+1}{2}} \right\} \tag{14}$$

where $\rho_3 = \frac{g_{thf}\Omega_2}{\gamma(b-ag_{thf})}$, $q_3 = \frac{\lambda d_{Rf}^n}{1-\mu_{Rf}}$, $u_3 = \frac{q_2 g_{thf}(\Omega_1\rho_3+\Omega_3)}{\gamma(b-ag_{thf})}$.

4 Simulation Results

In this section we present the simulation-based numerical analysis to evaluate the joint influence of erroneous CSI and SIC due to residual self-interference on the outage probability of HD-AF relay-based NOMA system. These numerical results will validate the probability of outage expressions of both near and far user presented in Sect. 3. Also we have compared outage probability results of this system with conventional OMA system. The key system parameters for AF relay assisted two-user downlink cooperative NOMA system is provided in Table 1.

In Fig. 2 we present the simulation results of outage probability for both near user and far user w.r.t. transmission SNR considering the joint impact of SIC error $\delta = 0.01$ and fading index $\lambda = \{1, 2\}$. We may observe that for medium and high SNR regime the simulation-based outage probability results for both the users have a perfect match with the results based on analytical expressions presented in Sect. 3. From figure it may also be seen that when λ is increased from 1 to 2, the outage performance is significantly improved but presence of SIC error in the system have a negative impact on the outage performance of both the users.

Figure 3 presents the simulation results of outage probability for both near user and far user w.r.t. CSI error index (μ) considering the effect of fading index $\lambda = \{1, 2\}$

Table 1 Parameters for numerical simulations

System parameters	Numerical values
Fixed gain of AF relay (α)	0.8
Target date rates (R_n, R_f)	1.5, 1
Power distribution coefficients (a, b)	0.15, 0.85
Base station power (P_B),	1 W
Relay power (P_R)	0.5 W
Direct link normalized distances (d_R, d_n)	0.4, 0.5
Hop link normalized distances (d_{Rn}, d_{Rf})	0.4, 0.8
Path los exponent (n)	2

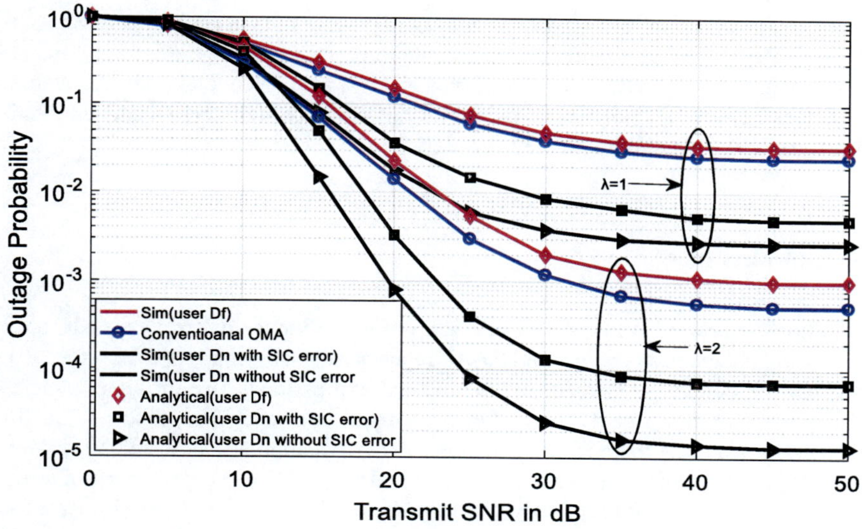

Fig. 2 Outage probability of D_n and D_f versus transmit SNR considering SIC error (δ) and fading index (λ)

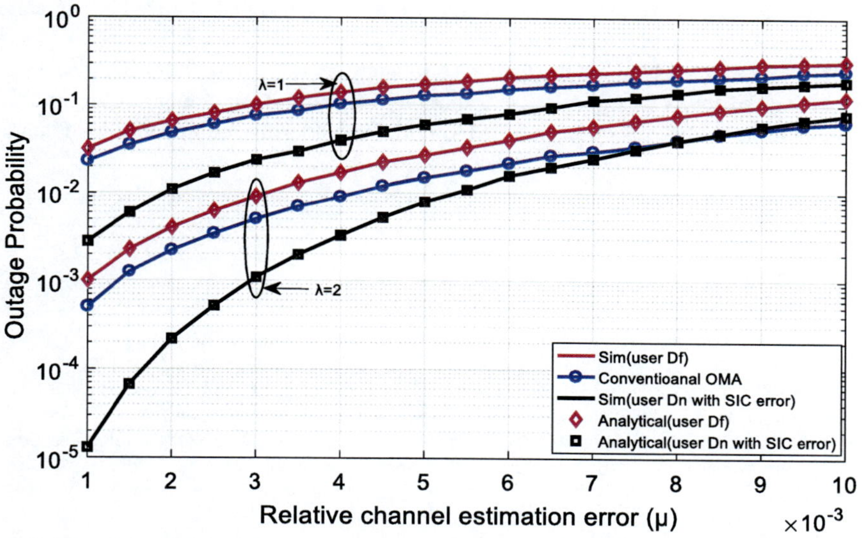

Fig. 3 Outage probability of D_n and D_f versus CSI error index (μ) considering fading index (λ)

with fixed transmit SNR of 40 dB. We may observe that outage performance of both the users degrades with the increase in estimation error but increase in fading index λ has positive effect on the outage performance. It is also evident that once the value of μ goes beyond 0.0083, outage probability of near user falls below conventional OMA. So CSI error should not be very high.

5 Conclusions

The aim of this work was to study and analyze the influence of erroneous SIC and CSI on the outage performance of HD-AF relay-based downlink NOMA system. To evaluate the outage performance mathematical expressions were presented where realistic assumptions of SIC and CSI errors for pair of downlink users were considered. To validate the mathematical analysis extensive simulations were carried out and it is observed that in the presence of residual IS ($\delta = 0.01$) outage performance for both the users is improved by increasing the multipath fading index (λ). Results of outage probability versus channel error (μ) shows that as μ increases outage performance of both the user degrades but overall the performance of near user in NOMA system outperforms the OMA user. Furthermore it has also been witnessed that in large SNR regime impact of residual interference signal and CSI errors was significant on the outage performance.

In future this study can be extended to full-duplex relay based NOMA systems. Also adaptive power allocation in combination with optimal relay selection algorithm may be investigated in two-user NOMA model considering the impact of SIC and CSI errors.

References

1. Liaqat M, Noordin KA, Abdul Latef T et al (2020) Power-domain non orthogonal multiple access (PD-NOMA) in cooperative networks: an overview. Wireless Netw 26:181–203
2. Islam SM, Zeng M, Dobre OA, Kwak K-S (2019) Nonorthogonal multiple access (NOMA): how it meets 5g and beyond. Wiley 5G Ref: the essential 5G reference, pp 1–2
3. Kim Y-B, Yamazaki K, Jung BC (2019) Virtual full-duplex cooperative NOMA: relay selection and interference cancellation. IEEE Trans Wireless Commun 18(12):5882–5893
4. Yang Z, Ding Z, Wu Y, Fan P (2017) Novel relay selection strategies for cooperative NOMA. IEEE Trans Veh Technol 66(11):10114–10123
5. Zhang J, Ge J, Ni Q, Wen M, Zhang Y (2018) Performance analysis of user ordering schemes in cooperative power-domain non-orthogonal multiple access network. IEEE Access 6: 47319–47331
6. Li Y, Li Y, Chen Y, Ye Y, Zhang H (2018) Performance analysis of cooperative NOMA with a shared AF relay. IET Commun 12(19):2438–2447
7. Kara F, Kaya H (2019) On the error performance of cooperative-NOMA with statistical CSIT. IEEE Commun Lett 23(1):128–131
8. Xu Y, Wang G, Zheng L, Jia S (2018) Performance of NOMA-based coordinated direct and relay transmission using dynamic scheme. IET Commun 12(18):2231–2242

9. Chu TMC, Zepernick H-J (2018) Performance of a non-orthogonal multiple access system with full-duplex relaying. IEEE Commun Lett 22(10):2084–2087
10. Manglayev T, Hee Kho Y, Kizilirmak R (2016) Optimum power allocation for non-orthogonal multiple access. In: IEEE 10th international conference on application of information and communication technologies (AICT), Baku, pp 1–5
11. Kara F, Kaya H (2020) Improved user fairness in decode-forward relaying non-orthogonal multiple access schemes with imperfect SIC and CSI. IEEE Access 8:97540–97556
12. Agrawal K, Flanagan MF, Prakriya S (2020) NOMA with battery-assisted energy harvesting full-duplex relay. IEEE Trans Veh Technol 69(11):13952–13957
13. Arzykulov S, Tsiftsis TA, Nauryzbayev G, Abdallah M (2019) Outage performance of cooperative underlay CR-NOMA with imperfect CSI. IEEE Commun Lett 23(1):176–179
14. Tian X, Li Q, Li X, Peng H, Zhang C, Rabie KM et al (2020) I/Q imbalance and imperfect SIC on two-way relay NOMA systems. Electronics 9(2):1–16
15. Li X, Liu M, Deng C, Mathiopoulos PT, Ding Z, Liu Y (2020) Full-duplex cooperative NOMA relaying systems with I/Q imbalance and imperfect SIC. IEEE Wirel Commun Lett 9(1):17–20
16. Im G, Lee JH (2019) Outage probability for cooperative NOMA systems with imperfect SIC in cognitive radio networks. IEEE Commun Lett 23(4):692–695
17. Do D, Le A, Lee BM (2020) NOMA in cooperative underlay cognitive radio networks under imperfect SIC. IEEE Access 8:86180–86195

Study the Impact of ZrO_2 High-k Dielectrics Gate Material on FD-SOI and PD-SOI MOSFET

Astha Srivastava, Narendra Yadava, Mangal Deep Gupta, and R. K. Chauhan

Abstract To overcome the problem of short-channel effect (SCE) and leakage current, a new device such as silicon-on-insulator (SOI) MOSFET comes into the picture. This paper studies the impact of ZrO_2 high-k dielectrics gate material on the performance of fully depleted (FD) SOI MOSFET and partially depleted (PD) SOI MOSFET. To simulate the proposed structure, the Silvaco TCAD tool has been used. We find that the better performance of FD-SOI MOSFET as compare to PD-SOI MOSFET in terms of on-state driving current and reduced off-state leakage current with better amplification in terms of transconductance.

Keywords Silicon-on-insulator SOI · PD-SOI MOSFET · FD-SOI MOSFET · High-k dielectrics · SOI floating body · Fringing electric field

1 Introduction

Today, with each day, microelectronics technical word is converting with the more advanced level that is of nanoelectronics. These devices are scaled more to meet the requirement of the growing demand of fabrication while maintaining the device performance factor in terms of gain, linearity, and reliability [1]. Reduced-size effects are becoming a limitation for the development of MOS devices [2–4]. Thus, FD-SOI MOSFET which electrical characteristics and its short-channel effect make a more considerable device [2, 5–9] to reduce the short-channel effect in SOI MOSFET as it degraded the performance of the device. To get a solution to this problem, a new technology came to light, i.e., silicon-on-insulator (SOI) [10]. To make the device more reliable and to get a better result in terms of lesser leakage current, a lightly doped drain structure is used nowadays to reduce the lateral drain electrical field [11].

A. Srivastava (✉) · N. Yadava · M. D. Gupta · R. K. Chauhan
Department of Electronics and Communication, Madan Mohan Malviya University of Technology, Gorakhpur, Uttar Pradesh, India
e-mail: suchi.astha1@gmail.com

B. Mishra and M. Tiwari (eds.), *VLSI, Microwave and Wireless Technologies*, Lecture Notes in Electrical Engineering 877,
https://doi.org/10.1007/978-981-19-0312-0_57

Today, in CMOS technology, using different dielectrics gives results in a decrement in terms of leakage current (I_{off}) and increment in ON-state current, i.e., driving current [12–14].

It is a fabrication technology of silicon semiconductor devices in a layered between silicon insulator and silicon substrate to decrease the parasitic capacitance [20] as for the improvement of device performance. In other words, we can also say that this fabrication technology or process gives the advantages of achieving higher performance and utilizes low power when compared with other conventional transistors [23].

As per the limitations to the CMOS bulk process design because of scaling. SOI technology is needed as its solution. SOI technology provides the facility of removing latch as no parasitic bipolar device due to SOI layer. As SOI technology provides following facilities i.e., firstly provide the facility of removing latch as no parasitic bipolar device due to SOI layer and secondly because of insulation layer above the substrate which helps the device to have smaller leakage current. SOI technology also provides high operational speed because of the presence of low capacitance between the device and substrate. Power dissipation is smaller as it operates at the smaller voltage and current level [24]·

A gate dielectric with a dielectric constant (k) is more than SiO_2 (k_{OX}) will have a smaller thickness value when compared with SiO_2 with larger physical thickness. Therefore, we can define equivalent thickness by the given formula, [26, 27]

$$T_{eq} = (k_{ox}/k)t_{phys.} \tag{1}$$

Thermal stability is one of the important factors to be considered as the high temperature is required to activate dopant in the source, drain, and polysilicon gate [27]. A more recent trend is focused on high-dielectric constant gate insulator's primary metal oxides. (eg—Ta_2O_5, TiO_2, ZrO_2, HfO_2, Y_2O_5, La_2O_3, Al_2O_3, and Gd_2O_3) [27].

The advantage of using a very high-k dielectric constant material is just to replace SiO_2 having a similar thickness which is limited due to the presence of two-dimensional electric fringing fields [29–30].

As in the case of SOI, we know that source, body, and a drain region of the device are insulated from the substrate. The body of the device is left unconnected and the outcome we get in form of a floating body. This floating body effect can charge/discharge freely due to the condition of switching which affects the threshold voltage (V_t) with other characteristics of the device [23].

SOI MOSFET is of having a similar structure to that bulk CMOS. The only difference between them is insulation layer is inserted underneath the device of the silicon substrate. Depending on the silicon thickness, SOI device are classified into two types - Partially depleted (PDSOI) and Fully depleted (FDSOI) [25].

In this PD-SOI MOSFET, here, keeping the thickness of the SOI layer is more when compared with the depletion width of the gate. This device also provides the same performance at low power and gives a better performance gain of 20–40%

when the device is compared with bulk CMOS technology [31], but the major issue with PD-SOI is the floating body effect.

Now, in the case of fully depleted SOI MOSFET, the thickness of the silicon layer is kept very less, and therefore, channel becomes fully depleted by the majority of carriers. Such that SOI layer becomes very thin when compared with the depletion width of the device [32].

In this paper, we compare both FD-SOI MOSFET and PD-SOI MOSFET using ZrO_2 high-k dielectric material to compare their performance in form of transfer characteristics, transconductance and leakage current.

2 Device Structure

For the analysis of the effect observed using different high-k dielectrics on SOI MOSFET, here, we are using schematic cross-section view of the SOI MOSFET is simulated using ATLAS TCAD device simulator shown in Figs. 1 and 2, respectively. Here, we are using the following data for the fabrication of the device, i.e., lightly doped concentration (1×10^{-14} cm^{-3}) to ignore threshold voltage variation and carrier mobility. Doping concentration for source/drain region is kept at (1×10^{-18} cm^{-3}). Gate length structure for both the device (i.e., FD-SOI MOSFET and PD-SOI MOSFET) is kept to be 20 nm; silicon film thickness is 10 nm; gate oxide thickness (SiO_2) is 0.3 nm, and box thickness is 25 nm, respectively. Here, we are considering the N-channel device for the simulation at metal gate work function of 4.5 eV (Work function of TiN) of SOI MOSFET. High-k dielectrics, i.e., ZrO_2 (*k*

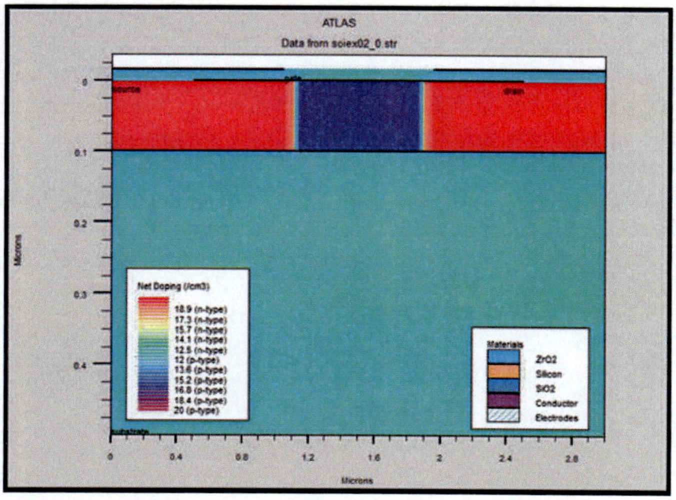

Fig. 1 Shows simulated structure of PD-SOI MOSFET using high-k dielectrics ZrO_2

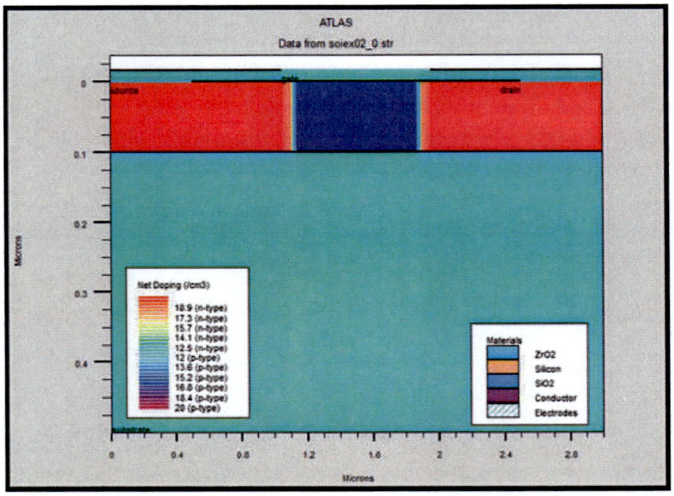

Fig. 2 Shows simulated structure of FD-SOI MOSFET using high-k dielectrics ZrO_2

= 25) are being used in this paper to evaluate its performance effect on both the structure and to compare both the device performance.

3 Simulation Results and Discussions

Deckbuild is used to design and simulate an FD-SOI MOSFET structure in this analysis. The Tonyplot displays the simulation result of the SOI MOSFET structure using ZrO2 high-k dielectrics material. The structure of the FD-SOI MOSFET is shown below and can be modified by selecting different in Tonyplot with contoured features. The color plot is called contour. The color plotting of impurity values is known as contours. Shown in Figs. 1 and 2.

It is very important for the device regarding the current capacity parameter which is also defined as the output characteristics of given device. Given below display the variation of drain current with respect to drain to source voltage in PD-SOI with FD-SOI MOSFET.

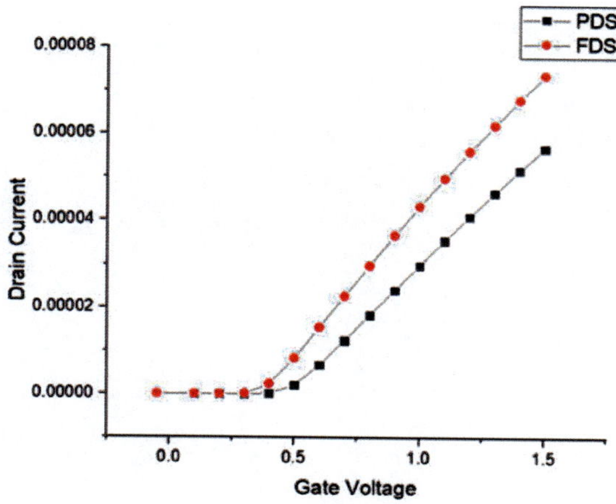

Fig. 3 Shows the comparative analysis of PD-SOI MOSFET with FD-SOI MOSFET

3.1 Characteristics Plot Between Drain Current (I_D) and Gate-Source Voltage (V_{GS})

Through Silvaco TCAD, we get a Tonyplot for both FD-SOI and PD-SOI MOSFET using dielectrics ZrO_2, Shown below in Fig. 3 which displays the transfer characteristics comparison of PD-SOI MOSFET and FD-SOI MOSFET in linear scale.

When both the transfer characteristics plot compares with each other, we obtained result in delayed in threshold voltage and because the floating body effect comes into the picture. To remove this effect, FD-SOI MOSFET is a better option in comparison with PD-SOI MOSFET.

In above, Fig. 4 gives the comparative analysis of transconductance (gm) for PD-SOI MOSFET and FD-SOI MOSFET. here using Eq. 2 for the calculation of transconductance in both PD-SOI and FD-SOI MOSFET,

$$gm = \partial I_D/\partial V_{GS} \quad (2)$$

on comparison of gm as shown in Fig. 4, we get a result of better gain in FD-SOI with PD-SOI MOSFET which means the FD-SOI gives better performance in terms of amplification when compared with PD-SOI MOSFET.

4 Conclusion

In this paper, see the impact of ZrO2 dielectrics gate material on FD-SOI and PD-SOI MOSFET. Through which, we can conclude that the switching effect is better

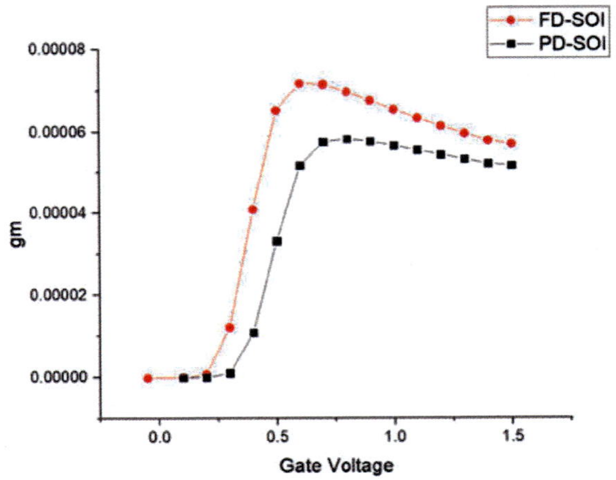

Fig. 4 Shows the comparative analysis of transconductance of PD-SOI MOSFET with FD-SOI MOSFET

in FD-SOI devices than in PD-SOI. Secondly, the leakage current is more in FD-SOI when compared with PD-SOI, i.e., OFF-state current (IOFF) is 1.01×10^{-10}A and 1.1×10^{-10}A corresponding FD and PD-SOI MOSFET and also obtained better amplification in FD-SOI with PD-SOI MOSFET in terms of transconductance (gm).

References

1. Ghosh S (2014) Modeling the performance of short-channel fully depleted silicon-on-insulator (SOI) metal oxide semiconductor field effect transistors (MOSFETs). Silicon-On-Insulator (SOI) Technol 77–101
2. El Dirani H, Fonteneau P, Solaro Y, Legrand C-A, Marin-Cudraz D, Ferrari P, Cristoloveanu S (2017) Sharp- switching band-modulation back-gated devices in advance FDSOI technology. Solid-State Electron 128:180
3. Shin M, Shi M, Mouis M, Cros A, Josse E, Kim GT, Ghibaudo G (2015) In depth characterization of electron transport in 14nm FD-SOI CMOS devices. Solid-State Electron 112:13–18
4. Xie Q, Xu J, Taur Y (2012) Review and critique of analytic models of MOSFET short-channel effects in subthreshold. IEEE Trans Electron Devices 59:1569
5. Coquand R, Barraud S, Casse M, Leroux P, Vizioz C, Comboroure C, Perreau P, Ernst E, Samson MP, Maffini-Alvaro V, Tabone C, Barnola S, Munteanu D, Ghibaudo G, Monfray S, Boeuf F, Poiroux T (2013) Scaling of high-k/metal-gate Trigate SOI nanowire transistor down to 10 nm width. Solid State Electron 88:32–36
6. Makovejev S, Planes N, Haonda M, Flandre D, Raskin JP, Kilchytska V (2015) Comparative study of parasitic elements on RF FoM in FD SOI and bulk technologies. Solid-State Electron
7. El Dirani H, Solaro Y, Fonteneau P, Legrand CA, Marin-Cudraz D, Golanski D, Ferrari P, Cristoloveanu S (2016) A sharp-switching gateless device (Z3-FET) in advanced FDSOI technology. Solid-State Electron 125:103
8. Morvan S, Andrieu F, Barbe JC, Ghibaudo G (2014) Study of an embedded buried SiGe structure as a mobility booster for fully—depleted SOI MOSFETs at the 10nm node, Solid-State Electron 98:50

9. Meel K, Gopal R, Bhatnagar D (2011) Three-dimensional analytic modelling of front and back gate threshold voltages for small geometry fully depleted SOI MOSFET's. Solid-State Electron 62:174
10. Choi YK, Asano K, Lindert N, Subramanian V, King T-J, Bokor J, Hu C (2000) Ultrathin-body SOI MOSFET for deep-sub -tenth micron era. IEEE Electron Device Lett 21(5):254
11. Ma M-W, Wu C-H, Yang T-Y, Kao K-H, Wu W-C, Wang S-J, Chao T-S. Lei T-F (2007) Impact of high-k offset spacer in 65nm node SOI devices. IEEE Electron Device Lett 28(3)
12. Tsuchiya R, Ohnishi K, Horiuchi M, Tsujikawa S, Shimamoto Y, Inada N, Yugami J, Ootsuka F, Onani T (2002) Femto-second CMOS technology with High-k offset spacer and SiN gate dielectric with oxygen enriched interface. VLSI Symp Tech Dig, Honolulu, HI, pp 150–151
13. Xiong Z, Liu H, Zhu C, Sin JKO (2004) Characteristics of high-k spacer offset-gated polysilicon TFTs. IEEE Trans Electron Devices 51(8):1304–1308
14. Kencke DL, Chen W, Wang H, Mudanai S, Ouyang Q, Tasch A, Banerjee SK (1999) Source-side barrier effects with very high-k dielectrics in 50nm Si MOSFETs. In: Proceedings DRC Digging, pp 22–23
15. Sentaurus Structure Editor User's Manual, Synopsys International
16. Husan S, Hameed S, Salehuddin F, Roslan AF, Maziah Mnin A, Mohd Duad N, Afifah Maheran AH, Zain ASM, Kaharudin KE (2019) Analysis of SOI PD-MOSFET device using different gate spacer materials, ResearchGate, August 2019
17. Rajneesh S, Rituraj SR, Ashwani KR (2018) Impact of high-k spacer on device performance of nanoscale underlap fully depleted SOI MOSFET. J Circuits Syst Comput
18. Kale S, Banchhor S, Kondekar P (2015) Performance study of high-k gate and spacer dielectrics Dopant segregated Schottky Barrier SOI MOSFET. In: Proceedings of international conference of electronics and communication system
19. Ke D-M, Wu D, Meng J, Yang F, Wan L-X, Yang J-G, Chang H (2018) Subthreshold characteristics analysis and modeling of fully depleted silicon-on- insulator MOSFETs with high-k SiO_2 stacked gate structure. Japanese J Appl Phys 57
20. Ranka D, Rana AK, Yadav RK, Giri D (2011) Performance analysis of FD-SOI MOSFET with different gate spacer dielectric. Int J Comput Appl 18(5)
21. Cellar K, Cristoloveanu S (2003) Frontiers of silicon-on-insulator. J Appl Phys 93(9):4955C
22. https://en.m.wikipedia.org/wiki/silicon-on-insulator
23. Vandana B (2013) A theoretical study of low power SOI technology. IOSR J VLSI Signal Process 2:30–37
24. Katti G, Das Gupta N, Das Gupta A (2004) IEEE T Electron Dev 51(7):1169
25. Rahou FZ, Guen Bouazza A, Rahou M, Global J (2013) Res Eng Electrical Electron Eng 13(1), Version 1.0
26. Buchen DA (1999) Scaling the gate dielectric: material, integration and reliability. IBM J Res Dev 43:245–264
27. Gusev EP, Lu HC, Garfunkel EL, Gustafsson T, Green ML (1999) Growth and characterization of ultrathin nitride silicon dioxide films. IBM J Res & Dev 43:265–286
28. Colinge J-P (2004) Multiple-gate SOI MOSFETs, solid-state electronics
29. Frank D, Dennard R, Nowak E, Solomon P, Taus Y, Wong HS (2001) Device scaling limits of Si MOSFETs and their application dependencies, proc. IEEE 89:259–288
30. Frank D, Wong HSP (2000) Analysis of the design space available for high-k gate dielectric in nanoscale MOSFETs. In: Proceeding of the IEEE silicon nanoelectronics workshop. pp 47–48
31. Shahidi GG et al (1999) In: IEEE International solid-state circuit conference, vol 426
32. Goel N, Tripathi A (2012) Int J Comput Appl 42(21):975
33. Mishra VK, Chauhan RK (2017) chapter-37 performance analysis of fully depleted SOI tapered body reduced source (FD-SOI TBRS) Mosfet for low power, digital application. Springer Science and Business Media LLC 2017
34. Essuir.sumdu.edu.ua.

Permanent Magnet Synchronous Motor (PMSM) Drive Using Multi-Objective Genetic Algorithm (MOGA) Technique

Deepti Yadav, Arunima Verma, and Frank Tittel

Abstract In this paper the performance of a permanent magnet synchronous motor (PMSM) drive is improved by using multi-objective genetic algorithm (MOGA) technique. Improve the performance of the PMSM by minimizing the speed and torque ripples that cause noise and vibrations. The performance is further enhanced by improving the transient response specifications. This work has provided an insight into the incorporation of the MOGA technique for tuning of a speed controller in the PMSM drive model.

Keywords Permanent magnet synchronous motor (PMSM) · Ziegler–Nichols (Z-N) · Multi-objective genetic algorithm (MOGA)

1 Introduction

Permanent magnet synchronous motor (PMSM) drive is particularly used in robotics, adjustable speed drives, and applications in electric vehicles. This is due to the ease of care, simple structure, and high efficiency [1–3]. The nonlinear behavior, resulting largely from load characteristics, motor dynamics and the uncertainties present in it make their control an exceptionally difficult task. Therefore, approach of speed controller should be effective and robust for industrial applications.

PID controller gains have been tuned using the conventional Zeigler–Nichols (Z-N) method [4]. The main criterion for tuning and designing of the speed controller is minimization of an overshoot, the peak time, the rise time, and the settling time. Algorithm using the Z-N method was developed to obtain optimized PID speed

D. Yadav (✉) · F. Tittel
Department of Electrical and Computer Engineering, Rice University, Houston, TX, USA
e-mail: deepti.yadav00@gmail.com

F. Tittel
e-mail: fkt@rice.edu

A. Verma
Department of Electrical Engineering, Institute of Engineering and Technology, Lucknow, India

B. Mishra and M. Tiwari (eds.), *VLSI, Microwave and Wireless Technologies*,
Lecture Notes in Electrical Engineering 877,
https://doi.org/10.1007/978-981-19-0312-0_58

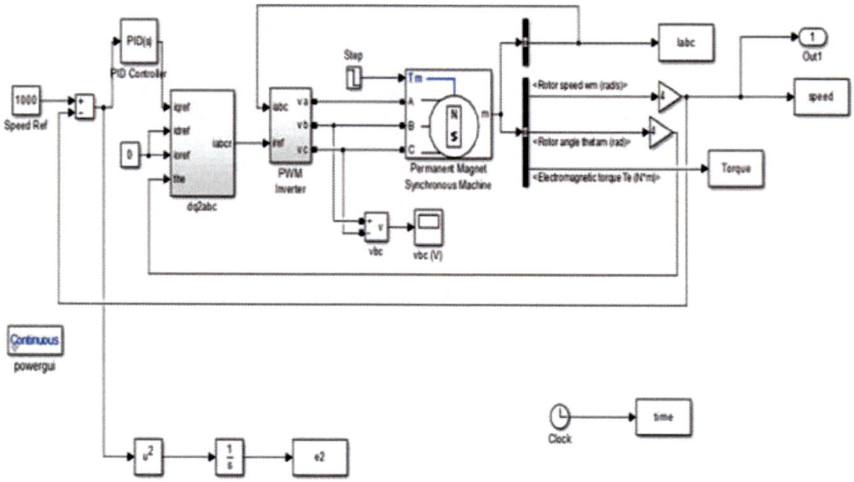

Fig. 1 PMSM drive with PID speed controller

controller gain parameters. These gain values are subsequently used in the drive model. The PMSM drive is analyzed different speed and load conditions. Main objective of this work is to carry out detailed investigation and the design of AI techniques based speed control methods of the PMSM drive system to improve and optimize transient response specifications [4, 5].

2 PMSM Drive Model

In MATLAB/SIMULINK, PMSM drive model is simulated using a field-oriented speed control scheme (FOC). Under different dynamic operating conditions, a 0.55 kW PMSM drive model fed from a 3-phase AC supply system is simulated. The proposed PMSM drive model composed of the PID speed controller is shown in Fig. 1 [4].

3 Ziegler–Nichols Method

This section addresses different steps involved in the algorithm for the Z-N method [5].

Step 1: Fig. 1 displays the plant model. The plant provides transfer function (TF) for the PMSM motor.
Step 2: TF and controller type are provided as input.

Table 1 PID controller gains using Z-N method

K_p	K_i	K_d
0.34584	3.460165	0.008641591

Step 3: Next, the plant transfer function estimates the gain margin (G_m), the phase margin (P_m), and the gain crossover frequency (W_{gc}).
Step 4: For this, K_{cr}, P_{cr}, must be calculated using the following equations (Table 1)

$$k_{cr} = G_m$$
$$p_{cr} = 2\pi / W_g$$

3.1 PMSM Drive Under a No-Load Condition

The performance of the 0.55 kW PMSM drive is analyzed during the starting period of the motor. The starting dynamics is shown in Fig. 2.

The reference speed is kept at 1000 rpm. The obtained result indicated following observations. First, the magnitude of the currents at the time of starting is high. At the moment, when the motor speeds up and attains its reference value, the currents acquire their no-load values along with their normal frequency, approximately at $t =$ 0.04 s; the electromagnetic torque developed decreases and reaches the load torque (zero value).

3.2 PMSM Drive at Speed Reversal Condition

The characteristic of the PMSM in reverse direction of the motor is presented in Fig. 3. The rated speed is reversed at $t = 0.08$ s to -1000 rpm. Furthermore, rotor speed attempts to achieve the reference value. The torque T_{em} increases in the reverse direction when the braking condition of the motor occurs. Thereafter, the PMSM drive achieves its no-load reference value.

3.3 PMSM Drive at Abrupt Load Condition

The dynamic reaction of the PMSM drive during abrupt load conditions is analyzed as depicted in Fig. 4. When a 6 Nm load is abruptly applied on the PMSM shaft at $t =$ 0.15 s, the rotor speed ω_m instantly decreases slightly. The PID controller increases the reference torque to be overcome by sudden load disturbance and increases the value of the torque produced. When there is an abrupt withdrawal of 6 Nm load and brought to the no-load value at $t = 0.35$ secs, the PID controller immediately responds

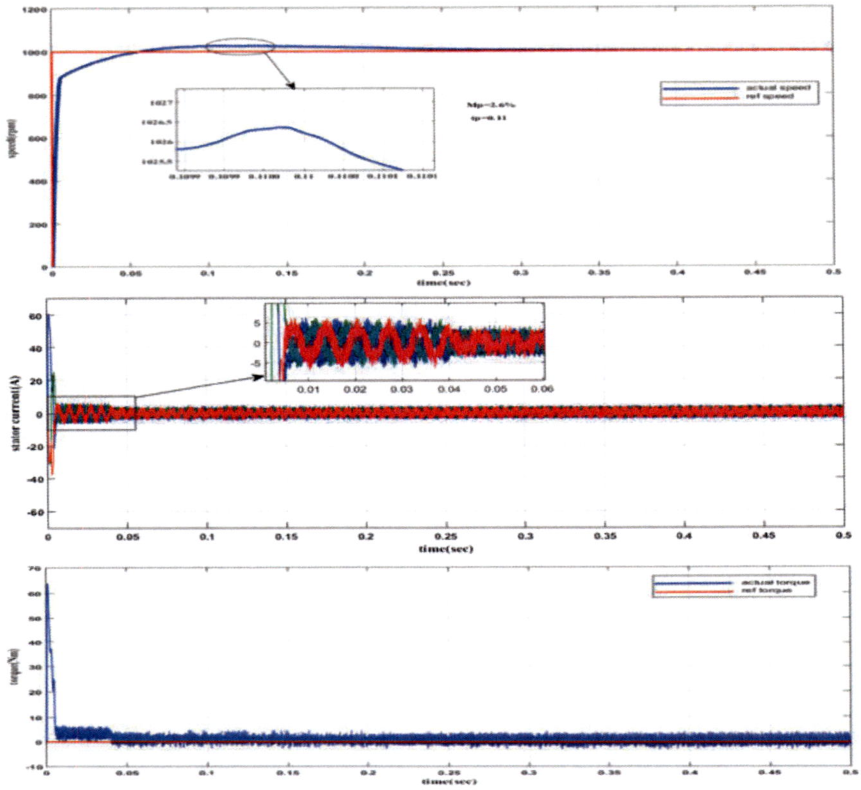

Fig. 2 No-load characteristics of a 0.55 kW Z-N method

by lowering the torque. As the load is increased, it draws heavy stator currents, and when it is removed, the stator currents regain their original rated value.

4 Methodology of Proposed MOGA

This section has carried out PID-tuned speed controller of PMSM using MOGA method. The work has matched the outcomes of both Z-N method and MOGA technique-tuned PID speed controller and has considered their performances for transient conditions like starting, speed reversal/braking, abrupt load changes [6–12]. The flowchart of the above algorithm implemented for tuning of the PID speed controller is shown in Fig. 5. PID controller gains in MOGA method have been obtained by min. of the four standard performance indices (PINs). These PINs are

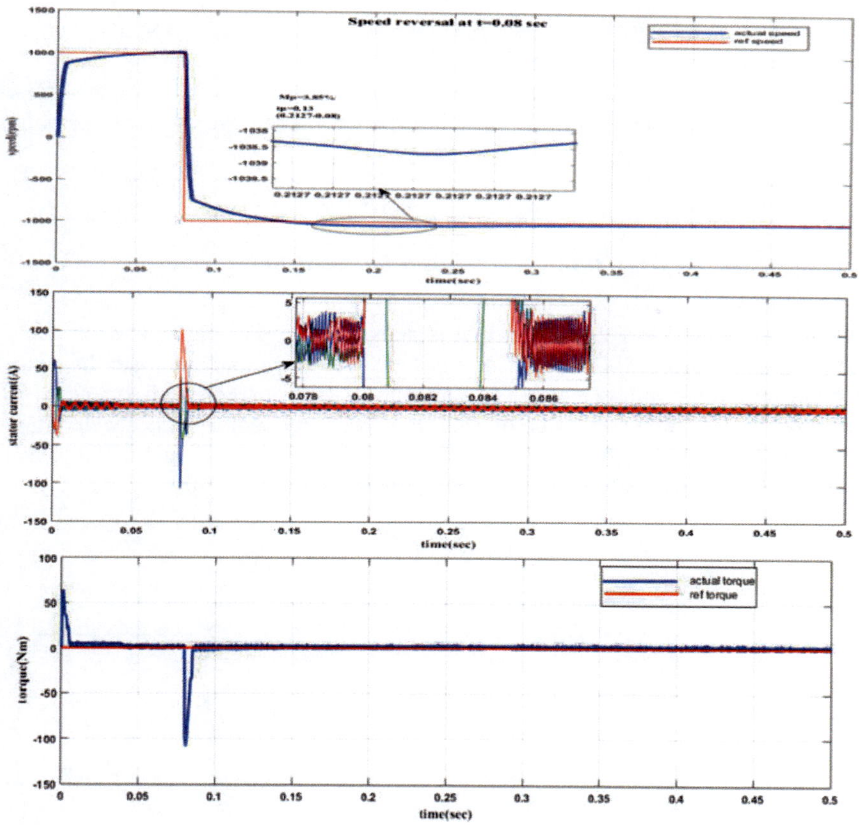

Fig. 3 Speed reversal characteristics of 0.55 kW PMSM drive using Z-N method

taken up as four different objective functions (OFs). The OF, in the multi-objective GA, is described as follows [13].

```
% pin calculation
% IAE-
F1=@(y), (sum(abc(e)*dt));
% ISE-
F2=@(y), (sum(e'*e*dt));
% ITSE-
F3=@(y), (sum((t.*e'*e)*dt));
% ITAE-
F4=@(y), (sum(t'.*abc(e)*dt));

PIN=@(y),[F1(y); F2(y); F3(y); F4(y)];
[y,fval_PIN,flag]=ga-multi-obj(PIN;2;[];[];[];[];[-2 -2];[2 2];options);
```

The K_p, K_i, and K_d gains of PID speed controller obtained from MOGA tuning method program are given in Table 2.

The modeling and simulation of the performance of 0.55 kW drive using MOGA method-tuned PID speed controller have been carried out under following operating conditions.

4.1 PMSM Drive Under a No-Load Condition

The starting dynamics of the PMSM drive is analyzed in detail in Fig. 6. The command value is maintained at 1000 rpm. Motor is switched on without any load; the stator winding currents stealthily increase to the desired value with reduced frequency. The speed reaches its reference value of 1000 rpm when the motor starts accelerating. Developed torque is high during gradual rise in the rotor speed, and as the speed achieves the reference value, it reaches the load torque of zero Nm.

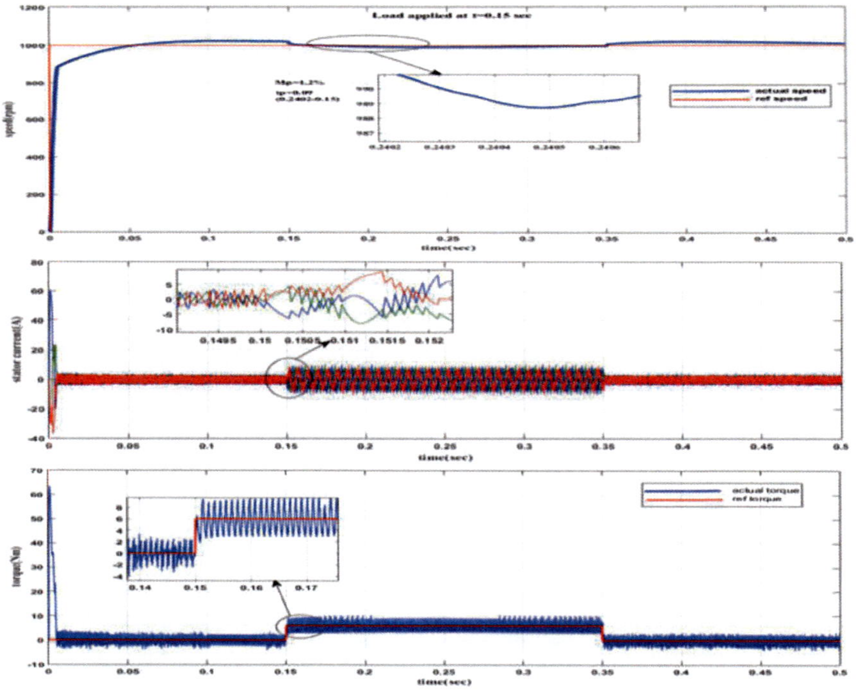

Fig. 4 Response of 0.55 kW PMSM with Z-N method for abrupt load disturbances

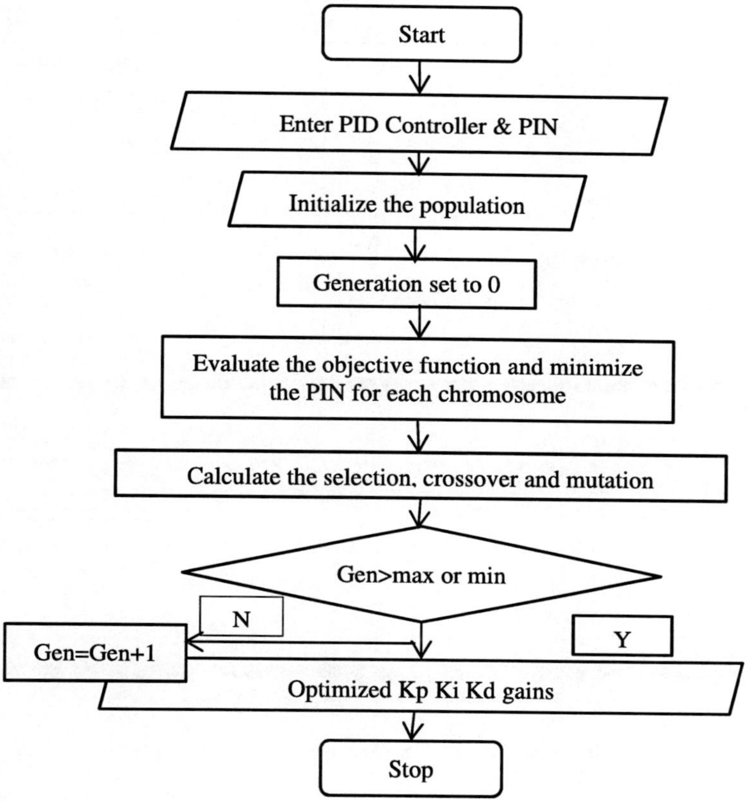

Fig. 5 Flowchart for MOGA algorithm [12]

Table 2 Values of PID gains for PMSM using MOGA method of tuning

	K_p	K_i	K_d
MOGA method-tuned PID speed controller	1.04587	3.00487	0.000741

4.2 *Speed Reversal Condition*

The behavior of the PMSM when the motor is met with sudden braking is shown in Fig. 7. The reference motor speed is reversed at $t = 0.08$ secs to -1000 rpm; the rotor speed tracks it and attains the rated value. The torque T_{em} decreases to get its reference value.

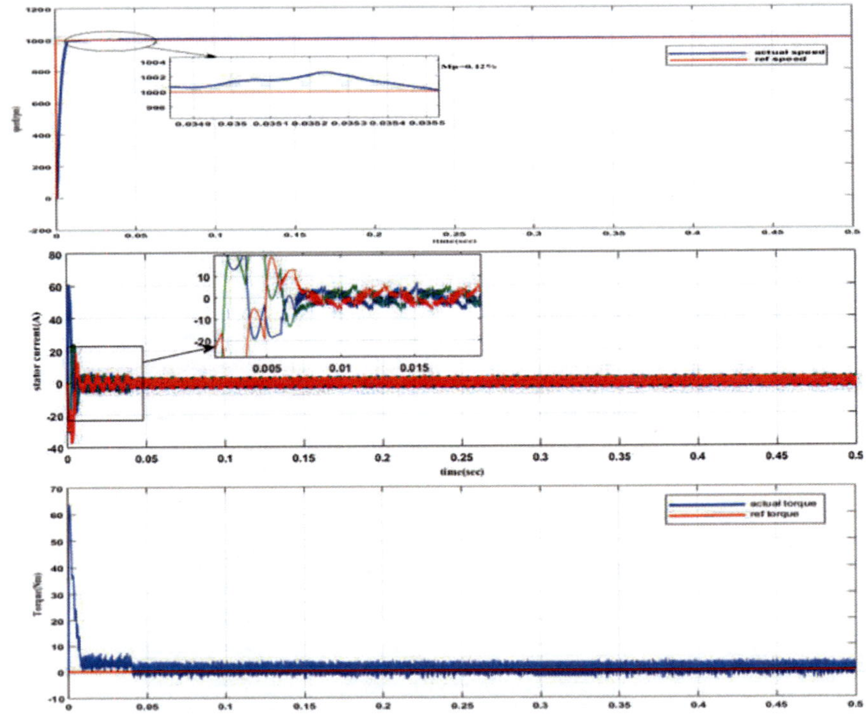

Fig. 6 MOGA-based PID speed controller PMSM drive: starting characteristics

4.3 Load Application and Load Removal

At $t = 0.15$ s, a 6 Nm load is realized. This load is suddenly removed at $t = 0.35$ secs. Load dynamics of PMSM drive for speed, torque, and currents are shown in Fig. 8. On the application of the load, T_{em}, i.e., the developed torque rises and settles at the reference load torque value of 6 Nm. It then reduces to zero Nm on the removal of the load. The rotor speed ω_m decreases as a consequence of the application of the load. On the removal of the load, the speed ω_m settles to 1000 rpm. Magnitude of stator winding currents increases to meet the increased load demand. When load is abruptly removed from motor drive, stator winding currents decrease in magnitude to acquire new values.

A comparative analysis of the PMSM using MOGA and Z-N-tuned PID controller has been reviewed in Table 3.

The observations inferred from Table 3 implemented in PMSM drive are as follows.

At the starting of the motor, MOGA tuning method has been successful in reducing peak overshoot by 95.38%. The MOGA method has also minimized t_s almost by 91.81% and t_r by 74%. At the speed reversal of the motor, MOGA in PMSM has

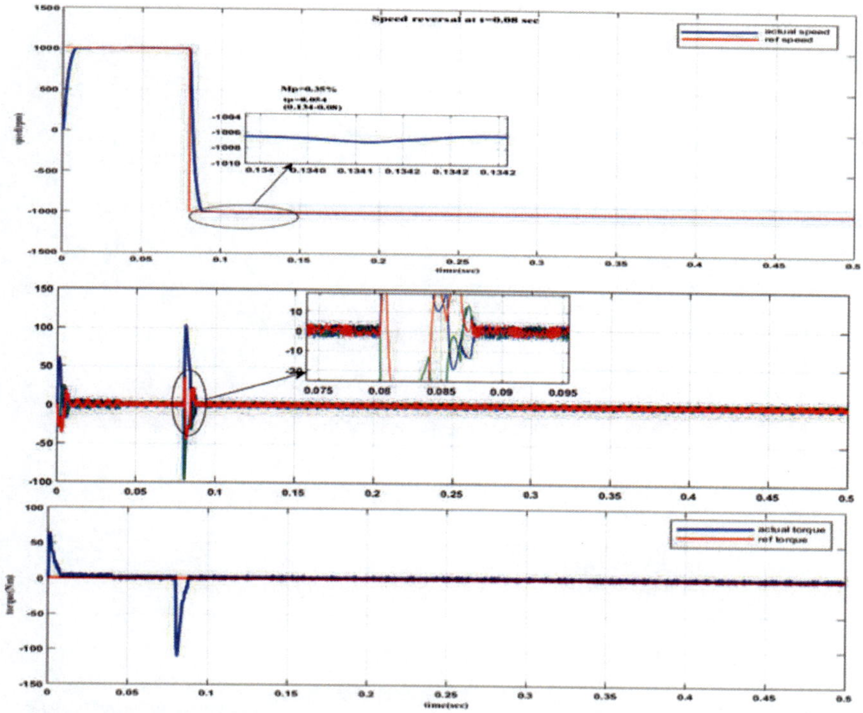

Fig. 7 Speed reversal characteristics of PMSM drive using MOGA

facilitated reduction of peak overshoot in speed by 90.9%; settling time is decreased by 15%. t_r is also decreased by 90%. At load application and removal, MOGA method has reduced M_p by 71.66 and 21.42% during load dynamics. It has taken less time to settle down to the reference value and has been able reduce t_s by 33.33 and 73.68% at the instant of abrupt load disturbances correspondingly.

5 Conclusion

The MOGA method in PMSM drive has been successful in meeting the various dynamic operating conditions. This has been achieved by decreasing the peak time, peak overshoot, settling time. The GA as an optimization tool has fairly faced all the operating conditions with the objective of enhancing the performance of the PMSM.

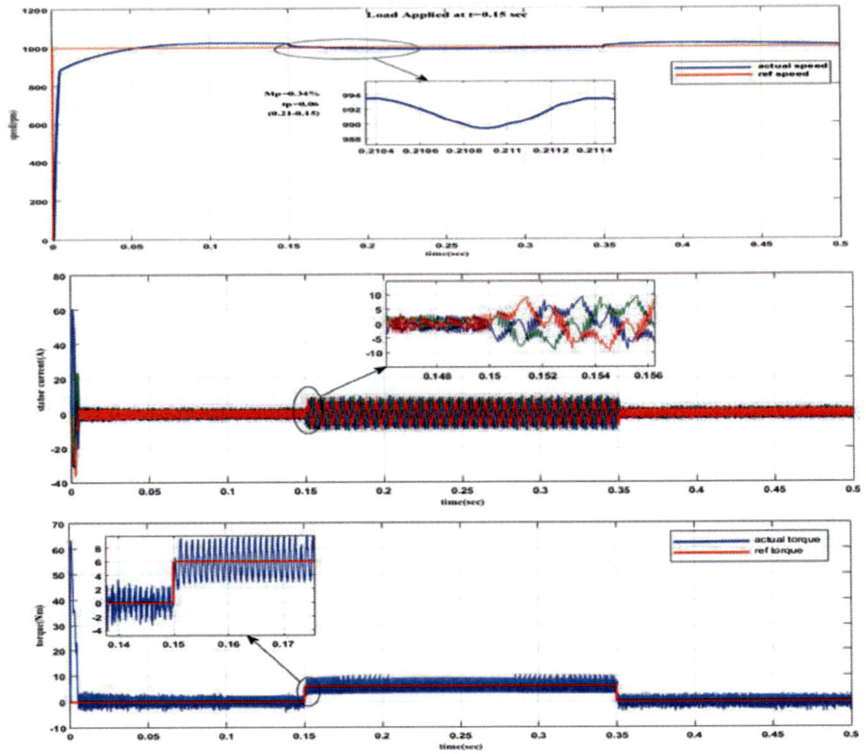

Fig. 8 MOGA-based PID controller in PMSM for abrupt load disturbances

Table 3 Dynamic response of the PMSM using MOGA and Z-N method-tuned PID controller

SN	Speed controller	Motor characteristics	Rise time (sec)	Peak overshoot (%)	Peak time (sec)	Settling time (sec)
1	Z-N method	Starting	0.01	2.6	0.11	0.44
		Speed reversal	0.03	3.85	0.13	0.40
		Load application	–	1.2	0.09	0.18
		Load removal	–	2.1	0.06	0.38
2	MOGA method	Starting	0.0026	0.12	0.026	0.036
		Speed reversal	0.003	0.35	0.054	0.34
		Load application	–	0.34	0.06	0.12
		Load removal	–	1.65	0.02	0.10

References

1. Krishnan R (2010) Permanent magnet synchronous and brushless DC motor drives. CRC Press Taylor & Francis Group, U.S.A.
2. Bose BK (2001) Modern power electronic and drives. The University of Tennessee, Knoxville, Upper Saddle River, NJ07458
3. Pan Tian-hong D, Shao-yuan LI (2009) Adaptive PID control for nonlinear systems based on lazy learning control theory and applications
4. Katsuhiko Ogata Z (2002) Modern control engineering, 4th (ed). Prentice Hall of India, New Delhi
5. Jain LC, Martin NM (1998) Fusion of neural networks, fuzzy systems and genetic algorithms: industrial applications. CRC Press
6. Cho DH, Jung HK, Chung TK, Lee CG (2000) Design of a short-time rating interior permanent magnet synchronous motor using a niching genetic algorithm. IEEE Trans Magn 36(4):1936–1940
7. Malekian K, Milimonfared J, Majidi B (2009) Performance improvement of direct torque controlled interior permanent magnet synchronous motor drives using artificial intelligence. Iranian J Electric Electron Eng 5(1):60–66
8. Ayyar KK, Ramesh K, Gurusamy G (2012) Design of speed controller for permanent magnet synchronous motor drive using genetic algorithm based lower order system modelling. J Comput Sci 8(10):1700–1710
9. Shamlou S, Mirsalim M (2012) Design, optimization, analysis and experimental verification of a new line-start permanent magnet synchronous shaded-pole motor. IET Electr Power Appl 7(1):16–26
10. Liu K, Zhu ZQ (2013) Quantum genetic algorithm based parameter estimation of PMSM under variable speed control accounting for system identifiability and VSI nonlinearity. IEEE Trans Ind Electron 0278–0046
11. Nakata T, Sanada M, Morimoto S (2017) Automatic design of IPMSM using a genetic algorithm combined with the coarse-mesh finite element method for enlarging the high-efficiency operation area. IEEE Trans Industr Electron 64(12):9721–9728
12. Yadav D, Verma A (2016) Performance analysis of PMSM drive using MPSO and MOGA technique. IEEE international conference on industrial electronics and applications conference (IEACon) 2016. Kota Kinabalu, Malaysia, pp 197–202
13. GA-Multiobjective Optimization Toolbox- User's Guide. The MathWorks. www.mathworks.com

Study of Structural and Electronic Properties of 2D WSe_2 Monolayer

Arunima Singh and Bramha P. Pandey

Abstract The structural and electronic properties of the monolayer WSe_2 are investigated in this paper. The ubiquitous behavior of WSe_2 monolayer is represented with two properties, structural and electronic. The lattice constant of the relaxed geometry is investigated and compared with other monolayers of TMD family as a structural parameter along with the bandgap of the monolayer as an electronic parameter. Further, the semiconducting nature is obtained from the monolayer band structure, and greater overlapping among W-5d and Se-4p orbitals shown in density of states (DOS) affirms the nature analyzed with the band structure.

Keywords Transition metal dichalcogenide (TMD) · Density functional theory (DFT) · PAW · WSe_2 · Bandgap

1 Introduction

In recent years, more research interest has been brought into the monolayer transition metal dichalcogenides (TMDs) for their novel actual properties and potential device applications. For instance, the bulk MX_2 (M = Mo, W; X = S, Se) has indirect bandgap and the monolayer MX_2 has direct bandgap, at the *K* points [1]. Consequently, it is exceptionally alluring to utilize DFT to mathematically compute the band structure of the TMDs and contrast the count results and the trial ones.

TMD monolayers are the atomically tenuous semiconductors of the MX_2 kind, where M belongs to transition metal atom (for example, Mo, W, etc.) and X belongs to a chalcogen atom (for example, S, Se, or Te) [1]. One layer containing M metal atoms is located between the dual layers of the X chalcogen atoms. They are member of the broad class of materials known as 2D materials, called because to emphasize their exceptionally tenuous width. The vital components of these materials are the communication of huge particles in the 2D structure as contrasted by first-row TMDs.

A. Singh (✉) · B. P. Pandey
Department of Electronics and Communication Engineering, Madan Mohan Malaviya University of Technology, Uttar Pradesh, Gorakhpur, India
e-mail: arunimas318@gmail.com

B. Mishra and M. Tiwari (eds.), *VLSI, Microwave and Wireless Technologies*, Lecture Notes in Electrical Engineering 877,
https://doi.org/10.1007/978-981-19-0312-0_59

The 2D semiconducting sheets have the base conceivable width and tunable bandgap (1−2 eV); from the perspective of usage, they can have applications in photovoltaic [1], detecting [1], biological medicines [1], solar cells [1], and catalysis [1]. The MX_2 compound comprising X-M-X sheets is roughly conjugated by solid bonds inside the layers. These materials have alluring as well as different physical properties; like, HfS_2 is an insulator, MoS_2 is a semiconductor, WTe_2 and TcS_2 have semi-metallic nature, and all $Nb(Ch)_2$ and $Ta(Ch)_2$ are superconductors [2]. Another favorable position within the nanoscale materials displayed another window for 2D TMDs. MX_2, for instance, MoS_2, $MoSe_2$, WS_2, and WSe_2, has sizeable bandgaps that convert from indirect to direct in single layer applications, for instance, semiconductors and photo-detectors.

Among TMD monolayers, the group VI-B TMDs (MoS_2, $MoSe_2$, WS_2, and WSe_2) are emanated as the compounds with several superior physical properties like enhanced mobility at normal conditions or controlling their electronic properties simply by using external field. The direct bandgap semiconductors are TMD monolayers such as MoS_2, $MoSe_2$, $MoTe_2$, WS_2, and WSe_2. It may be utilized in optical as emitter detector and electronic as semiconductors [2].

An inorganic compound with chemical formula WSe_2 is called as tungsten diselenide. The bond length between tungsten-selenium is 0.2526 nm, and interatomic spacing of selenium atoms is 0.334 nm. The heaviest transition metal in the class of basic TMDs is tungsten. Van der Waals force mediates the stacking of layers together [2]. WSe_2 is a truly steady semiconductor among TMDs of group VI. The 2D semiconducting sheets have the base conceivable width and wide band holes, from the perspective of use.

This paper is composed in four sections, in which Sect. 1 introduces about the TMDs, whereas Sect. 2 explains the methodology used along with Sect. 3 explaining the structural and electronic properties of WSe_2 monolayer, and Sect. 4 concludes with the scope of future work.

2 Computational Methods

The structural and electronic properties of WSe2 monolayer are computed utilizing GPAW simulation package [3] using plane augmented wave (PAW). The $4 \times 4 \times 1$ monolayer slab with lattice constant $a = 3.34$ Å and vacuum separation of 12 Å along z-direction is modeled using ASE simulation tool. The energy cut-off is 600 eV, and k-point of 15*15*1 is opted using Monkhorst k-point [4] scheme. The GGA is used as correlation parameter with Perdew–Burke–Ernzerhof (PBE) [5] as exchange function for accurately estimating the properties of monolayer. The geometry optimization of monolayer displayed in Fig. 1 is performed with convergence at 10^{-4} eV. The Fig. 1 represents the WSe2 surface slab consisting of 'W' as central atom surrounded by two 'Se' atoms constituted in a trigonal manner.

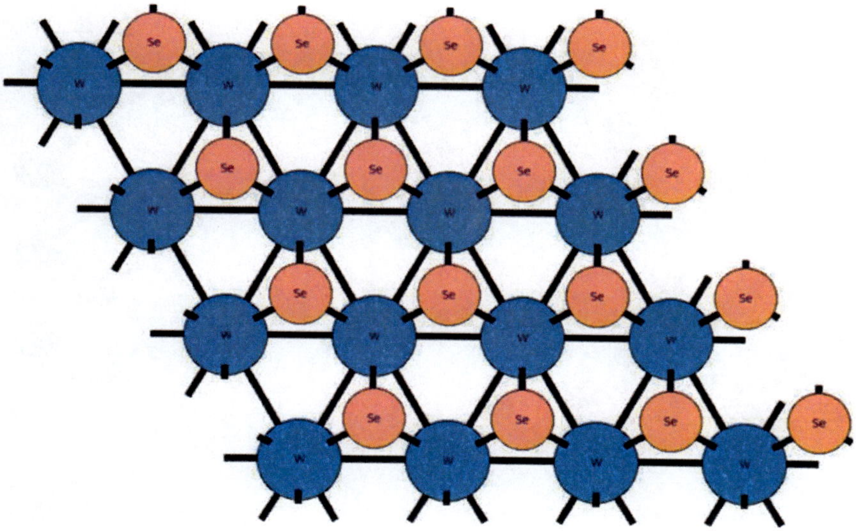

Fig. 1 Crystal structure of WSe2 monolayer

3 Results and Discussion

3.1 Structural Properties

The crystal structure of the WSe2 monolayer is a type of MX2 compound. The 'W' atoms involve a sub-cross section of the hexagonal plane sandwiched with two layers of Se atoms. In this paper, we get the value of lattice constant of WSe2 that is 3.34 Å which is in accordance with the reported value of 3.32 Å [6].

TMDs due to their layered pattern forming capability have already been explored for fullerenes-type nature. A solitary layer comprises of a tightly bonded X-M-X sandwich with numerous layers that are loosely stacked. The change in crystal structure occurs in each unit cell-based different counts of layers and their arrangement regarding each other [2]. WSe_2 is a layered material with strong and loose interactions in inside the plane and out of the plane, respectively, empowering exfoliation of thickness of a unit cell 2D layers. The TMDCs have overall rhombohedral or hexagonal evenness, with either trigonal or octahedral prismatic coordination of the metal atoms [5]. WSe_2 has a configuration like that of MoS_2; the only change is the distance between interlayers. The contrast with the nuclear size of 'W' brings about an adjustment in the interlayer dispersing. The monolayer WSe_2 is atomically thin with width of ~0.7 nm (Fig. 2).

These are the experimental value and have been acquired by different procedures like chemical vapor deposition (CVD) and mechanical peeling of mass WSe2.

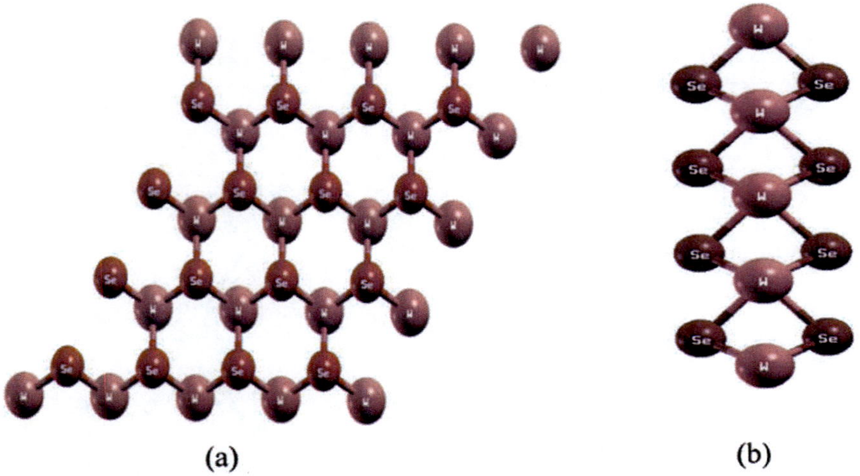

Fig. 2 Unit cell calculation **a** Front view, **b** Side view

3.2 Electronics Properties

WSe2 depicts various properties at the monolayer as well as at the levels existing in the bulk. While these materials have an indirect bandgap at the mass level, the bandgap rises for WSe2 at the monolayer level [5]. It is imperative to realize the bandgap of the material and transitions across the bandgap. The inspection of quantum–mechanical wave elements of a particular electron in vast atomic or molecular lattice atom structures is the reason behind choosing band structure [6].

To examine electronic properties of the WSe2 monolayer, we initially figure the band construction of the monolayer WSe2, as demonstrated in Fig. 3. We can infer the direct bandgap of monolayer WSe2. The TMDs are viewed as an immediate

Fig. 3 Band structure of WSe_2 monolayer

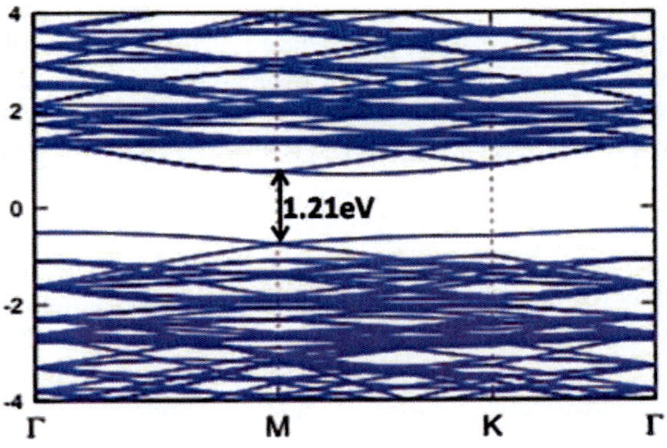

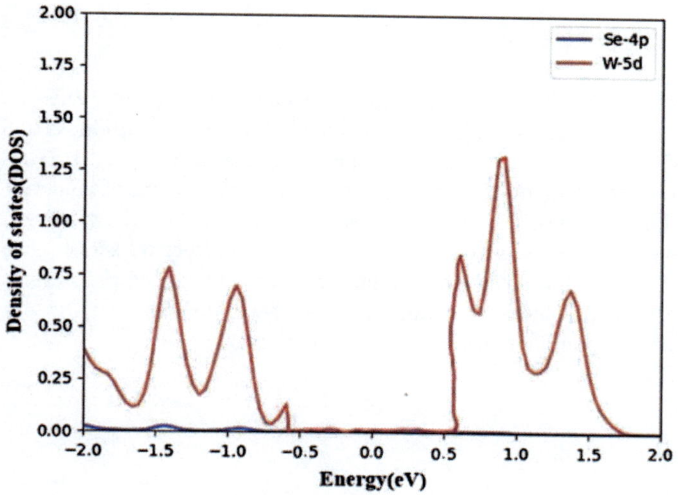

Fig. 4 Density of state (DOS) of WSe_2 monolayer

energy hole of monolayer created between the bands of conduction and valence. The gap of monolayer WSe2 is 1.21 eV [2]. This behavior is related to the density of states (DOS) as well. As observed from Fig. 4, the increased orbital overlapping of area between W and Se indicates the strong hybridization of 3d orbitals of W and 4p orbitals of Se atom, respectively. The flat states present close to the fermi level of DOS indicate the semiconducting behavior which is as confirmed from band structure.

Direct bandgap is involved in the band structure of monolayered WSe2; on the other hand, bulk WSe2 comprises as an indirect bandgap because of progress in enlarged layer width. The maxima and minima are located at 'M' the point in the valence band and conduction band, respectively. The circuitous direct bandgap changes are credited to the lost interlayer communications in monolayer-based MX2 [3]. The monolayer-based WSe2 has highest bandgap difference in comparison to all the other MX2 semiconductors [3]. The cause of it is the solid spin–orbit coupling and loss of inverse symmetry. A turn split construction emerges, consequently adding to a vital property of MX2 semiconductors [4]. The energy gap of semiconductors relies upon temperature. The rising temperature results in narrowed bandgap in most of the semiconductors. An expansion occurs in atomic spacing because of expansion in lattice vibrations on account of increment in thermal energy. As observed by the electrons within the material, the potential declines with subsequently inflating interatomic distance which ultimately reduces the expand of bandgap [7]. The spin–orbit coupling in tungsten is enormous, in contrast with that of molybdenum, and consequently, WSe2 (monolayer based) discovers its utility in the area of spintronics [5].

4 Conclusions

This paper explores the structural and electronic properties of WSe_2 monolayer utilizing first-principle method. The outcome shows the great solidness of the direct bandgap of monolayer. The bandgap of the monolayer WSe2 is 1.21 eV. The structure has 2H prototype with the trigonal prismatic arrangement of centered W atom surrounded by Se atoms. Therefore, the WSe2 monolayer band structure concludes the semiconducting nature as a result of indirect to direct transition. The tunable bandgap of WSe2 monolayer opens a new area with enhanced performance. Thus, different structural (electronic) properties can be extracted from the $4 \times 4 \times 1$ configuration of monolayer as future scope of work.

References

1. Hieu NN, Ilyasov VV, Vu TV, Poklonski NA, Phuc HV, Phuong LT, Nguyen CV (2018) First principles study of optical properties of molybdenum disulfide: from bulk to monolayer. Superlattices Microstruct 115:10–18
2. Khan F, Din HU, Khan SA, Rehman G, Bilal M, Nguyen CV, Amin B (2019) Theoretical investigation of electronic structure and thermoelectric properties of MX2 (M= Zr, Hf; X= S, Se) van der Waals heterostructures. J Phys Chem Solids 126:304–309
3. Wang QH, Kalantar-Zadeh K, Kis A, Coleman JN, Strano MS (2012) Electronics and optoelectronics of two-dimensional transition metal dichalcogenides. Nat Nanotechnol 7(11):699–712
4. Zhu ZY, Cheng YC, Schwingenschlögl U (2011) Giant spin-orbit-induced spin splitting in two-dimensional transition-metal dichalcogenide semiconductors. Phys Rev B 84(15):153402
5. Muoi D, Hieu NN, Phung HT, Phuc HV, Amin B, Hoi BD, Le PTT (2019) Electronic properties of WS2 and WSe2 monolayers with biaxial strain: a first-principles study. Chem Phys 519:69–73
6. Perdew JP, Burke K, Ernzerhof M (1996) Generalized gradient approximation made simple. Phys Rev Lett 77(18):3865
7. Mak KF, Lee C, Hone J, Shan J, Heinz TF (2010) Atomically thin MoS_2: a new direct-gap semiconductor. Phys Rev Lett 105(13):136805
8. Van Zeghbroeck B (2011) Principles of electronic devices. University of Colorado
9. Boochani A, Nowrozi B, Khodadadi J, Solaymani S, Jalali-Asadabadi S (2017) Novel graphene-like Co2VAl (111): case study on magnetoelectronic and optical properties by first-principles calculations. J Phys Chem C 121(7):3978–3986

Four-Port Compact MIMO Antenna for 5.8 GHz Applications

Akhilesh Kumar, Shadab Azam Siddique, Sanjay Kumar Soni, Kritika Singh, and Brijesh Mishra

Abstract In this article, an ultra-compact MIMO antenna resonating at 5.8 GHz is presented. The anticipated design consists of four electromagnetic radiating elements along with implanted square patch ground plane, having total size of antenna are $38 \times 38 \times 1.524\,\text{mm}^3$ and loss tangent is 0.025. Square-shaped stub is introduced between the four-port elements to improve the isolation of the proposed MIMO antenna design. The overall bandwidth of the presented MIMO antenna has 1.1 GHz (5.1–6.2) to measure the $-10\,\text{dB}$ impedance bandwidth. The anticipated design has been providing low mutual coupling in between the ports respectively, i.e., ($S_{31} < -13\,\text{dB}$), ($S_{41} < -15.57\,\text{dB}$), ($S_{32} < -18\,\text{dB}$) and ($S_{42} < -10\,\text{dB}$), and high peak gain 5.9 dB. The antenna performance is deliberate in terms of surface current distribution, reflection coefficient, radiation pattern, peak gain, envelop correlation coefficient, and diversity gain.

Keywords MIMO antenna · Diversity gain · Envelop correlation coefficient

1 Introduction

Wide bandwidth, high data rate, and consistency are the primary need for different wireless applications, such as video conferencing, online gaming, real-time application [1]. Meeting all these challenges is an essential for increasing the reliability of the wireless system. Fading is one of the prominent issue that reduces the overall transmission quality. To mitigate the effect of fading, various techniques are evolved in the literature, where space diversity plays significant role [2, 3]. Under this prin-

A. Kumar · S. A. Siddique · S. K. Soni · K. Singh
Department of Electronics and Communication, Madan Mohan Malaviya University of Technology, Gorakhpur, UP, India

B. Mishra (✉)
Department of Electronics and Communication, Shambhunath Institute of Engineering and Technology, Prayagraj, UP, India
e-mail: brijesh.mishra0933@gmail.com

B. Mishra and M. Tiwari (eds.), *VLSI, Microwave and Wireless Technologies*, Lecture Notes in Electrical Engineering 877,
https://doi.org/10.1007/978-981-19-0312-0_60

ciple, multiple antennas are placed at either transmitter (Tx), or receiver (Rx) and at both Tx and Rx.

Utilizing the earlier principal, various antenna systems exist in the open literature, namely single-input-single-output (SISO), multiple-input-single-output (MISO), single-input-multiple-output (SIMO), and multiple-input-multiple-output (MIMO). Among them, MIMO offers higher capacity and data rates [4]. Multiple antennas provides multiple channels for data transmission and reception and thus improves the capacity of channel [5]. The antenna size is very important in portable/compact devices due to restricted area. However, if the antenna elements have to be placed in closely, it produces considerable mutual coupling or isolation that can depreciate antenna performance [6].

The antenna independent and dependent decoupling techniques are preferably used for high isolation and compactness [7–10]. Authors in [7], proposed a generalized RF circuit that is further used to decouple 2 dual band antennas with the aid of T-stub circuit, where arbitrary and independent phase shift and impedance at dual frequencies is observed. In [8], author incorporated flag-shaped stub at ground plane center, to enhance isolation between elements and increase insulation between ports 1 and 2. Perhirin and Auffret in [9] presented a MIMO antenna with dual band that follows pattern diversity, for improving the port-to-port isolation between diagonal and adjacent components. With the help of orthogonal circularly polarized arranged elements, more than 30 dB separation is achieved, and the diagonal element has the same circularly polarized element [10]. Another antenna architecture proposed by Saurabh and Meshram in [11], where a T-shaped ground stub is incorporated to suppress the mutual coupling between the radiating elements. In order to receive improvement in isolation and impedance matching concurrently, an altered T-shaped ground stub is positioned between radiating components. In addition, the split U-shaped stub is mounted at the center of each MIMO antenna radiator to obtain improvement in impedance matching whereas resonant frequency decreases. Meander lines are used to enhance isolation between two elements of the microstrip antenna [12]. F-shaped stubs are added to generate several resonances and high isolation between the radiating elements of the ground plane [13]. Again in [14], with the aim to minimize the mutual coupling of the four-element MIMO antenna system, five distinct mushroom-like loadings are proposed, whereas a metamaterial absorber array is used between the two antenna elements [15].

In the proposed structure of antenna, we have designed a closely spaced compact four-element MIMO antenna with radiating elements at 5.8 GHz applications. Finally, a square-shaped stub is placed between the elements for achieving better isolation. The design is simulated by using electromagnetic tool and high-frequency structure simulator (HFSS). The proposed structure exhibits gain of 5.9 dB having impedance bandwidth 1.1 GHz (5.1–6.2 GHz). The paper organization is as follows: Sect. 2 provide detailed steps and methodology to design the antenna structure. In Sect. 3, elaborated discussion on the proposed geometry is provided with the help of surface current distribution (SCD), isolation, envelop correlation coefficient (ECC), and return loss, etc. Sect. 4, gives the final remarks.

The proposed work is novel in terms of compact geometry, four-port communication, and better performance like ECC, isolation, bandwidth (BW), and radiation efficiency. ECC's are 0.02544 (port 1–2), 0.015877 (port 1–3), 0.113464 (port 1–4), isolation between ports ($S_{31} = -13$ dB), ($S_{41} = -15.5798$ dB), ($S_{32} = -18.0122$ dB), and ($S_{42} = -10.3758$ dB), BW of 1.1 GHz and radiation efficiency $< 90\%$.

2 Antenna Geometry and Evolution

The complete structure of the anticipated 4 × 4 MIMO antenna having front view and back view is revealed in Fig. 1. The total size of the illustrated antenna is $38 \times 38\,\text{mm}^2$. The proposed antenna structure uses FR4-epoxy dielectric substrate with $\epsilon_r = 4.4$, thickness = 1.524 mm and loss tangent = 0.025. Figure 2 illustrates the systematic proposal steps (step-1 to step-4) of the suggested MIMO antenna. Figure 3 represents return loss of the antenna steps (1–4). In step-1, we have to design an I-shaped structure with ring ground plain. But this step is unable to cover −10 dB return loss as shown in Fig. 3. Secondly, I-shaped is modified and made it an inverted L-shaped structure with the ring ground plane has been designed as shown in step-2. Further in step-3, inverted L-shaped structure which is united with an I-shaped structure has been designed. In step-2, it covers −32.78289 dB resonate at 5.3 GHz, and in step-3, it covers −26.8 dB resonate at 6.3 GHz. Due to compact size of the proposed design, it realizes low isolation between the antenna ports which result in decrease the performance. In step-4, we have to design square-shaped patch isolator having dimensions $8 \times 8\,\text{mm}^2$, to minimize the inter-element coupling among closely packed radiators. Sizes of the anticipated antenna are in millimeter scale. It also illustrates that MIMO antenna exhibits 1.1 GHz impedance bandwidth as the proposed structure is having return loss below −10 dB between 5.1 and 6.2 GHz.

3 Result and Discussions

This section briefly explains the existence of the proposed structure at given frequency along with its favorable and unfavorable characteristics. Figure 4 illustrates the SCD of the presented antenna at radiating frequency of 5.8 GHz. The SCD on metallic surface reveals the behavior of structure isolation. Excitation is given to Port 1 and other ports are terminated by 50 Ω matched loads.

For SCD, we have to observe the maximum and minimum currents on the ports. At 5.8 GHz radiating frequency, we have seen that a small surface current (SC) exhibits on the ring ground structure of the corresponding elements. At other ports 2, 3, and 4, it shows a less current coupling due to the implanted square patch in between meander lines. The SC at port 2 and port 3 is having similar values due to equal distance from port 1. At port 4, the SC is minimum, due to the large separation from port 1. The SC lies the range of 0.0842–115 A/m at different ports. Similarly, equal current distribution (CD) can be attained when other port is excited and rests are terminated with 50 Ω load. The return loss of anticipated antenna is −33 dB resonate at 5.8 GHz frequency. Since all elements are similar, due to symmetry of the ports, their scattering parameters are also equivalent, i.e., S_{ij}, $(i = j)$ shown in Fig. 5.

Figure 6 represents the isolation between ports port (3–1), port (4–1), port (3–2), and port (4–2) of proposed antenna, i.e., ($S_{31} < -13$ dB), ($S_{41} < -15.5798$ dB), ($S_{32} < -18.0122$ dB) and ($S_{42} < -10.3758$ dB), respectively. Lower mutual coupling provides high isolation between the ports (3–1, 4–1, 3–2, and 4–2). To achieve high isolation, square-shaped stub introduces between the four-port MIMO antenna and reduced mutual coupling between antenna ports is observed.

The normalized values of the unidirectional far field (i.e., electric and magnetic fields) patterns are shown in Fig. 7. The electric field (for $\phi = 0$) radiation patterns for a single element MIMO antenna structure at a radiating frequency of 5.8 GHz... The obtained results show that the antenna element's predominant lobe direction is 180° and magnitude is 12.5 dBV/m. The MIMO antenna element has an angular 3 dB width of 112.4°. Equally, we observed that magnetic field (for = 90) shapes at

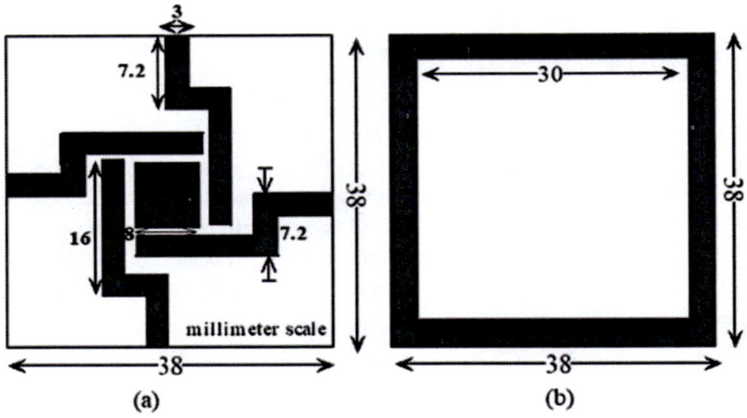

Fig. 1 Schematic of the 4 × 4 MIMO antenna. **a** Front view. **b** Back view

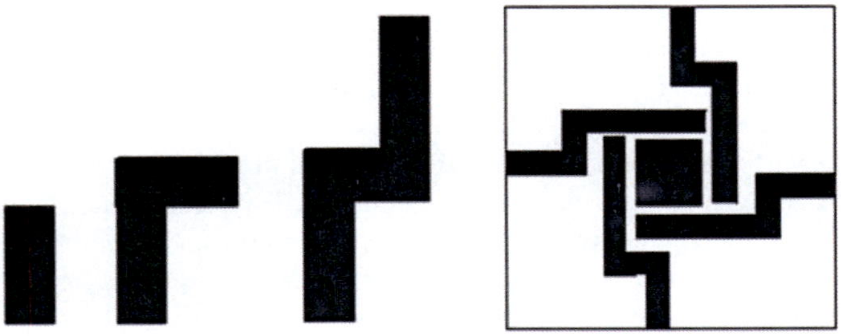

Fig. 2 Developments of antenna design

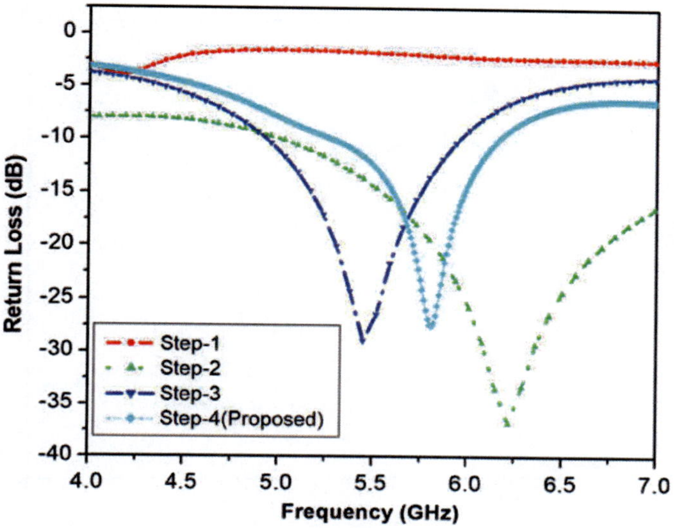

Fig. 3 S_{11} parameters comparison

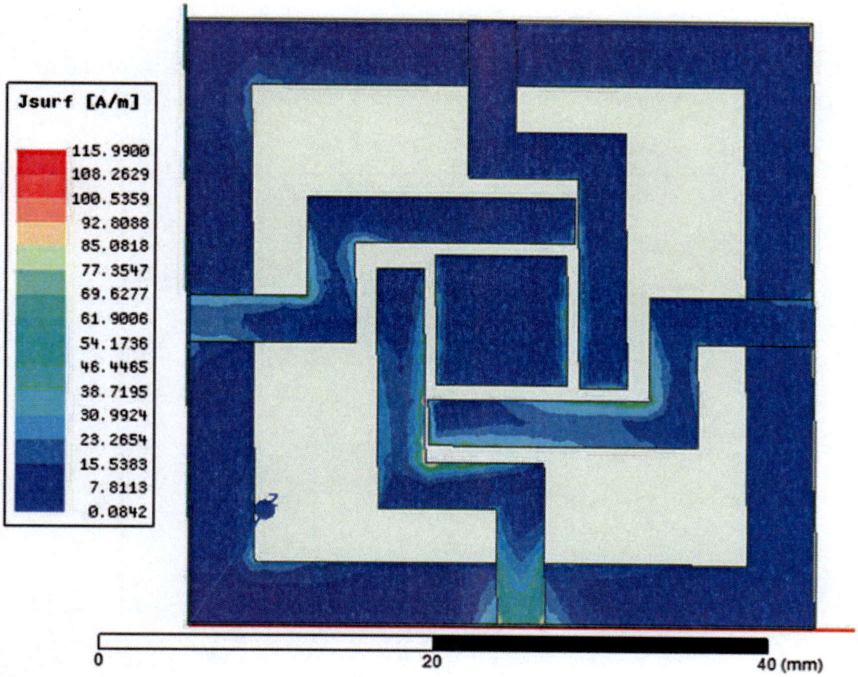

Fig. 4 Surface current distribution of the proposed antenna

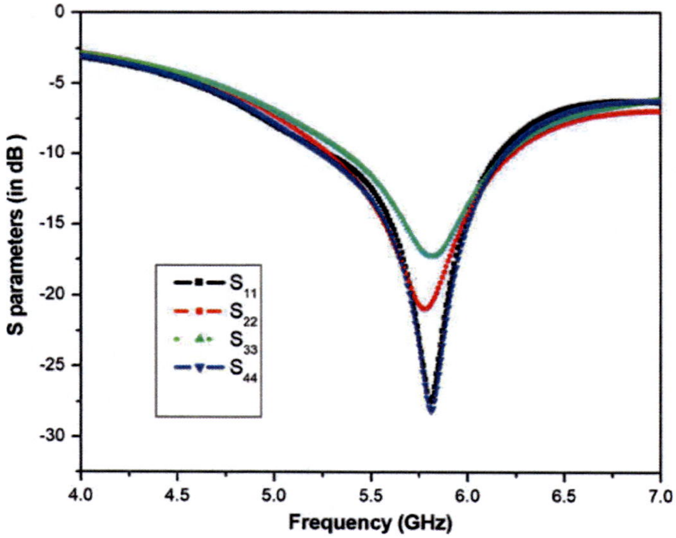

Fig. 5 Reflection coefficient of the proposed antenna

5.8 GHz, with a magnitude of 115.99 dBA/m, and the maximum value of the field seen at 165°.

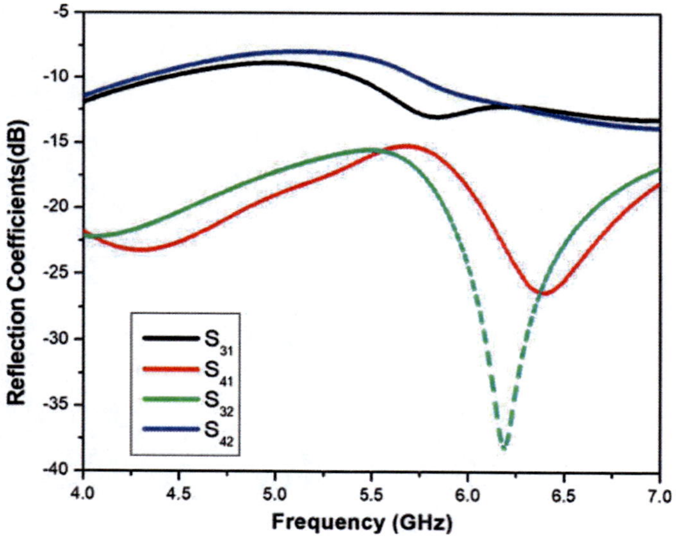

Fig. 6 *S*-parameters for the isolations (S_{31}, S_{41}, S_{32}, S_{42})

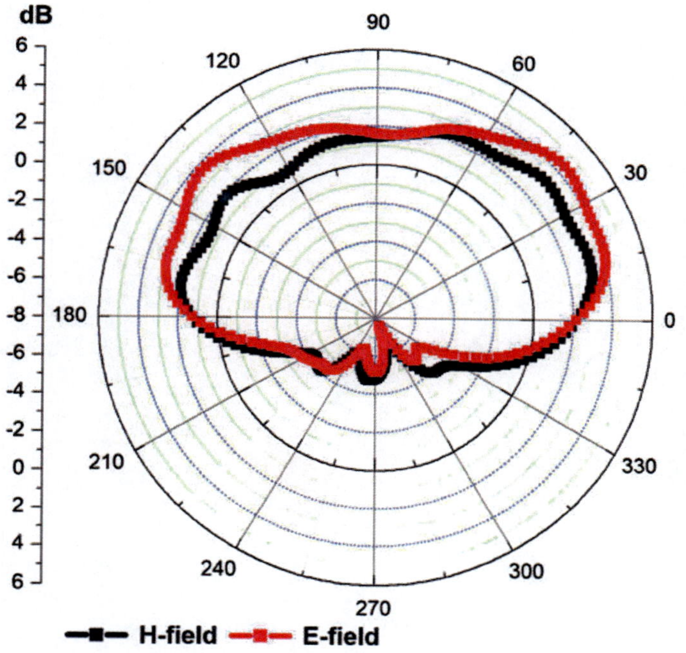

Fig. 7 E-field and H-field radiation pattern at 5.8 GHz

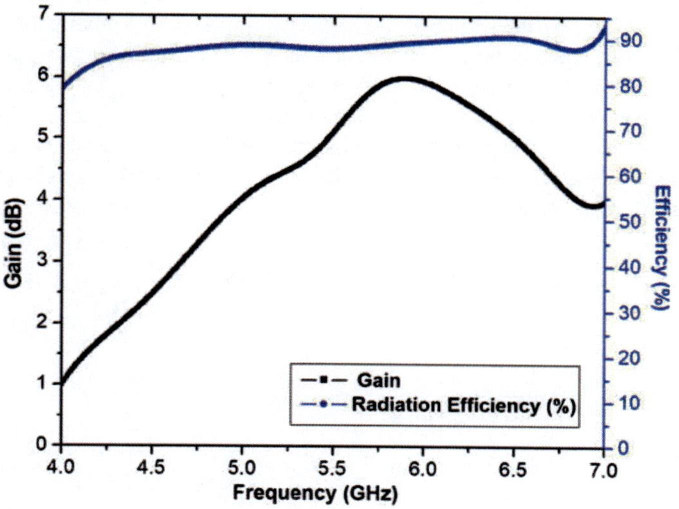

Fig. 8 Radiation efficiency and peak gain

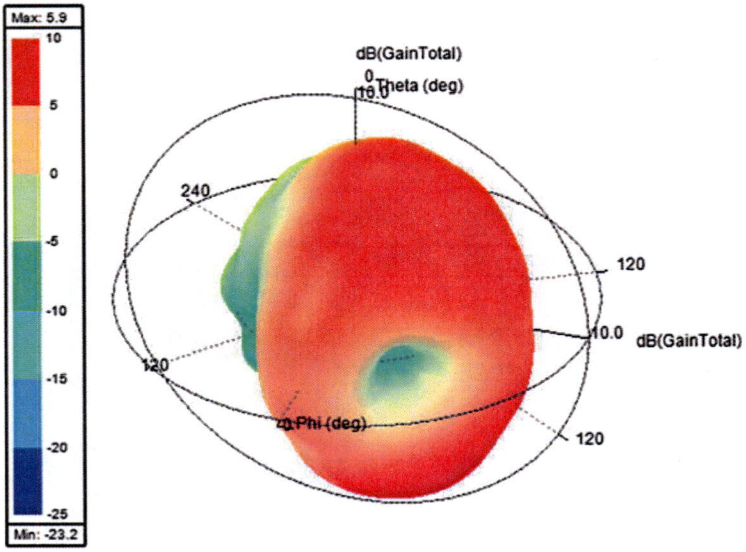

Fig. 9 3D radiation pattern for element-1 at 5.8 GHz and other elements are terminated

Radiation efficiency, 2D and 3D plot of antenna gain is shown in Figs. 8 and 9. Gain of anticipated antenna is close to 6 dB at resonating frequency as visible from Fig. 8, while observing Fig. 9, we witness simulated 3D peak gain 5.9 dB at resonate 5.8 GHz. Another important performance metric is the diversity gain, shown in Fig. 10. Computational value of diversity gain given by Eq. (5) provides a measure of improvement in the MIMO system, which should be greater than 9.95 dB [16]. With respect to proposed architecture, the radiation efficiency is less than 90% which satisfies the earlier given requirement and diversity gain is 10 dB as can be seen from Figs. 8 and 10, respectively.

$$\mathrm{ECC} = \frac{\left|\iint_{4\pi}[F_1(\theta,\phi).F_2(\theta,\phi)\mathrm{d}\Omega]\right|^2}{\iint_{4\pi}|F_1(\theta,\phi)|^2\mathrm{d}\Omega.\iint_{4\pi}|F_2(\theta,\phi)|^2\mathrm{d}\Omega]} \tag{1}$$

$$\mathrm{ECC}_{12} = \frac{|S_{11}*S_{12}+S_{12}*S_{22}+S_{13}*S_{32}+S_{14}*S_{42}|^2}{1-(|S_{11}|^2+|S_{12}|^2+|S_{13}|^2+|S_{14}|^2)} \tag{2}$$

$$\mathrm{ECC}_{13} = \frac{|S_{11}*S_{13}+S_{12}*S_{23}+S_{13}*S_{33}+S_{14}*S_{43}|^2}{1-(|S_{11}|^2+|S_{12}|^2+|S_{13}|^2+|S_{14}|^2)} \tag{3}$$

$$\mathrm{ECC}_{14} = \frac{|S_{11}*S_{14}+S_{12}*S_{24}+S_{13}*S_{34}+S_{14}*S_{44}|^2}{1-(|S_{11}|^2+|S_{12}|^2+|S_{13}|^2+|S_{14}|^2)} \tag{4}$$

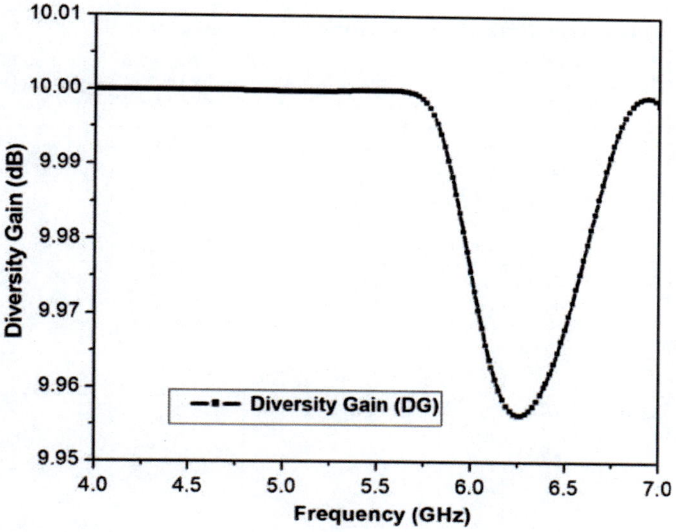

Fig. 10 Diversity gain

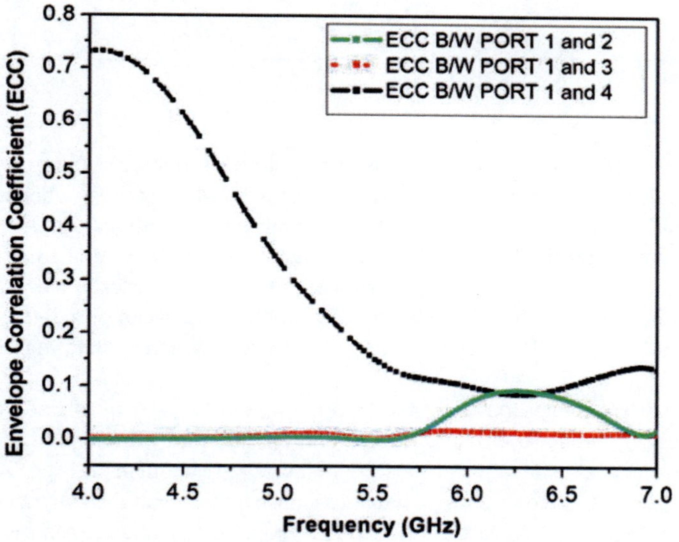

Fig. 11 Envelope correlation coefficients (ECC) of four-port compact MIMO antenna

$$DG = 10\sqrt{1 - ECC^2} \tag{5}$$

Figure 11 represents the plot of ECC versus frequency. It signifies amount of correlation exist between antenna radiation patterns. For example, if one antenna

Table 1 Performance comparison with previously published works

References	Year	Antenna size (mm^2)	Substrate used	BW (GHz)	Isolation (dB)	Peak gain (dB)	Radiation efficiency (dB)	3-D radiation pattern
[5]	2017	48 × 29 = 1392	$\epsilon_r = 4.4$ $\tan\delta = 0.02$ $h = 1.6$ mm	0.12	< -18	4.695	< 60%	4.695
[8]	2016	70 × 70 = 4900	$\epsilon_r = 4.4$ $\tan\delta = 0.025$ $h = 1.524$ mm	0.368 0.68	< -17	2.54 5.26	< 60% < 68%	NR
[10]	2016	26.69 × 97 = 2588.93	$\epsilon_r = 4.4$ $\tan\delta = 0.025$ $h = 1.524$ mm	0.534	< -33	5.34	NR	5.33
Proposed work	–	38 × 38 = 1444	$\epsilon_r = 4.4$ $\tan\delta = 0.025$ $h = 1.524$ mm	1.1	< -18	5.9	< 90%	5.9

NR not reported

is vertically polarized and other horizontally polarized then correlation is zero. If the ECC Value is ≤ 0.5, it is not considerable for MIMO antenna, whereas if it is ≤ 0.3 considered pretty good. In general, two well-known methods *S*-parameter and radiation pattern (far field) are employed for numerical calculation of ECC Eqs. (1), (2), (3) and (4). Mostly, radiation pattern method is used in order to calculate ECC values and is given in [16]. The value of ECC obtained between port 1–2, port 1–3, port 1–4 is 0.02544, 0.015877, and 0.113464, respectively, at the operating frequency 5.8 GHz.

Performance of the proposed work and previous works in terms of size, substrate, BW, peak gain, radiation efficiency, ECC and Isolation are enumerated in Table 1. Bandwidth and peak gain of the proposed antenna are greater than the reported papers [5, 8, 10]. Overall size of the proposed antenna is smaller than the antenna reported in [8, 10] by factor of 3.39 and 1.79, respectively, whereas greater than the antenna [5] by factor of 1.03. Isolations of the proposed antenna are better than [5, 8], i.e., < -18 dB [5], < -17 dB [8].

4 Conclusion

In this report, a compact four-port element MIMO antenna with a ring ground plane for 5.8 GHz applications has been designed. The anticipated antenna has simple geometry, and overall size is $38 \times 38\ \text{mm}^2$. We have to realize better isolation between the ports (3–1, 4–1, 3–2, 4–2) of MIMO elements which is achieved by introducing square-patch stub between the elements of the MIMO antenna. Maximum gain and efficiency obtained at resonate frequency are 5.9 dB and 88.63%. The ECC for ports (port 1–2), (port 1–3) and (port 1–4) is given as 0.02544, 0.015877, and 0.113464, respectively; the diversity gain is nearby to 10 dB. Other results of anticipated antenna design are radiation efficiency ($< 90\%$) and surface current distributions (115.99 A/m). The suggested antenna design can be modified further for energy harvesting applications as reported in [17].

Acknowledgements Research described in this paper was financially supported by The World Bank and National Project Implementation Unit (NPIU), MHRD, India, under the TEQIP III project scheme of Collaborative Research Scheme (CRS). Project ID: 1-5744809094, Entitled "Development of IoT controlled frequency/pattern reconfigurable MIMO antenna for harvesting systems".

References

1. Abdalla MA, Ibrahim AA (2013) Compact and closely spaced metamaterial MIMO antenna with high isolation for wireless applications. IEEE Antennas Wirel Propag Lett 12:1452–1455. https://doi.org/10.1109/LAWP.2013.2288338
2. Pant D et al (2019) Error probability and channel capacity analysis of wireless system over inverse-gamma shadowed fading channel with selection diversity. Int J Commun Syst e4083
3. Chauha PS et al (2018) Modelling and performance analysis of energy detector over Weibull shadowed fading channel with application to cooperative sensing. Wirel Pers Commun 103(4):2791–2809
4. Sarangi AK, Datta A (2018) Capacity comparison of SISO, SIMO, MISO MIMO systems. In: Proceedings of the 2nd international conference on computing methodologies and communication, ICCMC, pp 798–801. https://doi.org/10.1109/ICCMC.2018.8488147
5. Sharma Y, Sarkar D, Saurav K, Srivastava KV (2017) Three-element MIMO antenna system with pattern and polarization diversity for WLAN applications. IEEE Antennas Wirel Propag Lett 16:1163–1166. https://doi.org/10.1109/LAWP.2016.2626394
6. Sun XB, Cao MY (2017) Low mutual coupling antenna array for WLAN application. Electron Lett 53(6):368–370. https://doi.org/10.1049/el.2016.4563
7. Sui J, Wu KL (2017) A general T-stub circuit for decoupling of two dual-band antennas. IEEE Trans Microw Theory Tech 65(6):2111–2121. https://doi.org/10.1109/TMTT.2017.2647951
8. Malviya LD, Panigrahi RK, Kartikeyan MV (2016) A 2 × 2 dual-band MIMO antenna with polarization diversity for wireless applications. Prog Electromagn Res C 61:91–103. https://doi.org/10.2528/PIERC15110401
9. Perhirin S, Auffret Y (2013) A low consumption electronic system developed for a 10 km long all-optical extension dedicated to sea floor observatories using power-over-fiber technology and SPI protocol. Microw Opt Technol Lett 55(11):2562–2568. https://doi.org/10.1002/mop
10. Bhadade RS, Mahajan SP (2019) Circularly polarized 4 × 4 MIMO antenna for WLAN applications. Electromagnetics 39(5):325–342. https://doi.org/10.1080/02726343.2019.1619227

11. Saurabh AK, Meshram MK (2020) Compact sub-6 GHz 5G-multiple-input-multiple-output antenna system with enhanced isolation. Int J RF Microw Comput Eng 30(8):1–11. https://doi.org/10.1002/mmce.22246
12. Li Q et al (2018) Mutual coupling reduction between patch antennas using meander line. Int J Antennas Propag. https://doi.org/10.1155/2018/2586382
13. Iqbal A, Saraereh OA, Ahmad AW, Bashir S (2017) Mutual coupling reduction using F-shaped stubs in UWB-MIMO antenna. IEEE Access 6:2755–2759. https://doi.org/10.1109/ACCESS.2017.2785232
14. System FMA (2016) Mutual coupling reduction of a closely spaced using discrete mushrooms. IEEE Trans Antennas Propag 64(10):3060–3067
15. Agarwal M, Meshram MK (2016) Isolation improvement of 5 GHz WLAN antenna array using metamaterial absorber. In: 2016 URSI Asia-Pacific radio science conference URSI AP-RASC, pp 1050–1053. https://doi.org/10.1109/URSIAP-RASC.2016.7601144
16. Roy S, Chakraborty U (2020) Mutual coupling reduction in a multi-band MIMO antenna using meta-inspired decoupling network. Wirel Pers Commun 114:3231–3246
17. Kumar A, Kallollu Narayaswamy N, Venkatesh Kumar H, Mishra B, Azam Siddique S, Kumar Dwivedi A (2021) High-isolated WiFi-2.4 GHz/LTE MIMO antenna for RF-energy harvesting applications. AEU - Int J Electron Commun 141:153964. https://doi.org/10.1016/j.aeue.2021.153964

Energy Band Gap Tuning Using 633 nm Laser Exposed on $(GeS_2)_{85}(Sb_2S_3)_{15}$ Thin Films: An Optical Analysis Using Spectroscopic Techniques

Rajnish Raj, Pooja Lohia, and D. K. Dwivedi

Abstract Thin films of sulphur-based chalcogenide glasses have been deposited on ultra-clean glass substrate using thermal evaporation technique. Optical properties of the thin films have been studied using UV–Vis spectrophotometer with the help of absorption spectra recorded in the range 300–900 nm. In the present paper, optical constants such as the absorption coefficient (α), extinction coefficient (k) and optical band gap (E_g) are obtained. It has been found that the absorption coefficient increases with increase in photon energy (hν), and the extinction coefficient decreases with increases in wavelength for laser-exposed thin films at different time intervals.

Keywords Chalcogenide glass · Thin films · Laser illuminated · Optical band gap · Exposed films

1 Introduction

With the emergence of laser technology in 1960, researchers paid too much attention to the field of data communications through the optical system. More information could be sent by an optical device compared to microwaves as well as other electrical systems. D. Jones introduced the first multi-mode optical fibre scheme with losses of below 20 dB/km [1]. Currently designed optical fibre sensors are capable of transmitting gigabit (GB) information at light speeds. These optical fibre sensors are easy to move from one location to another having lightweight, compact in size, and are highly resistant to conventional sensors [2]. Due to its thermal, mechanical, electric and optoelectronic characteristics, the optical fibre sensor synthesised using non-oxide chalcogenide glasses has got a lot of interest from researchers due to its potential

R. Raj · P. Lohia
Department of Electronics and Communication Engineering, M. M. M. University of Technology, Gorakhpur 273010, India

D. K. Dwivedi (✉)
Photonics and Photovoltaic Research Lab, Department of Physics and Material Science, M. M. M. University of Technology, Gorakhpur 273010, India
e-mail: todkdwivedi@gmail.com

B. Mishra and M. Tiwari (eds.), *VLSI, Microwave and Wireless Technologies*,
Lecture Notes in Electrical Engineering 877,
https://doi.org/10.1007/978-981-19-0312-0_61

applications in industries [3–5]. Holography, xerography, far-infrared optical (IR) fibres, switching devices, X-ray imaging, photo-voltaices, electro-chemical sensors, bio-medical sensors, IR detectors, data storage devices have potential applications of chalcogen glasses, [6–8] IR biological sensor and gas sensor [9]. Chalcogen element doped with different materials can have varying characteristics. Alteration in the optical constant of such materials has enabled them to be used in storage devices for high-density information and high-resolution display [10, 11]. Many materials are used for optoelectronic applications, but due to their remarkable properties over traditional materials, glassy semiconductors also known as chalcogenide glasses alloys could be a good candidates for optical fibre sensors [12].

GeS2-Sb2S3 have excellent thermal stability and an advantageous optical properties [13]. These glasses are one of the most interesting metal chalcogenides, and thin films of Sb2S3 are of great interest for use in high-reflecting dielectric film, switching devices, diodes, detectors etc. [14, 15]. It was possible to reasonably optimise the glasses forming region, refractive index, optical non-linearity and other optical properties for their different applications in devices.

In the current paper, optical transmission of $(GeS_2)_{85}(Sb_2S_3)_{15}$thin films for 45, 60 and 90 min using He–Ne laser illumination. Thin films have been investigated for by UV–visible spectrophotometer in the 300–900 nm spectral range. For the design and analysis of various optical and optoelectronic devices, accurate value of the absorption coefficient, extinction coefficient and optical band gap of semiconducting films is important.

2 Experimental Analysis

The studied compound $(GeS_2)_{85}(Sb_2S_3)_{15}$ were synthesised by the conventional cost-effective process melt-quench technique. Electronic balance (WENSAR) weights the exact quantity of 5 N pure (99.999%) additive materials (Ge, S and Sb) with a minimum count of 0.0001 gm. Using acetone and distilled water, the quartz ampoule was cleaned. After that, all the materials are packed in an internal diameter 8-mm-quartz ampoule and a length of 7 cm held at a vacuum of 10^{-5} mbar. The ampoules are therefore placed in the muffle furnace for 12 h, held at a temperature of 800 °C. At the rate of 3–5°C/min, the temperature of the muffle furnace raised steadily. The quartz crystal ampoules are continuously rocked over an interval of 10 min during the heat treatment by rotating right and left having the support of ceramic rod hooked up to obtain its homogeneous nature of glass materials. The quartz ampoule holding the molten material was subsequently dipped rapidly in cold water with ice. After that, we separated the ampoule to collect the materials, and for thin film processing, bulk samples were grinded in powdered form. After that, the vacuum thermal evaporation method was used during the development of Ge–S–Sb glass alloy thin films. Thin film thickness was measured as 500 nm (nm) from the quartz crystal thickness monitor display throughout the vacuum thermal evaporation coating unit.

A UV–Vis computer-controlled spectrophotometer (Shimadzu, UV-2600) has been used to evaluate emission spectra for both as prepared and laser-illuminated (continuous wave He–Ne laser 633 nm) thin films at ambient temperature with 300–900 nm wavelength range.

3 Result and Discussion

3.1 Coefficient of Absorption (A) and Band Gap (E_g)

The optical absorption spectra to all of the deposited films were recorded between 300 and 900 nm employing UV–Vis spectrophotometer to study the existence of the electronic transitions for both the as prepared and laser-illuminated $(GeS_2)_{85}(Sb_2S_3)_{15}$ thin films. Optical constants such as the absorption coefficient (α), extinction coefficient (k) as well as the optical band gap (E_g) were obtained using transmittance spectra.

The absorption coefficient (α) [16–18] is dependent on the light absorbing capacity per unit thickness of the thin film. Observed value of an absorption coefficient (α) depends exponentially on photonic energy. Thus, coefficient of absorption is examined as

$$\alpha = \frac{2.303\ \mathrm{A}}{t} \tag{1}$$

where A defines the absorbance which is observed by optical transmission spectra $[A = 2 - \log T]$ whilst t denoting the thickness of thin films was recorded by quartz crystal display. The graph of as prepared sample, photon energy versus coefficient of absorption (α) reveals that α increases exponentially with the increasing photon energy which is shown in Fig. 1.

Both photons and phonons are absorbed by holes and electrons in the absorption process. Because of absorption, these phonon and photons acquire required energy and momentum, and these phenomenon can be described upon the basis of excitation absorption, basic absorption and acceptor absorption in valence band. The compared value of absorption coefficient of both unexposed and laser exposed thin films is given in Table 1.

The coefficient of absorption (α) has been used to determine to find the energy band gap (E_g) of the as prepared material and laser-illuminated films $(GeS_2)_{85}(Sb_2S_3)_{15}$ for 45, 60 and 90 min. Absorption edge is usually expressed as exponential tail present in amorphous semiconductor. The absorption coefficient just above the exponential tail obeys the following equation. In accordance with the Tauc plot [17].

$$(\alpha h\nu)^{1/n} = \mathrm{A}(h\nu - E_g) \tag{2}$$

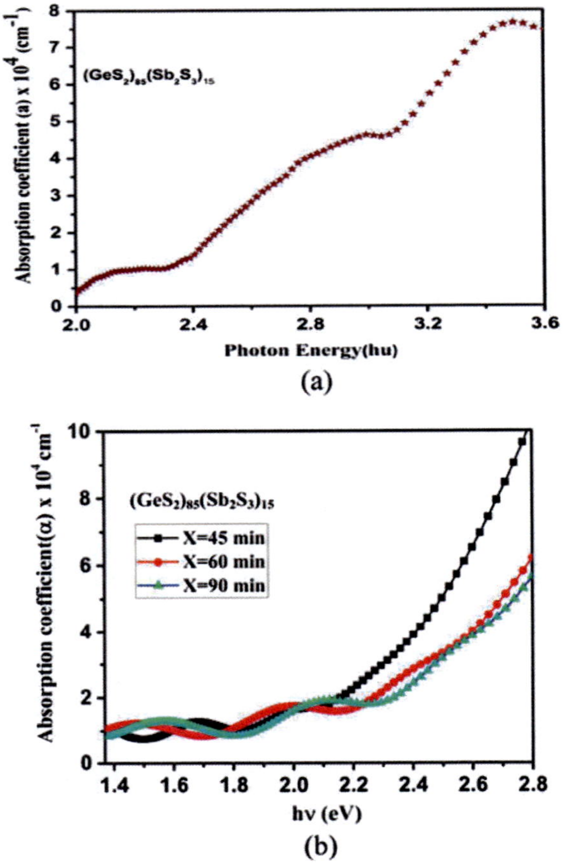

Fig. 1 Dependence of coefficient of absorption (α) on photon energy in $(GeS_2)_{85}(Sb_2S_3)_{15}$ **a** as prepared **b** laser-illuminated thin films

Table 1 Optical constant of as prepared $(GeS_2)_{85}(Sb_2S_3)_{15}$ and laser-annealed thin films different time durations

Composition $(GeS_2)_{85}(Sb_2S_3)_{15}$	As prepared	Laser exposed for 45 min	Laser exposed for 60 min	Laser exposed for 90 min
Absorption coefficient (at 2.5 eV) $\times 10^4$	2.10	5.04	3.34	3.16
Extinction coefficient (k) (at $\lambda = 395$ nm) $\times 10^{-3}$	15.68	54.03	35.15	32.83
Optical band gap (E_g) eV	2.40	2.06	2.16	2.23

where ν represents incident photons frequency and A represents a constant which relating the parameter of edge width and depends on the high probability of transition and is determined from the linear portion from the above equation. E_g denotes the optical bandgap; n in Eq. (2) seems to be a variable that accepts different values such as $n = 1/2$ and 2 that define direct and indirect allowable transitions alike, whereas 3 and $n = 3/2$ define indirect and direct and forbidden transitions, respectively.

In the mentioned study, the experimental result for $(GeS_2)_{85}(Sb_2S_3)_{15}$ thin films was found to be better suited at $n = 1/2$ which implies that direct transition is responsible for, as prepared and laser exposed (for 45, 60 90 min) thin films $(GeS_2)_{85}(Sb_2S_3)_{15}$. Non-oxide glassy materials optical band gap (Eg) energies were used to obtain at the edge of absorption by predicting the absorption through direct allowed transition by using classical Tauc equation [18]. All evaluated the observed value of optical band gap (E_g) of the unexposed and exposed thin films are given in Table 1. The comparison graph of unexposed and laser exposed with a time duration for the $(GeS_2)_{85}(Sb_2S_3)_{15}$ thin films is plotted between $(\alpha h\nu)^2$ and photonic energy $(h\nu)$ as shown in Fig. 2.

Fig. 2 Photon energy $(h\nu)$ versus of $(\alpha h\nu)^2$ for $(GeS_2)_{85}(Sb_2S_3)_{15}$ **a** as prepared **b** laser-illuminated thin films

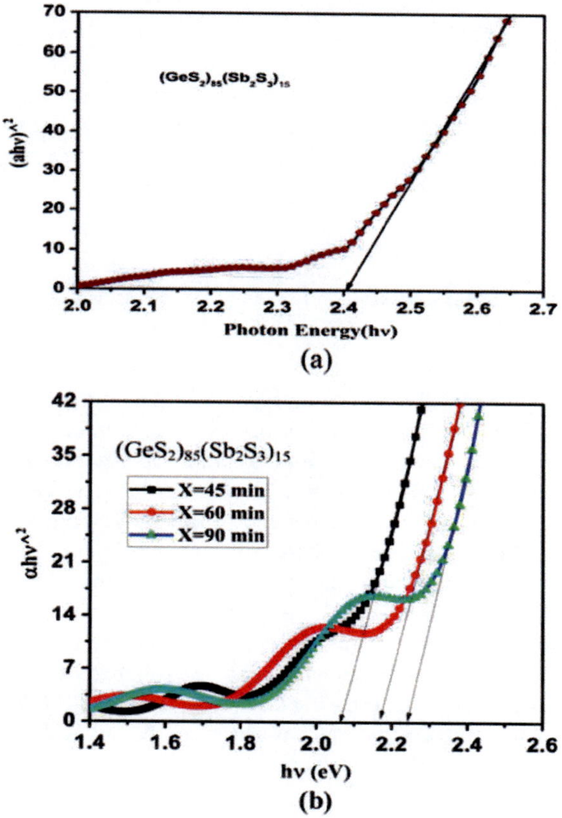

Where the direct allowed optical band gap (E_g) of non-oxide glassy material $(GeS_2)_{85}(Sb_2S_3)_{15}$ for the unexposed as prepared thin films is determined 2.4 eV by extrapolating from the linear part of the graphs down to zero, which is determined $(\alpha h\nu)^2 \rightarrow 0$ as $\alpha \rightarrow 0$. Increasing order of band gap of laser exposed thin films could be explained on the basis of band tail. Our outcomes of results are found in good agreement in accordance with the recently published findings of other researchers [19–21].

3.2 *Coefficient of Extinction* (k)

The coefficient of extinction (k) represents the decreasing rate of light transmission due to the absorption and scattering, and it could be obtained by [22–24]

$$k = \alpha\lambda/4\Pi \qquad (3)$$

where α is the coefficient of absorption, λ is the wavelength of light which emerges on the films. For $(GeS_2)_{85}(Sb_2S_3)_{15}$ thin film, where wavelength dependency of k is shown in Fig. 3.

It is investigated that, k decreases with wavelength (λ) linearly, which could be due to an exponential decrease in absorption of spectra at particular wavelength. The evaluated rate of extinction coefficient (k) of the unexposed and exposed thin films is given in Table 1.

4 Conclusion

In the purposed study $(GeS_2)_{85}(Sb_2S_3)_{15}$, fabrication of thin film of non-oxide glassy material mostly by vacuum evaporation technique coating unit has been examined for optical properties using spectroscopic techniques. Optical absorbance spectra have been used to determine different optical constants such as absorption coefficient (α), extinction coefficient (k) and optical band distance (E_g). With the incident photon energy ($h\nu$), the coefficient of absorption (α) increases linearly, whilst the extinction coefficient (k) decreases linearly with wavelength (λ) for as prepared as well as laser-exposed thin films. The observed value of optical band gap (E_g) of unexposed thin film is determined as 2.4 eV, whilst for laser exposed thin films for different time, optical band gap is observed to first decrease with respect to as prepared film, and thereafter, it increases which can be explained by Mott and Davis model. As E_g and α increase with increasing laser-exposed time, whilst the value of k decreases. Since the investigated material has a high coefficient of absorption (α), this could be a good applicant for the optoelectronic applications.

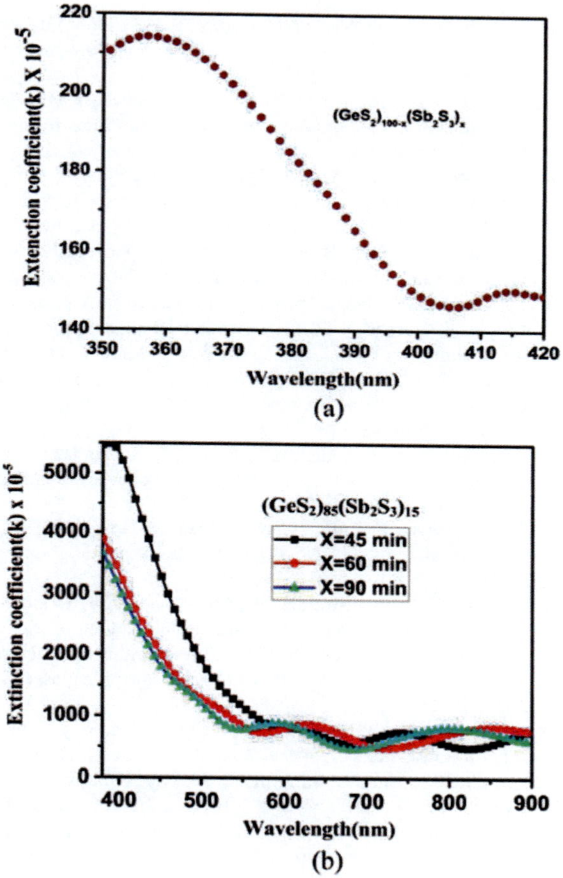

Fig. 3 Comparison graph for coefficient of extinction (k) on wavelength (nm) **a** as prepared **b** laser-illuminated $(GeS_2)_{85}(Sb_2S_3)_{15}$ thin films

References

1. Jones D (1998) Introduction to fiber optics, Naval Education and Training Professional Development and Technology Center
2. Girard S, Mescia L, Vivona M, Laurent A, Ouerdane Y, Marcandella C, Prudenzano F, Boukenter A, Robin T, Paillet P, Goiffon V, Gaillardin M, Cadier B, Pinsard E, Cannas M, Boscaino R (2013) Design of radiation-hardened rare-earth doped amplifiers through a coupled experiment/simulation approach. J Lightwave Technol 31:1247–1254
3. Seddon AB (1995) Chalcogenide glasses: a review of their preparation, properties and applications. J Non-Cryst Solids 184:44–50
4. Shiryaeva VS, Karaksina V, Churbanov MF, Kotereva TV, Stepanov BS, Ketkova LA, Evdokimov II, Koltashev VV, Plotnichenko VG, Filatov AI, Antonov IN (2018) Special pure germanium-rich Ga–Ge–As–Se glasses for active mid-IR fiber optics. Mater Res Bull 107:430–437
5. Zakery A, Elliott SR (2003) Optical properties and applications of chalcogenide glasses: a review. J Non-Cryst Solids 330(1–3):1–12

6. Tang Z, Furniss D, Neate NC, Benson TM, Seddon AB (2019) Low gallium-content, dysprosium III-doped, Ge–As–Ga–Se chalcogenide glasses for active mid-infrared fiber optics. J Am Chem Soc 102(1):195–206
7. Darwish AAA, Rashad M, Bekheet AE, El-Nahass MM (2017) Linear and nonlinear optical properties of GeSe2−xSnx ($0 \leq x \leq 0.8$) thin films for optoelectronic applications. J Alloys Compd 709:640–645
8. Darwish AAA, Ali HAM (2017) On annealing induced effect in optical properties of amorphous GeSeSn chalcogenide films for optoelectronic applications. J Alloys Compd 710:431–435
9. Lee B (2003) Review of the present status of optical fiber sensors. Optical Fiber Technol 9:57
10. Ryzhov VA, Melekh BT (2018) Optical properties of tellurium-based chalcogenide alloys in the far infrared region (λ>30 μm). Semiconductors 52(2):209–214
11. Savage JA (1982) Optical properties of chalcogenide glasses. J Non-Cryst Solids 47(1):101–115
12. Mehta N, Kumar A (20130 Recent advances in chalcogenide glasses for multifunctional applications in fiber optics. Recent Patents Mater Sci 6(1):59–67
13. Wang J, Yu X, Long N, Sun X, Yin G, Jiao Q, Liu X, Dai S, Lin C (2019) Spontaneous crystallization of PbC_{l2} nanocrystals in GeS_2-Sb_2S_3 basedchalcogenide glasses. J Non-Cryst Solids 521:119543
14. Petit L, Carlie N, Adamietz F, Couzi M, Rodriguez V, Richardson KC (2006) Correlation between physical, optical and structural properties of sulfide glasses in the system Ge–Sb–S. Mater Chem Phys 97:64–70
15. Fang X, Zhang X, Jing R, Chen G, Ma H, Adam JL (2010) Glass formation and crystallization behavior of novel GeS2–Sb2S3–PbS chalcogenide glass system. J Am Ceram Soc 89:2154–2157
16. Mohamed M, Abdel-Rahim MA (2014) Mater Sci Semicond Process 27:288–292
17. Tauc J (1974) Amorphous and liquid semiconductors. Plenum, New York, p 171
18. Raj R, Lohia P, Dwivedi DK (2020) Investigation of structural and optical properties of (GeS2) 100–x (Sb2S3) x (x = 15, 30, 45, 60) chalcogenide glasses for mid infrared applications. Optik 218:165041
19. Tauc J, Menth A (1972) States in the gap. J Non-Cryst Solids 8–10:569–585
20. Aparimita A, Naik R, Sripan C, Ganesan R. Laser-induced optical photo bleaching in Bi-doped Ge30Se70 amorphous thin films. Appl Phys A https://doi.org/10.1007/s00339-019-3194-4
21. Musgraves JD, Carlie N, Hub J, Petit L, Agarwal A, Kimerling LC, Richardson KA (2011) Comparison of the optical, thermal and structural properties of Ge–Sb–S thin films deposited using thermal evaporation and pulsed laser deposition techniques. Acta Mater 59:5032–5039
22. Guo Q, Xu Y, Guo H, Xiao X, Lin C, Cui X, Wang P, Gao F, Lu M, Peng B (2017) Effect of iodine (I2) on structural, thermal and optical properties ofGe-Sb-S chalcohalide host glasses and ones doped with Dy. J Non-Cryst Solids 464:81–88
23. Swanepoel RJ (1984) Phys E: Sci Instrum 17:896–903
24. Raj R, Lohia P, Dwivedi DK (2020) Structural and optical investigations of (GeS2)85(Sb2S3)15 chalcogenide glassy alloy: a material for IR devices. In: 2020 International Conference on Electrical and Electronics Engineering (ICE3), 14–15 Feb 2020

A Compact Circular Ultra-Wideband MIMO Antenna Sensor (CUMAS) Probe for Breast Cancer Detection

Ankit Kumar Gupta, Praveen Kumar Rao, and Rajan Mishra

Abstract In this paper, a compact ultra-wideband circular microstrip patch antenna is designed for the detection of breast cancerous tumors. The designed antenna is a four-port MIMO of 2 × 2 is proposed. The size of the proposed antenna is 20 × 30 × 1.6 mm. It is designed on a low-cost FR-4 epoxy substrate. The antenna is operated between the frequency ranges of 5.01–19.59 GHz. Isolation among antenna elements is greater than 22 dB. Further, the 3-D breast phantom model is also simulated for analysis of the effect of SAR. Due to the variation in the electrical properties of cancerous cells and healthy cells it is possible to identify the cancerous tumor using SAR analysis. The obtained maximum local SAR value without a cancerous tumor is 72.3 W/Kg and with a cancerous tumor is 113 W/Kg. Thus it shows the difference between the SAR value of a normal cell and a cancerous cell and helps to identify the tumor.

Keywords Multi-input multi-output (MIMO) antenna · Reflection co-efficient · Isolation co-efficient · Specific absorption rate (SAR)

1 Introduction

The most dangerous and common disease among all of the diseases is cancers. Cancer is of many types such as lung cancer, skin cancer, blood cancer, breast cancer, and many more. Breast cancer is the most frequent diagnosis form of cancer. The second most major cause of death among females is breast cancer. According to Global cancer statistics (GCS) 2020, there were over 2.3 million new cases in 2020, it is nearly 11.4% of all cancer [1].

The chance that women will die from cancer is 2.6%, however, from 2013 to 2018, the death rate decreased by 1% per year [2]. This decrease is possible due to early detection and Screening technology for better medical treatment. This motivates

A. K. Gupta · P. K. Rao (✉) · R. Mishra
Electronics and Communication Engineering, Madan Mohan Malaviya University of Technology, Gorakhpur, (U.P.), India
e-mail: rao.prabeen143@gmail.com

B. Mishra and M. Tiwari (eds.), *VLSI, Microwave and Wireless Technologies*, Lecture Notes in Electrical Engineering 877,
https://doi.org/10.1007/978-981-19-0312-0_62

the researcher to find a more efficient, cost-effective, comfortable, and non-invasive technique for the early detection of breast cancer.

Breast cancer is caused due to development of malignant cells in the breast which have different electrical properties than normal cells. Presently there are different types of early detection modalities are available for clinical examination of breast cancer [3]. These are X-Rays mammography, Ultrasonography, and Magnetic resonance imaging. Unfortunately, all previously mentioned techniques have some drawbacks such as false-negative results, high cost, patient comfort, and risk of other diseases due to ionization [4, 5].

The X-Ray mammography technique is conventional commonly and used for tumor detection. But due to the percentage of false results is more and it cannot find the accurate location of the malignant cells. It may also affect the healthier cells which are near to malignant cells because of ionization. Another alternative detection method is by ultrasound which gives less false-positive result than X-Ray mammography but not so accurate to detect the cancerous cells which are laying deep inside the breast [6]. Another technique is MRI it gives fewer false results by accurate identification between benign and malignant breast lesions. It is a time-consuming, expensive, and complex process. All above mention have their drawback which motives researcher to find another modality which can give more accurate, simple, less time, patient comfort, consuming, non-ionizing, and cost-effective technique.

Microwave imaging sensors specifically in the ultra-wideband region are an alternative to conventional methods to overcome the limitation of previously used techniques. The advantages of using microwave-based sensors antenna for identifying breast tumors are their wide bandwidth, compact size, repeatability, and no ionization effect [7]. The dielectric property (electrical property) of the different cells helps us to make difference between the cancerous cells and healthy cells.

In the last decade, UWB technology has received great attention and extraordinary development due to greater data rate, omnidirectional radiation, and channel capacity, but there is also some limitation of multipath fading in such communication [8]. MIMO antennas are used widely to overcome multipath fading degradation in wireless communication systems and enhanced system capabilities [9].

The electromagnetic sensing behavior of any biological tissue depends upon the dielectric property of that tissue. The two most relevant numerical models are the Debye model and the Cole–Cole model. In this paper the simulated breast phantom relative permittivity is calculated using the proposed antenna is evaluated by the Cole–Cole model as reported in [10].

Even though, much different design and size of antenna have been mentioned for malignant tumor screening in the literature; a breast tumor detection antenna setup was reported by Islam et al. [11], and antipodal Vivaldi antenna sensors were used as a transceiver. The breast phantom was positioned between these antennas. The proclaimed antenna sensor had operated between the frequencies ranges of 3.01–11 GHz. Simulation-based-SAR analysis using ultra-wideband antenna has also been reported in [12] and the antenna is operated within a 3–15 GHz frequency band. The SAR value of the healthy and cancerous phantom mainly depends on the electrical property (permittivity and conductivity) of matter. The tumor screening method was fixed which is based on the simulation-based average SAR and local SAR values. Electrochemical impedance spectroscopy was founded in [13] which is based on a Field-programmable array (FPGA) for bio-detection of different parts of the body but the experiment was performed only on the egg sample. Moreover, two vital functions of antenna sensors besides bio-detection were also carried out in [14, 15]. Regardless of huge advancement in microwave imaging still, there is the chance to design a much effective antenna for breast cancer screening.

In this paper, a circular ultra-wideband MIMO antenna sensor (CUMAS) is presented for the breast phantom diagnosis. The proposed antenna is formed in four steps. The CUMAS is designed with four circular-shaped having elliptical etch radiating elements over FR-4 epoxy substrate. Although it is lossy at higher frequency, it is cost-effective and easily available. The 3-D breast phantom has been simulated with the proposed antenna. The proposed antenna is fixed over the healthy and cancerous phantoms (having a single tumor) and different S-parameters are observed for further analysis. Even though multi-input multi-output antennas are many times used for communication fields as reported in [16, 17]. Here, the antenna is uniquely constructed for cancerous tumor identification. The designed antenna is to be considered as a replacement of antenna having a single port excited antenna. Only a single S-parameter ($S11$) can be calculated from the antenna having a single radiating element [18]. The tumor identification based on single specification analysis is very finite. However, the designed CUMAS probe can measure multiple S-parameters such as ($S11, S21, S31, S41$) and ($S12, S13$) despite one parameter. Therefore, more accurate analyses of the healthy and cancerous phantoms are achievable using the CUMAS. For more accurate identification between the healthy and cancerous phantom, SAR analysis is further done over healthy phantom and phantom containing single tumor of radius 5 mm. This antenna is specifically designed to operate on high SAR values to distinguish between healthy and cancerous phantom. The presented method of the 3-D phantom analysis for malignant tissue identification is much better than the formerly reported single antenna.

2 Antenna Design and Performance

The CUMAS for breast cancer detection is designed in five steps. In each step, there is a modification in the ground and patch of antenna for improving the operating bandwidth and isolation parameters. The presented antenna is designed on the FR4 substrate of 1.6 mm thickness having a dielectric constant of 4.4 and loss tangent of 0.02 due to easy availability and low-cost, the antenna cost decrease. Finally, a compact size of the 20×30 mm antenna is proposed. In step 1 the single monopole antenna with a circular patch of 4 mm radius with the partial ground of 5 mm and the uniform strip of the width of 1 mm providing impedance matching of 5 ohms. In next step 2 the elliptical slot, the rectangular notch is etched and the rectangular patch is united near the port in staircase form to enhance the bandwidth. Further in step 3, the monopole antenna is arranged side by side to form two port antenna. To improve the isolation parameter the T shape stub is introduced between them in the ground plane in step 4. The final step in which the antenna is arranged in a 2 × 2 array of 4 radiating elements to get all the other desired parameters that are $S13$, $S24$, and all of total 16 parameters.

The designed antenna is well resonated in the UWB frequency domain which has a bandwidth of 5.01–19.59 GHz.

The isolation between elements more than 20 dB is also obtained among the antenna elements shown in Fig. 4b. The final proposed antenna top view and bottom view are shown in Fig. 3. Now the designed antenna is ready for breast phantom analysis for a malignant cell inside the phantom (Figs. 1 and 2).

An antenna is a device that transforms the electrical energy received from the transmitter to the electromagnetic wave. How efficiently the antenna could perform this conversion, is called antenna radiation efficiency, which is shown in Fig. 4c. The plot of peak gain is also shown in Fig. 4d (Fig. 5).

The radiation patterns of the designed antenna are measured for E-field and H-field. The frequency at which the patterns are analyzed is 5.8 GHz, 9.5 GHz, and 13.3 GHz, respectively. The results regarding the same are shown above in the figure. The proposed CUMAS probe shows more directional E-field as compared to H-field at a lower frequency, i.e., 5.8 GHz. As we increase the frequency to 9.5 GHz we observe a distortion in the E-field. Further, if there is an increase in the frequency to

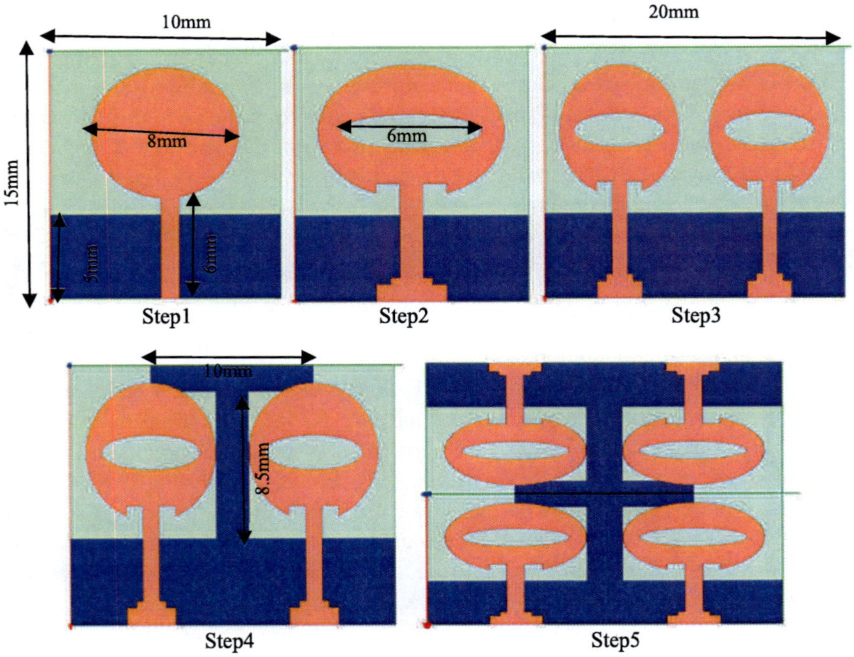

Fig. 1 Evolution step of the designed antenna

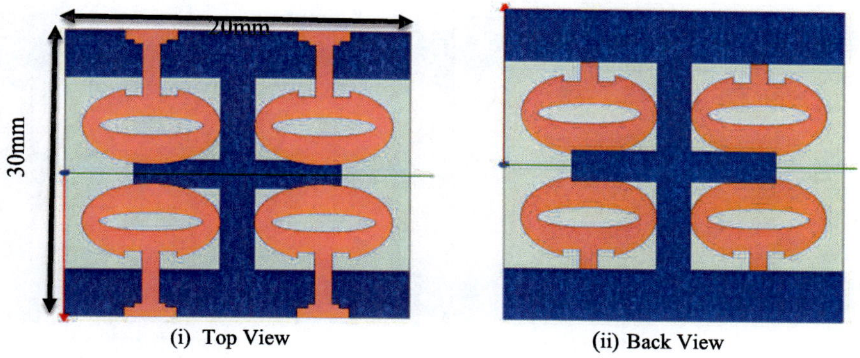

Fig. 2 (i) Top view (ii) Back view of designed antenna

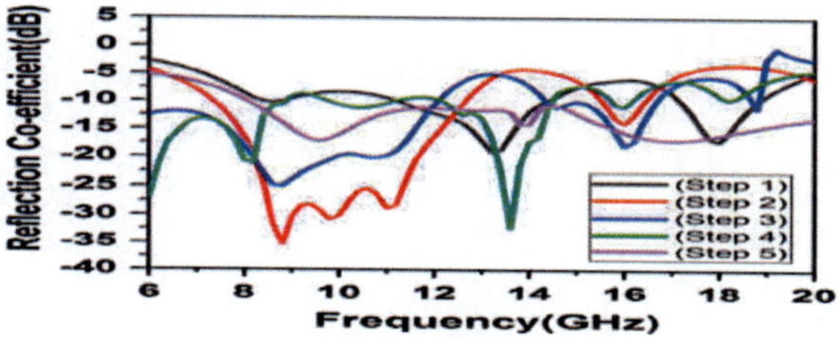

Fig. 3 The return loss of each step of evolution

13.3 GHz it is observed that the *E*-field is distorted omnidirectionally and H-field is reversed directionally having a high back lobe. So, from the following observations, we can conclude that at lover frequency, i.e., 5.8 GHz we attain a stable and directional *E*-field as compared to the higher frequencies (9.5 GHz and 13.3 GHz).

Now, the designed antenna is flawlessly verified and ready for analysis over a simulated breast phantom model, to identify the tumor embedded inside the phantom. A 3-D breast phantom is simulated for analysis purposes in the upcoming section.

3 Breast Cancer/tumor Identification

3.1 Breast Phantom

The goal of this section is to implement the proposed antenna on the breast phantom to identify a malignant tumor embedded inside the breast phantom. Phantom is structured in 3-D shape and replica of the original breast, the electrical property is maintained for a different layer of the breast such as skin, fat, glandular tissues, and malignant tumor. Considering the electrical property of breast tissue to be a Gaussian

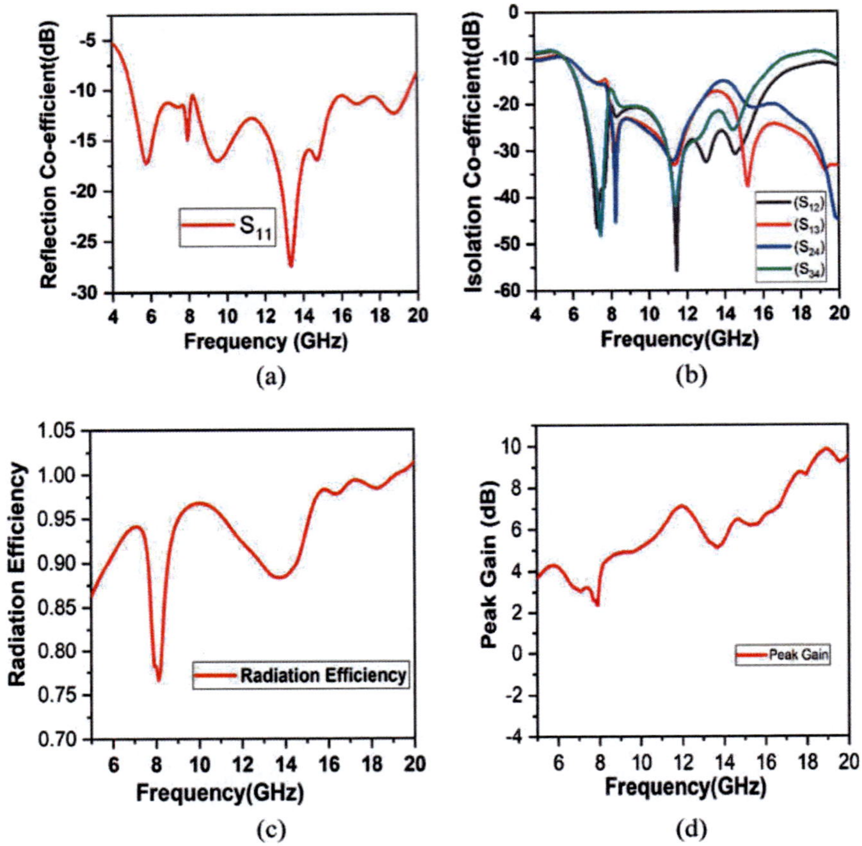

Fig. 4 **a** Reflection co-efficient **b** Isolation co-efficient **c** Radiation efficiency **d** Peak gain of proposed antenna

random variable with nearly ±10% approximation from normal data and this data value is mentioned in the article [19] (Fig. 6).

The above figure shows the 3-D model of breast phantom of 55 mm height. It is consisting of 3 layers of skin, fat, and glandular tissue. The outermost layer is formed by thin skin which has a thickness of 5 mm, σ 1.46 S/m having a mass density of 1010 kg/m^3, and dielectric permittivity of 38. The next layer is fat of thickness 10 mm, σ of 0.1 S/m, mass density 928 kg/m^3, and permittivity of 5.2. The inner part of the breast phantom is the glandular tissue having a thickness of 40 mm of electrical conductivity of 0.5 S/m and dielectric property of 20.1 constituting a mass density of 1035 kg/m^3.

For detection purposes, a tumor of radius 5 mm having a relative permittivity of 62 and a conductivity of 4.2 S/m is implanted inside the breast phantom.

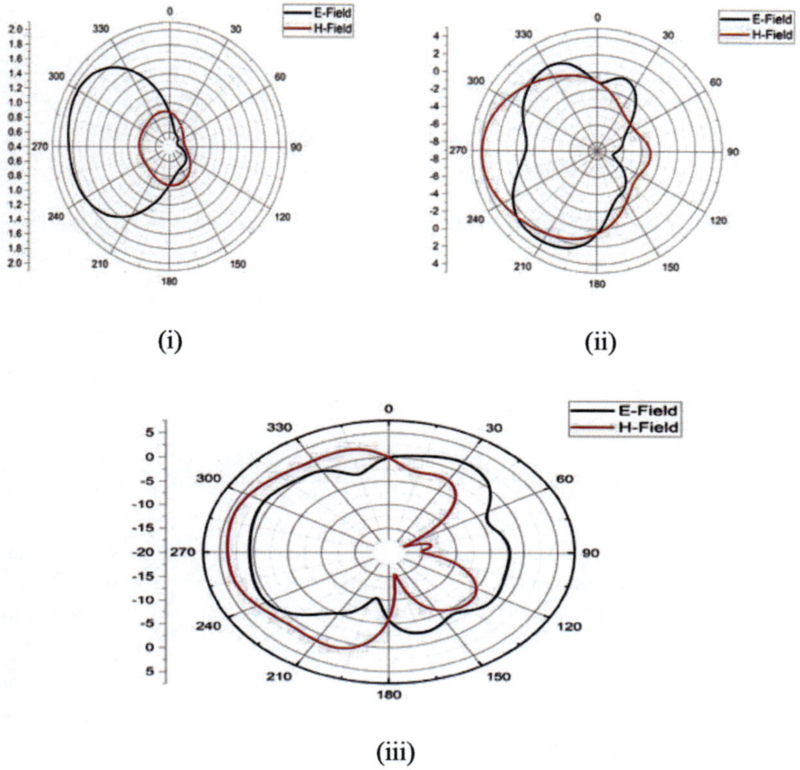

Fig. 5 The radiation pattern for *E* and *H*-Field of designed antenna sensor at (i) 5.8 GHz (ii) 9.5 GHz (iii) 13.3 GHz

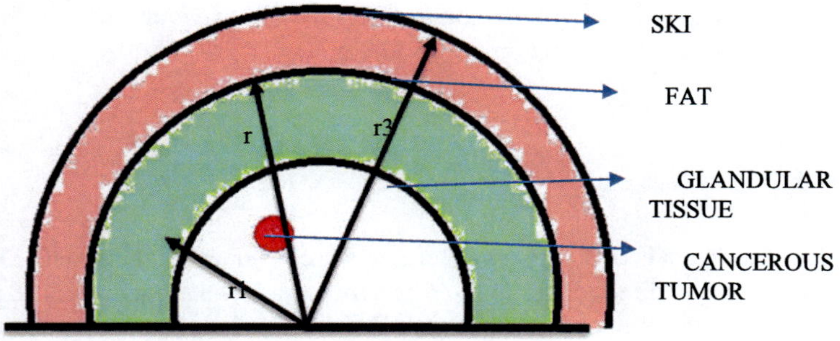

Fig. 6 3-D Breast phantom model ($r1 = 40$ mm, $r2 = 50$ mm, and $r3 = 55$ mm)

3.2 Specific Absorption Rate (SAR)

The SAR is defined as the quantity of energy absorbed by body tissues when brings in contact with an electromagnetic wave. The quantity of energy absorbed by any tissue is defined in [20]. SAR is calculated in watt per kilogram (W/Kg) which is calculated over the small amount of tissue mass of (1gm) or (10gm). SAR value is mainly depending upon the electric field within tissues [21].

The local SAR in (W/Kg) [30] at each point in tissue is evaluated by the formula:

$$\mathrm{SAR}_{\mathrm{local}}(r, \omega) = \frac{\sigma(r, \omega)|E(r, \omega)|^2}{2\rho(r)} \tag{1}$$

For computation average, SAR from the SAR of the local point is obtained by integrating over cube mass and divided by cube mass. It is given as:

$$\mathrm{SAR}_{\mathrm{average}}(r, \omega) = \frac{1}{v}\int \frac{\sigma(r, \omega)|E(r, \omega)|^2}{2\rho(r)} dr \tag{2}$$

where $E(r, \omega)$ is the electric field (V/m) at a distance of r, $\sigma(r, \omega)$ is the conductivity in (S/m) of the phantoms at an angular frequency (ω), (v) volume in (m^3), $\rho(r)$ is the tissue mass density in kilogram per unit volume in meter (kg/m3) and (r)is the position vector between the antenna and tissue.

3.3 Tumor Detection

The specific absorption rate value of the breast tissue without tumor and with cancerous tumor using the proposed antenna is simulated at 3 GHz. The maximum local specific absorption rate without tumor is 72.53 W/Kg and with a tumor of radius 5 mm at the location (15,15,30) mm is 113 W/Kg (Fig. 7).

4 Conclusion

The designed antenna is of a very compact size which is operated between 5.01 and 19.59 GHz. It shows directional radiation patterns at a lower frequency. The isolation between antenna elements is improved. The proposed antenna is also analyzed over a 3-D breast phantom model. The SAR analysis is carried over the breast phantom model consisting of a cancerous tumor of size 5 mm and without tumor. The specific absorption rate for a complete breast phantom without a cancerous tumor is 72.3 W/Kg and with a cancerous tumor is 113 W/Kg. From SAR analysis it is clear

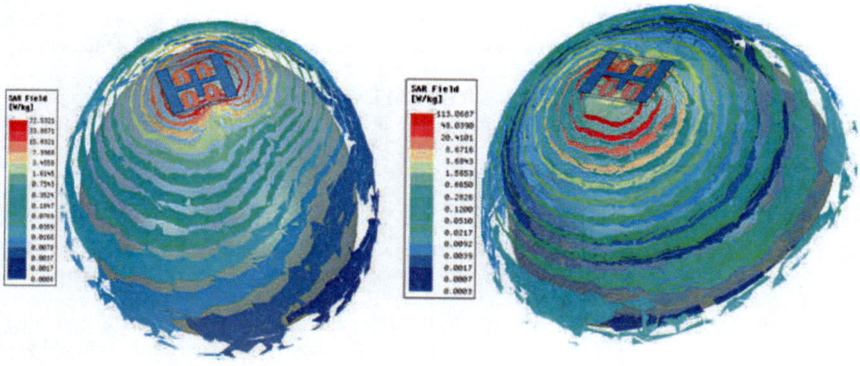

(a) Local SAR without tumor (b) Local SAR with tumor

Fig. 7 Simulated SAR results on 3-D breast phantom **a** Normal, **b** with a single tumor

that the SAR is higher for breast phantom containing the cancerous tumor. Hence, it is concluded that it possible is to detect the cancerous tumor in the breast.

References

1. Sung H et al (2021) Global cancer statistics 2020: GLOBOCAN estimates of incidence and mortality worldwide for 36 cancers in 185 countries. CA: a cancer journal for clinicians. 4 Feb 2021. https://doi.org/10.3322/caac.21660
2. American Cancer Society (2021) Cancer facts & figures 2021. American Cancer Society, Atlanta. Last Accessed 13 Mar 2021
3. Rao PK, Mishra R (2020) Elliptical shape flexible MIMO antenna with high isolation for breast cancer detection application. IETE J Res 24:1–9
4. Rao PK, Mishra R (2020) Elliptical shape CPW antenna for breast cancer detection applications. In: 2020 International conference on electrical and electronics engineering (ICE3), Gorakhpur, India, pp 489–494. https://doi.org/10.1109/ICE348803.2020.9122817
5. Rao PK, Mishra R (2020) UWB antenna for breast cancer detection using back scattering. In: Janyani V, Singh G, Tiwari M, Ismail T (eds) Optical and wireless technologies. Lecture Notes in Electrical Engineering, vol 648. Springer, Singapore. https://doi.org/10.1007/978-981-15-2926-942
6. Rao P, Yadav A, Mishra R (2020) AMC-based antenna sensor for breast tumors detection. Int J Microwave Technol 1–8. https://doi.org/10.1017/S1759078720001609
7. Rao PK, Mishra R (2021) Resonator based antenna sensor for breast cancer detection. Progress Electromagnet Res (PIER) M, 1–11 10 fed 2021
8. Liang C, Su R et al (2018) Compact printed MIMO antenna with 6.1GHz notched band for ultra-wide applications. Progress Electromagnet Res Lett 76:77–83
9. Parchin et al (2020) Ultra-wideband diversity MIMO antenna system for future mobile handsets. Sensors 20(8):2371
10. Said T, Varadan VV (2009) Variation of cole-cole model parameters with the complex permittivity of biological tissues. In: IEEE MTT-S international microwave symposium digest, Jun 2009, pp 1445–1448. https://doi.org/10.1109/MWSYM.2009.5165979

11. Islam MT, Samsuzzaman M, Islam MT, Kibria S, Singh MJ (2018) A homogeneous breast phantom measurement system with an improved modified microwave imaging antenna sensor. Sensors 18(9):2962. https://doi.org/10.3390/s18092962
12. Subramanian S, Sundarambal B, Nirmal D (2018) Investigation on simulation-based specific absorption rate in ultra-wideband antenna for breast cancer detection. IEEE Sensors J 18(24):10002–10009. https://doi.org/10.1109/JSEN.2018.2875621
13. Jiang Z et al (2019) Development of a portable electrochemical impedance spectroscopy system for bio-detection. IEEE Sensors J 19(15):5979–5987. https://doi.org/10.1109/JSEN.2019.2911718
14. Xavier GVR, da Costa EG, Serres AJR, Nobrega LAMM, Oliveira AC, Sousa HFS (2019) Design and application of a circular printed monopole antenna in partial discharge detection. IEEE Sensors J 19(10):3718–3725. https://doi.org/10.1109/JSEN.2019.2896580
15. Gargari AM, Zarifi MH, Markley L (2019) Passive matched mushroom structure for a high sensitivity low profile antenna-based material detection system. IEEE Sensors J 19(15):6154–6162. https://doi.org/10.1109/JSEN.2019.2908687
16. Sharma MK, Kumar M, Saini JP, Singh SP (2019) Computationally optimized MIMO antenna with improved isolation and extended bandwidth for UWB applications. Arabian J Sci Eng 45(3):1333–1343. https://doi.org/10.1007/s13369-019-03888-6
17. Gangwar D, Sharma A, Kanaujia BK, Singh SP, Lay-Ekuakille A Characterization and performance measurement of low RCS wideband circularly polarized MIMO antenna for microwave sensing applications. IEEE Trans Instrum Meas, to be published. https://doi.org/10.1109/tim.2019.2936707
18. Hussein M, Awwad F, Jithin D, El Hasasna H, Athamneh K, Iratni R (2019) Breast cancer cells exhibits specific dielectric signature in vitro using the open-ended coaxial probe technique from 200 MHz to 13.6 GHz. Sci Rep 9(1), Art. no. 4681. https://doi.org/10.1038/s41598-019-41124-1
19. Xie Y, Guo B, Xu L, Li J, Stoica P (2006) Multistatic adaptive microwave imaging for early breast cancer detection. IEEE Trans Biomed Eng 53(8):1647–1657
20. Kumari V, Sheoran G, Kanumuri T (2019) SAR analysis of directive antenna on anatomically real breast phantoms for microwave holography. Microw Opt Technol Lett 62(1):466–473
21. Michałowska-Samonek J, Miaskowski A, Wac-Włodarczyk A (2012) Numerical analysis of high frequency electromagnetic field distribution and specific absorption rate in realistic breast models. Electrotech Rev 88(12b):97–99

A Novel Framework for Recommendation of Various Communication Networks Using Multi-criteria

Pradeep Kr. Tiwari, Animesh Tripathi, Gaurav Srivastava, Shiv Prakash, and Narendra Kr. Shukla

Abstract In the modern world information sharing is an essential requirement for the users; therefore to fulfill these needs intelligent communication networks are required in an organized way. Moreover, in today's environment, a huge amount of data generated through diverse scenarios is passing through the communication networks. Accordingly, various communication network architectures are required for transmitting such data. In the past few decades, various such network architectures have been developed. Furthermore, a robust and proficient framework is required to provide communication in various regions. In this paper, first we discuss the network architectures (Internet core and delay tolerant network architecture) and then we propose a novel framework that handles the user requirements in an efficient manner with a multi-criteria recommendation. Under Internet architecture, optical networks are discussed while Mobile Ad hoc networks (MANETs), Vehicular Ad hoc networks (VANETs), Wireless Sensor Networks (WSNs), and Exotic Media Networks (EMNs), are discussed under the delay tolerant network architecture. In particular, we identify the key features, strengths and weaknesses of these networks and compared them by using various network parameters. The proposed future data-driven novel framework recommends the network architecture with the optimized result.

P. Kr. Tiwari (✉) · A. Tripathi · G. Srivastava · S. Prakash · N. Kr. Shukla
Department of Electronics and Communication Engineering, University of Allahabad, Prayagraj, UP, India
e-mail: pradeepjkald@gmail.com

A. Tripathi
e-mail: animeshjkald@gmail.com

G. Srivastava
e-mail: gaurav771980@gmail.com

S. Prakash
e-mail: shivprakash@allduniv.ac.in

N. Kr. Shukla
e-mail: nksjkiapt@gmail.com

B. Mishra and M. Tiwari (eds.), *VLSI, Microwave and Wireless Technologies*,
Lecture Notes in Electrical Engineering 877,
https://doi.org/10.1007/978-981-19-0312-0_63

Keywords Internet core architecture · Delay tolerant network architecture · Optical networks · QoS parameters

1 Introduction

In the modern era communication is an integral part of life. Therefore, the optimal and efficient platform is required to facilitate communication in various regions. The existing TCP/IP protocol-based service model often called Internet works on few key assumptions which are continuous and bidirectional host 2 host path, short round trip time, cumulative data rates, and low error rates can be occurred due to the overall performance criteria of the under-laying link. The standardization and mapping of the IP technique into network-related link layer data frames on every router by using a packet-switched service model supports interoperability. However, in certain network scenarios that suffer significant delay so-called challenged network, TCP/IP Internet techniques just do not work because of the violation of one or more assumptions of TCP/IP technique-based service model. So, if we have a set of interesting application scenarios for which TCP/IP simply cannot work, there is a need for some delay tolerant techniques. To address such challenged networks, the network architecture is developed known as delay tolerant network architecture and such challenged networks are commonly referred to as delay tolerant network [1, 2].

The structure of the paper is organized into four sections, Sect. 1 is an introduction in which we briefly introduce the existing network architectures. In Sect. 2, the problem and related work discussed. In Sect. 3, we focus on the comparative analysis and proposed framework, and finally, Sect. 4 will conclude with suggestions and highlights future work.

2 The Problem and Related Work

The quality of service (QoS) parameters such as delay, bandwidth, fault tolerance, reliability, availability, throughput, may be required for a multi-criteria recommendation. To fulfill the User requirements, bandwidth, fault tolerance, reliability, availability, throughput should be maximized and delay should be minimized. The equations of QoS parameters are as follows:

Bandwidth is a potential link measurement. The bandwidth is a very important quality of service parameter which may be used to facilitate different user requirements in a different format which is defined as-

$$\text{Bandwidth (BW)} = (f_{\max} - f_{\min}) \tag{1}$$

where $f_{\max}$ and $f_{\min}$ are maximum and minimum frequencies of the system, respectively.

Throughput is the actual amount of data sent/received successfully over the communication channel. So, it is an actual measurement of how fast we can send data through a network.

The total delay of the network can be calculated as-

$$\begin{aligned}\text{Total Delay} &= \text{Transmission Delay} + \text{Propagatin Delay}\\ &\quad + \text{Queuing delay} + \text{Processing Delay}\end{aligned} \tag{2}$$

$$\text{Transmission Delay} = \frac{\text{Message size}}{\text{Bandwidth}} \tag{3}$$

$$\text{Propagatin Delay} = \frac{\text{Distance}}{\text{Transmission Speed}} \tag{4}$$

$$\text{Queuing Delay} = \frac{(\text{no. of data packets } - 1)\text{ Message size}}{2\text{ Bandwidth}} \tag{5}$$

The already existing model of networks does not satisfy the user requirements of various regions. Thus, we have to propose a network framework that satisfies the user interest. Some existing network Architectures available in the literature are discussed in Sects. 2.1 and 2.2.

2.1 Internet Core Architecture

Internet architecture is developed by interconnecting computer systems. The Internet technique is designed for the use of packet-switched computer communication networks (PSCN) in interconnected systems. Figure 1 shows the complete sense of data transfer from one end to another by using Internet core architecture. If we want to send a datagram from one application module to another on the Internet, the transmitting application module prepares its data and dial-up its local Internet module for sending the data as a datagram and as the calling arguments pass the destination address and other parameters. The Internet module (IM) is prepared and appended the data to a datagram header. For this Internet address, the IM decides local network addresses. It transmits desired datagram and the local network (LN) address to the interface. The LN interface generates the header of LN and adds the datagram to the desired network, then transmits the results through the LN. The datagram reaches a gateway along with the LN header and the header gets stripped off at the LN interface and transfers the datagram to the IM. From the Internet address, the IM specifies that the datagram is to be forwarded to another host on a second network. An LN address for the destination host is determined by the IM. It calls on the LN interface to transfer the datagram to that network. This LN interface generates a header for the LN and adds a datagram that transfers the result to the target host. The datagram is extracted from the LN header by the LN interface at this destination host and handover to the

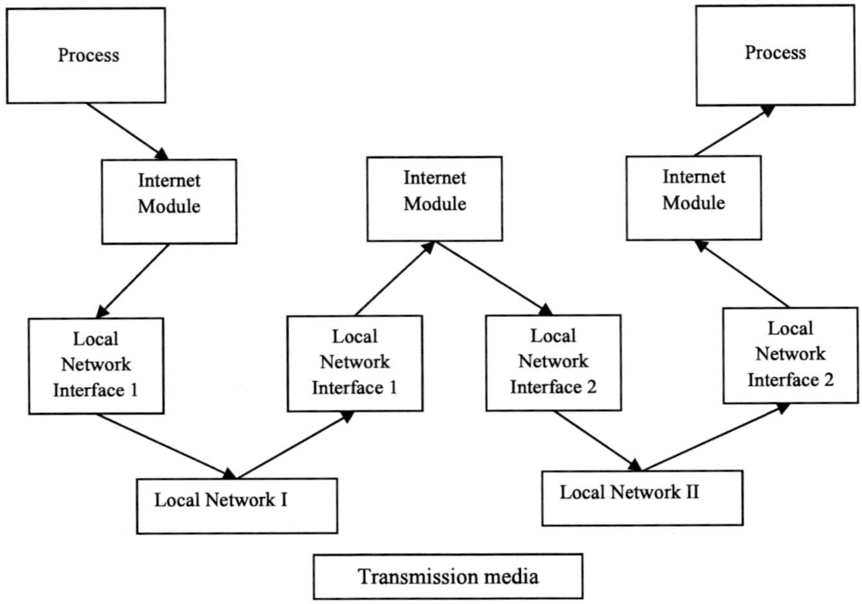

Fig. 1 Internet architecture

IM. The IM specifies that the datagram is for an application process in this hosting response to the system call, it transfers the data to the application module, passing the address of the source, and other parameters as call results.

The transmission path in this architecture can be wired or wireless. We use high-frequency EM wave signals for the transmission purpose for long-distance communication. If we use free space as a transmission path then this is wireless communication and this is done by baseband modulation/demodulation technique and if we use optical fiber cable as a transmission path then this comes into the category of wired communication which is based on intensity modulation and direct direction technique. Both type of communication is coming into the category of carrier wave communication. The optical fiber cable can handle only optical signals which are also high-frequency EM wave signals. The networks made by using optical fiber cable as transmission path are termed Optical Networks.

2.2 *Optical Networks*

Networks either retain signals in the optical form or at least use transmission channels that carry waves in the optical form. Typically, optical networks are organized into three major levels: access networks (ANs), metropolitan area networks (MANs), and backbone networks (BNs) [3].

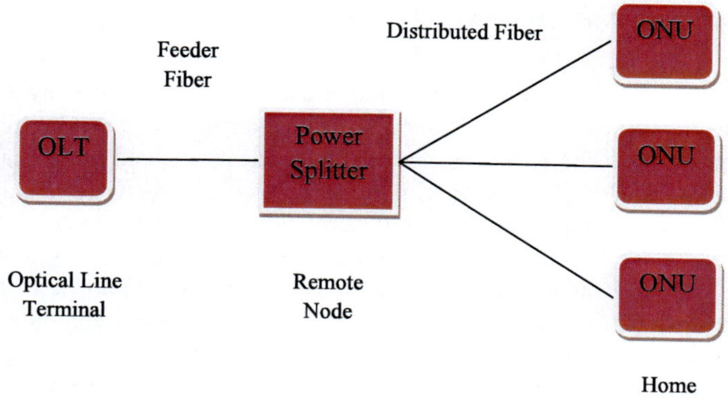

Fig. 2 Block diagram of passive optical network

Optical access networks (OANs) referred passive optical networks (PONs) usually have an inverse tree structure linking several distributed optical network units (ONUs; also known as optical network terminals (ONTs)) reside at home connected to remote node (power splitter) through distribution line and power splitter is connected to metro networks with a central office also known as optical line terminal (OLT) through feeder line. Figure 2 shows the basic block diagram of PON.

For a wide range of outlying networks, OANs can provide Internet connectivity [4–6]. Residential (home) wired or wireless local area networks interconnect individual end devices (hosts) and can link directly to an optical access network in a home or small business. Internet connectivity to a wide variety of mobile devices is provided by cellular wireless networks. Specialized cellular backhaul networks transmit wireless cellular network traffic from/to base stations to either wireless access networks or OANs. In addition, OANs are also used to link networks of the Data Center (DC) to the Internet. Highly specialized server units are interconnected by DC networks with specialized networking technologies that process and store huge amounts of data. DC provides "cloud" services for commercial and social media applications. [7–28].

Metropolitan optical area networks (MANs) interconnect the OANs with each other and with backbone (core) networks. Usually, ring or star topology is used for MANs and optical networking systems are widely used in it [29–34].

On a national or international scale, the optical BNs (wide area) link the individual MANs. Usually, BNs have a mesh structure and use very high-speed links for optical transmission.

Growing traffic on the Internet day by day continually challenges the advancements in communication and networking technology. In order to satisfy the client's enormous bandwidth requirements, the optical networks have many features, such as reliability, scalability, versatility, and performance. In today's scenario, elastic optical network (EON) or flex grid technology is one of the most sought-for technologies for high-speed communication [35].

2.3 Delay Tolerant Network Architecture

The most popular and widely used technique and architecture of today's Internet core architecture may not work in some challenging environment that provides long delay, frequent network partitions, intermittent connectivity, and no logical end to end connection. Such types of networks have their own architecture and they do not use TCP/IP protocol. The network architecture is proposed with some key features such as in network storage capability and retransmission, a coarse-grained class of service, interoperable naming and authenticated forwarding known as the delay tolerant network (DTN) to achieve interoperability between challenged networks. DTN nodes have permanent storage capabilities with large buffer sizes that follow the store-carry-forward message switching technology instead of globally adopted packet switching technology [1, 2].

VANETs, MANETs, WSNs, and EMNs are some popular examples of delay tolerant network, shown in Fig. 3.

VANETs allow the communication between any moving object with other adjacent vehicles and roadside units without infrastructure by transmitting data such as present direction, weather condition, speed, the position of vehicle, obstacle, and road condition to other nodes [36–38]. MANETs operate in such a condition where mobility, intentional jamming, or factors related to the environment are the main cause of the disruption. MANETs nodes such as Cellular Phones, GPSs, Laptops, PDAs, Tracking devices, mounted over continuously moving objects. The frequent

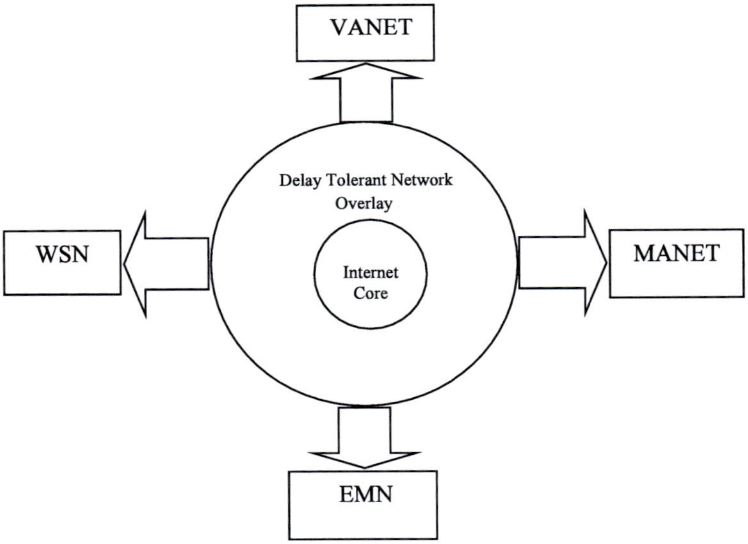

Fig. 3 Delay tolerant network overlay

interruption into these networks is mainly caused by nodes going out of coverage range of each other or node destruction.

WSNs are characterized by networks of small memory, CPU capability and end-node power to save energy by periodically switch between active and sleep mode.

EMNs include extra-terrestrial nodes like near-earth satellite, lander, deep space probes, orbiters, etc. These systems may suffer outage because of environmental circumstances or may have high latencies with predictable interruption [1, 2].

3 Comparative Analysis and Proposed Framework

Table 1 represents the comparative study of existing network architectures mainly Internet core and delay tolerant network architecture. This table also provides a brief idea about the user that how they will serve their near-optimal requirements.

Delay and disruption are very crucial parameters to decide the services. Figure 4 shows the future framework to facilitate users for the suggested networks for

Table 1 The comparative analysis of existing network architectures

Networks parameters	Internet core architecture	Delay tolerant network architecture
Topology	Fixed	Not fixed or changes frequently
End to end logical connection	Exist	Not exist
Delay	Very small (of the order of few milliseconds) delay	Very long delay
Reliability	Reliable	Not reliable
Switching technology	Packet switching	Store-carry-forward (message switching)
Protocol suite	TCP/IP	Bundle
Data delivery	Host to host or end to end	Hop to hop or node to node
Throughput	High	Low
In network storage capability	No	Yes
Examples	Access network, Backbone network	VANETs, MANETs, WSNs

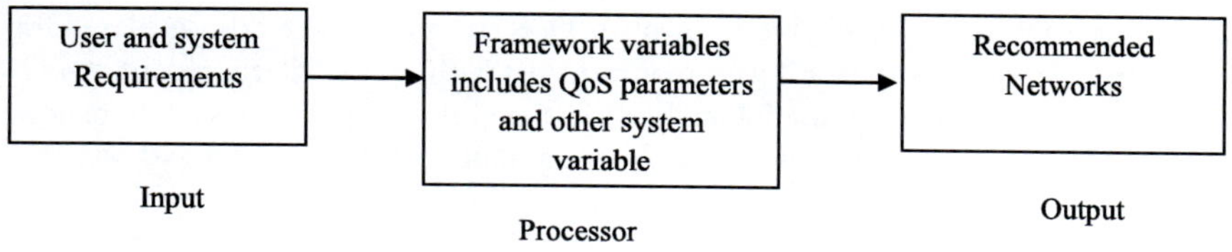

Fig. 4 Block diagram of proposed future framework

communication establishment. With the help of this proposed model, we can achieve suitable Internetwork architecture for all scenarios. The proposed framework is as follows-

The proposed framework is simulated by MATLAB using the required simulation environment of a network with different QoS parameters. Figure 5 shows the flow chart for the proposed novel framework. The framework is a simulated network environment for delay tolerant and other suggested fulfilling user requirements for a multi-criteria recommendation. The size of the simulated network with different parameters such as area (100 m $\times$ 100 m) and the number of nodes are 100.

The simulated framework for QoS parameters with the optimized results as follows for delay tolerant network because delay calculated into the network is 10 min. In this case, the calculated network delay is greater than two minutes so, the delay tolerant network architecture is the best choice suggested by the proposed framework to deliver the data from sender side to receiver because TCP connection cannot be maintained when the delay is produced by the network is greater than two minutes. The decision is taken based on network delay which is one of the QoS parameters to satisfy user requirements. So, by delay parameter, we can decide that either we follow the Internet core architecture or delay tolerant network architecture. After that, some other parameters like Buffer size and Node power are taken into the account to decide which network protocol is suited for that particular delay tolerant network. Similarly, different QoS parameters may be for the different environments to fulfill different user requirements with multi-criteria recommendations. Figure 6 shows that the decided network proposed by framework is having a large delay (10 min), small buffer capacity and low node power. Hence the proposed framework can decide which network is used to deliver the data from source to destination with the help of different QoS parameters.

4 Conclusion and Future Work

Due to user requirements, a well-defined model for communication networks is needed. Moreover, the passing of data generated from the various scenarios into the networks is one of the challenging issues using current existing network architectures. In this paper, a novel framework is proposed by using multi-criteria with comparative analysis of different types of networks and their architectures supported in various scenarios. The simulated framework uses Delay, Buffer size, and Node power as QoS parameters to decide the suggested network which fulfills the user requirements in an optimal manner. Similarly, other QoS parameters can be considered for different environments to fulfill user requirements with multi-criteria recommendations. Finally, the proposed model analyzes and optimizes the framework to handle the data generated from various scenarios using multi-criteria recommendations.

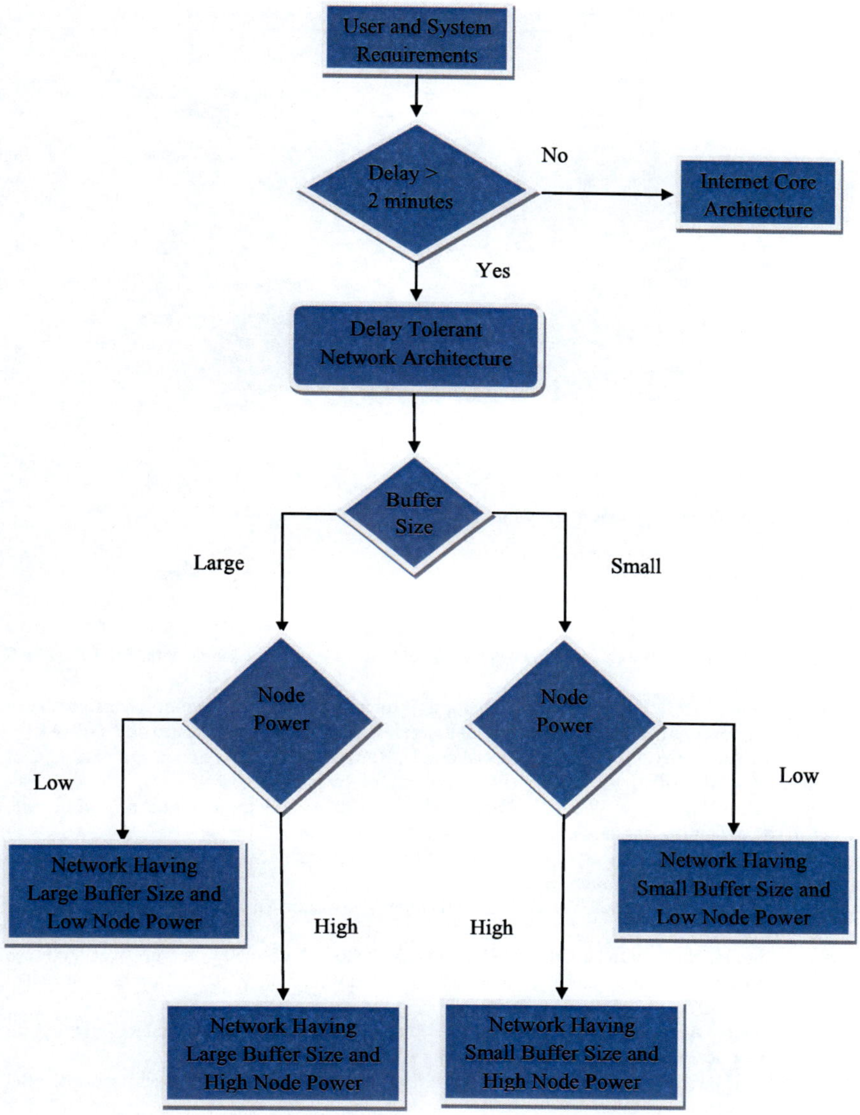

Fig. 5 Flow chart of proposed novel framework

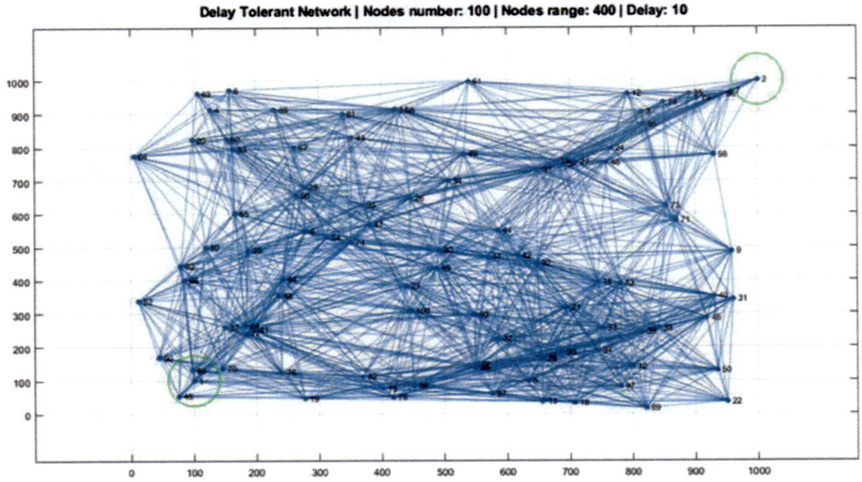

Fig. 6 Simulated result by using delay tolerant network architecture

References

1. Fall K (2003) A delay-tolerant network architecture for challenged internets. Intel Research Berkley
2. Khabbaz M, Assi C, Fawaz W (2012) Disruption-tolerant networking: a comprehensive survey on recent developments and persisting challenges. IEEE Commun Surv Tut 14(2):607–640
3. Talebi S, Alam F, Katib I, Khamis M, Salama R, Rouskas GN (2014) Spectrum management techniques for elastic optical networks: a survey. Opt Sw Netw 13:34–48
4. Yu X, Zhao Y, Zhang J, Wang X, Wang J, Zhang J (2014) Spectrum engineering in flexible grid data center optical networks. Opt Sw Netw 14:282–288
5. Gerstel O, Jinno M, Lord A, Yoo SJB (2012) Elastic optical networking: a new dawn for the optical layer? IEEE Commun Mag 50(2):S12–S20
6. Ramaswami R, Sivarajan K, Sasaki G (2009) Optical networks: a practical perspective. Morgan Kaufmann
7. Liu J, Guo H, Nishiyama H, Ujikawa H, Suzuki K, Kato N (2016) New perspectives on future smart FiWi networks: scalability, reliability and energy efficiency. IEEE Commun Surv Tut 18(2):1045–1072
8. Sarigiannidis AG et al (2015) Architectures and bandwidth allocation schemes for hybrid wireless-optical networks. IEEE Commun Surv Tut 17(1):427–468
9. Tsagklas T, Pavlidou F (2011) A survey on radio-and-fiber FiWi network architectures. J Sel Areas Telecommun March 18–24
10. Gurrola EI, McGarry MP, Luo Y, Effenberger F (2014) PON/xDSL hybrid access networks. Opt Sw Netw 14:32–42
11. Chen M, Wan J, Gonzalez S, Liao X, Leung V (2014) A survey of recent developments in home M2M networks. IEEE Commun Surv Tut 16(1):98–114
12. Capozzi F, Piro G, Grieco LA, Boggia G, Camarda P (2013) Downlink packet scheduling in LTE cellular networks: key design issues and a survey. IEEE Commun Surv Tut 15(2):678–700
13. Damnjanovic A et al (2011) A survey on 3GPP heterogeneous networks. IEEE Wirel Commun 18(3):10–21
14. Schwarz S, Ikuno JC, Simko M, Taranetz M, Wang Q, Rupp M (2013) Pushing the limits of LTE: a survey on research enhancing the standard. IEEE Access 1:51–62

15. Li J, Peng M, Cheng A, Yu Y, Wang C (2016) Resource allocation optimization for delay-sensitive traffic in fronthaul constrained cloud radio access networks. IEEE Syst J (in print)
16. Liu C, Zhang L, Zhu M, Wang J, Cheng L, Chang G-K (2013) A novel multi-service small-cell cloud radio access network for mobile backhaul and computing based on radio-over-fiber technologies. IEEE/OSA J Lightwave Tech 31(17):2869–2875
17. Park S, Simeone O, Sahin O, ShamaiShitz S (2014) Fronthaul compression for cloud radio access networks: signal processing advances inspired by network information theory. IEEE Signal Proc Mag 31(6):69–79
18. Peng M, Wang C, Lau V, Poor HV (2015) Fronthaul-constrained cloud radio access networks: insights and challenges. IEEE Wirel Commun 22(2):152–160
19. Raza H (2013) A brief survey of radio access network backhaul evolution: Part II. IEEE Commun Mag 51(5):170–177
20. Tipmongkolsilp O, Zaghloul S, Jukan A (2011) The evolution of cellular backhaul technologies: current issues and future trends. IEEE Commun Surv Tut 13(1):97–113
21. Yang M, Li Y, Jin D, Su L, Ma S, Zeng L (2013) OpenRAN: a software-defined RAN architecture via virtualization. ACM SIGCOMM Comput Commun Rev 43(4):549–550
22. Akyildiz IF, Wang X, Wang W (2005) Wireless mesh networks: a survey. Comput Netw 47(4):445–487
23. Alotaibi E, Mukherjee B (2012) A survey on routing algorithms for wireless *ad-hoc* and mesh networks. Comput Netw 56(2):940–965
24. Benyamina D, Hafid A, Gendreau M (2012) Wireless mesh networks design: a survey. IEEE Commun Surv Tut 14(2):299–310
25. Kuran MS, Tugcu T (2007) A survey on emerging broadband wireless access technologies. Comput Netw 51(11):3013–3046
26. Pathak PH, Dutta R (2011) A survey of network design problems and joint design approaches in wireless mesh networks. IEEE Commun Surv Tut 13(3):396–428
27. Vijayalayan KS, Harwood A, Karunasekera S (2013) Distributed scheduling schemes for wireless mesh networks: a survey. ACM Comp Surv 46(1):1–34
28. Cai Y, Yan Y, Zhang Z, Yang Y (2013) Survey on converged data center networks with DCB and FCoE: standards and protocols. IEEE Network 27(4):27–32
29. Kachris C, Tomkos I (2012) A survey on optical interconnects for data centers. IEEE Commun Surv Tut 14(4):1021–1036
30. Samadi P, Wen K, Xu J, Bergman K (2016) Software-defined optical network for metro-scale geographically distributed data centers. OSA Opt Expr 24(11):12310–12320
31. Yang H, Zhang J, Zhao Y, Han J, Lin Y, Lee Y (2016) SUDOI: software defined networking for ubiquitous data center optical interconnection. IEEE Commun Mag 54(2):86–95
32. Zhang J, Ren F, Lin C (2013) Survey on transport control in data center networks. IEEE Network 27(4):22–26
33. Bianco A, Bonald T, Cuda D, Indre R-M (2013) Cost, power consumption and performance evaluation of metro networks. IEEE/OSA J Opt Commun Netw 5(1):81–91
34. Bianco A, Cuda D, Finochietto JM (2015) Short-term fairness in slotted WDM rings. Comput Netw 83:235–248
35. Chatterjee BC, Ba S, Oki E (2018) Fragmentation problems and management approaches in elastic optical networks: a survey. IEEE Commun Surv Tut 20(1):183–210
36. Hiraishi H (2018) Passenger condition based route-planning for cognitive vehicle system. Int J Softw Sci Comput Intell (IJSSCI) 10(2):25–35
37. Hussain MDM, Beg MMS (2019) Using vehicles as fog infrastructures for transportation cyber-physical systems (T-CPS): fog computing for vehicular networks. Int J Softw Sci Comput Intell (IJSSCI) 11(1):47–69
38. Mirsadeghi F, Rafsanjani MK, Gupta BB (2020) A trust infrastructure based authentication method for clustered vehicular *ad hoc* networks. Peer-to-Peer Netw Appl

CDBA-Based MOS-C Quadrature Oscillator

Shekhar Suman Borah and Mourina Ghosh

Abstract In this article, MOS-C resistor-based quadrature sinusoidal oscillator configuration is proposed. The proposed structure is appropriate for the MOS-C implementation of a linear resistor that facilitates oscillation frequency electronically tunable. The three grounded/virtually grounded capacitors are ideal for the integration point of view. The sensitivity, non-ideality analysis, Monte-Carlo simulation, temperature analysis of the circuit is also provided. The proposed structure utilizes 900 μW power with 600 mV supply voltage and suitable for the Megahertz frequency range of applications. PSPICE simulation outcomes utilizing TSMC 0.18 μm CMOS process parameters have been incorporated to confirm the functionality of the offered design.

Keywords Active building block (ABB) · Current differencing buffered amplifier (CDBA) · Quadrature oscillators

1 Introduction

Quadrature oscillators exhibit two sinusoids producing 90° phase variations. The phase-locked sine–cosine connection of quadrature oscillator has beneficial utilization in the area of instrumentation, power electronics, and telecommunications [1, 2]. Throughout the years, various quadrature oscillator configurations have been implemented utilizing traditional operational amplifier to various voltage and current mode building blocks [3–8]. As an ABB, op-amp impersonated a dominant part in the last few years. But op-amp-based circuits show many shortcomings in their execution starting from the restricted bandwidth and slew rate. Hence, the current mode procedure has been placed to accomplish high-speed operations.

S. S. Borah (✉) · M. Ghosh
Department of ECE, IIIT, Guwahati, India
e-mail: borahsumanshekhar90@gmail.com

M. Ghosh
e-mail: mourina@iiitg.ac.in

B. Mishra and M. Tiwari (eds.), *VLSI, Microwave and Wireless Technologies*,
Lecture Notes in Electrical Engineering 877,
https://doi.org/10.1007/978-981-19-0312-0_64

The novel active device, CDBA was proposed in 1999 [9] to contribute additional potentialities in the circuit construction, analysis, and simulation. It offers a higher slew rate, extended range of frequency, and uncomplicated designs as the input ends are internally grounded [10–12].

In this report, CDBA-based second-order MOS-C quadrature oscillator configuration is presented and the circuit description is shown in Sect. 2. The oscillation condition and frequency are electronically tunable. Section 3 reports the non-ideality investigation of the proposed structure. Sensitivity analysis, Monte-Carlo simulation, and temperature analysis are also investigated. Simulation results shown in Sect. 4 verify that oscillator features are in good agreement with the theoretical prediction, and lastly, the paper ends in Sect. 5.

2 Proposed Circuit

The proposed quadrature oscillator circuit includes CDBA [13] as an active element. The circuit symbol of CDBA is displayed in Fig. 1 and the current–voltage properties are presented as:

$$V_p = V_n = 0, \quad I_z = I_p - I_n, \quad V_w = V_z \tag{1}$$

where p and n are the internally grounded low input impedance ports with voltage output port w and current output port z. Figure 2 shows the CMOS implementation of CDBA. The aspect ratios [14] are presented in Table 1.

Figure 3 presents the proposed MOS-C CDBA-based quadrature oscillator configuration. To make the proposed design resistor free, MOS-based active resistors are used in place of the conventional resistor and all the capacitors are grounded/virtually grounded. The resistance value may be tuned by the proper selection of gate voltages [15]. The quadrature signals with a phase variation of 90-degree are obtained from V_{01} and V_{02}. The characteristics equation for the proposed configuration is obtained as:

$$s^2 C_2 C_3 R_{\text{mos1}} R_{\text{mos2}} R_{\text{mos3}} + s(2C_3 R_{\text{mos1}} R_{\text{mos3}} - C_1 R_{\text{mos1}} R_{\text{mos2}}) + R_{\text{mos2}} = 0 \tag{2}$$

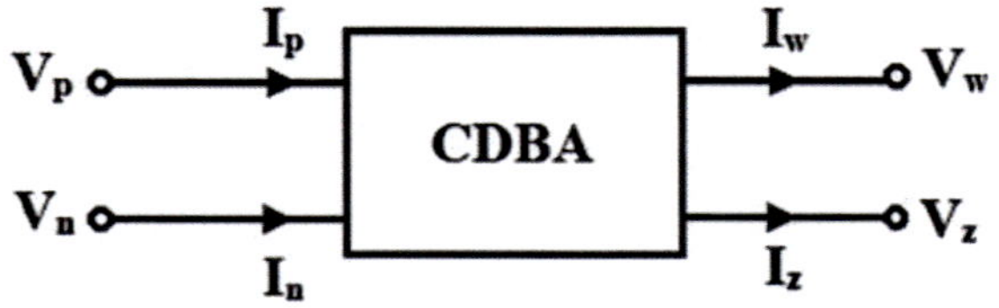

Fig. 1 Electrical symbol of CDBA [13]

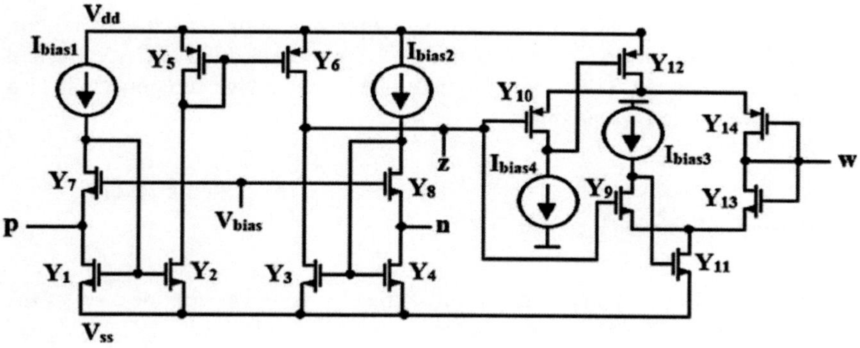

Fig. 2 CMOS realization of CDBA

Table 1 Aspects ratios of the transistors

Transistors	W/L (μm)
Y_1, Y_2, Y_3, Y_4	3.6/1.80
Y_5, Y_6, Y_7, Y_8	180/1.80
Y_9	45/0.36
Y_{10}, Y_{12}, Y_{14}	240/0.36
Y_{11}, Y_{13}	72/0.36

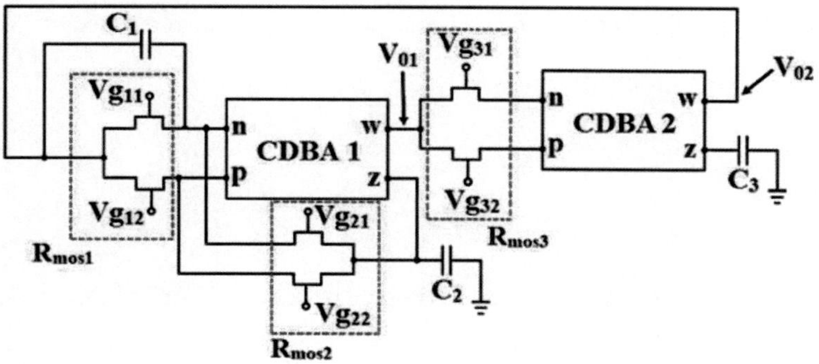

Fig. 3 Proposed MOS-C quadrature oscillator

The condition and frequency of oscillations for the circuit are given by:

$$C_3 R_{\text{mos3}} = \frac{C_1 R_{\text{mos2}}}{2} \quad (3)$$

$$\omega_0 = \sqrt{\frac{2}{R_{\text{mos1}} C_2 R_{\text{mos2}} C_1}} \quad (4)$$

It is observed from the above equations that the tuning of oscillation condition and frequency is independently established by the proper selection of C_3, R_{mos3}, and R_{mos1}, C_2, respectively. The sensitivity analysis for the passive components is also obtained and is found within unity.

3 Non-ideal Investigation

It is extremely crucial to study the non-idealities by using the non-ideal set of the equation [16] as:

$$V_p = V_n = 0, \quad I_z = e_p I_p - e_n I_n, \quad V_w = f_v V_z \tag{5}$$

where e_n and e_p are the current-tracking errors from terminal *n*/*p* to *z*. Also, the f_v is the voltage-tracking error from terminal *z* to *w*. Therefore, the ideal characteristics equation of the proposed quadrature oscillator is modified as:

$$s^2 C_2 C_3 R_{\text{mos1}} R_{\text{mos2}} R_{\text{mos3}} + s \begin{pmatrix} C_3 R_{\text{mos1}} R_{\text{mos3}} + e_{n1} C_3 R_{\text{mos1}} R_{\text{mos3}} \\ -e_{n1} e_{n2} f_{v1} f_{v2} C_1 R_{\text{mos1}} R_{\text{mos2}} \end{pmatrix} + e_{p1} e_{n2} f_{v1} f_{v2} R_{\text{mos2}} = 0 \tag{6}$$

Hence, the modified condition and frequency of oscillation are stated as:

$$C_3 R_{\text{mos3}} = \frac{e_{n1} e_{n2} f_{v1} f_{v2} C_1 R_{\text{mos2}}}{1 + e_{n1}} \tag{7}$$

$$\omega_0 = \sqrt{\frac{(1 + e_{n1}) e_{p1} e_{n2} f_{v1} f_{v2}}{e_{n1} e_{n2} f_{v1} f_{v2} R_{\text{mos1}} C_2 R_{\text{mos2}} C_1}} \tag{8}$$

It can be found that the modified condition and frequency of oscillation are effected by the non-ideal parameters.

4 Simulated Results

The simulations were conducted utilizing a CMOS-based CDBA. Where supply voltages, $V_{\text{dd}} = V_{\text{ss}} = \pm 600$ mV. The value of the MOS resistors and capacitors are chosen as $R_{\text{mos1}} = R_{\text{mos2}}/2 = R_{\text{mos3}} = 10\ \text{k}\Omega$ and $C_1 = C_2 = 10$ pF, $C_3 = 5.2$ pF which results in an oscillation frequency of 1.591 MHz. The simulated transient output voltage during the initial state is shown in Fig. 4. Figure 5 shows the steady-state response. Again, the frequency spectrum is displayed in Fig. 6a. The quadrature relation between the produced waveforms in the XY plane is presented in Fig. 6b.

Further, a performance inspection of the proposed oscillator is assessed utilizing Monte-Carlo simulation which yields the robustness of the circuit is displayed in Fig. 7. The steady-state output waveform (V_{02}) showing in Fig. 8 with temperature

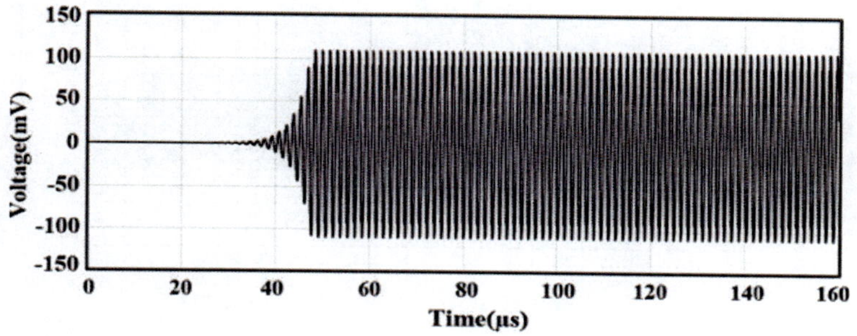

Fig. 4 Simulated transient response

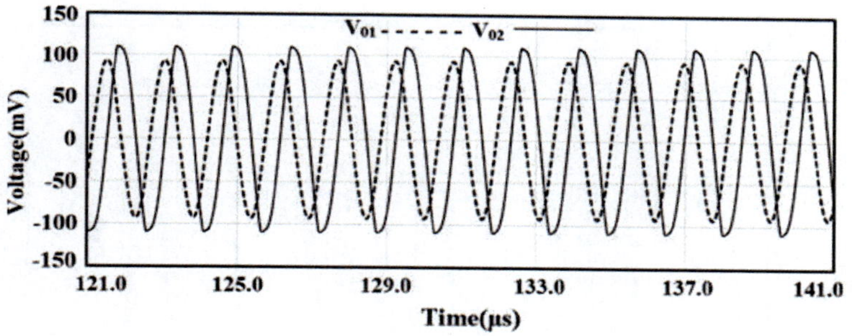

Fig. 5 Simulated steady-state response

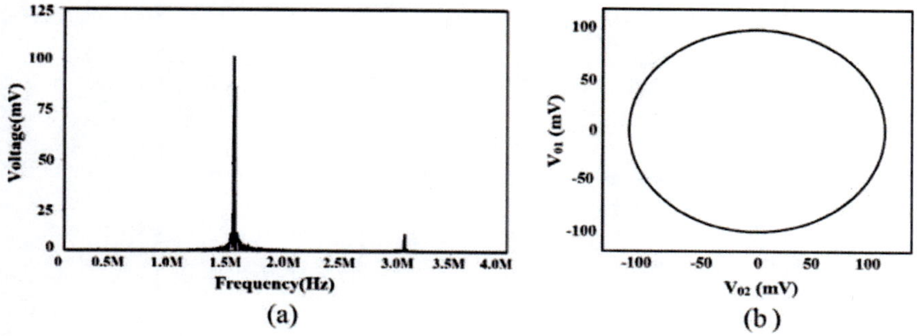

Fig. 6 **a** Simulated frequency spectrum, **b** plot of V_{01} versus V_{02}

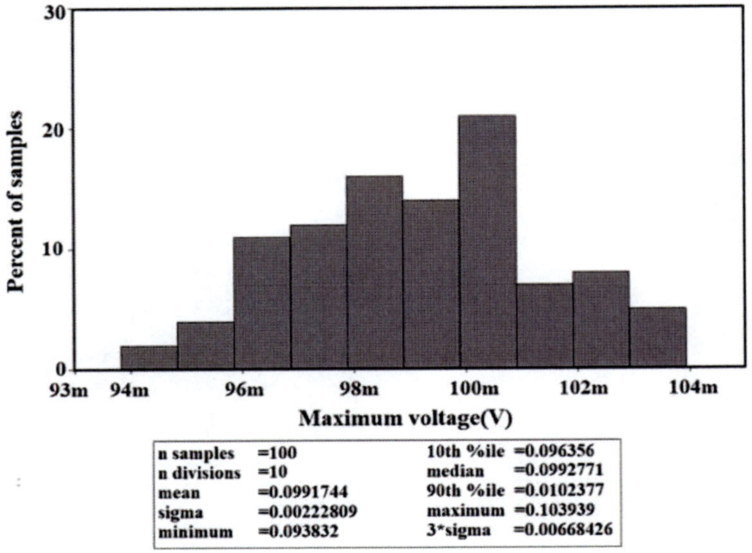

Fig. 7 Monte-Carlo analysis

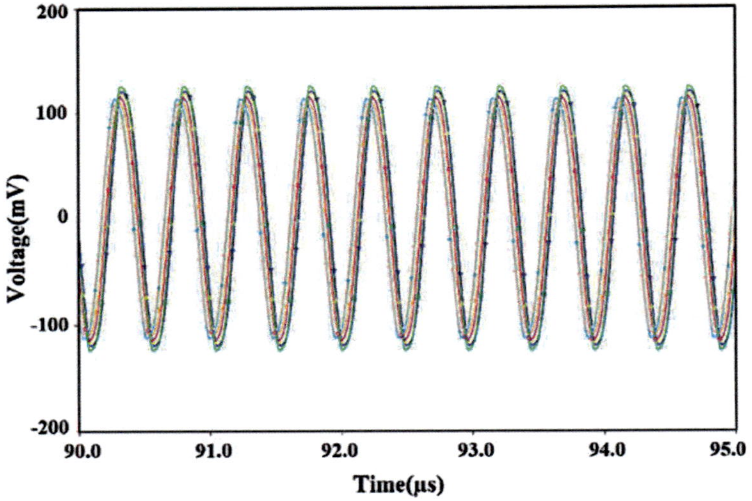

Fig. 8 Steady-state response (V_{02}) with respect to the variations in temperature

variations from −20 °C to 100 °C confirms that there is not any specific difference because of the fluctuations in temperature.

5 Conclusion

A novel topology of CDBA-based MOS-C quadrature oscillator configuration is presented in this paper. From the simulation results, it is witnessed that the outcomes are in close agreement with the analytical hypothesis. The sensitivity, non-ideal behavior, Monte-Carlo simulation, temperature analysis has also been incorporated in the stated work.

References

1. Horng JW, Hou CL, Chang CM, Chung WY, Tang HW, Wen YH (2005) Quadrature oscillators using CCIIs. Int J Electron 92:21–31
2. Sedra AS, Smith KC (2004) Microelectronic circuits. Oxford University Press, New York
3. Holzel R (1993) A simple wide-band sinewave quadrature oscillator. IEEE Trans Instrum Meas 42:758–760
4. Tangsrirat W, Pisitchalermpong S (2007) CDBA-based quadrature sinusoidal oscillator. Frequenz 61(3–4):102–104
5. Nandi R, Venkateswaran P, Das S, Kar M (2010) CDBA-based electronically tunable filters and sinusoid quadrature oscillator. J Telecommun 4:35–41
6. Tangsrirat W, Pukkalanun T, Surakampontorn W (2008) CDBA-based universal biquad filter and quadrature oscillator. Act Passive Electron Compon 2008:1–6
7. Sharma RK, Arora TS, Senani R (2016) The realization of canonic single-resistance-controlled oscillators using third generation current conveyors. IET Circuits Devices Syst 11:10–20
8. Sotner R, Jerabek J, Prokop R, Kledrowetz V (2016) Simple CMOS voltage differencing current conveyor-based electronically tunable quadrature oscillator. Electron Lett 52:1016–1018
9. Acar C, Ozoguz S (1999) A new versatile building block: current differencing buffered amplifier suitable for analog signal-processing filters. Microelectron J 30:157–160
10. Özoguz S, Toker A, Acar C (1999) A new current-mode continuous time fully integrated universal filters. Electron Lett 35:97–98
11. Toker A, Özoguz S, Çiçekoğlu O, Acar C (2000) Current mode allpass filters using CDBA and a new high Q bandpass filter configuration. IEEE Trans Circuits Syst 47:949–954
12. Borah SS, Singh A, Ghosh M, Ranjan A (2021) Electronically tunable higher-order quadrature oscillator employing CDBA. Microelectron J 108(104985):0026–2692
13. Sedef H, Acar C (2000) On the realization of voltage-mode filters using CDBA. Frequenz 54:198–202
14. Cakir C, Minaei S, Cicekoglu O (2009) Low voltage low power CMOS current differencing buffered amplifier. Analog Integr Circ Sig Process 62(2):237–244
15. Ozoguz S, Toker A, Acar C (1999) Current-mode continuous-time fully integrated universal filter using CDBAs. Electron Lett 35(2):97–98
16. Singh A, Borah SS, Ghosh M (2020) CDBA based quadrature sinusoidal oscillator with non-interactive control. In: International conference on communication and signal processing (ICCSP), Chennai, India, pp 0033–0037

Impact of Mixed Interlayer Thickness on Performance of Organic Light-Emitting Diodes

Pooja Maurya and Brijesh Kumar

Abstract Organic light-emitting diode (OLED) is very emerging research field due to low-cost solution for display applications. It also helps in providing the lightweight flexible displays for mobile phones and other electronic applications. Presently, lot of researchers from academia and industry are working in this field. They are looking for high mobility materials and novel device structures of OLEDs to enhance the performance. A necessary element to control of OLED performance is including the device parameters, operating voltage and stability, low-cost fabrication, and low-power consumption. These are the key merits of OLEDs that defined its wide applicability. In this paper, comparison among five devices is investigated, where the impact of thickness variation of mixed interlayer is analyzed. The optimized thickness, sequence of layer, and doping concentration of mixed interlayer improve the performance of multilayer OLEDs. The mixed interlayer is the ratio of the predominant CBP hole material to the predominant TPBi electron materials in the range of 3:2, respectively. Significant improvement is extracted that defines the outcomes in terms of current density, luminance, and luminescent power. The huge improvement in current density, luminance, and luminescence power efficiency is observed for Device A. The respective values for current density are 47 mA/cm^2, whereas luminance is 14269 cd/m^2, and extracted luminescence power efficiency value of 20.8 lm/W is observed, respectively. Improved current density, efficiency, luminance, and luminescence power efficiency are attributed to the mixed interlayer enlarge the charge carrier's transport, with optimized excitons. Improved OLED device can be further used in bio-medical applications as light source as well as light detector.

P. Maurya
Department of Electronics and Communication Engineering, Madan Mohan Malaviya University of Technology, Gorakhpur, UP, India

B. Kumar (✉)
Department of Information Technology, Indira Gandhi Delhi Technical University for Women, Delhi 110006, India
e-mail: brijeshkr@igdtuw.ac.in

B. Mishra and M. Tiwari (eds.), *VLSI, Microwave and Wireless Technologies*, Lecture Notes in Electrical Engineering 877,
https://doi.org/10.1007/978-981-19-0312-0_65

Keywords Organic light-emitting diode · Highest occupied molecular orbital · Lowest occupied molecular orbital · Electron block layer · Hole block layer · Mixed interlayer

1 Introduction

In the past decade, organic electronics has been the development of researchers to complement conventional technology based on silicon. These devices are selected because they are low-cost, thin-film, simple to fabrication, mechanically flexible, and lightweight manufacturing devices [1–3]. Due to this intensive research, several technologies have been seen recently. Dependent on organic transistors that have shown high performance, such as digital circuits, memory, and other applications like Radio-frequency identification (RFID) or organic display (OLEDs) and many more [4]. Organic light-emitting diode (OLED) is cutting-edge latest technology; it may have many applications and innovations in electronics fields, like solid-state lighting and flat panel display devices. OLED is excited because it is in various and functional features of wide viewing angle, flexibility, lower operating voltage, higher resolution, simple fabrication in room temperature, and accurate or faster response time. The OLED is an active and growing technology that can satisfy the need for smaller and better appliances. Because of this, extremely better characterization for OLEDs is actively used in televisions, smartphones, and display technologies. In determining the features of organic devices. Organic semiconductor (OSC) material or device structures have a play a central role [5]. OLED appears to be the perfect technology for all types of displays, but it also has some problems, short lifetime, manufacturing, high energy gap, and keeping color stability.

Several researchers are still attempting to further better their results. To enhance the performance of OLED, various techniques are being investigated. Scientists have used various techniques to improve their performance, such as a change in the architectural design of OLEDs. Various supporting layers are connected to the architecture (charge injection, transport, and blocking layers) to improve OLED performance. Negi et al. [6], and other parameters are used to hole block layer (HBL), electron block layer (EBL), double block layer (DBL), spacer layers, and mixed interlayer (MI) are used to highly enhance the performance of OLED. Mixed interlayer used because it is also important that the charge injection of a mixed layer with various HOMO or LUMO levels is used further. A very powerful solution is also the inclusion of emission layers, where a specialized layer has been used to harness a particular luminescence [7]. The mixed interlayer between the blue fluorescence-emissive layer and complementary yellow phosphorescence emissive layer performs a significant role in the brightness, efficiency, and lifetime [8]. A mixed interlayer is said to be able to mix hole and electron transport materials to attempt to control the location of the zone of exciton recombination, and by modifying the proportion of the two transport materials, the singlet and triplet excitons are controlled. To achieve better EL performance in the devices, the MI is an easy way of effectively balancing the carrier's property [8]. The various parameter is extracted such as current density, luminance, luminescent power efficiency, and internal analysis of structures. This is highly improved the performance of mixed interlayer OLED.

This paper is categorized into five parts with the involvement of introductory part in Sect. 1. Simulation setup and structure of OLED along with the variation in thickness of mixed interlayer and comparison of simulation results with fabricated data, extraction of luminance, or luminescent power part in Sect. 2. Analysis of difference thickness variation of a mixed interlayer part in Sect. 2.1. Internal analysis of mixed-layer OLED in Sect. 2.2. Result and discussion of different devices are discussed in Sect. 3. The conclusion of the paper is summarized in Sect. 4.

2 Simulation Setup and Device Structure

This article analyzes the fabricated device. It is used to introduce several different layers to the structure to improve the efficiency of OLED. For this reason, the tool used is Atlas by from Silvaco. It uses the versatility model of Poole and Frenkel. (ATLAS 2011) to examine the characteristics of organic devices. The Langevin recombination model, on the other hand, (ATLAS 2011) is used for carrier analysis in OLED, recombination, and exciton formation [9]. The OLED-simulated structure of the device is shown in Fig. 1. The indium tin oxide (ITO) electrode on a glass substrate, the resulting deposit of organic and metallic layers, was under a base pressure of thermal evaporation 2×10^{-6} Torr without the vacuum splitting. Al (Aluminum) is

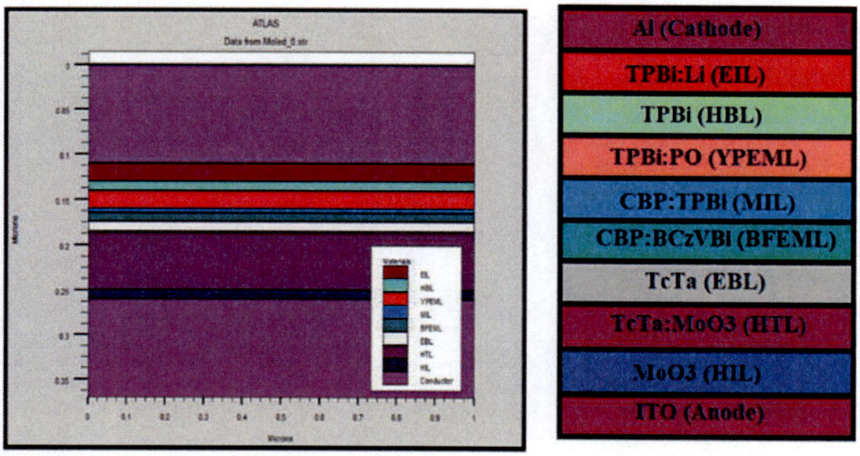

(a) (b)

Fig. 1 **a** Simulated device structure of mixed interlayer OLED and **b** structure of OLED with name of materials for different layers another electrode used as a cathode, and the different materials used like TPBi: Li 0.8 vol.% (20 nm) (2,2',2")-(1,3,5-Benzinetriyl)-tris(1-phenyl-1-H-benzimidazole): (Lithium) materials are used as an n-type doping electron injection layer (EIL). MoO_3 (10 nm) (molybdenum trioxide) has been used as a hole injection layer (HIL) of P-type-doped material. TcTa (10 nm) (4,4',4"-Tri (N-carbazolyl) triphenylamine) has been used as an electron block layer (EBL) is P-type material.

TcTa: MoO_3 (65 nm) has been used as a hole transport layer (HTL), or TPBi (10 nm) is used either as a self-triplet exciton-blocking layer and as a self-blocking (HBL) with N-type doped material. TPBi: PO (20 nm) ((2,2',2")-(1,3,5-Benzinetriyl)-tris(1-phenyl-1-H-benzimidazole): (Iridium -(III)-bis (4-phenylthieno [3,2-c] pyridinato-N, C2') acetylacetonate) in the PO is used as yellow phosphorescent dopant. The CBP: BCzVBi (10 nm) 4,4'-Bis (carbazol-9-yl) biphenyl: (4,4'-Bis (9-ethyl-3-carbazovinylene)-1,1'-biphenyl) in BCzVBi has been used as a blue, fluorescent doping agent. CBP: TPBi (5 nm) is used as a mixed interlayer, where CBP is used as host or interlayer spacer, and TPBi is used as electron transport layer host interlayer spacer, CBP: TPBi is (3:2) ratio used in this device [8]. The process of analysis begins with the taking of a fabricated device and simulated by using the Silvaco Atlas tool. Subsequently, devices are simulated for the current density versus anode voltage graph and compared with fabricated device result, and found that extracted result is 40 mA/cm^2 is shown in Fig. 2. The fabricated devices are used five devices with different ratios of mixed interlayer, but we have simulated only one green color device is the ratio of (3:2) (Fig. 3).

The numerous characteristic parameters are useful when the applicability of an OLED is calculated in many fields like luminance, luminescent power, and current

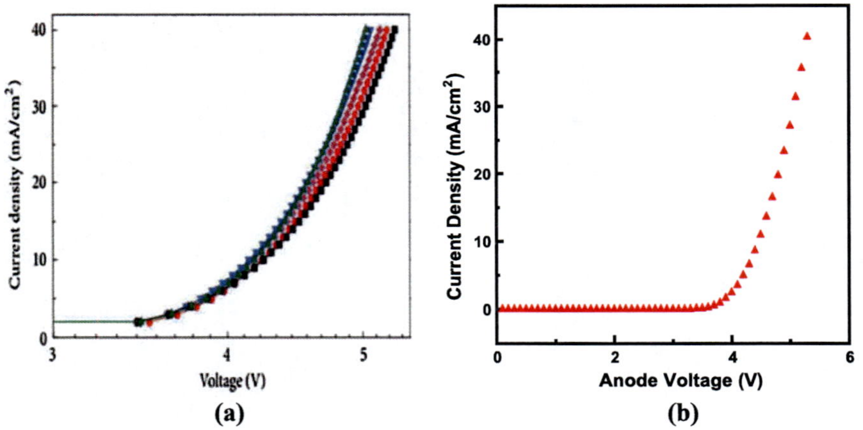

Fig. 2 Comparison of current density versus anode voltage characteristics with experimental data and simulation plots **a** experimental graph and **b** simulated with similar parameters as experimental

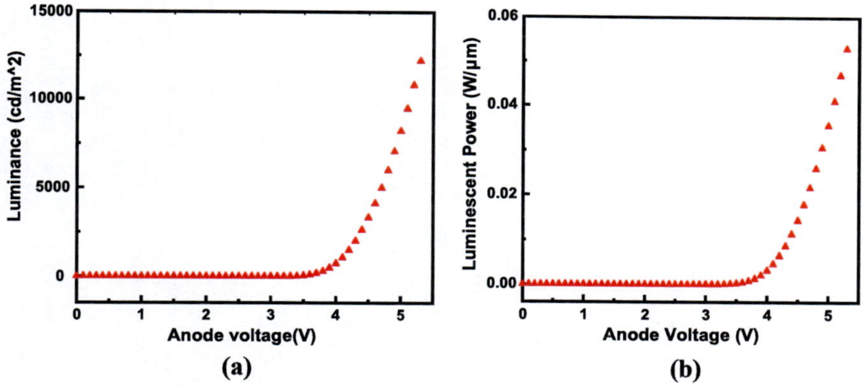

Fig. 3 **a** Luminance versus anode voltage graph. **b** Luminescent power versus anode voltage graph

density. The fabricated device did not calculate the value of luminance and luminescent power; we have calculated the luminance and luminescent power; the measured value is 12,200 cd/m^{2}and 0.0524 W/μm.

2.1 Analysis of Thickness Variation of Mixed Interlayer on OLEDs

In these devices, optimization of the device characteristics, such as the thickness of the emitting layers, the architecture of the interlayer, stacking to improve the sequence, and doping concentration, the efficiency and stability of colors have been important for OLED [10].

The thickness of a mixed interlayer and the order of collection of the emitting layer were beneficial in improving the light-emitting layer of the OLEDs performance. When the thickness is increased, the recombination rate will also increase, as extracted result current density and luminance may also enhance, and the performance of mixed-interlayer OLED is improved. The impact of thicknesses is investigated from 1 to 20 nm with step size of 5 nm for five (Device A to Device E) OLEDs is shown in Table 1. The parameters of OLED, properties, work function, materials, device dimensions, and HOMO (highest occupied molecular orbital), or LUMO (Lowest occupied molecular orbital) value is given in Table 1 [8].

Table 1 Specification of different thickness variation-proposed work OLED structure

Parameter	Properties	Device A	Device B	Device C	Device D	Device E
OSC active layer	Thickness of mixed interlayer	MI = 1 nm	MI = 5 nm	MI = 10 nm	MI = 15 nm	MI = 20 nm
Anode	Work function	5.8	5.8	5.8	5.8	5.8
	Material	ITO/Glass (110 nm)	ITO/Glass (110 nm)	ITO/Glass (110 nm)	ITO/Glass (110 nm)	ITO/Glass (110 nm)
Catode	Work function	2.8	2.8	2.8	2.8	2.8
	Material	Al (110 nm)	Al (110 nm)	Al (110 nm)	Al (110 nm)	Al (110 nm)
HIL	HOMO–LUMO	(5.83–2.30)	(5.83–2.30)	(5.83–2.30)	(5.83–2.30)	(5.83–2.30)
	Material	MoO_3 (10 nm)	MoO_3 (10 nm)	MoO_3 (10 nm)	MoO_3 (10 nm)	MoO_3 (10 nm)
EIL	HOMO–LUMO	(6.15–2.65)	(6.15–2.65)	(6.15–2.65)	(6.15–2.65)	(6.15–2.65)
	Material	TPBi: Li (20 nm)	TPBi: Li (20 nm)	TPBi: Li (20 nm)	TPBi: Li (20 nm)	TPBi: Li (20 nm)
HTL	HOMO–LUMO	(5.83–2.30)	(5.83–2.30)	(5.83–2.30)	(5.83–2.30)	(5.83–2.30)
	Material	TcTa: MoO_3(65 nm)	TcTa: MoO_3 (65 nm)	TcTa: MoO_3 (65 nm)	TcTa: MoO_3 (65 nm)	TcTa: MoO_3 (65 nm)
EBL	HOMO–LUMO	(5.49–2.41)	(5.49–2.41)	(5.49–2.41)	(5.49–2.41)	(5.49–2.41)
	Material	TcTa (10 nm)	TcTa (10 nm)	TcTa (10 nm)	TcTa (10 nm)	TcTa (10 nm)
YP-EML	HOMO–LUMO	(6.10–2.71)	(6.10–2.71)	(6.10–2.71)	(6.10–2.71)	(6.10–2.71)
	Material	TPBi: PO (20 nm)	TPBi: PO (20 nm)	TPBi: PO (20 nm)	TPBi: PO (20 nm)	TPBi: PO (20 nm)
BF-EML	HOMO–LUMO	(5.30–2.55)	(5.30–2.55)	(5.30–2.55)	(5.30–2.55)	(5.30–2.55)
	Material	CBP: BCzVBi (10 nm)	CBP: BCzVBi (10 nm)	CBP: BCzVBi (10 nm)	CBP: BCzVBi (10 nm)	CBP: BCzVBi (10 nm)
MIL	HOMO–LUMO	(5.75–2.55)	(5.75–2.55)	(5.75–2.55)	(5.75–2.55)	(5.75–2.55)
	Material	CBP: TBPi (1 nm)	CBP: TBPi (5 nm)	CBP: TBPi (10 nm)	CBP: TBPi (15 nm)	CBP: TBPi (20 nm)

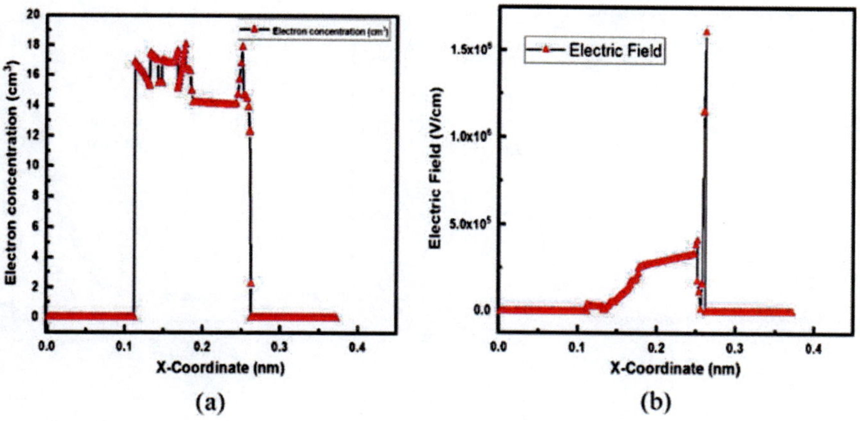

Fig. 4 Impact of thickness variation on **a** electric field and **b** electron concentration for device A

2.2 *Analysis of Internal Structure of Mixed Interlayer on OLED*

In the given section devices, which are stimulated and given result, and it is shown the internal analysis of device is build up the performance of OLED. The analysis is executed by drawing the cutline at the middle of the OLED to monitor the effect in a different layer on the internal parameter including electric field and electron concentration shown in Figures 4a, b. By drawing a cutline at the center of OLED, it may easily analyze the electric field and electron concentration. The measured maximum magnitude is 1.61×10^6 V/cm, 17.9 cm^2 (Fig. 4).

3 Result and Discussion

All-important parameters, like current density, luminance, and luminescence power efficiency, are extracted from ATLAS Silvaco Tool [11–14]. This is in Fig. 5a; we compared the current density with the applied voltage of the different thickness of the mixed interlayer. It is identified that as we increase the size of the thickness of mixed interlayers, the current density improves at a point, but it is decreased when increases the mixed interlayer dimension higher size; this is called by optimization of thickness of the mixed interlayer. When we examine the five highly current density devices it is given, comparing both Device A and Device C, the current density is much better in comparison to Device E. This is in Fig. 5b. We have been contrasting the luminance with a different voltage of anodes applied thickness of mixed interlayer. It is observed that as we increase the thickness dimension of layers, the luminance improves at a point, but it is decreased when increases the mixed interlayer dimension

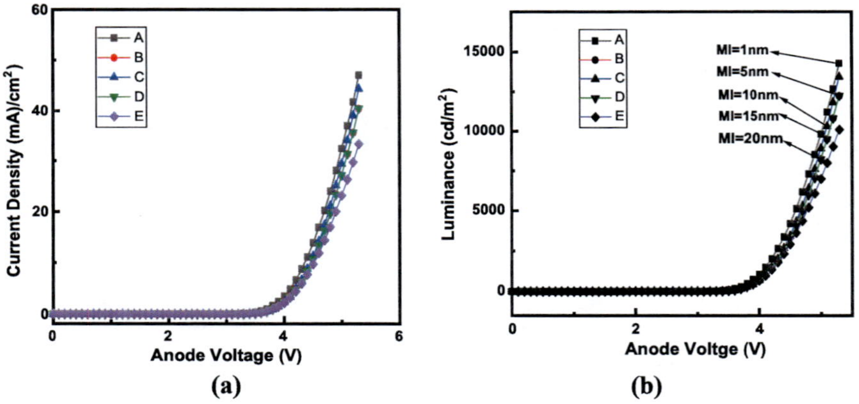

Fig. 5 **a** Combine graph of current density versus anode voltage. **b** Combine graph of luminance versus anode voltage with different thickness of mixed interlayer

highly. Among the five, highly luminance is given by is absorbed in case of Device A and Device C, and the worst current density obtained is Device E.

Device A to Device E is compared based on different performance characteristics at 5.32 V supply voltage. Characteristics such as current density, the luminance have been especially in comparison, and their values are being used in Table 2. We can see the Device A and Device C characterize the best performance, and Device E shows the worst characteristics. Device B is a fabricated device; this simulated value occurs as a current density is 40 mA/cm^2, and the proposed Device A and Device C value occurs as a current density of 47 and 42.2 mA/cm^2. In terms of current density, Device A is 12% better than the other Device B; Device C is 7% better than the other Device B, and Device D is 7.3% better than the other Device B. In terms, luminance device A is 13.42% better than the other device B; device C is 10.42% better than the other device B. The luminescence power efficiency of these OLEDs is extracted with the following formula of luminescence power efficiency [4].

$$\eta_p = L\pi / JV$$

Table 2 Comparison of performance parameter of different thickness of mixed interlayer for five devices

Parameter (at 5.32 V)	Device A	Device B	Device C	Device D	Device E
Current density (mA/cm^2)	47	40	44.2	44	33
Luminance (cd/m^2)	14,269	12,237	13,900	12,260	10,113
Luminescence power efficiency (lm/W)	20.8	17.7	20.6	17.9	14.0

where η_p is denoted by luminescence power efficiency, L is luminance, J is current density, and V is working voltage. The extracted values of luminescence power efficiency for devices A, B, C, D, and device E are summarized in Table 2. The highest device luminescence power efficiency is observed for device A. The extracted value for it is 20.8 lm/W.

4 Conclusion

In this paper, we have demonstrated the impact of variation in thickness of OLED mixed interlayer. This variation of a mixed interlayer structure by optimizing the mixing ratio (3:2) of hole-predominated material (CBP) and electron-predominated material (TPBi) is implemented and performance analyzed. The impact of thicknesses is investigated from 1 to 20 nm with a step size of 5 nm for five (Device A to Device E) of OLEDs. Device A has observed the best device with current density is 47 mA/cm^2, luminance is 14269 cd/m^2, and luminescence power efficiency value of 20.8 lm/W, respectively. Device A is very useful for bio-medical applications as a light source as well as a light detector.

References

1. Chou KY, Chao PCP, Chen CX, Wu CX, Huang SC (2017) Modeling and analysis of touch on flexible ultra-thin touch sensor panels for AMOLED displays employing finite element methods. Microcyst Technol 1(10):5211–5220
2. Kumar B, Kaushik BK, Negi YS (2014) Organic thin film transistors: structures, models, materials, fabrication, and applications. A review. Polym Rev 54(1):33–111
3. Fu Y, Tsai CC, Tsai JL (2016) Characterizing mechanical behaviors of a flexible AMOLED during the debonding process. Microcyst Technol 22(10):2397–2406
4. Negi S, Mittal P, Kumar B (2018) Impact of different layers on performance of OLED. Microsyst Technol 24:4981–4989
5. Negi S, Mittal P, Kumar B (2019) Analytical modeling and parameter extraction of multilayer OLED. IET Circuits Dev Syst 13:1255
6. Negi S, Mittal P, Kumar B (2020) Characteristic performance of OLED based on hole injection, transport and blocking layers. Rec Patent Eng 14(3)
7. Negi S, Mittal P, Kumar B (2020) In-depth analysis of structures, materials, models, parameters, and applications of organic light-emitting diodes. J Electron Mater 49(8)
8. Gao CH, Shi XB, Zhou DY, Zhang L, Wang ZK, Liao LS (2013) Highly efficient white organic light-emitting diodes with controllable excitons behavior by a mixed interlayer between fluorescence blue and phosphorescence yellow-emitting layers. Int J Photoenerg 2013(1–7)
9. (2011) ATLAS User's Manual device simulation software. Silvaco International Ltd, Sant Clara
10. Yook KS, Jeon SK, Joo CW, Lee JY (2009) Color stability and suppressed efficiency roll-off in white organic light-emitting diodes through management of interlayer and host properties. J Ind Eng Chem 15:420–422
11. Negi S, Mittal P, Kumar B (2020) Numerical modeling and parameters extraction of novel triple hole block layer based organic light emitting diode for display. J Soc Inform Display 28(12):956–964

12. Negi S, Mittal P, Kumar B (2019) Organic LED based light sensor for detection of ovarian cancer. Microelectron Eng 218:111154–111164
13. Kumar B, Kaushik BK, Negi YS (2013) Perspectives and challenges for organic thin film transistors: materials, devices, processes, and applications. J Mater Sci Mater Electron 25(1):1–30
14. Mittal P, Kumar B, Negi YS, Kaushik BK, Singh RK (2012) Channel length variation effect on performance parameters of organic field effect transistors. Microelectron J 43(12):985–994

Physical Properties Resemblance of Optical Material $ZnGeN_2$ with GaN Under Different Higher Pressures

Satish Chandra, Deepak Kumar, Rukmani Singh, Ritesh Kumar, and Virendra Kumar

Abstract In this paper, the physical properties of $ZnGeN_2$ and GaN compound semiconductors are calculated under different pressures using the density functional theory calculations. The lattice parameters and energy bandgap are studied at ambient conditions. Further, the energy bandgap of $ZnGeN_2$ and GaN under different pressures has been calculated. The bandgap values show that $ZnGeN_2$ is a direct bandgap up to 100 GPa and becomes an indirect bandgap at 110 GPa. GaN is a direct bandgap up to 150 GPa and turns out to be an indirect bandgap semiconductor at 160 GPa. The elastic parameters, i.e., elastic stiffness coefficients, have been estimated in the range of 0–190 GPa pressures. Results show that $ZnGeN_2$ and GaN are stable up to 180 and 150 GPa pressures, respectively. Comparative results show that the physical properties of $ZnGeN_2$ have a resemblance with GaN up to 100 GPa pressure and can be a potential candidate in place of GaN in various technological applications.

Keywords Gallium nitride · Density functional theory · $ZnGeN_2$ · Electronic properties · Elastic properties

1 Introduction

The gallium nitride (GaN) is a binary compound of III–V of family, which well-known direct bandgap semiconductor for its application as blue light-emitting diodes (LEDs) since 1990. $ZnGeN_2$ is a ternary nitride compound of $A^{II}B^{IV}N_2$ family with

S. Chandra (✉) · D. Kumar · R. Singh · R. Kumar
Electronics and Communication Engineering Department, Madan Mohan Malviya University of Technology, Gorakhpur, Uttar Pradesh 273010, India
e-mail: chandra.satish1989@gmail.com

R. Singh · R. Kumar · V. Kumar
Electronics Department, Indian Institute of Technology (Indian School of Mines), Dhanbad, Jharkhand 826004, India

D. Kumar
Electronics and Communication Engineering Department, National Institute of Technology, Patna, Bihar 800005, India

B. Mishra and M. Tiwari (eds.), *VLSI, Microwave and Wireless Technologies*, Lecture Notes in Electrical Engineering 877,
https://doi.org/10.1007/978-981-19-0312-0_66

a wide and direct bandgap, which is similar in properties with GaN. This makes $ZnGeN_2$ replace GaN in many applications such as lasers, LEDs, UV sensors, and other optoelectronic devices [1]. $ZnGeN_2$ and alloys of $ZnGeN_2$ have more bandgap flexibility, making them suitable for LEDs and photovoltaic absorbers [2]. It was first predicted by Maunaye and Lang in 1970 [3] that $ZnGeN_2$ has a less monoclinic symmetry within the wurtzite (WZ) structure. Using neutron-diffraction investigation, Wintenberger et al. [4] described the physical structure of $ZnGeN_2$ as an orthorhombic crystal. Further, many workers have used the MOCVD method [5, 6] to synthesize $ZnGeN_2$ and its alloy stated the energy bandgap of $ZnGeN_2$ in the range of 3.1−3.2 eV. For the computation of physical and nonlinear properties, Limpijumnong et al. [7] employed local density approximation (LDA). The electronic, optical, and elastic properties of $ZnGeN_2$ were studied by other workers using a plane-wave pseudopotential approach [8]. Recently, we have investigated the various physical properties of $ZnGeN_2$ [9], $LiGaS_2$ [10], $LiGaSe_2$ [10], graphene [11], AlN [12], and $A^{II}B_2^{III}C_4^{VI}$ group of semiconductors using first-principle method [13] and plasma oscillations theory of solids [14]. Increasing demand and use of GaN could not be able to fulfill the requirement alone. In this manner, we figured it would be remarkable to think about these compounds' characteristics at various higher pressures. In this paper, the physical properties of $ZnGeN_2$ and GaN are investigated at 0−190 GPa pressures employing first-principle calculations. The findings at 0 GPa pressure are contrasted with the known data reported by earlier workers. A sensibly great agreement has been acquired between them for both the compounds. On other pressures, calculated properties show the resemblance of $ZnGeN_2$ and GaN.

2 Computational Details

For optimization of $ZnGeN_2$ (space group Pbn21) and GaN (space group P63mc) with Cambridge sequential total energy package (CASTEP) code [15, 16], density functional theory within local density approximation (LDA) calculations is used. Under different pressures, the physical properties of these semiconductors are calculated. To obtain the optimized structure, a plane-wave basis set with ultrasoft pseudopotential (USP) at the cut-off of 350 eV [17] has been applied over the Monkhorst–Pack 2 × 3 × 3 mesh Brillouin zone. The Broyden, Fletcher, Goldfarb, and Shanno (BFGS) [18] system was used at the ultra-fine quality with "total energy difference of 5 × 10 – 6 eV/atom, the maximum displacement of 5 × 10 – 4 Å, the maximum stress of 0.02 GPa, and Hellmann–Feynman ionic force of 0.01 eV/Å for convergence threshold".

3 Results and Discussion

3.1 Electronics Properties

As a superstructure of the wurtzite structure, $ZnGeN_2$ crystallizes with orthorhombic symmetry under ambient conditions. Zn (0.620, 0.083, 0.000), Ge (0.125, 0.083, 0.000), N1 (0.115, 0.070, 0.365), and N_2 (0.640, 0.095, 0.385) [4] are the calculated atomic positions of $ZnGeN_2$, whereas GaN crystal atomic positions are Ga (0.333, 0.667. 0.000) and N (0.333, 0.667, 0.375) [19]. For $ZnGeN_2$ and WZ-GaN, the optimization curve, i.e., the total energy versus volume curves, is shown in Fig. 1, which shows that the minimum energy is attained for $ZnGeN_2$ is at −9464.98426

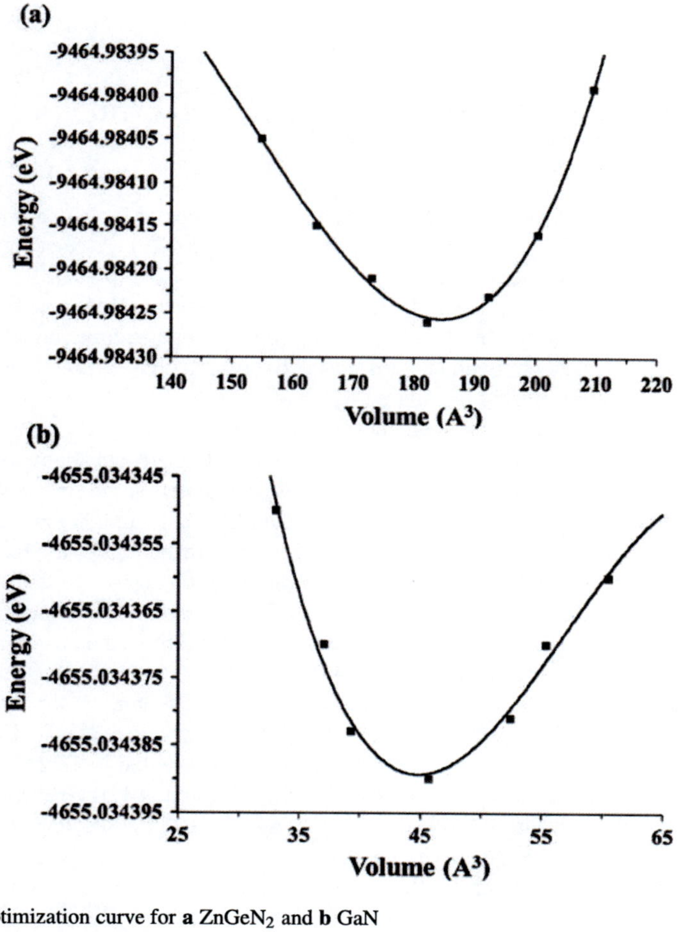

Fig. 1 Optimization curve for **a** $ZnGeN_2$ and **b** GaN

Table 1 Lattice parameter and energy bandgap of $ZnGeN_2$ and GaN semiconductors at ambient conditions

Compounds		Lattice parameters (Å)			E_g (eV)
$ZnGeN_2$		a	b	c	
	This work	6.291	5.345	5.119	2.548
	Expt. values	6.44[a]	5.45 [a]	5.19 [a]	3.40 ± 0.0[b], 3.3[c], 3.35*
	Theo. values	6.38[d], 6.42[d], 6.33[e]	5.45[d], 5.54[d], 5.36[e]	5.22[d], 5.27[d], 5.11[e]	1.66[d], 1.57[d], 1.615*
GaN		a		b	
	This work	3.170		5.113	2.027
	Expt. values	3.190[f]		5.189[f]	3.44[g], 3.50[h], 3.47*
	Theo. values	3.182[i]		5.189[i]	3.44[j], 3.50[k,l], 2.89[m], 3.0[n], 3.207*

[a][4], [b][7], [c][33], [d][8], [e][21], [f][22], [g][26], [h][27], [i][25], [j][28], [k][29], [l][30], [m][31], [n][34]
*Average value of theoretical and experimental data

and −4655.03439 eV for WZ-GaN. Table 1 lists the values of lattice parameters of $ZnGeN_2$ and GaN along with the known values [4, 8, 20–22]. A close agreement between them has been obtained.

The computed band structure of $ZnGeN_2$ and GaN is presented in Fig. 2. Figures 2a and b shows that $ZnGeN_2$ and GaN both are direct bandgap semiconductors with 2.548 eV and 2.027 eV energy bandgaps, respectively, at 0 GPa in the Γ–Γ direction (G-point). Table 1 presents the calculated values along with the earlier reported values [8, 23–31]. At 0 GPa, the calculated energy bandgap of $ZnGeN_2$ ($E_g = 2.54$ eV) is 24% lower than the average values of experimental data [23, 24] but about 50% better than the average known data of Eg. However, the calculated value for GaN is 42% lower than the average values of experimental data [25, 26] and 8% lower than the average values of theoretical data. These deviations are attributable to the recognized fact that the bandgap value is underestimated due to exchange–correlation energy discontinuity in the case of the LDA method [32]. It is essential to mention that, as reported in various papers, the experimental data of $ZnGeN_2$ changes from 3.3 to 3.40 eV, and for GaN, it changes from 3.44 to 3.50 eV.

In addition, the band structure is examined for $ZnGeN_2$ and GaN in the range 0–160 GPa under various pressures. The values of energy bandgap (E_g) of $ZnGeN_2$ and GaN have been calculated on these pressures and presented in Table 2. Table 2 shows that $ZnGeN_2$ is a direct bandgap up to 100 GPa, and at 110 GPa, it becomes an indirect bandgap, whereas GaN is a direct bandgap up to 150 GPa, and at 160 GPa it becomes an indirect bandgap. This indicates for $ZnGeN_2$ that the phase transition appears at 110 GPa. This implies that the electronic band structure is substantially affected by the change of pressure, resulting in an indirect bandgap at 110 GPa. The phase transition appears at 160 GPa pressure in the case of GaN. This demonstrates

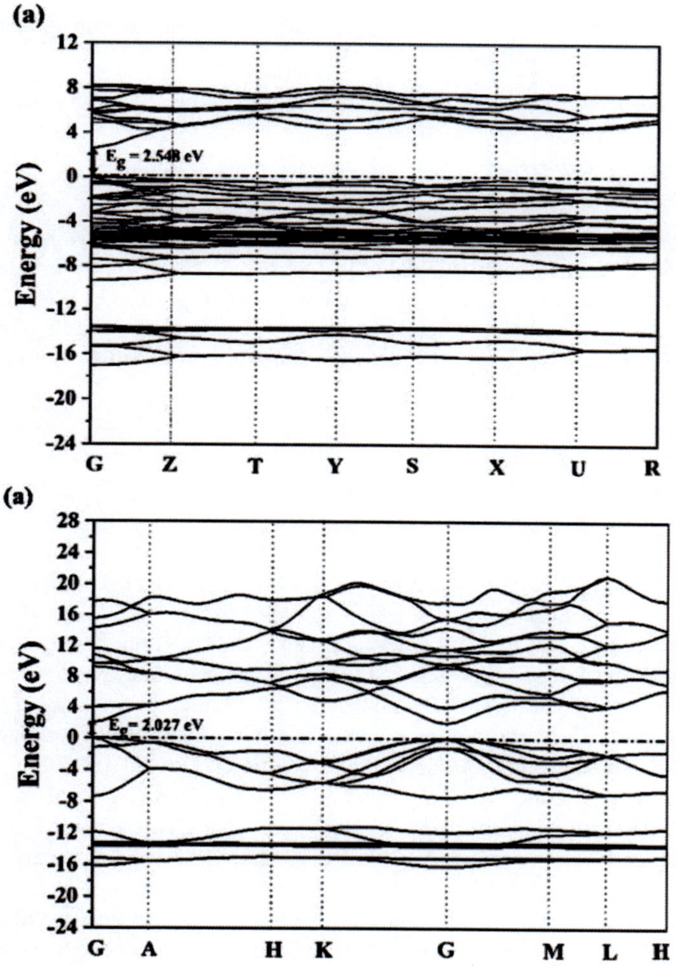

Fig. 2 Energy band structure of **a** $ZnGeN_2$ and **b** GaN

Table 2 Energy bandgap (E_g) of $ZnGeN_2$ and GaN semiconductors under different pressures

Compounds		At different pressures in GPa						
		0		50	100	110	150	160
$ZnGeN_2$	E_g (eV)	2.548	3.40 ± 0.0 [7], 3.3 [33]	3.679	4.284	4.359	0.192	0.222
	Type	Direct	Direct	Direct	Direct	Indirect	Indirect	Indirect
GaN	E_g (eV)	2.027	3.44 [26], 3.50 [27]	3.181	3.836	3.938	4.270	4.273
	Type	Direct	Direct	Direct	Direct	Direct	Direct	Indirect

that both semiconductors' energy bandgap up to 100 GPa, indicating a similarity between $ZnGeN_2$ and GaN of up to 100 GPa.

3.2 Elastic Properties

The elastic properties of the orthorhombic $ZnGeN_2$ and wurtzite GaN in the pressure range 0–190 GPa have been calculated in this paper. There are nine independent elastic stiffness constants of an orthorhombic crystal C_{ij}, i.e., C_{11}, C_{22}, C_{33}, C_{44}, C_{55}, C_{66}, C_{12}, C_{13}, and C_{23}, whereas there are six stiffness constants of WZ, i.e., C_{11}, C_{33}, C_{44}, C_{66}, C_{12}, and C_{13}. The orthorhombic crystal must meet the critical and sufficient requirements as follows [34]:

$$\begin{aligned} &C_{11} > 0;\ C_{11}C_{22} > C_{12}^2; \\ &C_{11}C_{22}C_{33} + 2C_{12}C_{13}C_{23} - C_{11}C_{23}^2 - C_{22}C_{13}^2 - C_{33}C_{12}^2 > 0; \\ &C_{44} > 0;\ C_{55} > 0;\ C_{66} > 0 \end{aligned} \tag{1}$$

$$\text{and } C_{11} > |C_{12}|;\ 2C_{13}^2\langle C_{33}(C_{11} + C_{12});\ C_{44}\rangle 0;\ C_{66} > 0\text{''} \tag{2}$$

Table 3 lists the calculated C_{ij} values of $ZnGeN_2$ and GaN, along with the available experimental and known data [35, 36]. Table 3 confirms that the elastic stiffness coefficient C_{ij} for $ZnGeN_2$ is in fair concurrence with the known data [35] and that the C_{ij} values for GaN are in good concurrence with the experimental data [36] at 0 GPa. The computed nine-elastic stiffness coefficients of $ZnGeN_2$, out of which six-constant coefficients, i.e., C_{11}, C_{22}, C_{33}, C_{12}, C_{13}, and C_{23}, increase with the increase in pressure, whereas the three-elastic constant C_{44}, C_{55}, and C_{66} decrease and subsequently get deviated at 120 GPa. At 190 GPa, the computed values of C_{ij} for $ZnGeN_2$ disregard the critical and necessary conditions specified in Eq. (1). This demonstrates that up to 180 GPa of $ZnGeN_2$ is stable. In the case of GaN, the values of C_{11}, C_{33}, C_{12}, and C_{13} increase with the increase of pressures, whereas C_{44} and C_{66} decrease with the increase of pressure up to 140 GPa and subsequently get deviated afterward. At 160 GPa, the computed value of C_{44} for GaN turns negative, which disregards the conditions specified in Eq. (2), and it is becoming unstable. This tells that $ZnGeN_2$ up to 180 GPa is mechanically stable, and phase transition begins at 190 GPa, while GaN up to 150 GPa is stable, and at 160 GPa, phase transition begins. Table 3 confirms that the six C_{ij} values for $ZnGeN_2$ and GaN, i.e., C_{11}, C_{33}, C_{44}, C_{66}, C_{12}, and C_{13}, follow a similar trend up to 120 GPa. This demonstrates the elastic stiffness coefficients of both semiconductors of up to 120 GPa, indicating a similarity between $ZnGeN_2$ and GaN of up to 120 GPa.

Table 3 Elastic stiffness constants C_{ij} (in GPa) of $ZnGeN_2$ and GaN under different pressures

Elastic constant	0 (GPa)				This work													
	$ZnGeN_2$		GaN															
	This work	Rep. [35]	This work	Rep. [36]	$ZnGeN_2$	GaN	$ZnGeN_2$	GaN	$ZnGeN_2$	GaN	$ZnGeN_2$	GaN	$ZnGeN_2$	GaN	$ZnGeN_2$	GaN	$ZnGeN_2$	$ZnGeN_2$
1	2	3	4	5	6	7	8	9	10	11	12	13	14	15	16	17	18	19
C_{11}	340.09	341	354.04	367	539.99	538.63	666.55	680.37	687.24	704.36	719.75	716.29	1395.75	2820.39	1470.1	3640.71	1240.07	1213.66
C_{33}	394.38	401	407.46	405	644.57	603.04	847.21	753.94	899.88	776.41	930.19	813.32	1397.9	907.15	1471.88	939.58	1580.52	1437.76
C_{44}	94.53	86	99.42	95	84.24	84.41	53.29	54.93	43.78	49.32	35.56	46.14	255.2	7.88	258.48	−1.23	277.78	246.99
C_{66}	110.33	105	111.66	116	99.07	103.94	76.92	84	56.85	85.34	77.79	62.59	264.72	288.8	266.49	396.92	252.44	205.12
C_{12}	132.33	136	130.72	115	336.46	330.76	521.87	512.37	562.12	533.67	589.66	591.11	498.37	2242.8	513.8	2846.87	699.64	515.42
C_{13}	87.52	103	104.07	92	292.32	296.05	517.88	497.22	555.1	530.34	596.58	563.64	501.04	924.53	519.9	1047.2	684.11	610.59
C_{22}	356.73	358			519.71		665.27		675.54		693.74		1391.6		1458.36		1248.07	2898.98
C_{55}	101.98	95			83.59		65.55		51.49		52.36		266.32		266.89		262.2	179.44
C_{23}	88.96	98			285.53		480.33		518.49		551.92		543.55		564.71		646.2	2785.53

4 Conclusion

For the physical properties of orthorhombic $ZnGeN_2$ and GaN compounds at 0−190 GPa pressures, the density functional theory within LDA method calculations is successfully performed. The calculated lattice parameter and energy bandgap (E_g) values in Table 1 are in fair concurrence with the known data. $ZnGeN_2$ energy bandgap values reveal a direct bandgap nature of up to 100 GPa and become an indirect bandgap at 110 GPa. GaN is a direct bandgap of up to 150 GPa, and at 160 GPa, it becomes an indirect bandgap. This broadly demonstrates that both are up to 100 GPa of the direct bandgap.

In addition, the elastic stiffness coefficient values of both semiconductors are computed in the 0–190 GPa range pressures and presented together with the known values in Table 3. In a few cases, they are reasonably concurring with the experimental and reported data at 0 GPa. The computed C_{ij} values show that $ZnGeN_2$ and GaN are stable, respectively, up to 180 and 150 GPa, and become unstable. Furthermore, elastic property analysis shows that six parameters of C_{ij}, i.e., C_{11}, C_{33}, C_{44}, C_{66}, C_{12}, and C_{13} of $ZnGeN_2$ and GaN, pursue the same pattern and have a resemblance of each other up to 120 GPa. Our reported values are in reasonable concurrence with the known data at 0 GPa in almost all instances. Accumulating the analysis of energy bandgap and elastic stiffness coefficients, it is noted that up to 100 GPa, $ZnGeN_2$ resembles GaN. Thus, in place of GaN and $ZnGeN_2$ can be used for up to 100 GPa in applications with greater chemical flexibility.

Acknowledgements For his continuous motivation and encouragement, we are grateful to Prof. R. K. Chauhan, Head, Electronics and Communication Engineering, M. M. M. Gorakhpur Technology University.

References

1. kachkov D, Jaroenjittichai AP, Huang LY, Lambrecht WRL (2016) Native point defects and doping in $ZnGeN_2$. Phys Rev B 93:1–15
2. Narang P, Chen S, Coronel NC, Gul S, Yano J, Wang L-W, Lewis NS, Atwater HA (2014) Bandgap tunability in Zn(Sn, Ge)N_2 semiconductor alloys. Adv Mater 26:1235–1241
3. Maunaye M, Lang J (1970) Preparation et proprietes de $ZnGeN_2$. Mater Res Bull 5:793–796
4. Wintenberger M, Maunaye M, Laurent Y (1973) Groupe spatial et ordre des atomes de zinc et de germanium dans $ZnGeN_2$. Mater Res Bull 8:1049–1053
5. Zhu LD, Maruska PH, Norris PE, Yip PW, Bouthillette LO (1999) GaN and related alloys. Materials Research Society, Warrendale, Pennsylvania
6. Muth J, Cai A, Osinsky A, Everitt H, Cook B, Avrutsky I (2004) Optical pProperties of II–IV-N_2 Semiconductors. MRS Proc 831:E11.45
7. Limpijumnong S, Rashkeev SN, Lambrecht WRL (1999) GaN and related alloys. Materials Research Society, Warrendale, Pennsylvania
8. Punya A, Lambrecht WRL, Van Schilfgaarde M (2011) Quasiparticle band structure of Zn-IV-N_2 compounds. Phys Rev B—Condens Matter Mater Phys 84:1–10
9. Chandra S, Kumar V (2019) Structural, electronic and elastic properties of $ZnGeN_2$ and WZ-GaN under different hydrostatic pressures: a first-principle study. Int J Mod Phys B 33:1950297

10. Kumar V, Chandra S, Santosh R (2018) First-principles calculations of the structural, electronic, elastic and optical properties of $LiGaS_2$ and $LiGaSe_2$ semiconductors under different pressures. J Electron Mater 47:1223–1231
11. Kumar V, Santosh R, Chandra S (2017) First-principle calculations of structural, electronic, optical and thermal properties of hydrogenated graphene. Mater Sci Eng, B 226:64–71
12. Kumar V, Singh BP, Chandra S (2017) Optical, elastic and thermal properties of ZB-AlN semiconductor from first-principle calculations. Indian J Phys 91:445–451
13. Chandra S, Sinha A, Kumar V (2019) Electronic and elastic properties of AIIB2IIIC4VI defect-chalcopyrite semiconductors. Int J Mod Phys B 33:1950340
14. Kumar V, Chandra S (2018) Second-order nonlinear optical tensor coefficients of $LiXTe_2$ (X = Al, Ga, In) chalcopyrite semiconductors. Cryst Res Technol 53:1800133
15. Perdew JP, Zunger A (1981) Self-interaction correction to density-functional approximations for many-electron systems. Phys Rev B 23:5048–5079
16. Segall MD, Lindan PJD, Probert MJ, Pickard CJ, Hasnip PJ, Clark SJ, Payne MC (2002) First-principles simulation: ideas, illustrations and the CASTEP code. J Phys: Condens Matter 14:2717–2744
17. Vanderbilt D (1990) Soft self-consistent pseudopotentials in a generalized eigenvalue formalism. Phys Rev B 41:7892–7895
18. Fischer TH, Almlöf J (1992) General methods for geometry and wave function optimization. J Phys Chem 96:9768–9774
19. Hanada T (2009) Basic properties of ZnO, GaN, and related materials
20. Paudel TR, Lambrecht WRL (2008) First-principles study of phonons and related ground-state properties and spectra in Zn-IV-N2 compounds. Phys Rev B 78:115204
21. Schulz H, Thiemann KH (1977) Crystal structure refinement of AlN and GaN. Solid State Commun 23:815–819
22. Brudnyi VN, Kosobutsky AV, Kolin NG (2011) Effect of pressure and mechanical stress on the electronic properties of AlN and GaN. Phys Solid State 53:679–688
23. Misaki T, Tsuchiya K, Sakai D, Wakahara A, Okada H, Yoshida A (2003) Growth and characterization of $ZnGeN_2$ by using remote-plasma enhanced metalorganic vapor phase epitaxy. Phys Stat Sol (c). 0:188–191
24. Du K, Bekele C, Hayman CC, Angus JC, Pirouz P, Kash K (2008) Synthesis and characterization of $ZnGeN_2$ grown from elemental Zn and Ge sources. J Cryst Growth 310:1057–1061
25. Yeo YC, Chong TC, Li MF (1998) Electronic band structures and effective-mass parameters of wurtzite GaN and InN. J Appl Phys 83:1429–1436
26. Monemar B (1974) Fundamental energy gap of gan from photoluminescence excitation spectra. Phys Rev B 10:676–681
27. Huang MZ, Ching WY (1985) A minimal basis semi-ab initio approach to the band structures of semiconductors. J Phys Chem Solids 46:977–995
28. Rezaei B, Asgari A, Kalafi M (2006) Electronic band structure pseudopotential calculation of wurtzite III-nitride materials. Physica B 371:107–111
29. Miwa K, Fukumoto A (1993) First-principles calculation of the structural, electronic, and vibrational properties of gallium nitride and aluminum nitride. Phys Rev B 48:7897–7902
30. Min BJ, Chan CT, Ho KM (1992) First-principles total-energy calculation of gallium nitride. Phys Rev B 45:1159–1162
31. Oussaifi Y, Said A, Ben Fredj A, Debbichi L, Ceresoli D, Said M (2012) Effect of pressure on the energy band gaps of wurtzite GaN and AlN and electronic properties of their ternary alloys AlxGa1-xN. Physica B 407:3604–3609
32. Tripathy SK, Kumar V (2014) Electronic, elastic and optical properties of $ZnGeP_2$ semiconductor under hydrostatic pressures. Mater Sci Eng, B 182:52–58
33. Misaki T, Wakahara A, Okada H, Yoshida A (2004) Epitaxial growth and characterization of $ZnGeN_2$ by metalorganic vapor phase epitaxy. J Cryst Growth 260:125–129
34. Mouhat F, Coudert FX (2014) Necessary and sufficient elastic stability conditions in various crystal systems. Phys Rev B 90:224104

35. Paudel TR, Lambrecht WRL (2009) First-principles calculations of elasticity, polarization-related properties, and nonlinear optical coefficients in Zn-IV-N_2 compounds. Phys Rev B 79:245205
36. Wright AF (1997) Elastic properties of zinc-blende and wurtzite AlN, GaN, and InN. J Appl Phys 82:2833–2839

Photonic Generation of Arbitrary Microwave Signal Based on Tunable Optoelectronic Oscillator

Shivangi Yadav, Ritesh Kumar, and Pooja Lohia

Abstract Optoelectronic oscillator (OEO) provides microwave signal with very high-frequency and low-phase noise. In this paper, design of OEO is proposed based on external modulator and dispersive components. Role of dispersive component is to provide frequency selection similar to the photonic filter. Here, non-linear chirp fiber Bragg grating (CFBG) having quadratic variation of grating period is used as dispersive component. It has fixed center frequency at 1552 nm. Effects of various parameters like variations in the index modulation, chirp function, and length of the grating have been investigate on the oscillating frequency. Also, frequency tunability in the oscillating frequency of OEO is demonstrated by changing the wavelength of optical carrier signal. The reflected signal from CFBG was passed through parallelly connected single mode fiber of 0.5, 0.6, and 0.8 km length. Fiber is used for the extra delay in the loop and for locking of oscillating frequency. Tunability of the design has been shown for the wavelength 1550 to 1555 nm, and corresponding frequency of oscillation was observed between 3.9 and 35 GHz. Observations in the oscillating frequency are made for variation of modulation index of the CFBG having no chirp as well as the linear chirp. The overall model is theoretically analyzed and verified with the simulated results.

Keywords Optoelectronic oscillator · Chirp fiber Bragg grating · Frequency tunable

S. Yadav · R. Kumar (✉) · P. Lohia
Department of Electronics and Communication Engineering, Madan Mohan Malaviya University of Technology, Gorakhpur, U.P. 273010, India
e-mail: ritesh.iitdhanbad@gmail.com

R. Kumar
Department of Electronics Engineering, Indian Institute of Technology (ISM), Dhanbad, Jharkhand 826004, India

B. Mishra and M. Tiwari (eds.), *VLSI, Microwave and Wireless Technologies*, Lecture Notes in Electrical Engineering 877,
https://doi.org/10.1007/978-981-19-0312-0_67

1 Introduction

Microwave waveforms are intensively deployed in defense, security navigation, space communications, and modern healthcare instruments. However, conventional approach for the generation of microwave frequencies has some limitations due to their low speed and sampling rate. Optically generated signals have inherent properties in terms of very high frequency and optical processing [1]. Some of the interesting features adoptable through photonic techniques are signal generation with frequency multiplying capability, chirping capability, optoelectronics oscillator (OEO)-based generation for low-phase noise, and design of microwave photonic filter (MPF). Photonically-assisted microwave waveform generation is generally based on, direct space-to-time pulse shaping, time domain processing of optical signal, external modulation techniques, and frequency selection by dispersive elements [2]. In addition to the ability to generate stable frequency RF signals, OEOs can be used to perform other task such as generation of arbitrary shaped electrical signal, optical detection for physical signal, low-power RF signal detection, clock recovery, frequency translation by performing up and down conversion of frequency. There are different photonic techniques proposed for microwave signal generation with very high frequencies with large bandwidth [3], frequency multiplying capability [4–6], chirping capability [7–10], and of arbitrary shape [11, 12]. However, in this technique, external modulating RF signal is required. But in case of OEO, no external function generator is required for modulating RF signal.

The OEO is characterized by having very high-quality factor (Q) and stability and ultralow-phase noise [13]. Its unique behavior results from the use of electro-optical (E/O) and microwave photonic filters, which are generally characterized with high efficiency, high speed, and low dispersion in the microwave frequency regime. Microwave photonic filters (MPFs) are photonic subsystem designed with the aim of carrying equivalent tasks to those of an ordinary microwave filter within a radio frequency system or link. Photonic link-based filter is now an effective solution for low-transmission loss, high frequency, and microwave signal processing [14]. It can be demonstrated by using time delay [15, 16] and non-time delay structures [17, 18]. MPFs are also realized using dispersive component—linear chirped fiber Bragg grating (LCFBG), dispersion compensator, or a single mode fiber (SMF) [19], rough selection of oscillation mode in optoelectronic oscillator is bring about by the dispersive element together with PolM and PBS, for fine tuning, double loop is used [20]. Recently, the author has also proposed external modulation-based photonic techniques for the generation of microwave signals and investigated their application in remote sensing area [7–11].

In the above discussion, existing method of OEO design is basically based on linear CFBG and phase shifted FBG. However, the system was found to have limited performance in terms of wavelength tuning, ripple in the reflection spectra as well as some delay fluctuation. So, these parameters can be further modified to have enhanced performance of the system. In our proposed system, non-linear chirp FBG with quadratic variation of grating period has been considered, and their various characteristics have been explored. The application of non-linear CFBG has also been investigated in the proposed methodology for tunable OEO.

In this work, we opt a dispersive element for frequency selection in optical domain. There is no requirement of electronic microwave filter for optoelectronic oscillator implementation. Three single mode fiber of different length is used to provide delay to signal splitted in three equal parts by the power splitter. The oscillation frequency is tuned by changing the wavelength of the light source or by changing the chromatic dispersion of the dispersive element. The central frequency of the microwave filter is function of optical wavelength of the laser source and the chromatic dispersion of the dispersive element. One of the key components is non-linear CFBG. The chirp function for the grating has quadratic variation with respect to grating length, and Gaussian apodization is used to remove the ripple from slope of the delay curve and to achieve the flat top reflection bandwidth. Here with Gaussian apodization, quadratic chirp variation is used to achieve the comparatively large reflection bandwidth of 2.49 nm which provides large tunable range of frequency in GHz range. Significance of proposed model is that it has simplified configuration as well as no electronic microwave filter is employed.

2 Theory

The schematic diagram of the proposed optoelectronic oscillator for photonic generation of microwave signal is shown in Fig. 1. The system consists of a tunable laser source (TLS), an external modulator, an optical amplifier (OA), an optical circulator (OC), a dispersive element which is a chirped fiber Bragg grating (CFBG), power splitter (PS) power coupler (PC), and a photo-detector (PD). An optical carrier signal has been taken for a continuous wave laser source. The laser output passes through an external modulator to perform phase modulation of the signal. The phase modulated signal is optically amplified by optical amplifier before processing through a dispersive component. The optically amplified signal is given to one port of circulator where it is circulated to CFBG. The reflected optical signal available at the output

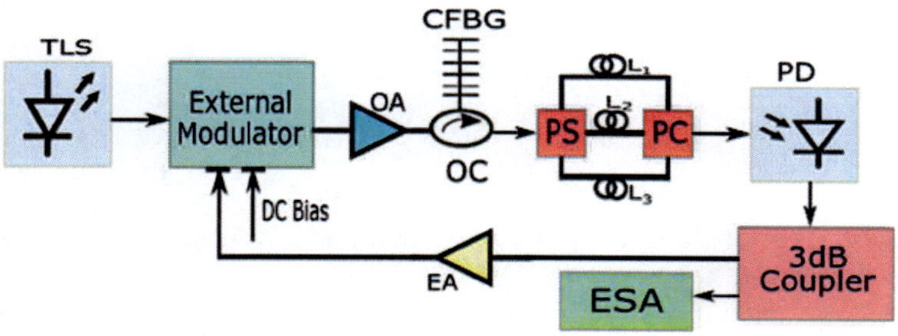

Fig. 1 Schematic diagram of proposed optoelectronic oscillator with dispersive element

port is further processed through combination of multiple fiber length. Here, three different lengths L_1,L_2, and L_3 have considered in parallel connection. The reflected signal power is divided into three arms by PS, where it experienced different delay due to three different lengths of dispersive fibers. These three differently delayed signals are combined through PC and detected by PD. Now, this signal is splitted in two parts, where one of the split parts is observed at electrical spectrum analyzer, and remaining part is feed to the RF port of external modulator.

The optical wave $E_o(t) = \sqrt{P_i(t)}e^{j\omega_o t}$ emitted from TLS acts as the source of energy for optoelectronic oscillator, where $P_i(t)$ is the input power, and ω_o is angular frequency of optical signal. Initial phase of the laser signal is assumed to be zero. The optically modulated signal at the end of external modulator can be presented as

$$E(t) = \sqrt{P_i(t)}\exp\{j[\omega_o t + \beta\cos(\omega_{osc}t)]\} \quad (1)$$

where, $\beta = \frac{\pi v_{Osc}}{2v_\pi}$ represents the modulation index of external modulator and v_π is its half wave voltage,ω_{osc} and v_{Osc} are respective angular frequency and amplitude of oscillation. When the modulated signal passes through the CFBG, the reflected signal will experience some phase change. Here, CFBG can be treated as optical band-pass filter whose transfer function is $H(f) = |H(f)|e^{j\emptyset}$, $|H(f)|$, and $\emptyset = \frac{\pi D f_{osc}^2}{c}$ represents magnitude and phase of transfer function in the reflection spectrum of CFBG. Here, c is the velocity of light in free space, and D is the dispersion coefficient.

Time domain representation of the reflected signal is given by

$$E'(t) = E(t)*h(t) \quad (2)$$

where $h(t)$ is the time domain representation of transfer function $H(f)$ of CFBG. When the loop is closed, the different length fiber L_1, L_2, and L_3 will form three comb filters. Such recombination has been taken for the oscillation frequency selection of the overall model. The transfer function of individual fiber length L_i is written as

$$H_i'(f) \propto \cos\left(\frac{2\pi f n_{\text{eff}} L_i}{c}\right) \quad (3)$$

The output of fiber in time domain is given by

$$E''(t) = E'(t)*\sum_{i=1}^{3} h'_i(t) \quad (4)$$

where $h_i'(t)$ is the time domain representation of transfer function $H_i'(f)$. The recovered signal at PD is given by

$$I \propto \frac{e^{j\omega_{osc}t}}{\left|1 - G.\left(\sum_{i=1}^{3} H_i'(f)\right)\right|} \quad (5)$$

where G is the overall gain provided by the loop due to the presence of optical and electrical components. The corresponding power at PD is given as

$$P \propto \frac{1}{1 + G^2\cos^2\left(\frac{\pi D f_{osc}^2}{c}\right) - 2G\cos\left(\frac{\pi D f_{osc}^2}{c}\right)\sum_{i=1}^{3} H_i'(f)} \tag{6}$$

In the above analysis, three different length of fiber is considered, which will provide oscillating frequencies with the phase increment of $\frac{2\pi f n_{\text{eff}} L_i}{c}$ with multiple of 2π. Here, three different length of fiber will generate three different frequencies. Least multiple of these frequencies will be locked, and a single frequency of oscillation can be selected by properly adjusting the fiber length.

3 Results and Discussion

The proposed design in Fig. 1 is implemented on OptiSystem software. Lightwave generated from tunable laser source is sent to the external modulator. The upcoming signal is firstly amplified by the optical amplifier and through circulator passed to the LCFBG for frequency selection, with the help of power splitter, the reflected signal power is equally splitted into three different arms containing unique length of fiber and fed to three single mode fibers having different length. The lengths of the single mode fiber in the loops are 0.5, 0.6, and 0.8 km. The non-linearity caused by optical fiber is compensated through CFBG. Signal is again combined by power combiner and fed to PD for optical to electrical conversion of the signal. A part of detected signal at the end of PD is sent as feedback to external modulator, and remaining part is used to observe through electrical spectrum analyzer where the generated electrical waveform can be seen. An electrical amplifier is used before feeding the split electrical signal to the RF port of external modulator. The operational frequency of optoelectronic oscillator is limited by frequency response of electro-optical devices and dispersion of CFBG. Large dispersion in short propagation length 5 cm is realized by CFBG. In the existing method of OEO design discussed in the literature have used linear CFBG, but in the proposed system, non-linear chirp FBG has been used which has the chirp function for the grating has quadratic variation with respect to grating length, and Gaussian apodization is used to remove the ripple from slope of the delay curve and to achieve the flat top reflection bandwidth of 2.49 nm. The index modulation is 0.0006, and radial photosensitivity index is real and constant at 1.46. This CFBG is investigated as microwave photonic filter for the generation of microwave signal realized on optoelectronic oscillator based on the set up shown in Fig. 1.

Reflectivity and delay characteristics of the considered CFBG have shown in Fig. 2. In Fig. 2a, reflectivity spectrum as function of propagating wavelength of CFBG is shown. The central wavelength of CFBG is 1552 nm. It can be seen that

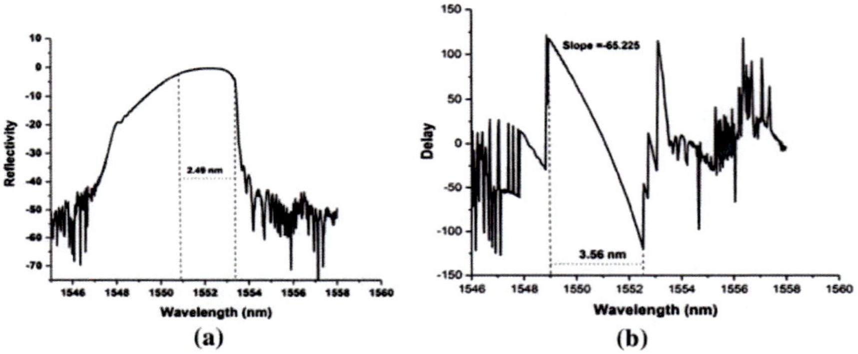

Fig. 2 **a** Reflection and **b** delay characteristics of quadratic chirp fiber Bragg grating with Gaussian apodization

maximum reflectivity occurs in the region 1550.87 nm to 1553.37 nm. Hence, reflection bandwidth of 2.49 nm gives good tunable range of CFBG. The delay curve of same is also showing Fig. 2b. It is found that delay is linear in the region 1548.95–1552.51 nm wavelengths with slope −65.23 ps/nm. The ripple effect in the obtained reflection bandwidth is negligible.

In Fig. 3a, a graph between wavelength and power is shown, taken from electrical spectrum analyzer. The system is tunable from 6 to 9 GHZ at wavelength 1552.52 to 1555 nm, 10dBm power is observed at wavelength 1552.50 nm. Maximum tunable frequency 9 GHz is observed at wavelength of 1552.52 nm at very low-absolute power –30dBm as shown in Fig. 3b.

Figure 4 shows the variation in oscillating frequency when laser wavelength is changed from 1552.52 nm to 1550 nm. In this case, CFBG of 1555 nm center wavelength with no chirp is considered, and remaining parameters are same as before.

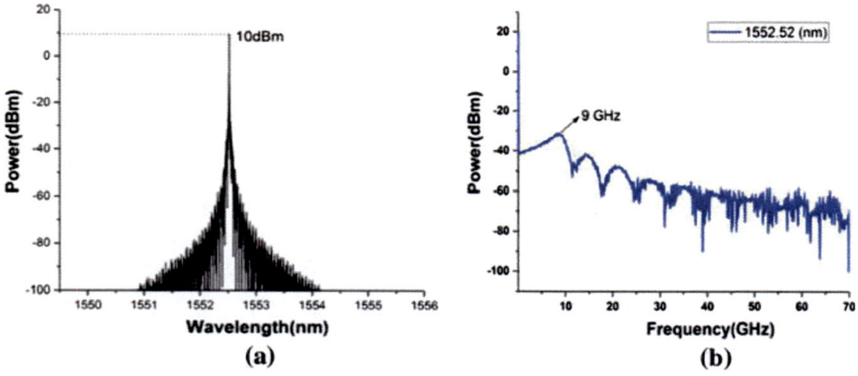

Fig. 3 **a** Reflected waveform after CFBG, **b** detected RF signal for optical carrier signal of 1552.52 nm

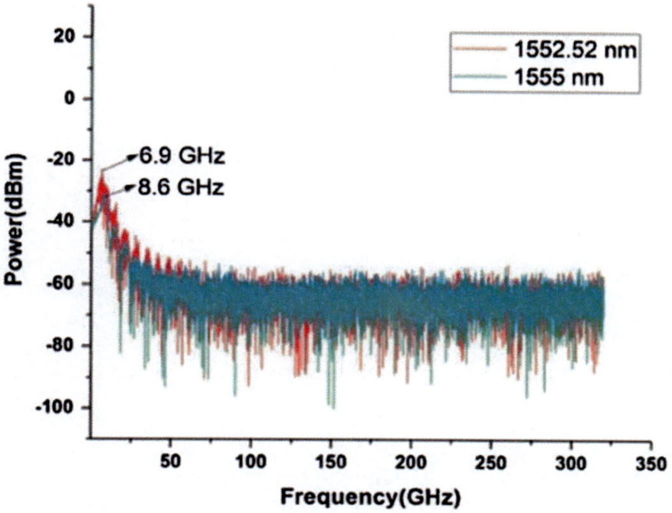

Fig. 4 Power of the detected signal at 1552.52 and 1555 nm laser wavelength for 1555 nm center wavelength CFBG

Here, for the same laser wavelength 1552.52 nm, the generated signal frequency has changed to 6.9 GHz. Again, frequency tunability is observed when laser wavelength is changed to 1555 nm. For this wavelength, oscillating frequency is 8.6 GHz. Result shows variation in oscillating frequencies with respect to center wavelength of CFBG as well as with the wavelength of laser source.

Again, effect of index modulation variation on oscillating frequency is investigated for no chirp and linear chirp of the fiber Bragg grating at laser wavelength of 1550 nm. Figure 5a shows the power of oscillating frequencies for different values

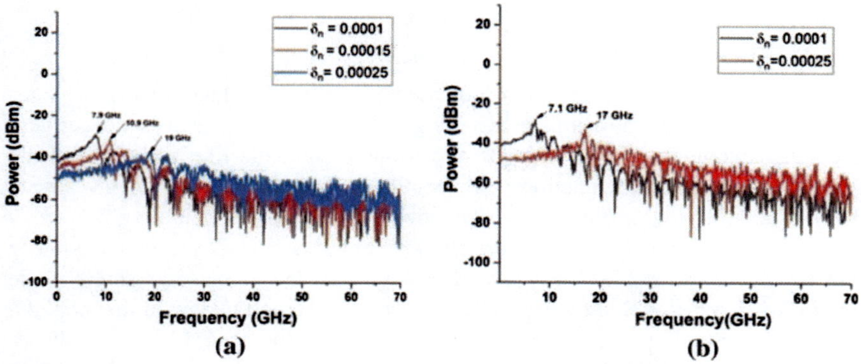

Fig. 5 **a** Index modulation variation for non-chirp fiber having length 20 mm. **b** Index modulation variation for linear chirp fiber with length 20 mm

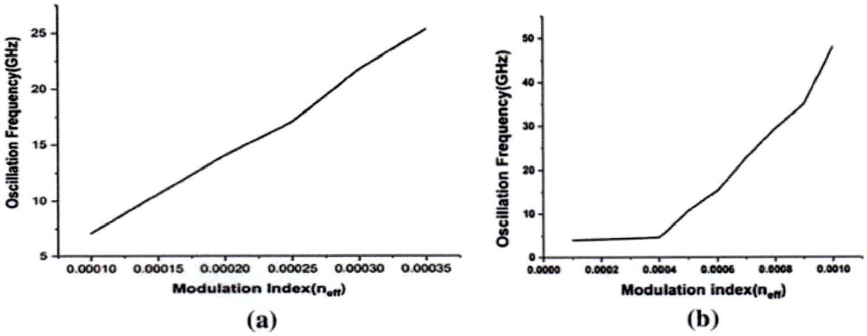

Fig. 6 Variation in oscillating frequency for different modulation index when **a** no chirp, **b** linear chirp of CFBG

of modulation index for no chirp. Oscillating frequencies are 7.9, 10.9, and 19 GHz, respectively, when index modulations are 1×10^{-4}, 1.5×10^{-4}, and 2.5×10^{-4}. In Fig. 5b, tunability in oscillating frequency is observed for index modulation variation for the case of linear chirp of CFBG. The oscillating frequencies are 7.1 and 17 GHz, respectively, for modulation index of 1×10^{-4} and 2.5×10^{-4}. In Fig. 5, it can be observed that oscillating frequency is increasing by increasing the index modulation of fiber Bragg grating. Next, effect of chirp can also be noticed which has been depicted in Fig. 6. Figure 6a represents the oscillating frequency variation with respect to index modulation when no chirp has taken, whereas Fig. 6b represents the same when linear chirp of CFBG is opted. A rapid variation in the oscillating frequency is observed in case of linear chirp compare to no chirp. However, smoothness in variation of oscillating frequency was found in case of no chirp compare to linear chirp of CFBG.

4 Conclusion

In this paper, design of OEO is proposed based on external modulator and chirp fiber Bragg grating for generation of signal having low-phase noise as well as independent of external modulating signal. Effect of various parameters like variations in the index modulation, chirp function, and length of the grating has been investigate on the oscillating frequency. Also, frequency tunability in the oscillating frequency of OEO is demonstrated by changing the wavelength of optical carrier signal. The reflected signal from CFBG was passed through parallelly connected single mode fiber of 0.5, 0.6, and 0.8 km length. Fiber is used for the extra delay in the loop and locking of oscillating frequency, CFBG. Tunability of the design has been shown for the wavelength 1550–1555 nm, and corresponding frequency of oscillation was observed between 3.9 and 35 GHz. Observations in the oscillating frequency are made for variation of modulation index of the CFBG having no chirp as well as the linear chirp. The overall model is theoretically analyzed and verified with the simulated results. The proposed technique does not require external function generator making the system less complex and useful in high-frequency application.

Acknowledgements This work is financially supported by TEQIP-III Collaborative Research Scheme (NPIU MHRD). The work is carried out under Project ID: 1-5744809099 and Project title: Generation of chirped arbitrary microwave waveforms with enhanced performance using photonic techniques for airborne application.

References

1. Seeds J, Williams KJ (2006) Microwave photonics. J Lightwave Technol 24(12):4628–4641
2. Yao JP (2011) Photonic generation of microwave arbitrary waveforms. Opt Comm 284:3723–3736
3. Xie Z, Li S, Yan H, Xiao X, Zheng X, Zhou B (2016) Tunable dual frequency optoelectronic oscillator with low intermodulation based on dual-parallel Mach-Zehnder modulator opticsexpress 30282:24(26)
4. Li W, Yao J (2010) Investigation of photonically assisted microwave frequency multiplication based on external modulation. IEEE Trans Microwave Theory Techniques 58(11)
5. Li W, Yao J (2010) Microwave generation based on optical domain microwave frequency octupling. IEEE Photon Technol Lett 22(1)
6. Li W, Yao J (2010) Microwave and terahertz generation based on photonically assisted microwave frequency twelvetupling with large tenability. IEEE Photon J 2(6)
7. Kumar R, Raghuwanshi RK (2019) Simultaneous photonic generation of multiple chirp and unchirp microwave waveform with frequency multiplying capability for optical beam forming system. Opt Quant Electron 51:344
8. Kumar R, Raghuwanshi SK (2017) A photonic scheme for the generation of dual linear chirp microwave waveform based on the external modulation technique and its airborne application. Opt Quant Electron 49:370
9. Raghuwanshi SK, Kumar R, Kumar M (2018) Generation of a dual nonlinear chirp microwave waveform in Ku-band and itscomparative study with dual linear chirping capability to distinguish range-Doppler coupling in radar application. In: Proceedigs SPIE 10531, Terahertz, RF, millimeter, and sub millimeter-wave technology and applications XI, 105311X
10. Raghuwanshi SK, Kumar R, Srivastava A, Srivastava NK (2016) Dual-chirp arbitrary microwave waveform generationby using a dual parallel mach-zehndermodulatorfeeding with RFchirpsignal. Progress In Electromagnet Res C 65:79–92
11. Kumar K, Raghuwanshi SK (2020) Photonic generation of a parabolic-shaped microwave signal and dual-linear-chirp microwave waveform. Appl Opt 59(20)
12. Li W, Wang WT, Zhu NH (2014) Photonic generation of radio-frequency waveforms based on dual-parallel mach–zehnder modulator. IEEE Photon J 6(3)
13. Yao XS, Maleki L (1996) Optoelectronic microwave oscillator. J Opt Soc Amer B 13 (8):1725–1735
14. Zhao J, Hu J, Deng P, Yu R, Liu R, Zhao M, Gu Y (2020) A tunable dual-passband microwave photonic filter based on optically injected distributed feedback semiconductor laser and dual-output mach-zehnder modulator. Appl Sci 10:3631
15. Capmany J (2003) Microwave photonic filters with negative coefficients based on phase inversion in an electro-optic modulator. Opt Lett 16:1415–1417
16. Zhang YM, Pan SL (2013) Complexcoeficient microwave photonic filter using a polarization-modulator-based phase shifter. IEEE Photon Technol Lett 25:187–189

17. Vidal B, Piqueras MA, Marti J (2007) Tunable and reconfigurable photonic microwave filter based on stimulatedBrillouin scattering. Opt Lett 32:23–25
18. Li WZ, Li M, Yao JP (2012) A narrow-passband and frequency-tunable microwave photonic filter basedon phase-modulation to intensity-modulation conversion using a phase-shifted fiber Bragg grating. IEEE Trans Microw Theory Tech 60:1287–1296
19. Tang Z, Pan S, Zhu D, Guo R, Zhao Y, Pan M, Ben D, Yao J (2012) Tunable optoelectronic oscillator based on a polarization modulator and a chirped FBG. IEEE Photon Technol Lett 24(17)
20. Li W, Yao JP (2010) An optically tunable optoelectronic oscillator. J Lightw Technol 28(18):2640–2645

Triple Band Semi-Fractal-Based Antenna for Wireless Applications in S, C and X-Bands

Vinay Singh, Brijesh Kumar Maurya, Praveen Kumar Rao, and Vipin Kumar Upaddhyay

Abstract In this paper, the proposed antenna is designed with U-shaped geometry. The proposed U-shaped antenna is designed with FR4 substrate having dimensions $26 \times 35.5\ \text{mm}^2$. This structure operates at three different frequencies for various wireless applications in *S*, *C* and *X*-bands. The multiband design covers bandwidth 2.45–2.75, 7.6–7.9 and 11.2–12.4 GHz. This diverse range of frequency bands is achieved by utilizing fractal configuration, such as the incorporation of semi-U and a rectangle slot at the bottom of the main U-slot. The antenna is simulated using HFSS software. This antenna is suitable for Bluetooth, radar and *C*-band applications, etc.

Keywords Patch antenna · *S*-band · *C*-band · *X*-band · FR4 epoxy · Slot

1 Introduction

The phenomenal increase in the use of wideband frequencies for wireless communication has opened a gateway for antennas operating in multiple frequency bands. These requirements have led the researchers to design antennas, which are compact in size, planar design along with low profile and additionally operates in the wideband region. Various techniques have been used to meet these design requirements for multiband applications such as semi or half ground substrate, tapered structure, split ring resonator, fractal geometry, [1–4]. Fractals are a family of complex geometric shapes in which the complete geometric shape is made up of several similarly shaped smaller parts, where each smaller part is the replica of the whole shape. The complex shapes possess the property of self-similarity and space-fitting which result in obtaining multiband antenna with compact geometry [5]. Apart from fractal

V. Singh
Department of Electronics and Communication Engineering, Institute of Technology and Management, Gida, Gorakhpur, India

B. K. Maurya · P. K. Rao · V. K. Upaddhyay (✉)
Department of Electronics and Communication Engineering, Madan Mohan Malviya University of Technology, Gorakhpur, India
e-mail: vipin08120@gmail.com

B. Mishra and M. Tiwari (eds.), *VLSI, Microwave and Wireless Technologies*, Lecture Notes in Electrical Engineering 877,
https://doi.org/10.1007/978-981-19-0312-0_68

geometry, there are various other shapes antennas such as metamaterial loaded [6, 7], notched band geometry, the pie-shaped quasi-complimentary antenna [8] are investigated in the open literature. Besides, triple band monopole antenna geometrical configured with compact disc shape design are proposed in [9]. The aforementioned shape is further modified by incorporating a crescent shape structure along with a circular disc, and additionally, the ground is also truncated into semi window shape form at the bottom [10]. This same structure helps to attain multiband operation in Bluetooth, WiMax, and Ku-band communication. In multiband antenna design positioning of different structures at optimized location plays a key role but it also leads to an increase in the antenna's dimensions. Following it, the authors in [11], shed the light on the importance of electric LC and EBG structure to achieve tri-band operations. Moreover, the importance of asymmetrical structure in antenna size miniaturization is discussed in [12] and triple circularly polarized bands using asymmetrical fractal curve structure are designed for the same. In [13], self-similarity fractal H shape antenna is designed for multiband operations. Ultra-wideband (bow-tie shape) is converted into Koch-like fractal antenna for multiband operations [14]. All the aforementioned discussions fail to meet the design requirements of a compact tri-band antenna. Motivated by this, we proposed to design a compact tri-band antenna based on fractal geometry.

2 Antenna Design

The realization of an antenna mostly depends on its dimensions, geometry and the property of its substrate. These design parameters are based on different characteristics like radiation pattern, resonant frequencies, directivity and return loss (Fig. 1).

The geometric structure of the proposed tri-band antenna is configured with two U-shaped joined with one lateral slot along with one additional rectangle slot at the bottom. This proposed configuration can be referred to as semi-fractal geometry. A co-axial feed is provided at the centre of the proposed tri-band design. The U-shape slot mounted on the FR-4 epoxy dielectric substrate with a dielectric loss tangent of 0.02 and a relative permittivity of 4.4. The dimensions of the substrate are 26, 35.5 and 1.6 mm as length, width and thickness, respectively. The additional rectangle slot added to the U-slot provides an extra frequency band operating in the ISM band between 2.4 and 2.8 GHz. The dimensions of all parameters tagged in the proposed design are tabulated in Table 1.

The second U-shaped slot partially radiated by primary U-shaped, which is mutually excited by Co-axial feed. The short slot added to U-shaped significantly compensates the impact of fringing effects of the primary slot, thus providing an additional band of operation. Moreover, the impedance matching between the proposed design and feed can be adjusted by varying the length of the short stub and the gap between the elements of the semi-fractal design.

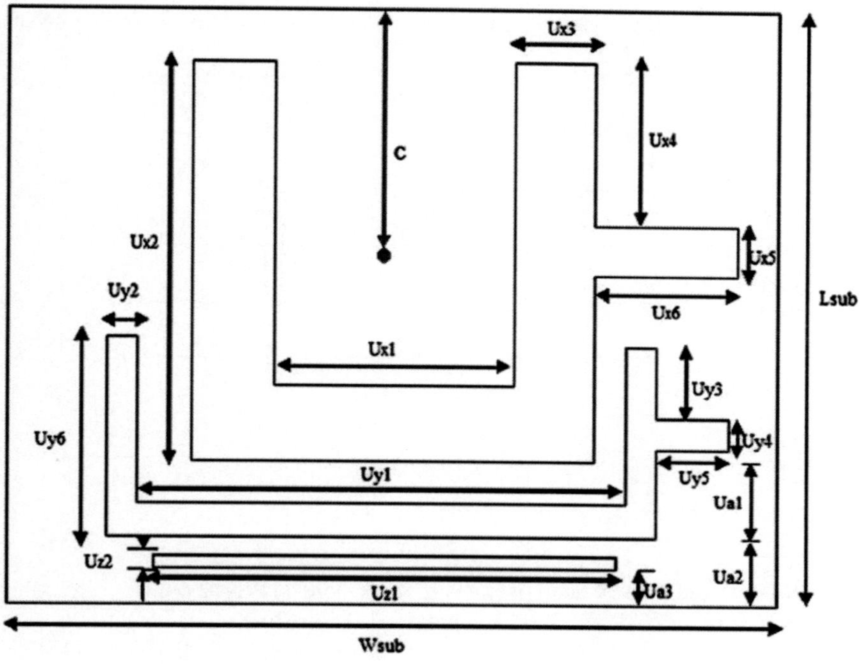

Fig. 1 Geometrical shape of the proposed antenna

Table 1 List of parameters

Parameters	Value (in mm)	Parameters	Value (in mm)	Parameters	Value (in mm)	Parameters	Value (in mm)
L_{sub}	26	W_{sub}	35.5	H	1.6	c	13.5
I_p	2.1	W_p	12	U_{x1}	7.8	U_{x2}	19.5
U_{x3}	2.1	U_{x4}	9.3	U_{x5}	1.1	U_{x6}	8
U_{y1}	20	U_{y2}	1	U_{y3}	2.5	U_{y4}	1
U_{y5}	4.0	U_{y6}	7	U_{z1}	16	U_{z2}	0.5
U_{a1}	4.8	U_{a2}	1.5	U_{a3}	0.5		

3 Result Analysis

The simulated result of the proposed antenna provides multiple frequency bands. The radiating patch of the proposed design is selected as a perfect electric conductor (PEC). The minimum return loss of this antenna is obtained as −33.8 dB in the *X*-band region. Figure 2 exhibits the reflection coefficient of the proposed antenna, with a minimum of −14 (dB) for all three bands of operation. The first band offers frequency operation in the *S*-band (2.45–2.8 GHz), the second band lies in the *C*-band

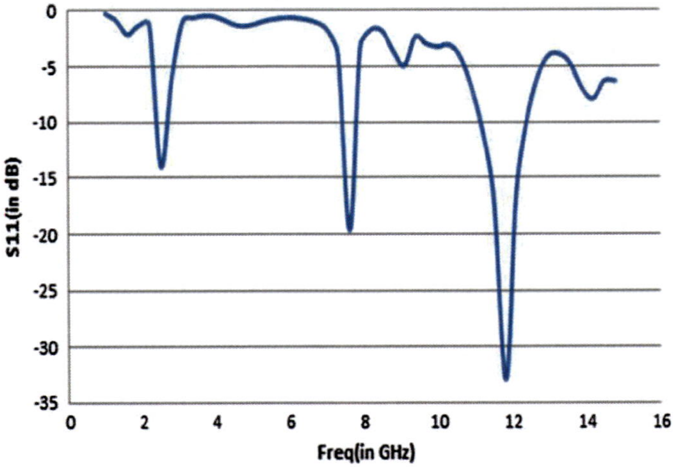

Fig. 2 Return loss of proposed antenna (dB)

(7.6–7.9 GHz) and the last band operates in the *X*-band (11.2–12.4 GHz). From the simulated return loss plot, the first band provides a return loss of −14 (dB) with −10 (dB) bandwidth of 350 GHz, second band's return loss lies at −20 (dB) with 300 MHz bandwidth. At last, the last *X*-band operates at a centre frequency of 11.5 GHz with a return loss of −33 (dB). The far-field radiation pattern in terms of *E* and *H* fields for the proposed Tri-bands design are provided in Figs. 3, 4 and 5, respectively. From Fig. 3, we can intuitively draw the inference that the far-field radiation pattern of the first band is in a doughnut shape and attains a Gain of approximately 8.5 (dB).

Figure 4 illustrates that the second band operating at 7.7 GHz offers a directional radiation pattern along with a gain of 4.3 (dB). Similarly, from Fig. 5, it is evident that the third's far-field radiation pattern is approximately directive in nature and achieved a directive gain of 7.2 (dB). Moreover, the stop bands between all three operating bands almost lie below −5 (dB). Specifically, the inter-band isolation between the first band and second band lies below –1 (dB), which exhibits excellent isolation. On the contrary, the isolation between second and third bands are compromised a little bit rendering to parasitic inductance and capacitance emerges due to third higher GHz operation.

In Fig. 6, the plot for radiation efficiency and gain versus frequency is exhibited concurrently. From the plot, it is evident that for all the triple bands, average 50% radiation efficiency is provided. Apart from that the gain at last frequency band is highest amongst all triple bands and lowest at first frequency band. Additionally, the gain offered at frequency bands other than the triple bands is very low that exhibits excellent inter-band rejection.

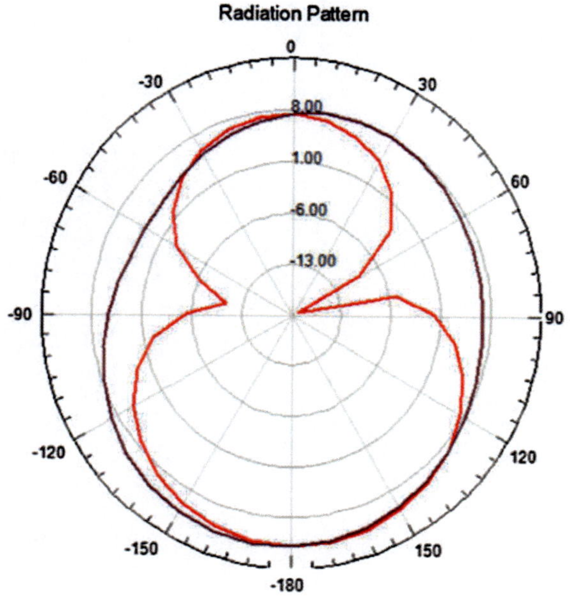

Fig. 3 Radiation pattern at 2.5 GHz

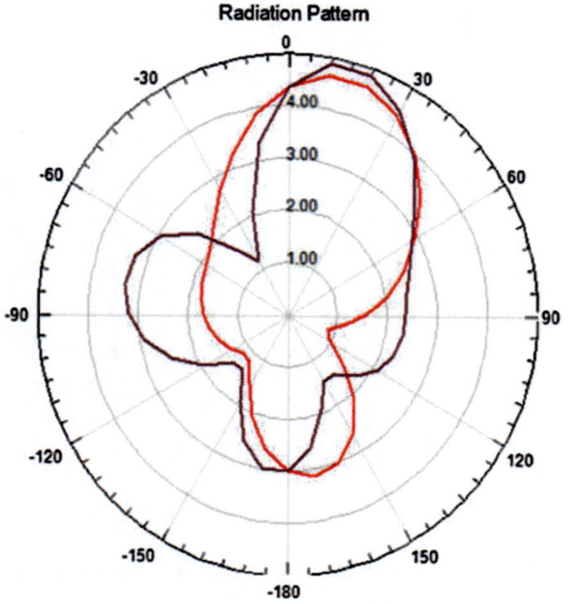

Fig. 4 Radiation pattern at 7.7 GHz

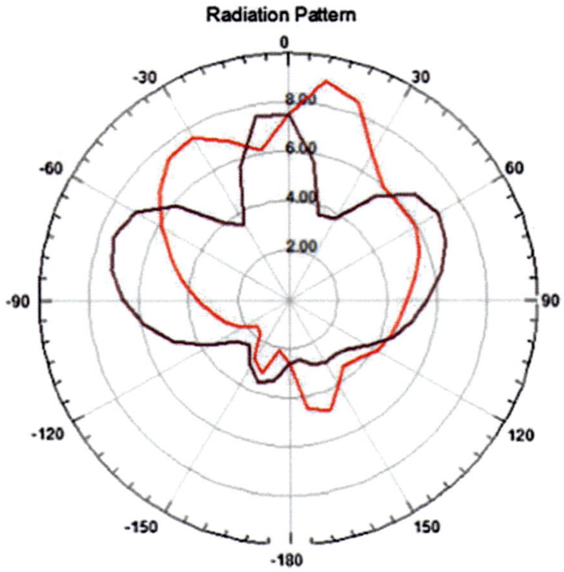

Fig. 5 Radiation pattern at 11.5 GHz

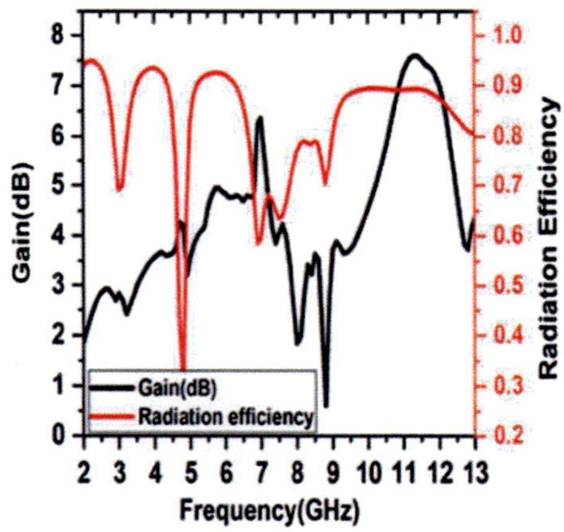

Fig. 6 Collective plot of gain and radiation efficiency

4 Conclusion

In this paper, a U-shaped slot with a rectangle shape added to the right-hand side of the slot in the middle is used which provides a notch in the frequency band. The use of multiple slots gives the multiple frequency bands that are used in different application for the *S*-band, *C*-band and *X*-band frequency regions. This antenna also gives a large bandwidth in the *X*-band frequency region. The *S*-band region bandwidth is used for

Bluetooth, microwave oven etc. The higher band applications of this antenna include weather forecasting, air traffic control and radar applications, etc.

References

1. Mrabet OE (2013) A compact split ring resonator antenna for wireless communication systems. Prog Electromagn Res Lett 36:201–207
2. Pujitha P (2017) Design of multiband antenna for wireless communications. Int J Sci Eng Res 8(6). ISSN 2229-55185
3. Kshetrimayum RS (2008) Printed monopole antennas for multiband applications. Int J Microw Opt Technol 3(4)
4. Werner DH, Ganguly S (2003) An overview of fractal antenna engineering research. IEEE Antennas Propag Mag 45(1):38–57
5. Balanis CA, Antenna theory, analysis and design, 3rd ed. John Wiley & Sons Inc., New Jersey
6. Antoniades MA, Eleftheriades GV (2009) A broadband dual-mode monopole antenna using NRI-TL metamaterial loading. IEEE Antennas Wireless Propag Lett 8:258–261
7. Zhu J, Antoniades MA, Eleftheriades GV (2010) A compact triband monopole antenna with single-cell metamaterial loading. IEEE Trans Antennas Propag 58:1031–1038
8. Lin CC, Huang CY, Chen GH (2013) Obtuse pie-shaped quasi-self-complementary antenna for WLAN applications. IEEE Antennas Wireless Propag Lett 12:353–355
9. Rao PK, Mishra R (2018) A disc shape monopole antenna with tri-band-notched 2.55/3.5/5.5 GHz characteristics. In: 2018 IEEE international students' conference on electrical, electronics and computer science (SCEECS). Bhopal, pp 1–6
10. Rao PK, Mishra R (2018) Multi band antenna with multi band notch 2.5/3.7/6.4/8.1/12.4/14.4GHz characteristics. In: 2018 5th IEEE Uttar Pradesh section international conference on electrical, electronics and computer engineering (UPCON). Gorakhpur, pp 1–5
11. Zhu K, Li C, Yuan-Ming L, Li C, Chang-Hong L (2013) Design of electrically small metamaterial antenna with ELC and EBG loading. IEEE Antennas Wireless Propag Lett 12:678–681
12. Reddy VV, Sarma NVSN (2014) Triband circularly polarized koch fractal boundary microstrip antenna. IEEE Antennas Wireless Propag lett 13
13. Chung WW, Hung CL (2014) An H-fractal antenna for multiband applications. IEEE Antennas Wireless Propag lett 13
14. Li D, Mao JF (2012) A koch-like sided fractal bow-tie dipole antenna. IEEE Trans Antennas Propag 60:2242–2251

Delay and Dispersion Investigation of Optical Components for Microwave Photonic Filter

Ritesh Kumar, Yadvendra Singh, Sanjeev Kumar Raghuwanshi, Satish Chandra, and Danish Nadeem

Abstract Photonic generation of microwave signal requires optical processing like filtering and shaping of light signal. For this purpose, main challenge is selection of optical delay components so that required microwave photonic filter can be designed. In this paper, theoretical and simulated analysis has been given for delay characterization of some most used optical components, like ring resonator, dispersive fiber, and fiber Bragg grating. Effect of coupling coefficient on delay characteristics and free spectral range has shown for single ring as well as the cascading of multiple rings. Then, group delay of optical fiber in a microwave photonic link has found for different dispersion coefficients. In the last, reflectivity, delay, and dispersion characteristics of fiber Bragg grating (FBG) have investigated for linearly chirped FBG with uniform, second order Gaussian, and fourth order Gaussian apodization function.

Keywords Microwave photonic filter · Ring resonator · Fiber · Chirp fiber Bragg grating

1 Introduction

Photonic generation of microwave signal imparts very high frequency and large bandwidth which is intensively used in airborne application [1, 2]. It allows various optical processing including frequency multiplying capability, pulse compression, and shaping of microwave signal [3, 4]. Convention electronic approaches for the same are limited by low frequency and bandwidth. One of the key components used for the optical signal processing is microwave photonic filter, which is very useful

R. Kumar (✉) · Y. Singh · S. K. Raghuwanshi · D. Nadeem
Department of Electronics Engineering, Indian Institute of Technology (ISM) Dhanbad, Jharkhand 826004, India
e-mail: ritesh.iitdhanbad@gmail.com

R. Kumar · S. Chandra
Department of Electronics and Communication, Madan Mohan Malaviya University of Technology, Gorakhpur, U.P. 273010, India

B. Mishra and M. Tiwari (eds.), *VLSI, Microwave and Wireless Technologies*, Lecture Notes in Electrical Engineering 877,
https://doi.org/10.1007/978-981-19-0312-0_69

as true time delay in optical beamforming network. It can also be used for dynamic phase shift in optically modulated signal to develop frequency multiplying capability in the generated signal. Microwave photonic filter has its large impact on the arbitrary shaping of microwave signal [5].

The photonic filter can be implemented based on different optical components. Some of the preferable optical components are dispersive fiber, optical ring resonator, fiber Bragg grating (FBG), and Mach–Zehnder delay interferometer. Optical fiber application can be recognized depending upon the delay and dispersion characteristics for various type of its dispersion coefficient. Photonic crystal fiber has very high-dispersion coefficient and provides a large delay even with very small length [6]. Periodic frequency response of optical ring resonator as true time delay element in optical beamforming network for beam steering of RF signal gives squint free operation [7]. FBG is widely used in beam steering application as well as in opto-electronic oscillator [8, 9]. There are various types of FBG which have been proven very much useful in remote sensing application [10–12]. These FBGs can be used for providing progressive phase shift and time delay in optical domain. In [8], linear chirp FBG is used for the conversion of intensity modulated optical signal to phase modulated optical signal. In paper [9], effect of ripple present in the group delay of chirp FBG is investigated on beam steering of RF signal. Recently, the author has also proposed external modulation-based photonic techniques for the generation of microwave signals and investigated their application in remote sensing area [4, 13–16].

In this paper, theoretical and simulated analysis has done for delay characterization of some most used optical components, like ring resonator, dispersive fiber, and fiber Bragg grating. Effect of coupling coefficient on delay characteristics and free spectral range has shown for single ring as well as the cascading of multiple rings. Then, group delay of optical fiber in a microwave photonic link has found for different dispersion coefficients. In the last, reflectivity, delay, and dispersion characteristics of fiber Bragg grating (FBG) have investigated for linearly chirped FBG with uniform, second order Gaussian, and fourth order Gaussian apodization function.

2 Group Delay Analysis of Optical Ring Resonator

Using complex scattering matrix of all pass network, transfer function of a ring resonator is given by following complex scattering matrix element,

$$S_{21} = \frac{a\sqrt{1-k} - ae^{-\gamma L}}{1 - a\sqrt{1-k}e^{-\gamma L}} \tag{1}$$

where a is the coupler loss, $\gamma = \alpha + j\beta$ is propagation constant, k is coupling coefficient of the ring. Assuming that there is no propagation loss in the resonator, i.e., all power coming to input port reaches to the output port. Then, α will be zero

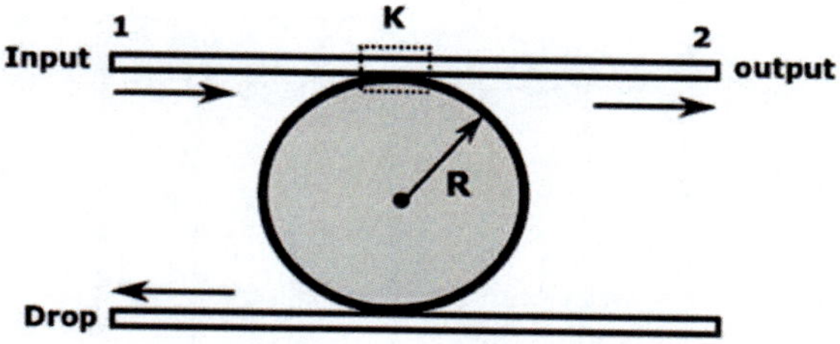

Fig. 1 Schematic diagram of four port ring resonator

and $\beta = \frac{2\pi n_{\text{eff}}}{\lambda}$ is the phase shift constant, which depends on the refractive index n_{eff} of the ring material and wavelength λ of the traveling light inside the ring. Delay and phase experienced by light inside waveguide depend on the round-trip length L of the ring. $L = 2\pi R$ is the circumference of resonator where R is the radius. Coupler loss $a = \sqrt{1-\gamma}$, where γ is fractional power loss of coupler. General expression of transfer function of ring resonator can be written by

$$H(f) = |H(f)|e^{j\emptyset} \tag{2}$$

where $|H(f)|$ is magnitude and $\emptyset$ is the phase of the transfer function. Four port representation of RR has shown in Fig. 1. In Fig. 1, optical signal is applied at the input port 1 and coupled to the RR through coupling coefficient K. Throughput output is considered at port 2 which suffer from the delay provided by RR. Phase induced in the output of signal is given by

$$\emptyset = \tan^{-1}\left(\frac{(1-a(1-k))\sin\beta L}{(1+a)\sqrt{1-k}-(1+a(1-k))\cos\beta L}\right) \tag{3}$$

Delay variation τ due to RR can be found by the phase-delay relation $\tau = -\frac{\lambda^2}{2\pi c}\cdot\frac{d\emptyset}{d\theta}$. If fractional loss of the coupler is neglected, then coupler loss will reduce to unity, i.e., $a = 1$, and magnitude of the transfer function will be unity. In this case, phase of transfer function $H(f)$ of the ring resonator and delay can be represented as in Eq. (4) and Eq. (5), respectively.

$$\emptyset = \tan^{-1}\left(\frac{k\sin\theta}{2\sqrt{1-k}-(2-k)\cos\theta}\right) \tag{4}$$

where $\theta = \beta L$. Group delay $\tau_g(f) = -\frac{d\emptyset}{d\omega}$ due to RR can be found by

$$\tau_g(f) = \frac{k\tau_r}{\left(2-k-2\sqrt{1-k}\cos(2\pi f\tau_r)\right)} \tag{5}$$

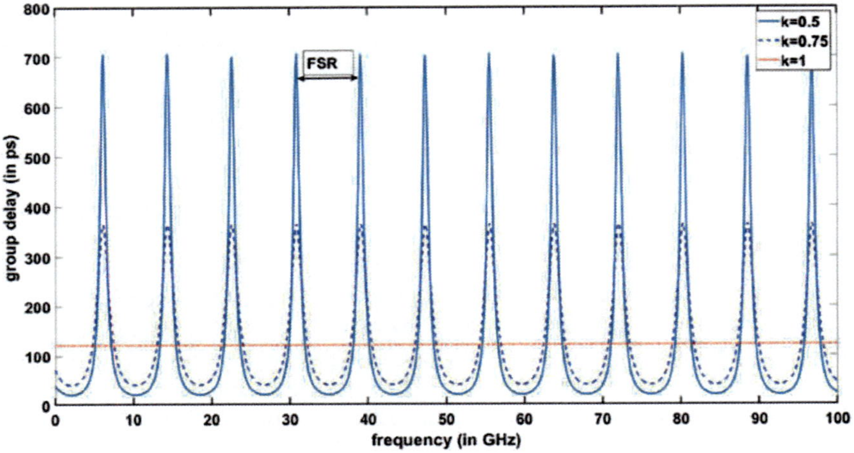

Fig. 2 Delay characteristics of ring resonator for different values of coupling coefficient

Here, $\tau_r = \frac{n_{\text{eff}}L}{c}$ is round-trip time of the ring. Delay characteristic of RR for different values of coupling coefficients has been shown in Fig. 2. It can be observed from Fig. 2, delay characteristic is periodic for specific values of coupling coefficient. Frequency spacing between two adjacent peaks of the delay is known as frequency spectral range (FSR). For the present case, FSR is found as 12.2 GHz. For the present case, coupling coefficients have been taken as $k = 0.5, 0.75$, and 1, which gives different delays without effecting the FSR. Group delay obtained in different cases is approximately 700 ps, 360 ps, and 120 ps for coupling coefficients 0.5, 0.75, and 1, respectively. This shows that the RR with tunable coupling coefficient can provide variable delay line. However, above suffers from frequency stability problem since the spectral width is much narrower and a particular frequency with spacing of $\pm$ FSR can only be utilized.

The bandwidth of peaks can be enhanced by cascading multiple ring resonators. Figure 3a shows group delay characteristics of three ring resonators in cascade configuration. Coupling coefficients of individual ring resonators have considered as 0.5, 0.75, and 1, respectively. In this case, FSR is found as 8.2 GHz with peak group delay of 654 ps. Figure 3b shows zoom in view of dotted region indicated in Fig. 3a. In Fig. 3b, it can be observed that spectral width of the delay characteristics has increased drastically. Spectral width of 1.4 GHz gives almost flat response, and approximately, 654 ps group delay with small deviation is found in that region.

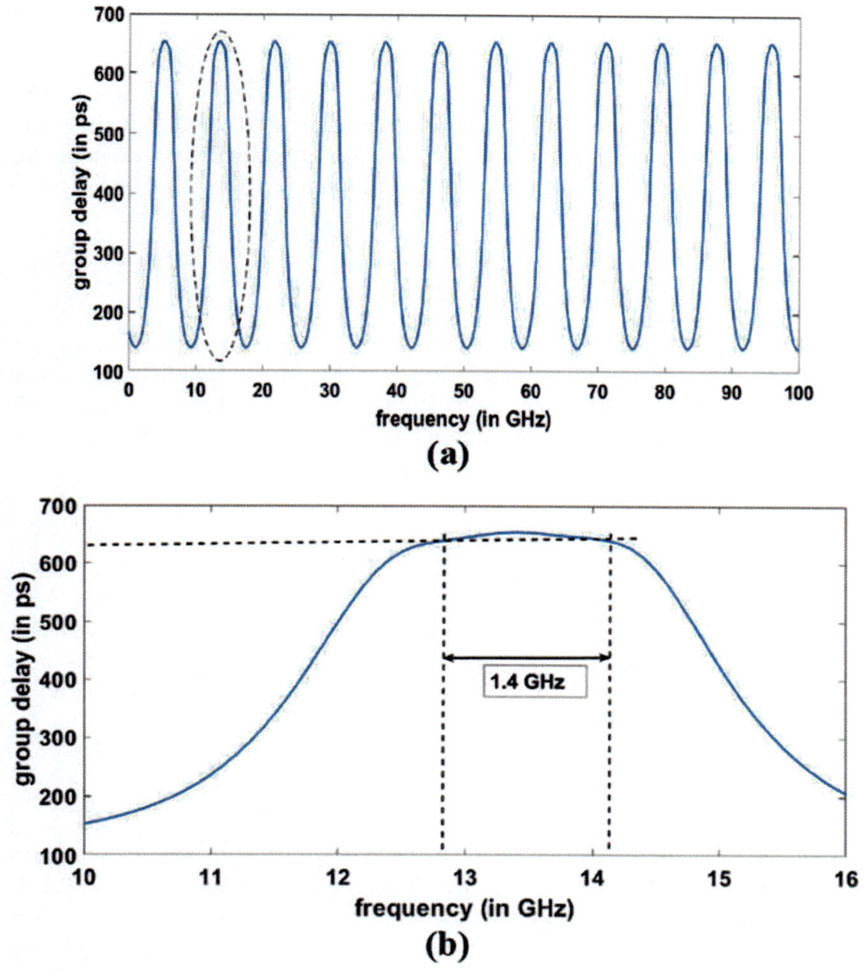

Fig. 3 **a** Group delay characteristics of three cascaded ring resonators, **b** zoom in view of group delay response in 10–16 GHz

3 Delay and Dispersion Analysis of Fiber

Let input optical signal provided by laser source is given by $E_i(t) = \sqrt{P_i(t)}e^{j\omega_0 t}$ where we have assumed zero phase, $P_i(t)$ is the input power, and ω_0 is the angular frequency of the optical signal. When above optical signal is passed through fiber of length 'L', induced phase is given by $e^{-j\beta L}$. β is the propagation constant may be expanded as

$$\beta = \beta(\omega_0) + \dot{\beta}(\omega_0)(\omega - \omega_0) + \frac{1}{2}\ddot{\beta}(\omega_0)(\omega - \omega_0)^2 + \ldots \tag{6}$$

where $\beta(\omega_0)$ is the zeroth order propagation constant, and $\dot{\beta}(\omega_0)$, $\ddot{\beta}(\omega_0)$ are higher order harmonics centered at optical frequency ω_0. Neglecting higher order harmonics in the delay expansion propagation constant can be approximated as

$$\beta \approx \beta(\omega_0) + \frac{n_{\text{eff}}\omega_{\text{RF}}}{c} - \frac{D\omega_{\text{RF}}^2\lambda^2}{4\pi c} \tag{7}$$

In Eq. (7), n_{eff} is the effective refractive index of fiber, c is the velocity of light, λ represents optical wavelength corresponding to ω_0, ω_{RF} shows shift in optical signal frequency, and D is the dispersion coefficient of the fiber used as delay element. D is function of operating wavelength and uniquely defined for different fibers. For example, dispersion of SMF-28 fiber is given by

$$D \propto \frac{S_0}{4}\left[\lambda - \frac{\lambda_0^4}{\lambda^3}\right] \tag{8}$$

where λ lies between 1200 and 1623 nm, λ_0 is zero dispersion wavelength lies between 1302 and 1322 nm, and s_0 is the zero dispersion slope less than 0.092 ps/nm^2.Km. Time domain representation of optical signal at the output of fiber can be represented as

$$E_0(t) = \sqrt{P_i(t - \tau L)}e^{j(\omega_0 t - \beta_0 L)}e^{-j\tau(\omega - \omega_0)L}e^{j\frac{D\omega_{RF}^2\lambda^2}{4\pi c}} \tag{8}$$

To demonstrate delay and dispersion induced in the optical signal when passed through a fiber, a microwave photonic link is considered shown in Fig. 4. In Fig. 4, a continuous wave laser source is considered to take an optical carrier signal. It is passed through Lithium Niobate Mach–Zehnder modulator to be modulated by RF signal. The RF signal has taken from function generator. The optically modulated signal is then passed through fiber and finally detected at the photodetector (PD). Delay and dispersion due to fiber element in the detected signal can be observed with respect to original RF signal.

Figure 5 represents the delayed RF signal when a fiber of 2 km length is used. In Fig. 5a, an RF signal of 32 GHz frequency from function generator has shown.

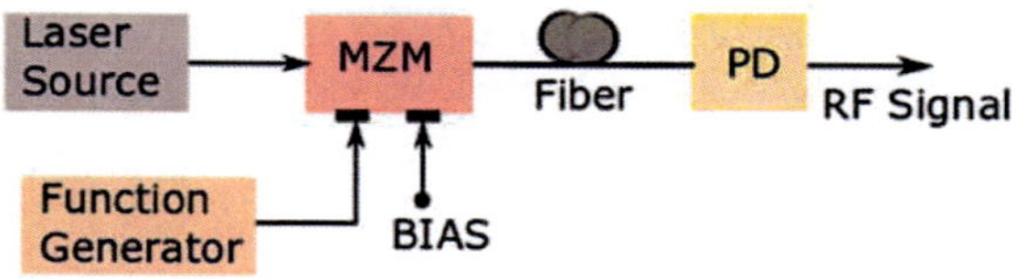

Fig. 4 Schematic diagram of the microwave photonic link

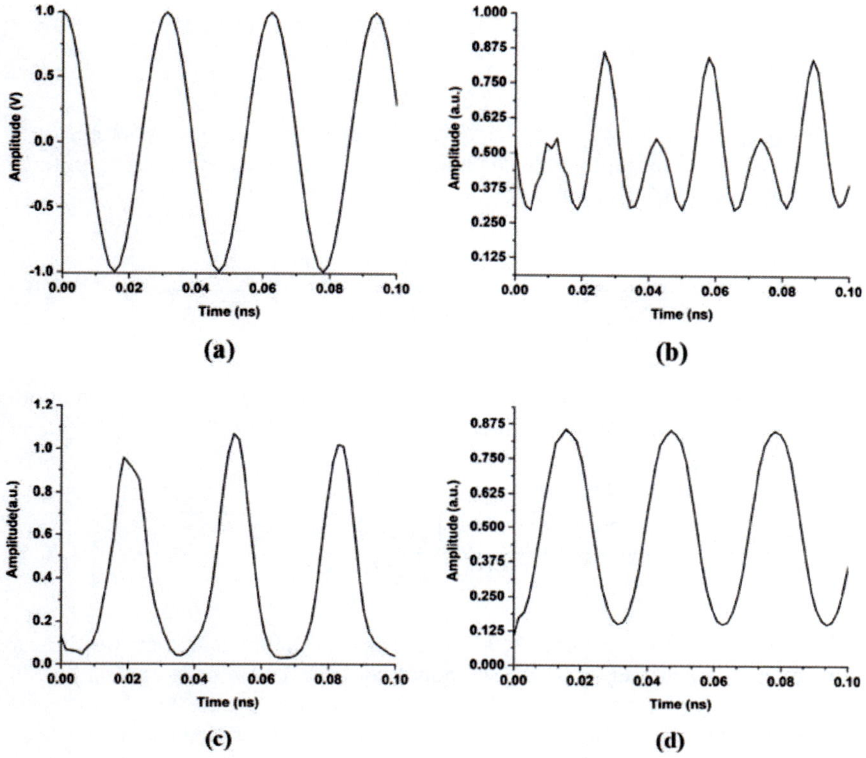

Fig. 5 **a** RF signal from function generator and detected RF signal for, **b** negative dispersive fiber, **c** positive dispersive fiber, **d** combination of positive and negative dispersive fiber

Figure 5b represents detected RF signal at the end of PD when a fiber with negative dispersion coefficient of magnitude 16.75 ps/nm Km is considered. The detected signal has delay of 15.6 ps but suffers from large dispersion due to negative dispersion coefficient. In Fig. 5c, effect of positive dispersion is investigated with the same length of fiber. In this case, same delay is obtained with reduced dispersion in the detected RF signal. In Fig. 5d, dispersion effect has shown, when two fibers of positive and negative dispersion coefficient of magnitude 16.75 ps/nm Km have taken. Length of each fiber element is 1 km. These two fibers are cascaded so that net length for propagation is maintained to 2 km. Opposite polarity of dispersion reduces the distortion very well.

4 Spectral Characteristics for Different Apodization of FBG

Fiber Bragg gratings are very much useful in the design of microwave photonic filter and delay elements for remote sensing application. Amplitude reflection of uniform fiber Bragg grating is given as [17],

$$\rho = \frac{-k\sinh\left(\sqrt{k^2 - \hat{\sigma}^2}L\right)}{\hat{\sigma}\sinh\left(\sqrt{k^2 - \hat{\sigma}^2}L\right) + i\sqrt{k^2 - \hat{\sigma}^2}\cosh\left(\sqrt{k^2 - \hat{\sigma}^2}L\right)} \tag{10}$$

where all parameters are well defined in [17] with the same symbolic representation. The delay expression from Eq. (10) is obtained by the relation $\tau = -\frac{\lambda^2}{2\pi c}\cdot\frac{d\theta_\rho}{d\lambda}$ and represented in Eq. (11).

$$\tau_g = -\frac{N\lambda_{\max}}{2Lc}\left[\frac{\mathrm{L.csch}^2\left(\sqrt{k^2 - \hat{\sigma}^2}L\right) - \frac{k^2}{\hat{\sigma}^2\sqrt{k^2-\hat{\sigma}^2}}\coth\left(\sqrt{k^2 - \hat{\sigma}^2}L\right)}{1 + \frac{k^2-\hat{\sigma}^2}{\hat{\sigma}^2}\coth^2\left(\sqrt{k^2 - \hat{\sigma}^2}L\right)}\right] \tag{11}$$

Due to large reflection bandwidth and linearity of delay characteristics, linear chirp fiber Bragg grating (LCFBG) is popularly used in airborne application. In this section, reflection and delay spectrum of LCFG are investigated for different apodization function. Average refractive index profile of the grating has taken whose grating shape is $\sin\left(\frac{2\pi z}{\Delta} - \frac{\pi\Delta_1 z^2}{2n_{\mathrm{eff}}\Delta^2}\right)$, where z is the distance along the length of grating, $\Delta = 2$ nm is total grating chirp, $\Delta_1 = 0.01$ nm chirp parameter of Bragg wavelength, and n_{eff}=1.46 is the effective refractive index.Δ_1 insures a linear variation of the local Bragg wavelength along the grating via the variation of pitch of the effective index. Effect of different apodization function has investigated and shown in Fig. 6. Figure 6 represents reflectivity (top), delay (middle), and dispersion characteristics (bottom) for uniform apodization (left most), second order Gaussian (center), and fourth order Gaussian apodization (right most) of CFBG. In case of uniform apodization, reflectivity is maximum for wavelength between 1547 and 1553 nm. The reflection bandwidth is 6 nm, but some ripple is present in this band. Also, the delay characteristics are linear in the same reflection band, but considerable ripple is present, which may cause deviation in the desired output when used in photonic filter. A large dispersion −2000–2000 ps/nm is observed in the wavelength range 1547–1553 nm for uniform apodization. When second order Gaussian apodization is used, maximum reflectivity has fewer ripples, but reflection bandwidth is reduced to 2 nm and contains the wavelength 1549–1551 nm. Again, the delay characteristics have same slope as in the previous case, but smoothness of the delay characteristics has improved. Dispersion is also reduced and lie between −200 and 200 ps/nm for the wavelength range 1549–1551 nm. So, it can be noticed that second order Gaussian apodization preferable compare to uniform apodization. However, still, there is

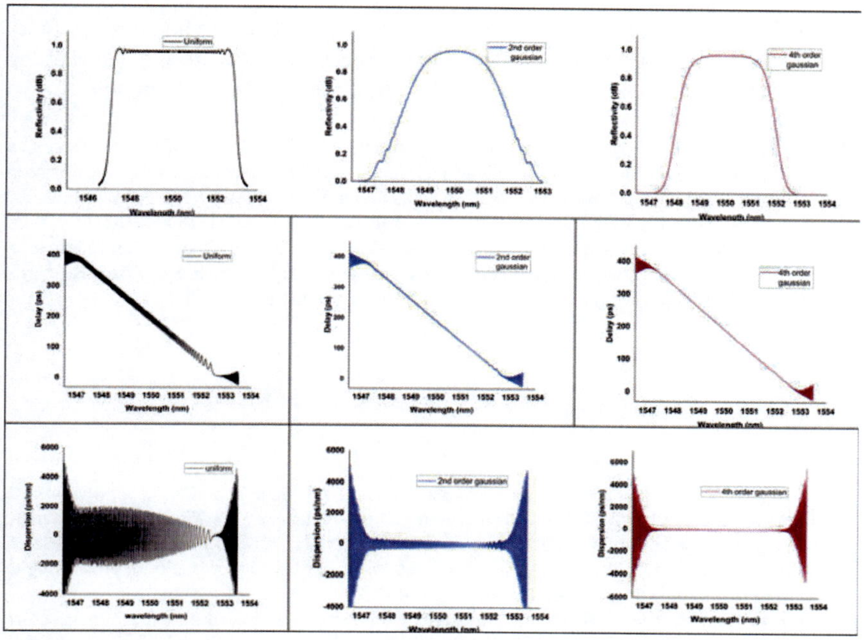

Fig. 6 Reflectivity (top), delay (middle), and dispersion (bottom) of chirp fiber Bragg grating for uniform, second order Gaussian function, and fourth order Gaussian function (left to right)

presence of considerable dispersion corresponding to the reflection bandwidth. In the right most column of Fig. 6, fourth order Gaussian apodization is investigated. In this case, maximum reflection occurs for the wavelength range approximately 1548.5–1551.5 nm. The reflection bandwidth is enhanced to 3 nm with no ripples. The delay characteristic is also linear with the slope of magnitude 66 ps/nm. There is no ripple observed in the linear region of delay characteristics. In the dispersion plot of fourth order Gaussian apodization, there is zero dispersion in wavelength range 1548.5–1551.5 nm as shown in the figure.

5 Conclusion

In this paper, theoretical and simulated analysis has been given for delay characterization of some most used optical components, like ring resonator, dispersive fiber, and fiber Bragg grating. Effect of coupling coefficient on delay characteristics and free spectral range has shown for single ring as well as the cascading of multiple rings. Then, group delay of optical fiber in a microwave photonic link has found for different dispersion coefficients. In the last, reflectivity, delay, and dispersion characteristics of fiber Bragg grating (FBG) have investigated for linearly chirped FBG with uniform, second order Gaussian, and fourth order Gaussian apodization function. These analyzes are beneficial for optical processing like filtering and shaping of light signal and can be applied for the design of microwave photonic filter.

Acknowledgements This work is financially supported by TEQIP-III Collaborative Research Scheme (NPIU MHRD) and Space Applications Center, Indian Space Research Organization (ISRO), Ahmedabad, India. The work is carried out under Project number: 1-5744809099 and ISRO/(10)/2018-2019/571/ECE (Grant No. DS-2B-13012(2)/13/2018 and ISRO/RES/3/775/19-20 dated 4 Feb 2020). Authors would like to thank Mr. R K Bahl, Head, Optical Communication Division, Satcom and Navigation Payload Area, Space Application Center, Indian Space Research Organization (ISRO), Ahmedabad 380053, for necessary experimental support.

References

1. Seeds AJ, Williams KJ (2006) Microwave photonics. J Lightwave Technol 24(12):4628–4641
2. Monsterleet A, Goldstein ST, Dolfi D, Huignard J, Sape P, Chazelas J (2005) Optically generated arbitrary waveforms for radar applications. Electron Lett 41(6):332–334
3. Loayssa A, Benito D, Garde MJ (2002) Applications of optical carrier Brillouin processing to microwave photonics. Opt Fiber Technol 8(1):24–42
4. Raghuwanshi SK, Kumar R, Kumar M (2018) Photonic generation of a dual nonlinear chirp microwave waveform in Ku-band and its comparative study with dual linear chirping capability to distinguish range doppler coupling in radar application. In: Terahertz, RF, millimeter, and submillimeter-wave technology and applications XI, vol 10531. International Society for Optics and Photonics, p 105311X
5. Weiner AM (2011) Ultrafast optical pulse shaping: a tutorial review. Optics Commun 284(15):3669–3692
6. Subbaraman H, Chen MY, Chen RT (2008) Photonic crystal fiber-based true-time-delay beamformer for multiple RF beam transmission and reception of an x-band phased-array antenna. J Lightwave Technol 26:2803–2809
7. Burla M, Marpaung AID, Zhuang L, Khan MR, Leinse A, Beeker W, Hoekman M, Heideman RG, Roeloffzen CGH (2014) Multiwavelength-integrated optical beamformer based on wavelength division multiplexing for 2-D phased array antennas. J Lightwave Technol 32(20):3509–3520
8. Li W, Yao J (2010) An optically tunable optoelectronic oscillator. J Lightwave Technol 28(18):2640–2645
9. Zhou B, Zheng X, Yu X, Zhang H, Guo Y, Zhou B (2008) Impact of group delay ripples of chirped fiber grating on optical beamforming networks. Opt Express 16(4):2398–2404
10. Dong X, Shum PP (2010) Tunable fiber Bragg grating filters realized by chirp rate tuning with a cantilever beam. Frontiers Optoelectron China 3(1):71–77
11. Hunter DB, Parker ME, Dexter JL (2006) Demonstration of a continuously variable true-time delay beamformer using a multichannel chirped fiber grating. IEEE Trans Microw Theory Tech 54(2):861–867
12. Korganbayev S, Orazayev Y, Sovetov S, Bazyl A, Schena E, Massaroni C, Gassino R, Vallan A, Perrone G, Saccomandi P, Caponero MA (2018) Detection of thermal gradients through fiber-optic Chirped Fiber Bragg Grating (CFBG): medical thermal ablation scenario. Opt Fiber Technol 41:48–55
13. Raghuwanshi SK, Kumar R, Srivastava A, Srivastava NK (2016) Dual-chirp arbitrary microwave waveform generation by using a dual parallel mach-zehnder modulator feeding with RF chirp signal. Progress Electromagnet Res C 65:79–92

14. Kumar R, Raghuwanshi SK (2017) A photonic scheme for the generation of dual linear chirp microwave waveform based on the external modulation technique and its airborne application. Opt Quant Electron 49(370):1–13
15. Kumar R, Raghuwanshi SK (2019) Simultaneous photonic generation of multiple chirp and unchirp microwave waveform with frequency multiplying capability for optical beam forming system. Opt Quant Electron 51(344):1–17
16. Kumar R, Raghuwanshi SK (2020) Photonic generation of a parabolic-shaped microwave signal and dual-linear-chirp microwave waveform. Appl Opt 59(20):6024–6029
17. Erdogan T (1997) Fiber grating spectra. J Lightwave Technol 15(8):1277–1294

Enhanced Performance of $MoSe_2$-CIGS Solar Cell with ZnSe Buffer Layer

Jyoti Singh, Uzair Alam, and Anupam Sahu

Abstract In this paper, the effect of $MoSe_2$ layer on the performance of thin-film CIGS solar cell has been investigated for different structural parameters. The proposed ZnO/ZnSe/CIGS/$MoSe_2$/Mo structure is analysed and compared with existing ZnO/CdS/CIGS/$MoSe_2$/Mo structure for varying thickness of CIGS absorber layer. The proposed ZnSe-based structure attains a maximum efficiency of 27.13%, with $MoSe_2$ layer in comparison to 23.38% without $MoSe_2$ layer. The remarkable enhanced efficiency of 27.13% is achieved with the proposed ZnSe-based structure compared to 22% of CdS-based structure.

Keywords Thin-film CIGS solar cell · $MoSe_2$ · Buffer layer ZnSe

1 Introduction

With the advent of the modernization of civilization, the growing energy demand can only be met by harnessing solar energy, which is in abundance and non-exhaustible. In this context, the thin-film copper-indium-gallium-selenide (CIGS) solar cell semiconductors have grabbed the limelight because of their numerable benefits such as high conversion efficiency and lower cost of production and stability. CIGS has a tunable bandgap energy that can be utilized for greater solar irradiance and has a high absorption coefficient in the visible solar spectrum.

In recent years, the investment in the domestic development of CIGS solar cells has surpassed 4 billion yuan. However, the recorded highest efficiency has achieved

J. Singh (✉) · U. Alam · A. Sahu
Department of Electronics and Communication Engineering, Madan Mohan Malaviya University of Technology, Gorakhpur, India
e-mail: replymejyoti@gmail.com

U. Alam
e-mail: uzairmzp@gmail.com

A. Sahu
e-mail: anupam.or.sahu@gmail.com

B. Mishra and M. Tiwari (eds.), *VLSI, Microwave and Wireless Technologies*,
Lecture Notes in Electrical Engineering 877,
https://doi.org/10.1007/978-981-19-0312-0_70

up to 22.6% [1]. It has been reported that the conversion efficiency can be enhanced by the reduction of series resistance at the back contact [2]. With the features such as higher absorption coefficient and tunable bandgap, CIGS has become as the most promising candidate for the absorber layers in thin-film photovoltaic devices [3]. The CIGS solar cell based on ZnSe buffer layer not only boosts the performance of ohmic contact but also decreases the contact resistance barrier and series resistance significantly.

In this paper, we proposed a CIGS solar cell based on ZnSe buffer and a thin $MoSe_2$ layer. The device is simulated in SCAPS-1D environment, and the result is investigated for different thickness of CIGS absorber layer.

2 Device Modelling and Simulation

The proposed structure of a device is ZnO/ZnSe/CIGS/$MoSe_2$/Mo solar cell which has been shown in Fig. 1. The proposed solar cell comprises of CIGS and $MoSe_2$ as a absorber layer with M_o deposited on the substrate. The properties of the intermediate $MoSe_2$ layer depend on the CIGS deposition method and improve adhesion and back surface field due to wider bandgap than that of CIGS [4]. The ZnSe buffer layer has been examined with the variation of CIGS layer thickness (1–3 μm). The absorber layer absorbs a considerable amount of photons incident on it due to its thicker thickness. Consequently, a significant number of e-h pairs are created from the absorber layer and ZnO acts as a window layer with a thickness of 0.08 μm. Buffer layers are typically highly resistive from an electronic viewpoint; they can protect shunting issues by acting as an intermediary between the TCO and absorber layer [5].

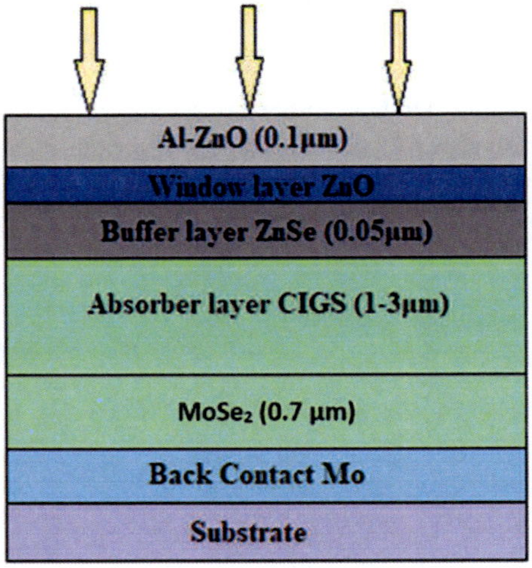

Fig. 1 Proposed device structure of $MoSe_2$ - CIGS solar cells

Table 1 Parameter used in simulation [2, 7]

Parameter	$MoSe_2$	CIGS	ZnSe	i-ZnO	Al:ZnO
Thickness (μm)	0.7	1–3	0.05	0.08	0.1
Bandgap (eV)	1.4	1.15	2.90	3.3	3.3
Electron affinity (eV)	4.32	4.5	4.09	4.45	4.45
Permittivity (ε)	7.29	13.6	10	9	9
Density of state at conduction band N_c(/cm^3)	9×10^{18}	2.2×10^{18}	1.5×10^{18}	2.2×10^{18}	2.2×10^{18}
Density of state at valance band N_v(/cm^3)	9×10^{18}	1.8×10^{19}	1.8×10^{18}	1.8×10^{19}	1.8×10^{19}
Electron mobility (cm^2/V-s)	100	100	100	100	100
Hole mobility (cm^2/V-s)	25	25	50	31	31
Electron thermal velocity V_{tn} (cm/s)	10^7	10^7	10^7	10^7	10^7
Hole thermal velocity V_{tp} (cm/s)	10^7	10^7	10^7	10^7	10^7
Acceptor density N_A (/cm^3)	10^{17}	5×10^{16}	0	0	0
Donar density N_D (/cm^3)	0	0	5.5×10^{17}	10^{18}	10^{20}
Defect density (/cm^3)	10^{14}	10^{14}	10^{18}	10^{17}	10^{17}
Cross-section electron σ_n (cm^2)	10^{-15}	10^{-15}	10^{-15}	10^{-15}	10^{-15}

The CIGS solar cell based on ZnSe buffer layer has been proven to be environment-friendly and non-toxic compound and achieves the second highest efficiency in thin-film market [6]. The optical properties of solar cells have been greatly enhanced by adding $MoSe_2$ at the fixed thickness of 2 μm of the absorption layer. $MoSe_2$ gives an additional tunnelling impact, which boosts performance and suppresses the compound effect near the back contact.

The simulation has been carried out to compute the efficiency of the proposed solar cell using SCAPS-1D. The parameters used in the simulation are listed in Table 1.

3 Result and Discussion

For the operating temperature at 300 K and incident solar power at $P = 100\,\text{mW/cm}^2$, the structure is analysed under the illumination of spectrum AM 1.5 G. The results in terms of efficiency (η), fill factor (FF), open circuit voltage (V_{OC}) and short circuit current density (J_{SC}) have been studied and plotted in Fig. 2a–d for thickness varying from 1–3 μm. The results of the proposed ZnSe-based structure (with and without $MoSe_2$) are compared with existing CdS-based structure.

The efficiency of the both the structure increases with the increase in thickness of CIGS absorber layer as shown in Fig. 2a. The increasing thickness provides the

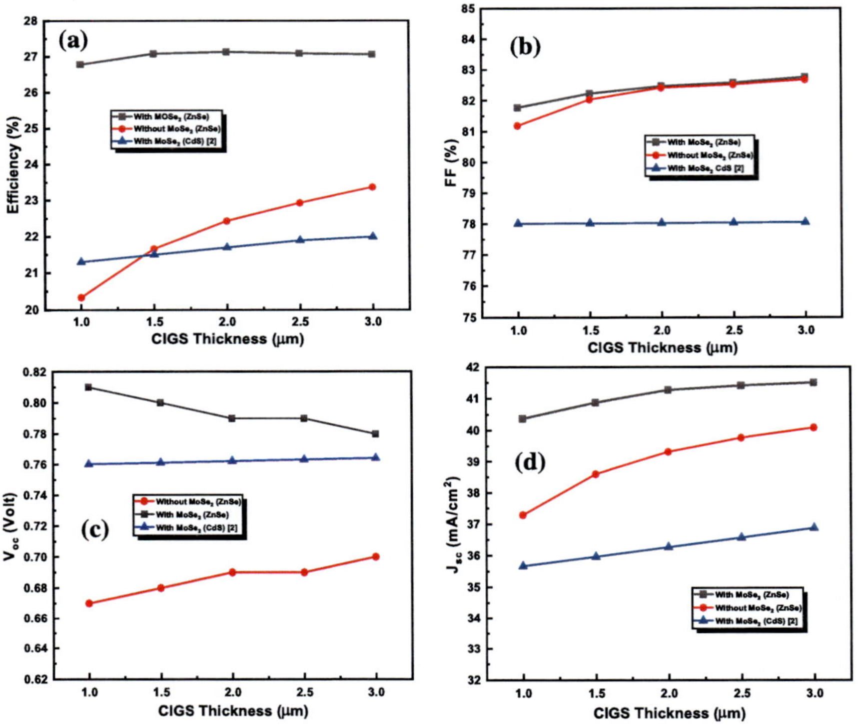

Fig. 2 Performance of CIGS with different thickness **a** Efficiency, **b** Fill factor, **c** Open circuit voltage, **d** Short circuit current density

larger surface to absorption of photons, which ultimately increases the efficiency of the solar cell. However, the efficiency of the proposed structure without $MoSe_2$ surpasses the CdS-based structure at a thickness of around 1.5 μm. Significantly, the major improvement in efficiency is achieved on the addition of thin $MoSe_2$ in the ZnSe-based solar cell. The bandgap of $MoSe_2$ is larger than CIGS and thus acts as a barrier for the electrons recombining at the back contact. This reduces the recombination of electrons and holes at the back contact and contributes more to the photocurrent. The thickness of buffer layer ZnSe has a major impact on the performance of CIGS solar cells as it creates a strong electric field between the junction and thus allows the absorption of larger number of photons.

For the proposed structure (with $MoSe_2$), the efficiency (η) varies from 26.78 to 27.13% in comparison to 21.3–22% of CdS-based structure as shown in Fig. 2a. Also, the proposed structure (with $MoSe_2$) achieves larger FF of 82.76% in comparison to 78.05% of CdS at CIGS thickness of 3 μm as shown in Fig. 2b. The maximum V_{OC} of 0.811 V and maximum J_{SC} of 41.50 mA/cm^2 and proposed structure (with $MoSe_2$) are obtained at CIGS thickness of 1 μm and 3 μm, respectively, as shown in Fig. 2c–d.

Table 2 **a** Performance of proposed optimized CIGS (with ZnSe buffer) solar cell without and with $MoSe_2$

S. No.	Parameters	Without $MoSe_2$	With $MoSe_2$
1	Efficiency (η) %	23.38	27.13
2	Fill factor (FF) %	82.88	82.47
3	Short circuit current (J_{sc}) mA/cm^2	40.09	41.27
4	Open circuit voltage (V_{oc}) Volts	0.70	0.79

Table 2 **b** Comparison of reported and proposed optimized CIGS solar cell parameters for CdS and ZnSe buffer layer

S. No.	Parameters	CdS (Reported) in Fig. 3 of Ref [2]	ZnSe (Proposed)
1	Efficiency (η) %	22.00	27.13
2	Fill Factor (FF) %	78.05	82.47
3	Short circuit current (J_{sc}) mA/cm^2	36.86	41.27
4	Open circuit voltage (V_{oc}) Volts	0.764	0.79

The detailed comparative analysis of solar cell parameters of the proposed structure (with and without $MoSe_2$) and with CdS and ZnSe buffer layer is shown in Table 2a, b, respectively.

4 Conclusion

The optical characteristics of thin-film CIGS-$MoSe_2$ has been analysed with different buffer layers. The enhanced characteristics of the proposed structure (with and without $MoSe_2$) is compared with the existing CdS-based structure for different thickness of CIGS absorber layer. The efficiency of proposed solar cell (ZnSe based) is enhanced on the application of thin $MoSe_2$. Thus, the $MoSe_2$ proves its fitness in creating better adhesion between the CIGS and M_o layers. The maximum efficiency of 27.13% is achieved at a CIGS thickness of 2 μm.

References

1. Jackson P, Wuerz R, Hariskos D, Lotter E, Witte W, Powalla M (2016) Physica status solidi (RRL)—rapid research letters. 10:583
2. Sun HY, Li PH, Xue YM, Qiao ZX, Sai LI (2019) Effect of $MoSe_2$ on the performance of CIGS solar cells. Optoelectron Lett 15(6)

3. Ong KH (2018) Review on substrate and molybdenum back contact in CIGS thin film solar cell. Int J Photoenergy
4. Hsiao KJ, Liu JD, Hsieh HH, Jiang TS (2013) Electrical impact of MoSe2 on CIGS thin-film solar cells. Phys Chem Chem Phys 15(41)
5. Oyedele SO, Aka B (2017) Numerical simulation of varied buffer layer of solar cells based on CIGS. Model Numer Simul Mater Sci 07(3)
6. Moradi M (2017) Buffer layer replacement: a method for increasing the conversion efficiency of CIGS thin film solar cells. Optik 136
7. Sadanand, Dwivedi DK (2019) Theoretical investigation on enhancement of output performance of CZTSSe based solar cell. Sol Energy 193:442–451

Study and Implementation of Smart Water Supply Management Model for Water Drain Region in India

Prashant Pandey, Ashish Ranjan Mishra, P. K. Verma, and Ravi P. Tripathi

Abstract In this paper, an Internet of Things (IoT)-based water supply management/monitoring model has been proposed which works in real time, real environment and guarantee sufficient water supply at each end user. It is also capable of monitoring its water supply network, auto detects required maintenance, calculates total consumption, measures supply flow and pressure using various sensors. In water drain region, water supply management becomes very important and needs specific requirement of sensors and actuator. After detailed analysis of water supply network at Rajkiya Engineering College (REC) Sonbhadra, a model has been developed and integrated with IoT applications. A number of sensors which have been interfaced with microcontroller, Radio Frequency (RF) device modules and communication software between nodes, form a Wireless Sensor Network (WSN). By using four layered architecture of IoT, these sensors will collect data at base station unit and collected data can be processed for various calculations. The water supply management is going to be an essential need in near future as water crisis is emerging as a major challenge. This paper leads toward innovative solution to water crisis problem in water drain region of India.

Keywords IoT · Water supply management · Monitoring model · Water drain

P. Pandey · P. K. Verma
Department of Electronics Engineering, Rajkiya Engineering College, Sonbhadra, UP 231206, India

A. R. Mishra (✉)
Department of Computer Science and Engineering, Rajkiya Engineering College, Sonbhadra, UP 231206, India
e-mail: ash.cs.recs@gmail.com

R. P. Tripathi
Department of Civil Engineering, Rajkiya Engineering College, Sonbhadra, UP 231206, India

B. Mishra and M. Tiwari (eds.), *VLSI, Microwave and Wireless Technologies*,
Lecture Notes in Electrical Engineering 877,
https://doi.org/10.1007/978-981-19-0312-0_71

1 Introduction

United Nations has expected that the world population will rise upto 9 billion by 2050 and water resources available will remain same. Water scarcity in water drain region is major issue in social development. Water is unequally distributed for different space and different time within a year. Most of water is wasted and unsustainably managed. Proper distribution of water needs proper monitoring at different levels and different locations. The motive behind developing a smart water management model is to achieve water safety at all levels (source, supply, distribution, storage and uses). Some of the primary features of water management system model include water flow management, smart leakage location detection techniques for water networks, maintenance of the water cycle, smart water meter technology and smart water quality monitoring.

In [1], different smart water supply projects and case study have been investigated and wireless connectivity models solution of total cost of ownership framework have been elaborated. In [2], a novel, very low cost and efficient water conductivity sensor has been developed and its use in different parameter characterization has been thoroughly explained. In [3], data obtained from Hall Effect sensor is stored, sent to cloud, processed by use of machine learning tools and need of water is modeled for each household. In [4], advanced metering system which is capable of remote connection has been explained. In [5], accurate prediction methods for water leakage and consumption for very complex water supply network have been proposed. In [6], a system, which can issue alerts to warn consumer about excess consumption of water providing excess monitoring and control capability to water supply system, have been discussed. In [7], different water supply network related key parameters can be detected using variation in pressure; temperature and conductivity of water have been discussed. In [8], an IoT technology having Wireless Fidelity (Wi-Fi) and web page accessibility is used in application to monitor the equally distributed water level and water in real time has been explained. In [9, 10], a portable, low cost, easy to configure and flexible system can solve problem of water wastage. Laser sensor with HC12 transceiver is used for detecting the leakage in the tanks and cloud platform Adafruit is used.

In proposed work, a model for water supply system has been implemented based on survey of water supply system at REC located in Sonbhadra district of Uttar Pradesh, India which is at 24.7°N 83.07°E. It has an average elevation of 1080 feet from sea level. It is located in the south-eastern ranges of the Vindhyanchal Mountain. Using different sensors and actuators, an IoT model for smart water supply management has been made for monitoring and controlling of water resources.

2 Field Survey and Modeling of Problem

Water Supply Management System of REC Sonbhadra has been modeled as according to specification and limitation. The location of REC Sonbhadra is suffering from water availability problems, as it is situated at high elevation due to presence of rock around college, it becomes too hard to dig for underground water. Figure 1 depicts the Water Supply Management System of REC Sonbhadra. The primary source of water supply in REC Sonbhadra is ground water. At a distance of 3.5 km away from the college, there are two pump houses; first pump house is located at Musahi which is equipped with generator and second is at Raup. Water is pumped and distributed through pipelines, and stored in overhead tanks inside the campus.

Pressure gages are used at both pump houses. In REC Campus, there is one motor and two reservoirs, first is overhead tank and second one is Clear Water Reservoir (CWR). When the pressure is sufficient enough, then water is sent directly to the overhead tanks without taking the help of motor but when the pressure is not enough then water stored in CWR and then with help of motor it is sent to overhead tank. The height of overhead tank is 18 m above the ground and capacity of tank is 5 lakh liters.

Field survey and analysis of different location and equipment installed in water supply system have been performed. Working of each component used in system has been thoroughly studied and issues associated in real world and real time has been investigated. Based on detailed analysis, potential issues have been identified and listed in Table 1.

Based on issues shown in Table 1, to tackle them efficiently, need of automation is identified at various locations. Sensors that need to be integrated with Wi-Fi module are decided. Based on interconnection and location of sensors A hardware model of water supply management system has been presented in Fig. 2.

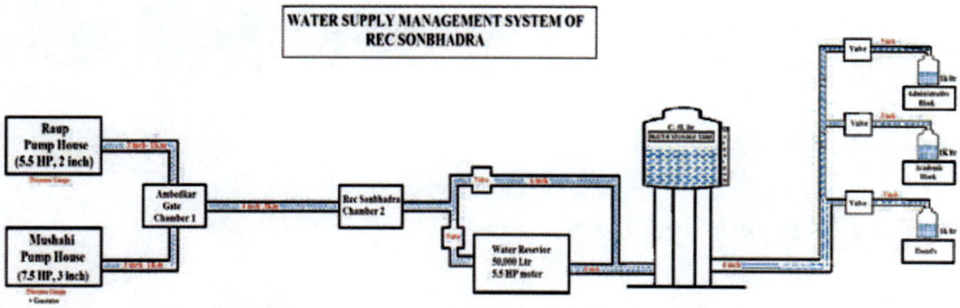

Fig. 1 Water supply management system of REC Sonbhadra

Table 1 Water management system of REC Sonbhadra

S. no	Device	Location	Specification	Distance (km)	Problems
1	Pumping Motor	Raup	5.5 HP, 2 inch	1 km	Electrical Supply, Motor Problem
		Mushahi	7.5 HP, 3 inch	1 km	
2	Pressure gage	At both pump house		–	Water Pressure
3	Champer Block 1	Ambedkar gate Mushahi basti	I/p-3 inch O/p-4 inch	3 km	leakage
4	Champer Block 2	REC Sonbhadra	I/p-4 inch O/p-6 inch	–	
5	Water reservoir tank		50,000 l I/p-6 inch O/p-6 inch 5.5HP Motor	–	Use in low pressure condition
6	Water over head tank		500,000 l I/p-6 inch O/p-6 inch	–	Water level sensor
7	Champer Block 3		I/p-6 inch O/p-3 inch	–	
8	Water level indicator	At all water tank		–	Water level sensor

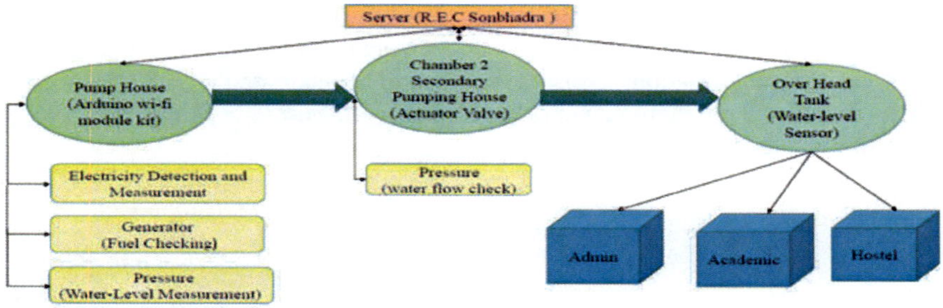

Fig. 2 Hardware used at various locations in model

3 Flow Diagram and Components

Overall flow diagram of water supply management system which we want to achieve has been explained in Fig. 3. The process starts at the pump house. With the help of the voltage sensor, we will detect whether there is a proper electricity supply given to the motor or not. If the water is flowing properly through pipeline then it will be stored in the reservoir. But if the water is not flowing in the pipeline then there occurs two conditions.

(1) There can be hindrance to the power supply.

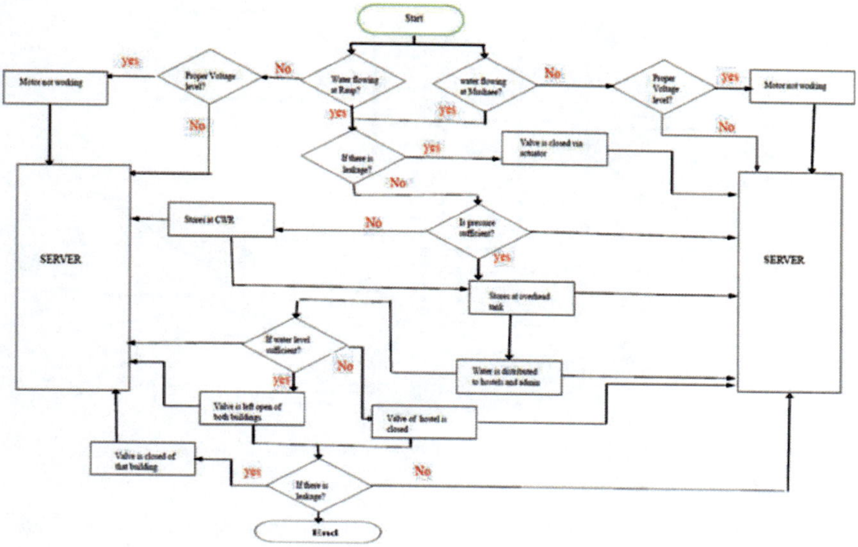

Fig. 3 Flow diagram of overall process

(2) Motor is not functioning well.

In order to tackle these problems efficiently, need of automation is identified at various location. Based upon requirement, sensors that need to be integrated with Wi-Fi module are decided. Based on interconnection and location of sensors block diagram for water supply management system is shown in Fig. 2.

In both cases, a notification will be sent to the operator about the improper functioning of the model by sending a message. The next step starts from the water that flows in the pipeline. The flow of the water is detected by the pressure sensor which is located at the end of the pipe. If there is leakage in the pipeline, it is detected by the pressure sensor. These sensors will collect data from the water which flows through it and send it to Arduino. The Arduino calculates this data and notify the operator.

A program, to detect whether there is a leakage in the pipeline or not, is written. Leakage will depend on data obtained from Hall Effect sensor at different locations. Microcontroller will process that data and obtain the location of the leakage. It will send data to monitoring system and with the help of relay and stepper motor the valve of that particular section will be closed automatically. After repair it will return back to normal condition. The whole information is sent from Wi-Fi to the server.

Based on the flow diagram shown in Fig. 3 and the hardware used an IoT-based wireless smart water supply management system shown in Fig. 4. In Fig. 4, different locations used in model as pump house, pipeline, overhead tank, different users (Hostel-1, Hostel-2 and Admin), reservoir have been shown. Different sensors used at different locations, their connectivity with Aurdino and Wi-Fi have been properly shown. Function at different locations, stage and connectivity with server is properly

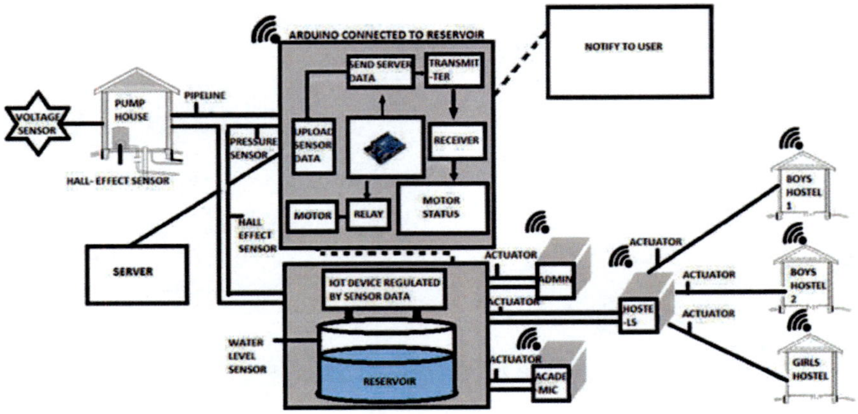

Fig. 4 Wireless smart water supply management system based on IoT

shown in diagram. Figure 4 explains formation of wireless sensor network. Water level sensors are equipped at each hostels and Admin with Wi-Fi and Aurdino. For optimum Utilization of water, Actuators have been installed so that valve can be opened and closed based on conditions observed at server. At water reservoir status of operation of motor, pressure in pipeline and water level is detected. Data is processed by Aurdino, Node MCU and sent to the server. Different hardware used are shown in detail in Table 2.

Figure 5 explains simplified flow diagram of model of water supply management system.

If electricity is available at pump house then the pump will start otherwise we need to on the generator for electrical supply, and check the water level at water reservoir at college premises if the tank is completely filled then pump should off otherwise continue the pumping, here Hall Effect sensor come into the picture which is placed at a regular distance at pipe joint to measure the water flow pressure and it also help in detecting leakage point. If the pressure of water flow of pump house is equal to the college point then no interruption is required otherwise it send an alert alarm to the pump operator and water store in reservoir. At peak demand time water should supply directly to the hostels and academic buildings but in normal time water can store in water overhead tank.

4 Working Methodology

Smart IoT-Based Water Supply Management System has hub as main part. One or multiple sensors can be connected to it. Hub can be connected to system router through Ethernet cable. Hub will control, transmit and receive data from smart IoT-based sensor. Hub will communicate to each node with help of Internet. So we

Table 2 Components used in model

Name	Specification	Quantity	Photo
Hall Effect Sensor	Model: YF-S201, I_{max}: 15 mA at 5 V Voltage range (V_R): 5 to 18 V DC Sensor type: Hall Effect, , Output Type: 5 V Working Flow Rate: 1 to 30 L/Minute, Max. Pressure of Flow Pr: 2.0 MPa, Pulse characteristics: Frequency (Hz) = 7.5 * Flow rate (L/min), Pulses per Lit.: 450(Pul/Lit.), Durability: min. 0.3 million cycles	2	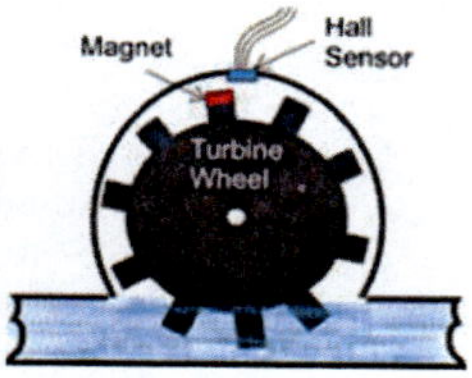
Fluid valve	Working Range: 5–12 V DC, Operates at 3 L/min of flow, ½″ BSP inlet and outlet :	1	
Water level sensor	Working Range: 5–12 V DC, Three Level Indication LEDs, Three transistors (BC 547), 7 Resistors (220 Ω), 1 switch, Accuracy: ±10%	1	
MSP-EXP430F5529LP	MSP430F5529, 16-bit MCU, System Clock cycle Up to 25-MHz, Voltage range: 1.8–V to 3.6-V, Flash memory: 128 kb, RAM: 8 kb, 4 serial interfaces, 12-bit ADC, Integrated USB, Analog Voltage comparator,	1	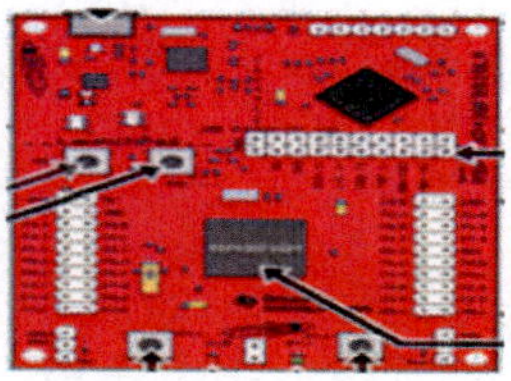

(continued)

Table 2 (continued)

Name	Specification	Quantity	Photo
MSP-EXP430G2ET	RISC architecture, 16-Bit, Clock System:16-MHz Flash memory: 16 kb SRAM: 512 b 10-bit ADC, 24 GPIOs 8-channel comparator 2 No. of 16-BitTimers with 3 capture/compare registers	1	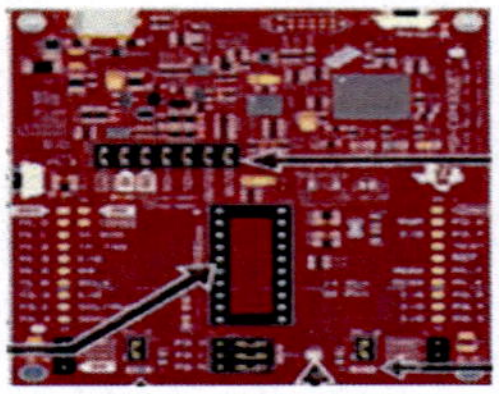
Node MCU	Developer ESP8266 Open source Community Type Single-board microcontroller Operating system XTOS CPUESP8266 (LX106) Memory 128 kb Storage 4 mb, Power USB	1	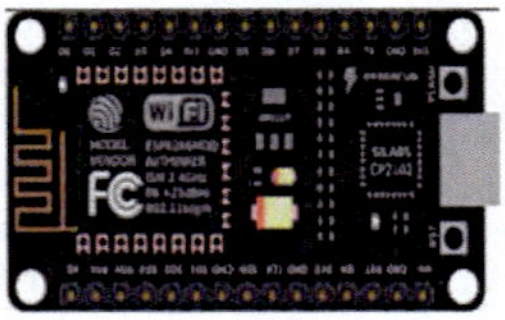
Arduino Uno	ATmega328P, Clock Speed: 16 MHz Voltage of operation: 5 V, V_{in} Range: 7–20 V, Digital I/O Pins: 14 (6 Pin PWM output), Flash Memory: 32 kb bootloader 0.5 kb EEPROM: 1 kb SRAM: 2 kb	3	

can connect and use system with smart phone having Internet facility from any remote location. We can connect to cloud any time. Widely used protocol for SMART WATER MANAGEMENT SYSTEM: CC3200, Wi-Fi, NODE MCU.

Several sensors for water flow and water level measurement are connected through common gateway. Tank is full or not it will be communicated through GUI. At the same time more than one sensor can communicate. Sensor repeaters can also be added. For long distance communication sensors hubs acting as repeaters Cloud is used to maintain data over different locations. Smart system can send information whenever requested by user and cloud will update data time to time. So water supply data is always available to user in real time to end user. GUI of model display is shown in Fig. 6 where pump on/off and Hall Effect Sensor data have been shown through GUI.

Real time monitoring and notifications is one of the key features of Smart IoT-based Water Management systems.

Hub can also control sensor by transferring data instantly to it. As it can start pump or start steeper motor so that valve can be rotated. It can also send alert notification whenever required.

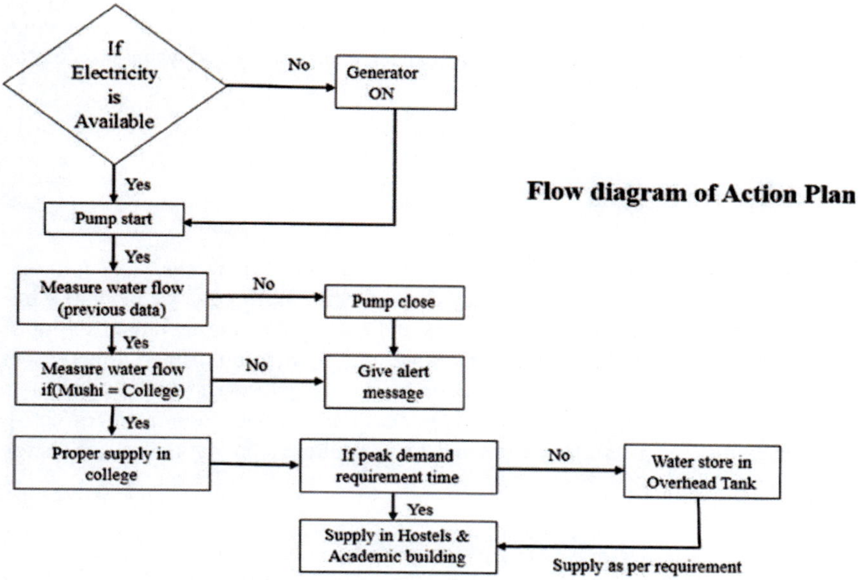

Fig. 5 Simplified flow diagram of model

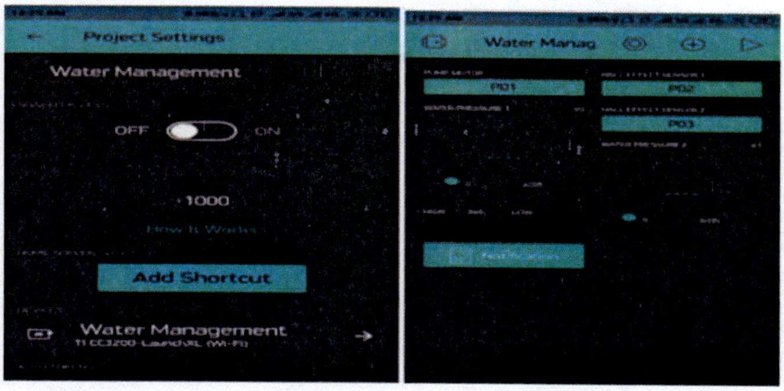

Fig. 6 IFTTT display view

5 Implemented Model and Observation

Over all projects is divided into three parts monitor, control and alert. By 2 Hall Effect sensor data of water flow is monitored. By stepper motor control action for pump on or valve on can be performed. Alert message can be sent to all interested. It is predicated that by 2025 almost 50% of the urban population will come in water-stressed areas. Thus, smart water management provides solution and pre-anticipated water crisis. This model with increased number of nodes will provide enough data which can be further processed for different kind of desired result. Such as peak hour, average flow, average requirement can be expected accurately and water sources can be activated accordingly.

6 Future Directions

Water is the most essential requirement for life of human and plants as well. Optimum use of water resources are present day requirement. With advancement in wireless technology application, smart sensors and data processing techniques a lot of advancement is expected in near future. Water quality measure can be directly sent to consumer.

Crisis can be managed and effective measures can be automated. Effective implementation can be further extended to agriculture application and sewerage treatment.

7 Conclusion

A multi node IoT-based system, ensures improved smart water management through various sensors and actuators which are installed at various locations. Smart IoT-based water supply management system requirements for water drain region have been studied and modeled successfully. In proposed model valve connected with stepper motor gives extra control on overall supply management. This model has been prepared using real time data and opens new direction for drinking water usage optimization.

References

1. Kulkarni P, Farnham T (2016) Smart city wireless connectivity considerations and cost analysis: lessons learnt from smart water case studies. IEEE Access 4:660–672. https://doi.org/10.1109/ACCESS.2016.2525041
2. Serra H, Bastos I, Melo JLAD, Oliveira JP, Paulino N, Nefzaoui E, Bourouina T (2019) A 0.9-V analog-to-digital acquisition channel for an IoT water management sensor node. IEEE Trans Circuits Syst II Express Briefs 66(10):1678–1682. https://doi.org/10.1109/TCSII.2019.2933276
3. Nasser AA, Rashad MZ, Hussein ES (2020) A two-layer water demand prediction system in urban areas based on micro-services and LSTM neural networks. IEEE Access 8:147647–147661. https://doi.org/10.1109/ACCESS.2020.3015655
4. Vinothini E, Suganya S (2014) Automated water distribution and performance monitoring system. Int J Eng Innov Technol (IJEIT) 3(8):30–32

5. Piasecki A, Jurasz J, Skowron R (2017) Forecasting surface water level fluctuations of lake Serwy (Northeastern Poland) by artificial neural networks and multiple linear regression. J Environ Eng Landsc Manag 25(4):379–388
6. Tharanyaa JPS, Jagadeesan A, Lavanya A (2013) Theft identification and automated water supply system using embedded technology. Int J Adv Res Electrical Electronics Instrumentation Eng 2(8):3727–3733
7. Proteus (2018) AdaPtative micROfluidic and nano-enabled smart systems for waTEr qUality Sensing (PROTEUS). H2020 grant No. 644852 (Online). Available: http://www.proteus-sensor.eu/
8. Patawala AH, Bansode N, Bhong YP, Zadbuke AS (2017) IoT based water management system for smart city. Int J Adv Res Ideas Innov Technol 3(2):379–383
9. Siregar B, Satria IS, Budiarto R, Rahmat RF (2017) Water pipeline monitoring and leak detection using flow liquid meter sensor. In: IOP conference series: materials science and engineering, vol 190. https://doi.org/10.1088/1757-899X/190/1/012036
10. Joseph J, Manju KM, Sajith MR, Sithara K, Sujith N, Vishnu PV, Sithara K (2018) Water management system using IoT. Int Res J Eng Technol (IRJET) 05(04):1887–1890

Improvement of Electrical Characteristics for Nanoscale Single-Gate FDSOI Using Gate Oxide Engineering

Aditya Kushwaha, Shailesh Shankar Upadhyay, Narendra Yadava, Mangal Deep Gupta, and R. K. Chauhan

Abstract In this paper, various heterodielectric gate oxide-based (HDGO) fully depleted silicon-on-insulator (FDSOI) arrangements are considered, i.e., dual heterodielectric gate oxide (DHDGO), triple heterodielectric gate oxide (THDGO), and quadruple heterodielectric gate oxide (QHDGO). Among all the proposed structures, QHDGO-FDSOI is a potential candidate. In this structure, the gate oxide region of C-FDSOI is divided into two equal parts along the x and y axes. To improve the ratio of gate-source capacitance (C_{gs}) to gate-drain capacitance (C_{gd}), low dielectric constant ($k = 7.5$, $k = 3.9$) in the lower part and high dielectric constant ($k = 9$, $k = 22$) in the upper part is used. The electrical characteristics of proposed HDGO-FDSOI are evaluated and compared with conventional FDSOI (C-FDSOI), find the improvements in ON to OFF current ratio (I_{ON}/I_{OFF}), intrinsic gain (A_V), drain-induced-barrier-lowering (DIBL), OFF current (I_{OFF}), and subthreshold slope (SS) value. The proposed structures are designed and simulated using the SILVACO ATLAS 2D simulator.

Keywords FDSOI · DHDGO · THDGO · QHDGO · DIBL · I_{ON} · I_{OFF}

1 Introduction

In recent decades, the scaling of planar MOSFETs has been carried out vigorously to improve integrated circuits' electrical efficiency. If this process continues on a nanoscale, SCEs are arising, which will degrade the device's performance. Due to the continuous decrease of channel length, it will affect the device's crucial parameters, i.e., there is an increase in OFF-state leakage currents, DIBL, and subthreshold slope values. To resolve these problems, a new FDSOI transistor is introduced. The detailed study of FDSOI metal oxide semiconductor (MOS) transistor is done in [1–3]. The leakage current is increased by the scaling down of the transistor. The thickness of conventional dielectric silicon dioxide (SiO_2) reaches its maximum

A. Kushwaha (✉) · S. S. Upadhyay · N. Yadava · M. D. Gupta · R. K. Chauhan
Madan Mohan Malaviya University of Technology, Gorakhpur, India
e-mail: kushawahaaditya@gmail.com

B. Mishra and M. Tiwari (eds.), *VLSI, Microwave and Wireless Technologies*,
Lecture Notes in Electrical Engineering 877,
https://doi.org/10.1007/978-981-19-0312-0_72

limit due to scaling. At this dimension of the gate oxide, the gate tunneling highly increases. And as a result, static power dissipation and OFF-state leakage current dramatically increase, which will reduce the device performance [4–6]. Therefore, to improve electrical efficiency, effective gate control over the channel is becoming an important issue. To resolve these problems, instead of conventional SiO_2 dielectric, high-k dielectrics are used directly over the silicon wafer or the SiO_2 layer [7, 8].

Recent research shows that direct use of high-k dielectrics over the silicon wafer will debase the short channel performance. Fringing fields lines either from gate to source/drain region or from source/drain region to channel region debase the device's performance [9, 10]. To get rid of these problems, to some extent, higher dielectric value material is stacked over the low dielectric value material [10–12].

In this research work, we examine analog and DC analysis of 50-nm channel length FDSOI [13], based on different HDGO structures [12]. The 2D SILVACO ATLAS tool has been used to simulate and design the proposed structure. In Sects. 2 and 3, device specification and results are discussed, respectively.

2 Device Specification and Simulation

A conventional FDSOI structure is shown in Fig. 1a. Figure 1b–d shows the different heterodielectric gate oxide FDSOI structures, i.e., DHGDO, THGDO, and QHGDO-FDSOI structures, respectively [12]. By improving the ratio of gate-source capacitance (C_{gs}) to gate-drain capacitance (C_{gd}), i.e., C_{gs}/C_{gd}, controlling the gate over the channel increases. So, decreasing C_{gd} will provide better control of the device over the channel. Consequently, we use low-value dielectric material on the drain side to reduce the C_{gd} value, calculated by Eq. 1 [14].

$$C = \frac{k\varepsilon_0 A}{t_{ox}} \tag{1}$$

where k is the relative dielectric constant value, A is the area of capacitance, ε_0 is the air dielectric constant, and t_{ox} is the thickness of gate oxide material. Proposed device specifications are mentioned in Table 1.

Equivalent dielectric constant (ε_{eq}) and its formula for different HDGO and C-FDSOI are shown in Table 2. Various models are in the simulation for mobility conmob and fldmob are used; SRH model and auger model are used for carrier recombination. Concentration-dependent and parallel model and perpendicular field-dependent mobility model are also used in simulations. The 2D SILVACO ATLAS tool has been used to simulate and design the proposed structure [15].

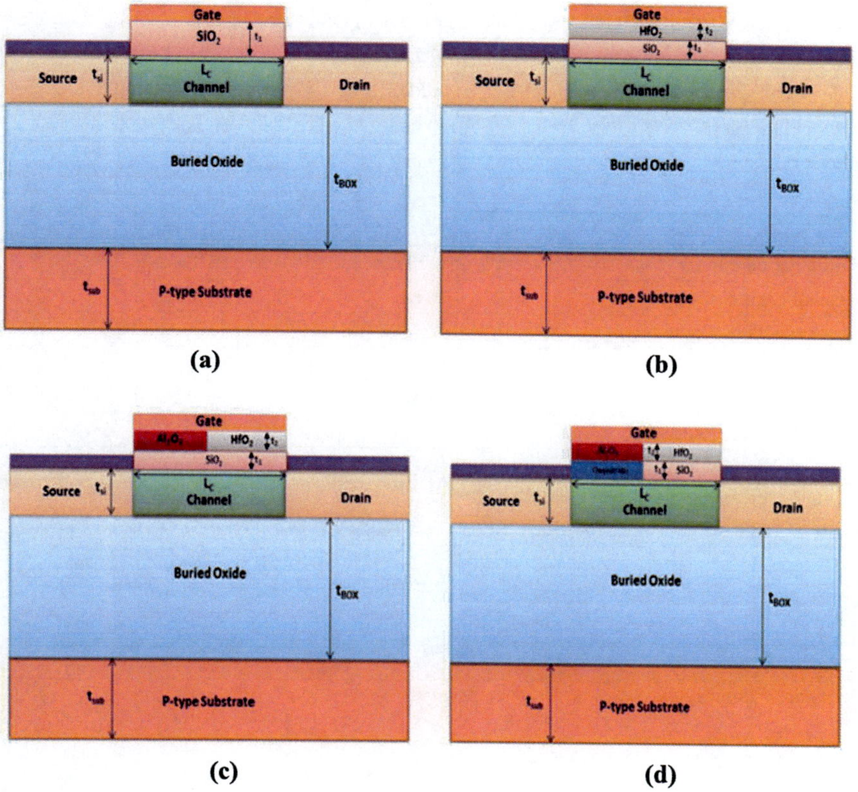

Fig. 1 **a** Conventional FDSOI structure. **b** DHDGO-FDSOI structure. **c** THDGO-FDSOI structure. **d** QHDGO-FDSOI structure

3 Results and Discussions

3.1 Transfer Characteristics

Transfer characteristics ($I_{DS}-V_{GS}$) show the variation of drain current with gate-source voltage (V_{GS}). In Fig. 2a, there is a comparative study of linear-scale transfer characteristics, and Fig. 2b shows log scale transfer characteristics for different proposed HDGO-FDSOI structures over C-FDSOI structure. As Fig. 2a and b show, QHDGO-FDSOI is a better candidate among overall FDSOI structures. Consequently, it will provide better gain and a low leakage current.

Table 1 Device description of conventional and HDGO-FDSOI structures

Device parameters		VALUES	
		C-FDSOI	HDGO-FDSOI
Channel length (L_c)		50 nm	50 nm
Oxide thickness	t_1	1 nm	0.5 nm
	t_2	–	0.5 nm
Channel doping (N_A)		1e16 cm^{-3}	1e16 cm^{-3}
Drain region doping (N_D)		1e20 cm^{-3}	1e20 cm^{-3}
Insulator thickness (t_{BOX})		20 nm	20 nm
Work function of gate (Φ)		4.77 eV (Cu)	4.77 eV (Cu)
Gate-source voltage (V_{GS})		1.0 V	1.0 V
Source region doping (N_D)		1e20 cm^{-3}	1e20 cm^{-3}
Silicon film thickness (t_{Si})		10 nm	10 nm
Total width of the device		100 nm	100 nm

Table 2 Equivalent dielectric constant for different HDGO and C-FDSOI

Structures	Formula	Equivalent dielectric constant (ε_{eq})
C-FDSOI	–	3.9
DHDGO-FDSOI ($\varepsilon_1 = 3.9$, $\varepsilon_2 = 22$)	$\varepsilon_{eq} = \frac{2\varepsilon_1\varepsilon_2}{(\varepsilon_1+\varepsilon_2)}$	6.6
THDGO-FDSOI ($\varepsilon_1 = 3.9$, $\varepsilon_2 = 22$, $\varepsilon_3 = 9$)	$\varepsilon_{eq} = \frac{2\varepsilon_1(\varepsilon_2+\varepsilon_3)}{(\varepsilon_1+\varepsilon_2+\varepsilon_3)}$	6.9
QHDGO-FDSOI ($\varepsilon_1 = 3.9$, $\varepsilon_2 = 22$, $\varepsilon_3 = 9$, $\varepsilon_4 = 7.5$)	$\varepsilon_{eq} = \frac{2(\varepsilon_1+\varepsilon_2)(\varepsilon_3+\varepsilon_4)}{(\varepsilon_1+\varepsilon_2+\varepsilon_3+\varepsilon_4)}$	20.1

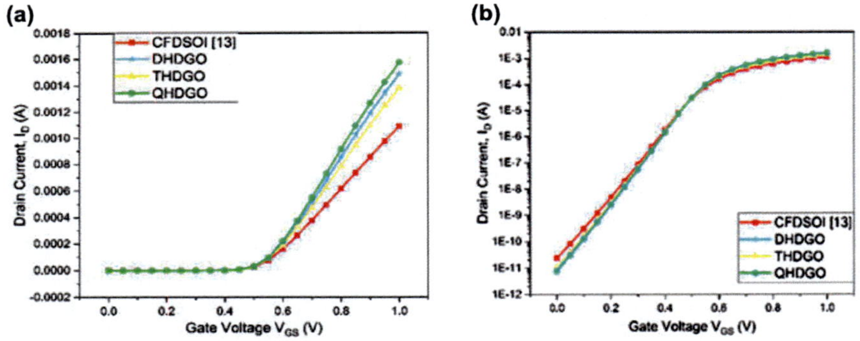

Fig. 2 **a** Linear transfer characteristics for different proposed HDGO and conventional FDSOI structures at $V_{DS} = 0.8$ V. **b** Linear transfer characteristics for different proposed HDGO and conventional FDSOI structures at $V_{DS} = 0.8$ V

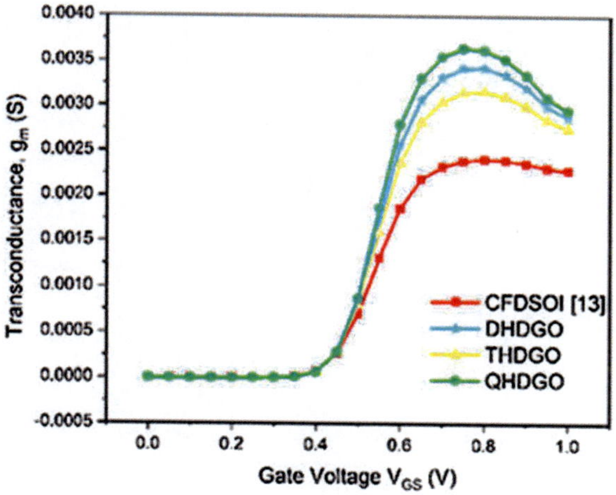

Fig. 3 Transconductance versus gate voltage characteristics for different proposed HDGO and conventional FDSOI structures at $V_{DS} = 0.8$ V

3.2 *Transconductance*

Transconductance (g_m) defines the variation of drain current on a small change in gate voltage, keeping drain-source voltage (V_{DS}) constant, calculated by Eq. 2. As we can observe from Fig. 3, QHDGO-FDSOI shows better transconductance performance over the proposed and conventional FDSOI structures. So, QHDGO-FDSOI is a potential candidate to provide better amplification.

$$g_m = \frac{\partial I_D}{\partial V_{GS}} \tag{2}$$

3.3 *Output Characteristics*

Output characteristics ($I_{DS}-V_{DS}$) show the variation of drain current with drain-source voltage. Figure 4 shows drain current as a function of drain-source voltage for different proposed HDGO and conventional FDSOI. As ON current (I_{ON}) is higher for QHDGO-FDSOI than the various proposed HDGO and C-FDSOI structures, the proposed QHDGO-FDSOI designs are useful for analog applications. The device can show a large voltage swing when used in the amplifier circuits [16, 17].

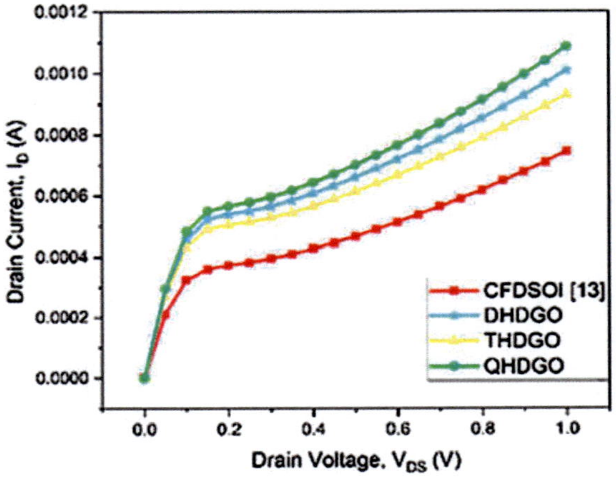

Fig. 4 Output characteristics for different proposed HDGO and conventional FDSOI structures at $V_{GS} = 0.8$ V

3.4 Output Conductance

Output conductance defines the change of drain current as a function of a small change in drain voltage by assuming constant gate voltage, and it can be calculated by Eq. 3. It is an important parameter to decide drive current. Figure 5 shows output characteristics for different proposed HDGO and C-FDSOI structures. Figure 5 output conductance is higher for QHDGO-FDSOI to some extent over the other proposed HDGO and conventional designs.

$$g_{\mathrm{d}} = \frac{\partial I_{\mathrm{D}}}{\partial V_{\mathrm{DS}}} \tag{3}$$

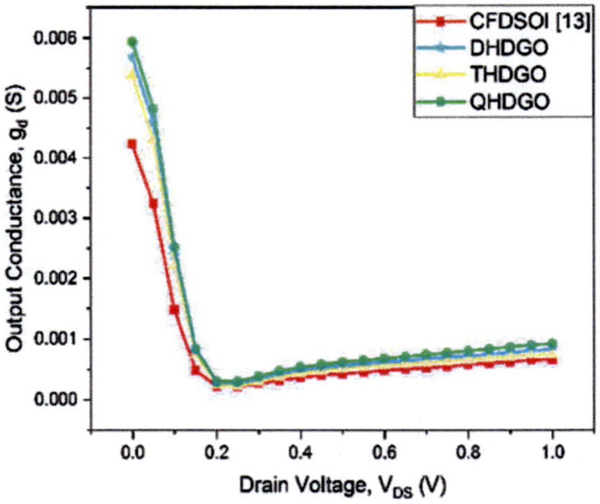

Fig. 5 Output conductance versus drain-voltage characteristics for different proposed HDGO and conventional FDSOI structures at $V_{GS} = 0.8$ V

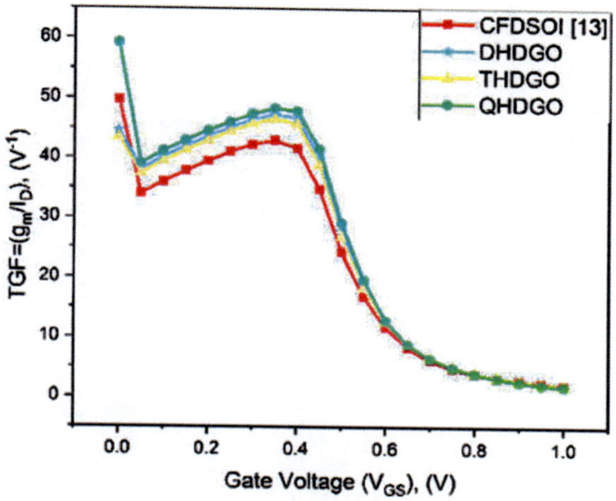

Fig. 6 Transconductance generation factor versus drain-voltage characteristics for different proposed HDGO and conventional FDSOI structures at $V_{GS} = 0.8$ V

3.5 *Transconductance Generation Factor*

The transconductance generation factor (TGF) is a crucial parameter to deciding the device's analog performance, calculated by using Eq. 4. As in Fig. 6, we found that QHDGO shows better TGF in comparison to the other proposed HDGO and conventional transistors.

$$\text{TGF} = \frac{g_{\text{m}}}{I_{\text{D}}} \tag{4}$$

3.6 *Static Power Dissipation*

As Fig. 7 shows, static power dissipation ($P_{\text{Diss.}}$) for QHDGO is lower than other FDSOI structures. So, proposed HDGO-FDSOI structures are the better option for digital applications.

Figure 8a–c shows OFF (I_{OFF}) current, ON (I_{ON}) current, and $I_{\text{ON}}/I_{\text{OFF}}$ current ratio for different structures. And these parameters are improved for QHDGO-FDSOI over the other structures. Figure 9 shows intrinsic gain variation against different FDSOI structures. THDGO shows the highest intrinsic gain among all the designs, a 15% improvement over conventional FDSOI. In Fig. 10, among all the structures, QHDGO exhibits the lowest subthreshold slope 71.1964 mV/decade another important electrical parameter, DIBL obtained for different FDSOI structures, is shown in Fig. 11. DIBL is reduced to a minimal value in QHDGO, while it is maximum for C-FDSOI structures. All these improvements help to overcome the short channel effects arising in nanoscaled FDSOI devices. QHDGO gate oxide in FDSOI has

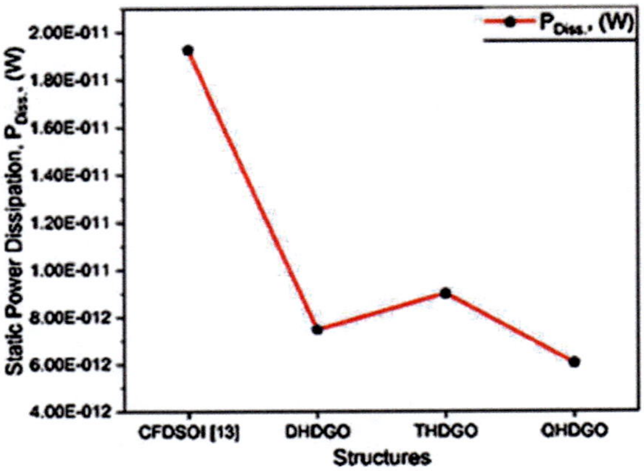

Fig. 7 Static power dissipation for different proposed HDGO and conventional FDSOI structures

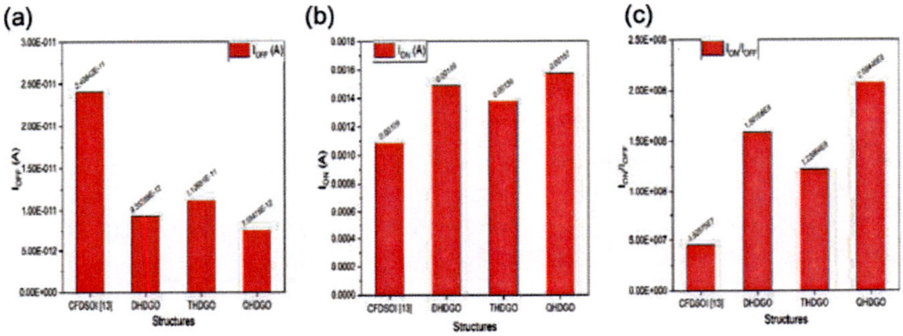

Fig. 8 **a** OFF-state current for different proposed HDGO and conventional FDSOI structures at $V_{DS} = 0.8$ V. **b** ON-state current for different proposed HDGO and conventional FDSOI structures at $V_{DS} = 0.8$ V **c** ON-state current for different proposed HDGO and conventional FDSOI structures at $V_{DS} = 0.8$ V

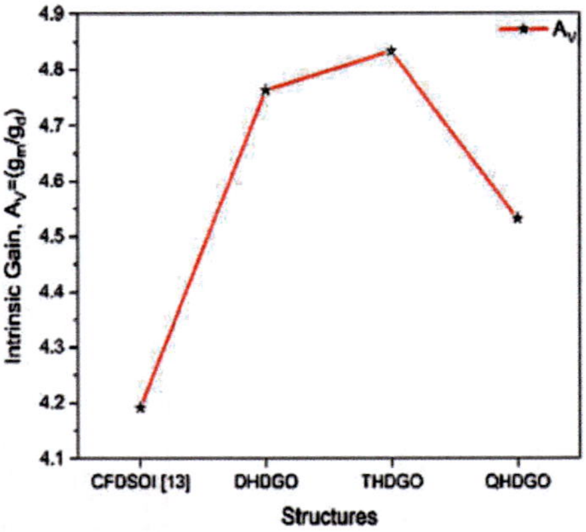

Fig. 9 Intrinsic gain value for different proposed HDGO and conventional FDSOI

Fig. 10 Subthreshold slope value for different HDGO and conventional proposed FDSOI

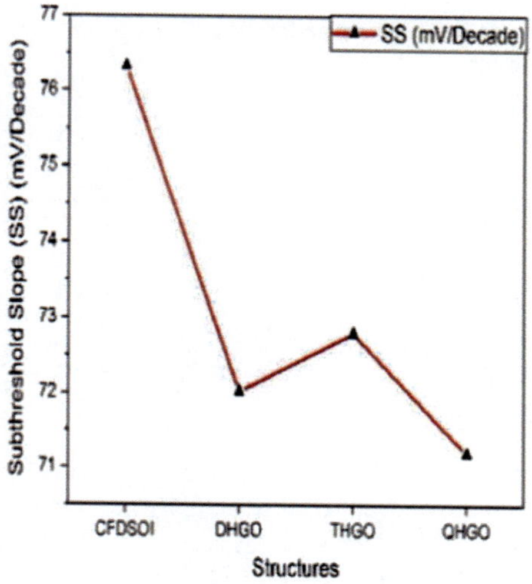

Fig. 11 DIBL value for different HDGO and conventional proposed FDSOI

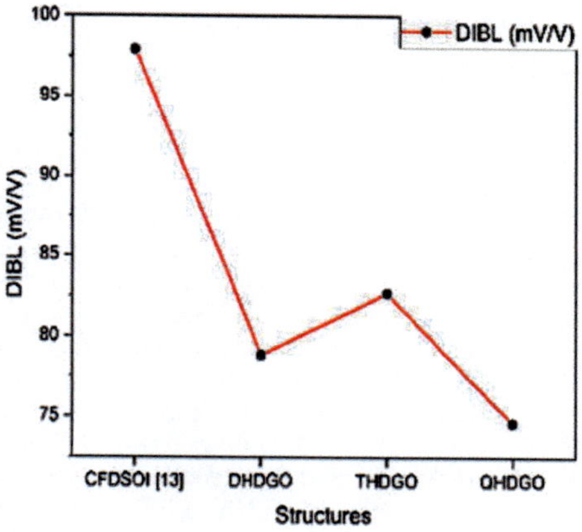

proven to be the best arrangement to mitigate the short channel effects. All the electrical parameters for different FDSOI structures are discussed in Table 3.

4 Conclusion

In this work, single-gate HDGO-FDSOI designs for different gate oxides are presented. Three separate structures, i.e., DHGDO, THGDO, QHGDO-FDSOI structures, are analyzed. Among all the structures, QHDGO-FDSOI is a better performer, which improves the short channel issues in a higher ratio. It improves the ON current

Table 3 Electrical performance comparison between C-FDSOI and HDGO-FDSOI on various parameters of the devices

Parameters	Structures			
	C-FDSOI	DHDGO-FDSOI	THDGO-FDSOI	QHDGO-FDSOI
DIBL (mV/V)	97.9267	78.832	82.676	74.6173
Subthreshold (mV/Decade)	76.3198	72.0185	72.7893	71.1964
I_{OFF} (pA)	24.0842	9.35359	11.2661	7.55475
I_{ON} (mA)	1.08999	1.48877	1.38195	1.57477
I_{ON}/I_{OFF}	4.52575e+007	1.59166e+008	1.22664e+008	2.08448e+008

by ~1.4 times, intrinsic gain by ~1.08 times, and ON–OFF current ratio by ~4.6 times over the conventional structure. And it reduces the OFF-state leakage current by ~3.2 times, DIBL by ~1.3 times, and subthreshold slope by ~2.4 times contrasted with C-FDSOI. So, QHDGO-FDSOI is a potential candidate for analog applications.

References

1. Cheng KG, Khakifirooz A (2016) Fully depleted SOI (FDSOI) technology. Sci China Inf Sci 59(6):061402
2. Skotnicki T, Hutchby A, J, King J, T, Wong PSH, Boeuf F (2005) The end of CMOS scaling—toward the introduction of new materials and structural changes to improve MOSFET performance. IEEE Circuits Dev Mag 21(1):16–26
3. Ferain I, Colinge A, C, Colinge P, J (2011) Multigate transistors as the future of classical f6lmetal-oxide-semiconductor field-effect transistors. Nature 479(7373):310–316
4. Kim BY (2010) Challenges for nanoscale MOSFETs and emerging Nanoelectronics. Trans Electr Electron Mater 11:93–105. https://doi.org/10.4313/TEEM.2010.11.3.093
5. Colinge PJ (2004) Multiple-gate SOI MOSFETs. Solid State Electron 48(6):897–905
6. Chaudhry A, Kumar JM (2004) Investigation of the novel attributes of a fully depleted dual-material gate SOI MOSFET. IEEE Trans Electron Dev 51(9):1463–1467
7. Duba A, Jaggi V, Verma A, Mishra A (2013) High K oxides as an alternative gate oxide in CMOS technology and its corresponding effects—survey approach. Int J Eng Res Tech (IJERT) 02(11)
8. Pradhan PK, Mohapatra S, Sahu KP, Behera KD (2014) Impact of high-k gate dielectric on analog and RF performance of nanoscale DG-MOSFET. Microelectron J 45:144–151
9. Lee HB, Song CS, Choi R, Kirsch P (2007) Metal electrode/high- k dielectric gate-stack technology for power management. IEEE Trans Electron Dev 55(1):8–20
10. Sharma KR, Gupta M, Gupta SR (2011) TCAD assessment of device design technologies for enhanced performance of nanoscale DG MOSFET. IEEE Trans Electron Dev 58(9):2936–2943
11. Nirmal D, Vijayakumar P, Samuel CPP, Jebalin KB, Mohankumar N (2013) Subthreshold analysis of nanoscale FinFETs for ultra low power application using high-k materials. Int J Electron 100(6):803–817
12. Bousari BN, Anvarifard KM., Haji-Nasiri S (2019) Improving the electrical characteristics of nanoscale triple-gate junctionless FinFET using gate oxide engineering. AEU-Int J Electron Commun 108:226–234

13. Mishra V, Chauhan RK (2017) Performance analysis of fully depleted SOI tapered body reduced source (FD-SOI TBRS) MOSFET for low power digital applications. Springer Nature Singapore Pte Ltd.
14. Chen Y, Xu R (2014) Analysis of the RF and noise performance of junctionless MOSFETs using monte carlo simulation. Int J Numer Model Electron Netw Dev Fields 27(5–6):822–833
15. ATLAS User's manual (2016) SILVACO, Santa Clara, CA, USA
16. Mishra V, Chauhan KR (2017) Performance analysis of modified source and tunnel diode body contact based fully-depleted silicon-on-insulator MOSFET for low power digital applications. J Nanoelectron Optoelectron 12:59–66
17. Ranka D, Rana KA, Yadav R, Giri D (2011) Performance analysis of FD-SOI MOSFET with different gate spacer dielectric. Int J Comput Appl 18:22–27. https://doi.org/10.5120/2280-2952

Improved Digital Performance of Modified Source-Drain FDSOI by Optimization of Buried Oxide Properties

Sudhir Rai, Mangal Deep Gupta, Saurabh Kumar, and R. K. Chauhan

Abstract In this work, the effect of buried oxide material and its thickness on the performance of modified source-drain fully depleted silicon on insulator (MS-MD FDSOI) is studied. The performance of the device has been analyzed at various buried oxide materials and thickness variation. We find that FDSOI with HfO_2 at 10 nm thickness provides better performance as compared to higher BOX thickness. The improvements in sub threshold slope (SS), drain induced barrier lowering (DIBL), Static power dissipation (SPD), Intrinsic gate delay (IGD), OFF current (I_{OFF}) and ON-to-OFF current ratio (I_{ON}/I_{OFF}) is observed to be 16.67%, 24.6%, 95.7%, 8.77%, 23.21 times and 19.49 times, respectively, for FDSOI with HfO_2 at 10 nm thickness as compared to FDSOI with SiO_2 BOX (10 nm thickness). The device structure is simulated using ATLAS 2-D device simulator.

Keywords FDSOI · Buried oxide · Modified source and modified drain · Short channel effects · ON-to-OFF current ratio

1 Introduction

The evolution of different technologies like Internet of things (IOT), information technology (IT) is possible due to advancement in MOS technologies. Requirement of highest speed with lowest power loss in MOS technology is in very much demand. To fulfill the requirement of high density ICs the nano scale devices are in great demand. MOSFETs are favorable among other technology beyond nanoscale due to its performance.

Decreasing the size of MOSFET will increase unpredictable adverse effects. Short-channel effect such as drain induced barrier lowering (DIBL), threshold voltage roll-off, hot carrier effect (HCEs) and leakage current comes in picture with decrease in size of MOS transistor [1–3]. This effect can limit the performance of these devices.

S. Rai (✉) · M. D. Gupta · S. Kumar · R. K. Chauhan
Madan Mohan Malaviya University of Technology, Gorakhpur, India
e-mail: sdece@mmmut.ac.in

B. Mishra and M. Tiwari (eds.), *VLSI, Microwave and Wireless Technologies*,
Lecture Notes in Electrical Engineering 877,
https://doi.org/10.1007/978-981-19-0312-0_73

To overcome this problem different type of advancement has been done to MOS transistors. Multi gate FET (FinFET) [4], fully depleted silicon on insulator (FDSOI) [5, 6] and Tunnel FET [7, 8] are the various technologies used to limit the adverse effects of decreasing the size of transistor [9]. FinFET having complex structure is used to overcome the various short channel effects [6].

Fully depleted silicon on insulator (FDSOI) is an excellent technology used for industrial purpose worldwide because of its excellent immunity over short channel effects (SCEs). FDSOI technology can also minimize the effect of channel density issue in Si MOSFET [8]. Early FDSOI MOSFET model is examined by Young [8] and at the interface of Si/SOI, electric field variation is examined by Suzuki and Pidin [10]. Nakajima et al. investigated of the thin BOX layer in the performance of FDSOI MOSFET [11]. The primary concern is the selection of appropriate material of buried oxide layer and its thickness in FDSOI-based MOS devices [11]. So that the electric field penetration of the device at SOI/BOX interfaces at sufficient drain bias is controlled [12]. Nowadays, the effect on performance of FDSOI MOSFET due to variation in back oxide thickness is discussed [12].

In this present work, the effect of different BOX material and its thickness variation on the performance of MS-MD FDSOI MOSFET has been analyzed. The main purpose of this work is to enhance the electrical performance of the MS-MD FDSOI. The performance of the present device has been analyzed based on the its DIBL, sub threshold slope, leakage current, ON-to-OFF current ratio, threshold voltage and static power dissipation, at different buried oxide materials and its thickness. The device structure is simulated using ATLAS 2-D device simulator.

2 Device Structure and Simulation

The simulated schematic of MS-MD FDSOI structures is shown in Fig. 1 with buried oxide material as (a) SiO_2 and (b) HfO_2, respectively. The source region is split vertically into two areas where the upper part is doped with a high concentration value while the bottom region is doped with a low concentration value [13]. The three buried oxide material with different dielectric constant (k_1, k_2, k_3) have been used here. The work function of gate electrode is 4.52 eV. The models used in simulation are BGN, SHOCKLEY-REED-HALL, FLDMOB, DRIFT-DIFFUSION MODEL. The device structure is simulated using ATLAS 2-D device simulator [14]. The device specification is listed in Table 1.

3 Result and Discussion

In Fig. 2 three different high-k dielectrics with fixed thickness of 10 nm are used to obtain I_d–V_g curve on log scale. Leakage current decreases to minimum value in case of HfO_2. If the value of dielectric constant is increased then the leakage current

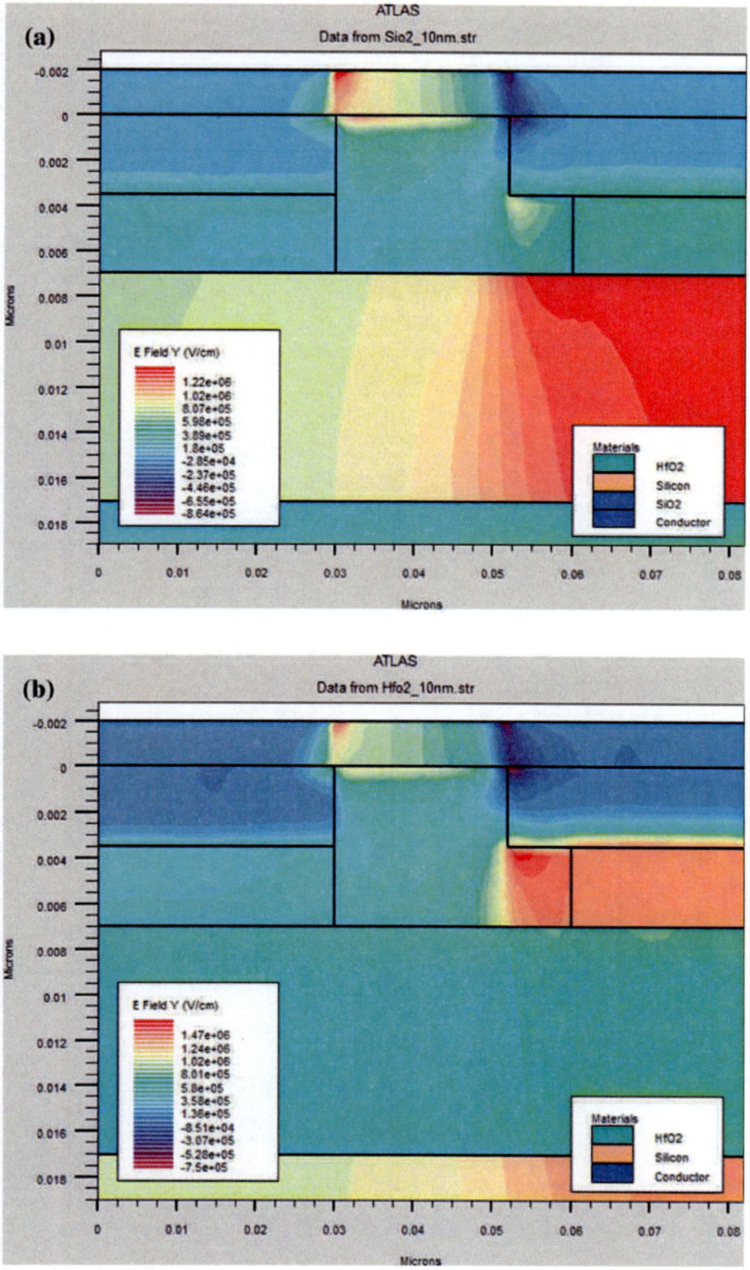

Fig. 1 **a** The MS-MD FDSOI structure with buried oxide material as SiO_2. **b** The MS-MD FDSOI structure with buried oxide material as HfO_2

Table 1 Device specifications used for simulation

Device parameters	MS-MD FDSOI devices [13]
Thickness of gate length	22 nm
Thickness of buried oxide (BOX)	0 nm, 20 nm, 30 nm, 40 nm
Thickness of gate oxide (T_{ox})	2 nm
Silicon thickness (T_{si})	7 nm
Doping concentration in substrate region	1e16
High doping concentration in drain (N^+ region)	1e20
Low doping concentration in drain (N^- region)	1e17
High doping concentration in drain N^+ region	1e20
Material of buried oxide	SiO_2, Si_3N_4, HfO_2
Gate work function	4.52
Dielectric constant of buried oxide	$k_1 = 3.9$, $k_2 = 7.5$, $k_3 = 22$
Material of gate oxide	HfO_2

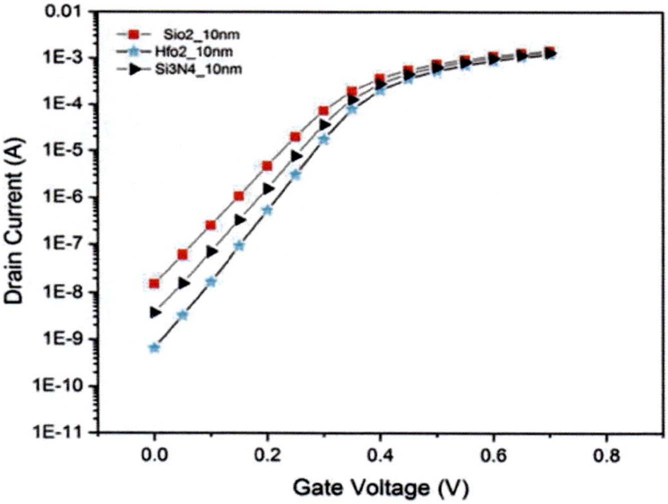

Fig. 2 I_d–V_g curve on log scale for different buried oxide material in MS-MD FDSOI

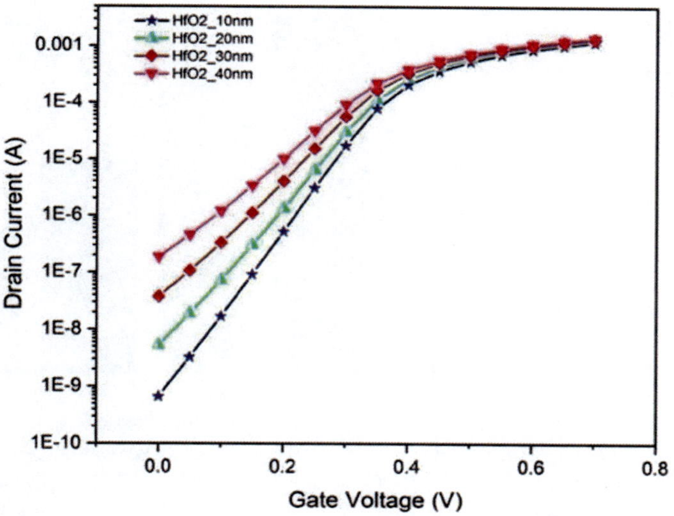

Fig. 3 I_d–V_g curve on log scale for different HfO_2 BOX thickness in MS-MD FDSOI

will decrease thus short channel effect will be reduced [15]. The highest leakage is observed for $SiO_2 = 1.539 \times 10^{-8}$ A and lowest value is observed for HfO_2 6.592 $\times 10^{-10}$ A.

Figure 3 shows the I_d–V_g variation in case of different HfO_2 BOX thickness (T_{BOX}). The BOX thickness is increased from 10 to 40 nm, a large variation is observed in the off state leakage current. The highest value of leakage current is observed as 1.84×10^{-7} A at BOX thickness of 40 nm and lowest value of leakage current is 6.59×10^{-10} A at BOX thickness of 10 nm. On the other hand ON current value is almost same for all buried oxide thickness.

The I_{ON}/I_{OFF} ratio behavior at different T_{BOX} and K_{BOX} are shown in Figs. 4 and 5, respectively. It can be seen from the Figs. 4 and 5 that the device gives negligible off state leakage current and better drive current. The ON-to-OFF current ratio is estimated to be 9.13×10^4 for10 nm SiO_2 BOX. Making the BOX more thinner will increases I_{ON}/I_{OFF} ratio which in turn leads to better controllability over the channel. If we increase the value of dielectric constant, the I_{ON}/I_{OFF} ratio will also increase. The ON-to-OFF current is estimated to be 1.79×10^6 in case of 10 nm HfO_2 BOX.

From Table 2, the threshold voltage has rolled up by reducing buried oxide thickness, which provides improved tolerance to the influences of short channel at low buried oxide thickness. Reduction in threshold voltage roll-off is obtained due to increase in T_{BOX} thickness [16]. So at thicker BOX layer, the consumption of less fraction of gate voltage is reduced.

Figure 6 shows variation of DIBL and sub threshold slope for different buried oxide materials. From Fig. 6 it is observed that HfO_2 gives least value of DIBL and sub threshold slope as compare to SiO_2 BOX material. The lower DIBL shows better

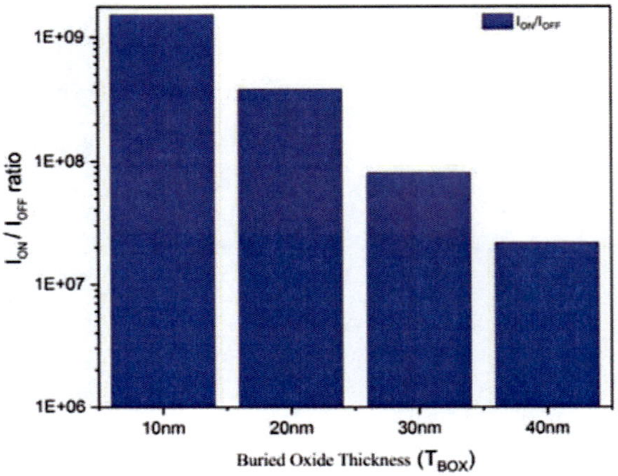

Fig. 4 Variation of I_{ON}/I_{OFF} ratio against HfO_2 buried oxide thickness

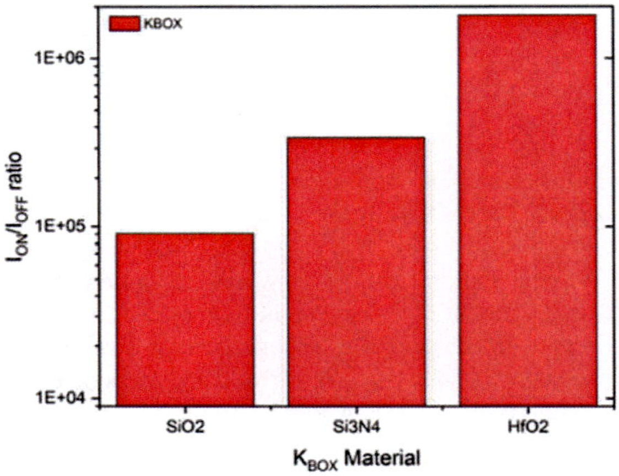

Fig. 5 Variation of I_{ON}/I_{OFF} ratio against buried oxide material

Table 2 Variation of T_{BOX} in MS-MD FDSOI

Buried oxide thickness (HfO_2)	OFF current	ON current	I_{ON}/I_{OFF}	Threshold voltage (V)	g_m (mS)	Sub threshold slope (mV/decade)	DIBL (mV/V)	Static power diss. (nW)
10 nm	6.59×10^{-10}	1.18×10^{-3}	1.79×10^6	0.03382	3.396	65.33	65.30	0.461
20 nm	5.26×10^{-9}	1.25×10^{-3}	2.37×10^5	0.03349	3.525	74.96	96.24	3.68
30 nm	3.71×10^{-8}	1.33×10^{-3}	3.58×10^4	0.03284	3.537	89.41	166.05	25.98
40 nm	1.84×10^{-7}	1.39×10^{-3}	7.55×10^3	0.02906	3.556	108.22	172.06	129.41

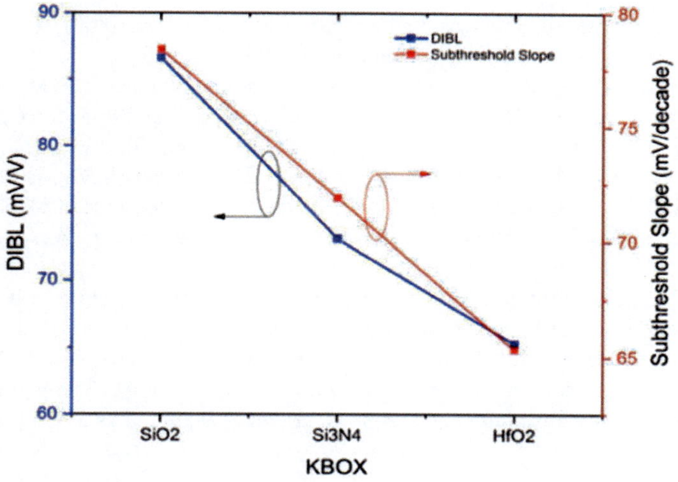

Fig. 6 Variation of DIBL and sub threshold slope against BOX material

immunity over the barrier lowering. Figure 7 shows the evaluation of sub threshold slope with variation in Buried oxide thickness for different BOX materials. At same BOX thickness (10 nm), the value of sub threshold slope decreases with increasing the high-*k* of BOX material. If we increase the thickness of BOX material, sub threshold

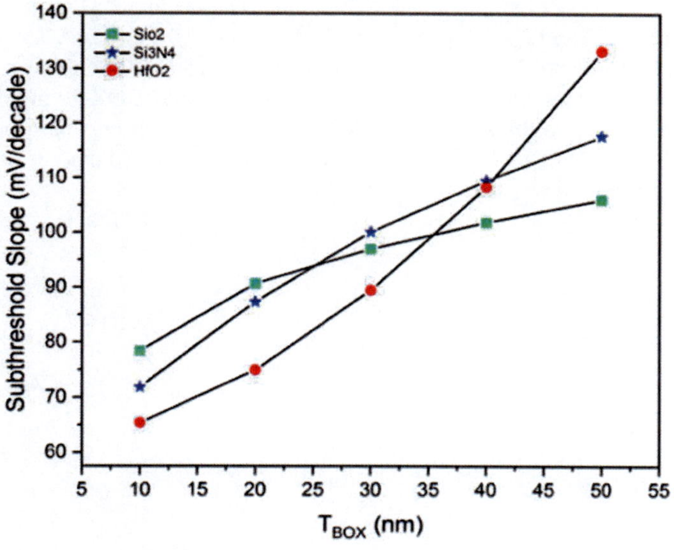

Fig. 7 Variation of sub threshold slope against BOX thickness

slope will be increased. The least value of SS is obtained in case of HfO_2 at 10 nm BOX thickness. All comparison is tabulated in Tables 2 and 3.

Figure 8 shows static power dissipation variation with increasing BOX thickness. With the reduction in BOX thickness the static power dissipation decreases. The highest power dissipation occurs at 40 nm box thickness, due to greater leakage current. The least value obtained in the high-*k* BOX material (HfO_2) is 0.461 nW at 10 nm BOX thickness and highest value of static power dissipation is 129.41 nW obtained at 40 nm BOX thickness. From Fig. 9 static power dissipation is obtained minimum in case of HfO_2.

From Fig. 10 it is seen that the intrinsic gate delay is lowest for HfO_2 and highest for SiO_2. Lower intrinsic gate delay confirms that a digital circuit can work at an enhanced speed. Improved transconductance for HfO_2 at 10 nm is shown in Fig. 11. An increment in the thickness of buried oxide increases transconductance. The highest value of transconductance is obtained at box = 40 nm and calculated as 3.556 mS.

4 Conclusion

In this present work, the effect of different BOX material and its thickness variation on the performance of MS-MD FDSOI MOSFET has been observed and analyzed. Replacing conventional SiO_2 BOX with high-*k* BOX (HfO_2) has helped to overcome the short channel effects. The improvement in digital performance of the MS-MD FDSOI has been obtained for HfO_2 BOX at 10 nm which greatly enhances immunity against the influences of short channel. The improvements in SS, DIBL, SPD, IGD, OFF current (I_{OFF}) and I_{ON}/I_{OFF} ratio are obtained to be 16.67%, 24.6%, 95.7% and 8.77%, 23.21 times and 19.49 times, respectively, for MS-MD FDSOI with HfO_2 at 10 nm thickness as compared to MS-MD FDSOI with SiO_2 BOX 10 nm thickness is observed. At 10 nm thickness HfO_2 BOX produces overall best performance as compared to 20, 30, 40 nm BOX thickness. Hence, it is suggested that the device can be suitable for low-power digital circuit applications.

Table 3 Variation of K_{BOX} in MS-MD FDSOI

Structure	OFF current	ON current	I_{ON}/I_{OFF}	Threshold voltage (V)	g_m (mS)	Sub threshold slope (mV/decade)	DIBL (mV/V)	Intrinsic delay (p (s))	Static power dissipation (nW)
SiO_2	1.53×10^{-8}	1.41×10^{-3}	9.13×10^{4}	0.03276	3.67587	78.40	86.61	0.58254	10.77657
Si_3N_4	3.74×10^{-9}	1.30×10^{-3}	3.48×10^{5}	0.03378	3.58372	71.88	73.06	0.5424	2.61909
HfO_2	6.59×10^{-10}	1.18×10^{-3}	1.79×10^{6}	0.03382	3.39685	65.33	65.30	0.5314	0.46147

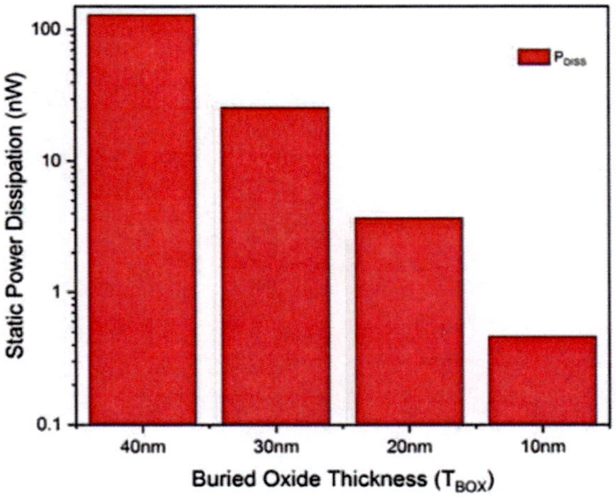

Fig. 8 Variation of static power dissipation versus buried oxide thickness in case of HfO_2

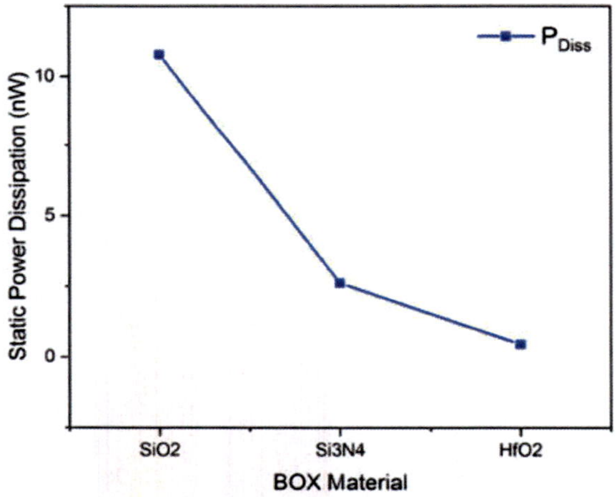

Fig. 9 Variation of static power dissipation versus buried oxide material

Fig. 10 Variation in intrinsic gate delay against K_{BOX} material

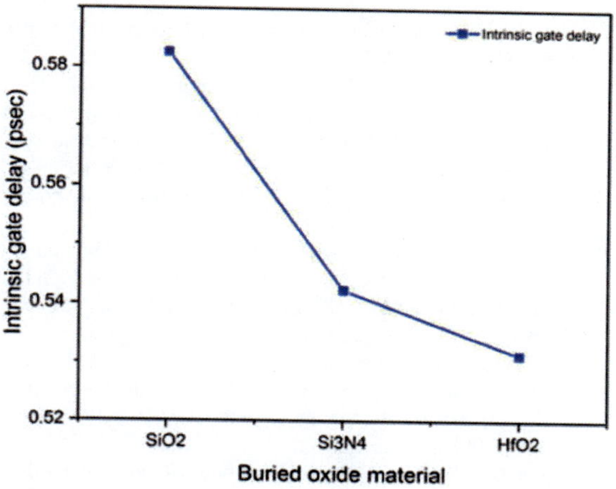

Fig. 11 Transconductance versus gate voltage plot with variation in T_{BOX} thickness

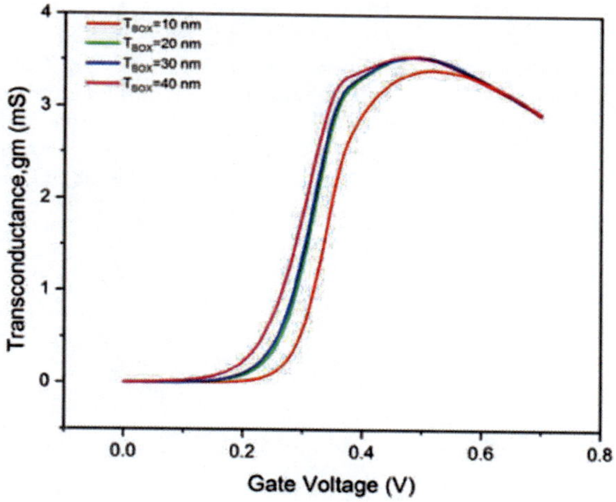

References

1. Dennard RH, Gaensslen FH, Yu H-N, Rideout VL, Bassous E, LeBlanc AR (1974) Design of ion-implanted MOSFET's with very small physical dimensions. IEEE J Solid-State Circuits 9(5):256–268
2. Davari B, Dennard RH, Shahidi G (1995) CMOS scaling for high performance and low power-the next ten years. Proc IEEE 83(4):595–606
3. (2008) 9th International conference on solid-state and integrated-circuit technology. IEEE Electron Device Lett 29(5):528
4. Narendar V, Mishra RA (2015) Analytical modeling and simulation of multigate FinFET devices and the impact of high-k dielectrics on short channel effects (SCEs). Superlattices Microstruct 85:357–436

5. Veeraraghavan S, Fossum JG (1989) Short-channel effects in SOI MOSFETs. IEEE Trans Electron Devices 36(3):522–528
6. Saremi M, Afzali-Kusha A, Mohammadi S (2012) Ground plane fin-shaped field effect transistor (GP-FinFET): a FinFET for low leakage power circuits. Microelectron Eng 95:74–82
7. Biswas K, Sarkar A, Sarkar CK (2018) Fin shape influence on analog and RF performance of junctionless accumulation-mode bulk FinFETs. Microsyst Technol 24(5):2317–2324
8. Young KK (1989) Short-channel effect in fully depleted SOI MOSFETs. IEEE Trans Electron Devices 36(2):399–402
9. Nowak EJ, Aller I, Ludwig T, Kim K, Joshi RV, Chuang C-T, Bernstein K, Puri R (2004) Turning silicon on its edge. IEEE Circ Devices Mag 20(1):20–31
10. Suzuki K, Pidin S (2003) Short-channel single-gate SOI MOSFET model. IEEE Trans Electron Devices 50(5):1297–1305
11. Nakajima Y, Tomita H, Aoto K, Ito N, Hanajiri T, Toyabe T, Morikawa T, Sugano T (2003) Characterization of trap states at silicon-on-insulator (SOI)/buried oxide (BOX) interface by back gate transconductance characteristics in SOI MOSFETs. Jpn J Appl Phys 42(4B), Part 1:2004–2008
12. Mukherjee S, Roy S, Dutta A, Sarkar CK (2016) Study on effect of back oxide thickness variation in FDSOI MOSFET on analogue circuit performance. IET Circuits Devices Syst 10(6):497–503
13. Mishra VK, Rao N (2020) Electrostatic investigation of intended source drain ultra-thin body FD-SOI MOSFET. SILICON 12(12):2819–2827
14. (2016) ATLAS user's manual. SILVACO, Santa Clara, CA, USA
15. Ranka DK, Rana A, Kumar Yadav R, Giri D (2011) Performance analysis of FD-SOI MOSFET with different gate spacer dielectric. Int J Comput Appl 18(5):22–27
16. Priya A, Srivastava NA, Mishra RA (2018) Perspective of buried oxide thickness variation on triple metal-gate (TMG) recessed-S/D FD-SOI MOSFET. Adv Electr Electron Eng 16(3)

Silicon on Insulator-Based Ultra-Small Micro-Ring Resonator for Temperature Sensing

Rukmani Singh, Rupam Srivastava, Y. K. Prajapati, and Dharmendra Kumar

Abstract In this paper, an ultra-small silicon on insulator (SOI)-based ring resonator has been rigorously analyzed for the temperature sensing. The computational analysis has been performed using finite element method (FEM)-based COMSOL Multiphysics. Temperature sensing has been done by measuring the variation in resonant wavelength due to alteration in temperature because of thermo-optic effect. The presented device is showing sensitivity of 106 pm/°C or 442 nm/RIU for 400 nm waveguide width on a very small footprint area with radius of 1.8785 μm. Performance of the device is also surveyed by varying the width of waveguide.

Keywords Optical temperature sensor · Ring resonator · Refractive index sensor · Silicon photonics

1 Introduction

In recent years, silicon photonics is emerging as very effective photonic integration platform. It is due to compatibility with CMOS fabrication process and high refractive index contrast. As a result, it provides advantages such as reliable fabrication, low-cost production, and high volume. The demand for sensors based on photonic integrated circuits has been increased from the last few years because these sensors offer several advantages such as high sensitivity, cost effectiveness, and reliability [1]. The sensors based on photonic integrated circuit are widely used in different fields

R. Singh (✉) · D. Kumar
Department of Electronics Engineering, Indian Institute of Technology (Indian School of Mines), Dhanbad 826004, Jharkhand, India
e-mail: rukmani.singh0@gmail.com

R. Singh
Department of Electronics and Communication Engineering, Madan Mohan Malviya University of Technology, Gorakhpur 273010, Uttar Pradesh, India

R. Srivastava · Y. K. Prajapati
Department of Electronics and Communication Engineering, Motilal Nehru National Institute of Technology Allahabad, Prayagraj 211002, Uttar Pradesh, India

B. Mishra and M. Tiwari (eds.), *VLSI, Microwave and Wireless Technologies*, Lecture Notes in Electrical Engineering 877,
https://doi.org/10.1007/978-981-19-0312-0_74

such as point-of-care applications, environmental monitoring systems, and food diagnostics along with temperature sensing and pressure sensing applications [2]. The mostly used efficient technology for photonic integrated circuits is the silicon on insulator (SOI). It gives the facilities of mass production and monolithic integration of electronic and photonic circuits, which are known as electronic photonic integrated circuits (EPIC) [3]. Due to the EPIC facility, detectors, sensors, light sources, and read-out electronics can be integrated in a single chip [4–6]. The photonic sensors based on SOI technology can be implemented mainly using interferometric or resonant structures. The interferometric structures are generally based on Mach–Zehnder interferometer configuration [7], and the resonant structures are realized using ring resonator [8]. Mach–Zehnder interferometer-based sensor gives output in terms of phase shift while ring resonator-based sensor works on the principle of resonant wavelength shift. Mach–Zehnder interferometer offers temperature stability more than ring resonator but provides lower sensitivity [9, 10]. Ring resonator provides facility for a dense sensor integration due to its advantage of high sensitivity and small footprint, compared to Mach–Zehnder interferometer [11–13].

Nowadays, accurate sensing of temperature has become a very important issue because there are many fields where accurate, fast, and compact device is required for temperature sensing [14]. Some of the important fields where accurate temperature sensing is required are medical, manufacturing and automobiles, environmental control, electronic chips, etc. Optical temperature sensors are being used frequently rather than conventional electrical sensors because they are immune to electromagnetic interference, highly sensitive, and also robust [15, 16]. Numerous optical methods have been presented in last few years to measure the temperature variation such as fiber Bragg grating [16] and Mach–Zehnder Interferometer [9]. Couples of optical temperature sensor based on the micro-ring resonator have been presented in last few years [15, 17]. Ring resonator-based temperature sensor can be fabricated using standard CMOS technology [17].

In this work, a photonic sensor based on ring resonator using SOI technology is demonstrated which is being used for temperature sensing due to thermo-optic effect. Ring resonator is formed with silicon waveguide on a buried oxide substrate of SiO_2. The dimensions of ring resonator are considered such that it can be implemented in very small area on the photonics chip with maintaining ease of fabrication. The radius of ring is chosen by performing the parametric optimization to satisfy the resonant condition on 400 nm width of waveguide and 1550 nm of resonant wavelength. When the ring resonator is subjected to temperature variations, there is change in the refractive index of silicon. The refractive index of silicon varies because of thermo-optic effect. Thus, by considering the change in refractive index due to temperature variation, shifting of resonant wavelength is measured to make the application of ring resonator effective in temperature sensing. The performance of device is also analyzed by varying the waveguide width, and it is obtained that the increment in waveguide width improves the device sensitivity.

It can also be implemented effectively as a biosensor, where temperature sensing is required to diagnose the diseases.

2 Theory

2.1 Operating Principle of Ring Resonator

A ring resonator consists of a bus waveguide and a ring waveguide, which are made of silicon on a silica substrate. When the light is given to the input port of bus/rectangular waveguide, a fraction of light is coupled to the ring resonator if the resonance condition is matched, while others are passed. The resonance peak at the output spectrum is observed, when the resonance condition meets and maximum light couples through ring waveguide. Generally, it acts like a notch filter. There are several important spectral characteristics of ring resonator which are used to describe its property such as resonant wavelength, FSR (free spectral range), finesse, FWHM (full width at half maximum), and quality factor [18, 19]. The resonant wavelength (λ_0) is the wavelength where light couples maximally to the ring waveguide from bus waveguide, and it can be given as Eq. (1).

$$\lambda_0 = \frac{L.n_{\text{eff}}}{k} \tag{1}$$

where 'n_{eff}' is effective refractive index of the waveguide, 'L' is length of cavity, which is $2\pi r$ for the ring cavity having 'r' as the radius, and 'k' is an integer which is used to define the resonance order.

The free spectral range can be defined as the spacing between two neighboring resonant wavelengths and expressed as Eq. (2).

$$\text{FSR} = \frac{\lambda_0^2}{L * n_g} \tag{2}$$

where 'n_g' is the group index.

The finesse is the ratio of FSR to full width at half maximum (FWHM). So, it can also be used to measure the resonance 'sharpness' relative to their spacing as shown in Eq. (3).

$$\text{Finesse} = \frac{\text{FSR}}{\text{FWHM}} \tag{3}$$

The quality factor can be defined as the ratio of resonant wavelength to the 3 dB bandwidth (BW) as shown in Eq. (4). It is very important parameter which affects the performance of ring resonator.

$$Q = \frac{\text{Resonant wavelength}}{\text{BW}} \tag{4}$$

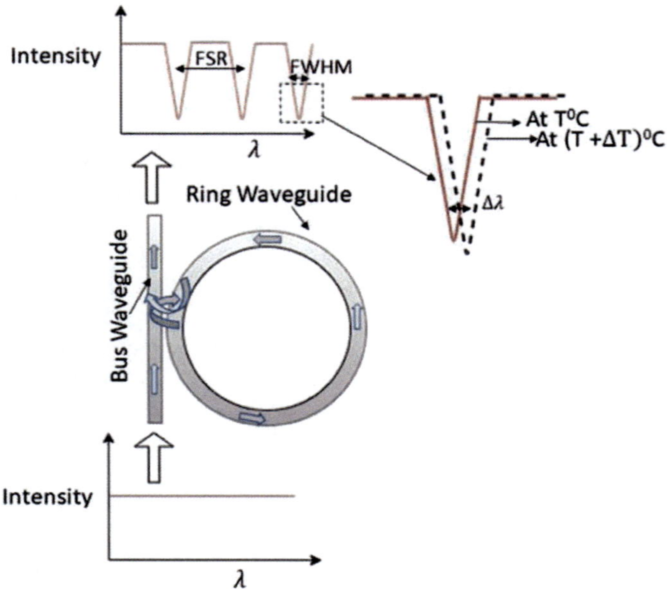

Fig. 1 Working principle of ring resonator as a temperature sensor

The working principle of SOI-based ring resonator to sense the temperature change is shown in Fig. 1. The dotted line shows the spectral response of ring resonator after alteration in temperature by ΔT°C.

2.2 Temperature Sensing Mechanism

The variation in resonant wavelength due to temperature caused by thermo-optic (TO) effect ($\Delta\lambda_T$) and thermal expansion effect ($\Delta\lambda_L$) is expressed by Eq. (5) [17]. TO effect changes the refractive index of the ring resonator, when temperature is varied, while thermal expansion effect expands the circumference of the ring with increment in temperature.

$$\Delta\lambda = \Delta\lambda_L + \Delta\lambda_T = \alpha_W \frac{n_{\text{eff}}}{n_g}\lambda\Delta T + \frac{\sigma_T}{n_g}\lambda\Delta T \quad (5)$$

$$\sigma_T = \frac{\partial n_{\text{eff}}}{\partial T} \quad (6)$$

where 'n_g' is the group index and 'n_{eff}' is the effective refractive index of the waveguide, 'ΔT' is the variation in temperature, 'α_W' is the coefficient of thermal expansion

(CTE), and 'σ_T' is the rate of change of n_{eff} with temperature. 'σ_T' depends on the thermo-optic coefficient of Si and SiO_2. The CTE of the silicon is ~2.6×10^{-6}/°C [17], and the thermo-optic coefficients of the Si and SiO_2 are 1.86×10^{-4}/°C and 1.0×10^{-5}/°C, respectively [17, 20, 21]. It can be observed that the thermo-optic coefficient of Si and SiO_2 is approximately 100 times than the CTE of Si. Thus, the variation in resonant wavelength is dominated by TO effect, i.e., thermal expansion effect can be ignored during the calculation of temperature sensitivity. TO effect is considered while thermal expansion effect is ignored in this paper for the analysis.

The relation between temperature variation and refractive index considering TO effect can be given by Eq. (7) [22].

$$n = n_0 + \alpha \Delta T \tag{7}$$

In the above equation, $n_0 = 3.4$ is the refractive index of silicon at 0 °C temperature, and 'α' is the thermo-optic coefficient, which is taken 2.4×10^{-4} for Si [22, 23].

The temperature sensitivity can be expressed as nm/°C as given in Eq. (8) in terms of change in resonant wavelength '$\Delta\lambda$' and the variation in temperature 'ΔT.'

$$S_T = \frac{\Delta\lambda}{\Delta T} \tag{8}$$

The refractive index sensitivity of device can be measured by Eq. (9), where Δn is change in refractive index.

$$S_n = \frac{\Delta\lambda}{\Delta n} \tag{9}$$

3 Computational Structure of Device

The structure of ring resonator based on SOI technology is analyzed here for temperature sensing. The computational cross-sectional structure of the device is shown in Fig. 2a. The ring resonator consists of a rectangular silicon waveguide and a circular silicon waveguide of width 'w_{core}' = 400 nm which is made on silica substrate layer having width 'w_{sub}' = 1000 nm and refractive index 1.44. From Eq. (1), it can be observed that radius of ring plays very important role in finding the resonance condition. The theoretical value of radius of the ring 'r_0' is calculated ~1.8864 μm for $k = 26$ and $\lambda_0 = 1550$ nm from Eq. (1). Since ring has non-zero width, this could not be exact value of ring radius for resonating condition. Hence, the resonant ring radius of the device is found out by parametric optimization with help of simulation using COMSOL. The radius is varied from 1870 to 1890 nm, and corresponding transmittance at output port is measured as shown in Fig. 2b. The radius of the ring 'r_0' is

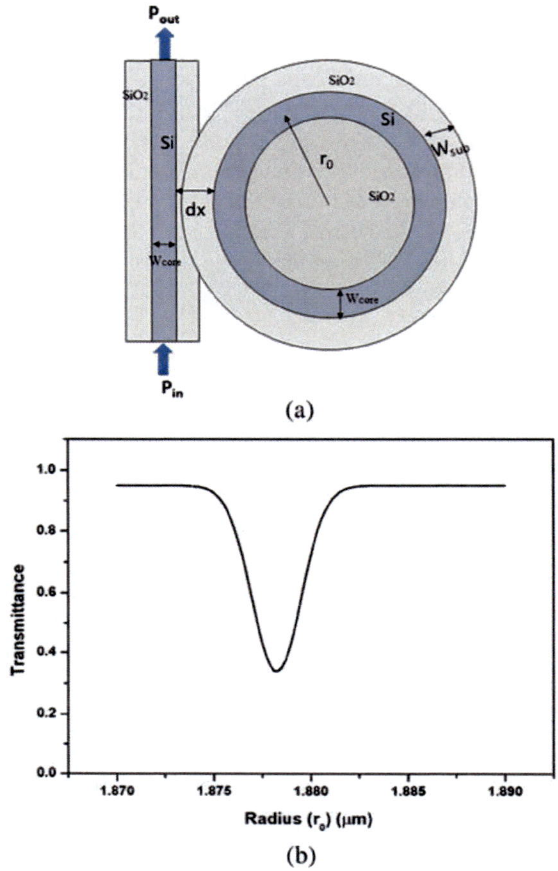

Fig. 2 **a** Computational model of ring resonator. **b** Parametric optimization of ring radius of ring resonator

chosen 1.8785 μm to provide maximum coupling of light from rectangular waveguide to circular ring waveguide. The separation between rectangular and circular waveguide '*dx*' is taken of 100 nm.

4 Result and Discussion

The ring resonator-based sensor is operating based on the variation of refractive index due to thermo-optic effect. Here, we have varied the temperature '*T*' from 0 °C to 20 °C in step of 5 °C, and corresponding refractive index of silicon is obtained from Eq. (7). Variation in refractive index of silicon changes the effective refractive index of the waveguide; hence, the coupling condition of ring resonator is changed according to Eq. (1); and consequently, resonant wavelength is shifted. The shifting in resonant wavelength is measured, and sensitivity of device is calculated by using Eqs. (8) and (9). Two-dimensional FEM-based COMSOL Multiphysics is used for the simulation. Scattering boundary condition and tetrahedral meshing are used with maximum size of 'w_{core}/10.' The coupling of light through the ring resonator is shown by the surface electric field propagation as displayed in Fig. 3a, which is obtained by simulation on COMSOL Multiphysics. Figure 3a shows maximum light couples from bus waveguide to ring resonator for the considered dimensional parameters. The shifting in spectral response of normalized transmission with variation in wavelength is shown in Fig. 3b. It can be noted that as temperature is increasing, the resonant wavelength is shifting toward right. The shifting takes place because of increment in temperature, which leads to rise in refractive index of silicon, and hence, resonant wavelength increases as given in Eq. (1). The measurement of shifting in resonant wavelength due to temperature or refractive index variation of silicon is shown by sensitivity. The sensitivity is calculated by using Fig. 3c and Eq. (8) in terms of nm/°C, and device sensitivity in terms of nm/RIU is calculated using Fig. 3d and Eq. (9).

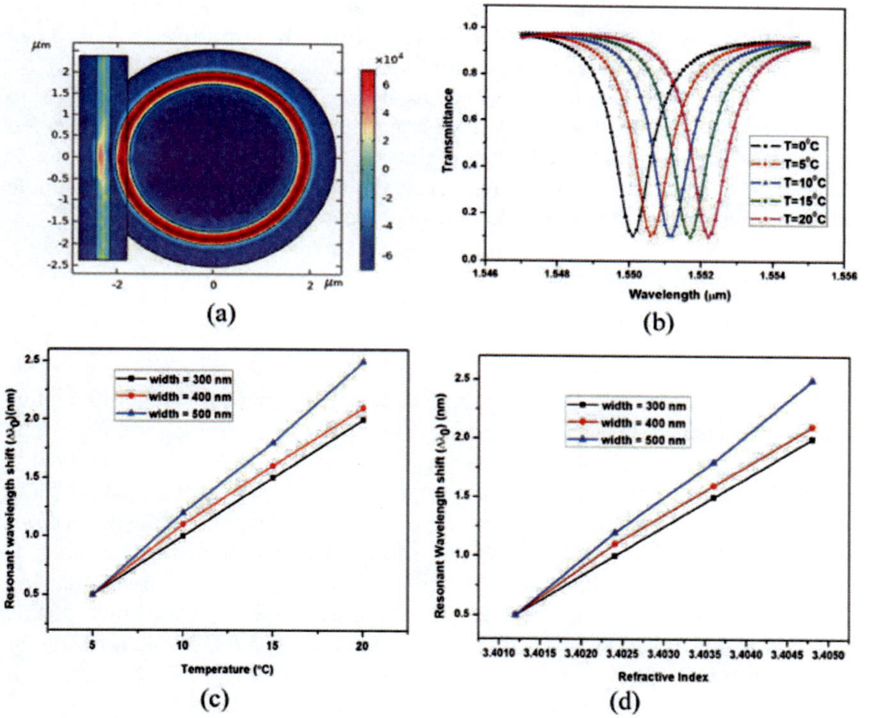

Fig. 3 Simulation and sensitivity calculation: **a** surface electric field propagation through the device. **b** Spectral response of device: normalized transmitted intensity versus wavelength (μm). **c** Temperature (°C) versus resonant wavelengths shift ($\Delta\lambda_0$) (nm). **d** Refractive index of silicon corresponding to different temperature versus resonant wavelength shift ($\Delta\lambda_0$) (nm)

Table 1 Comparison of characteristics with previous published works

Presented papers	Complexity of whole structure	Radius of ring	Sensitivity	Waveguide width
Ref. [17]	Simple	4 μm	83 pm/°C	500 nm
Ref. [24]	Very complex (10 rings)	5 μm	130 nm/°C	450 nm
Ref. [25]	Complex (2 rings)	271.94 μm of ring 1 232.23 μm of ring 2	229.6 pm/°C	350 nm of ring 1 450 nm of ring 2
Ref. [14]	Simple	3 μm	0.2 nm/°C	0.43 μm
Ref. [26]	Simple	4 μm	85 pm/°C	500 nm
This work	Simple	1.8785 μm	106 pm/°C	400 nm

The sensitivity of device is 106 pm/°C or 442 nm/RIU on small footprint area with radius of 1.8785 μm and waveguide core width of 400 nm on photonics chip. The shift in resonant wavelengths is also examined for variations in waveguide width of the ring. The device sensitivities are found 100 pm/°C or 416 nm/RIU and 125 pm/°C or 525 nm/RIU for 300 nm and 500 nm waveguide width, respectively. Thus, it can be concluded that increment in core width improves the device sensitivity.

Various optical temperature sensors based on SOI ring resonator have been proposed in the past few years, which are compared in Table 1. It can be observed that better sensitivity is achieved either at the cost of complexity of the structure or requirement of larger area. But in this paper, the demonstrated structure is simple with very small radius of the ring along with comparable sensitivity.

5 Conclusion

Here, we have demonstrated an ultra-small SOI micro-ring resonator in effective temperature sensing. Device is operating based on the thermo-optic effect. The variations in temperature influence the refractive index of device, consequently, the coupling condition of device is changed, and thus resonant wavelength shifts accordingly. The simulations have been performed using finite element method-based 2-dimensional COMSOL Multiphysics. A small radius of ring resonator is obtained by parametric optimization to achieve large coupling. Device is showing effective temperature sensing with very small radius of 1.8785 μm and waveguide width of 400 nm. Sensitivity of device is achieved as 106 pm/°C or 442 nm/RIU for 400 nm waveguide width, while it is 100 pm/°C or 416 nm/ RIU and 125 pm/°C or 525 nm/RIU for 300 nm and 500 nm waveguide widths, respectively. The presented device structure can be used in the applications like point of care, diagnosis, food industry, etc. wherever small chip area is required for the temperature sensing with requirement of accuracy and high sensitivity.

Acknowledgements This work was supported by National Project implementation Unit—Ministry of Human Resource Development (MHRD), Government of India via TEQIP Collaborative Research Scheme (ID-1-5764084542).

References

1. Thomson D, Zilkie A, Bowers JE, Komljenovic T, Reed GT, Vivien L, Marris-Morini D, Cassan E, Virot L, Fédéli J-M, Hartmann J-M, Schmid JH, Xu D-X, Boeuf F, O'Brien P, Mashanovich GZ, Nedeljkovic M (2016) Roadmap on silicon photonics. J Opt 18:073003. https://doi.org/10.1088/2040-8978/18/7/073003
2. Steglich P, Hülsemann M, Dietzel B, Mai A (2019) Optical biosensors based on silicon-on-insulator ring resonators: a review. Molecules 24:519. https://doi.org/10.3390/molecules24030519
3. Knoll D, Lischke S, Barth R, Zimmermann L, Heinemann B, Rucker H, Mai C, Kroh M, Peczek A, Awny A, Ulusoy C, Trusch A, Kruger A, Drews J, Fraschke M, Schmidt D, Lisker M, Voigt K, Krune E, Mai A (2015) High-performance photonic BiCMOS process for the fabrication of high-bandwidth electronic-photonic integrated circuits. In: 2015 IEEE international electron devices meeting (IEDM). IEEE, pp 15.6.1–15.6.4.https://doi.org/10.1109/IEDM.2015.7409706
4. Zhou Z, Yin B, Michel J (2015) On-chip light sources for silicon photonics. Light Sci Appl 4:e358–e358. https://doi.org/10.1038/lsa.2015.131
5. Sakib MNN, Sun J, Kumar R, Driscoll J, Rong H (2018) Demonstration of a 50 Gb/s all-silicon waveguide photodetector for photonic integration. In: Conference on lasers and electro-optics. OSA, Washington, D.C., p JTh5A.7. https://doi.org/10.1364/CLEO_AT.2018.JTh5A.7
6. Zang K, Zhang D, Huo Y, Chen X, Lu C-Y, Fei ET, Kamins TI, Feng X, Huang Y, Harris JS (2015) Microring bio-chemical sensor with integrated low dark current Ge photodetector. Appl Phys Lett 106:101111. https://doi.org/10.1063/1.4915094
7. Densmore A, Xu D-X, Janz S, Waldron P, Mischki T, Lopinski G, Delâge A, Lapointe J, Cheben P, Lamontagne B, Schmid JH (2008) Spiral-path high-sensitivity silicon photonic wire molecular sensor with temperature-independent response. Opt Lett 33:596. https://doi.org/10.1364/OL.33.000596
8. Steglich P, Villringer C, Pulwer S, Heinrich F, Bauer J, Dietzel B, Mai C, Mai A, Casalboni M, Schrader S (2017) Hybrid-waveguide ring resonator for biochemical sensing. IEEE Sens J 17:4781–4790. https://doi.org/10.1109/JSEN.2017.2710318
9. Zhang Y, Zou J, He J-J (2018) Temperature sensor with enhanced sensitivity based on silicon Mach-Zehnder interferometer with waveguide group index engineering. Opt Express 26:26057. https://doi.org/10.1364/OE.26.026057
10. Sakamoto H, Minpou Y, Sawai T, Enami Y, Suye S (2016) A novel optical biosensing system using Mach–Zehnder-type optical waveguide for influenza virus detection. Appl Biochem Biotechnol 178:687–694. https://doi.org/10.1007/s12010-015-1902-x
11. Rodriguez GA, Hu S, Weiss SM (2015) Porous silicon ring resonator for compact, high sensitivity biosensing applications. Opt Express 23:7111. https://doi.org/10.1364/OE.23.007111
12. Sarkaleh AK, Lahijani BV, Saberkari H, Esmaeeli A (2017) Optical ring resonators: a platform for biological sensing applications. J Med Signals Sens 7:185–191
13. Ajad AK, Islam MJ, Kaysir MR, Atai J (2021) Highly sensitive bio sensor based on WGM ring resonator for hemoglobin detection in blood samples. Optik 226:166009. https://doi.org/10.1016/j.ijleo.2020.166009
14. Singha S, Bhowmik BB (2018) On-chip photonic temperature sensor using micro ring resonator. In: 2018 fifteenth international conference on wireless and optical communications networks (WOCN). IEEE, pp 1–4.https://doi.org/10.1109/WOCN.2018.8556121

15. Xu H, Hafezi M, Fan J, Taylor JM, Strouse GF, Ahmed Z (2014) Ultra-sensitive chip-based photonic temperature sensor using ring resonator structures. Opt Express 22:3098. https://doi.org/10.1364/OE.22.003098
16. Klimov NN, Mittal S, Berger M, Ahmed Z (2015) On-chip silicon waveguide Bragg grating photonic temperature sensor. Opt Lett 40:3934. https://doi.org/10.1364/OL.40.003934
17. Kim G-D, Lee H-S, Park C-H, Lee S-S, Lim BT, Bae HK, Lee W-G (2010) Silicon photonic temperature sensor employing a ring resonator manufactured using a standard CMOS process. Opt Express 18:22215. https://doi.org/10.1364/OE.18.022215
18. Bogaerts W, De Heyn P, Van Vaerenbergh T, De Vos K, Kumar Selvaraja S, Claes T, Dumon P, Bienstman P, Van Thourhout D, Baets R (2012) Silicon microring resonators. Laser Photonics Rev 6:47–73. https://doi.org/10.1002/lpor.201100017
19. Xue C, Jin Y, Tong X, Cui D, Yan S, Zhang W (2011) Quality factor of silicon-on-insulator integrated optical ring resonator. Presented at the 8 Sept 2011.https://doi.org/10.1117/12.891814
20. Okada Y, Tokumaru Y (1984) Precise determination of lattice parameter and thermal expansion coefficient of silicon between 300 and 1500 K. J Appl Phys 56:314–320. https://doi.org/10.1063/1.333965
21. Teng J, Dumon P, Bogaerts W, Zhang H, Jian X, Han X, Zhao M, Morthier G, Baets R (2009) Athermal silicon-on-insulator ring resonators by overlaying a polymer cladding on narrowed waveguides. Opt Express 17:14627. https://doi.org/10.1364/OE.17.014627
22. Fu H, Zhao H, Qiao X, Li Y, Zhao D, Yong Z (2011) Study on a novel photonic crystal temperature sensor. Optoelectron Lett 7:419–422. https://doi.org/10.1007/s11801-011-0065-4
23. Tinker MT, Lee J-B (2005) Thermal and optical simulation of a photonic crystal light modulator based on the thermo-optic shift of the cut-off frequency. Opt Express 13:7174. https://doi.org/10.1364/OPEX.13.007174
24. Min SS, Hao DJ, Shu Bin Y (2018) Optical multi-ring cascade cavity temperature sensor with ultra-high sensitivity. Int J Sens Netw Data Commun 08. https://doi.org/10.4172/2090-4886.1000155
25. Kim H-T, Yu M (2016) Cascaded ring resonator-based temperature sensor with simultaneously enhanced sensitivity and range. Opt Express 24:9501. https://doi.org/10.1364/OE.24.009501
26. Kim G-D, Lee H-S, Lee S-S, Lim B-T, Bae H-K, Lee W-G (2009) Ultra small silicon resonator based temperature sensor. In: Frontiers in optics 2009/laser science XXV/fall 2009 OSA optics & photonics technical digest. OSA, Washington, D.C., p FMJ6.https://doi.org/10.1364/FIO.2009.FMJ6

Study of Sum-Rate Performance of Cooperative Cognitive Radio Networks Using MIMO

Champa Tanga, Joyatri Bora, Md. Anwar Hussain, and Karik Lego

Abstract As advancement in wireless communication became a new paradigm and challenges, this paper presents the advantages of multiple input multiple output (MIMO) in cooperative cognitive radio networks (CCRN), where primary users (PU) select some secondary users (SUs) with multiple antennas to transmit thePU data and simultaneously allowing the channel to transmit SU data. We have studied the average sum-rate performance of CCRN, which consists of MIMO secondary users (SU) network that operates in the range of multiple primary users (PUs). Next, the system average sum-rate performance of MIMO in CCRN with a highly scattering Rayleigh fading channel has been used in this study. This paper further analyzed the different linear precoding techniques such as SVD and ZF with varying the distance and signal to noise ratio power by keeping other parameters constant. It is found that SVD precoding techniques have given better sum-rate performance compared with the zero forcing beamforming precoding techniques.

Keywords Multiple inputs multiple output · Cooperative cognitive radio networks · Sum rate · SVD · ZF

1 Introduction

In recent wireless communication, the demand for the high-speed data is increasing very rapidly, and the capacity of the spectrum is becoming more deficits to fulfill the required demand of latest Internet of things [1, 2]. Furthermore, radio spectrum

C. Tanga (✉)
Department of Electronics and Communication, Rajiv Gandhi University, Doimukh 791102, India
e-mail: champa.tanga@gmail.com

J. Bora · Md. A. Hussain · K. Lego
Department of Electronics and Communication Engineering, NERIST, Nirjuli 791109, India
e-mail: jb@nerist.ac.in

Md. A. Hussain
e-mail: ah@nerist.ac.in

B. Mishra and M. Tiwari (eds.), *VLSI, Microwave and Wireless Technologies*,
Lecture Notes in Electrical Engineering 877,
https://doi.org/10.1007/978-981-19-0312-0_75

demand is increasing rapidly and making the deficient radio spectrum even more congested. As reported by the federal communication commission (FCC), most of the assigned spectra are underutilized. But lately, the FCC has been considering the more versatile and complete use of the available spectrum [3], by introducing the use of cognitive radio technology [4].

Cognitive radio improves the spectrum efficiency by making use of unused spectrum in dynamically changing environment. To enhance the efficiency of spectrum of wireless systems, it was found that the cognitive radio with the cooperative relay is a more potential area [5]. With the progress of cooperative communication, many complex issues were solved, viz., loss of rate to the cooperating mobile, average interference, and handoff in the network. In [6], a new paradigm has been coined as cooperative cognitive radio network. Here, the primary users recruit some secondary users to relay the primary data, and in return, some portions of channel access time are granted by which both the primary users and secondary users get benefited to transmit their own data. This process of sharing the channel access time creates a win–win situation. In CCRN framework, there are three phase. Where the SUs are confined to the third phase for transmission and PUs have to lease the channel completely for secondary users' transmission. Due to this, the throughput gain of both PUs and SUs is limited.

On the other side, MIMO is the latest technology that enables many antennas to be used for transmissions of data independently in the spatial domain to improve the channel capacity. In [7–10], the MIMO system was studied and found that the number of antennas will increase the capacity of the system. To overcome the problem in [3], a novel design was proposed, MIMO in CCRN in [11–14]. Multiple antenna SUs and primary users' recruit the secondary relays to improve channel capacity. The frame duration is divided in two phase. In phase one, the primary users will broadcast the data to secondary users. The secondary users also receive and transmit own data to other secondary users. In phase two, secondary users relay the primary data to primary receivers, and at the same time, secondary users transmit own data to other secondary users. In [15], the author has proposed a joint user antenna along with antenna selection algorithm and also considered different precoding techniques such as ZFBF, MMSE, and MRT for a range of SNR and studied the sum rate of the system.

In this paper, we have simulated the two primary users and four secondary users equipped with three MIMO antennas and studied the average sum-rate performance by varying the different parameters for two precoding techniques such as SVD and ZF. The rest of paper is presented as in Sect. 2, the system model is described, and in Sect. 3, the precoding techniques are discussed, and further in Sect. 4, the simulation results followed with conclusions.

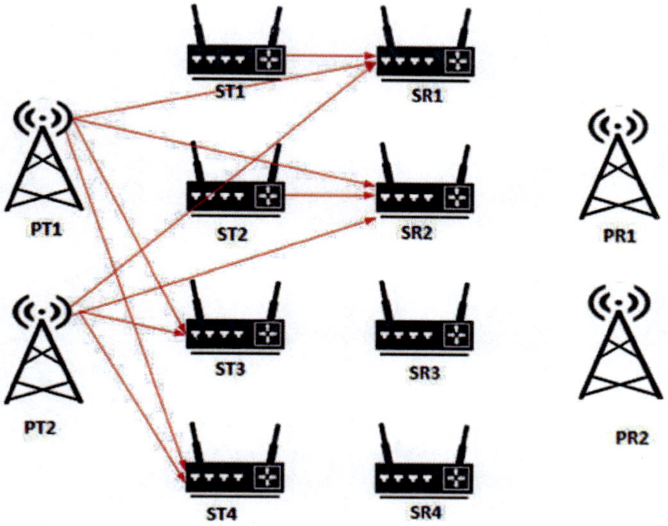

Fig. 1 First phase transmission

2 System Model

The system model consists of two transmitter–receiver pairs of primary user (PU). Between these primary transmitter receiver pairs, there are four transmitter–receiver pairs of secondary users (SU). The every secondary user is equipped with three pairs of MIMO antenna. There are two single antenna primary users. The complete data transmission is divided into two phases.

In first phase, primary users will select some secondary users as a cooperative relay and broadcast its data continuously to the selected relay. And at the same time in a time division multiple access (TDMA) fashion, the selected secondary relay will access the channel for its own transmission [12] (Fig. 1).

In second phase, the selected secondary user transmitter all together directs the PU signals to the respective primary receivers, and at the same time, the pairs in the subset of secondary users will use the channel spectrum in a TDMA fashion (Fig. 2).

3 Prerequisite Techniques

Precoding is a process of weighting the channel coefficient with a precoding vector and successively encoding at the transmitter. By weighting the phase and amplitude at the transmitter side, they can focus the signal to a desired user at the receiver. Precoding uses the basic idea of beamforming for the multiple antenna transmission in wireless communications [16].

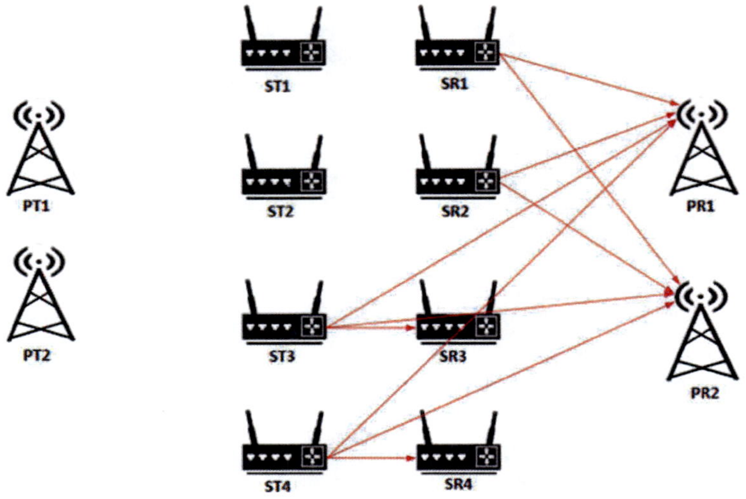

Fig. 2 Second phase transmission

3.1 Singular Value Decomposition (SVD)

In singular value decomposition, the channel coefficient matrix is decomposed to a singular value. To provide the possible signal path between the two transmitters and two receiver antennas, the channel coefficient numbers h_{11}, h_{12}, h_{21}, and h_{22} are there to show the possible data going through that path. If the path has larger value, that much portion of the total data is transmitted through that path. A channel information matrix is the matrix formed by the coefficient of these channel paths; this matrix represents the throughput of each of the possible path for transmission. The receiver and transmitter relationship are denoted as follows [16].

$$y = Hx$$
$$\begin{bmatrix} y_1 \\ y_2 \end{bmatrix} = \begin{bmatrix} h_{11} & h_{12} \\ h_{21} & h_{22} \end{bmatrix} \begin{bmatrix} x_1 \\ x_2 \end{bmatrix} \tag{1}$$

where y, H, and x are received data, channel information matrix, and the transmitted data, respectively. We have converted the channel information matrix into three matrices, namely U, S, and V with singular value decomposition. The operation of SVD is shown as follows.

$$
\begin{aligned}
y &= Hx \\
H &= \mathrm{USV}^H \\
y &= U^H(\mathrm{USV}^H)Vx \\
y &= \mathrm{ISI}x \\
y &= Sx
\end{aligned} \tag{2}
$$

The three matrix (USV^H) should be known to the receiver since the receiver can estimate it from channel matrix H of the received data and calculate these matrices from H.

3.2 Zero Forcing Beam Forming

The ZFBF techniques are used to nullify the undesired signals to avoid the interference at receiver side and increased the signal to noise ratio by forcing to form beams toward desired receivers. In transceiver, the beam forming can be done by proper coding on the signals [16].

4 Simulation Results

We have performed the simulation of the above described system model on MATLAB 2020a. In this study, simulation was performed for both uplink and downlink transmission by varying distance between the primary users and secondary users. Thereafter, analysis was performed between average sum rate (bits/Hz/s) versus signal to noise ratio (SNR dB) power. For downlink transmission, we have considered $n =$ 3.8 (path loss), and the thermal noise level is −96 dBm and also for observation, the downlink power is kept constant for the whole duration at 10 dB. Taking these parameters, we observe the result for two different type of precoding technique, i.e., SVD and ZF by varying the distance between the primary transmitter PT_1 and the secondary relay receiver in our system model. As we can see from Figs. 3 and 4, the graph shows that, if the distance between the primary transmitter and the secondary relay receiver is increased, the average sum rate decreases irrespective of precoding schemes. For this transmission, the SVD precoding gives the good performance as compared to ZF.

Figures 5 and 6 show the average sum rate of transmitters PT_1 to SR_1 and PT_1 to SR_2, respectively. It is observed that average sum rate is proportional to the transmission power. If the distance between the transmitter and receiver is fixed, the average sum rate increases with increase in power. Considering the channel noise level to be 1 and also assuming the distance between the primary transmitter and the secondary relay to be unity.

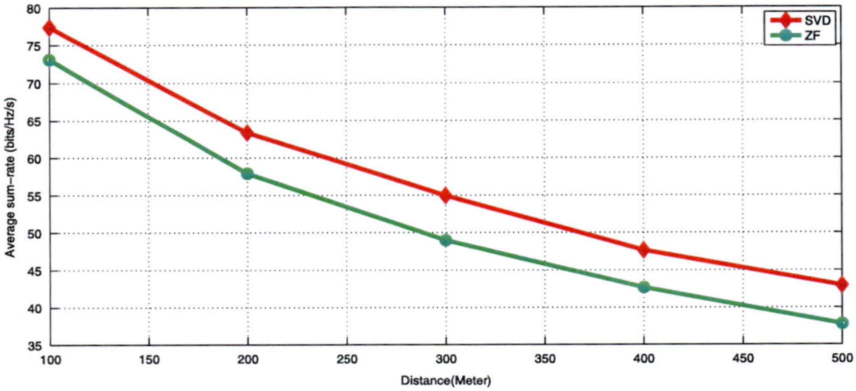

Fig. 3 Results show the average sum rate versus distance between PT_1 and SR_1

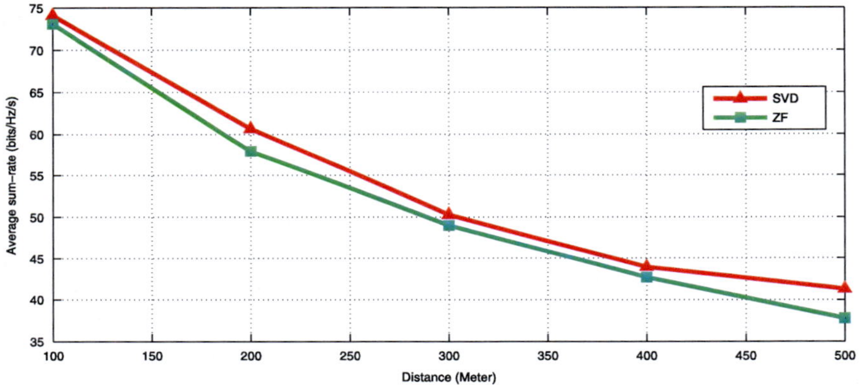

Fig. 4 Results of average sum rate versus distance between PT_1 and SR_2

For the uplink transmission, $n = 3.8$ (path loss) is taken, and the thermal noise level is kept at −96 dBm. From the Figs. 7 and 8, it can be observed that by increasing the distance between two secondary relays improves the average sum rate for SVD as compared to ZF precoding techniques. Also it can be seen that as the distance between primary transmitters and primary receivers increases, the average sum rate decreases proportionally.

Figures 9 and 10 show the comparison between the two precoding techniques and find average sum rate of primary transmitter 1 to the secondary relays (SR_1 to PR_2, SR_1 to PR_1) by varying the uplink power (SNR dB) from 5 to 25 dB for all precoding techniques and keep other parameters constant and also varying the distance between primary transmitter and primary receivers. The average sum rate

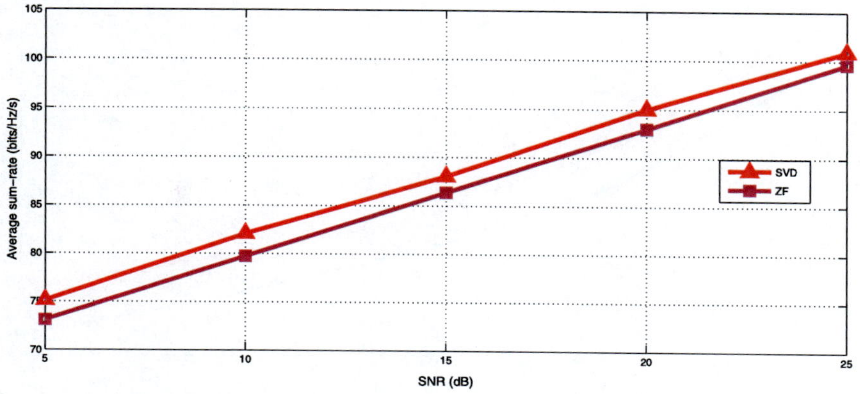

Fig. 5 Results of average sum rate versus SNR power PT_1 to SR_1

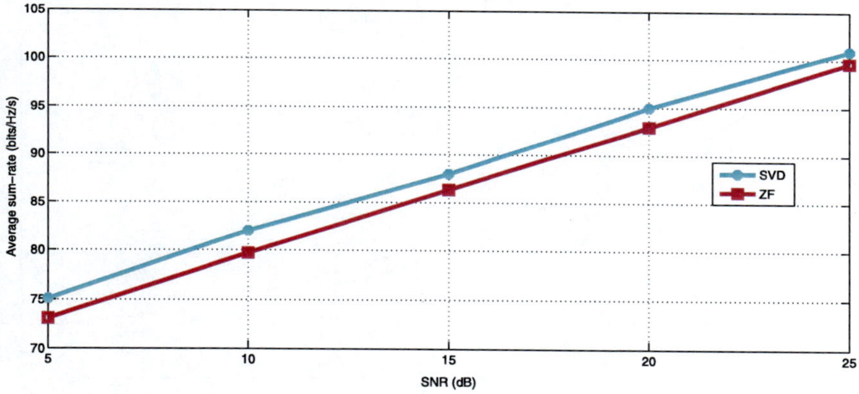

Fig. 6 Result of average sum rate versus SNR PT_1 to SR_2

increases as the power increases for all the precoding techniques but SVD gives the best performance in all the conditions.

5 Conclusion

In this paper, we have studied the average sum-rate performance of CCRN which consists of MIMO secondary users (SU) network that operate in the range of multiple primary users (PUs). The system average sum-rate performance of MIMO in CCRN with a highly scattering Rayleigh fading channel has been used in this study. This paper further analyzed the different linear precoding techniques such as SVD and ZF

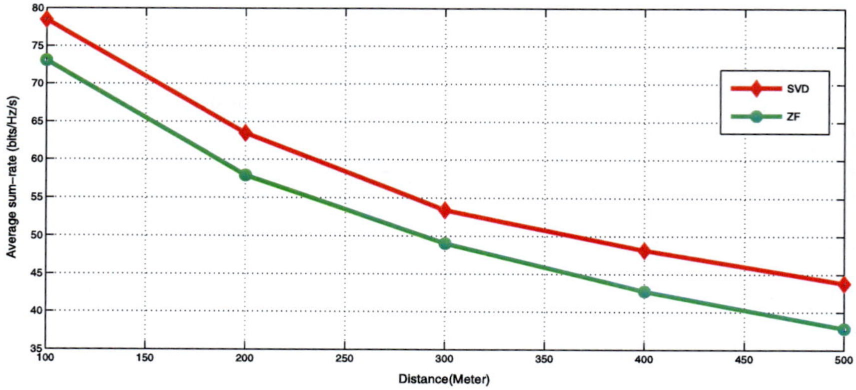

Fig. 7 Result of average sum rate versus distance between SR_1 and PR_1

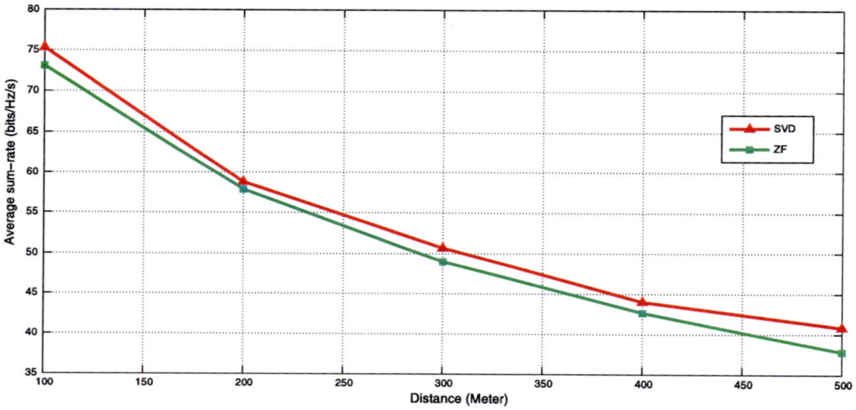

Fig. 8 Result of average sum rate versus distance between SR_1 and PR_2

with varying the distance and signal to noise ratio power by keeping other parameters constant. It is found that SVD precoding techniques have given the better average sum-rate performance compared with the zero forcing beam forming precoding techniques.

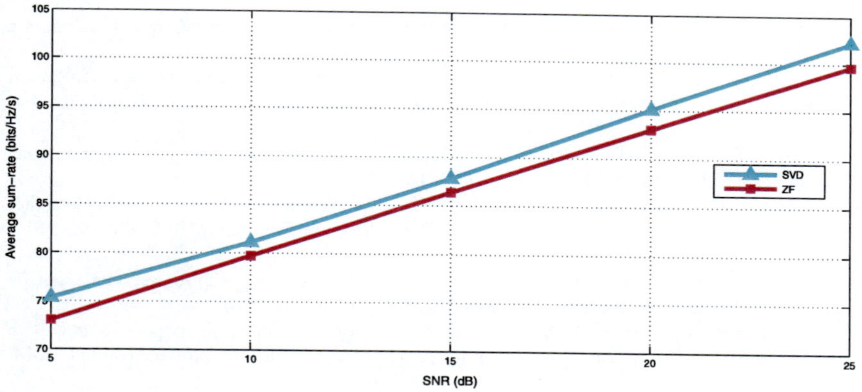

Fig. 9 Result of average sum rate versus SNR between SR_1 and PR_2

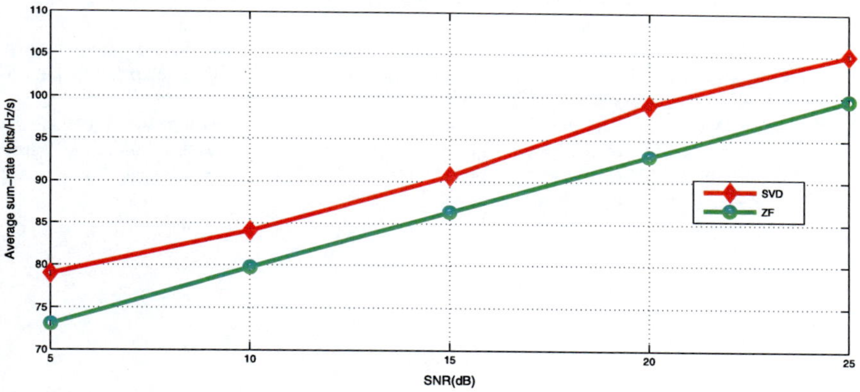

Fig. 10 Result of average sum rate versus SNR between SR_1 and PR_1

References

1. Cisco visual networking index: global mobile data traffic forecast update: 2013–2018, Cisco, Sanjose, CA, USA, February 2014
2. Mobile traffic forecasts 2010–2020 reports. UMTS Forum, Zurich, Switzerland, January 2011
3. Force S (2002) Spectrum policy task force report. Federal Communications Commission ET Docket 02, vol 135
4. Motila J (2000) Cognitive radio: an integrated agent architecture for software defined radio. P.Hd, Dissertation, KTH Royal Institute of Technology, Stockholm, Sweden
5. Akyildiz IF, Lo BF, Balakrishnan R (2011) Cooperative spectrum sensing in cognitive radio networks: a survey. Phys Commun 4(1):40–62. Elsevier, https://doi.org/10.1016/j.phycom.2010.12.003
6. Simeone O, Stanojev I, Savazzi S, Bar-Ness Y, Spagnolini, Pickholtz R (2008) Spectrum leasing to cooperating secondary ad hoc networks. IEEE J Sel Areas Commun 26:203–213

7. Telatar IE (1999) Capacity of multi-antenna Gaussian channels. Eur Trans Telecommun 10(6):585–595
8. Foschini G, Gans M (1998) On limits of wireless communications in a fading environment when using multiple antennas. Wirel Pers Commun 6:311–335. https://doi.org/10.1023/A:1008889222784
9. Zheng L, Tse DNC (2003) Diversity and multiplexing: a fundamental trade of multiple antenna channels. IEEE Trans Inf Theor 49(5):1037–1096
10. Mohamed E, Ahmad A (2015) Evaluation of MIMO system capacity over Rayleigh fading channel. Int J Comput Sci Eng Survey (IJCSES) 6:49–60. https://doi.org/10.5121/ijcses.2015.6305
11. Minchae J, Hwang K, Choi (2011) Interference minimization approach to precoding scheme in MIMO-based cognitive radio networks. IEEE Commun Lett 15(8)
12. Hua S, Liu H, Wu M, Panwar SS (2011) Exploiting multiple antennas in cooperative cognitive radio networks. IEEE Trans Veh Technol 63:2714–2722. https://doi.org/10.1109/INFCOM.2011.5935102
13. Hua S, Liu H, Zhuo X, Wu M, Pawar SS (2014) Exploiting multiple antennas in cooperative cognitive radio networks. IEEE Trans Veh Technol 63(7):3318–3330. https://doi.org/10.1109/TVT.2013.2297438
14. Ian, Lo F, Balakrishnan (2010) Cooperative spectrum sensing in cognitive radio networks : a survey. J Sel Topic Phys Commun 40–62. Elsevier
15. Sheikh TA, Bora J, Hussain MA Sum-rate performance of massive MIMO systems in highly scattering channel with semi-orthogonal and random user selection. Radioelectron Commun Syst 61:547. https://doi.org/10.3103/S0735272718120026
16. Busche H, Vanaev A, Rohling H (2009) SVD-based MIMO precoding and equalization schemes for realistic channel knowledge: design criteria and performance evaluation. Wirel Pers Commun 48:347–359. https://doi.org/10.1007/s11277-008-9526-x

Novel Power Gated (PG) and Sleep Body Bias (SBB) 6T CNTFET-Based SRAM Design for Ultra-Low-Power Application

Hemant Kumar, Balwinder Singh, Subodh Srivastava, Gulman Siddiqui, and Shekhar Yadav

Abstract The leakage power consumption accounts for progressively huge portion of average power consumption in nanometer regime. To limit the power dissipation, different low-power techniques are suggested. This article presents the design and performance investigation of 6T carbon nano-tube field effect transistor (CNTFET) static random access memory (SRAM) cell design by using power reduction technique, i.e., sleep approach and proposed SBB approach for ultra-low-power applications at 32 nm technology. Power reduction techniques can play vital role in significant improvement of performance of 6T CNTFET SRAM cell. By incorporating power-gated technique, in 6T CNTFET SRAM cell design, 84.43%, 79.48%, and 60.92% improvement in average power dissipation is achieved for sleep approach, sleep with header switch, and sleep with footer switch, respectively. Similarly, in 6T CNTFET SRAM memory cell, leakage power minimization of 25.34%, 13.18%, and 3.37% is observed for sleep approach, sleep with header, and sleep with footer techniques, respectively. Performance analysis for proposed sleep body bias (SBB) 6T CNTFET SRAM cell design shows that proposed design has significant improvements in delay (53.48%), average power consumption (56.86%), power delay product (PDP) (80%), and leakage power dissipation (4%) in comparison to PG CNTFET

H. Kumar (✉) · G. Siddiqui · S. Yadav
Department of Electronics, Banasthali Vidyapith, Vanasthali, Rajasthan, India
e-mail: hemantkumar@banasthali.in

G. Siddiqui
e-mail: mohdgulmansiddiqui@banasthali.in

S. Yadav
e-mail: yshekhar@banasthali.in

H. Kumar · S. Srivastava
Vivekananda Global University, Jaipur, Rajasthan, India
e-mail: srivastava.subodh@vgu.ac.in

B. Singh
Centre for Development of Advanced Computing, Mohali, Punjab, India
e-mail: balwinder@cdac.in

B. Mishra and M. Tiwari (eds.), *VLSI, Microwave and Wireless Technologies*,
Lecture Notes in Electrical Engineering 877,
https://doi.org/10.1007/978-981-19-0312-0_76

SRAM and basic 6T CNTFET SRAM cell. Stability, of memory cells is also considered one of the most important parameters, is also examined utilizing butterfly curve method for proposed CNTFET SRAM cell.

Keywords Average power consumption · CNTFET · Leakage power dissipation · Sleep approach and SRAM

1 Introduction

The unrelenting progression of electronics, semiconductor field has been primarily facilitated by regular development in silicon-based complementary metal-oxide semiconductor (CMOS) technology [1]. This unceasing evolution has been preserved essentially by its device scaling, which outcomes in aggressive prosperity in both packaging density and device performance. With the progressive development in silicon technology, it has faced the design and manufacturing-related issues such as leakage power dissipation, drain-induced barrier lowering (DIBL), and testing problems due to short-channel effects (SCEs) [2]. As per International Technology Roadmap for Semiconductors (ITRS), device feature size scaling can be done up to a certain limit; due to pugnacious device dimensions scaling, the performance of traditional MOS-based designs is altered and reliability, drain-induced barrier lowering current, hot-carrier degradation, gate oxide tunneling, etc., issues arise due to SCE [3]. So, Semiconductor Corporation is struggling to overwhelm the restrictions of conventional MOS technology by incorporating contemporary promising nanoscaled devices. Currently, nanoscale technologies have been proposed for realizing integrated circuits such as quantum-dot cellular automata (QCA), tunnel field effect transistor (TFET), fin-shaped field effect transistors (FinFET), nanowire (NW) transistors, graphene nanoribbon (GNR) transistor, single-electron transistor (SET), and memristor and carbon nanotube field effect transistor (CNTFET) to resolve such challenges [1, 3, 4]. Among these new nanoscale technologies, CNTFET technology, that have stronger performance, low off-current, high thermal strength, and admirable gate current attributes, has encouraging and optimal successor to reinstate current CMOS technology [5–8]. Current–voltage characteristics of the CNTFET device are same as of MOS device; similar to NMOS and PMOS, it has n-CNTFET and p-CNTFET. Hence, incurring to this resemblance and presence of both n-CNTFET and p-CNTFET, earlier circuits implemented on conventional MOS technology can be easily designed using CNTFET technology. CNTFETs' another noteworthy aspect incorporates 1-dimensional (1D) band design that has excellent electrostatic control over the channel in comparison to MOS devices and tremendous drive current due to huger current carrier mobility. Furthermore, CNTFET with its miniaturized size and low-power dissipating capability, it contributes the flexibleness to manage the threshold voltage by ratifying the CNT diameters of distinct values [8, 9].

Meanwhile the last few decades, memory has become the most remarkably component of digital system. The most remarkably section of any digital design is the memory cells that takes up higher than 80% of the chip size. So, it is the fundamental threat for inventors to invent memory cell which dissipates low power and has better performance to strengthen the digital system performance. SRAM memory cell is the affirming and inseparable part of any digital design. More previous SRAM memory cell design advancements were concentrated on memory size with high packaging density, while in the ongoing scenario, designers are attracted on immense clock speed with valuable perseverance of stored data and enormous storage proficiency with low-power consumption. In the modern attitude, memory block of digital design must be invented and investigated with regard to power, area, delay, and stability so that notable improvement can be observed in system performance. In this context, CNTFET-based SRAM cell has become good choice that has excellent speed, low-average power dissipation, and high stability [10, 11]. To make the CNTFET SRAM memory design more efficient in terms of power consumption and design metrics of cell, low-power reduction schemes must be employed so that all over performance of digital system can be enhanced. In this paper, sleep transistor or power gating technique is employed with 6T CNTFET-based SRAM cell, in addition to sleep transistor technique performance of the proposed SBB 6T CNTFET SRAM cell is also analyzed. Performance of PG CNTFET SRAM cell, proposed SBB CNTFET-based SRAM cell is also compared with basic 6T CNTFET-based SRAM cell, and the result shows that proposed SRAM cell has noteworthy advancement in power utilization, leakage current, and delay. Data preserving capacity of SRAM cells that has an important design metrics of any memory cell, that is also examined using popular method, i.e., butterfly curve analysis [12].

The organization of this paper is as follows. Section 2 specifies the overview of emerging CNTFET device with its parameters, CNTFET-based SRAM cell and its operations explained in Sect. 3. Section 4 describes the power gating technique with CNTFET SRAM memory cell and proposed SBB CNTFET-based SRAM cell is discussed in Sect. 5. In Sect. 6, simulation results and performance analysis is stated; then after, finally, conclusion is formulated.

2 The CNTFET

Carbon nano-tubes (CNTs) were discovered by Ljima in Japan 1991; CNTs with nanostructure are hexagonal sheets of graphene wrapped into hollow cylinder. Chirality of CNTs decides either it is semiconducting or metallic, chiral vector (*n*, *m*) of CNTs define its structure, is given by

$$C_n = na_1 + ma_2 \tag{1}$$

$$n, m = \text{Chirality Parameter}$$

$$a_1 = a\left(\sqrt{3}, 0\right) \tag{2}$$

$$a_2 = a\left(\sqrt{3}/2, \sqrt{3}/2\right) \tag{3}$$

where $a = 0.142$ A° bond length between two carbon atoms if $n = m$ then structure is armchair, and if nor m is zero, then structure is zigzag. Nanotube diameter and chiral angles are given as

$$d_t = \frac{\sqrt{3}a_{cc}\sqrt{(m^2 + mn + n^2)}}{\pi} \tag{4}$$

$$\theta = \tan^{-1}\frac{\sqrt{3}n}{2m + n} \tag{5}$$

Study shows that with increasing diameter of CNTs delay will get reduced, while power consumption will get increased; gate current is also dependent on number of CNT tubes, with increasing tubes gate current is also increasing [10]. Basic difference between MOSFET and CNTFET is that CNTFET has one CNT or more CNTs as channel material instead of Si in conventional MOS devices. Top-gated view of CNTFET is shown in Fig. 1, and internal view of CNTFET can be shown as Fig. 2.

So, we can see from top-gate view CNTFET has CNT tube as channel material as conventional MOS has bulk Si, that is why CNTFET has improved electrical characteristics. Gate terminal is controlling terminal to control the flow of control between drain and source. By choosing appropriate diameter and chiral vector, improved

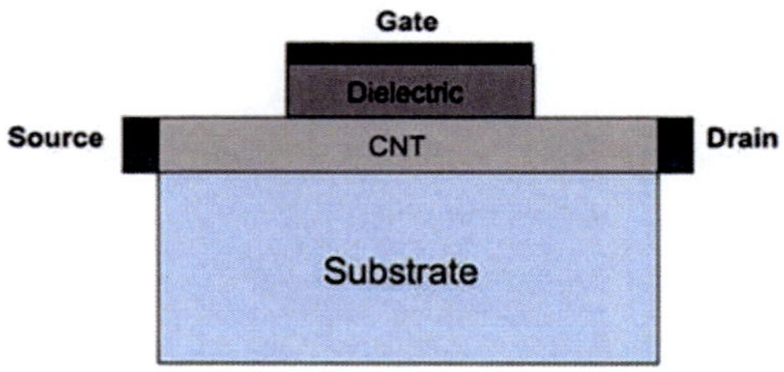

Fig. 1 Top-gate view of CNTFET

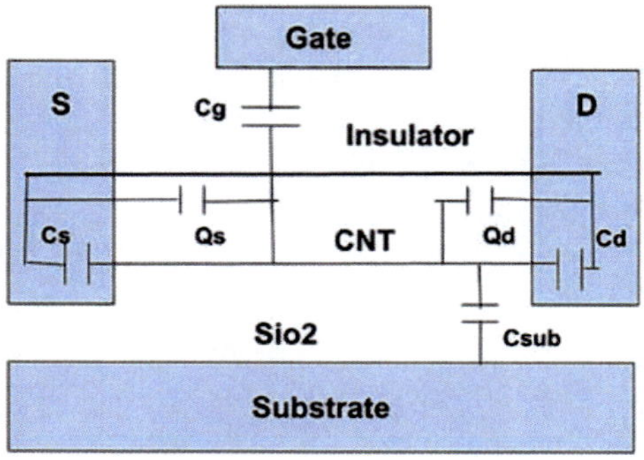

Fig. 2 Internal view of CNTFET

characteristics can be obtained. In internal view of structure, IDS denotes current between drain and source terminal, and Cg, C_D, C_S, C_{Sub} denote the capacitance of gate, drain, source, and substrate, respectively. CNTFET has also P-CNTFET and N-CNTFET as MOS has NMOS and PMOS having electron/holes as majority charge carriers.

3 6T CNTFET-Based SRAM Cell

SRAM memory is a volatile semiconductor memory that has read and write capability, and contents of memory can be accessed in random manner. There are different types of structure available of SRAM cell with depending of application. In this paper, 6T SRAM cell structure is implemented with using emerging CNTFET technology. Schematic of 6T CNTFET-based SRAM cell is represented in Fig. 3, i.e., composed of six transistors, in which N1, N2, P1, and P2 four transistors are connected in cross-coupled form and two transistors N4 and N5 transistors called access transistors associated with bit_bar (BLB)/bit (BL) lines, respectively.

SRAM memory cell design has basically three different mode of operation, i.e., standby or idle mode, read mode, and write mode [10]. In read mode, data can be read from node or cell; in write mode, contents are modified, and in idle mode, read/write operation is not performed. It stays in idle condition for WL = 0, and we cannot write the data; this means that same data will be hold by SRAM memory cell until word line signal is activated. SRAM cell must be robust, i.e., read and write operation must be performed carefully with choosing proper W/L ratio.

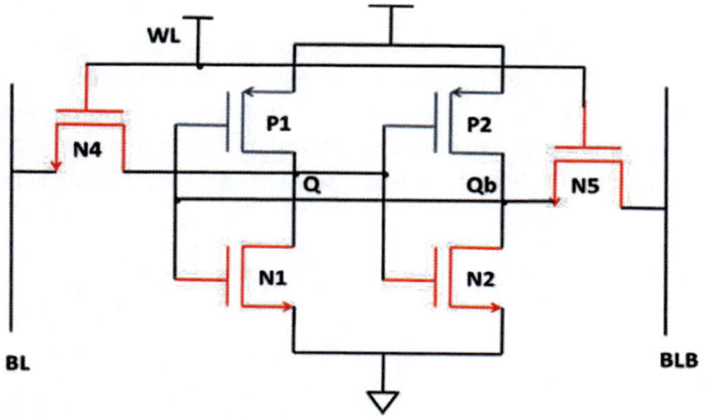

Fig. 3 Schematic of 6T CNTFET SRAM cell

4 Power-Gated (PG) CNTFET SRAM Cell

Power utilization is one of the dominant things for the very large-scale integration (VLSI) circuit designers. Nowadays, designers have aimed at low-power devices due to contemporary thriving requirements of IoT-based digital portable devices like mobile and laptop [13, 14]. To settle the power dissipation issue, many experts have come up with distinct schemes from the different level of designs, i.e., architecture level, device level. Though there is no ubiquitous practice to evade tradeoff among area, power and delay, and hence, inventors are obliged to adopt convenient low-power schemes that accomplish the need of user or application [15]. Power utilization in digital VLSI circuits can be segregated as static and dynamic power consumption. Static power dissipation is observed in the circuit when it is in an idle state, and dynamic power is because of the switching of the transistors [13]. It has been observed that by incorporating many power reduction techniques, dynamic power dissipation has minimized at a significant level, but there is still issue with static power dissipation. Static power consumption is primarily owed to the leakage current components flowing in circuits when device is in standby mode [16, 17]. The leakage current components in 6T CNTFET-based SRAM cell is shown in Fig. 4.

Figure 4 shows that two assertive sub-threshold leakage components paths in a 6T CNTFET-based SRAM cell exist, i.e., power supply VDD to ground named as cell leakage paths, and another one is from BL/BLB to ground path through the N3/N4 named as bit line leakage paths [6]. Leakage current components get increased in an exponential way with reducing threshold voltage; hence, leakage power consumption will get increased [17]. To minimize the leakage power consumption in a CNTFET-based SRAM cell using intrusive threshold-voltage scaling, this article introduces the power gating approach, i.e., sleep transistor scheme [18–23] and new proposed body bias sleep transistor technique.

Mostly, whenever portable gadgets are not in use, they move in standby mode. For example, although the cell phone is in idle mode, it exhausts power from battery; hence, it lowers battery life. So, key idea is that battery life can be enhanced by forcing downward the supply voltage from the circuit, when it is not in active state. This can be done in 6T CNTFET-based SRAM memory cell as shown in Fig. 5.

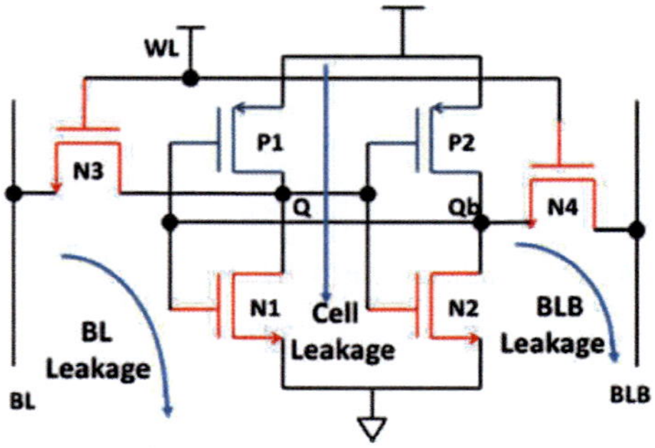

Fig. 4 Leakage components in CNTFET-based SRAM cell

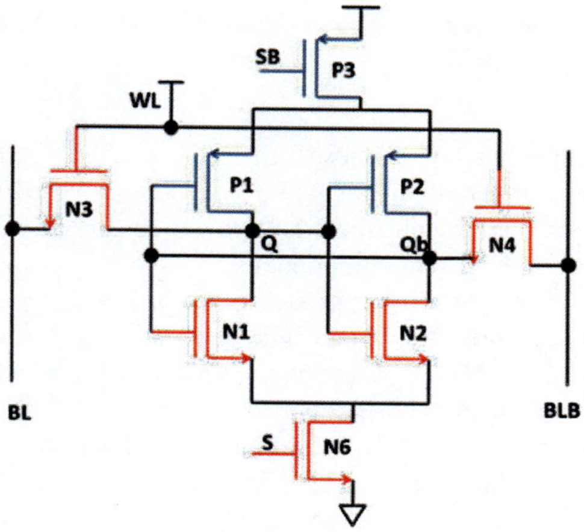

Fig. 5 Schematic of PG 6T CNTFET SRAM cell

Extra transistors say sleep transistors N6 (called footer switch) and P3 (called header switch) are introduced in the VDD and GND of the SRAM cells, i.e., controlled by S and SB signal. Power dissipation can be minimized by controlling the switches when circuit is in active mode or idle mode [22]. Variation of sleep approach using header switch and footer switch with 6T CNTFET based SRAM memory cell as presented in Fig. 6, these schemes have less area overhead in comparison of sleep transistor approach.

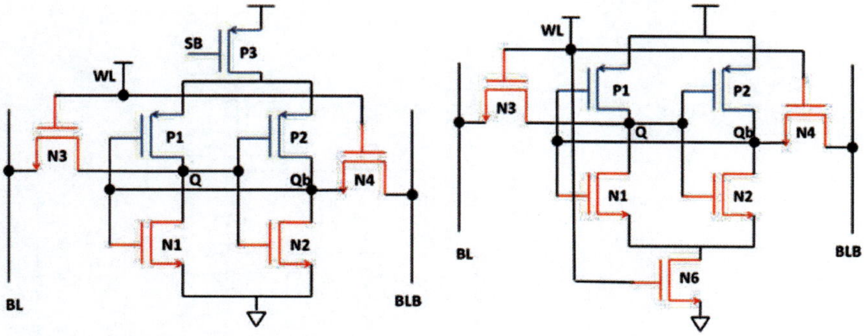

Fig. 6 6T CNTFET SRAM cell with header switch and footer switch

5 Proposed Sleep Body Bias (SBB) 6T CNTFET SRAM Cell

Schematic of proposed SBB 6T CNTFET SRAM cell is revealed in Fig. 7; it is comprised of body bias effect-empowered CNTFET SRAM memory cell that designed to contribute low-power consumption and the substantial minimization in leakage current components.

The primitive approach of body or substrate bias effect is to provide biasing at body end, rather than joining it through ground or supply voltage for N-CNFET and P-CNFET, respectively. Negative body biasing is applied to N-CNTFET, and positive body biasing is applied to P-CNTFET. As the body or substrate bias voltage is increased, depletion region will get widens, and hence, extraneous gate voltage is applied to compensate it. In this way, threshold voltage of CNTFET device will get enhanced; this concept is called substrate or body bias effect. Increased threshold voltage by providing body biasing can be represented by [17]

$$V_t = V_{t_0} + \gamma\left(\surd(V_{SB} - 2\emptyset f)\right) - \surd|2\emptyset f|) \quad (6)$$

Because of body or substrate bias effect, threshold voltage of the device is raised, and hence, leakage current component will get lowered when memory cell is in hold mode. Negative biasing is provided to body terminal of N-CNTFET device; hence, power consumption will get declined as substrate bias voltage is further increased, as displayed in Fig. 8. Positive biasing is provided at the body terminal for P-CNTFET device, and power consumption will be reduced with increasing the substrate bias voltage as observed in Fig. 8.

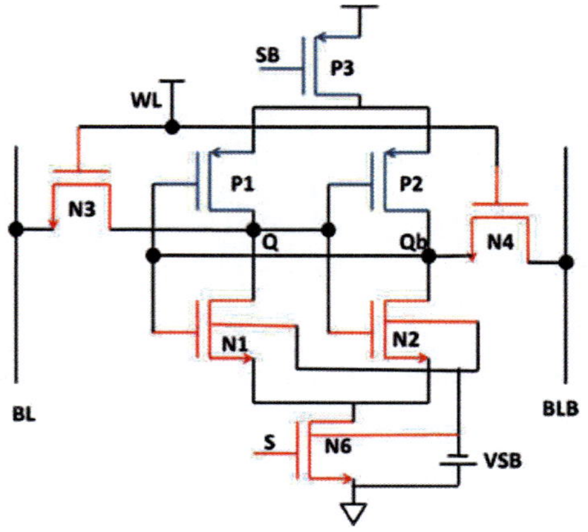

Fig. 7 Schematic of proposed SBB CNTFET SRAM cell

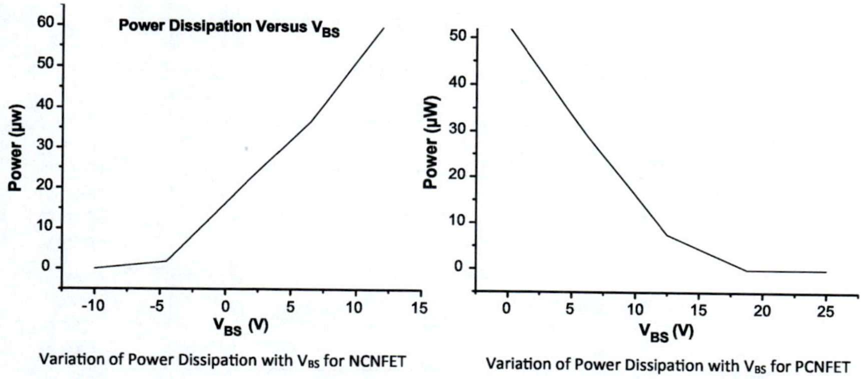

Fig. 8 Variation of power consumption with substrate bias voltage V_{BS} for N-CNTFET and P-CNTFET

6 Simulation Results and Discussions

This section describes the simulation results and analysis of performance metrics of 6T CNTFET based SRAM memory cell, PG CNTFET based SRAM and proposed SBB CNTFET based SRAM memory cell, those are implemented at 32 nm technology with power supply of 0.7 V. Simulation results as waveform of 6T CNTFET-based SRAM and proposed CNTFET-based SRAM cell validate the authenticity of this work, i.e., shown in Fig. 9. Transient analysis is performed as simulation result with the help of Cadence Virtuoso tool. For CNTFET device, Verilog-A model, represented by Stanford Compact Model [24], is used to generate symbol of the device. Different parameters used for this work are as follows: average-free path for CNT 200 nm, average-free path for doped CNT 15 nm doped region length 32 nm, work function 4.5 eV, number of tubes 1, and chiral vector (19,0) [25]. Comparative performance investigation is listed in Table 1.

Comparative performance analysis of CNTFET SRAM cells shows that incorporation of power gating technique and proposed sleep body bias technique results into power-efficient CNTFET SRAM cell; hence, performance of SRAM cells will get enhanced. It can be observed from Table 1 that there is 84.43%, 79.48%, and 60.92% improvement in average power dissipation achieved for sleep approach, with header switch, and with footer switch, respectively, in comparison of 6T CNTFET-based SRAM memory cell. Similarly, in 6T CNTFET SRAM cell design, leakage power reduction of 25.34%, 13.18%, and 3.37% is observed for sleep approach, with header, and with footer techniques, respectively.

Proposed SBB CNTFET SRAM cell is more effective in comparison to PG CNTFET SRAM and shows excellent performance design metrics, i.e., 854 pw average power dissipation, 10.07 ns delay, 0, 27.67 ps delay 1, and 98.84 pw leakage power dissipation. Analysis of performance parameters of CNTFET SRAM cell can

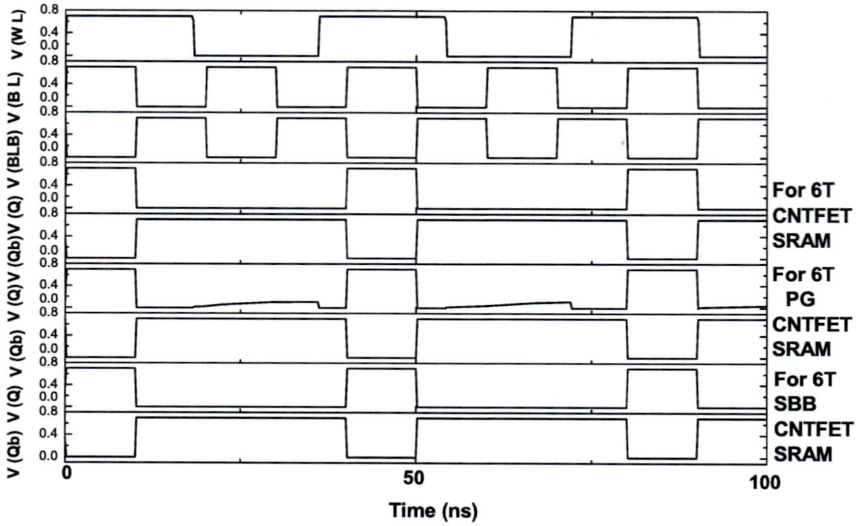

Fig. 9 Simulation waveform of proposed CNTFET SRAM cell

Table 1 Comparative performance analysis of CNTFET SRAM cells

Parameters/Structure (6T CNTFET SRAM)	6T CNTFET SRAM	PG CNTFET SRAM	With header switch	With footer switch	SBB CNTFET SRAM
Average power consumption (w)	12.72n	1.98n	2.61n	4.97n	854p
Delay0 (s)	10.07n	17.47p	21.3p	24.15p	10.07n
Delay1 (s)	96.06p	59.49p	59.51p	72.03p	27.67p
PDP0 (s)	128.26a	34.72z	55.61z	120.2z	8.599a
PDP1(s)	1.2237a	118.2z	155.4z	358.3z	23.63z
Leakage power consumption (w)	136.5p	101.9p	118.5p	131.9p	98.84p

be shown in Figs. 10, 11, and 12 for average power utilization, delay, and leakage power consumption.

Memory cell stability is described as the maximal endures the voltage of extraneous turbulence or DC interruptions without altering the initial cell state [12]. There are different ways to examine the cell stability of SRAM cell, but the method employed in this work for the same is butterfly curve. Butterfly technique is the most acceptable practice to evaluate the stability of SRAM memory cell that is computed in form of static noise margin (SNM). Cell stability can be probed by delineating a butterfly curve, and next, the length of the peak feasible square is evaluated that can be enclosed in it; this is defined as SNM that is read SNM for read mode, write SNM

Fig. 10 Average power dissipation for CNTFET SRAM cells

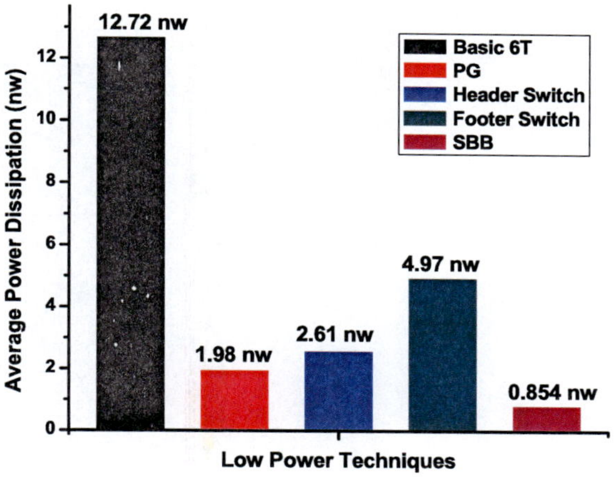

Fig. 11 Delay for CNTFET SRAM cells

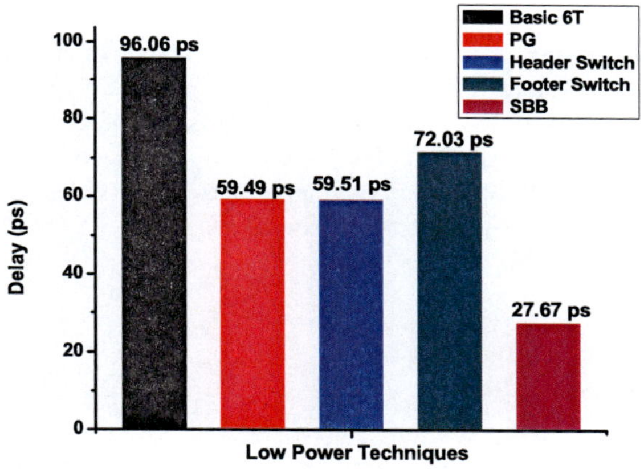

Fig. 12 Leakage power dissipation for CNTFET SRAM cells

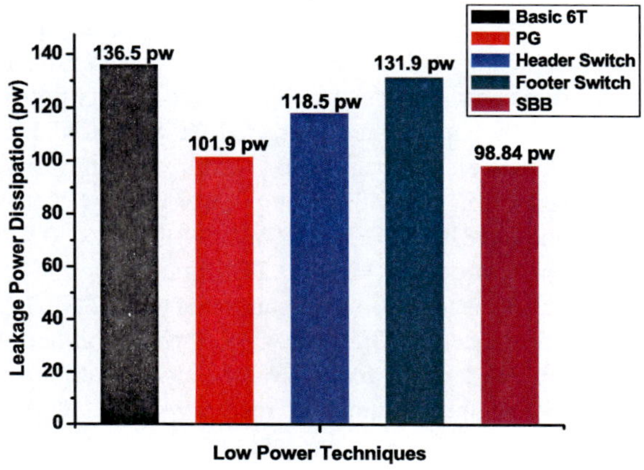

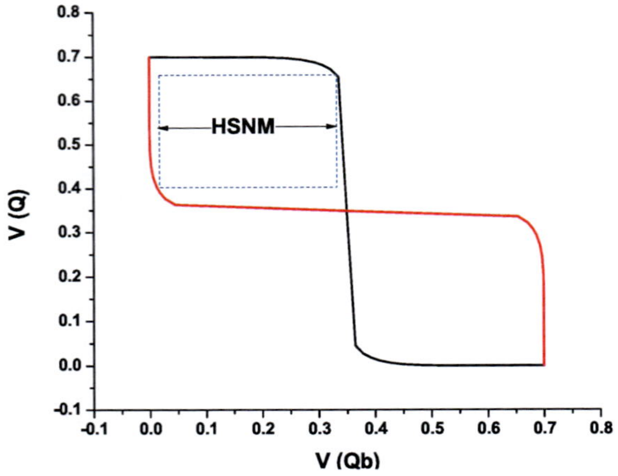

Fig. 13 HSNM curve of proposed CNTFET SRAM cell

for write mode, and hold SNM for hold mode of operation of SRAM cell. Figures 13 and 14 show the SNM plot for the proposed CNTFET SRAM cell.

Comparative analysis of stability is done for 6T SRAM Cell, PG CNTFET, and proposed SBB CNTFET SRAM cell, evaluated values of SNM are listed in Table 2. Stability design metrics can be varied with respect to supply voltage. Variation of stability design metrics can be shown in Fig. 15 for proposed SBB CNTFET-based SRAM cell.

7 Conclusion

In this paper, power minimization schemes are applied in 6T CNTFET-based SRAM memory cell to make it more efficient so that system performance can be enhanced. Proposed SRAM cell design surpasses the other designs of SRAM cells those are included in this work, in terms of power consumption, leakage power dissipation, and stability design metrics. The simulation results notify legitimacy of the suggested technique as phenomenal proposed 6T SBB CNTFET SRAM cell, exclusively in terms of power dissipation and stability in comparison of other design those are discussed in this paper. By using power gating (PG CNTFET SRAM cell) and SBB techniques (SBB CNTFET SRAM cell), 0.84×, 0.93× times reduction in average power dissipation and 0.25×, 0.27× times reduction in leakage power dissipation are observed, respectively, in comparison of 6T CNTFET SRAM cell, without degrading the performance. CNTFET SRAM cell with modified sleep approach (power gating), i.e., with header and footer switch, also exhibits the significant improvement in performance specifications of memory cell. Cell stability is also observed in this work by using most popular method of stability, i.e., butterfly curve, which shows

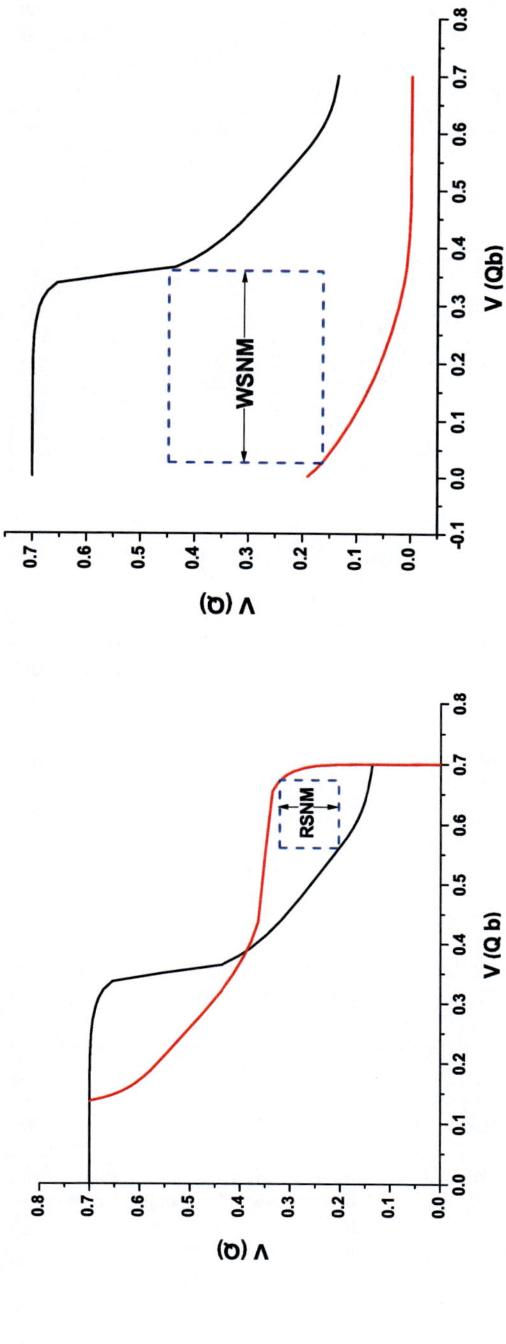

Fig. 14 Butterfly curve of proposed CNTFET SRAM cell for RSNM and WSNM

Table 2 Comparative stability analysis of CNTFET SRAM cells

Parameters/Structure (6T CNTFET SRAM)	6T CNTFET SRAM	PG CNTFET SRAM	With header switch	With footer switch	SBB CNTFET SRAM
HSNM (mv)	300	313.63	321.2	320	321
RSNM (mv)	130	101	127	112	125
WSNM (mv)	340	378	391	392	348

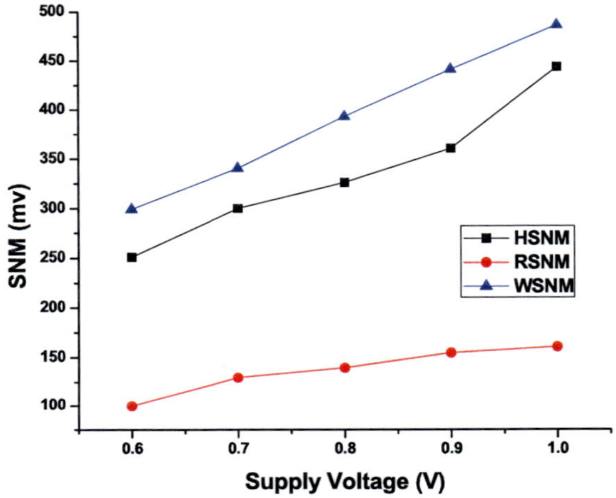

Fig. 15 SNM parametric variation with respect to supply voltage for SBB CNTFET SRAM cell

that by incorporating low power reduction schemes stability design metrics can be enhanced.

References

1. Moaiyeri MH, Rahi A, Sharifi F, Navi K (2017) Design and evaluation of energy-efficient carbon nanotube FET-based quaternary minimum and maximum circuits. J Appl Res Technol 15:233–241. https://doi.org/10.1016/J.JART.2016.12.006
2. Gupta D (2020) Low-voltage analog integrated circuit design. In: Dhiman R, Chandel R (eds) Nanoscale VLSI. Energy systems in electrical engineering. Springer, Singapore. https://doi.org/10.1007/978-981-15-7937-0_1
3. Kim Y-B (2010) Challenges for nanoscale MOSFETs and emerging nanoelectronics. Trans Electr Electron Mater 11(3):93–105. https://doi.org/10.4313/TEEM.2010.11.3.093
4. Singh A, Khosla M, Raj B (2016) Comparative analysis of carbon nanotube field effect transistor and nanowire transistor for low power circuit design. J Nanoelectron Optoelectron 11(3):388–393. https://doi.org/10.1166/jno.2016.1913
5. Saha P, Jain A, Sarkar SK (2013) A comparative study of CMOS and carbon nanotube field effect transistor based inverter at 32 nm technology node., Asian J Chem 25:424–426

6. Rajendra Prasad S, Madhavi BK, Kishore KL (2012) Leakage-power SRAM cell design using CNTFETs at 32 nm technology. In: Das VV, Stephen J (eds) Advances in communication, network, and computing. CNC 2012. Lecture notes of the Institute for Computer Sciences, Social Informatics and Telecommunications Engineering, vol 108. Springer, Berlin. https://doi.org/10.1007/978-3-642-35615-5_24
7. Jafarzadehpour F, Keshavarzian P (2016) Low-power consumption ternary full adder based on CNTFET. IET Circuits Devices Syst 10(5):365–374. https://doi.org/10.1049/iet-cds.2015.0264
8. Sharma T, Kumre L (2019) CNTFET-based design of ternary arithmetic modules. Circuits Syst Signal Process 38:4640–4666. https://doi.org/10.1007/s00034-019-01070-9
9. Sinha SK, Chaudhury S (2014) Comparative study of leakage power in CNTFET over MOSFET device. J Semicond 35(11):114002 1–6. https://doi.org/10.1088/1674-4926/35/11/114002
10. Kumar H, Srivastava S, Singh B (2020) Low power, high-performance reversible logic enabled CNTFET SRAM cell with improved stability. Mater Today: Proc 1–7. https://doi.org/10.1016/j.matpr.2020.06.475
11. Sayed SI, Abutaleb MM, Nossair ZB (2016) Optimization of CNFET parameters for high performance digital circuits. 1–9.https://doi.org/10.1155/2016/6303725
12. Jadav S, Chandel R (2018) High performance 9T adiabatic SRAM and novel stability characterization using pole zero placement. Analog Integr Circ Sig Process 98(2):347–355. https://doi.org/10.1007/s10470-018-1308-0
13. Padmavathi B, Geetha BT, Bhuvaneshwari K (2017) Low power design techniques and implementation strategies adopted in VLSI circuits. In: IEEE international conference on power, control, signals and instrumentation engineering (ICPCSI), pp 1764–1767. https://doi.org/10.1109/ICPCSI.2017.8392017
14. Mandi BC (2020) Low power design techniques for integrated circuits. In: Dhiman R, Chandel R (eds) Nanoscale VLSI. Energy systems in electrical engineering. Springer, Singapore. https://doi.org/10.1007/978-981-15-7937-0_4
15. Pandit S (2020) Design methodology for ultra-low-power CMOS analog circuits for ELF-SLF applications. In: Dhiman R, Chandel R (eds) Nanoscale VLSI. Energy systems in electrical engineering. Springer, Singapore. https://doi.org/10.1007/978-981-15-7937-0_2
16. Yadav R, Yadav R (2020) Dynamic power reduction techniques for CMOS logics using 45 nm technology. In: Singh Tomar G, Chaudhari N, Barbosa J, Aghwariya M (eds) International conference on intelligent computing and smart communication 2019. Algorithms for intelligent systems. Springer, Singapore. https://doi.org/10.1007/978-981-15-0633-8_112
17. Pal A (2015) Low-power VLSI circuits and systems. Springer, New Delhi. https://doi.org/10.1007/978-81-322-1937-8
18. Kumar H, Saun S (2019) Power gated technique to improve design metrics of 6t SRAM memory cell for low power applications. ICTACT J Microelectron 5:3
19. Muralidharan V, Sathish Kumar N (2020) Design and implementation of low power and high speed multiplier using quaternary carry look-ahead adder. Microprocess Microsyst 75:103054. https://doi.org/10.1016/j.micpro.2020.103054
20. Nag A, Ruchira Reddy K, Majumder N, Debbarma E, Pradhan SN (2021) A novel NOR gate-based dynamic power gating technique in SRAM. In: Nath V, Mandal JK (eds) Proceedings of the fourth international conference on microelectronics, computing and communication systems. Lecture notes in electrical engineering, vol 673. Springer, Singapore. https://doi.org/10.1007/978-981-15-5546-6_28
21. Saranya L, Arvind C, Karthigaikumar P, Balachander K (2020) A 4 × 4 modified 8T SRAM cell array using power gating technique. Mater Today: Proc 1–7 (2020) (Article in Press). https://doi.org/10.1016/j.matpr.2020.08.746
22. Geetha Priya M, Baskaran K, Krishnaveni D (2011) Leakage power reduction techniques in deep submicron technologies for VLSI applications. In: International conference on communication technology and system design. Procedia Eng 30:1163–1170. https://doi.org/10.1016/j.proeng.2012.01.976

23. Metaj R, Stopjakova V, Arbet D (2017) Design techniques for low voltage analog integrated circuits. J Electr Eng 68(4):245–255. https://doi.org/10.1515/jee-2017-0036
24. Stanford CNTFET Model. Available: https://nano.stanford.edu/stanford-cnfet-model. Accessed on 5 Feb 2021
25. Hemant K, Subodh S, Balwinder S (2019) Comparative analysis of 6T, 7T conventional CMOS and CNTFET based SRAM cell design. Adv Sci Eng Med 11(1/2):3–10. https://doi.org/10.1166/asem.2019.2301

Tuning of White-Colour Gamut Using Strain Adapting Interfacial Layer in 'Giant' Colloidal Core–shell Quantum Dot

Anupam Sahu and Dharmendra Kumar

Abstract In this paper, the 'giant' colloidal core–shell quantum dot with strain adapting interfacial layer (g-CISQD) has been theoretically investigated for the application in light-emitting diode (LED). The CdSe/$CdSe_xS_{1-x}$/CdS g-CISQD is modeled in the effective mass approximation considering the strain at the core-interfacial-shell interface. The optical gain characteristics have been investigated considering the transition between the highest occupied molecular orbit (HOMO) and lowest unoccupied molecular orbit (LUMO). The emission of different colours of white light has been achieved using the efficient structural and material compositions.

Keywords Core–shell · Interfacial · Fermi level · Colloidal quantum dot

1 Introduction

The carriers in the low-dimensional semiconductor structure such as quantum well, wire and dot confine are confined in one, two and three dimensions, respectively. The 3D confinement of carriers in quantum dots (QDs) leads to the formation of discrete energy levels and delta-like density of states. The discretization is mainly because of two effects: the quantum confinement and columbic interaction between the charged particles. The electronic properties of QDs exist between those of bulk semiconductors and discrete molecules.

Further, the surface engineering of QDs with some semiconductor materials having different bandgaps forms a core–shell quantum dots (CSQDs) [1]. This ensures chemical stability and improved photoluminescence efficiency over conventional QDs [2]. The increasing shell thickness over the core materials favors the development of quasi type-II band alignment [3, 4]. Such 'giant' CSQD (g-CSQD)

A. Sahu (✉) · D. Kumar
Department of Electronics and Communication Engineering, Madan Mohan Malaviya University of Technology, Gorakhpur, India
e-mail: anupam.or.sahu@gmail.com

D. Kumar
e-mail: dharmendra0127@gmail.com

B. Mishra and M. Tiwari (eds.), *VLSI, Microwave and Wireless Technologies*, Lecture Notes in Electrical Engineering 877,
https://doi.org/10.1007/978-981-19-0312-0_77

provides enhanced quantum yield and has been explored by researchers for several optoelectronic devices [5–7].

The strain arises at the core–shell interface of g-CSQD is responsible for the development of unwanted carrier confinement. Therefore, the use of strain adapting interfacial layer to form core-interfacial-shell (g-CISQD) helps in minimizing these anomalies. Also, the use of an intermediate interfacial layer provides the tailored emission and absorption spectra suitable for device applications.

In this paper, we have theoretically investigated the optical gain spectrum for $CdSe/CdSe_xS_{1-x}/CdS$ g-CISQD for different structural parameters and optimised the structure for application in light-emitting diode (LED) applications. The $CdSe_xS_{1-x}$ is the alloyed interfacial layer that provides the intermediate potential step between the core and shell material. The structure is modeled in the effective mass approximation, considering the significant strain at the core-interfacial-shell interface. We reported the tuning capability of peak gain wavelength in the spectrum of white-colour gamut using appropriate alloying 'x' and core radius.

2 Mathematical Modelling

Consider the spherical semiconductor material (CdSe) as a core with radius r_1, encapsulated with the interfacial layer ($CdSe_xS_{1-x}$) with radius r_2 and shell material (CdS) with the radius R. Thus, the total thickness of the interfacial and shell layer is designated as h_1 and h_2, respectively. The strain adapting $CdSe_xS_{1-x}$ at the interface forms a multi-step potential for the carriers, as shown in Fig. 1.

Here, the V_c and V_v are the confinement potential at the core–shell interface for conduction band (CB) and valence band (VB), respectively.

In the framework of effective mass approximation, the envelope of electron and hole wave functions can be obtained using the Schrodinger equation shown below [8]:

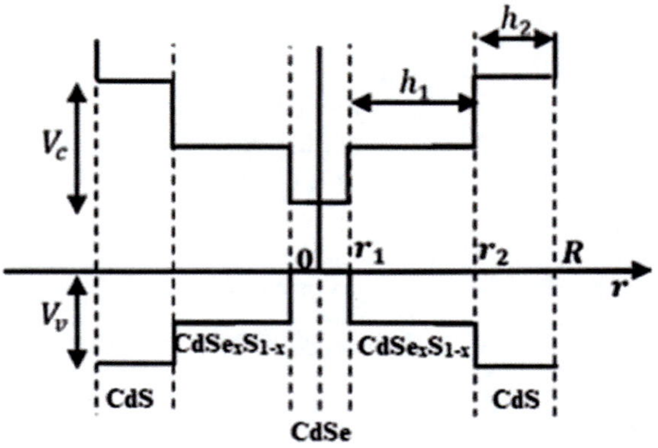

Fig. 1 Represents the band structure of the 'giant' core-interfacial-shell quantum dot

$$H_i \Psi_i(\overrightarrow{r_i}) = E_i \Psi_i(\overrightarrow{r_i}) \quad (i = e, h) \tag{1}$$

where the suffix i stands for carriers in the region. Considering the Ben Daniel–Duke boundary condition, the Hamiltonian can be written as:

$$H_i = \left(-\frac{\hbar^2}{2m_i^*(r_i)} \nabla^2 + V_i(r_i) \right) \tag{2}$$

The detailed modeling description of strained g-CISQD has been shown in Ref. [9].

2.1 Optical Gain Characteristics

The linear optical gain of the QD system under adequate current injection is expressed as [10, 11]:

$$g(E) = \frac{2\pi e^2 \hbar N_d}{c n_r \varepsilon_o m_o^2} \sum_{c,v} \frac{|P_{c,v}|^2}{E_{cv}} \int_{-\infty}^{+\infty} [f_c(E') - f_v(E')] G(E' - E_{cv}) B_{cv}(E - E') dE' \tag{3}$$

where c and v denote the discrete energy state of CB and VB, respectively, N_d is the inverse of the QD volume (V), n_r is the refractive index of the QD material, E_{cv} denotes the interband transition, $\hbar$ is the reduced Plank's constant, c is the speed of light, and $G(E)$ is the Dirac delta function. $f_c(E')$ and $f_v(E')$ are the Fermi–Dirac distribution function of the CB and VB, respectively, and are expressed as:

$$f_i = \frac{1}{\exp\left(\frac{E_i - E_{fi}}{k_B T}\right) + 1} \quad i = c, v \tag{4}$$

where E_{fi} is the quasi-Fermi level and T is the temperature in Kelvin. The values of E_{fc} and E_{fv} can be obtained using the relations [12]:

$$N = \sum_i \frac{2}{\left[1 + \exp\left(\frac{E_i - E_{fc}}{k_B T}\right)\right] V} \tag{5a}$$

$$P = \sum_i \frac{2}{\left[1 + \exp\left(\frac{E_{fv} - E_i}{k_B T}\right)\right] V} \tag{5b}$$

where $N \approx P$. The B_{cv} represents the Lorentzian distribution functions and is expressed as [10]:

$$B_{cv}(E - E') = \frac{\frac{\hbar\Gamma_{cv}}{\pi}}{(E - E')^2 + (\hbar\Gamma_{cv})^2} \tag{6}$$

where $\hbar\Gamma_{cv}$ is the inhomogeneous broadening due to the interband scattering relaxation time Γ. The transition matrix element $|P_{c,v}|^2$ for interband transition is expressed as:

$$|P_{c,v}|^2 = |I_{c,v}|^2 M^2 \tag{7}$$

where $I_{c,v}$ denotes the overlap of electron and hole wave functions and M is the element dipole moment.

3 Results and Discussion

The optical properties for interband transition $1s_e$-$1s_h$ ($l = m = 0, n = 1$) in 'giant' colloidal CdSe/$CdSe_xS_{1-x}$/CdS CISQD have been investigated for different alloying 'x' (0, 0.25, 0.50, 0.75) and varying core radius (1–2.25 nm). The dimension of interfacial, outermost shell and values of other parameters are taken from [9].

Figure 1 represents the interband transition energy with a varying core radius for different alloying 'x' of the structure. Evidently, the transition energy decreases for increasing core radius, majorly due to the quantum confinement effect. As the core radius increases, the energy eigenvalue for electrons in CB and holes in VB decreases. This, in turn, decreases the resultant transition energy, which is in concurrence with the literature. On the other hand, on increasing 'x', the interband transition energy decreases. This is due to the decrement in the intermediate step potential at the interfacial layer, which decreases the confinement of the carriers in the core region.

Figure 3 represents the variation of peak gain wavelength with the variation of core radius and different alloying 'x' of the structure. In Fig. 3a, as the core radius increases, the peak wavelength increases. The result is attributed to Fig. 2, where on the increasing core radius, the transition energy (E_{cv}) decreases, and hence, peak wavelength increases. In Fig. 3b, the range of peak wavelength for different alloying 'x' is plotted for core radius varying from 1–2.25 nm. For '$x = 0$', the g-CSQD (i.e., without interfacial layer) offers the blue, green and yellow spectrum of white light. The redshift in the range of peak gain wavelength has been observed on the 'x' (g-CISQD), and finally, the red colour spectrum is achieved at $x = 0.75$.

Figure 4 represents the normalised gain spectrum for a different colour of white light obtained via tuning of core radius considering the different 'x' of the structure.

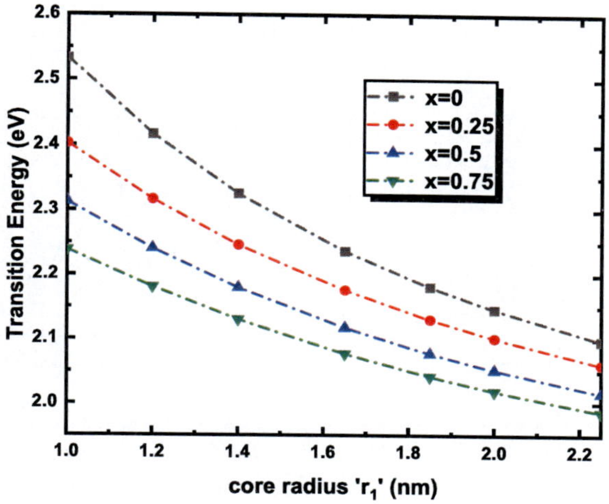

Fig. 2 Represents the interband transition energy with varying core radius in g-CISQD for different alloying 'x'

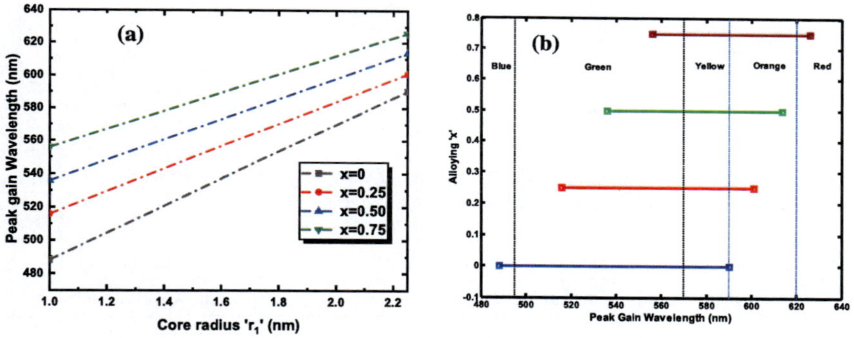

Fig. 3 Represents **a** the peak gain wavelength as a function of core radius for different alloying 'x' and **b** the range of peak gain wavelength as a function of alloying 'x' with core radius varying from $1 - 2.25$ nm in g-CISQD

For the carrier injection of $4 \times 10^{24}/\text{m}^3$, the detailed tuning parameters for the emission spectrum have been shown in Tables 1a and 1b, respectively:

4 Conclusion

The optical transition energy between the HOMO and lowest LUMO in strained g-CISQD has been computed for different structural parameters. The transition energy

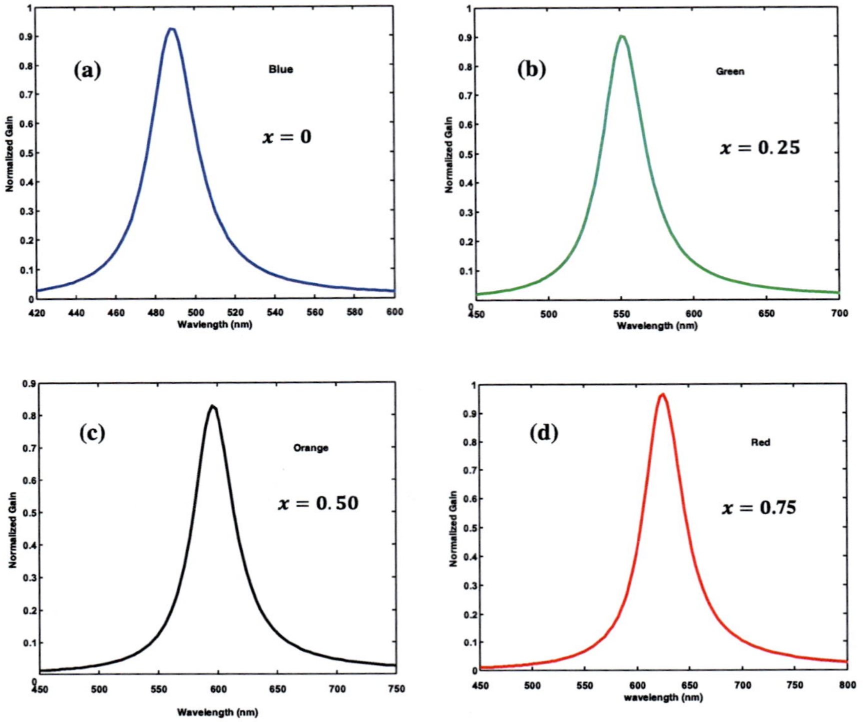

Fig. 4 Represents the normalised gain as a function of wavelength for a different colour of white light obtained at different alloying 'x' with varying core radius

Table 1a Structural dimensions and material composition for white-light emission

S. no	Alloying 'x'	Core radius 'r_1' (nm)	Colour	Peak gain wavelength (nm)
1.	$x = 0$	1.0	Blue	488
2.	$x = 0.25$	1.4	Green	551
3.	$x = 0.50$	1.85	Orange	596
4.	$x = 0.75$	2.25	Red	626

Table 1b Comparison of proposed g-CISQD results with the literature

S. no	Colour	g-CISQD		Refs. [11, 13]	
		Peak gain wavelength (nm)	Energy (eV)	Energy (eV)	Peak gain wavelength (nm)
1.	Blue	488	2.5360	2.6683	470
2.	Green	551	2.2462	2.3171	530
3.	Orange	596	2.0781	–	–
4.	Red	626	1.9828	2.0213	580

decreases for increasing core radius and alloying 'x'. Further, the noticeable redshift in the peak gain wavelength has been observed for increasing the alloying 'x' of the interfacial layer. The emission of the blue, green, yellow, orange, and red colours of light has been achieved on varying core radius for a particular alloying 'x'. The results of g-CISQD in terms of peak gain wavelength render the promising candidate for the emission of different colour of white in LED.

References

1. Haus JW, Zhou HS, Honma I, Komiyama H (1993) Quantum confinement in semiconductor heterostructure nanometer-size particles. Phys Rev B. https://doi.org/10.1103/PhysRevB.47.1359
2. Vasudevan D, Gaddam RR, Trinchi A, Cole I (2015) Core-shell quantum dots: properties and applications. https://doi.org/10.1016/j.jallcom.2015.02.102
3. Meinardi F, Colombo A, Velizhanin KA, Simonutti R, Lorenzon M, Beverina L, Viswanatha R, Klimov VI, Brovelli S (2014) Large-area luminescent solar concentrators based on stokes-shift-engineered nanocrystals in a mass-polymerized PMMA matrix. Nat Photonics 8:392–399. https://doi.org/10.1038/nphoton.2014.54
4. Coropceanu I, Bawendi MG (2014) Core/shell quantum dot based luminescent solar concentrators with reduced reabsorption and enhanced efficiency. Nano Lett 14(7):4097–4101. https://doi.org/10.1021/nl501627e
5. Huang B, Xu R, Zhuo N, Zhang L, Wang H, Cui Y, Zhang J (2016) "Giant" red and green core/shell quantum dots with high color purity and photostability. Superlattices Microstruct 91:201–207. https://doi.org/10.1016/j.spmi.2016.01.015
6. Adhikari R, Jin L, Navarro-Pardo F, Benetti D, AlOtaibi B, Vanka S, Zhao H, Mi Z, Vomiero A, Rosei F (2016) High efficiency, Pt-free photoelectrochemical cells for solar hydrogen generation based on "giant" quantum dots. Nano Energy 27:265–274. https://doi.org/10.1016/j.nanoen.2016.07.010
7. Navarro-Pardo F, Zhao H, Wang ZM, Rosei F (2018) Structure/property relations in "giant" semiconductor nanocrystals: opportunities in photonics and electronics. Acc Chem Res 51:609–618. https://doi.org/10.1021/acs.accounts.7b00467
8. Sahu A, Kumar D (2020) Computation of intersubband transition energy and absorption spectra of binary capped GaAs/Al0.42Ga0.58As core-shell quantum dot. J Nanophotonics. https://doi.org/10.1117/1.jnp.14.026003
9. Sahu A, Kumar D (2021) Theoretical modeling of a "giant" colloidal core–shell quantum dot with an alloyed interfacial layer for solar cell applications. J Opt Soc Am B. https://doi.org/10.1364/josab.414664
10. Sakamoto A, Sugawara M (2000) Theoretical calculation of lasing spectra of quantum-dot lasers: effect of homogeneous broadening of optical gain. IEEE Photonics Technol Lett. https://doi.org/10.1109/68.823485
11. Ranjbaran A (2012) Temperature effects on output characteristics of quantum dot white light emitting diode. Front Optoelectron.https://doi.org/10.1007/s12200-012-0275-9
12. Asada M, Miyamoto Y, Suematsu Y (1986) Gain and the threshold of three-dimensional quantum-box lasers. IEEE J Quantum Electron. https://doi.org/10.1109/JQE.1986.1073149
13. Qian L, Yang Y, Han C, Cheng Y, Chen L, Zhang T, Xue W (2015) White light emission from quantum well structured quantum dots. Mater Sci Eng B Solid-State Mater Adv Technol 193:1–3. https://doi.org/10.1016/j.mseb.2014.10.013

The Effect of Finite Ground Plane on Wideband Slim Rectangular Dielectric Resonator Antenna

Sovan Mohanty and Baibaswata Mohapatra

Abstract In this paper, a slim profile microstrip-fed wideband rectangular dielectric resonator antenna is proposed, and the effect of the finite ground plane on its performance is studied. Due to the availability of specified finite dimensions and fixed dielectric constant material, wide range of frequency tuning is one of the key challenges in the design of RDRA. Stabilized frequency tuning cannot be achieved without significant variation in the impedance matching. Therefore, to achieve broader frequency tuning, the loading effect of the radiator can be altered by varying the dimensions of the finite ground plane. In this proposed radiator, wideband operations can be achieved by carving a notch on the designed structure. The influence of the finite ground plane is altered for practical characterization and to test the overall performance of the antenna. The proposed antenna provides an impedance bandwidth of more than 80% and radiation efficiency of 99.9%. It is achieved with the optimized finite ground plane of length 24 mm. The RDRA provides a peak gain of 5 dBi at 11.4 GHz. This proposed RDRA is found to be suitable for short-range radio links in X and Ku band applications.

Keywords Rectangular dielectric resonator antenna (RDRA) · Wideband antenna · Dielectric waveguide model (DWM)

1 Introduction

RF and microwave technology have undergone a tremendous surge in the last decade in the quest for high-speed communication and navigational applications. The modern system demands ultra-wide bandwidth, polarization purity, consistent gain, and high radiation efficiency with temperature stability in the operating frequency intervals. The miniaturized design stimulates physically, electrically, and functionally small embeddable antenna as the front-end device with considerable power handling

S. Mohanty (✉) · B. Mohapatra
School of Electrical, Electronics and Communication Engineering, Galgotias University, Greater Noida, U.P., India
e-mail: mohanty.sovan@gmail.com

B. Mishra and M. Tiwari (eds.), *VLSI, Microwave and Wireless Technologies*, Lecture Notes in Electrical Engineering 877,
https://doi.org/10.1007/978-981-19-0312-0_78

capabilities [1]. The operating bandwidth should be high for effective output radiation performance, input characteristics like impedance and phase response while handling digital impulse signals. So far, traditional microstrip patch antenna is a dominating commercial technology, but its operation is restricted to narrowband due to its intrinsic capacitive behavior [2–4]. The DRA reduces the reactive and distributive effects at microwave and mm-wave frequencies. It provides numerous advantages like high radiation efficiency, wide bandwidth, miniaturized design, and stable output characteristics [5–8]. Though most of the antenna design is based on an infinite ground plane, on a practical note, all the ground planes are finite. It is due to the presence of various circuit components that are fitted along with the antenna. If the dimension of the ground plane altered and the source is positioned close to the ground plane, then there will be diffraction of the incident wave from the edge of the finite ground plane. Predominantly, through the effect of diffraction, electromagnetic waves follow a path around the obstacles. The current at the edge of the ground plane will excite the radiation outward. It makes the edge to behave like a line source [9]. The characteristics of the diffracted field and the diffracted coefficient will be calculated by the geometric theory of diffraction [10]. The radiation and impedance characteristics are proportional to the location and separation from the ground plane [11]. The dimension of the ground plane influences the directivity, beam pointing angle, and input resistance of the radiator [12]. Therefore, the ground plane radius has to be chosen carefully to optimize the performance of the antenna [13].

In this paper, a stepped microstrip line-fed slim wideband-operated rectangular dielectric resonator antenna with a finite ground plane is being proposed. A notch is carved on the RDRA to reduce the relative permittivity. Therefore, effective Q-factor can be lowered to enhance the bandwidth [14, 15]. Each portion of the radiator will operate in the same modes but with slightly different resonating frequencies. This cumulative response enhances the impedance bandwidth. A cut out of dimension $2.5 \times 4.94 \times 9$ m m^3 is made on the DRA which reduces the effective dielectric constant and surface current density to obtain a multiband or wideband responses. Multi-functionality and multiband operation can be achieved by exciting the ground plane through a suitable matching network. The matching network along with the ground plane can excite the fundamental modes according to the characteristics of the mode theory. To optimize the proposed design, parametric analysis was carried out by employing the finite element method. The antenna design methodology is explained in Sect. 2 followed up by the design approach is in Sect. 3. Thorough result analyses are presented in Sect. 4. At the end, conclusions are given in Sect. 5.

2 Antenna Design

Figures 1 and 2 show the detailed geometry of the designed slim profile wideband antenna. The antenna is mounted on a piece of a substrate (Roger TMM4 (tm), $\varepsilon_r = 4.5$, $\tan\delta = 0.002$) having an area of 50×50 mm^2 and thickness of 1.6 mm. Here, Roger thermostat microwave material is chosen as substrate because of its high

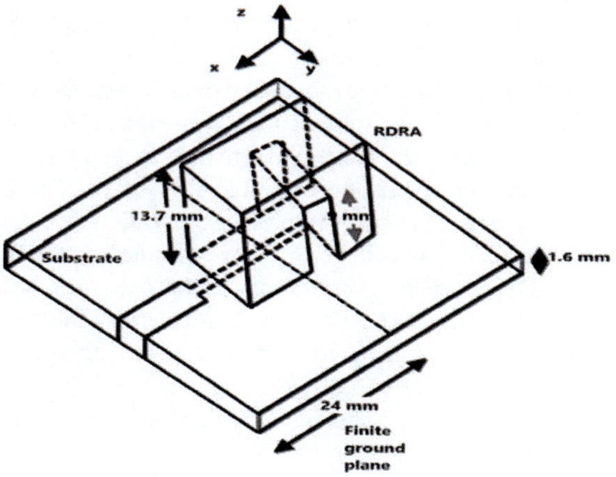

Fig. 1 Geometry of the proposed antenna with finite ground plane

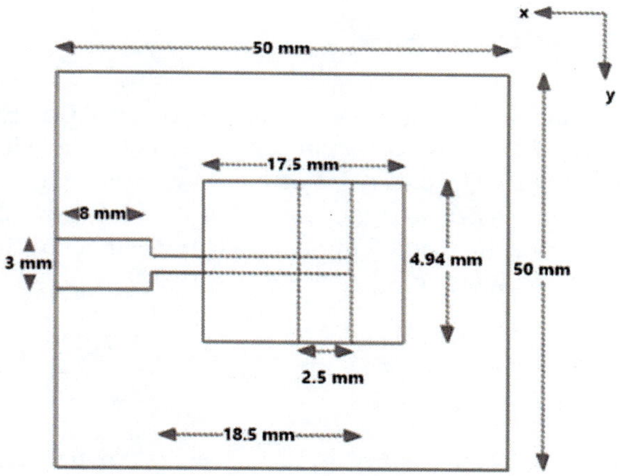

Fig. 2 Top view of the proposed antenna

electrical strength and easy availability. Its isotropic coefficient of thermal expansion is similar to that of copper; therefore, it can support the microstrip feed. Finite ground plane initiates scattering from the edges under excitation which results in the generation of multiple modes. These multiple modes can be controlled by proper impedance matching [16]. The finite floating ground plane provides zero potential reference system is placed at the backside of the substrate. The length of the ground plane controls the electrical characteristics and output patterns of the antenna. Out of the various experiments, optimal performance can be realized when the dimension of the ground plane is set to $24 \times 50\ \text{mm}^2$. The orientation is based on the Cartesian coordinate system such that the bottom portion of the substrate aligned along the x–y plane. The feeding network is composed of two steps-based microstrip line. The optimum length and width of the microstrip patch are $8 \times 3\ \text{mm}^2$, and the terminated patch strip is $18.5 \times 2\ \text{mm}^2$.

Dielectric waveguide model and modified dielectric waveguide model [1] provide a simplified approach to calculate the resonant frequency and the radiation quality factor. The dielectric waveguide model can be applied when RDRA is placed on an infinite ground plane and placed in free space. The finite element method is used where the entire arbitrary volume of the RDRA will be discretized [16]. A simplified and fast approach will be to determine the approximate dimension with the help of the DWM method and compare the outcome with the commercial simulation software HFSS. It is based on the finite element method which is a frequency domain differential technique. After a thorough investigation, the dimension of the RDRA (Roger RT/duroid 6010/6010 LM (tm)), $\varepsilon_r = 10.2$, $\tan\delta = 0.0023$ is taken as $17.5 \times 4.94 \times 13.7\ \text{mm}^3$. A notch of dimension $2.5 \times 4.94 \times 9\ \text{mm}^3$is placed on the RDRA.

3 Design Approach

The infinite ground plane having infinite conductivity will reflect all the energies directed toward it, but an imperfect ground plane will absorb a certain amount of electromagnetic energy incident upon it. According to image theory [16], radiation from the image after the removal of the finite ground plane will be attenuated in amplitude and shifted in phase. The reflection coefficient for the reflected ray is given by Holzman [9]

$$\Gamma_r = \frac{\varepsilon_c \cos\theta - \sqrt{\varepsilon_c - \sin^2\theta}}{\varepsilon_c \cos\theta - \sqrt{\varepsilon_c - \sin^2\theta}} \quad (1)$$

where $\varepsilon_c = \varepsilon_o(\varepsilon - j\sigma/\omega\varepsilon_0)$ and $\mu = \mu_0$. So, the radiated field over the imperfect ground plane will be:

$$E_{\theta r} = S_\theta(r, \theta, \phi)\left(e^{\mathrm{jkh}\cos\theta} + \Gamma_r e^{-\mathrm{jkh}\cos\theta}\right) \quad (2)$$

It has been observed that there will be either total cancellation or a doubling of the field at certain angles. But, for the imperfect ground plane, there will be the presence of no null, and the field strength is less than double for some angles. This is due to the complex nature of the reflection coefficient, arbitrary phase shift, and depolarization of waves. Because of the presence of the finite ground plane, DWM method is unsuitable for calculating the resonant frequency and the radiation Q-factor. However, because of the severity of the numerical technique with rectangular DRA and finite ground plane, RDRA is designed based on DWM. A minor approximate modification is carried out to optimize the desired parameters. The resonant frequency of $\mathrm{TE}^x{}_{\delta 11}$ can be obtained through the solution of the transcendental equation [1].

$$k_x \tan\left(\frac{k_x d}{2}\right) = \sqrt{(\varepsilon_r - 1)k_0^2 - k_x^2} \tag{3}$$

where $k_0 = \frac{2\pi}{\lambda_0} = \frac{2\pi f_0}{c}$, $k_y = \frac{\pi}{w}$, $k_z = \frac{\pi}{b}$, and $k_x^2 + k_y^2 + k_z^2 = \varepsilon_r k_0^2$. The transcendental equation can be solved for selected ratio of w/b as a function of d/b with respect to normalized frequency F. The normalized frequency F can be defined as

$$F = \frac{2\pi \mathrm{wf}_0\sqrt{\varepsilon_r}}{c} \tag{4}$$

Here, f_0 is the resonant frequency. The value of the normalized frequency F can be determined from the F versus (d/b) plot for different value of (w/b). The above equation can be effectively written as

$$f_{\mathrm{GHz}} = \frac{15F}{w_{\mathrm{cm}}\pi\sqrt{\varepsilon_r}} \tag{5}$$

where resonant frequency is represented in GHz scale and w is in cm.

4 Result Analysis

Figure 3 shows the return loss $|S_{11}|$ versus frequency for the various dimensions of the ground plane. It shows a comparative analysis of the $|S_{11}|$ for the finite ground plane of length 22, 23, 24, 25, 26, 27 mm, and the infinite ground plane with respect to 50 mm width. It is observed that −10 dB impedance bandwidth of more than 80% with a peak return loss of 37.5 dB is realized when the dimension of the finite ground plane is 24 × 50 mm^2. It covers both X and Ku band applications. Figure 4 implies the real and imaginary part of Z (1, 1).

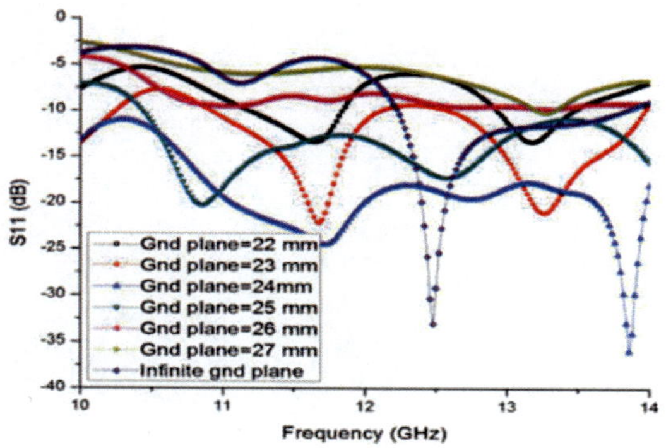

Fig. 3 Plot of return loss ($|S_{11}| < 10$dB) versus frequency

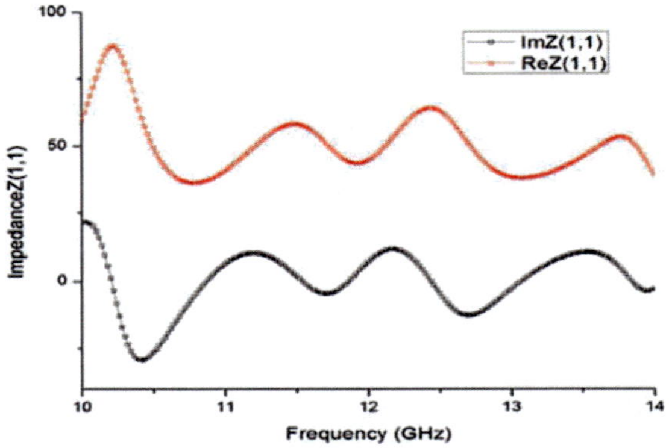

Fig. 4 Plot of impedance versus frequency

Figures 6 and 7 show the radiation pattern indicating E and H-plane co- and cross-polarization of the antenna. The plane of the antenna parallel to the electric field is the E-plane, and similarly, the plane of the antenna parallel to the magnetic field is called an H-plane. Due to radiation from the corner edges of the ground plane, the difference between cross- and co-polarization is around 8–10 dB in the major radiating plane. The plot of gain versus frequency for different dimensions of the ground plane is shown in Fig. 7.

Figure 8 shows the total gain versus frequency under the different sizes of the finite ground plane. It is observed that a maximum peak gain of 5 dB is observed when optimized size of the ground plane is 24 mm. The total gain of the antenna

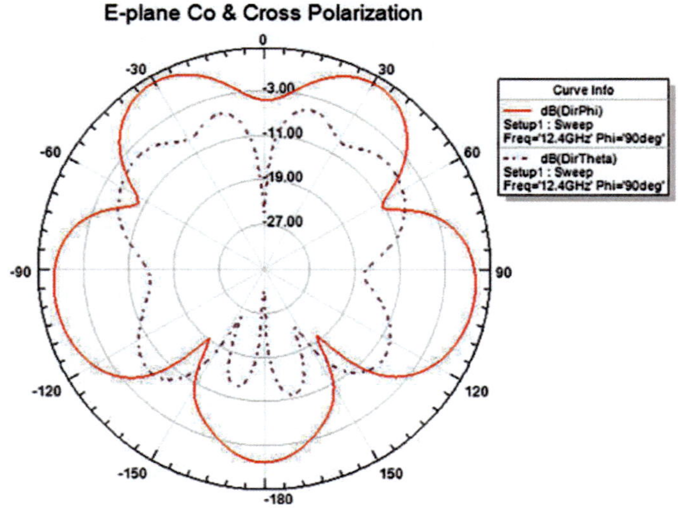

Fig. 5 Radiation pattern indicating E-plane co and cross-polarization

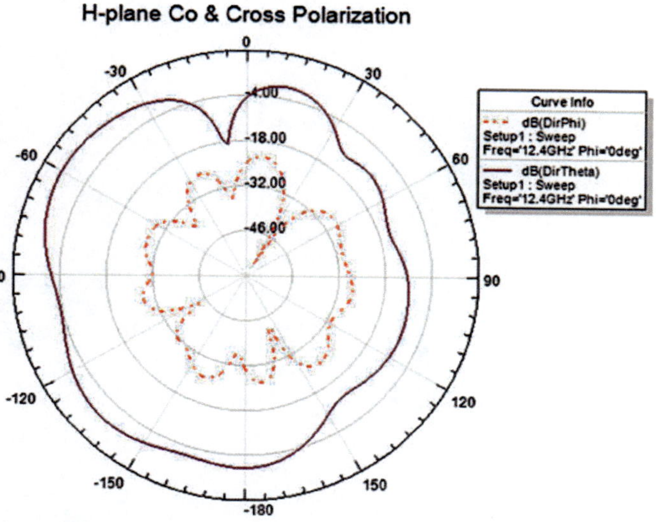

Fig. 6 Radiation pattern indicating H-plane co and cross-polarization

Fig. 7 Plot of gain versus frequency

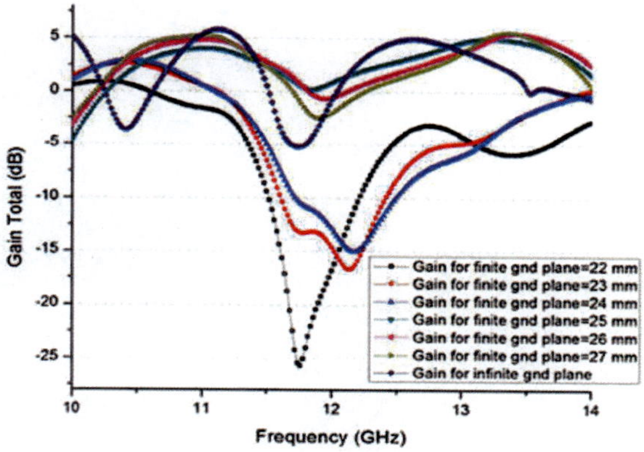

can be defined as the sum of the E_θ and E_ϕ components. It shows comparatively high-gain stability for the entire wide band of frequencies.

Figures 8 and 9 indicate three-dimensional gains of the designed antenna with respect to azimuth and elevation angles. It has been observed that due to high bandwidth response, there is a relatively low-gain bandwidth product.

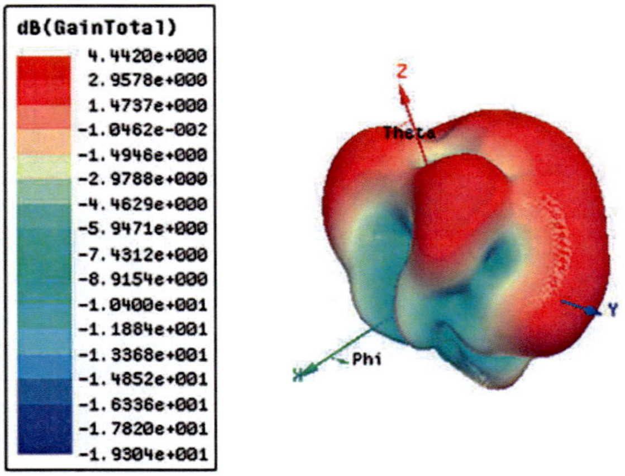

Fig. 8 3D plot of gain along azimuth and elevation

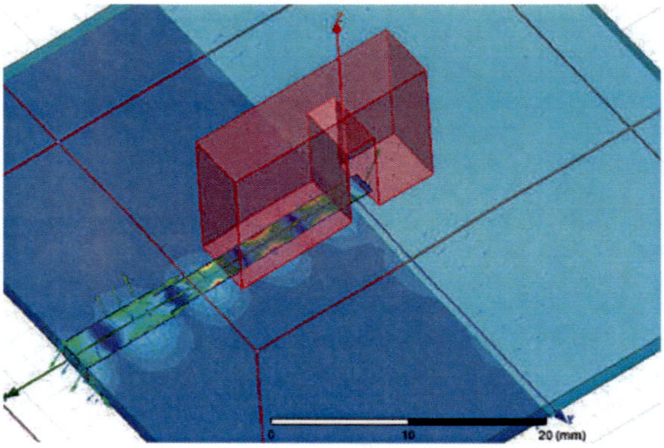

Fig. 9 Formation of surface current on the antenna

5 Conclusions

In this paper, a slim profile, microstrip-fed wideband rectangular dielectric resonator antenna is designed, and the influence of the finite ground plane for its characterization is investigated. The ground plane plays a significant role in deciding the radiation patterns, impedance, and potential of the antenna. The designed structure shows a wideband response with an impedance bandwidth of 80% and a radiation efficiency of 99.9%. A peak radiation gain of 5 dBi at 11.4 GHz makes it suitable for short-range radio links in X and Ku band applications. The multi-functionality and multiband operation can be obtained by exciting the ground plane through the matching network. A genetic algorithm can be used to optimize its performance.

References

1. Petosa A (2007) Dielectric resonator antenna handbook. Artech House, Norwood, MA
2. Pozar DM, Antenna MP (1992) 1st ed. IEEE Press, pp 203–259
3. Guha D, Antar YMM (2011) Microstrip and printed antennas new trends, techniques and applications, 1st ed. John Wiley and Sons Ltd, United Kingdom, pp 305–340
4. Waterhouse RB (2003) Microstrip patch antennas: a designer's guide. Khuwer Academic Publishers, Massachusettes, USA
5. Luk KM, Leung KW (2003) Dielecric resonator antennas. Research Studies Press, Baldock, Herfordshire, UK, pp 1–4
6. Liu B, Qiu J, Chen L, Li G (2019) Dual band-notched rectangular dielectric resonator antenna with tunable characteristic. Electronics 8:472. https://doi.org/10.3390/electronics8050472
7. Mashhadia SHH, Jiaob YC, Chena J, Wide 3-dB beamwidth step-walled rectangular dielectricresonator antenna. J Electromagn Waves Appl. https://doi.org/10.1080/09205071.2020.1713224.
8. Mohanty S, Mohapatra B (2020) Leaky wave guide based dielectric resonator antenna for millimeter-wave applications. Trans Electr Electron Mater. https://doi.org/10.1007/s42341-020-00240-w
9. Holzman E (2006) Essentials of RF and microwave grounding. Artech House, Norwood
10. Huang J (1983) The finite ground plane effect on the microstrip antenna radiation patterns. IEEE Trans Antennas Propag AP-31(4)
11. Ott HW (1979) Ground—a path for current flow. In: Proceedings of IEEE international symposium on electromagnetic compatibility, pp 167–170
12. Abedian M, Oraizi H, Rahim SKA, Danesh S, Ramli MR, Jamaludin MH (2018) Wideband rectangular dielectric resonator antenna for low-profile application 12(1):115–119
13. Noghanian S, Shafai L (1998) Control of microstrip antenna radiation characteristics by ground plane size and shape. IEE Proc-Microwave Antennas Propag. 145(3):207–212
14. Denidni TA, Weng Z (2009) Rectangular dielectric resonator antenna for ultra wideband applications. Electron let 19, 45(24)
15. Wang L, En YF (2021) Broadband circularly polarized dielectric resonator antenna with L-shaped dielectric resonator strip. Int J RF Microwave Comput Aided Des. https://doi.org/10.1002/mmce.22569
16. Harrington RF (2001) Time harmonic electromagnetic fields. In: IEEE press series on electromagnetic wave theory

Design and Computational Simulation of CZTS-Based Solar Cell Using SCAPS-1-D Software

Avijeet Rai and B. K. Panday

Abstract The direct energy band gap of around 1.4–1.6 eV and greater absorption coefficient of the order of 10^4 cm^{-1} make CZTS as one of the best materials for solar cell. Here we have performed the numerical modelling and computational simulation with the help of SCAPS-1-D software to investigate the performance of solar cell. In the present work we have studied the effect of variation of the absorber layer (CZTS) thickness from 0.5 to 2 μm with buffer layer thickness constant at 0.05 μm at temperature of 300 K and under "AM 1.5 spectrum" on electrical parameters such as open circuit voltage (V_{OC}), short circuit current (I_{SC}), fill factor (FF) and efficiency (eta). It is observed that the electrical parameters are greatly affected by the variation in the thickness of absorber layer. The efficiency of solar cell gets decreased with increase in temperature. At best simulated parameter (2 μm absorber layer, 0.05 buffer layer, 300 K operating temperature and at "AM 1.5 sun spectrum") the efficiency was found to be 28.54%. This result can be considered extremely useful in fabrication of thin film CZTS solar cell.

Keywords CZTS · SCAPS-1-D · Solar cell · Thin film · Simulation

1 Introduction

The amount of solar energy we are getting cannot be neglected when we are looking for renewable sources of energy. By an estimate, the total solar energy incident on earth surface is near about 7.45×10^{11} Kwh annually and total energy consumed by humans on earth is near about 0.5×10^{14} Kwh annually [1]. The conversion of solar energy into electricity has great technological and social advantages, the practical fabrication and manufacturing of solar cell is very expansive due to the material, machinery and technique used. Also, the cost of maintenance cannot be neglected, hence it becomes mandatory for us to look for the materials which are available

A. Rai · B. K. Panday (✉)
Department of Physics and Material Science, M.M.M. University of Technology, Gorakhpur, India
e-mail: bkpmmmec11@gmail.com

B. Mishra and M. Tiwari (eds.), *VLSI, Microwave and Wireless Technologies*, Lecture Notes in Electrical Engineering 877,
https://doi.org/10.1007/978-981-19-0312-0_79

in large amount on earth surface and whose extraction can be easily and cheaply done. Beside the above there are many other ways of reducing the cost of solar cell which are being looked by researchers and scientists. The manufacturing of thin film solar cell is expanding their base in the field of solar cell. There are various type of thin film solar cell which are available to us such as CIGS, CdTe which have efficiency around 21%–24%, respectively, which are very efficient considering the cost of manufacturing, but there are some problem associated with them such as they use rare earth metal and are toxic in nature, which are harmful to our nature. Hence, we have to look for materials which are non-toxic in nature and available in abundant amount on earth. One of them is copper, zinc, tin, sulphide (CZTS). The most promising factor of CZTS is their direct and suitable band gap having the order of 1.4–1.6 eV and a large absorption coefficient of 10^4 cm^{-1}.There are many techniques to fabricate CZTS solar cell such as sputtering thermal evaporation, deposition using pulse laser, sol–gel, and spray pyrolysis. To manufacture a solar cell having high efficiency, a deep study of mechanism of synthesis is essential including the study of the outcome of solar cell is also required. To do this we use numerical simulation and computational technique that allow our researchers of the field of solar cell to look deep into characteristic of solar cells [1, 2].

2 The Modelling and Simulation Programme SCAPS-1-D

Computational simulation is important technique, which is used heavily these days, to study deep into the mechanisms. Simulation is a fast, easy and reliable option to study various parameters of solar cells. There is various simulation programme available such as (AFORS HET, SILVACO, AMPS, SCAPS-1-D, and MATLAB) to study the mechanism and characteristics of solar cell [3–6]. Here we have used SCAPS-1-D in 1-dimensional software. In SCAPS a maximum of 7 layer can be made, i.e. a thin solar cell of 7 layer can be numerically modelled. In this software first we define structure MO/CZTS/ZnS/ZnO, for each layer we define their thickness, band gap, electron affinity, their impurity concentration, then we define the external parameters temperature of 300 K, "AM 1.5 solar spectrum", external series and shunt resistance. After doing all this we define the voltage upto which the simulation is to be performed on defined structure, then we click on stop after V_{OC}, so that unnecessary simulation after V_{OC} is not performed by software, then we click at single shot. After this we made batch of parameters for which we want to optimize our solar cell. At last, we made a recorder setup to see the electrical parameters variations with batch we have made earlier. The various parameter for simulation is shown in Table 1 [7–10].

Table 1 Different parameters used in this simulation

Parameters	ZnO	ZnS	CZTS	Mo
Thickness (μm)	0.1	0.05	0.5–2	0.15
Band gap (eV)	3.3	3.68	1.50	1.7
Electron affinity (eV)	4.60	4.3	4.5	4.20
Dielectric permittivity	9.0	8.3	10.0	13.6
CB effective density of state (cm^{-3})	2.2×10^{18}	2.2×10^{18}	2.2×10^{18}	2.2×10^{18}
VB effective density of state (cm^{-3})	1.8×10^{19}	1.8×10^{19}	1.8×10^{19}	1.8×10^{19}
Thermal velocity of electron (cm/s)	1.0×10^{7}	1.0×10^{7}	1.0×10^{7}	1.0×10^{7}
Thermal velocity of holes (cm/s)	1.0×10^{7}	1.0×10^{7}	1.0×10^{7}	1.0×10^{7}
Electron mobility (cm^2/V_s)	100	165	100	100
Hole mobility (cm^2/V_s)	25	5	25	25
Shallow uniform donor density N_D (cm^{-3})	1.0×10^{18}	1.0×10^{17}	10	0
Shallow uniform acceptor density N_A (cm^{-3})	0	0	2.0×10^{14}	1.0×10^{16}

3 Results and Discussions

3.1 Effect of the Variation of Absorber Layer's Thickness (CZTS)

The total cost of solar cell depends on the width of absorber layer here we have taken CZTS as absorber layer. For thin film solar such as CZTS whose cost is quite high as compared to the other traditional PV cells. For observing the various parameters (V_{OC}, J_{SC}, FF, eta), the width of CZTS is tuned from 0.5 to 2 μm during all simulation process, the band gap the CZTS layer was kept constant at 1.5 eV [12], with a fixed Zns layer of width of 0.05 μm. It can be seen from Fig. 1 that there is an increase in efficiency (eta) and a slight increase in V_{OC}, J_{SC}. The small increase in V_{OC}, J_{SC} may be considered because of the process of recombination of the electron hole pair at the MO layer. From Fig. 1, the fill factor first increases from 0.4 to 0.8 μm, then a continuous decrease from 0.8 to 2 μm. As the width of CZTS is increased, the number of photons absorbed by it also increased and create many electron–hole pairs, due to which to efficiency (eta) increases, V_{OC}, and J_{SC}. The increase in eta is also because of V_{OC}, J_{SC}, and FF.

3.2 Effect of Variation of Temperature

Many literatures published earlier shows that there is a negative effect of increase in temperature at electrical parameters of solar cell. From the above Fig. 2 it is very much clear that there is a linear decrease in V_{OC}, eta, FF and there is no change in

Fig. 1 Schematic structural view of CZTS solar cells

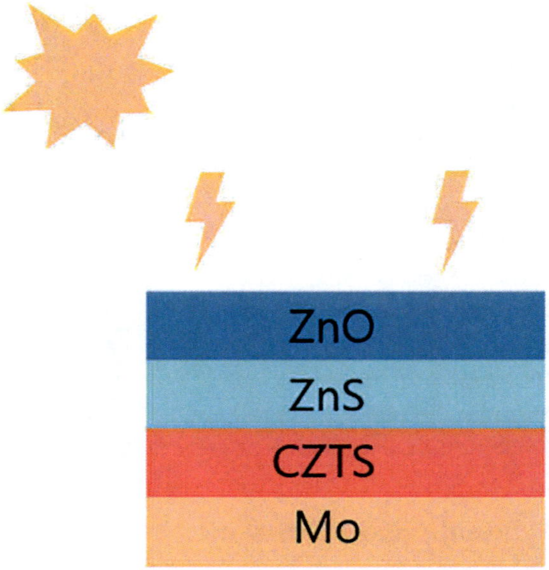

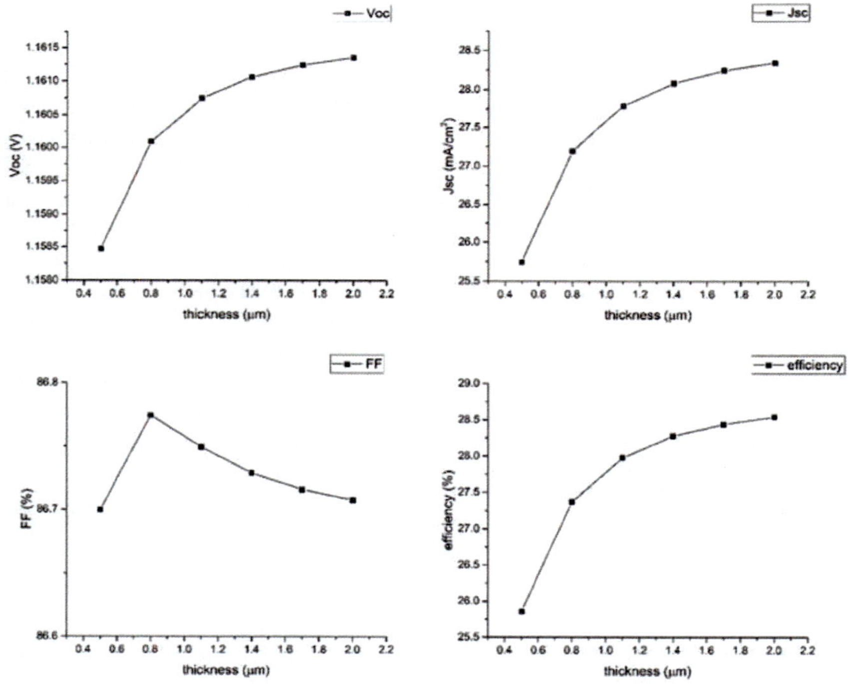

Fig. 2 Effect of various thickness of CZTS absorber layer

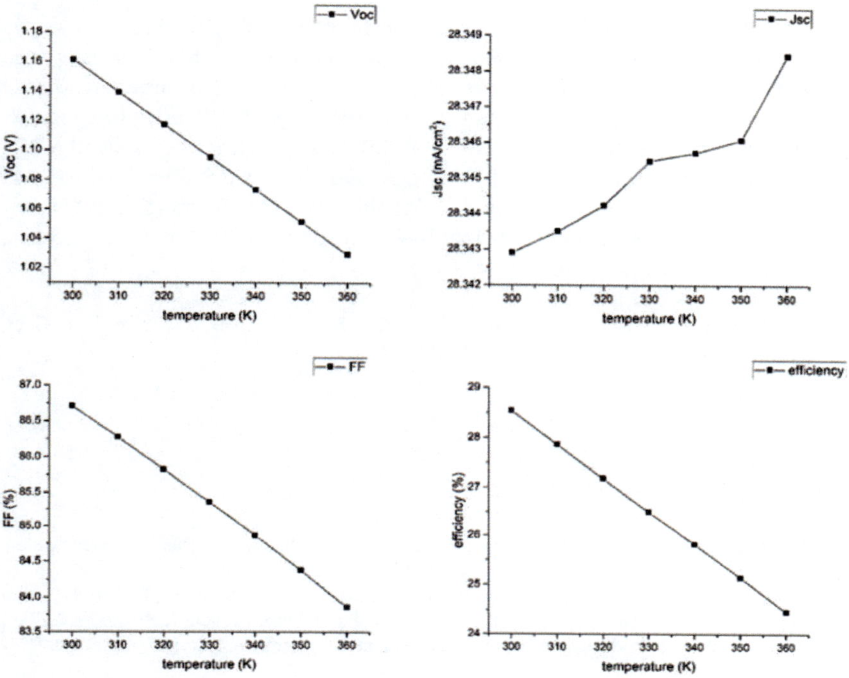

Fig. 3 Effect of temperature variation on solar cell parameters

J_{SC}. The decrease in eta is mainly considered due to decrease in V_{OC} as J_{SC} remains constant. From our simulated results as shown in Fig. 2 the maximum efficiency is observed at 300 K, after increase in temperature there is decrease in efficiency due to decrease in current with rise in temperature [11–13] (Fig. 3).

4 Conclusion

The modelling of CZTS photovoltaic solar cell has been done in our present work which shows that optimization of the absorber layer (p-CZTS) gives a solar cell of improved efficiency, it was also absorbed that there is decrease in efficiency with increase in temperature above 300 K. Both (CZTS) and (ZnS) are non-toxic. The results obtained in simulation also satisfying the experimental work at desired temperature. Since CZTS is thin film solar cell hence its thickness is of very important, here the optimized thickness is 2 μm and band gap of energy 1.5 eV gives an efficiency of 28.54% at 300 K.

Comparing with other experimental works done in recent times, we see in experimental work done by Haghighi et al. [14]. SnS_2 is used as buffer layer which is deposited by pulsed layer deposition (PLD) method on CZTS absorber layer shows an efficiency of 12.3%. Experimental work done by Sinha et al. [15] uses atomic layer deposition (ALD) to deposit various buffer layer on CZTS absorber layer. ZnO is deposited by ALD method which gives an efficiency of 2.46% and deposition of Zn-based ternary materials such as (Sn + Zn) shows an efficiency around 6.2%. In a computer simulation work [16] CdS, ZnS, and $Cd_{0.6}Zn_{0.6}S$ were optimized as buffer layer with CZTS absorber layer shows that $Cd_{0.4}Zn_{0.6}S$ shows an efficiency of 11.20%. Hence, we can conclude that in CZTS solar cell ZnS may be used as a buffer layer for its better performance.

Acknowledgements The authors are very grateful to Marc Burgelman, (ELIS), University of Gent, Belgium, for giving the free access of above 1D-SCAPS SOFTWARE.

References

1. Chopra KL, Das SR (1983) Thin film solar cells. Springer Science + Business Media, New York
2. Ito K (ed) (2015) Copper zinc tin sulphide-based thin-film solar cells. Wiley, United Kingdom
3. Stangl R, Leendertz C, Haschke J (2010) Numerical simulation of solar cells and solar cell characterization Methods: the open-source on demand program AFORS-HET. ISBN 978-953-307-052-0
4. Michael S (2005) The design and optimization of advanced multijunction solar cells using the silvaco ATLAS software package. Solar Energy Mater Solar Cells 87:785–794
5. Liu Y, Sun Y, Rockett A (2012) A new simulation software of solar cells—wxAMPS. Solar Energy Mater Solar Cells 98:124–128
6. Edouard M, Njomo D (2013) Mathematical modeling and digital simulation of PV solar panel using MATLAB software. Int J Emerg Technol Adv Eng 3(9)
7. Yasin S, Waar ZA, Al Zoubi T (2020) Development of high efficiency CZTS solar cell through buffer layer parameters optimization using SCAPS-1D. https://doi.org/10.1016/j.matpr.2020.05.064
8. Belarbia F, Rahalb W, Rachedc D, benghabrita S, Adnanea M (2020) A comparative study of different buffer layers for CZTS solar cell using Scaps-1D simulation program. https://doi.org/10.1016/j.ijleo.2020.164743
9. Zhou X, Han J (2020) Design and simulation of C2N based solar cell by SCAPS-1D software. https://doi.org/10.1088/2053-1591/abcdd6
10. Rai N, Dwivedi DK (2020) Numerical modelling for enhancement of output performance of CIGS based thin film solar cell using SCAPS 1-D simulation software. https://doi.org/10.1063/5.0001233
11. Dauodia AK, El Hassouani Y, Benami A (2016) Investigation of the effect of thickness, bandgap and temperature on the efficiency of CIGS Solar cells through SCAPS 1-D. (IJETR) 6(2) ISSN: 2321-0869(0) 2454-4698(P)
12. Chelvanathan P et al (2010) Current Appl Phys 10:S387–S391
13. Henriche H et al (2017) New ultra-thin CIGS structure solar cells using SCAPS simulation program. Int J Hydrogen Energy. https://doi.org/10.1016/j.ijhydene.2017.02.099
14. Haghighi M, Minbashi M, Taghavinia N, Kim DH, Mahdavi SM, Kordbacheh AA (2018) A modelling study on utilizing SnS_2 the buffer layer of CZT(S, Se) solar cells. Solar Energy 167:165–171

15. Sinha S, Nandi DK, Kim SH, Heo J (2018) Atomic layer—deposited buffer layers for thin film solar cells using earth abundant absorber materials: a review. Solar Energy Mater Solar Cells 176:49–68
16. Jhuma FA, Shaily MZ, Rashid MJ (2019) Mater Renewable Sustain Energy. https://doi.org/10.1007/s40243-019-0144-1